Handbook of
X-Ray Spectrometry

PRACTICAL SPECTROSCOPY
A SERIES

Edited by Edward G. Brame, Jr.

The CECON Group
Wilmington, Delaware

ADDITIONAL VOLUMES IN PREPARATION

Handbook of X-Ray Spectrometry

Methods and Techniques

edited by

René E. Van Grieken

Department of Chemistry
University of Antwerp
Antwerp, Belgium

Andrzej A. Markowicz

Institute of Physics and Nuclear Techniques
Academy of Mining and Metallurgy
Cracow, Poland

Marcel Dekker, Inc.　　　　**New York • Basel • Hong Kong**

CHEM
seplae

Library of Congress Cataloging-in-Publication Data

Handbook of X-ray spectrometry : methods and techniques / edited by
René E. Van Grieken, Andrzej A. Markowicz.
    p.   cm. -- (Practical spectroscopy series ; v. 14.)
    Includes bibliographical references and index.
    ISBN 0-8247-8483-9
    1. X-ray spectroscopy.   I. Grieken, R. Van (René)   II. Markowicz,
Andrzej.   III. Series.
QD96.H36   1992
543'.08586--dc20

            92-25554
             CIP

This book is printed on acid-free paper.

MARCEL DEKKER, INC.
270 Madison Avenue, New York, New York 10016

Current printing (last digit):
10 9 8 7 6 5 4 3 2 1

PRINTED IN THE UNITED STATES OF AMERICA

Preface

Scientists in recent years have been somewhat ambivalent regarding the role of x-ray emission spectrometry in analytical chemistry. Whereas no radically new and stunning developments have been seen, there has been remarkably steady progress, both instrumental and methodological, in the more conventional realms of x-ray fluorescence. For the more specialized approaches—for example, x-ray emission induced by synchrotron radiation, radioisotopes and polarized x-ray beams, and total-reflection x-ray fluorescence—and for advanced spectrum analysis methods, exponential growth and/or increasing acceptance has occurred. Contrary to previous books on x-ray emission analysis, these latter approaches make up a large portion of the present *Handbook of X-Ray Spectrometry.*

The major milestone developments that shaped the field of x-ray spectrometry and now have widespread applications all took place more than twenty years ago. After wavelength-dispersive x-ray spectrometry had been demonstrated and a high-vacuum x-ray tube had been introduced by Coolidge in 1913, the prototype of the first modern commercial x-ray spectrometer with a sealed x-ray tube was built by Friedmann and Birks in 1948. The first electron microprobe was successfully developed in 1951 by Castaing, who also outlined the fundamental concepts of quantitative analysis with it. The semiconductor or Si(Li) detector, which heralded the advent of energy-dispersive x-ray fluorescence, was developed around 1965 at Lawrence Berkeley Laboratory. Accelerator-based particle-induced x-ray emission analysis was developed just before 1970, mostly at the University of Lund. The various popular matrix correction methods by Lucas-Tooth, Traill and Lachance, Claisse and Quintin, Tertian, and several others, were all proposed in the 1960s. One may thus wonder whether the more conventional types of x-ray fluorescence analysis have reached a state of saturation and consolidation, typical for a mature and routinely applied analysis technique. Reviewing the state of the art and describing recent progress for wavelength- and energy-dispersive x-ray fluorescence, electron and heavy charged-particle-induced x-ray emission, quantification, and sample preparation methods is the purpose of the remaining part of this book.

Chapter 1 reviews the basic physics behind the x-ray emission techniques, and refers to the appendixes for all the basic and generally applicable x-ray physics constants. Chapter 2 outlines established and new instrumentation and discusses the performances of wavelength-dispersive x-ray fluorescence analysis, which, with probably 14,000 units in operation worldwide today, is still the workhorse of x-ray analysis with applications in a wide range of disciplines including process control, materials analysis, metallurgy, mining, and almost every other major branch of science. Chapter 3 discusses the basic principles, background, and recent advances in the tube-excited energy-dispersive mode, which, after hectic growth in the 1970s, has now apparently leveled off to make up approximately 20% of the x-ray fluorescence market; it is invoked frequently in research on environmental and biological samples. Chapter 4 reviews in depth the available alternatives for spectrum evaluation and qualitative analysis; techniques for deconvolution of spectra have enormously increased the utility of energy-dispersive x-ray analysis, but deconvolution is still its most critical step. Chapters 5 and 6 review the quantification problems in the analysis of samples that are infinitely thick and of intermediate thickness, respectively. Chapter 7 is a very practical treatment of radioisotope-induced x-ray analysis, which is now rapidly acquiring wide acceptance for dedicated instruments and field applications. Chapter 8 reviews synchrotron-induced x-ray emission analysis, the youngest branch, with limited accessibility but an exponentially growing literature due to its extreme sensitivity and microanalysis potential. Although its principles have been known for some time, it is only since the advent of powerful commercial units that total reflection x-ray fluorescence has been rapidly introduced, mostly for liquid samples and surface layer characterization; this is the subject of Chapter 9.

Polarized beam x-ray fluorescence is outlined in Chapter 10. Particle-induced x-ray emission analysis is available at many accelerator centers worldwide; the number of annual articles on it is growing and it undergoes a revival in its microversion; Chapter 11 treats the physical backgrounds, instrumentation, performance, and applications of this technique. Although the practical approaches to electron-induced x-ray emission analysis, now a standard technique with wide applications in all branches of science and technology, are often quite different from those in other x-ray analysis techniques, a separate treatment of its potential for quantitative and spatially resolved analysis is given in Chapter 12. Finally, Chapter 13 briefly reviews the sample preparation techniques that are invoked most frequently in combination with x-ray fluorescence analysis.

This book is a multi-authored effort. We believe that having scientists who are actively engaged in a particular technique covering those areas for which they are particularly qualified and presenting their own points of view and general approaches outweighs any advantages of uniformity and homogeneity that characterize a single-author book. Three chapters were written by the editors and a coworker. For all the other chapters, we were fortunate enough to have the co-operation of eminent specialists. The editors wish to thank all the contributors for their efforts.

We hope that novices in x-ray emission analysis will find this book useful and instructive, and that our more experienced colleagues will benefit from the large amount of readily accessible information available in this compact form, some of it for the first time. This book is not intended to replace earlier works, some of which were truly excellent, but to supplement them. Some overlap is inevitable, but an effort has been made to emphasize the fields and developments that have come into prominence lately and have not been treated in a handbook before.

René E. Van Grieken
Andrzej A. Markowicz

Contents

Contributors

J. L. de Vries *Eindhoven, The Netherlands* (retired)

Robert D. Giauque *Lawrence Berkeley Laboratory, University of California, Berkeley, California*

J. A. Helsen *Catholic University of Leuven, Leuven, Belgium*

Jasna Injuk *University of Antwerp, Antwerp, Belgium*

Joseph M. Jaklevic *Lawrence Berkeley Laboratory, University of California, Berkeley, California*

Koen H. A. Janssens *University of Antwerp, Antwerp, Belgium*

Keith W. Jones *Brookhaven National Laboratory, Upton, New York*

Joachim Knoth *GKSS Forschungszentrum, Geesthacht, Germany*

Andrzej Kuczumow *Maria Sklodowska-Curie University, Lublin, Poland*

Willy Maenhaut *University of Gent, Gent, Belgium*

Klas G. Malmqvist *University of Lund and Lund Institute of Technology, Lund, Sweden*

Andrzej A. Markowicz *Academy of Mining and Metallurgy, Cracow, Poland*

Richard W. Ryon *Lawrence Livermore National Laboratory, Livermore, California*

Heinrich Schwenke *GKSS Forschungszentrum, Geesthacht, Germany*

John A. Small *National Institute of Standards and Technology, Gaithersburg, Maryland*

Pierre J. M. Van Espen *University of Antwerp, Antwerp, Belgium*

René E. Van Grieken *University of Antwerp, Antwerp, Belgium*

Bruno A. R. Vrebos *Philips Analytical Research Laboratories, Almelo, The Netherlands*

John S. Watt *Commonwealth Scientific and Industrial Research Organization, Sydney, New South Wales, Australia*

John D. Zahrt *Los Alamos National Laboratory, Los Alamos, New Mexico*

1

X-ray Physics

Andrzej A. Markowicz   *Academy of Mining and Metallurgy, Cracow, Poland*

I. INTRODUCTION

X-rays, or Röntgen rays, are electromagnetic radiations having wavelengths roughly within the range from 0.05 to 100 Å. At the short-wavelength end, they overlap with γ-rays, and at the long-wavelength end they approach ultraviolet radiation.

II. HISTORY

X-rays were discovered in 1895 by Wilhelm Conrad Röntgen at the University of Würzburg, Bavaria. He noticed that some crystals of barium platinocyanide near a discharge tube completely enclosed in black paper became luminescent when the discharge occurred. By examining the shadows cast by the rays, Röntgen traced the origin of the rays to the walls of the discharge tube. In 1896, Campbell-Swinton introduced a definite target (platinum) for the cathode rays to hit; this target was called the anticathode.

For his work with x-rays, Röntgen received the first Nobel Prize in physics, in 1901. It was the first of six to be awarded in the field of x-rays by 1927.

The obvious similarities with light led to the crucial tests of established wave optics: polarization, diffraction, reflection, and refraction. With limited experimental facilities, Röntgen and his contemporaries could find no evidence of any of these, hence the designation ''*x*'' (unknown) of the rays, generated by the stoppage at anode targets of the cathode rays, identified by Thomson in 1897 as electrons.

The nature of x-rays was the subject of much controversy. In 1906, Barkla found evidence in scattering experiments that x-rays could be polarized and must therefore be waves, but W. H. Bragg's studies of the produced ionization indicated that they were corpuscular. The essential wave nature of x-rays was established in 1912 by Laue, Friedrich, and Knipping, who showed that x-rays could be diffracted by a crystal (copper sulfate pentahydrate) that acted as a three-dimensional diffraction grating. W. H. and W. L. Bragg (father and son) found the law for the selective reflection of x-rays. In 1908, Barkla and Sadler deduced, by scattering experiments, that x-rays contained components characteristic of the material of the target, and they called these components K and L radiations. That these radiations had sharply defined wavelengths was shown by the diffraction experiments of W. H. Bragg in 1913. These experiments demonstrated clearly the existence of a line spectrum superimposed upon a continuous ("white") spectrum. In 1913, Moseley showed that the wavelengths of the lines were characteristic of the element of which the target was made and, further, showed that they had the same sequence as the atomic numbers, thus enabling atomic numbers to be determined unambiguously for the first time. The characteristic K absorption was first observed by de Broglie and interpreted by W. L. Bragg and Siegbahn. The effect on x-ray absorption spectra of the chemical state of the absorber was observed by Bergengren in 1920. The influence of the chemical state of the emitter on x-ray emission spectra was observed by Lindh and Lundquist in 1924. The theory of x-ray spectra was worked out by Sommerfeld and others. In 1919, Stenström found the deviations from Bragg's law and interpreted them as the effect of refraction. The anomalous dispersion of x-rays was discovered by Larsson in 1929, and the extended fine structure of x-ray absorption spectra was qualitatively interpreted by Kronig in 1932.

Soon after the first primary spectra excited by electron beams in an x-ray tube were observed, it was found that secondary fluorescent x-rays were excited in any material irradiated with beams of primary x-rays and that the spectra of these fluorescent x-rays were identical in wavelengths and relative intensities with those excited when the specimen was bombarded with electrons. Beginning in 1932, Hevesy, Coster, and others investigated in detail the possibilities of fluorescent x-ray spectroscopy as a means of qualitative and quantitative elemental analysis.

III. GENERAL FEATURES

The properties of x-rays, some of which are discussed in detail in this chapter, are summarized as follows:

- Invisible
- Propagated in straight lines with a velocity of 3×10^8 m s^{-1}, as is light
- Unaffected by electrical and magnetic fields
- Differentially absorbed in passing through matter of varying composition, density, or thickness
- Reflected, diffracted, refracted, and polarized

- Capable of ionizing gases
- Capable of affecting electrical properties of liquids and solids
- Capable of blackening a photographic plate
- Able to liberate photoelectrons and recoil electrons
- Capable of producing biological reactions, for example to damage or kill living cells and to produce genetic mutations
- Emitted in a continuous spectrum whose short-wavelength limit is determined only by the voltage on the tube
- Emitted also with a line spectrum characteristic of the chemical elements
- Found to have absorption spectra characteristic of the chemical elements

IV. EMISSION OF CONTINUOUS RADIATION

Continuous x-rays are produced when electrons, or other high-energy charged particles, such as protons or α particles, lose energy in passing through the Coulomb field of a nucleus. In this interaction the radiant energy (photons) lost by the electron is called *Bremsstrahlung* (from the German *bremsen,* to brake, and *Strahlung,* radiation; this term sometimes designates the interaction itself). The emission of continuous x-rays finds a simple explanation in terms of classic electromagnetic theory, since according to this theory the acceleration of charged particles should be accompanied by emission of radiation. In the case of high-energy electrons striking a target, they must be rapidly decelerated as they penetrate the material of the target, and such a high negative acceleration should produce a pulse of radiation.

The continuous x-ray spectrum generated by electrons in an x-ray tube is characterized by a short-wavelength limit λ_{min}, corresponding to the maximum energy of the exciting electrons:

$$\lambda_{min} = \frac{hc}{eV_0} \tag{1}$$

where h is Planck's constant, c is the velocity of light, e is the electron charge, and V_0 is the potential difference applied to the tube. This relation of the short-wavelength limit to the applied potential is called the Duane-Hunt law.

The probability of radiative energy loss (bremsstrahlung) is roughly proportional to $q^2 Z^2 T/M_0^2$, where q is the particle charge in units of the electron charge e, Z is the atomic number of the target material, T is the particle kinetic energy, and M_0 is the rest mass of the particle. Because protons and heavier particles have large masses compared to the electron mass, they radiate relatively little; for example, the intensity of continuous x-rays generated by protons is about four orders of magnitude lower than that generated by electrons.

The ratio of energy lost by bremsstrahlung to that lost by ionization can be approximated by

$$\left(\frac{m_0}{M_0}\right)^2 \frac{ZT}{1600 m_0 c^2} \tag{2}$$

where m_0 is the rest mass of the electron.

A. Spectral Distribution

The continuous x-ray spectrum generated by electrons in an x-ray tube (thick target) is characterized by the following features:

1. Short-wavelength limit, λ_{min} [Eq. (1)]; below this wavelength no radiation is observed,
2. Wavelength of maximum intensity λ_{max}, approximately 1½ times λ_{min}; however, the relationship between λ_{max} and λ_{min} depends to some extent on voltage, voltage waveform, and atomic number,
3. Total intensity nearly proportional to the square of the voltage and the first power of the atomic number of the target material.

The most complete empirical work on the overall shape of the energy distribution curve for a thick target has been that of Kulenkampff [1,2], who found the following formula for the energy distribution:

$$I(v)\, dv = i[aZ(v_0 - v) + bZ^2]\, dv \tag{3}$$

where $I(v)\, dv$ is the intensity of the continuous x-rays within a frequency range $(v, v + dv)$, i is the electron current striking the target, Z is the atomic number of the target material, v_0 is the cutoff frequency $(= c/\lambda_{min})$ above which the intensity is zero, and a and b are constants independent of atomic number, voltage, and cutoff wavelength. The second term in Eq. (3) is usually small compared to the first and is often neglected.

The total integrated intensity at all frequencies is

$$I = i(a'ZV_0^2 + b'Z^2V_0) \tag{4}$$

in which $a' = a(e^2/h^2)/2$ and $b' = b(e/h)$. An approximate value for b'/a' is 16.3 V; thus

$$I = a'iZV_0(V_0 + 16.3Z) \tag{5}$$

The efficiency Eff of conversion of electric power input to x-rays of all frequencies is given by

$$\text{Eff} = \frac{I}{V_0 i} = a'Z(V_0 + 16.3Z) \tag{6}$$

where V_0 is in volts. Experiments give $a' = (1.2 \pm 0.1) \times 10^{-9}$ [3].

The most complete and successful efforts to apply quantum theory to explain all features of the continuous x-ray spectrum are those of Kramers [4] and Wentzel [5]. By using the correspondence principle, Kramers found the following formulas for the energy distribution of the continuous x-rays generated in a thin target:

$$
\begin{aligned}
I(v)\, dv &= \frac{16\pi^2 AZ^2 e^5}{3\sqrt{3}m_0 V_0 c^3}\, dv & v &< v_0 \\
I(v)\, dv &= 0 & v &> v_0
\end{aligned}
\tag{7}
$$

where A is the atomic mass of the target material. When the decrease in velocity of the electrons in a thick target was taken into account by applying the Thomson-

Whiddington law [6], Kramers found, for a thick target,

$$I(v) \, dv = \frac{8\pi e^2 h}{3\sqrt{3}lm_0c^3} Z(v_0 - v) \, dv \tag{8}$$

where l is approximately 6. The efficiency of production of the x-rays calculated via Kramers' law is given by

$$\text{Eff} = 9.2 \times 10^{-10} \, ZV_0 \tag{9}$$

which is in qualitative agreement with the experiments of Kulenkampff [7], for example:

$$\text{Eff} = 15 \times 10^{-10} \, ZV_0 \tag{10}$$

It is worth mentioning that the real continuous x-ray distribution is described only approximately by Kramers' equation. This is related, inter alia, to the fact that the derivation ignores self-absorption of x-rays and electron backscattering effects.

Wentzel [5] used a different type of the correspondence principle than Kramers, and he explained the spatial distribution asymmetry of the continuous x-rays from thin targets.

An accurate description of continuous x-rays is crucial in all x-ray spectrometry. The spectral intensity distributions from x-ray tubes are of great importance for applying fundamental mathematical matrix correction procedures in quantitative x-ray fluorescence (XRF) analysis. A simple equation for accurate description of the actual continuum distributions from x-ray tubes was recently proposed by Tertian and Broll [8]. It is based on a modified Kramers' law and a refined x-ray absorption correction. Also, a strong need to model the spectral bremsstrahlung background exists in electron probe x-ray microanalysis (EPXMA). First, fitting a function through the background portion of an EPXMA spectrum on which the characteristic x-rays are superimposed is not easy; several experimental fitting routines and mathematical approaches, such as the simplex method, have been proposed in this context lately. Second, for bulk multielement specimens, the theoretical prediction of the continuum bremsstrahlung is not trivial; indeed, it has been known for several years that the commonly used Kramers' formula with Z directly substituted by the average $\bar{Z} = \sum_i W_i Z_i$ (W_i and Z_i are weight fraction and atomic number of the ith element, respectively) can lead to significant errors. In this context some improvements are offered by several modified versions of Kramers' formula developed for a multielement bulk specimen [9–12]. Recently, a new expression for the continuous x-rays emitted by thick composite specimens was proposed [13,14]; it was derived by introducing the compositional dependence of the continuum x-rays already in the elementary equations. The new expression has been combined with known equations for the self-absorption of x-rays [15] and electron backscattering [16] to obtain an accurate description of the detected continuum radiation. A third problem is connected with the description of the x-ray continuum generated by electrons in specimens of thickness smaller than the continuum x-ray generation range. This problem arises in the analysis of both thin films and particles by EPXMA. A theoretical model for the shape of the continuous x-rays generated in multielement specimens of finite thickness was recently developed [17]; both composition and thickness dependence have been considered. Further refinements of the theoretical approach are hampered by the lack of knowledge concerning the shape of the electron interaction volume, the distribution of the electrons within the interaction volume, and the anisotropy of the continuous radiation for different x-ray energies and for different film thicknesses.

B. Spatial Distribution and Polarization

The spatial distribution of the continuous x-rays emitted by thin targets has been investigated by Kulenkampff [18]. The author made an extensive survey of the intensity at angles between 22° and 150° to the electron beam in terms of dependence upon wavelength and voltage. The target was a 0.6 μm thick Al foil. Figure 1 shows the continuous x-ray intensity observed at different angles for voltages of 37.8, 31.0, 24.0, and 16.4 kV filtered by 10, 8, 4, and 1.33 mm of Al, respectively

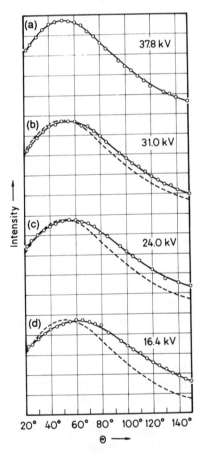

Figure 1 Intensity of continuous x-rays as a function of direction for different voltages. (Curve I is repeated as a dotted line.) (From Ref. 7. Reprinted by permission of the author and Springer-Verlag.)

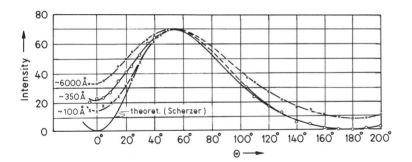

Figure 2 Intensity of continuous x-rays as a function of direction for different thicknesses of the Al target together with a theoretical prediction. (From Ref. 7. Reprinted by permission of the author and Springer-Verlag.)

[7]. Curve I is repeated as a dotted line near each of the other curves. The angle of the maximum intensity varied from 50° for 37.8 kV to 65° for 16.4 kV. Figure 2 illustrates the intensity of the continuous x-rays observed in the Al foil for different thicknesses as a function of the angle [7] for a voltage of 30 kV. The theoretical curve is from the theory of Scherzer [19]. The continuous x-ray intensity drops to zero at 180°, and although it is not zero at 0° as the theory of Scherzer predicts, it can be seen from Fig. 2 that for a thinner foil, a lower intensity at 0° is obtained. Summarizing, it appears that the intensity of the continuous x-rays emitted by thin foils has a maximum at about 55° relative to the incident electron beam and becomes zero at 180°.

The continuous radiation from thick targets is characterized by a much smaller anisotropy than that from thin targets. This is because in thick targets the electrons are rarely stopped in one collision and usually their directions have considerable variation. The use of electromagnetic theory predicts a maximum energy at right angles to the incident electron beam at low voltages, with the maximum moving slightly away from perpendicularity toward the direction of the electron beam as the voltage is increased. In general, an increase in the anisotropy of the continuous x-rays from thick targets is observed at the short-wavelength limit and for low-Z targets [6].

Continuous x-ray beams are partially polarized only from extremely thin targets; the angular region of polarization is sharply peaked about the photon emission angle $\theta = m_0c^2/E_0$, where E_0 is the energy of the primary electron beam. Electron scattering in the target broadens the peak and shifts the maximum to larger angles. Polarization is defined by [20]

$$P(\theta, E_0, E_\nu) = \frac{d\sigma\perp(\theta, E_0, E_\nu) - d\sigma\parallel(\theta, E_0, E_\nu)}{d\sigma\perp(\theta, E_0, E_\nu) + d\sigma\parallel(\theta, E_0, E_\nu)} \tag{11}$$

where an electron of energy E_0 radiates a photon of energy E_ν at angle θ; $d\sigma\perp(\theta, E_0, E_\nu)$ and $d\sigma\parallel(\theta, E_0, E_\nu)$ are the cross sections for generation of the continuous radiation with the electric vector perpendicular $\perp$ and parallel $\parallel$ to the plane defined by the incident electron and the radiated photon, respectively. Polarization is difficult to observe, and only thin, low-yield radiators give evidence for this effect. When the electron is relativistic before and after the radiation, the electrical vector is most probably in the ⊥ direction. Practical thick-target bremsstrahlung shows no polarization effects whatever [6,7,20].

V. EMISSION OF CHARACTERISTIC X-RAYS

The production of characteristic x-rays involves transitions of the orbital electrons of atoms in the target material between allowed orbits, or energy states, associated with ionization of the inner atomic shells. When an electron is ejected from the K shell by electron bombardment or by the absorption of a photon, the atom becomes ionized and the ion is left in a high-energy state. The excess of energy the ion has over the normal state of the atom is equal to the energy (the binding energy) required to remove the K electron to a state of rest outside the atom. If this electron vacancy is filled by an electron coming from an L level, the transition is accompanied by the emission of an x-ray line known as the $K\alpha$ line. This process leaves a vacancy in the L shell. On the other hand, if the atom contains sufficient electrons, the K shell vacancy might be filled by an electron coming from an M level that is accompanied by the emission of the $K\beta$ line. The L or M state ions that remain may also give rise to emission if the electron vacancies are filled by electrons falling from further orbits.

A. Inner Atomic Shell Ionization

As already mentioned, the emission of characteristic x-rays is preceded by ionization of inner atomic shells, which can be accomplished either by charged particles (e.g., electrons, protons, and α particles) or by photons of sufficient energy. The cross section for ionization of an inner atomic shell of element i by electrons is given by [21–23]

$$Q_i = \pi e^4 n_s b_s \frac{\ln U}{U E_{c,i}^2} \tag{12}$$

where $U = E/E_{c,i}$ is the overvoltage, defined as the ratio of the instantaneous energy of the electron at each point of the trajectory to that required to ionize an atom of element i, $E_{c,i}$ is the critical excitation energy, and n_s and b_s are constants for a particular shell:

$$s = K: \quad n_s = 2, b_s = 0.35$$
$$s = L: \quad n_s = 8, b_s = 0.25$$

The cross section for ionization Q_i is a strong function of the overvoltage, which shows a maximum at $U \simeq 3$–4 [24,25].

The probability (or cross section) of ionization of an inner atomic shell by a charged particle is given by [26]

$$\sigma_s = \frac{8\pi r_0^2 q^2 f_s}{Z^4 \eta_s} \tag{13}$$

where r_0 is the classic radius of the electron equal to 2.818×10^{-15} m, q is the particle charge, Z is the atomic number of the target material, f_s is a factor depending on the wave functions of the electrons for a particular shell, and η_s is a function of the energy of the incident particles.

In the case of electromagnetic radiation (x or γ) the ionization of an inner atomic shell is a result of the photoelectric effect. This effect involves the disappearance of a radiation photon and the photoelectric ejection of one electron from the absorbing atom, leaving the atom in an excited level. The kinetic energy of the ejected photoelectron is given by the difference between the photon energy $h\nu$ and the atomic binding energy of the electron E_c (called also ionization energy). Critical absorption wavelengths [27] related to the critical absorption energies [28] via the equation $\lambda(\text{Å}) = 12.4/E(\text{keV})$ are presented in App. I.

For energies far from the absorption edge and in the nonrelativistic range, the cross section τ_K for the ejection of an electron from the K shell is given by [29]

$$\tau_K = \frac{32\sqrt{2}}{3} \pi r_0^2 \frac{Z^5}{(137)^4} \left(\frac{m_0 c^2}{h\nu}\right)^{7/2} \tag{14}$$

Equation (14) is not fully adequate in the neighborhood of an absorption edge; in this case Eq. (14) should be multiplied by a correction factor $f(X)$ [30]:

$$f(X) = 2\pi \left(\frac{D}{h\nu}\right)^{1/2} \frac{e^{-4X \operatorname{arccot} X}}{1 - e^{-2\pi X}} \tag{15}$$

where

$$X = \left(\frac{D}{h\nu - D}\right)^{1/2} \tag{15a}$$

with

$$D \simeq \frac{1}{2}(Z - 0.3)^2 \frac{m_0 c^2}{(137)^2} \tag{15b}$$

When the energy of the incident photon is of the order of $m_0 c^2$ or greater, relativistic cross sections for the photoelectric effect must be used [31].

B. Spectral Series in X-rays

The energy of an emission line can be calculated as the difference between two terms, each term corresponding to a definite state of the atom. If E_1 and E_2 are the term values representing the energies of the corresponding levels, the frequency of an x-ray line is given by the relation

$$\nu = \frac{E_1 - E_2}{h} \tag{16}$$

Using the common notations one can represent the energies of the levels E by means of the atomic number and the quantum numbers $n, l, s,$ and j [32]:

$$\begin{aligned}
\frac{E}{Rh} &= \frac{(Z - S_{n,l})^2}{n^2} + \alpha^2 \frac{(Z - d_{n,l,j})^2}{n^3} \left(\frac{1}{l + \frac{1}{2}} - \frac{3}{4n}\right) \\
&\quad - \alpha^2 \frac{(Z - d_{n,l,j})^4}{n^3} \frac{j(j + 1) - l(l + 1) - s(s + 1)}{2l\,(l + \frac{1}{2})(l + 1)}
\end{aligned} \tag{17}$$

where $S_{n,l}$, $d_{n,l,j}$ are screening constants that must be introduced to correct for the effect of the electrons on the field in the atom, R is the universal Rydberg constant valid for all elements with $Z > 5$ or throughout nearly the whole x-ray region, and α is the fine structure constant given by

$$\alpha = \frac{2\pi e^2}{hc} \tag{17a}$$

The theory of x-ray spectra reveals the existence of a limited number of allowed transitions; the rest are "forbidden." The most intense lines create the electric dipole radiation. The transitions are governed by the selection rules for the change of quantum numbers:

$$\Delta l = \pm 1 \qquad \Delta j = 0 \qquad \text{or} \qquad \pm 1 \tag{18}$$

The j transition $0 \to 0$ is forbidden.

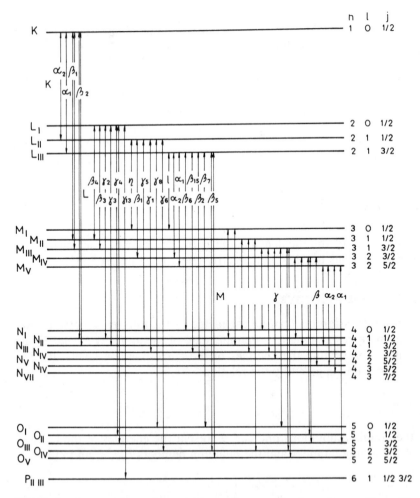

Figure 3 Commonly used terminology of energy levels and x-ray lines. (From Ref. 32. Reprinted by permission of the author and Springer-Verlag.)

According to Dirac's theory of radiation [33], transitions that are forbidden as dipole radiation can appear as multipole radiation, for example as electric quadrupole, and magnetic dipole transitions. The selection rules for the former are

$$\Delta l = 0 \quad \text{or} \quad \pm 2 \quad \Delta j = 0 \quad \pm 1 \quad \text{or} \quad \pm 2 \tag{19}$$

The j transition $0 \to 0$, $\frac{1}{2} \to \frac{1}{2}$, and $0 \rightleftarrows 1$ are forbidden.

The selection rules for magnetic dipole transitions are

$$\Delta l = 0 \quad \Delta j = 0 \quad \text{or} \quad \pm 1 \tag{20}$$

The j transition $0 \to 0$ is forbidden.

The commonly used terminology of energy levels and x-ray lines is shown in Fig. 3.

A general expression relating the wavelength of an x-ray characteristic line with the atomic number of the corresponding element is given by Moseley's law [34]:

$$\frac{1}{\lambda} = k(Z - \sigma)^2 \tag{21}$$

where k is a constant for a particular spectral series and σ is a screening constant for the repulsion correction due to other electrons in the atom. Moseley's law plays an important role in the systematizing of x-ray spectra. Appendix II tabulates the energies and wavelengths of the principal x-ray emission lines for K, L, and M series with their approximate relative intensities, which can be defined either by means of spectral line peak intensities or by the area below their intensity distribution curve. In practice, the relative intensities of spectral lines are not constant because they depend not only on the electron transition probability but on the specimen composition as well.

Considering one x-ray peak of a series, $K\alpha$, the $K\alpha$ fraction of the total K spectrum is defined by the transition probability $p_{K\alpha}$, which is given by [35]

$$p_{K\alpha} = \frac{I(K\alpha_1 + K\alpha_2)}{I(K\alpha_1 + K\alpha_2) + I(K\beta_1 + K\beta_2)} \tag{22}$$

Wernisch [23] recently proposed a useful expression for the calculation of the transition probability $p_{K\alpha}$ for different elements:

$$p_{K\alpha,i} = \begin{cases} 1.052 - 4.39 \times 10^{-4}Z_i^2 & 11 \leq Z_i \leq 19 \\ 0.896 - 6.575 \times 10^{-4}Z_i & 20 \leq Z_i \leq 29 \\ 1.0366 - 6.82 \times 10^{-3}Z_i & \\ + 4.815 \times 10^{-5}Z_i^2 & 30 \leq Z_i \leq 60 \end{cases} \tag{23}$$

For the L level, split into three subshells, several electron transitions exist. The transition probability $p_{L\alpha}$, defined as the fraction of the transitions resulting in $L\alpha_1$ and $L\alpha_2$ radiation from the total of possible transitions into the L_{III} subshell, can be calculated by the expression [23]

$$p_{L\alpha,i} = \begin{cases} 0.944 & 39 \leq Z_i \leq 44 \\ -4.461 \times 10^{-1} + 5.493 \times 10^{-2}Z_i & \\ -7.717 \times 10^{-4}Z_i^2 + 3.525 \times 10^{-6}Z_i^3 & 45 \leq Z_i \leq 82 \end{cases} \tag{24}$$

Radiative transition probabilities for various K and L x-ray lines [36] are presented in detail in App. III.

C. X-ray Satellites

A large number of x-ray lines have been reported that do not fit into the normal energy level diagram [37,38]. Most of the x-ray lines, called satellites or nondiagram lines, are very weak and are of rather little consequence in analytical x-ray spectrometry. By analogy to the satellites in optical spectra, it was supposed that the origin of the nondiagram x-ray lines is double or manyfold ionization of an atom through electron impact. Following the ionization, a multiple electron transition results in emission of a single photon of energy higher than that of the characteristic x-rays. In practice the most important nondiagram x-ray lines occur in the K_α series; they are denoted the $K\alpha_3, \alpha_4$ doublet, and their origin is a double-electron transition. The probability of a multiple-electron transition resulting in the emission of satellite x-ray lines is considerably higher for low-Z elements than for heavy and medium elements. For instance, the intensity of the Al$K\alpha_3$ satellite line is roughly 10% of that of the Al$K\alpha_1, \alpha_2$ characteristic x-rays.

Appendix IV tabulates wavelengths of the K satellite lines.

A new class of satellites that are inside the natural width of the parent lines was recently observed by Kawai and Gohshi [38]. The origin of these satellites, called parasites or hidden satellites, is multiple ionization in nonadjacent shells.

D. Soft X-ray Emission-Band Spectra

In the soft x-ray region, the characteristic emission spectra of solid elements include continuous bands of width varying from 1 to 10 electronvolts (eV); the same element in vapor form produces only the usual sharp spectral lines. The bands occur only when an electron falls from the outermost or valency shell of the atom, the levels of which are broadened into a wide band when the atoms are packed in a crystal lattice. Investigation of the emission band spectra is of great significance in understanding the electronic structure of solid metals, alloys, and complex coordination compounds.

E. Auger Effect

It has already been stated that the excess of energy an atom possesses after removing one electron from an inner shell by whatever means may be emitted as characteristic radiation. Alternatively, however, an excited atom may return to a state of lower energy by ejecting one of its own electrons from a less tightly bound state. The radiationless transition is called the Auger effect, and the ejected electrons are called Auger electrons [39,40]. Generally, the probability of the Auger effect increases with a decrease in the difference of the corresponding energy states, and it is the highest for the low-Z elements.

Since an excited atom already has one electron missing, for example in the K shell, and another electron is ejected in an Auger process, for example from the L shell, the atom is left in a doubly ionized state in which two electrons are missing. This atom may return to its normal state by single- or double-electron jumps with the emission of diagram or satellite lines, respectively. Alternatively, another Auger process may occur in which a third electron is ejected from the M shell.

The Auger effect also occurs after capture of a negative meson by an atom. As the meson changes energy levels in approaching the nucleus, the energy released may be either emitted as a photon or transferred directly to an electron that is emitted as a high-energy Auger electron (in the keV range for hydrogen and MeV range for heavy elements).

Measurements of the energy and intensity of the Auger electrons are applied extensively in surface physics studies (Auger electron spectroscopy).

F. Fluorescence Yield

An important consequence of the Auger effect is that the actual number of x-ray photons produced from an atom is less than expected, since a vacancy in a given shell might be filled through a nonradiative transition. The probability that a vacancy in an atomic shell or subshell is filled through a radiative transition is called the fluorescence yield. The application of this definition to the K shell of an atom is straightforward, and the fluorescence yield of the K shell is

$$\omega_K = \frac{I_K}{n_K} \tag{25}$$

where I_K is the total number of characteristic K x-ray photons emitted from a sample and n_K is the number of primary K shell vacancies.

The definition of the fluorescence yield of higher atomic shells is more complicated, for the following two reasons:

1. Shells above the K shell consist of more than one subshell; the average fluorescence yield depends on how the shells are ionized,
2. Coster-Kronig transitions, which are nonradiative transitions between the subshells of an atomic shell having the same principal quantum number [41,42].

In the case of the absence of Coster-Kronig transitions, the fluorescence yield of the ith subshell of a shell, whose principal quantum number is indicated by X ($X = L, M, \ldots$), is given as

$$\omega_i^X = \frac{I_i^X}{n_i^X} \tag{26}$$

An average or mean fluorescence yield $\overline{\omega_X}$ for the shell X is defined as

$$\overline{\omega_X} = \sum_{i=1}^{k} N_i^X \omega_i^X \tag{27}$$

where N_i^X is the relative number of primary vacancies in the subshell i of shell X:

$$N_i^X = \frac{n_i^X}{\sum\limits_{i=1}^{k} n_i^X} \qquad \sum_{i=1}^{k} N_i^X = 1 \tag{28}$$

The summations in Eqs. (27) and (28) extend over all k subshells of shell X. For the definition of the average fluorescence yield, the primary vacancy distribution must be fixed; that is, Coster-Kronig transitions must be absent. It is noteworthy that $\overline{\omega_X}$ generally is not a fundamental property of the atom but depends both on the atomic subshell fluorescence yields $\omega_i{}^X$ and on the relative number of primary vacancies $N_i{}^X$ characteristic of the method used to ionize the atoms.

In the presence of Coster-Kronig transitions, which modify the primary vacancy distribution by the transfer of ionization from one subshell with a given energy to a subshell with less energy, the average fluorescence yields can be calculated by using two alternative approaches. In the first, the average fluorescence yield $\overline{\omega_X}$ is regarded as a linear combination of the subshell fluorescence yields $\omega_i{}^X$ with a vacancy distribution modified by Coster-Kronig transitions:

$$\overline{\omega_X} = \sum_{i=1}^{k} V_i^X \omega_i^X \qquad \sum_{i=1}^{k} V_i^X > 1 \tag{29}$$

where V_i^X is the relative number of vacancies in the subshell i of shell X, including vacancies shifted to each subshell by Coster-Kronig transitions. The V_i^X values can be expressed in terms of the relative numbers N_i^X of primary vacancies and the Coster-Kronig transition probability for shifting a vacancy from a subshell X_i to a higher subshell X_j denoted as f_{ij}^X [42]

$$\begin{aligned}
V_1^X &= N_1^X \\
V_2^X &= N_2^X + f_{12}^X N_1^X \\
V_3^X &= N_3^X + f_{23}^X N_2^X + (f_{13}^X + f_{12}^X f_{23}^X) N_1^X
\end{aligned} \tag{30}$$

$$\vdots$$

In an alternative approach, the mean fluorescence yield $\overline{\omega_X}$ is a linear combination of the relative numbers of primary vacancies N_i^X:

$$\overline{\omega_X} = \sum_{i=1}^{k} N_i^X \nu_i^X \tag{31}$$

where ν_i^X represents the total number of characteristic x-rays that result per primary vacancy in the X_i subshell. The transformation relations between the coefficients ν_i^X and the subshell fluorescence yields ω_i^X follow from Eqs. (29) through (31) and are given in Refs. 41 and 42.

Between the fluorescence yield ω_i^X, the Auger yield a_i^X, and the Coster-Kronig transition probabilities f_{ij}^X, the following relationship must hold:

$$\omega_i^X + a_i^X + \sum_{i=1}^{k} f_{ij}^X = 1 \tag{32}$$

The mean Auger yield $\overline{a_X}$ is given by

$$\overline{a_X} = \sum_{i=1}^{k} V_i^X a_i^X = 1 \tag{33}$$

The values of the K, L, and M shell fluorescence yields as well as of the Coster-Kronig transition probabilities are given in App. V.

Although in principle the K shell fluorescence yield ω_K can be calculated theoretically, experimental data are applied in practice. The following semiempirical equation, due to Burhop [40], gives values correct to a few percent between $Z = 23$ and $Z = 57$ and less accurate values outside these limits:

$$\left(\frac{\omega_K}{1 - \omega_K}\right)^{1/4} = 6.4 \times 10^{-2} + 3.40 \times 10^{-2}Z - 1.03 \times 10^{-6}Z^3 \qquad (34)$$

The fluorescence yield for the K series can also be calculated from a different equation:

$$\left(\frac{\omega_K}{1 - \omega_K}\right)^{1/4} = -0.217 + 0.03318Z - 1.14 \times 10^{-6}Z^3 \qquad (35)$$

which gives quite good agreement with the experimental values for almost all elements. Other useful expressions for the calculation of the fluorescence yields ω_K ($12 \leq Z \leq 42$) and ω_{LIII} ($38 \leq Z \leq 79$) have recently been found, based on literature and experimental data [43]:

$$\omega_K = 3.3704 \times 10^{-1} - 6.0047 \times 10^{-2}Z + 3.3133 \times 10^{-3}Z^2$$
$$- 3.9251 \times 10^{-5}Z^3 \qquad (36)$$

$$\omega_{LIII} = 4.41 \times 10^{-2} - 4.7559 \times 10^{-3}Z + 1.1494 \times 10^{-4}Z^2$$
$$- 1.8594 \times 10^{-7}Z^3 \qquad (37)$$

It should be emphasized that the K shell fluorescence yield below $Z = 20$ becomes so small that the experimental values sometimes disagree by a factor of 3–4.

For the M lines the fluorescence effects are generally ignored, and therefore the average M shell fluorescence yields are less important; this is fortunate, since they are almost entirely unknown [41,42].

A comparison of the total x-ray yields for bulk samples (including both the probability of ionization and the fluorescence yield) in terms of photons per steradian per incident quantum for electrons, protons, and x-ray photons is shown in Fig. 4.

G. Fine Features of X-ray Emission Spectra (Valence or Chemical Effects)

Since characteristic x-ray emission is a process in which the innermost electrons in the atom are concerned, it is reasonable to suppose that the external, or valence, electrons have little or no effect upon the x-ray emission lines. However, this is not fully true for K lines of low-Z elements and L or M lines of higher Z elements, where the physical state and chemical combination of the elements affect the characteristic x-rays [37]. The changes in fine features of x-ray emission spectra with chemical combination can be classified into three groups: (1) shifting in wavelength, (2) distortion of line shape, and (3) intensity changes. Wavelength shifts to both longer and shorter wavelengths result from energy level changes due to electrical shielding or screening of the electrons when the valence electrons are drawn into a bond. Generally, the so-called last or highest energy member of a given series is most affected by chemical combination; maximum energy shifts

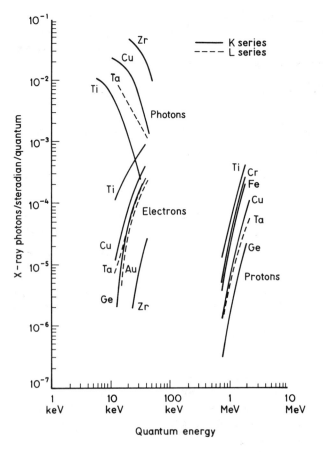

Figure 4 Total x-ray yields for excitation by electrons, protons, and primary x-ray photons as a function of energy of the exciting quantum. (From Ref. 44. Reprinted by permission of John Wiley and Sons, Inc.)

are of the order of a few electronvolts. Distortion of an x-ray emission line shape gives some indication of the energy distribution of the electrons occupying positions in or near the valence shell. The changes in the characteristic x-ray intensity are a result of alteration in excitation probabilities of the electrons undergoing transitions. Certain x-ray lines or bands appear or disappear with chemical combinations. In the case of the K series the most noticeable chemical effects on x-ray emission are seen in spectra from low-Z elements ($4 \leq Z \leq 17$). The L series shows as large or even larger changes with chemical combination of the elements than K series. The valence effects in L spectra have been observed for elements of the first transition series and others nearby in the periodic table.

Since the fine features of x-ray emission spectra may be applied to determine how each element is chemically combined in the sample (speciation), the valence effects found numerous applications in such fields as physics of solids and surface or near surface characterization.

VI. INTERACTION OF PHOTONS WITH MATTER

Interactions of photons with matter, by which individual photons are removed or deflected from a primary beam of x or γ radiation, may be classified according to

- The kind of target, such as electrons, atoms, or nuclei, with which the photon interacts
- The type of event, such as absorption, scattering, or pair production, that takes place

These interactions are thought to be independent of the origin of the photon (nuclear transition for γ-rays versus electronic transition for x-rays); hence we use the term "photon" to refer here to both γ- and x-rays.

Possible interactions are summarized in Table 1 [45], where τ is the total photoelectric absorption cross section per atom ($\tau = \tau_K + \tau_L + \cdots$) and σ_R and σ_C are Rayleigh and Compton collision cross sections, respectively.

The probability of each of these many competing independent processes can be expressed as a collision cross section per atom, per electron, or per nucleus in the absorber. The sum of all these cross sections, normalized to a per atom basis, is then the probability σ_{tot} that the incident photon will have an interaction of some kind while passing through a very thin absorber that contains one atom per cm^2 of area normal to the path of the incident photon:

$$\sigma_{tot} = \tau + \sigma_R + \sigma_C + \cdots \tag{38}$$

The total collision cross section per atom σ_{tot}, when multiplied by the number of atoms per cm^3 of absorber, is then the linear attenuation coefficient μ^* per centimeter of travel in the absorber:

$$\mu^* \frac{1}{cm} = \sigma_{tot} \frac{cm^2}{atom} \cdot \rho \frac{g}{cm^3} \cdot \frac{N_0}{A} \frac{atoms}{g} \tag{39}$$

where ρ is the density of the medium and N_0 is Avogadro's number (6.02252×10^{23} atoms/g atom).

The mass attenuation coefficient $\mu(cm^2/g)$ is the ratio of the linear attenuation coefficient and the density of the material.

It is worth mentioning that the absorption coefficient is a much more restricted concept than the attenuation coefficient. Attenuation includes the purely elastic process in which the photon is merely deflected and does not give up any of its initial energy to the absorber; in this process only a scattering coefficient is involved. In a photoelectric interaction, the entire energy of the incident photon is absorbed by an atom of the medium. In the Compton effect, some energy is absorbed and appears in the medium as kinetic energy of a Compton recoil electron; the balance of the incident energy is not absorbed and is present as a Compton scattered photon. Absorption, then, involves the conversion of incident photon energy into the kinetic energy of a charged particle (usually an electron), and scattering involves the deflection of incident photon energy.

For narrow, parallel, and monochromatic beams, the attenuation of photons in homogeneous matter is described by the exponential law

$$I = I_0 e^{-\mu^* t} \tag{40}$$

Table 1 Classification of Photon Interactions

Type of interaction	Absorption	Scattering		
		Elastic (coherent)	Inelastic (incoherent)	Multiphoton effects
Interaction with atomic electrons	Photoelectric effect[a] τ $\sim Z^4$ low energy $\sim Z^5$ high energy	Rayleigh scattering[a] $\sigma_R \sim Z^2$	Compton scattering[a] $\sigma_C \sim Z$	Two-photon Compton scattering $\sim Z$
Interaction with nucleus or bound nucleons	Nuclear photoelectric effect: reactions (γ, n) (γ, p), photofission $\sim Z$ $(E \geq 10 \text{ MeV})$	Nuclear coherent scattering (γ, γ) $\sim Z^2$	Nuclear Compton scattering (γ, γ') $\sim Z$	
Interaction with electrical field surrounding charged particles	1. Electron-positron pair production in field of nucleus, $\sim Z^2$ $(E \geq 1.02 \text{ MeV})$ 2. Electron-positron pair production in electron field $\sim Z$ $(E \geq 2.04 \text{ MeV})$ 3. Nucleon-antinucleon pair production $(E \geq 3 \text{ GeV})$	Delbrück scattering $\sim Z^4$		
Interaction with mesons	Photomeson production $(E \geq 150 \text{ MeV})$	Coherent resonant scattering (γ, γ)		

[a] Major effects of photon attenuation in matter, which are of great importance in practical x-ray spectrometry.
Source: From Ref. 45. Reprinted by permission of the author.

where I is the transmitted intensity, I_0 is the incident intensity, and t is the absorber thickness in centimeters.

If the absorber is a chemical compound or a mixture, its mass attenuation coefficient can be approximately evaluated from the coefficients μ_i for the constituent elements according to the weighted average

$$\mu = \sum_{i=1}^{n} W_i \mu_i \tag{41}$$

where W_i is the weight fraction of the ith element and n is the total number of the elements in the absorber. The "mixture rule" [Eq. (41)] ignores changes in the atomic wave function resulting from changes in the molecular, chemical, or crystalline environment of an atom. Above 10 keV, errors from this approximation are expected to be less than a few percent (except in the regions just above absorption edges), but at very low energies (10–100 eV) errors of a factor of 2 can occur [46].

For situations more complicated than the narrow-beam geometry, the attenuation is still basically exponential but is modified by two additional factors. The first of these, sometimes called a geometry factor, depends on the source absorber geometry. The other factor, often called the buildup factor, takes into account secondary photons produced in the absorber, mainly as the result of one or more Compton scatters, which finally reach the detector. The determination of the buildup factor, defined as the ratio of the observed effect to the effect produced only by the primary radiation, constitutes a large part of γ-ray transport theory [47].

In subsequent sections only major effects of photon attenuation are discussed in detail.

A. Photoelectric Absorption

In the photoelectric absorption described partially in Sec. V.A, a photon disappears and an electron is ejected from an atom. The K shell electrons, which are the most tightly bound, are the most important for this effect in the energy region considered in x-ray spectrometry. If the photon energy drops below the binding energy of a given shell, however, an electron from that shell cannot be ejected. Hence, a plot of τ versus photon energy exhibits the characteristic "absorption edges."

The mass photoelectric absorption coefficient $\tau N_0/A$ at the incident energy E (keV) can approximately be calculated based on Walter's equations [48]

$$\frac{\tau N_0}{A} = \begin{cases} \dfrac{30.3Z^{3.94}}{AE^3} & \text{for } E > E_K \\[2mm] \dfrac{0.978Z^{4.30}}{AE^3} & \text{for } E_{LI} < E < E_K \\[2mm] \dfrac{0.78Z^{3.94}}{AE^3} & \text{for } E_{MI} < E < E_{LIII} \end{cases} \tag{42}$$

Based on available experimental and theoretical information for approximately 10,000 combinations of Z and E covering 87 elements and the energy range 1 keV to 1 MeV, the following ln-ln polynomials for the photoeffect cross section τ_j have been fitted in incident photon energy between each absorption-edge region [49]:

$$\ln \tau_j = \sum_{i=0}^{1,2,\text{ or }3} A_{ij} \{\ln [E \text{ (keV)}]\}^i \tag{43}$$

In this polynomial, the total photoeffect cross section τ_j represents one of the sums:

$$\tau_1 = \tau_M + \tau_N + \tau_O + \cdots \quad \text{or } \tau_N + \tau_O + \cdots$$

$$E_{MI} < E < E_{LIII} \quad \text{or } E < E_{MV} \tag{44}$$

$$\tau_2 = \tau_L + \tau_M + \tau_N + \tau_O + \cdots \quad E_{LI} < E < E_K$$

$$\tau_3 = \tau_K + \tau_L + \tau_M + \tau_N - \tau_O + \cdots \quad E > E_K$$

The values of the fitted coefficients A_{ij} for the ln-ln representation are given in App. VI [50]. In multiple-edge regions, for example between L_I and L_{III} edge energies, the photoelectric absorption cross sections are also obtained via Eq. (43) by using the following constant "jump ratios" j (τ just above an absorption edge divided by τ just below that absorption edge):

$$j = 1.6 \qquad\qquad\qquad \text{for } E_{LII} < E < E_{LI}$$

$$j = 1.64 = (1.16 \times 1.41) \qquad \text{for } E_{LIII} < E < E_{LII}$$

$$j = 1.1 \qquad\qquad\qquad \text{for } E_{MII} < E < E_{MI} \tag{45}$$

$$j = 1.21 = (1.1 \times 1.1) \qquad \text{for } E_{MIII} < E < E_{MII}$$

$$j = 1.45 = (1.1 \times 1.1 \times 1.2) \qquad \text{for } E_{MIV} < E < E_{MIII}$$

$$j = 2.18 = (1.1 \times 1.1 \times 1.2 \times 1.5) \qquad \text{for } E_{MV} < E < E_{MIV}$$

Simple expressions for calculating the values of the energies of all photoabsorption edges are given in Sec. VI.D.

The experimental ratio of the total photoelectric absorption cross section τ to the K shell component τ_K can be fitted with an accuracy of $\sim$2–3% by the equation [45]

$$\frac{\tau}{\tau_K} \simeq 1 + 0.01481 \ln^2 Z - 0.000788 \ln^3 Z \tag{46}$$

Based on the tables of McMaster et al. [50], Poehn et al. [51] recently found a useful approximation for the calculation of the jump ratios (called also jump factors) for the K shell (j_K) and L_{III} subshell (j_{LIII}):

$$j_K = 1.754 \times 10 - 6.608 \times 10^{-1}Z + 1.427 \times 10^{-2}Z^2$$

$$- 1.1 \times 10^{-4}Z^3 \quad \text{for } 11 \le Z \le 50 \tag{47}$$

$$j_{LIII} = 2.003 \times 10 - 7.732 \times 10^{-1}Z + 1.159 \times 10^{-2}Z^2$$

$$- 5.835 \times 10^{-5}Z^3 \quad \text{for } 30 \le Z \le 83$$

As already mentioned [Eq. (45)], the values of the jump factors at the L_{II} and L_I absorption edges are constant for all elements and equal to 1.41 and 1.16, respectively.

When the apparently sharp x-ray absorption discontinuities are examined at high resolution, they are found to contain a fine structure that extends in some cases to about a few hundred electronvolts above the absorption edge. The fine structure very close to an absorption edge (less than ≈ 30 eV above the edge) is generally referred to as a Kossel structure. Peaks and trenches in this region, which can differ by a factor of 2 or more from the smoothly extrapolated data, can be described in terms of transitions of the (very low energy) ejected electrons to unfilled discrete energy states of the atom (or molecule), rather than to the continuum of states beyond a characteristic energy [32]. Superimposed on the Kossel structure is the so-called Kronig structure (the extended fine structure), which usually extends to about 300 eV above the absorption edge (occasionally to nearly 1 keV above an edge). The Kronig structure can be described in terms of interference effects on the de Broglie waves of the ejected electrons by the molecular or crystalline spatial ordering of neighboring atoms. The oscillations of the absorption coefficient are of the order of 50% in the energy region 50–60 eV above an absorption edge and of the order of 15% in the region beyond 200 eV above the edge.

Modulations of the absorption coefficient in the energy region above an absorption edge can be described theoretically in terms of the electronic parameters [52]. Through a Fourier transform relationship, the modulations are closely related to the radial distribution function around the element of interest [53]. Since both the Kossel and the Kronig fine structures can vary in magnitude and in energy displacement of the features, depending on the molecular, crystalline, or thermal environment of the atom, they can be applied for local structural analysis (extended x-ray absorption fine structure) [54].

B. Compton Scattering

Compton scattering [55,56] is the interaction of a photon with a free electron that is considered to be at rest. The weak binding of electrons to atoms may be neglected provided the momentum transferred to the electron greatly exceeds the momentum of the electron in the bound state. Considering conservation of momentum and energy leads to the following equations:

$$h\nu = \frac{h\nu_0}{1 + \gamma(1 - \cos\theta)} \tag{48}$$

$$T = h\nu_0 - h\nu = h\nu_0 \frac{\gamma(1 - \cos\theta)}{1 + \gamma(1 - \cos\theta)} \tag{49}$$

$$\tan\phi = \frac{1}{1 + \gamma} \cot\frac{\theta}{2} \tag{50}$$

with

$$\gamma = \frac{h\nu_0}{m_0c^2}$$

where $h\nu_0$ and $h\nu$ are the energies of the incident and scattered photon, respectively, θ is the angle between the photon directions of travel before and following a scattering interaction, and T and ϕ are the kinetic energy and scattering angle of the Compton recoil electron, respectively.

For $\phi = 180°$, Eqs. (48) and (49) reduce to

$$(h\nu)_{min} = h\nu_0 \frac{1}{1 + 2\gamma} \tag{51}$$

and

$$T_{max} = h\nu_0 \frac{2\gamma}{1 + 2\gamma} \tag{52}$$

The differential Klein-Nishina collision cross section $d\sigma_{KN}/d\Omega$ (defined as the ratio of the number of photons scattered in a particular direction to the number of incident photons) for unpolarized photons striking unbound, randomly oriented electrons is given by [57]

$$\frac{d\sigma_{KN}}{d\Omega} = \frac{r_0^2}{2} \left(\frac{h\nu}{h\nu_0}\right)^2 \left(\frac{h\nu_0}{h\nu} + \frac{h\nu}{h\nu_0} - \sin^2 \theta\right) \qquad \frac{cm^2}{electron \cdot sr} \tag{53}$$

where sr is an abbreviation for steradian.

Substitution of Eq. (48) for Eq. (53) gives the differential cross section as a function of the scattering angle θ:

$$\frac{d\sigma_{KN}}{d\Omega} = \frac{r_0^2}{2} \frac{1 + \cos^2 \theta}{[1 + \gamma(1 - \cos \theta)]^2}$$
$$\times \left\{1 + \frac{\gamma^2(1 - \cos \theta)^2}{(1 + \cos^2 \theta)[1 + \gamma(1 - \cos \theta)]}\right\} \qquad \frac{cm^2}{electron \cdot sr} \tag{54}$$

For very small energies $h\nu_0 \ll m_0 c^2$, the expression reduces to the classic Thomson scattering cross section for electromagnetic radiation on an electron:

$$\frac{d\sigma_{Th}}{d\Omega} = \frac{r_0^2}{2} (1 + \cos^2 \theta) \qquad \frac{cm^2}{electron \cdot sr} \tag{55}$$

For low energies of incident photons (approximately less than a few tens keV), the angular distribution of Compton scattered photons is symmetrical about $\theta = 90°$; at higher incident photon energies, the Compton scattering becomes predominantly forward.

The differential Klein-Nishina scattering cross section $d\sigma_{KN}^s/d\Omega$ for unpolarized radiation, defined as the ratio of the amount of energy scattered in a particular direction to the energy of incident photons, is given by

$$\frac{d\sigma_{KN}^s}{d\Omega} = \frac{h\nu}{h\nu_0} \frac{d\sigma_{KN}}{d\Omega} \qquad \frac{cm^2}{electron \cdot sr} \tag{56}$$

The average (or total) collision cross section σ_{KN} gives the probability of any Compton interaction by one photon while passing normally through a material containing one electron per cm²:

$$\sigma_{KN} = \int_0^\pi \frac{d\sigma_{KN}}{d\Omega} 2\pi \sin\theta \, d\theta = 2\pi r_0^2 \left\{ \frac{1+\gamma}{\gamma^2} \left[\frac{2(\gamma+1)}{1+2\gamma} - \frac{\ln(1+2\gamma)}{\gamma} \right] \right.$$

$$\left. + \frac{\ln(1+2\gamma)}{2\gamma} - \frac{1+3\gamma}{(1+2\gamma)^2} \right\} \qquad \frac{cm^2}{electron} \qquad (57)$$

Again, at the low-energy limit this cross section reduces to the classic Thomson cross section

$$\sigma_{Th} = \frac{8}{3}\pi r_0^2 = 0.6652 \times 10^{-24} \qquad \frac{cm^2}{electron} \qquad (58)$$

At extremely high energies $h\nu_0 \gg m_0 c^2$, Eq. (57) reduces to

$$\sigma_{KN} = \pi r_0^2 \frac{1}{\gamma} (\ln 2\gamma + \frac{1}{2}) \qquad \frac{cm^2}{electron} \qquad (59)$$

The average (or total) scattering cross section, defined as the total scattered energy in photons of various energies $h\nu$, is given by

$$\sigma_{KN}^s = \int_0^\pi \frac{d\sigma_{KN}^s}{d\Omega} 2\pi \sin\theta \, d\theta = \pi r_0^2 \left[\frac{\ln(1+2\gamma)}{\gamma^3} \right.$$

$$\left. + \frac{2(1+\gamma)(2\gamma^2 - 2\gamma - 1)}{\gamma^2(1+2\gamma)^2} + \frac{8\gamma^2}{3(1+2\gamma)^3} \right] \qquad \frac{cm^2}{electron} \qquad (60)$$

The usual Klein-Nishina theory that assumes the target electron is free and at rest cannot directly be applicable in some cases. Departures from it occur at low energies because of electron binding effects and, at high energies, because of the possibility of emission of an additional photon (double Compton effect) and radiative corrections associated with emission and reabsorption of virtual photons; these corrections are discussed in Ref. 45.

The total incoherent (Compton) collision cross section per atom σ_C, involving the binding corrections by applying the so-called incoherent scattering function $S(x, Z)$, can be calculated according to

$$\sigma_C = \frac{1}{2} r_0^2 \int_{-1}^1 \{[1 + \gamma(1 - \cos\theta)]^{-2} \left[1 + \cos^2\theta \right.$$

$$\left. + \frac{\gamma^2(1-\cos\theta)^2}{1+\gamma(1-\cos\theta)} \right] ZS(x, Z)\}2\pi d(\cos\theta) \qquad \frac{cm^2}{atom} \qquad (61)$$

where $x = \sin(\theta/2)/\lambda$ is the momentum transfer parameter and λ is the photon wavelength in angstroms.

The values of the incoherent scattering function $S(x, Z)$ and the incoherent collision cross section σ_C are given in Ref. 58.

The incoherent collision cross sections σ_C can also be calculated by using ln-ln polynomials already defined by Eq. (43) (by simply substituting τ_j with σ_C and taking $i = 3$). The values of the fitted coefficients for the ln-ln representation for σ_C valid in the photon energy range 1 keV to 1 MeV are given in App. VII.

To complete this section, it is worth mentioning the Compton effect for polarized radiation. The differential collision cross section $(d\sigma_{KN}/d\Omega)_{pp}$ for the plane-polarized radiation scattered by unoriented electrons has also been derived by Klein and Nishina. It represents the probability that a photon, passing through a target containing one electron per cm^2, will be scattered at an angle θ into a solid angle $d\Omega$ in a plane making an angle β with respect to the plane containing the electrical vector of the incident wave:

$$\left(\frac{d\sigma_{KN}}{d\Omega}\right)_{pp} = \frac{r_0^2}{2}\left(\frac{h\nu}{h\nu_0}\right)^2 \left(\frac{h\nu_0}{h\nu} + \frac{h\nu}{h\nu_0} - 2\sin^2\theta\cos^2\beta\right) \quad \frac{cm^2}{electron\cdot sr} \quad (62)$$

The cross section has its maximum value for $\beta = 90°$, indicating that the photon and electron tend to be scattered at right angles to the electrical vector of the incident radiation.

The scattering of circularly polarized (cp) photons by electrons with spins aligned in the direction of the incident photon is described by

$$\left(\frac{d\sigma_{KN}}{d\Omega}\right)_{cp} = r_0^2 \left(\frac{h\nu}{h\nu_0}\right)^2 \left[\left(\frac{h\nu_0}{h\nu} + \frac{h\nu}{h\nu_0} - \sin^2\theta\right)\right.$$
$$\left. \pm \left(\frac{h\nu_0}{h\nu} - \frac{h\nu}{h\nu_0}\right)\cos\theta\right] \quad \frac{cm^2}{electron\cdot sr} \quad (63)$$

The first term is the usual Klein-Nishina formula for unpolarized radiation. The + sign for the additional term applies to right circularly polarized photons.

C. Rayleigh Scattering

Rayleigh scattering is a process by which photons are scattered by bound atomic electrons and in which the atom is neither ionized nor excited. The incident photons are scattered with unchanged frequency and with a definite phase relation between the incoming and scattered waves. The intensity of the radiation scattered by an atom is determined by summing the amplitudes of the radiation coherently scattered by each of the electrons bound in the atom. It should be emphasized that, in Rayleigh scattering, the coherence extends only over the Z electrons of individual atoms. The interference is always constructive, provided the phase change over the diameter of the atom is less than one-half a wavelength; that is, whenever

$$\frac{4\pi}{\lambda} r_a \sin\frac{\theta}{2} < 1 \quad (64)$$

where r_a is the effective radius of the atom.

Rayleigh scattering occurs mostly at the low energies and for high-Z materials, in the same region where electron binding effects influence the Compton scattering cross section.

The differential Rayleigh scattering cross section for unpolarized photons is given by [59]

$$\frac{d\sigma_R}{d\Omega} = \frac{1}{2} r_0^2 (1 + \cos^2\theta) \cdot |F(x, Z)|^2 \quad \frac{cm^2}{atom\cdot sr} \quad (65)$$

where $F(x, Z)$ is the "atomic form factor,"

$$F(x, Z) = \int_0^\infty \rho(r)4\pi r \frac{\sin[(2\pi/\lambda)rs]}{(2\pi/\lambda)rs} dr \tag{66}$$

where $\rho(r)$ is the total electron density, r the distance from the nucleus, and $s = 2\sin(\theta/2)$. The atomic form factor has been calculated for $Z < 26$ using the Hartree electronic distribution [59] and for $Z > 26$ using the Fermi-Thomas distribution [48].

At high photon energies, Rayleigh scattering is confined to small angles; at low energies, particularly for high-Z materials, the angular distribution of the Rayleigh-scattered radiation is much broader. A useful simple criterion for judging the angular spread of Rayleigh scattering is given by [60]

$$\theta_R = 2\arcsin \frac{0.0133Z^{1/3}}{E} \frac{}{\text{MeV}} \tag{67}$$

where θ_R is the opening half-angle of a cone containing at least 75% of the Rayleigh-scattered photons. In the forward direction, $|F(x, Z)|^2 = Z^2$, so that Rayleigh scattering becomes appreciable in magnitude and must be accounted for in any γ- or x-ray scattering experiments.

The total coherent (Rayleigh) scattering cross section per atom σ_R can be calculated from

$$\begin{aligned}
\sigma_R &= \frac{1}{2} r_0^2 \int_{-1}^1 (1 + \cos^2\theta) |F(x, Z)|^2 2\pi d(\cos\theta) \\
&= \frac{3}{8} \sigma_{\text{Th}} \int_{-1}^1 (1 + \cos^2\theta) |F(x, Z)|^2 d(\cos\theta) \quad \frac{\text{cm}^2}{\text{atom}}
\end{aligned} \tag{68}$$

The values of the atomic form factor $F(x, Z)$ and the coherent scattering cross section σ_R are given in Ref. 58.

The simplest method for calculating the coherent scattering cross section σ_R consists in applying the ln-ln representation (see Eq. (43) with σ_R instead of τ_j and $i = 3$). The values of the fitted coefficients for ln-ln polynomials for calculating σ_R in the photon energy range 1 keV to 1 MeV are given in App. VIII.

D. Total Mass Attenuation Coefficient

An extensive review of current tabulations of x-ray attenuation coefficients has been given by Hubbell [61]. Differences between various compilations of total mass attenuation coefficients result from uncertainties in our knowledge of partial cross sections for the interaction of photons with matter as a function of elemental atomic number Z and photon energy E. Present discrepancies are disturbing, to say the least, frequently amounting to 5–10% in the photon energy region below 10 keV and rising to as much as 30% near an absorption edge.

Hubbell [62] recently tabulated mass attenuation coefficients and mass energy absorption coefficients for photon energies from 1 keV to 20 MeV for 40 elements ranging from hydrogen ($Z = 1$) to uranium ($Z = 92$) and for 45 mixtures and compounds of dosimetric interest.

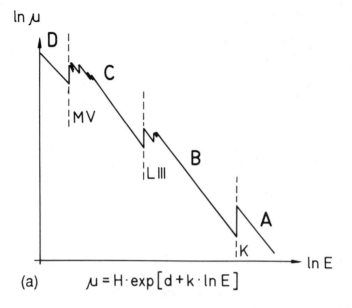

(a) $\mu = H \cdot \exp[d + k \cdot \ln E]$

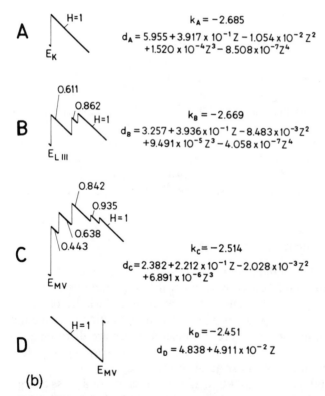

A $k_A = -2.685$
$d_A = 5.955 + 3.917 \times 10^{-1} Z - 1.054 \times 10^{-2} Z^2$
$+ 1.520 \times 10^{-4} Z^3 - 8.508 \times 10^{-7} Z^4$

B $k_B = -2.669$
$d_B = 3.257 + 3.936 \times 10^{-1} Z - 8.483 \times 10^{-3} Z^2$
$+ 9.491 \times 10^{-5} Z^3 - 4.058 \times 10^{-7} Z^4$

C $k_C = -2.514$
$d_C = 2.382 + 2.212 \times 10^{-1} Z - 2.028 \times 10^{-3} Z^2$
$+ 6.891 \times 10^{-6} Z^3$

D $k_D = -2.451$
$d_D = 4.838 + 4.911 \times 10^{-2} Z$

(b)

Figure 5 Definition of the energy ranges (a) and values of the parameters *H, d,* and *k* (b) applied for the calculation of the total mass attenuation coefficients according to Eq. (69). (From Ref. 63. Reprinted by permission of John Wiley and Sons, Ltd.)

The uncertainty ranges for the total mass attenuation coefficient values in the tabulation of McMaster et al. [50] have been estimated by Hubbell et al. [49]. These ranges of uncertainties fall into four categories. Category I (uncertainty below 2%) applies over the energy region 6–40 keV (except near absorption edges) for the following elements: C, Mg, Al, Ti, Fe, Ni, Cu, Zn, Zr, Mo, Pd, Ag, Cd, Sn, La, Gd, Ta, W, Pt, Au, Pb, Th, and U. In this category the photon energy region above 100 keV is also included for all elements in which incoherent scattering comprises more than 90% of the total cross section. Category II (uncertainty of 2–5%) applies to the energy region 2–6 keV for all elements, 6–40 keV for elements not specified in category I, and above 40 keV except for the scattering-dominated region specified in category I. In category III (uncertainty of 5–15%), the authors [49] include (1) the elements hydrogen, helium, and lithium, (2) the energy region 1–2 keV for all elements, and (3) the regions containing *K, L, M,* and *N* absorption edges, and the fine structure regions extending 200 eV to 1 keV above each of these regions. The experimental uncertainties in some cases 1 and 2 greatly exceed 15%. Category IV (uncertainty above 15%) applies to the photon energy region about 200 eV above an absorption edge (Kossel and Kronig fine structure regions) for all elements.

Based on the tables published by McMaster et al. [50], Wernisch et al. [63] recently developed an algorithm for the calculation of the total mass attenuation coefficient valid for the photon energy range from 1 to 50 keV and for 73 elements ($11 \leq Z \leq 83$). The authors have applied the simple expression

$$\mu \, \frac{cm^2}{g} = H e^{d + k \ln E} \tag{69}$$

Values of H, d, and k have been obtained from least-squares fits applied to the data published by McMaster et al. [50] and are given in Fig. 5.

The values of the edge energies E_K, E_{LI}, E_{LII}, E_{LIII}, E_{MI}, E_{MII}, E_{MIII}, E_{MIV}, and E_{MV} can simply be calculated from [63]

$$E_i = r_i + s_i Z + t_i Z^2 + n_i Z^3 \qquad keV \tag{70}$$

Table 2 Values of the Parameters r_i, s_i, t_i, and n_i Applied for Calculating the Energies of Absorption Edges via Eq. (70)

i	r_i	s_i	t_i	n_i	%[a]	Z_{min}	Z_{max}
K	-1.304×10^{-1}	-2.633×10^{-3}	9.718×10^{-3}	4.144×10^{-5}	3.5	11	63
L_I	-4.506×10^{-1}	1.566×10^{-2}	7.599×10^{-4}	1.792×10^{-5}	2.2	28	83
L_{II}	-6.018×10^{-1}	1.964×10^{-2}	5.935×10^{-4}	1.843×10^{-5}	2.3	30	83
L_{III}	3.390×10^{-1}	-4.931×10^{-2}	2.336×10^{-3}	1.836×10^{-6}	1.9	30	83
M_I	-8.645	3.977×10^{-1}	-5.963×10^{-3}	3.624×10^{-5}	0.4	52	83
M_{II}	-7.499	3.459×10^{-1}	-5.250×10^{-5}	3.263×10^{-5}	0.4	55	83
M_{III}	-6.280	2.831×10^{-1}	-4.117×10^{-3}	2.505×10^{-5}	0.4	55	83
M_{IV}	-4.778	2.184×10^{-1}	-3.303×10^{-3}	2.115×10^{-5}	0.4	60	83
M_V	-2.421	1.172×10^{-1}	-1.845×10^{-3}	1.397×10^{-5}	0.4	61	83

[a] Standard deviation of calculated energies [Eq. (70)] relative to the energies from the tables of McMaster et al. [50].
Source: From Ref. 63. Reprinted by permission of John Wiley and Sons, Ltd.

where the parameters r_i, s_i, t_i, and n_i for various absorption edges i are given in Table 2.

Most of the available tabulations of x-ray attenuation coefficients do not include the photon energy region below 1 keV that corresponds to the energies of characteristic K x-rays of light elements ($Z < 11$). Experimental data in this energy region are incomplete, and it should not be assumed that the accuracy of the available tabulated values is better than ~15% (Ref. 64 and App. IX).

E. Diffraction, Refraction, and Dispersion

When a beam of monochromatic x-rays falls onto a crystal lattice, a regular periodic arrangement of atoms, a diffracted beam only results in definite directions. The phenomenon of x-ray diffraction at an ordered array of atoms (or molecules) can also be interpreted as a reflection of an incident x-ray beam by the interior planes of a crystal (Bragg reflection). By elementary calculation of the difference in path between two coherent rays, W. L. Bragg found the reinforcement condition for reflection (known as Bragg's equation or law):

$$n\lambda = 2d \sin \frac{\theta_n}{2} \tag{71}$$

where n is the order of reflection, d is the interplanar spacing, and $\theta_n/2$ is the angle of reflection (or Bragg angle) defined as the angle between the reflecting plane of the crystal and the incident or reflected beam. The first-order reflection ($n = 1$) is normally strongest, and the reflected intensity decreases as n increases.

Bragg's law as given in Eq. (71) is only a first approximation, as the refraction in the crystal interferes with the angle of reflection. Since the refractive index of x-rays is slightly less than unity, the deviations from Bragg's law, Eq. (71), were not observed in the early years until methods were found for precise measurements of x-ray wavelengths. The refraction is accounted for by ascribing a slightly different value d_n of the lattice constant to each order of reflection; the simple Bragg's law [Eq. (71)] can thus be written as [32]

$$n\lambda = 2d\left(1 - \frac{4d^2}{n^2} \frac{\delta}{\lambda^2}\right) \sin \frac{\theta_n}{2} \tag{72}$$

where $\delta = 1 - n'$ for small photon absorption ($\lambda < 1$ Å), and n' is the refractive index for x-rays. δ is a small positive number of the order 10^{-5} for heavy elements and 10^{-6} for light elements at $\lambda = 1$ Å and is proportional to λ^2.

Since the value of δ is positive, total reflection occurs back into air when an x-ray beam meets a surface at a large enough angle of incidence. Provided no absorption occurs, the critical glancing angle θ_{tr} (tr, total reflection) is defined by [32]

$$\sin \theta_{tr} = \sqrt{2\delta} \tag{73}$$

For $\lambda = 1$ Å, the value of θ_{tr} is of the order 10^{-3} for light elements and 5×10^{-3} for heavy elements, increasing in proportion to λ.

The refractive index n' for a medium containing one type of atom can be calculated from [65]

$$n' = 1 - \frac{N\lambda^2}{2\pi} \frac{e^2}{m_0 c^2} F(0) \tag{74}$$

where N is the number of atoms per unit volume and $F(0)$ is the atomic scattering factor at zero scattering angle. Equation (74) shows that n' depends on the wavelength; this phenomenon is called dispersion.

The anomalous dispersion causes the quantity δ/λ^2, Eq. (72), to vary slightly with the wavelength. The variations become important only in the neighborhood of the absorption edges of the constituents of the crystal. Dividing the quantity δ/λ^2 into one normal part $(\delta/\lambda^2)_n$ and one anomalous part $(\delta/\lambda^2)_a$, the theory of anomalous dispersion leads to an expression of Bragg's law that can be written as [32]

$$n\lambda = 2d \left[1 - \frac{4d^2}{n^2} \left(\frac{\delta}{\lambda^2} \right)_n - \frac{4d^2}{n^2} \left(\frac{\delta}{\lambda^2} \right)_a \right] \sin \frac{\theta_n}{2} \tag{75}$$

Combined with Bragg's law in its uncorrected form, Eq. (71), this expression turns into [32]

$$\lambda = 2d \left[1 - \frac{4d^2}{n^2} \left(\frac{\delta}{\lambda^2} \right)_n \right] \frac{\sin \theta_n/2}{n} - \frac{4d^2}{n^2} \left(\frac{\delta}{\lambda^2} \right)_a \lambda \tag{76}$$

The values of $(\delta/\lambda^2)_a$ giving a correction for anomalous dispersion can be determined experimentally.

The theory of anomalous dispersion has been applied by Sparks [66,67] to explain the inelastic angular-independent scattering from elements having an absorption-edge energy just above the energy of the incident x-rays. The observed intensity of the inelastically scattered radiation was found to be dependent on the nearness of the energy of an absorption edge to the energy of the incident x-rays. The energy of the inelastic peaks is shifted from the incident energy by the binding energy of the most tightly bound shell from which electrons could be photoejected by the incident radiation.

F. X-ray Raman Scattering

Immediately after the discovery of Raman scattering in the visible wavelength region, a similar effect concerning x-ray radiation was experimentally examined [68,69]. X-ray Raman scattering appears as a band spectrum having a short-wavelength edge corresponding to a definite energy loss equal to the K electron binding energy E_K of the element. This inelastic effect was observed, for example, when $CrK\alpha$ and $CuK\alpha$ radiation was scattered by solids of light elements, such as lithium, beryllium, boron, and graphite [70–72]. The shape of the Raman band is similar to that of the soft x-ray K absorption spectrum of the solids.

Mizuno and Ohmura [73] have found the following two conditions for x-ray Raman scattering:

$$\frac{4\pi\bar{a}}{\lambda_0} < 1 \tag{77}$$

and

$$h\nu_0 \gg E_K \tag{78}$$

where $\bar{a}$ is the mean radius of charge distribution of the K electrons and λ_0 is the wavelength of the incident x-rays. The intensity of the x-ray Raman scattering $I(\theta, \lambda)$ is given by [72]

$$I(\theta, \lambda) \simeq (1 + \cos^2 \theta) \left[T_1(\lambda) \left(\frac{4\pi\bar{a}}{\lambda_0} \sin \frac{\theta}{2} \right)^2 + T_2(\lambda) \left(\frac{4\pi\bar{a}}{\lambda_0} \sin \frac{\theta}{2} \right)^2 \right] \qquad (79)$$

where $T_1(\lambda)$ and $T_2(\lambda)$ factors are related to the dipole and multipole transitions, respectively. Although according to this equation a slight displacement of the peak position of the Raman band is expected with scattering angle, the peak position does not shift in most experiments.

In general, x-ray Raman scattering gives information about the unoccupied states above the Fermi level of the relevant solids. Moreover, this kind of inelastic scattering by electrons in solids should sometimes be taken into account in x-ray spectrum evaluation.

VII. INTENSITY OF CHARACTERISTIC X-RAYS

This section provides some of the necessary background information for subsequent chapters dealing with quantitative x-ray analysis. Derivation of any relationship between excitation source intensity and measured characteristic x-rays is sometimes complex and is presented in detail in many relevant books, such as those by Jenkins et al. [74] and Tertian and Claisse [75] on x-ray fluorescence analysis and by Goldstein et al. [25] and Heinrich [24] on electron probe x-ray microanalysis.

A. Photon Excitation

When continuous (polychromatic) radiation is used to excite the characteristic x-rays of element i in a completely homogeneous sample of thickness T (cm) and when enhancement effects are neglected, the intensity of the fluorescent radiation $I_i(E_i)$ is described by

$$I_i(E_i) \, d\Omega_1 \, d\Omega_2 = \frac{d\Omega_1 \, d\Omega_2}{4\pi} \frac{\epsilon(E_i)}{\sin \Psi_1} \int_{E_{c,i}}^{E_{max}}$$

$$\times a_i(E_0) \frac{1 - \exp[-\rho T(\mu(E_0) \csc \Psi_1 + \mu(E_i) \csc \Psi_2)]}{\mu(E_0) \csc \Psi_1 + \mu(E_i) \csc \Psi_2} I_0(E_0) \, dE_0 \qquad (80)$$

with

$$a_i(E_0) = W_i \tau_i'(E_0) \omega_i p_i \left(1 - \frac{1}{j_i} \right) \qquad (81)$$

where $d\Omega_1$ and $d\Omega_2$ are the differential solid angles for the incident (primary) and emerging (characteristic) radiation, respectively; $\epsilon(E_i)$ is the intrinsic detector efficiency for recording a photon of energy E_i; $E_{c,i}$ and E_{max} are the critical absorption energy of element i and the maximum energy in the excitation spectrum; ρ is the density of the specimen in g cm^{-3}; Ψ_1 and Ψ_2 are the effective incidence and takeoff angles, respectively; $\mu(E_0)$ and $\mu(E_i)$ are the total mass attenuation coefficients in cm^2/g for the whole specimen [Eq. (41)] at energies E_0 and E_i, respectively; $I_0(E_0)\,dE_0$ is the number of incident photons per second per steradian in the energy interval E_0 to $E_0 + dE_0$; W_i is the weight fraction of the ith element; and $\tau_i'(E_0)$ is the total photoelectric mass absorption coefficient for the ith element at the energy E_0 in cm^2/g.

Since in practical spectrometers large solid angles Ω_1 and Ω_2 are used for the excitation and characteristic x-rays, Eq. (80) should also be integrated over these finite solid angles. Such calculations can often be omitted, however, and for a given measurement geometry, an experimentally determined geometry factor G can be applied.

As seen from Eq. (80), the intensity of characteristic x-rays is modified by the effects of primary ($\mu(E_0)$) and secondary ($\mu(E_i)$) absorption in the specimen; this is a major source of the so-called matrix effects in XRF analysis.

If the excitation source is monochromatic (emits only one energy), Eq. (80) simplifies to

$$I_i(E_i) = G\,\frac{\epsilon(E_i)a_i(E_0)I_0(E_0)]}{\sin \Psi_1}\,\frac{1 - \exp[-\rho T(\mu(E_0)\csc \Psi_1 + \mu(E_i)\csc \Psi_2)]}{\mu(E_0)\csc \Psi_1 + \mu(E_i)\csc \Psi_2}$$

(82)

The enhancement effect, consisting of an extra excitation of the element of interest by the characteristic radiation of some matrix elements, modifies the equations for the intensity $I_i(E_i)$. In the case of monochromatic photon excitation, a factor $1 + H_i$ should be included in Eq. (82), where H_i is the enhancement term defined as

$$H_i = \frac{1}{2\mu_i(E_0)}\sum_{k=1}^{m} W_k\omega_k\left(1 - \frac{1}{j_k}\right)\mu_i(E_k)\mu_k(E_0)$$

$$\times \left[\frac{\ln(1 + \mu(E_0)/[\mu(E_k)\sin \Psi_1])}{\mu(E_0)/\sin \Psi_1} + \frac{\ln(1 + \mu(E_i)/[\mu(E_k)\sin \Psi_2])}{\mu(E_i)/\sin \Psi_2}\right]$$

(83)

where $\mu_i(E_0)$ and $\mu_i(E_k)$ are the total mass attenuation coefficients for the ith element at the energy of incident radiation (E_0) and characteristic radiation of the element k (E_k), respectively, $\mu_k(E_0)$ is the total mass attenuation coefficient for the element k at the energy E_0, and $\mu(E_k)$ is the total mass attenuation coefficient for the whole specimen at the energy E_k.

1. Thin-Sample Technique

If the total mass per unit area of a given sample $m = \rho T$ is small, Eq. (82) simplifies to

$$I_i^{thin}(E_i) = \frac{G}{\sin \Psi_1}\,\epsilon(E_i)a_i(E_0)I_0(E_0)m$$

(84)

The relative error resulting from applying Eq. (84) instead of Eq. (82) does not exceed 5% when the total mass per unit area satisfies the condition

$$m_{thin} \leq \frac{0.1}{\mu(E_0)\csc \Psi_1 + \mu(E_i)\csc \Psi_2}$$

(85)

A major feature of the thin-sample technique is that the intensity of the characteristic x-rays, I_i^{thin}, depends linearly on the concentration of the ith element (or
on its mass per unit area); it is equivalent to the fact that, in the thin-sample
technique, matrix effects can safely be neglected.

2. Thick-Sample Technique

By the term "thick sample" we mean a sample whose mass per unit area (thickness) is greater than the so-called saturation mass (thickness). The saturation
thickness is defined as a limiting value above which practically no further increase
in the intensity of the characteristic radiation is observed as the sample thickness
is increased.

  If the total mass per unit area of a given sample is sufficiently large, Eq. (82)
simplifies to

$$I_i^{thick}(E_i) = \frac{G\epsilon(E_i)a_i(E_0)I_0(E_0)}{\mu(E_0) + \sin\Psi_1/\sin\Psi_2\,\mu(E_i)} \tag{86}$$

The relative error resulting from applying Eq. (86) instead of Eq. (82) does not
exceed 1% when the total mass per unit area satisfies the condition

$$m_{thick} \geq \frac{4.61}{\mu(E_0)\csc\Psi_1 + \mu(E_i)\csc\Psi_2} \tag{87}$$

B. Electron Excitation

For a thin foil (defined as a foil of thickness such that the beam electron undergoes
only one scattering act), the intensity of the characteristic x-rays I_i^{tf} (in photons
per second) is given by

$$I_i^{tf} = K_iI_0Q_i\omega_iN_0\frac{1}{A_i}\rho TW_i \tag{88}$$

where K_i is a factor depending on the measurement geometry and detection efficiency and the other symbols have the following dimensions: I_0 = electrons/s,
Q_i = ionizations per electron for 1 atom/cm^2, ω_i = photons/ionization, N_0 =
atoms/mol, $1/A_i$ = mol/g, ρ = g/cm^3, and T = cm.

  For a bulk target, Eq. (88) must be integrated over the electron path, taking
into account the loss of energy by the electron beam:

$$I_i^{bulk} = I_0\omega_iK_iN_0\frac{W_i}{A_i}\int_{E_0}^{E_{c,i}}\frac{Q_i(E)}{dE/d(\rho x)}\,dE \tag{89}$$

where $dE/d(\rho x)$ is the electron energy loss per unit of distance travelled in a
material, given by the Bethe equation [24,76]:

$$\frac{dE}{d(\rho x)} = -78,500\frac{1}{E}\sum_{j=1}^{n}\frac{W_jZ_j}{A_j}\ln 1.166\frac{E}{J_j} \qquad keV/g\,cm^{-2} \tag{90}$$

where J_j is the mean ionization potential of element j (J_j = $9.76Z_j$ +
$58.5Z_j^{-0.19}$ eV) [25]. Strictly speaking, the Bethe equation is only valid for the
electron energy $E > 6.3J_j$. Below this limit, the energy loss of the electrons should

be described either by the modification of Rao-Sahib and Wittry [77], Eq. (91), or by the modification of Love et al. [78], Eq. (92):

$$\frac{dE}{d(\rho x)} = -0.6236 \times 10^5 \frac{1}{\sqrt{E}} \sum_{j=1}^{n} \frac{W_j Z_j}{A_j \sqrt{J_j}} \qquad \text{keV/g cm}^{-2} \tag{91}$$

$$\frac{dE}{d(\rho x)} = -\sum_{j=1}^{n} \frac{W_j Z_j}{A_j J_j} \frac{1}{1.18 \times 10^{-5}(E/J_j)^{1/2} + 1.47 \times 10^{-6} E/J_j} \qquad \text{keV/g cm}^{-2} \tag{92}$$

However, to describe the intensity of characteristic x-rays emitted, three effects should additionally be considered [24,25,79]:

1. The absorption of characteristic x-rays within the specimen
2. Electron backscattering
3. Secondary fluorescence by characteristic x-rays and/or bremsstrahlung continuum produced by an electron beam

C. Particle Excitation

For a thin, uniform, homogeneous target, the intensity of characteristic x-rays I_i^{tt} is given by the simple formula

$$I_i^{tt} = K_i I_0 N_0 \sigma_{s,i}(E_0) \omega_i \frac{1}{A_i} m_i \tag{93}$$

where m_i is the areal density of the element with atomic number Z_i and atomic mass A_i.

For a thick, homogeneous target the intensity of characteristic x-rays, I_i^{thick}, from the element i of concentration W_i can be calculated by [80]

$$I_i^{thick} = \frac{K_i \omega_i N_0}{A_i} I_0 W_i \sum_{E_0}^{0} \frac{\sigma_{s,i}(E) T_i(E)}{S(E)} dE \tag{94}$$

where $S(E)$ is the stopping power and $T_i(E)$ is the photon attenuation factor. The latter is given by

$$T_i(E) = \exp\left(-\mu(E_i) \frac{\sin \Psi_1}{\sin \Psi_2} \int_{E_0}^{E} \frac{dE}{S(E)}\right) \tag{95}$$

More details on various x-ray analytical techniques are provided in subsequent chapters.

VIII. IUPAC NOTATION FOR X-RAY SPECTROSCOPY

The nomenclature currently used in x-ray spectroscopy to describe x-ray emission spectra was introduced by M. Siegbahn in the 1920s and is based upon the relative intensity of lines from different series. Recently, a new and more systematic notation for x-ray emission lines and absorption edges, based upon the energy level designation, was developed by the International Union of Pure and Applied Chemistry [83]. Since the new notation, called the IUPAC notation, replaces (or will replace) the existing Siegbahn notation, some characteristic features of the new nomenclature must be mentioned. In general, the x-ray level notation follows earlier conventions, except for a minor difference. The IUPAC notation prescribes Arabic numerals for subscripts; the original notation uses Roman numerals (e.g., L_2 and L_3, instead of L_{II} and L_{III}). In the IUPAC notation, states with double or multiple vacancies should be denoted by, for example, K^2, KL_1, and $L_{2,3}^n$, which correspond to the electron configurations $1s^{-2}$, $1s^{-1}2s^{-1}$, and $2p^{-n}$, respectively. X-ray transitions and x-ray emission diagram lines are denoted by the initial (placed first) and final x-ray levels separated by a hyphen. To conform with the IUPAC notation of x-ray spectra, the hyphen separating the initial and final state levels should also be introduced into the notation for Auger electron emission process. The IUPAC notation is compared with the Siegbahn notation in App. X.

APPENDIX I CRITICAL ABSORPTION WAVELENGTHS AND CRITICAL ABSORPTION ENERGIES

Table 1

Atomic number	Element	K edge Å	K edge keV	L_I edge Å	L_I edge keV	L_{II} edge Å	L_{II} edge keV	L_{III} edge Å	L_{III} edge keV	M_{IV} edge Å	M_{IV} edge keV	M_V edge Å	M_V edge keV
1	H	918	0.014										
2	He	504	0.025										
3	Li	226.953	0.055										
4	Be	106.9	0.116										
5	B	64.6	0.192										
6	C	43.767	0.283										
7	N	31.052	0.399										
8	O	23.367	0.531										
9	F	18.05	0.687										
10	Ne	14.19	0.874	258	0.048	564	0.022	564	0.022				
11	Na	11.48	1.08	225	0.055	365	0.034	365	0.034				
12	Mg	9.512	1.303	197	0.063	248	0.050	253	0.049				
13	Al	7.951	1.559	143	0.087	170	0.073	172	0.072				
14	Si	6.745	1.837	105	0.118	125	0.099	127	0.098				
15	P	5.787	2.142	81.0	0.153	96.1	0.129	96.9	0.128				
16	S	5.018	2.470	64.2	0.193	75.6	0.164	76.1	0.163				
17	Cl	4.397	2.819	52.1	0.238	61.1	0.203	61.4	0.202				
18	Ar	3.871	3.202	43.2	0.287	50.2	0.247	50.6	0.245				
19	K	3.437	3.606	36.4	0.341	41.8	0.297	42.2	0.294				
20	Ca	3.070	4.037	30.7	0.399	35.2	0.352	35.5	0.349				
21	Sc	2.757	4.495	26.8	0.462	30.2	0.411	30.8	0.402				
22	Ti	2.497	4.963	23.4	0.530	27.0	0.460	27.3	0.454				
23	V	2.269	5.462	20.5	0.604	23.9	0.519	24.2	0.512				
24	Cr	2.070	5.987	18.3	0.679	21.3	0.583	21.6	0.574				
25	Mn	1.896	6.535	16.3	0.762	19.1	0.650	19.4	0.639				
26	Fe	1.743	7.109	14.6	0.849	17.2	0.721	17.5	0.708				
27	Co	1.608	7.707	13.3	0.929	15.6	0.794	15.9	0.779				
28	Ni	1.488	8.329	12.22	1.015	14.2	0.871	14.5	0.853				
29	Cu	1.380	8.978	11.27	1.100	13.0	0.953	13.3	0.933				
30	Zn	1.283	9.657	10.33	1.200	11.87	1.045	12.13	1.022				
31	Ga	1.196	10.365	9.54	1.30	10.93	1.134	11.10	1.117				
32	Ge	1.117	11.100	8.73	1.42	9.94	1.248	10.19	1.217				
33	As	1.045	11.860	8.107	1.529	9.124	1.358	9.39	1.32				
34	Se	0.980	12.649	7.506	1.651	8.416	1.473	8.67	1.43				
35	Br	0.920	13.471	6.97	1.78	7.80	1.59	8.00	1.55				
36	Kr	0.866	14.319	6.46	1.92	7.21	1.72	7.43	1.67				
37	Rb	0.816	15.197	5.998	2.066	6.643	1.865	6.89	1.80				
38	Sr	0.770	16.101	5.583	2.220	6.172	2.008	6.387	1.940				
39	Y	0.728	17.032	5.232	2.369	5.755	2.153	5.962	2.079				
40	Zr	0.689	17.993	4.867	2.546	5.378	2,304	5.583	2.220				
41	Nb	0.653	18.981	4.581	2.705	5.026	2.467	5.223	2.373				
42	Mo	0.620	<u>19.996</u>	4.298	<u>2.883</u>	4.718	<u>2.627</u>	4.913	<u>2.523</u>				
43	Tc	0.589	21.045	4.060	3.054	4.436	2.795	4.632	2.677				
44	Ru	0.561	22.112	3.83	3.24	4.180	2.965	4.369	2.837				
45	Rh	0.534	23.217	3.626	3.418	3.942	3.144	4.130	3.001				
46	Pd	0.509	24.341	3.428	3.616	3.724	3.328	3.908	3.171				
47	Ag	0.486	25.509	3.254	3.809	3.514	3.527	3.698	3.351				
48	Cd	0.464	26.704	3.085	4.018	3.326	3.726	3.504	3.537				
49	In	0.444	27.920	2.926	4.236	3.147	3.938	3.324	3.728				
50	Sn	0.425	29.182	2.777	4.463	2.982	4.156	3.156	3.927				
51	Sb	0.407	30.477	2.639	4.695	2.830	4.380	3.000	4.131				
52	Te	0.390	31.800	2.511	4.937	2.687	4.611	2.855	4.340				
53	I	0.374	33.155	2.389	5.188	2.553	4.855	2.719	4.557				
54	Xe	0.359	34.570	2.274	5.451	2.429	5.102	2.592	4.780				
55	Cs	0.345	35.949	2.167	5.719	2.314	5.356	2.474	5.010				

Table 1 *Continued*

Atomic number	Element	K edge Å	K edge keV	L_I edge Å	L_I edge keV	L_II edge Å	L_II edge keV	L_III edge Å	L_III edge keV	M_IV edge Å	M_IV edge keV	M_V edge Å	M_V edge keV
56	Ba	0.331	37.399	2.068	5.994	2.204	5.622	2.363	5.245	15.56	0.7967	15.89	0.7801
57	La	0.318	38.920	1.973	6.282	2.103	5.893	2.258	5.488				
58	Ce	0.307	40.438	1.889	6.559	2.011	6.163	2.164	5.727				
59	Pr	0.295	41.986	1.811	6.844	1.924	6.441	2.077	5.967	13.122	0.9448	13.394	0.9257
60	Nd	0.285	43.559	1.735	7.142	1.843	6.725	1.995	6.213	12.459	0.9951	23.737	0.9734
61	Pm	0.274	45.207	1.665	7.448	1.767	7.018	1.918	6.466				
62	Sm	0.265	46.833	1.599	7.752	1.703	7.279	1.845	6.719	11.288	1.0983	11.552	1.0732
63	Eu	0.256	48.501	1.536	8.066	1.626	7.621	1.775	6.981	10.711	1.1575	11.013	1.1258
64	Gd	0.247	50.215	1.477	8.391	1.561	7.938	1.710	7.250				
65	Tb	0.238	51.984	1.421	8.722	1.501	8.256	1.649	7.517				
66	Dy	0.231	53.773	1.365	9.081	1.438	8.619	1.579	7.848				
67	Ho	0.223	55.599	1.317	9.408	1.390	8.918	1.535	8.072				
68	Er	0.216	57.465	1.268	9.773	1.338	9.260	1.482	8.361	8.601	1.4415	8.847	1.4013
69	Tm	0.209	59.319	1.222	10.141	1.288	9.626	1.433	8.650			8.487	1.4609
70	Yb	0.202	61.282	1.182	10.487	1.243	9.972	1.386	8.941				
71	Lu	0.196	63.281	1.140	10.870	1.199	10.341	1.341	9.239				
72	Hf	0.190	65.292	1.100	11.271	1.155	10.732	1.297	9.554				
73	Ta	0.184	67.379	1.061	11.681	1.114	11.128	1.255	9.874	6.87	1.804	7.11	1.743
74	W	0.178	69.479	1.025	12.097	1.075	11.533	1.216	10.196	6.59	1.880	6.83	1.814
75	Re	0.173	71.590	0.990	12.524	1.037	11.953	1.177	10.529	6.33	1.958	6.560	1.890
76	Os	0.168	73.856	0.956	12.968	1.001	12.380	1.140	10.867	6.073	2.042	6.30	1.967
77	Ir	0.163	76.096	0.923	13.427	0.967	12.817	1.106	11.209	5.83	2.126	6.05	2.048
78	Pt	0.158	78.352	0.893	13.875	0.934	13.266	1.072	11.556	5.59	2.217	5.81	2.133
79	Au	0.153	80.768	0.863	14.354	0.903	13.731	1.040	11.917	5.374	2.307	5.584	2.220
80	Hg	0.149	83.046	0.835	14.837	0.872	14.210	1.008	12.3	5.157	2.404	5.36	2.313
81	Tl	0.415	85.646	0.808	15.338	0.843	14.695	0.979	12.655	4.952	2.504	5.153	2.406
82	Pb	0.141	88.037	0.782	15.858	0.815	15.205	0.950	13.041	4.757	2.606	4.955	2.502
83	Bi	0.137	90.420	0.757	16.376	0.789	15.713	0.923	13.422	4.572	2.711	4.764	2.603
84	Po	0.133	93.112	0.732	16.935	0.763	16.244	0.897	13.817				
85	At	0.130	95.740	0.709	17.490	0.739	16.784	0.872	14.215				
86	Rn	0.126	98.418	0.687	18.058	0.715	17.337	0.848	14.618				
87	Fr	0.123	101.147	0.665	18.638	0.693	17.904	0.825	15.028				
88	Ra			0.645	19.229	0.671	18.478	0.803	15.439				
89	Ac	0.116	106.759	0.625	19.842	0.650	19.078	0.782	15.865				
90	Th	0.113	109.741	0.606	20.458	0.630	19.677	0.761	16.293	3.557	3.485	3.729	3.325
91	Pa	0.110	112.581	0.588	21.102	0.611	20.311	0.741	16.731	3.436	3.608	3.618	3.436
92	U	0.108	115.610	0.569	21.764	0.592	20.938	0.722	17.160	3.333	3.720	3.497	3.545
93	Np	0.105	118.619	0.553	22.417	0.574	21.596	0.704	17.614				
94	Pu	0.102	121.720	0.537	23.097	0.557	22.262	0.686	18.066				
95	Am	0.099	124.816	0.521	23.793	0.540	22.944	0.669	18.525				
96	Cm	0.097	128.088	0.506	24.503	0.525	23.640	0.653	18.990				
97	Bk	0.094	131.357	0.491	25.230	0.509	24.352	0.637	19.461				
98	Cf	0.092	134.683	0.477	25.971	0.494	25.080	0.622	19.938				
99	Es	0.090	138.067	0.464	26.729	0.480	25.824	0.607	20.422				
100	Fm	0.088	141.510	0.451	27.503	0.466	26.584	0.593	20.912				

Source: From Ref. 27, and 28.

APPENDIX II CHARACTERISTIC X-RAY WAVELENGTHS (Å) AND ENERGIES (keV)

Table 1 K Series Diagram Lines (Å)[a]

Line		$\alpha_{1,2}$	α_1	α_2	β_1	β_3	β_2	β_4
Approximate intensity		150	100	50	15		5	<1
Li	3	230						
B	4	113						
B	5	67						
C	6	44						
N	7	31.603						
O	8	23.707						
F	9	18.307						
Ne	10	14.615			14.460			
Na	11	11.909			11.574	11.726		
Mg	12	9.889			9.559	9.667		
Al	13	8.339	8.338	8.341	7.960	8.059		
Si	14	7.126	7.125	7.127	6.778			
P	15	6.155	6.154	6.157	5.804			
S	16	5.373	5.372	5.375	5.032			
Cl	17	4.729	4.728	4.731	4.403			
Ar	18	4.192	4.191	4.194	3.886			
K	19	3.744	3.742	3.745	3.454			
Ca	20	3.360	3.359	3.362	3.089			
Sc	21	3.032	3.031	3.034	2.780			
Ti	22	2.750	2.749	2.753	2.514			
V	23	2.505	2.503	2.507	2.285			
Cr	24	2.291	2.290	2.294	2.085			
Mn	25	2.103	2.102	2.105	1.910			
Fe	26	1.937	1.936	1.940	1.757			
Co	27	1.791	1.789	1.793	1.621			
Ni	28	1.659	1.658	1.661	1.500		1.489	
Cu	29	1.542	1.540	1.544	1.392	1.393	1.381	
Zn	30	1.437	1.435	1.439	1.296		1.284	
Ga	31	1.341	1.340	1.344	1.207	1.208	1.196	
Ge	32	1.256	1.255	1.258	1.129	1.129	1.117	
As	33	1.177	1.175	1.179	1.057	1.058	1.045	
Se	34	1.106	1.105	1.109	0.992	0.993	0.980	
Br	35	1.041	1.040	1.044	0.933	0.933	0.921	
Kr	36	0.981	0.980	0.984	0.879	0.879	0.866	0.866
Rb	37	0.927	0.926	0.930	0.829	0.830	0.817	0.815
Sr	38	0.877	0.875	0.880	0.783	0.784	0.771	0.770
Y	39	0.831	0.829	0.833	0.740	0.741	0.728	0.727
Zr	40	0.788	0.786	0.791	0.701	0.702	0.690	0.689
Nb	41	0.748	0.747	0.751	0.665	0.666	0.654	0.653
Mo	42	0.710	0.709	0.713	0.632	0.633	0.621	0.620
Tc	43	0.676	0.675	0.679	0.601		0.590	
Ru	44	0.644	0.643	0.647	0.572	0.573	0.562	0.561
Rh	45	0.614	0.613	0.617	0.546	0.546	0.535	0.534

Table 1 *Continued*

Line		$\alpha_{1,2}$	α_1	α_2	β_1	β_3	β_2	β_4
Approximate intensity		150	100	50	15		5	<1
Pd	46	0.587	0.585	0.590	0.521	0.521	0.510	
Ag	47	0.561	0.559	0.564	0.497	0.498	0.487	0.486
Cs	48	0.536	0.535	0.539	0.475	0.476	0.465	
In	49	0.514	0.512	0.517	0.455	0.455	0.445	0.444
Sn	50	0.492	0.491	0.495	0.435	0.436	0.426	0.425
Sb	51	0.472	0.470	0.475	0.417	0.418	0.408	0.407
Te	52	0.453	0.451	0.456	0.400	0.401	0.391	
I	53	0.435	0.433	0.438	0.384	0.385	0.376	
Xe	54	0.418	0.416	0.421	0.369		0.360	
Cs	55	0.402	0.401	0.405	0.355	0.355	0.346	
Ba	56	0.387	0.385	0.390	0.341	0.342	0.333	
La	57	0.373	0.371	0.376	0.328	0.329	0.320	0.319
Ce	58	0.359	0.357	0.362	0.316	0.317	0.309	0.307
Pr	59	0.346	0.344	0.349	0.305	0.305	0.297	
Nd	60	0.334	0.332	0.337	0.294	0.294	0.287	
Pm	61	0.322	0.321	0.325	0.283			
Sm	62	0.311	0.309	0.314	0.274	0.274	0.267	
Eu	63	0.301	0.299	0.304	0.264	0.265	0.258	
Gd	64	0.291	0.289	0.294	0.255	0.256	0.249	
Tb	65	0.281	0.279	0.284	0.246	0.246	0.239	
Dy	66	0.272	0.270	0.275	0.237	0.238	0.231	
Ho	67	0.263	0.261	0.266	0.230	0.231		
Er	68	0.255	0.253	0.258	0.222	0.223	0.217	
Tm	69	0.246	0.244	0.250	0.215	0.216		
Yb	70	0.238	0.236	0.241	0.208	0.208	0.203	
Lu	71	0.231	0.229	0.234	0.202	0.203	0.197	
Hf	72	0.224	0.222	0.227	0.195	0.196	0.190	
Ta	73	0.217	0.215	0.220	0.190	0.191	0.185	0.184
W	74	0.211	0.209	0.213	0.184	0.185	0.179	0.179
Re	75	0.204	0.202	0.207	0.179	0.179	0.174	0.174
Os	76	0.198	0.196	0.201	0.173	0.174	0.169	0.168
Ir	77	0.193	0.191	0.196	0.168	0.169	0.164	0.163
Pt	78	0.187	0.158	0.190	0.163	0.164	0.159	0.159
Au	79	0.182	0.180	0.185	0.159	0.160	0.155	0.154
Hg	80	0.177	0.175	0.180	0.154	0.155	0.150	0.150
Tl	81	0.172	0.170	0.175	0.150	0.151	0.146	0.146
Pb	82	0.167	0.165	0.170	0.146	0.147	0.147	0.141
Bi	83	0.162	0.161	0.165	0.142	0.143	0.138	0.138
Po	84	0.185	0.156	0.161	0.138		0.133	
At	85		0.152	0.157	0.134	0.135		
Rn	96		0.148	0.153	0.131	0.132		
Fr	87		0.144	0.149	0.127	0.128		
Ra	88		0.144	0.149	0.127	0.128		
Ac	89		0.140	0.145	0.124	0.125		
Th	90	0.135	0.133	0.138	0.117	0.118	0.114	0.114
Pa	91		0.131	0.136	0.115	0.116		
U	92	0.128	0.126	0.131	0.111	0.112	0.108	0.108

[a] Conversion equation: $E \text{ keV} = 12.4/\lambda \text{ Å}$.
Source: From Ref. 27.

Table 2 L Series Diagram Lines (Å)

Line		α	α_1	α_2	β_1	β_2	β_3	β_4	β_6	β_7	β_9
Approximate intensity		110	100	10	50	20	6	4	<1	<1	<1
Na	11										
Mg	12										
Al	13										
Si	14										
P	15										
S	16										
Cl	17										
Ar	18										
K	19										
Ca	20	36.393			36.022						
Sc	21	31.393			31.072						
Ti	22	27.445			27.074						
V	23	24.309			23.898						
Cr	24	21.713			21.323		19.429				
Mn	25	19.489			19.158		17.575				
Fe	26	17.602			17.290		15.742				
Co	27	16.000			15.698		14.269				
Ni	28	14.595			14.308		13.167				
Cu	29	13.357			13.079		12.115				
Zn	30			12.282	12.009		11.225				
Ga	31			11.313	11.045						
Ge	32			10.456	10.194						
As	33			9.671	9.414		8.930				
Se	34			8.990	8.735						
Br	35			8.375	8.126						
Kr	36										
Rb	37		7.318	7.325	7.075		6.788	6.821	6.984		
Sr	38		6.863	6.870	6.623		6.367	6.403	6.519		
Y	39		6.449	6.456	6.211		5.983	6.018	6.094		
Zr	40		6.070	6.077	5.836	5.586	5.632	5.668	5.710		
Nb	41		5.725	5.732	5.492	5.238	5.310	5.346	5.361		

Table 2 *Continued*

Line		β_{10}	γ_1	γ_2	γ_3	γ_4	γ_5	l	η
Approximate intensity		<1	10	1	2	<1	<1	3	1
Na	11							410	
Mg	12							260	
Al	13							180	
Si	14								
P	15								
S	16								
Cl	17							67.84	67.25
Ar	18							56.212	56.813
K	19							47.835	47.325
Ca	20							41.042	40.542
Sc	21							36.671	35.200
Ti	22							31.423	30.942
V	23							27.826	27.375
Cr	24							24.840	24.339
Mn	25							22.315	21.864
Fe	26							20.201	19.73
Co	27							18.358	17.86
Ni	28							16.693	16.304
Cu	29							15.297	14.940
Zn	30							14.081	13.719
Ga	31							12.976	12.620
Ge	32							11.944	11.608
As	33							11.069	10.732
Se	34							10.293	9.959
Br	35							9.583	9.253
Kr	36								
Rb	37			6.045			6.754	8.363	8.042
Sr	38			5.644			6.297	7.836	7.517
Y	39			5.283			5.875	7.356	7.040
Zr	40		5.384	4.953			5.497	6.918	6.606
Nb	41		5.036	4.654			5.151	6.517	6.210
Mo	42		4.726	4.380			4.387	6.150	5.847

Table 2 *Continued*

Line		α	α₁	α₂	β₁	β₂	β₃	β₄	β₆	β₇	β₉
Approximate intensity		110	100	10	50	20	6	4	<1	<1	<1
Mo	42		5.406	5.414	5.176	4.923	5.013	5.048	5.048		
Tc	43										
Ru	44		4.846	4.854	4.620	4.372	4.487	4.532	4.487		
Rh	45		4.597	4.605	4.374	4.130	4.253	4.289	4.242		
Pd	46		4.368	4.376	4.146	3.909	4.034	4.071	4.016		3.792
Ag	47		4.154	4.162	3.935	3.703	3.834	3.870	3.808		3.605
Cd	48		3.956	3.965	3.739	3.514	3.644	3.681	3.614		3.430
In	49		3.752	3.781	3.555	3.339	3.470	3.507	3.436		3.268
Sn	50		3.600	3.609	3.385	3.175	3.306	3.344	3.270	3.155	3.115
Sb	51		3.439	3.448	3.226	3.023	3.152	3.190	3.115	3.005	2.973
Te	52		3.290	3.299	3.077	2.882	3.009	3.046	2.971	2.863	2.839
I	53		3.148	3.157	2.937	2.751	2.874	2.912	2.837	2.730	2.713
Xe	54										
Cs	55		2.892	2.902	2.683	2.511	2.628	2.666	2.593	2.485	2.478

Line		β₁₀	γ₁	γ₂	γ₃	γ₄	γ₅	l	η
Approximate intensity		<1	10	1	2	<1	<1	3	1
Tc	43								
Ru	44		4.182	3.897			4.288	5.503	5.204
Rh	45		3.944	3.685			4.045	5.217	4.922
Pd	46	3.799	3.725	3.489			3.822	4.952	4.660
Ag	47	3.611	3.523	3.307			3.616	4.707	4.418
Cd	48	3.437	3.336	3.137			3.426	4.480	4.193
In	49	3.274	3.162	2.980		2.026	3.249	4.269	3.983
Sn	50	3.121	3.001	2.835		2.778	3.085	4.071	3.789
Sb	51	2.979	2.852	2.695		2.639	2.932	3.888	3.607
Te	52	2.847	2.712	2.567		2.511	2.790	3.716	3.438
I	53	2.720	2.582	2.447		2.391	2.657	3.557	3.280
Xe	54								
Cs	55	2.492	2.348	2.237	2.233	2.174	2.417	3.267	2.994

Table 2 *Continued*

Line		α_1	α_2	β_1	β_2	β_3	β_4	β_5	β_6	β_7
Approximate intensity		100	10	50	20	6	4	1	<1	<1
Ba	56	2.776	2.785	2.567	2.404	2.516	2.555		2.482	2.382
La	57	2.665	2.674	2.458	2.303	2.410	2.449		2.379	2.275
Ce	58	2.561	2.570	2.356	2.208	2.311	2.349		2.282	2.180
Pr	59	2.463	2.473	2.259	2.119	2.216	2.255		2.190	2.091
Nd	60	2.370	2.382	2.166	2.035	2.126	2.166		2.103	2.009
Pm	61	2.283		2.081						
Sm	62	2.199	2.210	1.998	1.882	1.962	2.000	1.779	1.946	1.856
Eu	63	2.120	2.131	1.920	1.812	1.887	1.926		1.875	1.788
Gd	64	2.046	2.057	1.847	1.746	1.815	1.853		1.807	1.723
Tb	65	1.976	1.986	1.777	1.682	1.747	1.785	1.577	1.742	1.659
Dy	66	1.909	1.920	1.710	1.623	1.681	1.720		1.681	1.599
Ho	67	1.845	1.856	1.647	1.567	1.619	1.658		1.622	
Er	68	1.785	1.796	1.587	1.514	1.561	1.601		1.567	1.494
Tm	69	1.726	1.738	1.530	1.463	1.505	1.544		1.515	
Yb	70	1.672	1.682	1.476	1.416	1.452	1.491	1.387	1.466	1.395
Lu	71	1.619	1.630	1.424	1.370	1.402	1.441	1.342	1.419	1.350
Hf	72	1.569	1.580	1.374	1.327	1.353	1.392	1.298	1.374	1.306
Ta	73	1.522	1.533	1.327	1.285	1.307	1.346	1.256	1.331	1.264
W	74	1.476	1.487	1.282	1.245	1.263	1.302	1.215	1.290	1.224
Re	75	1.433	1.444	1.238	1.206	1.220	1.260	1.177	1.252	1.186
Os	76	1.391	1.402	1.197	1.169	1.179	1.218	1.140	1.213	1.149
Ir	77	1.352	1.363	1.158	1.135	1.141	1.179	1.106	1.179	1.115
Pt	78	1.313	1.325	1.120	1.102	1.104	1.142	1.072	1.143	1.082
Au	79	1.277	1.288	1.083	1.070	1.068	1.106	1.040	1.111	1.050
Hg	80	1.242	1.253	1.049	1.040	1.034	1.072	1.010	1.080	1.019

Table 2 *Continued*

Line		β_9	β_{10}	β_{15}	β_{17}	γ_1	γ_2	γ_3	γ_4	γ_5
Approximate intensity		<1	<1	<1	<1	10	1	2	<1	<1
Ba	56	2.376	2.387			2.442	2.138	2.134	2.075	2.309
La	57	2.282	2.290			2.141	2.046	2.041	1.983	2.205
Ce	58	2.188	2.195			2.048	1.960	1.955	1.899	2.110
Pr	59	2.100	2.107			1.961	1.879	1.874	1.819	2.020
Nd	60	2.016	2.023			1.878	1.801	1.797	1.745	1.935
Pm	61									
Sm	62	1.862	1.870			1.726	1.659	1.655	1.606	
Eu	63	1.792	1.800			1.657	1.597	1.591	1.544	1.708
Gd	64		1.731			1.592	1.534	1.529	1.485	
Tb	65		1.667			1.530	1.477	1.471	1.427	
Dy	66					1.473	1.423	1.417	1.374	1.518
Ho	67					1.417	1.371	1.364	1.323	1.462
Er	68	1.485	1.494			1.364	1.321	1.315	1.276	1.406
Tm	69					1.316	1.274	1.268		1.355
Yb	70	1.384	1.392			1.268	1.228	1.222	1.185	1.307
Lu	71	1.336	1.343	1.372		1.222	1.185	1.179	1.143	1.260
Hf	72	1.291	1.299	1.328	1.437	1.179	1.144	1.138	1.103	1.215
Ta	73	1.247	1.254	1.287		1.138	1.105	1.099	1.065	1.173
W	74	1.204	1.212	1.247	1.339	1.098	1.068	1.062	1.028	1.132
Re	75	1.165	1.172	1.208	1.293	1.061	1.032	1.026	0.993	1.057
Os	76	1.126	1.133	1.171		1.025	0.998	0.992	0.959	1.057
Ir	77	1.090	1.097	1.137		0.991	0.966	0.959	0.928	1.022
Pt	78	1.054	1.062		1.166	0.958	0.934	0.928	0.897	0.988
Au	79	1.021	1.028	1.072	1.128	0.927	0.905	0.898	0.867	0.956
Hg	80	0.986	0.996	1.041	1.090	0.897	0.876	0.869	0.839	0.925

Table 2 *Continued*

Line	γ_6	γ_8	l	η	s	t
Approximate intensity	<1	<1	3	1	<1	<1
Ba 56		2.222	3.135	2.862		
La 57			3.006	2.740		
Ce 58		2.023	2.892	2.620		
Pr 59		1.936	2.784	2.512		
Nd 60	1.855		2.675	2.409		
Pm 61						
Sm 62			2.482	2.218		
Eu 63		1.632	2.395			
Gd 64			2.312	2.049		
Tb 65			2.234			
Dy 66			2.158	1.898		
Ho 67			2.086	1.826		
Er 68			2.019	1.757		
Tm 69			1.955	1.695		
Yb 70	1.243	1.250	1.894	1.635		1.831
Lu 71	1.198	1.204	1.836	1.478		1.776
Hf 72	1.155	1.161	1.782	1.523	1.663	1.723
Ta 73	1.114	1.120	1.728	1.471	1.612	1.672
W 74	1.074	1.081	1.678	1.421		
Re 75	1.037	1.044	1.630	1.374		
Os 76	1.001	1.008	1.585	1.328		
Ir 77	0.967	0.974	1.541	1.285		
Pt 78	0.934	0.941	1.499	1.243		
Au 79	0.903	0.910	1.460	1.202	1.352	1.414
Hg 80	0.873	0.880	1.422	1.164		

Line	α_1	α_2	β_1	β_2	β_3	β_4	β_5	β_6	β_7
Approximate intensity	100	10	50	20	6	4	1	<1	<1
Tl 81	1.207	1.218	1.015	1.010	1.001	1.039	0.981	1.050	0.990
Pb 82	1.175	1.186	0.982	0.983	0.969	1.007	0.953	1.021	0.962
Bi 83	1.144	1.155	0.952	0.955	0.939	0.977	0.926	0.993	0.935
Po 84	1.114	1.126	0.921	0.929	0.908	0.948	0.900	0.967	
At 85									
Rn 86									
Fr 87	1.030		0.840	0.858					
Ra 88	1.005	1.017	0.814	0.836	0.803	0.841	0.807	0.871	0.817
Ac 89									
Th 90	0.956	0.968	0.766	0.794	0.755	0.793	0.765	0.828	0.775
Pa 91	0.933	0.945	0.742	0.774	0.732	0.770	0.746	0.803	0.755
U 92	0.911	0.923	0.720	0.755	0.710	0.748	0.726	0.789	0.736
Np 93	0.890	0.901	0.698	0.735					
Pu 94	0.868	0.880	0.678	0.719	0.669	0.707	0.691		
Am 95	0.849	0.860	0.658	0.701					

Table 2 *Continued*

Line		β_9	β_{10}	β_{15}	β_{17}	γ_1	γ_2	γ_3	γ_4	γ_5
Approximate intensity		<1	<1	<1	<1	10	1	2	<1	<1
Tl	81	0.957	0.964	1.012	1.056	0.868	0.848	0.842	0.812	0.895
Pb	82	0.927	0.934	0.984	1.022	0.840	0.822	0.815		0.867
Bi	83	0.898	0.905	0.957	0.989	0.814	0.796	0.790	0.761	0.840
Po	84			0.931		0.786	0.771	0.764		
At	85									
Rn	86									
Fr	87					0.716				
Ra	88	0.769	0.776	0.838	0.844	0.694	0.682	0.675	0.649	0.717
Ac	89									
Th	90	0.723	0.730			0.653	0.642	0.635	0.611	0.675
Pa	91	0.701	0.708			0.634	0.624	0.617	0.594	0.655
U	92	0.681	0.687			0.615	0.605	0.598	0.577	0.635
Np	93					0.597				
Pu	94					0.579				
Am	95					0.562				

Line		γ_6	γ_8	l	η	s	t
Approximate intensity		<1	<1	3	1	<1	<1
Tl	81	0.845	0.852	1.385	1.127	1.279	1.342
Pb	82	0.817	0.824	1.350	1.092	1.244	1.308
Bi	83	0.791	0.799	1.317	1.058	1.210	
Po	84	0.765		1.283			
At	85						
Rn	86						
Fr	87						
Ra	88	0.673	0.680	1.167	0.908		
Ac	89						
Th	90	0.632	0.640	1.115	0.855	1.011	1.080
Pa	91	0.613		1.091	0.830		
U	92	0.595	0.601	1.067	0.806	0.964	1.035
Np	93						
Pu	94						
Am	95						

Source: From Ref. 27.

Table 3 M Series Diagram Lines (Å)

Line		α_1	α_2	β	γ	l
K	19					680
Cu	29					170
Ru	44				26.85	
Rh	45				25.00	
Pd	46					
Ag	47				21.80	
Cd	48				20.46	
In	49					
Sn	50				17.94	
Sb	51				16.92	
Te	52				15.93	
Ba	56				12.700	
La	57		14.88	14.51	12.064	
Ce	58		14.06	13.78	11.534	18.38
Pr	59				10.997	
Nd	60		12.675	Band	10.504	
Sm	62		Band	Band	9.599	
Eu	63		Band	10.744	9.211	14.22
Gd	64		Band	10.253	8.844	13.57
Tb	65		Band	9.792	8.485	12.98
Dy	66		Band	9.364	8.144	12.43
Ho	67		Band	8.965	7.865	11.86
Er	68		Band	8.593	7.545	11.37
Tm	69		8.460	8.246		
Yb	70	8.139	8.155	7.909	7.023	10.48
Lu	71		7.840	7.600	6.761	10.07
Hf	72	7.539	7.546	7.304	6.543	9.69
Ta	73	7.251	7.258	7.022	6.312	9.32
W	74	6.983	6.990	6.756	6.088	8.96
Re	75	6.528		6.504	5.887	8.63
Os	76		6.490	6.267	5.681	
Ir	77	6.261	6.275	6.037	5.501	8.02
Pt	78	6.046	6.057	5.828	5.320	7.74
Au	79	5.840	5.854	5.623	5.145	7.47
Hg	80		5.666	5.452		
Tl	81	5.461	5.472	5.250	4.825	6.97
Pb	82	5.285	5.299	5.075	4.674	6.74
Bi	83	5.118	5.129	4.909	4.531	6.52
Th	90	4.138	4.151	3.942	3.679	5.24
Pa	91	4.022	4.035	3.827	3.577	5.08
U	92	3.910	3.924	3.715	3.480	4.95

Source: From Ref. 27.

Table 4 Energies of Principal K and L X-ray Emission Lines (keV)

Atomic number	Element	$K\beta_2$	$K\beta_1$	$K\alpha_1$	$K\alpha_2$	$L\gamma_1$	$L\beta_2$	$L\beta_1$	$L\alpha_1$	$L\alpha_2$
3	Li			0.052						
4	Be			0.110						
5	B			0.185						
6	C			0.282						
7	N			0.392						
8	C			0.523						
9	F			0.677						
10	Ne			0.851						
11	Na		1.067	1.041						
12	Hg		1.297	1.254						
13	Al		1.553	1.487	1.486					
14	Si		1.832	1.740	1.739					
15	P		2.136	2.015	2.014					
16	S		2.464	2.309	2.306					
17	Cl		2.815	2.622	2.621					
18	Ar		3.192	2.957	2.955					
19	K		3.589	3.313	3.310					
20	Ca		4.012	3.691	3.688				0.344	0.341
21	Sc		4.460	4.090	4.085				0.399	0.395
22	Ti		4.931	4.510	4.504				0.458	0.492
23	V		5.427	4.952	4.944				0.519	0.510
24	Cr		5.946	5.414	5.405				0.581	0.571
25	Mn		6.490	5.898	5.887				0.647	0.636
26	Fe		7.057	6.403	6.390				0.717	0.704
27	Co		7.647	6.930	6.915				0.790	0.775
28	Ni	8.328	8.264	7.477	7.460				0.866	0.849
29	Cu	8.976	8.904	8.047	8.027				0.943	0.928
30	Zn	9.657	9.571	8.638	8.615				1.032	1.009
31	Ga	10.365	10.263	9.251	9.234				1.122	1.096
32	Ge	11.100	10.981	9.885	9.854				1.216	1.166
33	As	11.863	11.725	10.543	10.507				1.517	1.282
34	Se	12.651	12.495	11.221	11.181				1.419	1.379
35	Br	13.465	13.290	11.923	11.877				1.526	1.480
36	Kr	14.313	14.112	12.648	12.597				1.638	1.587
37	Rb	15.184	14.960	13.394	13.335			1.752	1.694	1.692
38	Sr	16.083	15.834	14.164	14.097			1.872	1.806	1.805
39	Y	17.011	16.736	14.957	14.882			1.996	1.922	1.920
40	Zr	17.969	17.666	15.774	15.690	2.302	2.219	2.124	2.042	2.040
41	Nb	18.951	18.621	16.614	16.520	2.462	2.367	2.257	2.166	2.163
42	Mo	19.964	19.607	17.478	17.373	2.623	2.518	2.395	2.293	2.290
43	Tc	21.012	20.585	18.410	18.328	2.792	2.674	2.538	2.424	2.420
44	Ru	22.072	21.655	19.278	19.149	2.964	2.836	2.683	2.558	2.554
45	Rh	23.169	22.721	20.214	20.072	3.144	3.001	2.834	2.696	2.692
46	Pd	24.297	23.816	21.175	21.018	3.328	3.172	2.990	2.838	2.833
47	Ag	25.454	24.942	22.162	21.988	3.519	3.348	3.151	2.984	2.978
48	Cd	26.641	26.093	23.172	22.982	3.716	3.528	3.316	3.133	3.127
49	In	27.859	27.274	24.207	24.000	3.920	3.713	3.487	3.287	3.279
50	Sn	29.106	28.483	25.270	25.042	4.131	3.904	3.662	3.444	3.436
51	Sb	30.387	29.723	26.357	26.109	4.347	4.100	3.543	3.605	3.595
52	Te	31.698	30.993	27.471	27.200	4.570	4.301	4.029	3.769	3.758
53	I	33.016	32.292	28.610	28.315	4.800	4.507	4.220	3.937	3.926
54	Xe	34.446	33.644	29.802	29.485	5.036	4.720	4.422	4.111	4.098

Table 4 *Continued*

Atomic number	Element	$K\beta_2$	$K\beta_1$	$K\alpha_1$	$K\alpha_2$	$L\gamma_1$	$L\beta_2$	$L\beta_1$	$L\alpha_1$	$L\alpha_2$
55	Cs	35.819	34.984	30.970	30.623	5.280	4.936	4.620	4.286	4.272
56	Ba	37.255	35.376	32.191	31.815	5.531	5.156	4.828	4.467	4.451
57	La	38.728	37.799	33.440	33.033	5.789	5.384	5.043	4.651	4.635
58	Ce	40.231	39.255	34.717	34.276	6.052	5.613	5.262	4.840	4.823
59	Pr	41.772	40.746	36.023	35.548	6.322	5.850	5.489	5.034	5.014
60	Nd	43.298	42.269	37.359	36.845	6.602	6.090	5.722	5.230	5.208
61	Pm	44.955	43.945	38.649	38.160	6.891	6.336	5.956	5.431	5.408
62	Sm	46.553	45.400	40.124	39.523	7.180	6.587	6.206	5.636	5.609
63	Eu	48.241	47.027	41.529	40.877	7.478	6.842	6.456	5.846	5.816
64	Gd	49.961	48.718	42.983	42.280	7.788	7.102	6.714	6.039	6.027
65	Tb	51.737	50.391	44.470	43.737	8.104	7.368	6.979	6.275	6.241
66	Dy	53.491	52.178	45.985	45.193	8.418	7.638	7.249	6.495	6.457
67	Ho	55.292	53.934	47.528	46.686	8.748	7.912	7.528	6.720	6.680
68	Er	57.088	55.690	49.099	48.205	9.089	8.188	7.810	6.948	6.904
69	Tm	58.969	57.576	50.730	49.762	9.424	8.472	8.103	7.181	7.135
70	Yb	60.959	59.352	52.360	51.326	9.779	8.758	8.401	7.414	7.367
71	Lu	62.946	61.282	54.063	52.959	10.142	9.048	8.709	7.654	7.604
72	Hf	64.936	63.209	55.757	54.579	10.514	9.346	9.021	7.898	7.843
73	Ta	66.999	65.210	57.524	56.270	10.892	9.649	9.341	8.145	8.087
74	W	69.090	67.233	59.310	57.973	11.283	9.959	9.670	8.396	8.333
75	Re	71.220	69.298	61.131	59.707	11.684	10.273	10.008	8.651	8.584
76	Os	73.393	71.404	62.991	61.477	12.094	10.596	10.354	8.910	8.840
77	Ir	75.605	73.549	64.886	63.278	12.509	10.918	10.706	9.173	9.098
78	Pt	77.866	75.736	66.820	65.111	12.939	11.249	11.069	9.441	9.360
79	Au	80.165	77.968	68.794	66.980	13.379	11.582	11.439	9.711	9.625
80	Hg	82.526	80.258	70.821	68.894	13.828	11.923	11.823	9.987	9.896
81	Tl	84.904	82.558	72.860	70.320	14.288	12.268	12.210	10.266	10.170
82	Pb	87.343	84.922	74.957	72.794	14.762	12.620	12.611	10.549	10.448
83	Bi	89.833	87.335	77.097	74.805	15.244	12.977	13.021	10.836	10.729
84	Po	92.386	89.809	79.296	76.868	15.740	13.338	13.441	11.128	11.014
85	At	94.976	92.319	81.525	78.956	16.248	13.705	13.873	11.424	11.304
86	Rn	97.616	94.877	83.800	81.080	16.768	14.077	14.316	11.724	11.597
87	Fr	100.305	97.483	86.119	83.243	17.301	14.459	14.770	12.029	11.894
88	Ra	103.048	100.136	88.485	85.446	17.845	14.839	15.233	12.338	12.194
89	Ac	105.838	102.846	90.894	87.681	18.405	15.227	15.712	12.650	12.499
90	Th	108.671	105.592	93.334	89.942	18.977	15.620	16.200	12.966	12.808
91	Pa	111.575	108.408	95.851	92.271	19.559	16.022	16.700	13.291	13.120
92	U	114.549	111.289	98.428	94.648	20.163	16.425	17.218	13.613	13.438
93	Np	117.533	114.181	101.005	97.023	20.774	16.837	17.740	13.945	12.758
94	Pu	120.592	117.146	103.653	99.457	21.401	17.254	18.278	14.279	14.082
95	Am	123.706	120.163	106.351	101.932	22.042	17.677	18.829	14.618	14.411
96	Cm	126.875	123.235	109.098	104.448	22.699	18.106	19.393	14.961	14.743
97	Bk	130.101	126.362	111.896	107.023	23.370	18.540	19.971	15.309	15.079
98	Cf	133.383	129.544	114.745	109.603	24.056	18.980	20.562	15.661	15.420
99	Es	136.724	132.781	178.646	112.244	24.758	19.426	21.166	16.018	15.764
100	Fm	140.122	136.075	120.598	114.926	25.475	19.879	21.785	16.379	16.113

APPENDIX III RADIATIVE TRANSITION PROBABILITIES

Table 1 Radiative Transition Probabilities for K X-ray Lines

Atomic number	Element	$K\alpha_2/K\alpha_1$	$K\beta_3/K\beta_1$	$(K\beta_1 + K\beta_3)/K\alpha_1$	$K\beta_1'^{a}/K\alpha_1$	$K\beta_2'^{b}/K\alpha_1$	$K\beta^{c}/K\alpha^{d}$
20	Ca	0.505		0.116	0.116		0.069
22	Ti	0.505		0.137	0.137		0.095
24	Cr	0.506		0.155	0.156		0.114
26	Fe	0.506		0.172	0.171		0.128
28	Ni	0.507		0.189	0.187		0.133
30	Zn	0.509		0.202	0.202		0.137
32	Ge	0.511		0.215	0.215		0.142
34	Se	0.513		0.225	0.225	0.006	0.153
36	Kr	0.515		0.235	0.235	0.013	0.164
38	Sr	0.518		0.244	0.244	0.022	0.175
40	Zr	0.520		0.251	0.252	0.034	0.185
42	Mo	0.523		0.258	0.259	0.043	0.193
44	Ru	0.526		0.264	0.265	0.048	0.201
46	Pd	0.528		0.270	0.271	0.051	0.209
48	Cd	0.531		0.275	0.277	0.054	0.216
50	Sn	0.533	0.516	0.280	0.282	0.056	0.222
52	Te	0.536	0.517	0.285	0.287	0.060	0.226
54	Xe	0.537	0.518	0.290	0.292	0.064	0.232
56	Ba	0.542	0.519	0.294	0.297	0.070	0.240
58	Ce	0.545	0.521	0.298	0.301	0.076	0.244
60	Nd	0.549	0.522	0.303	0.306	0.082	0.247
62	Sm	0.551	0.523	0.307	0.311	0.085	0.250
64	Gd	0.556	0.525	0.310	0.314	0.088	0.253
66	Dy	0.560	0.526	0.314	0.318	0.089	0.256
68	Er	0.565	0.527	0.317	0.322	0.090	0.259
70	Yb	0.568	0.529	0.320	0.325	0.090	0.261
72	Hf	0.572	0.531	0.324	0.329	0.091	0.263
74	W	0.576	0.532	0.326	0.332	0.092	0.267
76	Os	0.580	0.534	0.330	0.336	0.094	0.270
78	Pt	0.585	0.535	0.333	0.339	0.097	0.274
80	Hg	0.590	0.537	0.336	0.343	0.100	0.277
82	Pb	0.595	0.539	0.339	0.346	0.103	0.282
84	Po	0.600	0.541	0.342	0.350	0.106	0.285
86	Rn	0.605	0.542	0.345	0.353	0.110	0.288
88	Ra	0.612	0.544	0.348	0.356	0.113	0.291
90	Th	0.619	0.546	0.351	0.360	0.118	0.295
92	U	0.624	0.548	0.354	0.363	0.123	0.299
94	Pu	0.631	0.550	0.356	0.366	0.125	0.301
96	Cm	0.638	0.552	0.359	0.370	0.130	0.305
98	Cf	0.646	0.554	0.362	0.374	0.134	0.309
100	Em	0.652	0.556	0.364	0.377	0.138	0.312

[a] $K\beta_1' = KM_{II} + KM_{III} + KM_{IV,V}$.
[b] $K\beta_2' = KN_{II,III} + KO_{II,III}$.
[c] $K\beta = K\beta_1' + K\beta_2'$.
[d] $K\alpha = K\alpha_1 + K\alpha_2$.

Source: From Ref. 36. Reprinted by permission of the author and Academic Press, Inc.

Table 2 Radiative Transition Probabilities for L_I X-ray Lines Normalized to $L\beta_3 = 100$

Atomic number	Element	$L\beta_3$	$L\beta_4$	$L\gamma_3$	$L\gamma_2$	Atomic number	Element	$L\beta_3$	$L\beta_4$	$L\gamma_3$	$L\gamma_2$
36	Kr	100	—	10.8	—	66	Dy	100	67.8	31.4	—
38	Sr	100	—	14.7	—	68	Er	100	67.6	31.4	—
40	Zr	100	—	18.1	—	70	Yb	100	67.5	31.4	18.5
42	Mo	100	71.0	21.0	—	72	Hf	100	67.6	31.4	20.0
44	Ru	100	64.9	23.4	—	74	W	100	68.3	31.6	21.5
46	Pd	100	61.9	25.5	—	76	Os	100	69.6	31.9	23.2
48	Cd	100	61.0	27.2	—	78	Pt	100	71.7	32.3	25.0
50	Sn	100	61.4	28.5	—	80	Hg	100	74.8	33.0	26.9
52	Te	100	62.6	29.5	—	82	Pb	100	78.8	34.0	28.8
54	Xe	100	64.1	30.3	—	84	Po	100	83.9	35.2	30.9
56	Ba	100	65.5	30.8	—	86	Rn	100	89.8	36.8	33.1
58	Ce	100	66.7	31.2	—	88	Ra	100	95.0	38.7	35.3
60	Nd	100	67.5	31.4	—	90	Th	100	102.0	41.0	37.7
62	Sm	100	69.7	31.5	—	92	U	100	110.0	43.8	40.2
64	Gd	100	68.0	31.5	—	94	Pu	100	120.0	47.5	44.0

Table 3 L_{II} X-ray Lines Normalized to $L\beta_1 = 100$

Atomic number	Element	$L\beta_1$	$L\eta$	$L\gamma_1$	$L\gamma_6$	Atomic number	Element	$L\beta_1$	$L\eta$	$L\gamma_1$	$L\gamma_6$
30	Zn	100	12.3	—	—	64	Gd	100	2.10	16.1	—
32	Ge	100	10.4	—	—	66	Dy	100	2.10	16.5	—
34	Se	100	8.75	—	—	68	Er	100	2.10	17.0	—
36	Kr	100	7.40	—	—	70	Yb	100	2.12	17.6	—
38	Sr	100	6.25	—	—	72	Hf	100	2.13	18.4	—
40	Zr	100	5.25	0.91	—	74	W	100	2.16	19.2	0.375
42	Mo	100	4.35	6.71	—	76	Os	100	2.20	20.1	1.73
44	Ru	100	3.63	10.6	—	78	Pt	100	2.23	20.9	2.42
46	Pd	100	3.0	13.1	—	80	Hg	100	2.28	21.7	2.98
48	Cd	100	2.6	14.5	—	82	Pb	100	2.33	22.3	3.45
50	Sn	100	2.35	15.3	—	84	Po	100	2.40	22.8	3.88
52	Te	100	2.25	15.4	—	86	Rn	100	2.45	23.2	4.29
54	Xe	100	2.2	15.6	—	88	Ra	100	2.50	23.4	4.74
56	Ba	100	2.16	15.6	—	90	Th	100	2.60	23.7	5.25
58	Ce	100	2.12	15.7	—	92	U	100	2.80	24.0	5.88
60	Nd	100	2.10	15.8	—	94	Pu	100	2.30	24.2	6.65
62	Sn	100	2.10	15.9							

Table 4 L_{III} X-ray Lines Normalized to $L\alpha_1 = 100$

Atomic number	Element	$L\alpha_1$	$L\beta_{2,15}$	$L\alpha_2$	$L\beta_5$	$L\beta_6$	Ll
22	Ti	100	—	—	—	—	40.37
24	Cr	100	—	—	—	—	26.13
26	Fe	100	—	—	—	—	15.35
28	Ni	100	—	—	—	—	10.29
30	Zn	100	—	—	—	—	7.56
32	Ge	100	—	—	—	—	5.96
34	Se	100	—	—	—	—	4.98
36	Kr	100	—	—	—	—	4.36
38	Sr	100	—	—	—		3.98
40	Zr	100	2.43	—	—	—	3.75
42	Mo	100	6.40	12.5	—	—	3.65
44	Ru	100	9.55	12.2	—	—	3.58
46	Pd	100	12.1	12.1	—	—	3.55
48	Cd	100	13.9	11.9	—	—	3.56
50	Sn	100	15.4	11.7	—	—	5.59
52	Te	100	16.4	11.5	—	—	3.62
54	Xe	100	17.2	11.3	—	—	3.67
56	Ba	100	17.8	11.2	—	—	3.73
58	Ce	100	18.2	11.1	—	—	3.79
60	Nd	100	18.5	11.1	—	—	3.86
62	Sm	100	18.8	11.0	—	—	3.92
64	Gd	100	19.2	11.1	—	—	3.99
66	Dy	100	19.6	11.1	—	—	4.07
68	Er	100	20.0	11.2	—	—	4.15
70	Yb	100	20.5	11.2	—	—	4.23
72	Hf	100	21.2	11.3	—	1.15	4.32
74	W	100	21.9	11.3	0.242	1.28	4.42
76	Os	100	22.7	11.4	0.873	1.38	4.53
78	Pt	100	23.5	11.4	1.74	1.46	4.65
80	Hg	100	24.4	11.5	2.62	1.55	4.78
82	Pb	100	25.3	11.5	3.24	1.59	4.93
84	Po	100	26.2	11.4	3.85	1.65	5.09
86	Rn	100	26.8	11.4	4.28	1.70	5.27
88	Ra	100	27.3	11.3	4.69	1.75	5.46
90	Th	100	27.5	11.1	4.94	1.80	5.69
92	U	100	27.5	11.0	5.20	1.85	5.93
94	Pu	100	27.0	10.5	5.40	1.89	6.18

Source: From Ref. 36. Reprinted by permission of the author and Academic Press, Inc.

APPENDIX IV WAVELENGTHS OF *K* SATELLITE LINES (Å)

Atomic number	Element	α^{II}	α^{I}	α_3	α_3^{I}	α_3^{II}	α_4	α_4^{I}	α_5	α_6	α_7	β_0	β^{I}
11	Na		11.860	11.830	11.810				11.742	11.711			
12	Mg		9.845	9.820			9.804		9.749	9.724			
13	Al		8.302	8.283			8.267		8.226	8.206			
14	Si	7.100	7.095	7.082			7.073	7.069	7.035	7.017	7.026		6.816
15	P	6.140	6.131	6.117			6.109		6.075	6.063			6.838
16	S	5.370	5.354	5.341	5.339	5.344	5.334						
17	Cl	4.725											4.415
18	Ar												
19	K	3.739	3.728	3.721	3.718	3.724	3.717					3.496	
20	Ca	3.356	3.347	3.340	3.338	3.343	3.337					3.133	3.101
21	Sc	3.029	3.021	3.015	3.013	3.018	3.013					2.819	2.789
22	Ti	2.747	2.740	2.733	2.731	2.737	2.731					2.551	2.522
23	V	2.502	2.496	2.491	2.489	2.493	2.490					2.320	2.291
24	Cr		2.282	2.279	2.277	2.280	2.278					2.118	2.090
25	Mn		2.095	2.093	2.091	2.094	2.091						1.914
26	Fe		1.930	1.928	1.926	1.929	1.927					1.783	1.760
27	Co		1,784	1.782	1.780	1.783	1.781					1.645	1.623
28	Ni		1.653	1.651	1.650	1.652	1.650					1.522	1.502
29	Cu		1.536	1.535	1.533		1.534						1.394
30	Zn		1.431	1.430	1.428		1.429						1.296
31	Ga		1.338	1.335	1.334		1.335						1.209
32	Ge		1.250	1.250	1.248		1.249						
33	As		1.173	1.172	1.170		1.171					1.061	
34	Se			1.101	1.100		1.101					0.9958	
35	Br			1.036	1.035		1.036					0.9367	
36	Kr											0.8832	
37	Rb											0.8835	
38	Sr			0.8727	0.8646		0.8721					0.7881	
39	Y											0.7457	
40	Zr			0.7836	0.7828		0.7832					0.7056	
41	Nb				0.7432		0.7436					0.6706	
42	Mo				0.7065		0.7070					0.6369	
43	Tc												
44	Ru											0.5766	
45	Rh											0.5493	
46	Pd				0.5833		0.5837						
47	Ag											0.5007	

Continued

Atomic number	Element	β_{1x}	β_x	β^{V}	β^{II}	β^{II}_{III}	β^{III}	β^{IV}	β_5	β_6	β_7	β_8	β_9	β_{10}
11	Na													
12	Mg													
13	Al													
14	Si		6.753											
15	P	5.800	5.792		5.712		5.691							
16	S	5.028	5.023											
17	Cl			4.400	4.395									
18	Ar			3.882										
19	K			3.449	3.441		3.412	3.404						
20	Ca			3.087	3.082		3.054	3.048						
21	Sc				2.772		2.749	2.744						
22	Ti				2.506		2.489							
23	V				2.277		2.262							
24	Cr				2.078		2.066							
25	Mn						1.895							
26	Fe				1.749		1.742							
27	Co				1.614		1.607							
28	Ni				1.494		1.487							
29	Cu				1.391		1.380							
30	Zn				1.295		1.282							
31	Ga				1.207									
32	Ge				1.128		1.116							
33	As					1.042	1.044		1.064	1.050	1.054	1.047		
34	Se					0.9770	0.9786		0.9996	0.9854	0.9889			
35	Br					0.9177	0.9194		0.9403	0.9269	0.9297	0.9228		
36	Kr					0.8745								
37	Rb					0.8135	0.8155		0.8365	0.8234	0.8259	0.8193		
38	Sr					0.7681	0.7699		0.7911	0.7778	0.7802	0.7736		
39	Y					0.7261	0.7218		0.7492	0.7362	0.7382	0.7315		
40	Sr					0.6870	0.6890		0.7105	0.6973	0.6993	0.6924		
41	Nb					0.6510	0.6531		0.6741	0.6619	0.6636	0.6568		
42	Mo									0.6288	0.6303	0.6235	0.6256	0.6519
43	Tc													
44	Ru					0.5589			0.5808	0.5693	0.5708	0.5639		0.5607
45	Rh									0.5427	0.5439	0.5374	0.5399	0.5452
46	Pd										0.5189			
47	Ag									0.4943	0.4954	0.4902	0.4920	

Source: From Ref. 27.

APPENDIX V FLUORESCENCE YIELDS AND COSTER-KRONIG TRANSITION PROBABILITIES

Table 1 K Shell Fluorescence Yield ω_K

Atomic number	Element	ω_K	Atomic number	Element	ω_K
6	C	0.0009	45	Rh	0.81
7	N	0.0015	46	Pd	0.82
8	O	0.0022	47	Ag	0.83
10	Ne	0.0100	48	Cd	0.84
11	Na	0.020	49	In	0.85
12	Mg	0.030	50	Sn	0.86
13	Al	0.040	51	Sb	0.87
14	Si	0.055	52	Te	0.875
15	P	0.070	53	I	0.88
16	S	0.090	54	Xe	0.89
17	Cl	0.105	55	Cs	0.895
18	Ar	0.125	56	Ba	0.90
19	K	0.140	57	La	0.905
20	Sc	0.165	58	Ce	0.91
21	Ca	0.190	59	Pr	0.915
22	Ti	0.220	60	Nd	0.92
23	V	0.240	61	Pm	0.925
24	Cr	0.26	62	Sm	0.93
25	Mn	0.285	63	Eu	0.93
26	Fe	0.32	64	Gd	0.935
27	Co	0.345	65	Tb	0.94
28	Ni	0.375	66	Dy	0.94
29	Cu	0.41	67	Ho	0.945
30	Zn	0.435	68	Er	0.945
31	Ga	0.47	69	Tm	0.95
32	Ge	0.50	70	Yb	0.95
33	As	0.53	71	In	0.95
34	Se	0.565	72	Hf	0.955
35	Br	0.60	73	Ta	0.955
36	Kr	0.635	74	W	0.96
37	Rb	0.665	75	Re	0.96
38	Sr	0.685	76	Os	0.96
39	Y	0.71	77	Ir	0.96
40	Zr	0.72	78	Pd	0.965
41	Nb	0.755	79	Au	0.965
42	Mo	0.77	80	Hg	0.965
43	Tc	0.785	82	Pb	0.97
44	Ru	0.80	92	U	0.97

Source: From Refs. 81 and 82. Reprinted by permission of John Wiley and Sons, Inc., and CRC Press, Inc.

Table 2 Experimental L Subshell Fluorescence Yields ω_i

Atomic number	Element	ω_1	ω_2	ω_3
54	Xe			0.10 ± 0.01
56	Ba	0.06		0.05 ± 0.01
65	Tb	0.18	0.165 ± 0.018	0.188 ± 0.016
67	Ho			0.22 ± 0.03
			0.170 ± 0.055	0.169 ± 0.030
68	Er			0.21 ± 0.03
			0.185 ± 0.060	0.172 ± 0.032
70	Yb			0.20 ± 0.02
			0.188 ± 0.011	0.183 ± 0.011
71	Lu			0.22 ± 0.03
				0.251 ± 0.035
72	Hf			0.22 ± 0.03
				0.228 ± 0.025
73	Ta		0.25 ± 0.02	0.27 ± 0.01
			0.257 ± 0.013	0.25 ± 0.03
				0.191
				0.228 ± 0.013
				0.254 ± 0.025
74	W			0.207
				0.272 ± 0.037
75	Re			0.284 ± 0.043
76	Os			0.290 ± 0.030
77	Ir			0.244
				0.262 ± 0.036
78	Pt		0.331 ± 0.021	0.262
				0.31 ± 0.04
				0.317 ± 0.029
79	Au			0.291 ± 0.018
				0.276
				0.31 ± 0.04
80	Hg		0.39 ± 0.03	0.317 ± 0.025
			0.319 ± 0.010	0.40 ± 0.02
				0.32 ± 0.05
				0.367 ± 0.050
81	Tl	0.07 ± 0.02	0.319 ± 0.010	0.300 ± 0.010
			0.373 ± 0.025	0.37 ± 0.07
				0.386 ± 0.053
				0.306 ± 0.010
82	Pb	0.07 ± 0.02	0.363 ± 0.015	0.330 ± 0.021
		0.09 ± 0.02		0.337
				0.315 ± 0.013
				0.32
				0.35 ± 0.05
83	Bi	0.12 ± 0.01	0.32 ± 0.04	0.354 ± 0.028
		0.095 ± 0.005	0.38 ± 0.02	0.367
				0.36
				0.37 ± 0.05
				0.362 ± 0.029
				0.40 ± 0.05
				0.340 ± 0.018

Table 2 *Continued*

Atomic number	Element	ω_1	ω_2	ω_3
90	Th			0.42
				0.517 ± 0.042
91	Pa			0.46 ± 0.05
92	U			0.44
				0.500 ± 0.040
96	Cm	0.28 ± 0.06	0.552 ± 0.032	0.515 ± 0.034
			0.55 ± 0.02	0.63 ± 0.02

Source: From Ref. 42. Reprinted by permission of the author and the American Physical Society.

Table 3 Measured L Shell Coster-Kronig Yields

Atomic number	Element	f_{12}	f_{13}	f_{23}
56	Ba	0.66 ± 0.07		
65	Tb	0.41 ± 0.36	0.43 ± 0.28	0.066 ± 0.014
67	Ho			0.205 ± 0.034
68	Er			0.225 ± 0.025
70	Yb			0.142 ± 0.009
73	Ta	<0.14	0.19	0.148 ± 0.010
			<0.36	0.20 ± 0.04
74	W		0.27 ± 0.03	
75	Re		0.30 ± 0.04	
77	Ir		0.46 ± 0.06	
78	Pt		0.50 ± 0.05	
79	Au	0.25 ± 0.13	0.51 ± 0.13	0.22
			0.61 ± 0.07	
80	Hg	0.74 ± 0.04		0.22 ± 0.04
				0.08 ± 0.02
				0.188 ± 0.010
81	Tl	0.17 ± 0.05	0.76 ± 0.10	0.25 ± 0.13
		0.14 ± 0.03	0.57 ± 0.10	0.169 ± 0.010
			0.56 ± 0.07	0.159 ± 0.013
			0.56 ± 0.05	
82	Pb	0.15 ± 0.04	0.57 ± 0.03	0.164 ± 0.016
		0.17 ± 0.05	0.61 ± 0.08	0.156 ± 0.010
83	Bi	0.19 ± 0.05	0.58 ± 0.05	$0.06 \begin{smallmatrix} +0.14 \\ -0.06 \end{smallmatrix}$
		0.18 ± 0.02	0.58 ± 0.02	0.164
92	U			0.23 ± 0.12
93	Np	0.10 ± 0.04	0.55 ± 0.09	$0.02 \begin{smallmatrix} +0.05 \\ -0.02 \end{smallmatrix}$
94	Pu			0.22 ± 0.08
				0.24 ± 0.08
96	Cm	0.038 ± 0.022	0.68 ± 0.04	0.188 ± 0.019

Source: From Ref. 42. Reprinted by permission of the author and the American Physical Society.

Table 4 Theoretical L Subshell Fluorescence Yields ω_i and Coster-Kronig Yields f_{ij}[a]

Atomic number	Element	ω_1	ω_2	ω_3	f_{12}	f_{13}	$f_{12} + f_{13}$	f_{23}
13	Al	3.05−6		2.40−3			0.982	
14	Si	9.77−6		1.08−3			0.975	
15	P	2.12−5		4.1−4			0.971	
16	S	3.63−5		2.9−4			0.968	
17	Cl	5.60−5		2.3−4			0.964	
18	Ar	8.58−5		1.9−4			0.965	
19	K	1.15−4		2.1−4			0.962	
20	Ca	1.56−4		2.1−4			0.955	
22	Ti	2.80−4		1.18−3	0.313	0.629		
24	Cr	2.97−4		3.29−3	0.317	0.636		
26	Fe	3.84−4	1.43−3	5.59−3	0.302	0.652		7.24−2
				1.49−3				
28	Ni	4.63−4	2.69−3	8.02−3	0.325	0.622		9.97−2
29	Cu		3.57−3	3.83−3				0.109
30	Zn	5.23−4		1.08−2	0.322	0.624		
32	Ge	7.70−4	7.72−3	1.44−2	0.266	0.671		2.49−2
33	As	1.40−3	8.85−3	9.74−3	0.282	0.547		4.13−2
34	Se	1.30−3	9.94−3	1.78−2	0.302	0.616		5.95−2
35	Br		1.09−2					7.64−2
36	Kr	1.85−3	2.20−2	2.36−2	0.230	0.686		8.97−2
		2.19−3	1.19−2	1.23−2	0.225	0.585		9.22−2
37	Rb	1.32−2						0.107
38	Sr	3.00−3	2.24−2	2.43−2	0.249	0.646		0.115
40	Zr	3.97−3	2.94−2	2.95−2	0.236	0.648		0.118
		3.96−3	1.89−2	2.01−2	0.271	0.522		0.123
42	Mo	5.75−3	3.50−2	3.73−2	0.166	0.689		0.124
		6.34−3	2.45−2	2.59−2	0.048	0.692		0.126
44	Ru	7.74−3	4.18−2	4.50−2	0.057	0.779		0.136
47	Ag	1.02−2	5.47−2	6.02−2	0.052	0.786		0.152
		1.01−2	4.30−2	4.49−2	0.064	0.695		0.130
50	Sn	1.30−2	6.56−2	7.37−2	0.052	0.784		0.162
		1.30−2	5.67−2		0.072	0.693		0.136
51	Sb	3.11−2	6.16−2	6.33−2	0.164	0.316		0.138
54	Xe	5.84−2	9.12−2	9.70−2	0.179	0.274		0.173
56	Ba	4.46−2	9.07−2	8.99−2	0.168	0.336		0.151
60	Nd	7.46−2	0.133	0.135	0.207	0.303		0.141
		6.00−2	0.120	0.120	0.165	0.332		0.142
65	Tb		0.166	0.160				0.131
67	Ho	0.112	0.203	0.201	0.202	0.309		0.138
		0.094			0.178	0.317		
70	Yb	0.112			0.180	0.316		
74	W	0.115	0.287	0.268	0.195	0.332		0.123
		0.138	0.271	0.253	0.160	0.324		0.117
79	Au	0.105	0.357	0.327	0.083	0.644		0.132
80	Hg	0.098	0.352	0.321	0.101	0.618		0.108
83	Bi	0.120	0.417	0.389	0.069	0.656		0.101
85	At	0.129	0.422	0.380	0.082	0.612		0.100
90	Th	0.197	0.529	0.461	0.069	0.575		0.102
93	Np		0.460	0.472				0.209

[a] Figures following a sign indicate powers of 10. For example, $3.05 - 6$ means 3.05×10^{-6}.
Source: From Ref. 42. Reprinted by permission of the author and the American Physical Society.

Table 5　Measured M Shell Fluorescence Yields and Coster-Kronig Probabilities

Atomic number	Element	$\bar{\omega}_M$	$\omega_{LM}{}^a$	$\omega_{LM}{}^b$	$\omega_1 + f_{12}\omega_2$	ν_i	ω_i
76	Os		0.013 ± 0.0024	0.016 ± 0.003			
79	Au	0.023 ± 0.001					
79	Au		0.024 ± 0.005	0.030 ± 0.006			
82	Pb	0.029 ± 0.002					
82	Pb		0.026 ± 0.005	0.032 ± 0.006			
83	Bi	0.037 ± 0.007					
83	Bi	0.035 ± 0.002					
83	Bi		0.030 ± 0.006	0.037 ± 0.005			
92	U	0.06					
93	Np				$0.002 \begin{smallmatrix}+\,0.003\\-\,0.002\end{smallmatrix}$	$\nu_1 = 0.065 \pm 0.014$	
						$\nu_2 = 0.080 \pm 0.029$	
						$\nu_3 = 0.062 \pm 0.005$	
						$\nu_4 = 0.065 \pm 0.012$	$\omega_5 = 0.06 \pm 0.012$
						$\nu_4 = 0.065 \pm 0.012$	
96	Cm				$0.0075 \begin{smallmatrix}+\,0.0089\\-\,0.0075\end{smallmatrix}$	$\nu_{4,5} = 0.081 \pm 0.016$	
						$\nu_1 = 0.068 \pm 0.023$	$\omega_2 = 0.0046 \begin{smallmatrix}+\,0.0051\\-\,0.0046\end{smallmatrix}$
						$\nu_2 = 0.062 \pm 0.019$	
						$\nu_3 = 0.080 \pm 0.006$	
						$\nu_{4,5} = 0.075 \pm 0.012$	
							$\omega_5 = 0.075 \pm 0.012$

[a] Corrected for a 20% contribution from double M shell vacancies.
[b] Uncorrected values.
Source: From Ref. 42. Reprinted by permission of the author and the American Physical Society.

APPENDIX VI COEFFICIENTS FOR CALCULATING THE PHOTOELECTRIC ABSORPTION CROSS SECTIONS τ (BARNS/ATOM) VIA LN-LN REPRESENTATION

Table 1 $E > E_K{}^a$

Atomic number	Element	Atomic weight	A_0	A_1	A_2	A_3
1	H	1.008	2.44964	-3.34953	$-4.71370-2$	$7.09962-3$
2	He	4.003	6.06488	-3.29055	$-1.07256-1$	$1.44465-2$
3	Li	6.940	7.75370	-2.81801	$-2.41738-1$	$2.62542-2$
4	Be	9.012	9.04511	-2.83487	$-2.10021-1$	$2.29526-2$
5	B	10.811	9.95057	-2.74173	$-2.15138-1$	$2.27845-2$
6	C	12.010	$1.06879+1$	-2.71400	$-2.00530-1$	$2.07248-2$
7	N	14.008	$1.12765+1$	-2.65400	$-2.00445-1$	$2.00765-2$
8	O	16.000	$1.17130+1$	-2.57229	$-2.05893-1$	$1.99244-2$
9	F	19.000	$1.20963+1$	-2.44148	$-2.34461-1$	$2.19537-2$
10	Ne	20.183	$1.24485+1$	-2.45819	$-2.12591-1$	$1.96489-2$
11	Na	22.997	$1.26777+1$	-2.24521	$-2.74873-1$	$2.50270-2$
12	Mg	24.320	$1.28793+1$	-2.12574	$-2.99392-1$	$2.67643-2$
13	Al	26.970	$1.31738+1$	-2.18203	$-2.58960-1$	$2.22840-2$
14	Si	28.086	$1.32682+1$	-1.98174	$-3.16950-1$	$2.73928-2$
15	P	30.975	$1.33735+1$	-1.86342	$-3.39440-1$	$2.88858-2$
16	S	32.066	$1.37394+1$	-2.04786	$-2.73259-1$	$2.29976-2$
17	Cl	35.457	$1.36188+1$	-1.71937	$-3.54154-1$	$2.90841-2$
18	Ar	39.944	$1.39491+1$	-1.82276	$-3.28827-1$	$2.74382-2$
19	K	39.102	$1.37976+1$	-1.54015	$-3.94528-1$	$3.23561-2$
20	Ca	40.080	$1.42950+1$	-1.88644	$-2.83647-1$	$2.26263-2$
21	Sc	44.960	$1.39664+1$	-1.40872	$-4.14365-1$	$3.34355-2$
22	Ti	47,900	$1.43506+1$	-1.66322	$-3.31539-1$	$2.62065-2$
23	V	50.942	$1.47601+1$	-1.88867	$-2.71861-1$	$2.15792-2$
24	Cr	51.996	$1.48019+1$	-1.82430	$-2.79116-1$	$2.17324-2$
25	Mn	54.940	$1.48965+1$	-1.79872	$-2.83664-1$	$2.22095-2$
26	Fe	55.850	$1.43456+1$	-1.23491	$-4.18785-1$	$3.21662-2$
27	Co	58.933	$1.47047+1$	-1.38933	$-3.86631-1$	$3.03286-2$
28	Ni	58.690	$1.42388+1$	$-9.67736-1$	$-4.78070-1$	$3.66138-2$
29	Cu	63.540	$1.45808+1$	-1.18375	$-4.13850-1$	$3.12088-2$
30	Zn	65.380	$1.44118+1$	$-9.33083-1$	$-4.77357-1$	$3.62829-2$
31	Ga	69.720	$1.36182+1$	$-3.18459-1$	$-6.11348-1$	$4.58138-2$
32	Ge	72.590	$1.39288+1$	$-4.79613-1$	$-5.72897-1$	$4.31277-2$
33	As	74.920	$1.34722+1$	$-7.73513-2$	$-6.60456-1$	$4.92177-2$
34	Se	78.960	$1.30756+1$	$1.83235-1$	$-6.94264-1$	$5.02280-2$
35	Br	79.920	$1.32273+1$	$1.37130-1$	$-6.83203-1$	$4.95424-2$
36	Kr	83.800	$1.35927+1$	$-3.05214-2$	$-6.51340-1$	$4.77616-2$
37	Rb	85.480	$1.30204+1$	$3.82736-1$	$-7.32427-1$	$5.29874-2$
38	Sr	87.620	$1.35888+1$	$2.20194-3$	$-6.38940-1$	$4.60070-2$
39	Y	88.905	$1.34674+1$	$1.91023-1$	$-6.86616-1$	$4.97356-2$
40	Zr	91.220	$1.27538+1$	$6.97409-1$	$-7.89307-1$	$5.64531-2$
41	Nb	92.906	$1.33843+1$	$2.81028-1$	$-6.86607-1$	$4.86607-2$
42	Mo	95.950	$1.39853+1$	$-1.17426-1$	$-5.91094-1$	$4.17843-2$
43	Tc	99.000	$1.28214+1$	$7.51993-1$	$-7.87006-1$	$5.58668-2$
44	Ru	101.070	$1.26658+1$	$8.85020-1$	$-8.11144-1$	$5.73759-2$
45	Rh	102.910	$1.21760+1$	1.19682	$-8.66697-1$	$6.06931-2$
46	Pd	106.400	$1.39389+1$	$1.64528-1$	$-6.62170-1$	$4.76289-2$
47	Ag	107.880	$1.33926+1$	$4.41380-1$	$-6.93711-1$	$4.82085-2$

Table 1 *Continued*

Atomic number	Element	Atomic weight	A_0	A_1	A_2	A_3
48	Cd	112.410	1.15254 + 1	1.07714	− 8.31424 − 1	5.79120 − 2
49	In	114.820	1.18198 + 1	1.45768	− 8.88529 − 1	6.05982 − 2
50	Sn	118.690	1.30323 + 1	7.90788 − 1	− 7.62349 − 1	5.27872 − 2
51	Sb	121.760	9.06999	3.28791	− 1.26203	8.53470 − 2
52	Te	127.600	1.16656 + 1	1.71052	− 9.48281 − 1	6.53213 − 2
53	I	126.910	1.21075 + 1	1.43635	− 8.82038 − 1	6.03575 − 2
54	Xe	131.300	1.10857 + 1	2.08356	− 1.01209	6.90310 − 2
55	Cs	132.910	1.13757 + 1	1.94161	− 9.83232 − 1	6.71986 − 2
56	Ba	137.360	1.02250 + 1	2.67835	− 1.12648	7.62669 − 2
57	La	138.920	1.09780 + 1	2.23814	− 1.03549	7.02339 − 2
58	Ce	140.130	1.02725 + 1	2.74562	− 1.14174	7.74162 − 2
59	Pr	140.920	1.10156 + 1	2.22056	− 1.02216	6.90465 − 2
60	Nd	144.270	1.17632 + 1	1.79481	− 9.36661 − 1	6.35332 − 2
61	Pm	147.000	1.13864 + 1	2.05593	− 9.88180 − 1	6.69106 − 2
62	Sm	150.350	1.19223 + 1	1.79546	− 9.42902 − 1	6.44202 − 2
63	Eu	152.000	1.16168 + 1	1.97533	− 9.70901 − 1	6.58459 − 2
64	Gd	157.260	9.91968	3.03111	− 1.17520	7.86751 − 2
65	Tb	158.930	1.13818 + 1	2.14447	− 9.99222 − 1	6.75569 − 2
66	Dy	162.510	1.14845 + 1	2.10451	− 9.89870 − 1	6.69382 − 2
67	Ho	164.940	8.75203	3.71822	− 1.29273	8.55026 − 2
68	Er	167.270	1.20195 + 1	1.84815	− 9.39582 − 1	6.38106 − 2
69	Tm	168.940	1.25613 + 1	1.57523	− 8.90467 − 1	6.09779 − 2
70	Yb	173.040	7.42791	4.28955	− 1.35167	8.66136 − 2
71	Lu	174.990	1.26387 + 1	1.55476	− 8.81094 − 1	6.02036 − 2
72	Hf	178.500	7.58160	4.47037	− 1.42808	9.39044 − 2
73	Ta	180.950	8.65271	3.73117	− 1.26359	8.23539 − 2
74	W	183.920	7.57541	4.28874	− 1.34998	8.65200 − 2
75	Re	186.200	1.36944	7.79444	− 1.99822	1.26225 − 1
76	Os	190.200	1.37534 + 1	1.02122	− 7.77126 − 1	5.38811 − 2
77	Ir	192.200	1.25506 + 1	1.63090	− 8.75676 − 1	5.92011 − 2
78	Pt	195.090	1.27882 + 1	1.63605	− 8.98523 − 1	6.18550 − 2
79	Au	197.200	4.96352	5.79212	− 1.61842	1.02911 − 1
80	Hg	200.610	1.97594 + 1	− 1.97990	− 2.76981 − 1	2.68856 − 2
81	Tl	204.390	1.52879 + 1	2.73664 − 1	− 6.38890 − 1	4.57495 − 2
82	Pb	207.210	8.63374	3.69400	− 1.21312	7.74601 − 2
83	Bi	209.000	9.44293	3.44965	− 1.19886	7.83484 − 2
86	Rn	222.000	1.51782 + 1	3.49021 − 1	− 6.37638 − 1	4.51377 − 2
90	Th	232.000	1.34336 + 1	1.34805	− 8.13282 − 1	5.55664 − 2
92	U	238.070	1.37951 + 1	1.23983	− 8.01545 − 1	5.53596 − 2
94	Pu	239.100	1.82787 + 1	− 1.17371	− 3.68344 − 1	2.98738 − 2

[a] Notation abbreviated as in App. V, Table 4.

Table 2 $E_{LI} < E < E_K{}^a$

Atomic number	Element	Atomic weight	A_0	A_1	A_2
11	Na	22.997	$1.02355 + 1$	-2.55905	$-1.19524 - 1$
12	Mg	24.310	$1.05973 + 1$	-2.89818	$2.34506 - 1$
13	Al	26.970	$1.08711 + 1$	-2.77860	$1.75853 - 1$
14	Si	28.086	$1.12237 + 1$	-2.73694	$1.27557 - 1$
15	P	30.975	$1.15508 + 1$	-2.92200	$2.54262 - 1$
16	S	32.066	$1.18181 + 1$	-2.64618	$-9.68049 - 2$
17	Cl	35.457	$1.20031 + 1$	-2.41694	$-2.40897 - 1$
18	Ar	39.944	$1.22960 + 1$	-2.63279	$-7.36600 - 2$
19	K	39.102	$1.24878 + 1$	-2.53656	$-1.04892 - 1$
20	Ca	40.080	$1.27044 + 1$	-2.55011	$-9.43195 - 2$
21	Sc	44.960	$1.28949 + 1$	-2.40609	$-1.77791 - 1$
22	Ti	47.900	$1.31075 + 1$	-2.53576	$-9.57177 - 2$
23	V	50.942	$1.32514 + 1$	-2.49765	$-1.06383 - 1$
24	Cr	51.996	$1.34236 + 1$	-2.51532	$-1.01999 - 1$
25	Mn	54.040	$1.35761 + 1$	-2.49761	$-1.05943 - 1$
26	Fe	55.850	$1.36696 + 1$	-2.39195	$-1.37648 - 1$
27	Co	58.933	$1.38699 + 1$	-2.50669	$-8.69945 - 2$
28	Ni	58.690	$1.39848 + 1$	-2.48080	$-8.88115 - 2$
29	Cu	63.540	$1.42439 + 1$	-2.58677	$-6.67398 - 2$
30	Zn	65.380	$1.43221 + 1$	-2.62384	$-2.64926 - 2$
31	Ga	69.720	$1.44795 + 1$	-2.54469	$-7.57204 - 2$
32	Ge	72.590	$1.46813 + 1$	-2.69285	$-2.08355 - 2$
33	As	74.920	$1.46431 + 1$	-2.48397	-7.9
34	Se	78.960	$1.47048 + 1$	-2.38853	-1.0
35	Br	79.920	$1.48136 + 1$	-2.42347	-9.1
36	Kr	83.800	$1.49190 + 1$	-2.42418	-8.7
37	Rb	85.480	$1.49985 + 1$	-2.39108	-9.5
38	Sr	87.620	$1.50114 + 1$	-2.28169	-1.2
39	Y	88.905	$1.51822 + 1$	-2.38946	-8.8
40	Zr	91.220	$1.52906 + 1$	-2.38703	-9.12
41	Nb	92.906	$1.52088 + 1$	-2.20278	-1.36
42	Mo	95.950	$1.53494 + 1$	-2.26646	-1.16
43	Tc	99.000		733	-9.87
44	Ru	101.070		080	-1.19
45	Rh	102.910		976	-1.133
46	Pd	106.400		229	$-1.27652 - 1$
47	Ag	107.880		636	$-1.12223 - 1$
48	Cd	112.410		363	$-8.01104 - 2$
49	In	114.820		838	$-5.49951 - 2$
50	Sn	118.690		010	$-1.13539 - 1$
51	Sb	121.760		460	$-1.40745 - 1$
52	Te	127.600		7876	$-9.29405 - 2$
53	I	126.910		8214	$-5.07179 - 2$
54	Xe	131.300		1679	$-8.54498 - 2$
55	Cs	132.910		6363	$-5.42849 - 2$
56	Ba	137.360	$1.66217 + 1$	-2.48972	$-4.49623 - 2$
57	La	138.920	$1.63134 + 1$	-2.20156	$-9.80569 - 2$

Table 2 *Continued*

Atomic number	Element	Atomic weight	A_0	A_1	A_2
58	Ce	140.130	1.65862 + 1	− 2.36288	− 6.54708 − 2
59	Pr	140.920	1.67179 + 1	− 2.40326	− 6.12619 − 2
60	Nd	144.270	1.65964 + 1	− 2.26073	− 8.72426 − 2
61	Pm	147.000	1.68368 + 1	− 2.38881	− 6.45041 − 2
62	Sm	150.350	1.68725 + 1	− 2.39051	− 6.01080 − 2
63	Eu	152.000	1.70692 + 1	− 2.48046	− 4.47055 − 2
64	Cd	157.260	1.71159 + 1	− 2.47838	− 4.37107 − 2
65	Tb	158.930	1.71499 + 1	− 2.45507	− 4.71370 − 2
66	Dy	162.510	1.73446 + 1	− 2.54821	− 3.17606 − 2
67	Ho	164.940	1.76583 + 1	− 2.72523	− 8.19409 − 4
68	Er	167.270	1.77988 + 1	− 2.74671	− 2.87580 − 3
69	Tm	168.940	1.74250 + 1	− 2.51103	− 3.29454 − 2
70	Yb	173.040	1.69795 + 1	− 2.22577	− 7.32557 − 2
71	Lu	174.990	1.72638 + 1	− 2.37189	− 4.95994 − 2
72	Hf	178.500	1.64329 + 1	− 1.82851	− 1.32268 − 1
73	Ta	180.950	1.72410 + 1	− 2.30313	− 5.91006 − 2
74	W	183.920	1.72533 + 1	− 2.23874	− 7.27338 − 2
75	Re	186.200	1.78750 + 1	− 2.61051	− 1.36093 − 2
76	Os	190.200	1.73525 + 1	− 2.28550	− 5.88047 − 2
77	Ir	192.200	1.65270 + 1	− 1.76315	− 1.35232 − 1
78	Pt	195.090	1.73636 + 1	− 2.21112	− 7.30934 − 2
79	Au	197.200	1.74240 + 1	− 2.23911	− 6.63720 − 2
80	Hg	200.610	1.71857 + 1	− 2.08470	− 8.53294 − 2
81	Tl	204.390	1.77379 + 1	− 2.37745	− 4.33223 − 2
82	Pb	207.210	1.77963 + 1	− 2.37691	− 4.55883 − 2
83	Bi	209.000	1.75348 + 1	− 2.23353	− 5.96161 − 2
86	Rn	222.000	1.75028 + 1	− 2.13876	− 7.24638 − 2
90	Th	232.000	1.85481 + 1	− 2.61281	− 7.90574 − 3
92	U	238.070	1.75258 + 1	− 2.07237	− 7.23932 − 2
94	Pu	239.100	1.75519 + 1	− 2.02162	− 8.22940 − 2

[a] Notation abbreviated as in App. V, Table 4.

Table 3 $E_{MI} < E < E_{LIII}$[a]

Atomic number	Element	Atomic weight	A_0	A_1
31	Ga	69.720	1.22646 + 1	− 2.68965
32	Ge	72.590	1.24133 + 1	− 2.53085
33	As	74.920	1.25392 + 1	− 2.41380
34	Se	78.960	1.26773 + 1	− 2.39750
35	Br	79.920	1.27612 + 1	− 2.37730
36	Kr	83.800	1.28898 + 1	− 2.26021
37	Rb	85.480	1.30286 + 1	− 2.38693
38	Sr	87.620	1.31565 + 1	− 2.36655
39	Y	88.905	1.32775 + 1	− 2.43174
40	Zr	91.220	1.34508 + 1	− 2.50201
41	Nb	92.906	1.35434 + 1	− 2.50135
42	Mo	95.950	1.36568 + 1	− 2.48982
43	Tc	99.000	1.37498 + 1	− 2.44737
44	Ru	101.070	1.38782 + 1	− 2.48066
45	Rh	102.910	1.40312 + 1	− 2.61303
46	Pd	106.400	1.41392 + 1	− 2.57206
47	Ag	107.880	1.41673 + 1	− 2.48078
48	Cd	112.410	1.43497 + 1	− 2.52756
49	In	114.820	1.44115 + 1	− 2.49401
50	Sn	118.690	1.45572 + 1	− 2.56792
51	Sb	121.760	1.46268 + 1	− 2.55562
52	Te	127.600	1.47125 + 1	− 2.54324
53	I	126.910	1.47496 + 1	− 2.48179
54	Xe	131.300	1.47603 + 1	− 2.45068
55	Cs	132.910	1.49713 + 1	− 2.53145
56	Ba	137.360	1.50844 + 1	− 2.56341
57	La	138.920	1.51863 + 1	− 2.58287
58	Ce	140.130	1.52693 + 1	− 2.58174
59	Pr	140.920	1.53379 + 1	− 2.57086
60	Nd	144.270	1.54353 + 1	− 2.59006
61	Pm	147.000	1.55131 + 1	− 2.59623
62	Sm	150.350	1.56006 + 1	− 2.61328
63	Eu	152.000	1.57063 + 1	− 2.63481
64	Gd	157.260	1.57159 + 1	− 2.60843
65	Tb	158.930	1.58415 + 1	− 2.64040
66	Dy	162.510	1.59225 + 1	− 2.65289
67	Ho	164.940	1.60140 + 1	− 2.67903
68	Er	167.270	1.60672 + 1	− 2.67587
69	Tm	168.940	1.61269 + 1	− 2.67886
70	Yb	173.040	1.61794 + 1	− 2.67715
71	Lu	174.990	1.62289 + 1	− 2.67128
72	Hf	178.500	1.62758 + 1	− 2.66622
73	Ta	180.950	1.63068 + 1	− 2.66148
74	W	183.920	1.62613 + 1	− 2.60672
75	Re	186.200	1.63564 + 1	− 2.62453
76	Os	190.200	1.64233 + 1	− 2.63163
77	Ir	192.200	1.65144 + 1	− 2.64832

Table 3 *Continued*

Atomic number	Element	Atomic weight	A_0	A_1
78	Pt	195.090	1.67024 + 1	−2.71631
79	Au	197.200	1.64734 + 1	−2.57834
80	Hg	200.610	1.65903 + 1	−2.60670
81	Tl	204.390	1.66564 + 1	−2.61593
82	Pb	207.210	1.67131 + 1	−2.61538
83	Bi	209.000	1.67078 + 1	−2.58648
86	Rn	222.000	1.69000 + 1	−2.60945
90	Th	232.000	1.70483 + 1	−2.58569
92	U	238.070	1.70353 + 1	−2.56903
94	Pu	239.100	1.72953 + 1	−2.62164

[a] Notation abbreviated as in App. V, Table 4.

Table 4 $E < E_{MV}$[a]

Atomic number	Element	Atomic weight	A_0	A_1
61	Pm	147.000	1.55131 + 1	−2.59623
62	Sm	150.350	1.56006 + 1	−2.61328
63	Eu	152.000	1.57063 + 1	−2.63481
64	Gd	157.260	1.57159 + 1	−2.60843
65	Tb	158.930	1.58415 + 1	−2.64040
66	Dy	162.510	1.59225 + 1	−2.65289
67	Ho	164.940	1.60140 + 1	−2.67903
68	Er	167.270	1.60672 + 1	−2.67587
69	Tm	168.940	1.61269 + 1	−2.67886
70	Yb	173.040	1.39111 + 1	−2.40380
71	Lu	174.990	1.39813 + 1	−2.40841
72	Hf	178.500	1.40548 + 1	−2.42829
73	Ta	180.950	1.41313 + 1	−2.47214
74	W	183.920	1.42536 + 1	−2.32582
75	Re	186.200	1.42392 + 1	−2.35326
76	Os	190.200	1.42795 + 1	−2.21971
77	Ir	192.200	1.43422 + 1	−2.40183
78	Pt	195.090	1.43785 + 1	−2.34834
79	Au	197.200	1.44398 + 1	−2.32838
80	Hg	200.610	1.45195 + 1	−2.33016
81	Tl	204.390	1.45473 + 1	−2.26773
82	Pb	207.210	1.45771 + 1	−2.25279
83	Bi	209.000	1.46832 + 1	−2.30940
86	Rn	222.000	1.47243 + 1	−2.12905
90	Th	232.000	1.47730 + 1	−1.91192
92	U	238.070	1.49036 + 1	−2.12148
94	Pu	239.100	1.48535 + 1	−1.87733

[a] Notation abbreviated as in App. V, Table 4.
Source: From Ref. 50. Reprinted by permission of the U.S. Department of Energy.

APPENDIX VII COEFFICIENTS FOR CALCULATING THE INCOHERENT COLLISION CROSS SECTIONS σ_C (BARNS/ATOM) VIA THE LN-LN REPRESENTATION

Atomic number	Element	A_0	A_1	A_2	A_3
1	H	− 2.15772	1.32685	− 3.05620 − 1[a]	1.85025 − 2
2	He	− 2.56357	2.02536	− 4.48710 − 1	2.79691 − 2
3	Li	− 1.08740	1.03368	− 1.90377 − 1	7.79955 − 3
4	Be	− 6.90079 − 1	9.46448 − 1	− 1.71142 − 1	6.51413 − 3
5	B	− 7.91177 − 1	1.21611	− 2.39087 − 1	1.17686 − 2
6	C	− 9.87878 − 1	1.46693	− 2.93743 − 1	1.56005 − 2
7	N	− 1.23693	1.74510	− 3.54660 − 1	1.98705 − 2
8	O	− 1.73679	2.17686	− 4.49050 − 1	2.64733 − 2
9	F	− 1.87570	2.32016	− 4.75412 − 1	2.80680 − 2
10	Ne	− 1.75510	2.24226	− 4.47640 − 1	2.55801 − 2
11	Na	− 9.67717 − 1	1.61794	− 2.87191 − 1	1.31526 − 2
12	Mg	− 5.71611 − 1	1.35498	− 2.22491 − 1	8.30141 − 3
13	Al	− 4.39322 − 1	1.30867	− 2.11648 − 1	7.54210 − 3
14	Si	− 4.14971 − 1	1.34868	− 2.22315 − 1	8.41959 − 3
15	P	− 4.76903 − 1	1.46032	− 2.51331 − 1	1.07202 − 2
16	S	− 6.56419 − 1	1.65408	− 2.98623 − 1	1.42979 − 2
17	Cl	− 7.18627 − 1	1.74294	− 3.19429 − 1	1.58429 − 2
18	Ar	− 6.82105 − 1	1.74279	− 3.17646 − 1	1.56467 − 2
19	K	− 3.44007 − 1	1.49236	− 2.54135 − 1	1.07684 − 2
20	Ca	− 9.82420 − 2	1.32829	− 2.13747 − 1	7.73065 − 3
21	Sc	− 1.59831 − 1	1.39055	− 2.25849 − 1	8.51954 − 3
22	Ti	− 2.30573 − 1	1.45848	− 2.39160 − 1	9.38528 − 3
23	V	− 3.08103 − 1	1.52879	− 2.52768 − 1	1.02571 − 2
24	Cr	− 3.87641 − 1	1.59727	− 2.66240 − 1	1.11523 − 2
25	Mn	− 2.47059 − 1	1.49722	− 2.38781 − 1	8.93208 − 3
26	Fe	− 3.42379 − 1	1.57245	− 2.53198 − 1	9.85822 − 3
27	Co	− 4.28804 − 1	1.64129	− 2.66013 − 1	1.06512 − 2
28	Ni	− 5.04360 − 1	1.70040	− 2.76443 − 1	1.12628 − 2
29	Cu	− 5.70210 − 1	1.75042	− 2.84555 − 1	1.16930 − 2
30	Zn	− 4.20535 − 1	1.63400	− 2.53646 − 1	9.27233 − 3
31	Ga	− 3.58218 − 1	1.60050	− 2.44908 − 1	8.61898 − 3
32	Ge	− 3.34383 − 1	1.60237	− 2.45555 − 1	8.71239 − 3
33	As	− 3.39189 − 1	1.62535	− 2.50783 − 1	9.09103 − 3
34	Se	− 4.32927 − 1	1.72833	− 2.77138 − 1	1.11735 − 2
35	Br	− 4.48001 − 1	1.76082	− 2.85099 − 1	1.17865 − 2
36	Kr	− 3.91810 − 1	1.73010	− 2.76824 − 1	1.11280 − 2
37	Rb	− 1.28039 − 1	1.53044	− 2.27403 − 1	7.39033 − 3
38	Sr	7.99161 − 2	1.38397	− 1.92225 − 1	4.78611 − 3
39	Y	6.29057 − 2	1.41577	− 1.99713 − 1	5.33312 − 3
40	Zr	3.66697 − 2	1.45207	− 2.08122 − 1	5.95139 − 3
41	Nb	2.02289 − 4	1.49347	− 2.17419 − 1	6.62245 − 3
42	Mo	− 5.62860 − 2	1.55778	− 2.33341 − 1	7.85506 − 3
43	Tc	7.57616 − 2	1.44950	− 2.04890 − 1	5.64745 − 3
44	Ru	− 4.24981 − 2	1.54639	− 2.26470 − 1	7.18375 − 3

Continued

Atomic number	Element	A_0	A_1	A_2	A_3
45	Rh	$-1.60399-1$	1.64861	$-2.50238-1$	$8.93818-3$
46	Pd	$-2.67564-1$	1.73740	$-2.69883-1$	$1.03248-2$
47	Ag	$-1.66475-1$	1.65794	$-2.48740-1$	$8.66218-3$
48	Ce	$-5.16701-2$	1.57426	$-2.27646-1$	$7.05650-3$
49	In	$-8.17283-3$	1.55865	$-2.24492-1$	$6.85776-3$
50	Sn	$1.42151-2$	1.55754	$-2.24736-1$	$6.91395-3$
51	Sb	$1.56362-2$	1.57175	$-2.28753-1$	$7.26386-3$
52	Te	$-4.07579-2$	1.64267	$-2.47897-1$	$8.80567-3$
53	I	$-4.04420-2$	1.65596	$-2.51067-1$	$9.04874-3$
54	Xe	$-2.82407-3$	1.64039	$-2.47642-1$	$8.82144-3$
55	Cs	$1.84861-1$	1.50030	$-2.13333-1$	$6.24264-3$
56	Ba	$3.44376-1$	1.38742	$-1.86356-1$	$4.24917-3$
57	La	$4.09104-1$	1.33075	$-1.70883-1$	$3.04111-3$
58	Ce	$4.39881-1$	1.30925	$-1.64548-1$	$2.52641-3$
59	Pr	$4.49124-1$	1.30351	$-1.61841-1$	$2.27394-3$
60	Nd	$4.37283-1$	1.31370	$-1.62866-1$	$2.29377-3$
61	Pm	$4.05823-1$	1.33837	$-1.67229-1$	$2.55570-3$
62	Sm	$3.55383-1$	1.37733	$-1.74941-1$	$3.06213-3$
63	Eu	$2.80316-1$	1.44016	$-1.88641-1$	$4.01226-3$
64	Gd	$2.73133-1$	1.43842	$-1.86137-1$	$3.75240-3$
65	Tb	$2.57539-1$	1.45064	$-1.87591-1$	$3.79932-3$
66	Dy	$2.42685-1$	1.46266	$-1.89102-1$	$3.85628-3$
67	Ho	$2.28493-1$	1.47438	$-1.90559-1$	$3.90903-3$
68	Er	$2.15233-1$	1.48545	$-1.91908-1$	$3.95645-3$
69	Tm	$2.02656-1$	1.49625	$-1.93234-1$	$4.00233-3$
70	Yb	$2.02248-1$	1.48804	$-1.89143-1$	$3.62264-3$
71	Lu	$1.97176-1$	1.50264	$-1.92474-1$	$3.85751-3$
72	Hf	$1.99469-1$	1.50233	$-1.91385-1$	$3.74011-1$
73	Ta	$1.96871-1$	1.50623	$-1.91396-1$	$3.70889-3$
74	W	$1.91015-1$	1.51240	$-1.91922-1$	$3.71450-3$
75	Re	$1.89644-1$	1.50867	$-1.89570-1$	$3.49584-3$
76	Os	$1.16448-1$	1.57615	$-2.05532-1$	$4.66731-3$
77	Ir	$7.19908-2$	1.61204	$-2.13186-1$	$5.20497-3$
78	Pt	$4.20186-2$	1.63611	$-2.17964-1$	$5.52670-3$
79	Au	$1.56916-2$	1.65406	$-2.20982-1$	$5.70751-3$
80	Hg	$1.14587-1$	1.58076	$-2.02968-1$	$4.35692-3$
81	Tl	$1.47052-1$	1.56695	$-2.00347-1$	$4.20901-3$
82	Pb	$1.82167-1$	1.54661	$-1.95793-1$	$3.90772-3$
83	Bi	$1.89860-1$	1.56125	$-2.00932-1$	$4.36768-3$
86	Rn	$1.96619-1$	1.60080	$-2.13800-1$	$5.51717-3$
90	Th	$1.70890-1$	1.65561	$-2.29702-1$	$6.92516-3$
92	U	$1.08277-1$	1.74158	$-2.54104-1$	$8.95056-3$
94	Pu	$3.88791-2$	1.82229	$-2.76009-1$	$1.07392-2$

[a] Notation as in App. V, Table 4.
Source: From Ref. 50. Reprinted by permission of the U.S. Department of Energy.

APPENDIX VIII COEFFICIENTS FOR CALCULATING THE COHERENT SCATTERING CROSS SECTIONS σ_R (BARNS/ATOM) VIA THE LN-LN REPRESENTATION

Atomic number	Element	A_0	A_1	A_2	A_3
1	H	− 1.19075 − 1[a]	− 9.37086 − 1	− 2.00538 − 1	1.06587 − 2
2	He	1.04768	− 8.51805 − 2	− 4.03527 − 1	2.69398 − 2
3	Li	1.34366	1.81557 − 1	− 4.23981 − 1	2.66190 − 2
4	Be	2.00860	− 4.61920 − 2	− 3.37018 − 1	1.86939 − 2
5	B	2.61862	− 2.07916 − 1	− 2.86283 − 1	1.44966 − 2
6	C	3.10861	− 2.60580 − 1	− 2.71974 − 1	1.35181 − 2
7	N	3.47760	− 2.15762 − 1	− 2.88874 − 1	1.51312 − 2
8	O	3.77239	− 1.48539 − 1	− 3.07124 − 1	1.67303 − 2
9	F	4.00716	− 5.60908 − 2	− 3.32017 − 1	1.87934 − 2
10	Ne	4.20151	4.16247 − 2	− 3.56754 − 1	2.07585 − 2
11	Na	4.26374	1.34662 − 1	− 3.70080 − 1	2.14467 − 2
12	Mg	4.39404	1.37858 − 1	− 3.59540 − 1	2.02380 − 2
13	Al	4.51995	1.40549 − 1	− 3.52441 − 1	1.93692 − 2
14	Si	4.64678	1.62780 − 1	− 3.58563 − 1	1.96926 − 2
15	P	4.76525	1.68708 − 1	− 3.60383 − 1	1.97155 − 2
16	S	4.92707	1.65746 − 1	− 3.59424 − 1	1.95505 − 2
17	Cl	5.07222	1.49127 − 1	− 3.52858 − 1	1.89439 − 2
18	Ar	5.21079	1.35618 − 1	− 3.47214 − 1	1.84333 − 2
19	K	5.25587	1.88040 − 1	− 3.59623 − 1	1.93085 − 2
20	Ca	5.32375	2.06685 − 1	− 3.61664 − 1	1.93328 − 2
21	Sc	5.43942	2.00174 − 1	− 3.59064 − 1	1.91027 − 2
22	Ti	5.55039	1.97697 − 1	− 3.57694 − 1	1.89866 − 2
23	V	5.65514	1.99533 − 1	− 3.57487 − 1	1.89691 − 2
24	Cr	5.77399	2.03858 − 1	− 3.59699 − 1	1.92225 − 2
25	Mn	5.84604	2.13814 − 1	− 3.59718 − 1	1.91459 − 2
26	Fe	5.93292	2.25048 − 1	− 3.61748 − 1	1.93024 − 2
27	Co	6.01478	2.37959 − 1	− 3.64056 − 1	1.94754 − 2
28	Ni	6.09204	2.52277 − 1	− 3.66568 − 1	1.96586 − 2
29	Cu	6.17739	2.73123 − 1	− 3.72360 − 1	2.01638 − 2
30	Zn	6.23402	2.84312 − 1	− 3.72143 − 1	2.00525 − 2
31	Ga	6.28298	2.91334 − 1	− 3.69391 − 1	1.97029 − 2
32	Ge	6.33896	2.91512 − 1	− 3.65643 − 1	1.92896 − 2
33	As	6.39750	2.88866 − 1	− 3.61747 − 1	1.88788 − 2
34	Se	6.45637	2.86737 − 1	− 3.58794 − 1	1.85618 − 2
35	Br	6.51444	2.86324 − 1	− 3.57027 − 1	1.83557 − 2
36	Kr	6.57129	2.87711 − 1	− 3.56311 − 1	1.82470 − 2
37	Rb	6.59750	3.02389 − 1	− 3.56755 − 1	1.81706 − 2
38	Sr	6.62203	3.24559 − 1	− 3.61651 − 1	1.84800 − 2
39	Y	6.67096	3.25075 − 1	− 3.60613 − 1	1.83325 − 2
40	Zr	6.72275	3.23964 − 1	− 3.59463 − 1	1.81890 − 2
41	Nb	6.79013	3.11282 − 1	− 3.55233 − 1	1.78231 − 2
42	Mo	6.84600	3.02797 − 1	− 3.51131 − 1	1.74403 − 2
43	Tc	6.87599	3.26165 − 1	− 3.58969 − 1	1.80482 − 2

Continued

Atomic number	Element	A_0	A_1	A_2	A_3
44	Ru	6.93136	$3.34794 - 1$	$-3.63497 - 1$	$1.84429 - 2$
45	Rh	6.97547	$3.46394 - 1$	$-3.67794 - 1$	$1.87885 - 2$
46	Pd	7.03216	$3.49838 - 1$	$-3.70099 - 1$	$1.89983 - 2$
47	Ag	7.06446	$3.63456 - 1$	$-3.73597 - 1$	$1.92478 - 2$
48	Cd	7.09856	$3.72199 - 1$	$-3.75345 - 1$	$1.93481 - 2$
49	In	7.12708	$3.82082 - 1$	$-3.76855 - 1$	$1.94151 - 2$
50	Sn	7.16085	$3.85512 - 1$	$-3.76481 - 1$	$1.93305 - 2$
51	Sb	7.19665	$3.85543 - 1$	$-3.75054 - 1$	$1.91608 - 2$
52	Te	7.23464	$3.82493 - 1$	$-3.72715 - 1$	$1.89194 - 2$
53	I	7.27415	$3.77223 - 1$	$-3.69728 - 1$	$1.86280 - 2$
54	Xe	7.31469	$3.70315 - 1$	$-3.66280 - 1$	$1.83025 - 2$
55	Cs	7.33490	$3.76825 - 1$	$-3.65713 - 1$	$1.81843 - 2$
56	Ba	7.35812	$3.79361 - 1$	$-3.64099 - 1$	$1.79817 - 2$
57	La	7.39532	$3.69895 - 1$	$-3.59376 - 1$	$1.75406 - 2$
58	Ce	7.44255	$3.71328 - 1$	$-3.59642 - 1$	$1.75852 - 2$
59	Pr	7.48347	$3.68431 - 1$	$-3.57689 - 1$	$1.74099 - 2$
60	Nd	7.52334	$3.66462 - 1$	$-3.56048 - 1$	$1.72620 - 2$
61	Pm	7.56222	$3.65055 - 1$	$-3.54511 - 1$	$1.71214 - 2$
62	Sm	7.60020	$3.64134 - 1$	$-3.53086 - 1$	$1.69894 - 2$
63	Eu	7.63711	$3.63957 - 1$	$-3.51909 - 1$	$1.68783 - 2$
64	Gd	7.66938	$3.59752 - 1$	$-3.48899 - 1$	$1.65890 - 2$
65	Tb	7.70798	$3.65345 - 1$	$-3.50031 - 1$	$1.66927 - 2$
66	Dy	7.74188	$3.67107 - 1$	$-3.49433 - 1$	$1.66273 - 2$
67	Ho	7.77470	$3.69722 - 1$	$-3.49132 - 1$	$1.65862 - 2$
68	Er	7.80643	$3.73226 - 1$	$-3.49147 - 1$	$1.65710 - 2$
69	Tm	7.83711	$3.77547 - 1$	$-3.49441 - 1$	$1.65780 - 2$
70	Yb	7.86662	$3.82933 - 1$	$-3.50126 - 1$	$1.66173 - 2$
71	Lu	7.89137	$3.86034 - 1$	$-3.49756 - 1$	$1.65480 - 2$
72	Hf	7.91803	$3.87021 - 1$	$-3.48881 - 1$	$1.64406 - 2$
73	Ta	7.94534	$3.87299 - 1$	$-3.47926 - 1$	$1.63299 - 2$
74	W	7.97266	$3.87704 - 1$	$-3.47155 - 1$	$1.62372 - 2$
75	Re	7.99940	$3.88739 - 1$	$-3.46726 - 1$	$1.61751 - 2$
76	Os	8.02574	$3.90458 - 1$	$-3.46658 - 1$	$1.61455 - 2$
77	Ir	8.05150	$3.93143 - 1$	$-3.47052 - 1$	$1.61573 - 2$
78	Pt	8.08084	$3.95790 - 1$	$-3.48032 - 1$	$1.62345 - 2$
79	Au				
80	Hg				
81	Tl				
82	Pb				
83	Bi				
86	Rn				
90	Th				
92	U				
94	Pu				

[a] Notation as in App. V, Table 4.

Source: From Ref. 50. Reprinted by permission of the U.S. Department of Energy.

APPENDIX IX TOTAL MASS ATTENUATION COEFFICIENTS FOR LOW-ENERGY $K\alpha$ LINES

Absorber		Emitter wavelength (Å) energy (keV)								
		B	C	N	O	F	Ne	Na	Mg	Al
Atomic		$6.67+1^a$	$4.48+1$	$3.16+1$	$2.36+1$	$1.83+1$	$1.46+1$	$1.19+1$	$9.89+0$	$8.34+0$
number	Element	$1.83-1$	$2.77-1$	$3.92-1$	$5.25-1$	$6.77-1$	$8.49-1$	$1.04+0$	$1.25+0$	$1.49+0$
1	H	$1.81+3$	$4.89+2$	$1.51+2$	$5.85+1$	$2.62+1$	$1.22+1$	$6.32+0$	$3.76+0$	$2.23+0$
2	He	$1.20+4$	$4.06+3$	$1.25+3$	$5.07+2$	$2.23+2$	$1.06+2$	$5.58+1$	$3.15+1$	$1.77+1$
3	Li	$2.98+4$	$1.08+4$	$3.67+3$	$1.60+3$	$7.57+2$	$3.84+2$	$2.09+2$	$1.20+2$	$6.82+1$
4	Be	$5.82+4$	$2.29+4$	$8.40+3$	$3.78+3$	$1.86+3$	$9.59+2$	$5.44+2$	$3.26+2$	$1.89+2$
5	B	$3.93+3$	$3.75+4$	$1.53+4$	$7.22+3$	$3.63+3$	$1.93+3$	$1.10+3$	$6.73+2$	$3.94+2$
6	C	$7.32+3$	$2.75+3$	$2.44+4$	$1.22+4$	$6.29+3$	$3.43+3$	$1.98+3$	$1.21+3$	$7.22+2$
7	N	$1.19+4$	$4.71+3$	$1.83+3$	$1.73+4$	$9.23+3$	$5.14+3$	$3.02+3$	$1.90+3$	$1.14+3$
8	O	$1.74+4$	$6.90+3$	$2.68+3$	$1.27+3$	$1.21+4$	$6.94+3$	$4.17+3$	$2.67+3$	$1.63+3$
9	F	$2.45+4$	$9.73+3$	$3.73+3$	$1.79+3$	$9.12+2$	$8.54+3$	$5.23+3$	$3.44+3$	$2.11+3$
10	Ne	$3.72+4$	$1.36+4$	$5.70+3$	$2.71+3$	$1.39+3$	$7.54+2$	$7.11+3$	$4.55+3$	$2.84+3$
11	Na	$4.25+4$	$1.80+4$	$7.26+3$	$3.52+3$	$1.83+3$	$1.01+3$	$5.93+2$	$5.65+3$	$3.53+3$
12	Mg	$5.62+4$	$2.44+4$	$1.00+4$	$4.89+3$	$2.55+3$	$1.42+3$	$8.37+2$	$5.06+2$	$4.39+3$
13	Al	$6.44+4$	$2.92+4$	$1.23+4$	$6.10+3$	$3.22+3$	$1.80+3$	$1.07+3$	$6.63+2$	$4.07+2$
14	Si	$7.48+4$	$3.47+4$	$1.58+4$	$7.93+3$	$4.22+3$	$2.37+3$	$1.41+3$	$8.93+2$	$5.52+2$
15	P	$7.49+4$	$4.19+4$	$1.93+4$	$9.74+3$	$5.19+3$	$2.92+3$	$1.74+3$	$1.09+3$	$6.78+2$
16	S	$8.18+4$	$4.95+4$	$2.38+4$	$1.21+4$	$6.52+3$	$3.69+3$	$2.21+3$	$1.39+3$	$8.64+2$
17	Cl	$7.44+3$	$5.17+4$	$2.67+4$	$1.39+4$	$7.50+3$	$4.27+3$	$2.57+3$	$1.63+3$	$1.01+3$
18	Ar	$9.51+3$	$5.24+4$	$3.00+4$	$1.59+4$	$8.59+3$	$4.89+3$	$2.93+3$	$1.85+3$	$1.15+3$
19	K	$1.14+4$	$5.92+3$	$3.51+4$	$1.94+4$	$1.07+4$	$6.13+3$	$3.71+3$	$2.36+3$	$1.47+3$
20	Ca	$1.31+4$	$7.15+3$	$3.49+4$	$2.21+4$	$1.24+4$	$7.19+3$	$4.38+3$	$2.75+3$	$1.74+3$
21	Sc	$1.43+4$	$7.80+3$	$3.97+3$	$2.34+4$	$1.34+4$	$7.79+3$	$4.74+3$	$2.99+3$	$1.90+3$
22	Ti	$1.54+4$	$8.45+3$	$4.35+3$	$2.20+4$	$1.45+4$	$8.59+3$	$5.27+3$	$3.34+3$	$2.13+3$
23	V	$1.68+4$	$9.22+3$	$4.77+3$	$1.66+4$	$1.57+4$	$9.40+3$	$5.81+3$	$3.70+3$	$2.36+3$
24	Cr	$2.09+4$	$1.11+4$	$5.61+3$	$3.13+3$	$1.58+4$	$1.09+4$	$6.75+3$	$4.30+3$	$2.74+3$
25	Mn	$2.23+4$	$1.20+4$	$6.13+3$	$3.46+3$	$1.68+4$	$1.16+4$	$7.27+3$	$4.67+3$	$3.00+3$
26	Fe	$2.61+4$	$1.40+4$	$7.10+3$	$4.00+3$	$2.31+3$	$1.31+4$	$8.22+3$	$5.31+3$	$3.42+3$
27	Co	$2.86+4$	$1.54+4$	$7.85+3$	$4.41+3$	$2.58+3$	$1.22+4$	$8.84+3$	$5.73+3$	$3.71+3$
28	Ni	$3.34+4$	$1.81+4$	$9.19+3$	$5.25+3$	$3.01+3$	$1.80+3$	$1.02+4$	$6.58+3$	$4.26+3$
29	Cu	$3.65+4$	$1.96+4$	$9.96+3$	$5.94+3$	$3.20+3$	$2.01+3$	$9.47+3$	$7.04+3$	$4.52+3$
30	Zn	$4.03+4$	$2.26+4$	$1.17+4$	$6.56+3$	$3.73+3$	$2.25+3$	$5.26+3$	$7.51+3$	$4.96+3$
31	Ga	$4.18+4$	$2.42+4$	$1.26+4$	$7.10+3$	$4.08+3$	$2.47+3$	$1.52+3$	$6.92+3$	$5.23+3$
32	Ge	$4.41+4$	$2.64+4$	$1.40+4$	$7.89+3$	$4.56+3$	$2.75+3$	$1.70+3$	$4.72+3$	$5.63+3$
33	As	$4.41+4$	$2.86+4$	$1.55+4$	$8.80+3$	$5.10+3$	$3.08+3$	$1.95+3$	$1.30+3$	$5.32+3$
34	Se	$4.49+4$	$3.03+4$	$1.69+4$	$9.64+3$	$5.61+3$	$3.38+3$	$2.17+3$	$1.53+3$	$4.86+3$
35	Br	$4.19+4$	$3.22+4$	$1.86+4$	$1.07+4$	$6.28+3$	$3.87+3$	$2.44+3$	$1.68+3$	$1.01+3$
36	Kr	$3.94+4$	$3.28+4$	$2.05+4$	$1.19+4$	$6.98+3$	$4.23+3$	$2.66+3$	$1.72+3$	$1.14+3$
37	Rb	$3.52+4$	$3.56+4$	$2.17+4$	$1.28+4$	$7.57+3$	$4.63+3$	$2.92+3$	$1.89+3$	$1.25+3$
38	Sr	$2.83+4$	$3.47+4$	$2.29+4$	$1.37+4$	$8.21+3$	$5.04+3$	$3.20+3$	$2.08+3$	$1.38+3$
39	Y	$2.04+4$	$3.10+4$	$2.34+4$	$1.51+4$	$9.06+3$	$5.58+3$	$3.55+3$	$2.32+3$	$1.52+3$
40	Zr	$6.41+3$	$3.06+4$	$2.46+4$	$1.61+4$	$9.77+3$	$6.03+3$	$3.86+3$	$2.53+3$	$1.66+3$
41	Nb	$4.28+3$	$2.96+4$	$2.67+4$	$1.78+4$	$1.08+4$	$6.69+3$	$4.27+3$	$2.80+3$	$1.84+3$
42	Mo	$4.66+3$	$2.23+4$	$2.29+4$	$1.85+4$	$1.14+4$	$7.09+3$	$4.55+3$	$2.99+3$	$1.97+3$
43	Tc	$4.99+3$	$1.41+4$	$2.29+4$	$1.78+4$	$1.19+4$	$7.45+3$	$4.90+3$	$3.57+3$	$2.12+3$
44	Ru	$5.49+3$	$4.18+3$	$2.47+4$	$1.95+4$	$1.30+4$	$8.16+3$	$5.37+3$	$3.92+3$	$2.32+3$
45	Rh	$5.87+3$	$4.64+3$	$2.48+4$	$1.94+4$	$1.38+4$	$8.79+3$	$5.80+3$	$4.23+3$	$2.51+3$
46	Pd	$6.09+3$	$5.22+3$	$2.21+4$	$1.80+4$	$1.46+4$	$9.42+3$	$6.23+3$	$4.54+3$	$2.69+3$
47	Ag	$6.34+3$	$5.48+3$	$1.10+4$	$1.91+4$	$1.47+4$	$1.00+4$	$6.66+3$	$4.86+3$	$2.90+3$

Continued

Absorber		Emitter wavelength (Å) energy (keV)								
		B	C	N	O	F	Ne	Na	Mg	Al
Atomic		$6.67+1^a$	$4.48+1$	$3.16+1$	$2.36+1$	$1.83+1$	$1.46+1$	$1.19+1$	$9.89+0$	$8.34+0$
number	Element	$1.83-1$	$2.77-1$	$3.92-1$	$5.25-1$	$6.77-1$	$8.49-1$	$1.04+0$	$1.25+0$	$1.49+0$
48	Cd	$6.45+3$	$5.75+3$	$4.10+3$	$2.01+4$	$1.47+4$	$1.04+4$	$7.09+3$	$5.84+3$	$4.48+3$
49	In	$6.61+3$	$6.06+3$	$4.42+3$	$2.23+4$	$1.36+4$	$1.10+4$	$7.51+3$	$6.20+3$	$4.75+3$
50	Sn	$6.64+3$	$6.27+3$	$4.70+3$	$9.12+3$	$1.38+4$	$1.08+4$	$7.88+3$	$6.50+3$	$4.99+3$
51	Sb	$6.71+3$	$6.50+3$	$4.94+3$	$3.36+3$	$1.48+4$	$1.13+4$	$8.26+3$	$6.82+3$	$5.25+3$
52	Te	$6.65+3$	$6.58+3$	$5.09+3$	$3.53+3$	$1.75+4$	$1.13+4$	$8.35+3$	$6.90+3$	$5.32+3$
53	I	$5.77+3$	$6.99+3$	$5.52+3$	$3.88+3$	$1.18+4$	$1.02+4$	$8.24+3$	$6.93+3$	$5.39+3$
54	Xe	$5.79+3$	$7.27+3$	$5.88+3$	$4.17+3$	$2.84+3$	$1.08+4$	$8.72+3$	$6.75+3$	$5.30+3$
55	Cs	$5.91+3$	$7.52+3$	$6.13+3$	$4.40+3$	$3.02+3$	$1.15+4$	$9.79+3$	$6.67+3$	$5.29+3$
56	Ba	$4.09+3$	$7.64+3$	$6.21+3$	$4.52+3$	$3.13+3$	$1.16+4$	$7.51+3$	$6.36+3$	$5.21+3$
57	La	$3.76+3$	$7.74+3$	$6.32+3$	$4.66+3$	$3.26+3$	$4.21+3$	$7.89+3$	$6.56+3$	$5.22+3$
58	Ce	$6.79+3$	$7.98+3$	$7.14+3$	$5.16+3$	$3.57+3$	$2.45+3$	$8.81+3$	$7.12+3$	$5.43+3$
59	Pr	$8.74+3$	$8.77+3$	$7.60+3$	$5.45+3$	$3.77+3$	$2.59+3$	$9.49+3$	$6.10+3$	$5.26+3$
60	Nd	$1.05+4$	$9.43+3$	$7.92+3$	$5.64+3$	$3.89+3$	$2.68+3$	$8.29+3$	$6.19+3$	$5.47+3$
61	Pm	$1.26+4$	$1.04+4$	$8.40+3$	$5.92+3$	$4.07+3$	$2.79+3$	$2.51+3$	$6.59+3$	$5.73+3$
62	Sm	$1.48+4$	$1.13+4$	$8.95+3$	$6.26+3$	$4.27+3$	$2.93+3$	$2.01+3$	$6.94+3$	$5.84+3$
63	Eu	$1.58+4$	$1.20+4$	$9.42+3$	$6.58+3$	$4.49+3$	$3.08+3$	$2.12+3$	$7.32+3$	$4.88+3$
64	Gd	$1.49+4$	$1.12+4$	$9.39+3$	$6.60+3$	$4.53+3$	$3.12+3$	$2.17+3$	$4.57+3$	$5.01+3$
65	Tb	$1.87+4$	$1.24+4$	$9.93+3$	$7.38+3$	$4.99+3$	$3.39+3$	$2.34+3$	$1.61+3$	$5.49+3$
66	Dy	$2.04+4$	$1.37+4$	$1.06+4$	$7.83+3$	$5.26+3$	$3.56+3$	$2.47+3$	$1.73+3$	$5.68+3$
67	Ho	$2.16+4$	$1.47+4$	$1.13+4$	$8.29+3$	$5.56+3$	$3.76+3$	$2.61+3$	$1.87+3$	$5.70+3$
68	Er	$2.32+4$	$1.60+4$	$1.21+4$	$8.86+3$	$5.94+3$	$4.00+3$	$2.77+3$	$2.01+3$	$2.84+3$
69	Tm	$2.43+4$	$1.72+4$	$1.29+4$	$9.49+3$	$6.36+3$	$4.27+3$	$2.96+3$	$2.14+3$	$1.56+3$
70	Yb	$2.28+4$	$1.79+4$	$1.35+4$	$9.94+3$	$6.67+3$	$4.47+3$	$3.10+3$	$2.32+3$	$1.47+3$
71	Lu	$2.15+4$	$1.75+4$	$1.36+4$	$1.01+4$	$6.86+3$	$4.62+3$	$3.22+3$	$2.42+3$	$1.54+3$
72	Hf	$2.13+4$	$1.80+4$	$1.27+4$	$9.96+3$	$7.20+3$	$4.86+3$	$3.38+3$	$2.54+3$	$1.62+3$
73	Ta	$2.04+4$	$1.82+4$	$1.32+4$	$1.04+4$	$7.54+3$	$5.11+3$	$3.53+3$	$2.45+3$	$1.77+3$
74	W	$1.93+4$	$1.85+4$	$1.37+4$	$1.09+4$	$7.90+3$	$5.36+3$	$3.70+3$	$2.57+3$	$1.86+3$
75	Re	$1.76+4$	$1.84+4$	$1.39+4$	$1.12+4$	$8.18+3$	$5.57+3$	$3.87+3$	$2.69+3$	$1.95+3$
76	Os	$1.58+4$	$1.67+4$	$1.39+4$	$1.12+4$	$8.36+3$	$5.74+3$	$4.02+3$	$2.92+3$	$1.99+3$
77	Ir	$1.38+4$	$1.61+4$	$1.44+4$	$1.15+4$	$8.36+3$	$6.09+3$	$4.27+3$	$3.11+3$	$2.12+3$
78	Pt	$1.13+4$	$1.57+4$	$1.50+4$	$1.12+4$	$8.83+3$	$6.43+3$	$4.51+3$	$3.28+3$	$2.24+3$
79	Au	$9.09+3$	$1.49+4$	$1.51+4$	$1.16+4$	$9.23+3$	$6.74+3$	$4.68+3$	$3.28+3$	$2.29+3$
80	Hg	$6.62+3$	$1.47+4$	$1.50+4$	$1.17+4$	$9.30+3$	$6.95+3$	$4.77+3$	$3.31+3$	$2.36+3$
81	Tl	$5.42+3$	$1.29+4$	$1.40+4$	$1.21+4$	$9.55+3$	$7.24+3$	$5.13+3$	$3.88+3$	$2.50+3$
82	Pb	$4.19+3$	$1.14+4$	$1.38+4$	$1.23+4$	$8.93+3$	$7.10+3$	$5.34+3$	$4.04+3$	$2.61+3$
83	Bi	$3.17+3$	$1.01+4$	$1.35+4$	$1.26+4$	$9.23+3$	$7.35+3$	$5.55+3$	$4.20+3$	$2.72+3$
84	Po	$2.54+3$	$8.62+3$	$1.34+4$	$1.33+4$	$9.68+3$	$7.72+3$	$5.82+3$	$4.42+3$	$2.88+3$
85	At	$2.28+3$	$6.67+3$	$1.31+4$	$1.23+4$	$1.02+4$	$8.14+3$	$5.70+3$	$4.57+3$	$3.04+3$
86	Rn	$2.52+3$	$4.13+3$	$1.21+4$	$1.16+4$	$1.02+4$	$8.26+3$	$5.72+3$	$4.46+3$	$3.04+3$
87	Fr	$2.69+3$	$3.02+3$	$1.10+4$	$1.17+4$	$1.05+4$	$7.52+3$	$5.99+3$	$4.57+3$	$3.18+3$
88	Ra	$2.66+3$	$2.42+3$	$9.31+3$	$1.17+4$	$1.04+4$	$7.75+3$	$6.22+3$	$4.68+3$	$3.30+3$
89	Ac	$2.49+3$	$2.59+3$	$8.17+3$	$1.15+4$	$1.01+4$	$7.88+3$	$6.35+3$	$4.49+3$	$3.40+3$
90	Th	$1.99+3$	$2.12+3$	$6.44+3$	$1.11+4$	$8.83+3$	$7.85+3$	$6.22+3$	$4.64+3$	$3.48+3$
91	Pa	$2.34+3$	$2.38+3$	$3.44+3$	$1.13+4$	$9.45+3$	$8.35+3$	$5.96+3$	$4.97+3$	$3.71+3$
92	U	$2.57+3$	$2.30+3$	$2.70+3$	$1.09+4$	$9.39+3$	$8.36+3$	$6.03+3$	$4.96+3$	$3.76+3$
93	Np	$2.93+3$	$2.41+3$	$2.62+3$	$9.65+3$	$9.74+3$	$8.59+3$	$6.34+3$	$5.21+3$	$3.74+3$
94	Pu	$3.73+3$	$2.39+3$	$2.72+3$	$6.59+3$	$1.01+4$	$8.24+3$	$6.66+3$	$5.40+3$	$3.89+3$

^{a} Notation as in App. V, Table 4.

Source: From Ref. 64. Reprinted by permission of CRC Press, Inc.

APPENDIX X CORRESPONDENCE BETWEEN OLD SIEGBAHN AND NEW IUPAC NOTATION X-RAY DIAGRAM LINES

Siegbahn	IUPAC	Siegbahn	IUPAC	Siegbahn	IUPAC
$K\alpha_1$	$K\text{-}L_3$	$L\alpha_1$	$L_3\text{-}M_5$	$L\gamma_1$	$L_2\text{-}N_4$
$K\alpha_2$	$K\text{-}L_2$	$L\alpha_2$	$L_3\text{-}M_4$	$L\gamma_2$	$L_1\text{-}N_2$
$K\beta_1$	$K\text{-}M_3$	$L\beta_1$	$L_2\text{-}M_4$	$L\gamma_3$	$L_1\text{-}N_3$
$K\beta_2^{I}$	$K\text{-}N_3$	$L\beta_2$	$L_3\text{-}N_5$	$L\gamma_4$	$L_1\text{-}O_3$
$K\beta_2^{II}$	$K\text{-}N_2$	$L\beta_3$	$L_1\text{-}M_3$	$L\gamma_4'$	$L_1\text{-}O_2$
$K\beta_3$	$K\text{-}M_2$	$L\beta_4$	$L_1\text{-}M_2$	$L\gamma_5$	$L_2\text{-}N_1$
$K\beta_4^{I}$	$K\text{-}N_5$	$L\beta_5$	$L_3\text{-}O_{4,5}$	$L\gamma_6$	$L_2\text{-}O_4$
$K\beta_4^{II}$	$K\text{-}N_4$	$L\beta_6$	$L_3\text{-}N_1$	$L\gamma_8$	$L_2\text{-}O_1$
$K\beta_{4x}$	$K\text{-}N_4$	$L\beta_7$	$L_3\text{-}O_1$	$L\gamma_8'$	$L_2\text{-}N_{6(7)}$
$K\beta_5^{I}$	$K\text{-}M_5$	$L\beta_7'$	$L_3\text{-}N_{6,7}$	$l\eta$	$L_2\text{-}M_1$
$K\beta_5^{II}$	$K\text{-}M_4$	$L\beta_9$	$L_1\text{-}M_5$	Ll	$L_3\text{-}M_1$
		$L\beta_{10}$	$L_1\text{-}M_4$	Ls	$L_3\text{-}M_3$
		$L\beta_{15}$	$L_3\text{-}N_4$	Lt	$L_3\text{-}M_2$
		$L\beta_{17}$	$L_2\text{-}M_3$	Lu	$L_3\text{-}N_{6,7}$
				Lv	$L_2\text{-}N_{6(7)}$
				$M\alpha_1$	$M_5\text{-}N_7$
				$M\alpha_2$	$M_5\text{-}N_6$
				$M\beta$	$M_4\text{-}N_6$
				$M\gamma$	$M_3\text{-}N_5$
				$M\zeta$	$M_{4,5}\text{-}N_{2,3}$

Source: From Ref. 83.

REFERENCES

1. H. Kulenkampff, *Ann. Phys.* 69:548 (1922)
2. H. Kulenkampff, *Handbuch der Physik,* Geiger-Scheel, Berlin, *Bd.* 23: p. 433 (1933).
3. E. U. Condon, in *Handbook of Physics,* Part 7 (E. U. Condon and H. Odishaw, Eds.), McGraw-Hill, New York, 1958, pp. 7–126.
4. H. A. Kramers, *Philos. Mag.* 46:836 (1923).
5. G. Wentzel, *Z. Phys.* 27:257 (1924).
6. N. A. Dyson, *X-rays in Atomic and Nuclear Physics,* Longman Group, London, 1973, p. 42.
7. S. T. Stephenson, in *Encyclopedia of Physics,* Vol. XXX (S. Flügge, Ed.), Springer-Verlag, Berlin, 1957, p. 337.
8. R. Tertian and N. Broll, *X-Ray Spectrom.* 13:134 (1984).
9. P. J. Statham, *X-Ray Spectrom.* 5:154 (1976).
10. L. Lifshin, *Adv. X-Ray Anal.* 19:113 (1976).
11. W. M. Sherry and J. B. Vander Sande, *X-Ray Spectrom.* 6:154 (1977).
12. D. G. W. Smith and S. J. B. Reed, *X-Ray Spectrom.* 10:198 (1981).
13. A. A. Markowicz and R. E. Van Grieken, *Anal. Chem.* 56:2049 (1984).
14. A. Markowicz, H. Storms, and R. Van Grieken, *X-ray Spectrom.* 15:131 (1986).
15. N. G. Ware and S. J. B. Reed, *J. Phys. E.* 6:286 (1973).
16. P. J. Statham, *Proc. Annu. Conf. Microbeam Anal. Soc.* 14:247 (1979).
17. A. A. Markowicz, H. M. Storms, and R. E. Van Grieken, *Anal. Chem.* 57:2885 (1985).
18. H. Kulenkampff, *Ann. Phys.* 87:597 (1928).
19. O. Scherzer, *Ann. Phys.* 13:137 (1932).

20. R. W. Kenney, in *Encyclopedia of Physics* (R. M. Besançon, Ed.), Reinhold Publishing, New York, 1966, p. 86.
21. H. A. Bethe, *Ann. Phys.* (*Leipzig*) *5*:325 (1930).
22. M. Green and V. E. Cosslett, *Proc. Phys. Soc.* *78*:1206 (1961).
▷23. J. Wernisch, *X-Ray Spectrom.* *14*:109 (1985).
24. K. F. J. Henrich, *Electron Beam X-Ray Microanalysis,* Van Nostrand-Reinhold, New York, 1981, p. 234.
25. J. I. Goldstein, D. E. Newbury, P. Echlin, D. C. Joy, C. Fiori, and E. Lifshin, *Scanning Electron Microscopy and X-ray Microanalysis,* Plenum Press, New York, 1981, p. 105.
26. E. Merzbacher and H. W. Lewis, in *Encyclopedia of Physics,* Vol. 34 (S. Flügge, Ed.), Springer-Verlag, Berlin, 1958, p. 166.
27. G. L. Clark (Ed.), *Encyclopedia of X-Rays and Gamma Rays,* Reinhold Publishing, New York, 1963, p. 1124.
28. A. Burr, *Handbook of Spectroscopy,* Vol. I (J. W. Robinson, Ed.), CRC Press, Cleveland, Ohio, 1974, p. 25.
29. W. Heitler, *The Quantum Theory of Radiation,* 3rd ed., Oxford University Press, London, 1954, p. 207.
30. M. Stobbe, *Ann. Phys.* *7*:661 (1930).
31. F. Sauter, *Ann. Phys.* *9*:217 (1931).
32. A. E. Sandström, in *Encyclopedia of Physics,* Vol. XXX (S. Flügge, Ed.), Springer-Verlag, Berlin, 1957, p. 78.
33. P. A. M. Dirac, *Principles of Quantum Mechanics,* 3rd ed., Oxford, 1947.
34. H. G. J. Moseley, *Philos. Mag.* *27*:703 (1914).
35. T. P. Schreiber and A. M. Wims, *X-Ray Spectrom.* *11*:42 (1982).
36. R. C. West (Ed.), *Handbook of Chemistry and Physics,* 53rd ed., Chemical Rubber Co., Cleveland, 1972–73, p. E-183.
37. G. L. Clark, *Applied X-Rays,* McGraw-Hill, New York, 1955, p. 144.
38. J. Kawai and Y. Gohshi, *Spectrochim. Acta 41B*:265 (1986).
39. P. Auger, *J. Phys.* *6*:205 (1925).
40. E. H. S. Burhop, *The Auger Effect and Other Radiationless Transitions,* Cambridge University Press, New York, 1952.
41. R. W. Fink, *Handbook of Spectroscopy* (J. W. Robinson, Ed.), CRC Press, Cleveland, 1974, p. 219.
42. W. Bambynek, B. Crasemann, R. W. Fink, H. U. Freund, H. Mark, C. D. Swift, R. E. Price, and P. Venugopala Rao, *Rev. Modern Phys.* *44*:716 (1972).
43. W. Hanke, J. Wernisch, and C. Pöhn, *X-Ray Spectrom.* *14*:43 (1985).
44. L. S. Birks, *Electron Probe Microanalysis,* Wiley-Interscience, New York, 1971, p. 51.
45. J. H. Hubbell, *Photon Cross Sections, Attenuation Coefficients, and Energy Absorption Coefficients from 10 keV to 100 GeV,* NSRDS-NBS 29, Nat. Bur. Stand. (U.S.), Washington, D.C., 1969.
46. R. D. Deslattes, *Acta Crystallogr. A25*:89 (1969).
47. R. D. Evans, in *American Institute of Physics Handbook* (D. E. Gray, Ed.), McGraw-Hill, New York, 1963, pp. 8–81.
48. A. H. Compton and S. K. Allison, *X-Rays in Theory and Experiment,* Van Nostrand, New York, 1935, p. 537.
▷49. J. H. Hubbell, W. H. McMaster, N. Kerr Del Grande, and J. H. Mallett, in *International Tables for X-Ray Crystallography,* Vol. 4 (J. A. Ibers and W. C. Hamilton, Eds.), Kynoch Press, Birmingham, England, 1974, p. 47.
50. W. H. McMaster, N. Kerr Del Grande, J. H. Mallett, and J. H. Hubbell, *Compilation of X-Ray Cross Sections,* Lawrence Radiation Laboratory (Livermore) Report UCRL-50174, Sec. II, Rev. 1, University of California, 1969.

51. C. Poehn, J. Wernisch, and W. Hanke, *X-Ray Spectrom.* *14*:120 (1985).
52. P. A. Lee and J. B. Pendry, *Phys. Rev. B11*:2795 (1975).
53. D. Sayers, W. Lytle, and E. Stern, *Adv. X-Ray Anal.* *13*:248 (1970).
54. P. Lagarde, *Nucl. Instrum. Methods 208*:621 (1983).
55. A. H. Compton, *Phys. Rev. 21*:15 (1923).
56. A. H. Compton, *Phys. Rev. 22*:409 (1923).
57. O. Klein and Y. Nishina, *Z. Phys. 52*:853 (1929).
58. J. H. Hubbell, V. J. Veigele, E. A. Briggs, R. T. Brown, D. T. Cromer, and R. J. Howerton, *J. Phys. Chem. Reference Data 4*:471 (1975).
59. M. H. Pirenne, *The Diffraction of X-Rays and Electrons by Free Molecules,* Cambridge University Press, New York, 1946.
60. R. D. Evans, in *Encyclopedia of Physics,* Vol. XXXIV (S. Flügge, Ed.), Springer-Verlag, Berlin, 1958, p. 218.
61. J. H. Hubbell, Proceedings of the Workshop on New Directions in Soft X-Ray Photoabsorption, Asilomar Conference Center, Pacific Grove, CA, April 8–11, 1984.
62. J. H. Hubbell, *Int. J. Appl. Rad. Isot. 33*:1269 (1982).
63. J. Wernisch, C. Pöhn, W. Hanke, and H. Ebel, *X-Ray Spectrom. 13*:180 (1984).
64. W. J. Veigele, in *Handbook of Spectroscopy,* Vol. I (J. W. Robinson, Ed.), CRC Press, Cleveland, 1974, p. 155.
65. J. Thewlis (Ed.), *Encyclopaedic Dictionary of Physics,* Pergamon Press, Oxford, 1962, p. 800.
66. C. L. Sparks, Jr., in *Proc. Inter-Congress Conference on Anomalous Scattering* (S. Ramaseshan, S. C. Abrahams, and D. Hodgkin, Eds.), Munksgaard, Copenhagen, 1974, p. 175.
67. C. J. Sparks, Jr., *Phys. Rev. Lett. 33*:262 (1974).
68. B. Davis and D. P. Mitchell, *Phys. Rev. 32*:331 (1928).
69. K. S. Krishnan, *Nature 122*:961 (1928).
70. T. Suzuki, *J. Phys. Soc. Japan 21*:2087 (1966).
71. T. Suzuki, T. Kishimoto, T. Kaji, and T. Suzuki, *J. Phys. Soc. Japan 29*:730 (1970).
72. T. Suzuki and H. Nagasawa, *J. Phys. Soc. Japan 39*:1579 (1975).
73. Y. Mizuno and Y. Ohmura, *J. Phys. Soc. Japan 22*:445 (1967).
74. R. Jenkins, R. W. Gould, and D. Gedcke, *Quantitative X-Ray Spectrometry,* Marcel Dekker, New York, 1981.
75. R. Tertian and F. Claisse, *Principles of Quantitative X-Ray Fluorescence Analysis,* Heyden, London, 1982.
76. H. A. Bethe and J. Ashkin, *Experimental Nuclear Physics,* Vol. 1, Wiley, New York, 1953, p. 252.
77. T. S. Rao-Sahib and D. B. Wittry, Proceedings of the 6th International Conference on X-Ray Optics and Microanalysis, University of Tokyo Press, Tokyo, 1972, p. 131.
78. G. Love, M. G. Cox, and V. D. Scott, *J. Phys. D 11*:7 (1978).
79. G. Love and V. D. Scott, *Scanning 4*:111 (1981).
80. J. I. Campbell and J. A. Cookson, *Nucl. Instrum. Meth. Phys. Res. B3*:185 (1984)
81. L. S. Birks, *Electron Probe Microanalysis,* Wiley-Interscience, New York, 1971, p. 172.
82. R. W. Fink, *Handbook of Spectroscopy,* Vol. 1 (J. W. Robinson, Ed.), CRC Press, Cleveland, 1974, p. 222.
83. R. Jenkins, R. Manne, J. Robin, and C. Senemaud, Part VIII. Nomenclature system for X-ray spectroscopy, *Pure Appl. Chem., 63*(5):735 (1991).

2

Wavelength-Dispersive X-ray Fluorescence

J. A. Helsen   *Catholic University of Leuven, Leuven, Belgium*

Andrzej Kuczumow   *Maria Sklodowska-Curie University, Lublin, Poland*

I. INTRODUCTION

Moseley's law, formulated soon after the discovery of x-rays by Röntgen, represented the direct and definite onset of the use of x-ray spectrometry in chemistry. The first successful period was completed in the beginning of the 1920s by creating order in the periodic table of the elements and by the discovery of the missing elements [1,2]. X-ray fluorescence spectrometry (XRFS and XRF) has some unique features so that it became an irreplaceable tool for the analyst. Not many techniques had such a brilliant start to their analytical careers!

The first commercial instrument became available around 1940 and was derived from the x-ray goniometer of a diffraction instrument. A reasonable estimation of the number of wavelength-dispersive instruments working today is 14,000, of which some 20% are multichannel spectrometers. Some 2000 energy-dispersive instruments are working independently, and some 1500–2000 others are attached to electron and other microprobes; a few tens of instruments are connected to synchrotrons and linear accelerators as excitation sources. The commercial value of the instruments running today may be estimated at over $3 billion U.S. [2,3]. An important body of scientific literature relates to x-ray spectrometry [3–8].

About 10% of the instruments are applied in geological research and prospecting (sequential and simultaneous spectrometers). Most of the simultaneous instruments are installed in industry, and the sequential instruments are distributed over industry and research institutes in universities and public services.

XRF is unique in element analysis of inanimate matter: in qualitative analysis it has unbeaten selectivity for all elements between boron and uranium and also for transuranium elements and an extremely wide dynamic range in quantitative analysis (ppm to 100%). Although the plasma emission spectrometry (DCP or ICP), atomic absorption spectrophotometry, and neutron activation analysis have lower or much lower detection limits, the wide dynamic range remains a unique feature of XRF. The detection limits (without preconcentration about ppm level) and the limits to the maximum precision and accuracy that can be obtained (connected to sample homogeneity and counting statistics) determine the boundaries of its application.

Earlier books describing the technique are mentioned in the reference list [9–15]. The purpose here is to describe the basic principles that allow wavelength dispersion of x-rays and to describe the anatomy of a spectrometer.

Insofar as the fundamental principles are concerned, we may recall the law of Moseley [16,17], which is the basis of the qualitative application of XRF:

$$\sqrt{\nu} = \sqrt{\frac{c}{\lambda}} = k(Z - \sigma) \tag{1}$$

where ν = frequency in Hz, c = speed of light in m/s, λ = wavelength in m, k = constant for a given series, and σ = screening constant. The relationship between the wavelength of analogous lines and atomic number is illustrated by Fig. 1.

On increasing Z the spectra become more complex and full exploitation for qualitative analysis of mixtures is not always simple. Practical unraveling of spectra is dealt with in *Sec. IV*.

The intensity of any spectral line is proportional to the number of atoms emitting photons of energies attributed to this line. Simple linear proportionality is unfortunately not the rule. From the concepts of mass attenuation coefficients

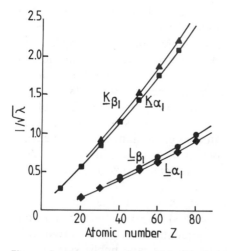

Figure 1 Representation of Moseley's law for K and L spectral series.

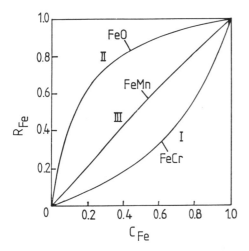

Figure 2 Absorption phenomena in heavy matrix (curve I); in light matrix (curve II); and in neutral matrix (curve III).

introduced in *Chap. 1*, it should be clear that the intensity of an analyte line of one species is attenuated by atoms of the same species and by any other atoms present in the *matrix. Matrix* refers to all elements present in a sample except the analyte element. Attenuation by absorption represents a first complication compromising the proportionality.

If the matrix contains elements with an absorption edge of higher wavelength than the analyte line, strong absorption may occur. This results in a further attenuation of the analyte line, on one hand, and, on the other hand, in enhancement of the spectral line related to the said absorption edge of the other matrix element. This phenomenon is known as secondary fluorescence. The process may be repeated with respect to all matrix elements with an absorption edge at a higher wavelength than a fluorescence line emitted by another matrix element. Secondary and higher order fluorescence is the major complication deteriorating the simple proportionality between intensity and concentration.

In Fig. 2 three typical kinds of relationships between relative intensity (measured intensity divided by the intensity obtained for the pure element) versus fractional concentration are represented. Curve I is obtained when the intensity of the analyte line is attenuated by absorption only. If secondary enhancement occurs, the analyte line intensity is higher than expected from primary excitation only and the curve is situated above the diagonal. A similar curve, however, can be obtained when the mass attenuation coefficient of the matrix is lower than the related coefficient of the element for its own radiation, as shown in the case for Fe in FeO (curve II). In an exceptional case the mutual relation between attenuation coefficients and/or enhancement may be such that both effects cancel precisely and then the diagonal is approximated (curve III).

The effects just discussed are all wavelength dependent. This means that any calculation for converting intensities into concentration must take into account attenuation and higher order fluorescence effects and must integrate over all wavelengths present in the exciting primary beam below the absorption edge and over all fluorescence lines. This is an *ab initio* approach for conversion of intensities into concentration implemented by several authors. We mention here the papers by Gillam and Heal [18], Sherman [19,20], and Shiraiwa and Fujino [21,22], who in the 1950s and the 1960s proposed suitable equations. Recent papers by Sparks [23] and Zhang Li-Xing [24] implemented small corrections and remarks concerning the details of these expressions. Likewise, Gardner and Hawthorne [25] obtained similar results by Monte Carlo simulation of x-ray excitation. For the complete treatment of those equations we refer to *Chap. 5.*

II. FUNDAMENTALS OF WAVELENGTH DISPERSION

The crystal monochromator is the heart of a wavelength-dispersive spectrometer. Wavelength dispersion of electromagnetic radiation in the x-ray region cannot be performed as a rule by normal gratings but only by diffraction on crystals or, for the long-wavelength x-ray region, on multilayers. We briefly explain the principles because the construction features of the monochromator are directly derived from them.

   Consider a monochromatic beam of x-rays of wavelength λ with their electrical vectors of equal amplitude in phase along any point of the direction of propagation. Assume further that the beam is parallel and is incident on a crystal at an angle θ between a given crystal plane (and the parallel planes) and the incident beam. The beam is scattered and diffracted rays results of equal λ but interfering constructively only in those directions for which the phase relationship is conserved. This happens at an angle θ for the scattered rays 1 and 2 (Fig. 3) for which the path difference *ABC* of ray 2 with ray 1 is equal to an integral number *n* of wavelengths. From Fig. 3 it is clear that $AB + BC = d \sin \theta + d \sin \theta = n\lambda$ or, according to W. L. Bragg and W. H. Bragg [26], who first formulated this relation, written as

$$n\lambda = 2d \sin \theta \tag{2}$$

where *n* denotes the number of wavelength difference between the rays scattered

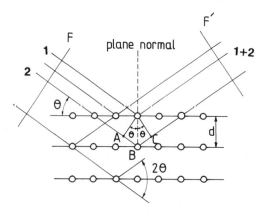

Figure 3   Derivation of Bragg's law. *ABC* is the path difference.

by the adjacent planes. As recommended by IUPAC [27], the order is denoted by n. If $n = 1$, the difference is one wavelength and the diffraction is said to be first order. If $n = 2$, the difference is two wavelengths and the diffraction is second order, and so on.

All x-rays emitted under angles different from θ cancel because they are out of phase and destructive interference occurs. From Fig. 3 one notices that in Bragg scattering the incident and exit angles are equal and in this sense there is some analogy with mirror reflection in classical optics. However, diffraction is by no means equal to reflection in classical optics, because the diffraction is a volume, not a surface process. It is less efficient (with great loss of intensity) and is performed only under particular angular conditions, that is, according to Bragg's law. Full analogy between reflection of x-rays and reflection of optical rays happens only at grazing angles (less than the Brewster angle). This is intrinsically a consequence of the refraction index, which is about equal to 1 for x-rays.

Thus the static condition for obtaining a diffraction of a monochromatic x-ray beam in some direction in the volume surrounding the analyzing crystal is given. What happens in the case of a polychromatic beam of x-rays? For a crystal, one set of planes is chosen (for different reasons) and d is constant. If only first-order diffraction is considered and constructive interference must be realized for all λ present in the incident beam, then θ is the only variable:

$$\lambda = \text{constant} \sin \theta \qquad (2a)$$

The diffraction angle θ is detected by a detector placed on a goniometer arm. The detector rotates around an axis through the macroscopic plane of the analyzing crystal. For a source at a fixed position, the detector rotates over an angle 2θ and the analyzing crystal rotates over an angle θ. The wavelength is calculated from the constant $= 2d$ and $\sin \theta$. Notice here that this holds only for first-order diffraction. If second-order diffraction is used, the wavelength is equal to half that value. The maximum wavelength λ_{max} that can be diffracted in first-order diffraction by a crystal is equal to $2d$ because $\sin \theta \leq 1$.

Because wavelength and energy are related, one more equation must be given, allowing us to convert wavelengths into energy units:

$$E = \frac{12398.5}{\lambda} \qquad (3)$$

where energy E is in electronvolts (eV) and λ in angstroms (Å). It is common practice that the units used by x-ray spectroscopists are still eV or keV for energy and Å for wavelength. When *Joules* and *nanometers* are used, as required by the international rules, the numerical value of the constant becomes $E = 1.98645 \times 10^{-16}$ J.

The presence of different wavelengths of different order on the same goniometer position has a particular consequence for wavelength-dispersive spectrometry. For a given position of goniometer (and detector), one may have a first-order wavelength of 0.6 Å, second-order diffraction of $\lambda = 0.3$ Å, third-order diffraction of $\lambda = 0.2$ Å, or 20.66, 41.32, or 61.98 keV, respectively. Figures 4 and 5 [28] demonstrate a practical situation and Table 1 reprints a fragment from x-ray tables from Ref. 29. To sort out these wavelengths, a pulse amplitude selector or detector is needed that can make the distinction between energy levels.

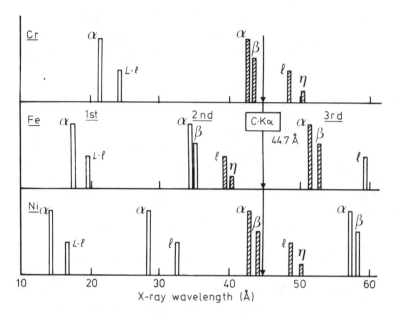

Figure 4 Higher order overlapping of L series x-rays of Cr, Fe, and Ni to $K\alpha$ of carbon. (From Ref. 28. Reprinted by permission of the author and Plenum Publ. Corp.)

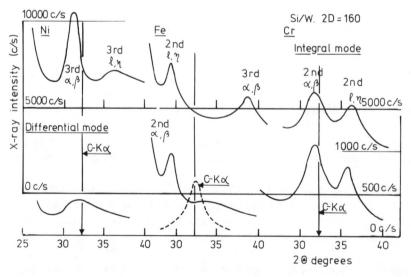

Figure 5 Overlapping of C $K\alpha$ by higher order x-rays of Ni, Fe, and Cr. (From Ref. 27. Reprinted by permission of the author and Plenum Publ. Corp.)

Table 1 A Fragment of the Detailed Tables of X-ray Lines

Raies de diagramme: 1251,56–1270,35

Notations, p. 3*i*   Notes p. 254. Références bibliographiques, p. 299

$\lambda/\mu X$	$\lambda/m\text{Å}$	Elément	Transition	Notation usuelle	Ordre	E/keV	ν/R	$(\nu/R)^{1/2}$	Bibl.
1251,56	1254,19	42 Mo	KM_{IV}	$K\beta_5^{II}$	2				(35.7)
1251,64 (^{a})	1254,27	58 Ce	KM_{IV}	$K\beta_5^{II}$	4				(36.8)
1252,192	1254,824	62 Sm	KL_{II}	$K\alpha_2$	4				(59.3)
1252,58	*1255,21*	73 Ta	abs. L_{III}	—	1	9,8776	725,99	26,9441	(41.4)
1252,6	1255,2	90 Th	$L_I N_{IV}$	—	2				(40.5)
1252,82	*1255,45*	73 Ta	$L_{III}O_{IV,V}$	$L\beta_5$	1	9,8758	725,85	26,9416	(41.4)
1255,17	*1257,81*	73 Ta	$L_{III}N_{VI,VII}$	$L\beta_7$	1	9,8573	724,49	26,9163	(41.4)
1255,43	*1258,07*	32 Ge	KL_{II}	$K\alpha_2$	1	9,8552	724,34	26,9135	(S.177)
1255,50	*1258,14*	69 Tm	KL_I	—	5				(70.22)
1256,60	*1259,24*	75 Re	$L_I M_{II}$	$L\beta_4$	1	9,8460	723,66	26,9010	(37.6)
1256,71	*1259,35*	73 Ta	$L_{III}O_{III}$	—	1	9,8452	723,60	26,8998	(42.4)
1257,1 (^{a})	*1259,7*	71 Lu	$L_{II}N_I$	$L\gamma_5$	1	9,842	723,4	26,8956	(42.5)
1257,12	1259,76	90 Th	abs. L_{II}	—	2				(40.5)
1257,66	*1260,30*	73 Ta	$L_{III}O_{II}$	—	1	9,8377	723,05	26,8897	(42.4)
1258,74	1261,39	54 Xe	KL_{II}	$K\alpha_2$	3				(33.5)
1259,210	1261,856	68 Er	KL_{III}	$K\alpha_1$	5				(59.3)
1259,8	1262,4	90 Th	$L_{II}P_{II,III}$	—	2				(40.5)
1260,30	*1262,95*	74 W	$L_I M_{III}$	$L\beta_3$	1	9,8171	721,54	26,8615	(39.4)
1260,6	1263,2	90 Th	$L_{II}P_I$	—	2				(40.5)
1260,620 (^{a})	1263,269	58 Ce	KM_{III}	$K\beta_1$	4				(36.8)
1261,15 (^{a})	1263,80	64 Gd	$KM_{IV,V}$	$K\beta_5$	5				(58.14)
1261,23	*1263,88*	73 Ta	$L_{III}O_I$	$L\beta_7$	1	9,8099	721,01	26,8516	(41.4)
1261,956	1264,608	42 Mo	KM_{III}	$K\beta_1$	2				(S.179)
1262,68	1265,33	90 Th	$L_{II}O_{IV}$	$L\gamma_6$	2				(40.5)
1263,086	1265,741	42 Mo	KM_{III}	$K\beta_3$	2				(S.179)
1263,428 (^{a})	1266,083	58 Ce	KM_{II}	$K\beta_3$	4	9,784	719,1	26,8158	(36.8)
1264,6	*1267,3*	74 W	$L_{III}N_{III}$	—	1				(34.3)
1264,99 (^{a})	*1267,65*	70 Yb	$L_{II}N_{IV}$	$L\gamma_1$	1	9,7807	718,86	26,8116	(42.5)
1265,0	1267,7	91 Pa	$L_{II}N_{IV}$	$L\gamma_1$	2				(S.202)
1265,24 (^{a})	*1267,90*	69 Tm	$L_I N_{III}$	$L\gamma_3$	1	9,7788	718,72	26,8090	(53.5)
1268,520 (^{s})	1271,186	92 U	$L_{II}N_I$	$L\gamma_5$	2				(58.15)
1268,66	1271,33	90 Th	$L_I N_{III}$	$L\gamma_3$	2				(40.5)
1268,68 (^{a})	*1271,35*	68 Er	$L_I O_{IV,V}$	—	1	9,7523	716,77	26,7726	(68.4)
1270,2	*1272,9*	74 W	$L_{II}M_V$	—	1	9,741	715,9	26,7566	(34.3)
1270,35 (^{a})	*1273,02*	64 Gd	KM_{III}	$K\beta_1$	5				(58.14)

Source: From Ref. 29, p. 67. Reprinted by permission of Pergamon Books, Ltd.

A detector is a device of which the principle is explained by the particle character of electromagnetic radiation. This is one reason that the energy of the impacting photons is often expressed in terms of energy, not wavelength.

There are also other differences between wavelength-dispersive (WD) and energy-dispersive (ED) XRF:

1. The brightness of a WD spectrometer is very low, the crystal being responsible for an important part of the losses. This problem may be overcome by the use of radiation sources of significant intensity.

2. The crystal is the dispersive device only, not the detecting device. The situation is different in EDXRF, where detectors play a double role, as the dispersive device and as the detector at the same time.
3. Since Bragg's law is of geometric character, the conditions for the collimation of primary and secondary beams are very severe for WDXRF, not the same as for EDXRF.
4. An attractive aspect of EDXRF is the simultaneous collection of the whole spectrum; the action of a typical WDXRF device is exclusively sequential. The maximum count rate for an EDXRF instrument, however, is 30 kcps for the whole spectrum, which severely limits the total number of accumulated counts and consequently limits the precision (counting statistics). Simultaneous WD instruments are composed of a series of individual crystal spectrometers (channels) operating simultaneously, but the number of channels is limited.

III. LAYOUT OF A SPECTROMETER

In its simplest form the main parts of a spectrometer can be represented by Fig. 6. The analyzing crystal is the central point of a wavelength-dispersive instrument. On the left side of the crystal we find (1) the source of excitation, (2) the filters and devices for shaping the fluorescence radiation (collimators and masks), and (3) the sample. On the right side we find (4) devices for shaping the diffracted beam (collimators) and (5) the detector. Signals from the detector are fed into the electronic circuitry for shaping the signals and subsequently manipulated by the computer software for data analysis. This arrangement can be reduced or made more complicated according to demand. However, all possible reductions lead to less flexible devices, and complication does not necessarily create a qualitatively new instrument.

A. Sources

A variety of radiation sources of sufficient energy, emitting either particles or γ- or x-rays, are potential candidates as sources for exciting some or all elements of the periodic table and some or most of the spectral lines of analytical interest. Other chapters deal with excitation by protons (proton-induced x-ray emission, PIXE, *Chap. 11*), by electrons (electron microprobe, *Chap. 12*), or by x-rays

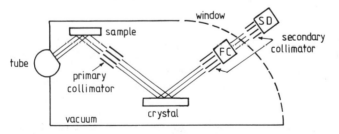

Figure 6 A wavelength-dispersive spectrometer: FC = proportional flow counter; SD = scintillation detector. (Reprinted by permission of Siemen AG.)

emitted from secondary targets or from synchrotron radiation (*Chap. 8*). Excitation by x-rays or soft γ-rays from radioisotopes and x-rays from low-power tubes is mainly restricted to energy-dispersive spectrometers (*Chaps. 3* and *7*). The most ideal source would be a tunable x-ray laser (monochromatic and intense, allowing the best choice of exciting wavelength and often selective excitation), but this cannot be expected before the end of the twentieth century [30]. From the practical point of view, vacuum x-ray tubes are the overwhelming choice among other potential excitation sources. High-power tubes are the only ones dealt with in detail in this chapter; low-power tubes are discussed in *Chap. 3*.

All modern tubes owe their existence to Coolidge's hot cathode x-ray tube as presented in *Physical Review* some 75 years ago [31]. It consists essentially of a sealed tube containing a hot tungsten filament for the production of electrons, a cooled anode, and a beryllium window. From a variety of modifications proposed over more than half a century, two geometries have emerged as the most suitable for all practical purposes: the end-window tube (EWT) and the side-window tube (SWT), both having their own merits and limitations.

The general requirements of an x-ray source are as follows:

1. Sufficient photon flux over a wide spectral range, with increasing emphasis on the intensity of the long-wavelength tail of the white spectrum. The actual vivid interest in low-Z element analysis will certainly activate research in this direction.
2. Good stability of the photon flux (<0.1% at least). Long-term stability reduces the frequency of recalibration in routine analysis; short-term stability is an absolute requirement for obtaining an acceptable precision.
3. Switchable tube potential (10–100 kV) allowing the creation of the most effective excitation conditions for each element. The intensity of the analyte lines varies considerably with excitation conditions. An extreme example is given by Vrebos and Helsen for simulated Al-Mo alloys [32].
4. Freedom from too many interfering lines from the characteristic spectrum of the tube anode.

Requirement 3 can only be realized in SWT, but the advantage of SWT remains a topic of discussion between manufacturers that has not yet been settled. Freedom from interfering lines is important because although the scattered characteristic lines of the anode may spectroscopically interfere with analyte lines, the accuracy of the conversion of intensity to concentration becomes more complicated. This is found in the results obtained after correction by some algorithms as well as by fundamental parameter calculations.

An x-ray tube is characterized by its anode element (a single element or two elements as in dual-anode tubes), its input power (expressed in watts, W or kW, typically between 0.2 and 5 kW for high-power tubes), the voltage range between anode and cathode (10–100 kV for SWT and limited to 75 kV for EWT), tube current (milliampere, mA; typically up to 60 mA or sometimes to 100 mA for high-power tubes), and an open or closed anode cooling circuit. The photon output of the tube or, more importantly, the photon flux hitting the sample (expressed in counts per second per watt input power), is determined by α, the incidence angle of the electron beam on the anode, the takeoff angle β (for SWT), the distance t to the beryllium window, the thickness d of the beryllium window, and

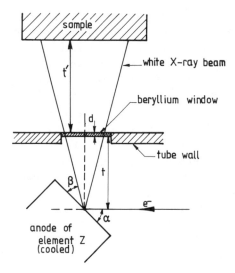

Figure 7 Detail of the geometry of a side-window x-ray tube and sample.

t' the distance between window and sample (Fig. 7). The radiation output power is rather poor with respect to the input power and is of the order of 1%, making the device a very inefficient transformer of electron current to electromagnetic radiation.

The impact of the electrons creates an excitation volume from which the white radiation and the characteristic radiation of the target must escape. This phenomenon represents the first distinct difference between SWT and EWT. As shown in Fig. 8a, the "escape" path of the photons in SWT geometry is on an average less than in EWT geometry. An immediate consequence of this characteristic is that only in SWT are dual anodes used: a light element (e.g., scandium) covering a heavier element (e.g., molybdenum). By switching the excitation voltage, x-rays are produced either in the upper (lighter) element layer or in the substrate (higher Z element), resulting in two distinct tube spectra with different yields in low- and high-wavelength regions (Fig. 8b).

The smaller the distance t, the higher is the output. A decrease in t to $0.5t$ increases the intensity roughly by a factor of 4. The reduction of t is limited in the classical setup, however. The smaller the value of t, the more intense is the bombardment by electrons and subsequent heating of the window by scattered electrons (Fig. 7). In SWT this bombardment is rather intensive because in this geometry the anode and window are both electrically at ground potential. In EWT, on the contrary, the filament and the window are at the same potential, and the heating of the window is negligible. For EWT, however, t cannot be reduced at will because it faces the anode at high potential (Fig. 9).

1. Alternative Configurations

To overcome the absorption of the low-wavelength tail of the continuum, a windowless configuration was considered. This is an obvious solution but it requires

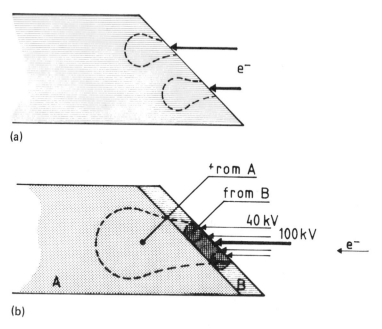

Figure 8 (a) Principle of excitation in a side-window x-ray tube. (b) Principle of excitation in a dual-anode x-ray tube (side window); Mo, W, or Au (for A) and Sc or Cr (for B) are commercially available as anode elements. The spectral feature of the tube spectrum depends on the energy of the impacting electrons.

that the whole spectrometer with sample, collimators, crystals, and flow counter be evacuated to the low pressure suitable for certifying an acceptably long life for the tube filament.

Nordfors advocated a dual-anode tube with two anodes physically separated [33]. They were excited by deviating the electron beam to either of the two anodes. Lack of stability prevented this solution from gaining commercial interest (Fig. 10).

Tubes with exchangeable anodes are another possibility to meet the problem of comfortable switching between different anodes. To the best of our knowledge this solution is not commercially available because of the tedious exchange operation.

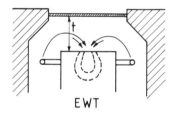

Figure 9 Principle of end-window x-ray tubes.

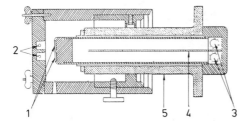

Figure 10   X-ray tube with two separate anodes: (1) anodes; (2) filaments; (3) cooling; (4) separating brass wall; (5) insulator. (After Nordfors, Ref. 32. Reprinted by permission of the Royal Swedish Academy of Sciences.)

Using the sample as anode is another solution, but then the application is bound to conducting materials. In a modified form this solution is implemented in scanning electron microprobes. Because this "anode" cannot be adequately cooled, the input power must necessarily be restricted to very low values, entailing low count rates and consequently reducing the precision that can be obtained within reasonably low counting times.

Another possibility is to make use of an existing 18 kW rotating anode x-ray tube generator. Sufficient intensity is available so that use can be made of the microdiffraction beam collimator. This collimates the beam to about 30 μm and thus allows microfluorescence [34,35].

2. Optimization

Because the takeoff angle is very important, the anode surface can be executed in steps in such a way that the yield of low-wavelength radiation is optimized. As is clear from Fig. 11, the step-shaped surface decreases the escape depths.

In a conventional EWT the reduction in t (Fig. 7) has limits. According to an old idea of Thordarson [36], Botden et al. [37] developed in 1952 an x-ray tube with a gold anode evaporated directly on the beryllium window. The tube was used for surface radiotherapy. After many years Philips returned to Botden's idea and again reduced t to zero by sputtering the anode target element directly on the beryllium window (Fig. 12). The anode is in SWT at ground potential, which considerably facilitates and simplifies the safety measurements. This geometry enhances the intensity by a factor of about 10, allowing a much lower input power:

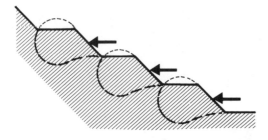

Figure 11   X-ray tube with stepwise configuration of the anode.

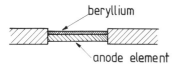

Figure 12 Anode elements deposited on the beryllium window.

200 W for an intensity equivalent to the earlier 3 kW EWT. As a result of the low input power, the heat produced in the target by the electron beam is easily dissipated and no supplementary water cooling is required. The tube construction is reduced to its most basic components. The commercial name for this development is T^3 or TTT (target transmission tube; Fig. 13c). It is used exclusively for simultaneous instruments, especially for the new version of the PW1660 [38].

3. End-Window Tube

The essential parts of an EWT are displayed in Fig. 13a. The ring-shaped cathode and the cooling circuit of the anode are typical features of EWT construction. The latter consists of a closed circuit filled with deionized water. Deionized water is necessary to reduce the conductivity and enhance the safety because the circuit is an anode potential! Electrons are diverted to the anode surface by electron optics. Because filament and window are at equal potential, no electrons hit the window. Heating is mainly by radiation from the anode, and an external cooling circuit is provided, which may be connected to normal water supplies because it is not in contact with a voltage source.

4. Side-Window Tube

An exploded view of a SWT is shown in Fig. 13b. The essential differences from an EWT are the distance between cathode and anode, an earthed anode, and a single cooling circuit for the anode only. This circuit may be directly connected to the normal water supply because it is at ground potential. The geometric arrangement of the cathode supports higher potentials of the cathode with respect to the anode. This is the basic reason that tube potentials of 100 kV are allowed. Another consequence of switchable anode potentials is the use of dual anodes. The available dual-anode tubes are summarized later in Table 10.

5. High-Voltage Generator

When EWT or SWT is used, a high-voltage generator supplying up to 60 kV for EWT or up to 100 kV for SWT and an output power of 3 kW (or much lower for T^3 tubes) is required. A special generator is needed if one wants the high-power rotating anode tube from Rigaku. The conventional power supplies were very cumbersome because of the size of the transformer. The new generation belongs to the so-called switched power supplies. Such a power supply is basically a dc/dc converter in which a dc voltage is electronically switched at high frequencies (several kHz) and fed into an inductance-capacitance network. The output power is not continuously regulated but is regulated through pulse-width modulation. One possible scheme of a switched power supply is given in Fig. 14. The mains

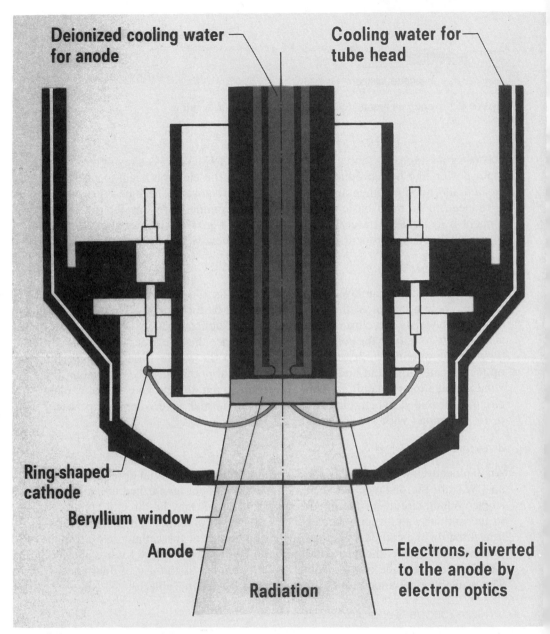

(a)

Figure 13 Different types of construction of commercial x-ray tubes: (a) end-window tube (EWT) (courtesy of Siemens AG); (b) side-window tube (SWT) (courtesy of Philips Analytical); (c) target transmission tube (TTT) (courtesy of Philips Analytical).

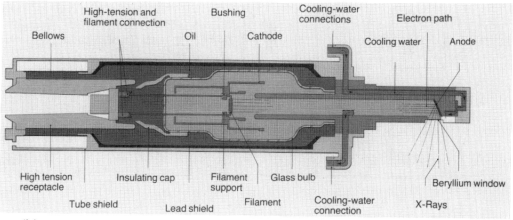

(b)

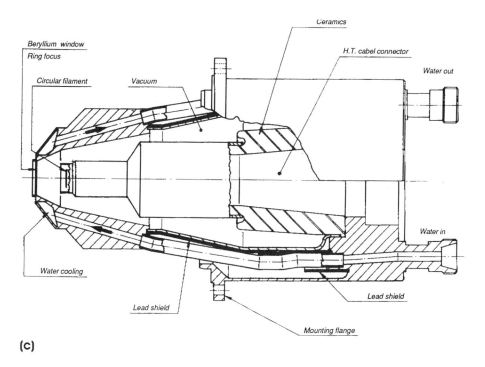

(c)

are rectified (a), high-frequency switched (b), transformed (c), and rectified and smoothed (d). The output voltage is sensed (e) and compared to a reference voltage (f), and this signal monitors the pulse-width modulator (g), which in turn commands the switching circuit (b). In the high-frequency transformer a ferrite core is used; this is a lightweight component. The whole setup allows the size of the generator to be reduced to a fraction of the volume of classical generators.

The whole system is generally monitored by microprocessor. Switching the tube to high voltage, as is possible and useful for SWT with, for example, a

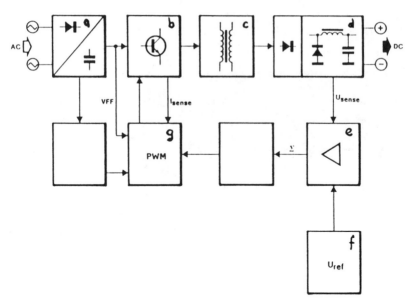

Figure 14   Principle of switched power supply: dc/dc converter.

rhodium anode, is done along the isowatt curve (constant input power), all under microprocessor control. The system has a number of safety switches on flow and temperature controllers of cooling water with microswitches on all panels, which when removed could expose the customer either to the electrical circuitry (high voltage of the tube supply or the mains voltage) or to unacceptable radiation levels.

B. X-ray Tube Spectrum

The first attempts to propose a quantitative description of the x-ray tube intensity as a function of wavelength were made by Kulenkampff [39] and Kramers [40]. The quantitative treatment of intensities of fluorescent x-rays (obtained by poly-chromatic excitation) by fundamental parameter programs requires knowledge of the intensity function of the exciting beam. This explains the continued interest in the rigorous numerical description of the tube spectrum.

Description of the tube spectrum is divided in two parts, the first concerning the continuum spectrum (Bremsstrahlung) and the second concerning the characteristic lines of the anode material superimposed on the continuum. The relative importance of those two contributions depend on the target element: for a tungsten target the characteristic L lines constitute the minority of emitted radiation, about 24%, and the K lines from the copper tube represents about 60% of the emitted radiation at 45 kV.

In continuum research the main effort was focused on the accuracy of Kramers' law. Some deviations of this law from experimental results were corrected by the introduction of a nonintegral exponent m of the wavelength in Kramers' law [41]:

$$N_\lambda \sim \frac{Z^r}{\lambda^2}\left(\frac{\lambda}{\lambda_0} - 1\right) \qquad \text{Kramers' law} \qquad (4a)$$

$$N_\lambda \sim \frac{Z^r}{\lambda^2}\left(\frac{\lambda}{\lambda_0} - 1\right)^m \qquad \text{Brunetto's result} \qquad (4b)$$

where λ_0 is the wavelength at the short wavelength end of the continuum and r and m are constant nonintegral exponents obtained by fitting to experimental results. Brunetto's result was confirmed later by Tertian and Broll [42]. To Eq. (4b) more corrections must be added: the absorption correction (often this of Philibert; see Pella et al. [43] and Markowicz and Van Grieken [44]), the electron backscatter loss term (for example, that of Statham; see Markowicz and Van Grieken [44]), and the exponential correction term taking into account the absorption of the beryllium window.

Furthermore, in the paper by Pella et al. [43], an interesting algorithm was proposed for taking the characteristic line intensities into account. For this purpose these authors made use of the equation proposed by Green and Cosslett [45]. The ratio of peak to background intensity was found to be a function of overvoltage U_0 and atomic number Z:

$$\frac{N_{\text{char}}}{N_{\text{con}}} = \exp\left[0.5\left(\frac{U_0 - 1}{1.17U_0 + 3.2}\right)^2\right]\left(\frac{a}{b + Z^4} + d\right)\left(\frac{U_0 \cdot ln U_0}{U_0 - 1} - 1\right) \qquad (5)$$

Overvoltage U_0 is the ratio of the initial electron energy E_0 to the critical excitation energy of a given shell E_q. The symbols a, b, and d are experimentally determined constants. The intrinsic merit of this formula is its dependence on the physical constant U_0 and on Z (see also the paper by Murata and Shibahara [46] with Monte Carlo estimations of the penetration depth of electrons).

Apart from all efforts concentrating on the numerical estimation of x-ray tube emission, experimentally recorded spectra may always be used (collected in Refs. 47–49).

C. Collimators and Masks

The wavelength-dispersive mode of operation depends strongly on rigorous geometric constraints on goniometer construction and analyzing crystal. Ideally only a parallel beam of fluorescence radiation should be diffracted and stray radiation should be absent. Stray radiation enhances the background. Diverging beams result in worsening the spectral resolution.

Masks, which reduce the shape of the fluorescent beam in front of the collimator, cut off x-rays emitted or scattered by an area larger than the sample area. Practical implementation of masks is discussed in *Sec. VI*.

Three positions for collimators in the optical path of the spectrometer are possible: (1) between sample and crystal, (2) between crystal and detector, and sometimes (3) an auxiliary position between two detectors working in tandem. Collimators are not ideal devices because they have their own angle of divergence:

$$\alpha = \arctan \frac{a}{l} \tag{6}$$

where a is the spacing between the blades and l the length of the blades. Besides divergence, there is also a substantial loss of counts while passing a collimator; for example, for radiation of about 10 keV passing by a typical 100×0.25 mm Soller collimator, the transmission is close to 10^{-5}. On the bonus side of the use of collimators it must be said that the resolution is improved very substantially (see Fig. 15). Thus, one must find a compromise between better resolution and higher brightness. In the analysis of light elements, for which the intensity is generally relatively low (excitation and attenuation) but with relatively well spaced spectral lines, intensity prevails over resolution and, when possible, a coarse collimator (or none at all?) is largely sufficient. In all focusing spectrometers the use of collimators is superfluous and the action of collimators is taken over by pinholes or slits.

D. Dispersive Elements

Both the primary photon beam derived from the x-ray tube (or another source of radiation) and the secondary or fluorescence beam derived from the sample can be monochromatized. The monochromatization of the primary beam is not essential but may be made for more efficient excitation of the sample by photons at chosen wavelengths or for radical simplification of calculation procedures (the fundamental parameter calculation is then transformed to simple Lachance-Traill-Tertian correction). Monochromatization may be compared to the action of a narrow bandpass filter by which only a small band of the whole spectrum is transmitted. The width of the bandpass is related to the spectral resolution. A number of parameters characterize this process.

The ratio $\Delta E/E$, where ΔE denotes the energy width of transmitted radiation band and E the energy of photons to be transmitted, is called the relative spectral resolution of the monochromator. ΔE is generally expressed as full width at half-maximum (FWHM) of the transmitted band (a peak in intensity-energy representation). The reciprocal of spectral resolution $E/\Delta E$ is called the resolving

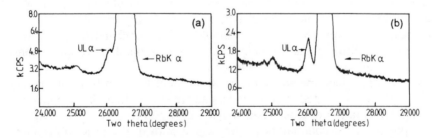

Figure 15 Two spectra from the same sample demonstrate how an auxiliary collimator reduces line overlap when measuring heavy elements (a) without and (b) with auxiliary collimator. (Courtesy of Philips Analytical. From Ref. 50.)

power. The same can be expressed in wavelength units as $\Delta\lambda/\lambda$ by transformation through relation (3). Both expressions are numerically equivalent but of opposite sign. For the wavelength-dispersive spectrometer the angular dispersion (Bragg's dispersion) is important and is obtained by differentiation of Bragg's law:

$$\frac{d\theta}{d\lambda} = \frac{n}{2d\cos\theta} \tag{7}$$

The monochromatization of the primary beam, if required, is made by the use of filters or secondary targets. The absorption edge of the elements present in the filter should be at lower wavelengths than the wavelength domain to be transmitted. Secondary targets emit their own characteristic radiation without white background (except the scattered part of the primary beam) and with a population density of peaks depending on the atomic number of the secondary target. In both these cases the monochromatization is far from perfect. The criteria for the monochromatization of the primary beam become very sharp in some methods of research with synchrotron radiation excitation.

The monochromatization of the secondary beam is the key problem. Dispersion of radiation is needed for the specific detection of characteristic lines. There are, however, a number of analytical problems in which the requisite dispersion is low. In limited cases it may be sufficient to subdivide the spectrum into two spectral windows, from which only one is to be detected. In other cases the cutting off of a relatively wide spectral window may be sufficient. In practice this selection of wide spectral domains is easily performed by energy-dispersive systems and pulse-height selection, by filters, or by a set of balanced filters. A spectrometer for general use over the widest range of elements, however, should provide a good compromise for the overall resolution, as is obtained by crystal diffraction.

1. Crystals

According to the crystal plane parallel to the macroscopic plane of the crystal, different values for d are preferentially diffracting, and accordingly, a different resolution results for a given wavelength. A few values have been calculated for LiF in different orientations and these are listed in Table 2. From the dispersion it can be expected that the $K\alpha$ line of manganese can be separated from the chromium $K\beta$ by LiF (220), not by LiF (200). The values in Table 2 are calculated for ideal crystals and a parallel beam. For real crystals, dispersion and resolving

Table 2 LiF: d Spacings for Different
Orientations and Angular Dispersion[a]

Crystal	$2d$ (Å)	Dispersion
LiF (422)	1.652	—
LiF (420)	1.800	—
LiF (220)	2.848	0.5154
LiF (200)	4.027	0.2902

[a] Calculated for θ angles corresponding to Mn
$K\alpha$ (2.102 Å) and $n = 1$.

Table 3 Currently Available Crystals

Crystal	$2d$ (nm)	Element range	Remarks
LiF (420)	0.180	Ni–U	High resolution, special applications
LiF (220)[a]	0.285	V–U	High resolution
LiF (200)[a]	0.402	K–U	General purpose
Si(111)	0.626	P, S, Cl	Suppresses even orders
Ge (111)[a,b]	0.653	P, S, Cl	Suppresses even orders
PG (002)[a]	0.671	P, S, Cl	
InSb (111)[a,b]	0.784	Si	
PE (002)[b]	0.874	Al–Cl	
EddT (020)	0.880	Al–Cl	
ADP (101)	1.064	Mg	
TlAP (100)	2.575	O–Mg	Especially F, Na, Mg
PX-1[a]	5.1	O–Mg	
PX-2[a]	12.0	B–C	
PX-3[a]	20.0	B	
PX-4[a]	12.0	C–(N, O)	
PbSt	10.0	F, C	
OVO 55	5.5	Mg, Na, F	
OVO 100	10.0	C, O	
OVO 160	16.0	B, C	

[a] Also available curved in simultaneous instruments.
[b] Also available transversely curved for sequential instruments.

power are less favorable, and some real values for a complete spectrometer are provided later. In Table 3 a list is presented for crystals currently available for spectrometers: lithium fluoride (LiF), silicon (Si), germanium (Ge), pyrolytic graphite (PG), indium antimonide (InSb), pentaerythritol (PE or PET), ethylenediamine-d-tartrate (EDDT), ammoniumdihydrogen phosphate (ADP), thallium hydrogen phosphate (TlAP), and multilayers with their commercial designations. Important characteristics other than interplanar distances codetermine the ultimate usefulness of an analyzing crystal: spectral resolution, mosaicity, reflectivity, stability, the thermal expansion coefficient, and the spectral range.

a. Resolution

The spectral resolution of crystals oscillates about the value $\Delta E/E = 10^{-2}$, with few exceptions. The data for resolution of different crystals are included later in Fig. 22, a summarizing figure on resolution. The spectral resolution for crystals is, generally speaking, better than that of any other dispersive devices, such as detectors or filters, but the difference with respect to energy-dispersive spectrometers becomes less favorable for the shorter wavelength range.

b. Mosaicity and Reflectivity

Real crystals have all kind of imperfections, and mosaicity is one of them. Mosaicity refers to the existence of "blocks" within the crystal with sizes of the order of magnitude of about 100 Å, which have slightly different orientations and lead

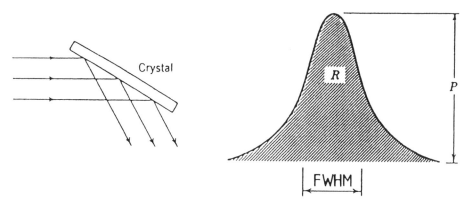

Figure 16 Rocking curve that would be obtained from a real crystal-diffracting parallel monochromatic radiation (rocking curves are actually measured on double-crystal spectrometers): P = the peak diffraction coefficient; R = the integral reflection coefficient. (From Ref. 10. Reprinted by permission of John Wiley and Sons, Inc.)

to widening of the diffracted peaks. This happens to widely differing degrees: topaz, EDDT, ADP, and gypsum exhibit low, quartz and LiF a little higher, and PET a significantly higher mosaicity. The effect is a slight angular widening of the diffracted peak on one hand, but on the other hand, the intensity of the peak is generally enhanced. The widening is smaller than the widening due to other geometric factors of the whole spectrometer and can be tolerated with respect to intensity. Some mechanical treatments enhance the mosaicity, and LiF crystals are easily treated in this respect. The angular intensity distribution of a diffracted peak, the rocking curve, can be determined by double-crystal spectrometers. In Fig. 16 the peak height of the represented curve obtained by a high-resolution spectrometer is determined by atomic scatter factors and the space group of the analyzing crystal; the width at half-maximum is determined by the mosaicity of the analyzing crystal and the divergence of the incident and diffracted beams. The surface under the curve is the integral reflection or, when the curve is normalized with respect to the intensity of the incident beam, the integral reflection coefficient. This coefficient is very much dependent on mosaicity. Abraded LiF has a 10-fold increase in the reflection coefficient with respect to freshly cleaved LiF $(3 \times 10^{-5} - 4 \times 10^{-4}$ rad), measured using CuK'_α radiation [10]. LiF crystals also have the great advantage of low absorption coefficients by their constituting atoms. Topaz and quartz, otherwise good crystals, have bad reflectivity properties, but PET is very good in this respect. Reflection constants as a function of wavelength are given in Fig. 17.

c. Stability and Temperature

The mechanical stability of most crystals is as a rule satisfactory, but there are exceptions. Gypsum can effloresce (especially in high vacuum); PET has a tendency to change phases on aging, and it is soft so that it is easily damaged when manipulated. The temperature inside the spectrometer is kept as constant as necessary (≤ 1 or $\leq 0.01°C$ if chemical peak shifts have to be measured). In these

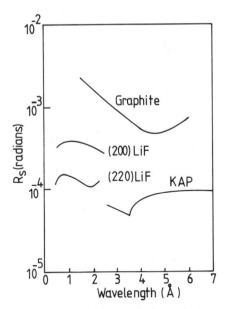

Figure 17 Single-crystal integral reflection coefficients of graphite, LiF (200), LiF (220), and KAP. (From Ref. 48. Reprinted by permission of *Analytical Chemistry*. Copyright © American Chemical Society.)

circumstances damage of the organic crystals by high temperature is not likely to occur. When very precise measurements of peak position are to be made for speciation purposes, however, the thermal expansion coefficient is important. PET is very disadvantageous in this respect, but while topaz is in the opposite situation. Examples are given in *Sec. V.*

It should be noted here that characteristic spectral lines of elements of the crystal can be superimposed on the spectrum, making it obscure or even unfit for use. Crystals of excellent resolving features, Be and Ge (111), have their own lines in the soft x-ray region.

d. Spectral Range

The useful spectral range covered by one crystal is limited. The maximum wavelength is of trigonometric origin and is imposed by Bragg's law, namely by the condition that $\sin \theta \leq 1$ and $\lambda_{max} \leq 2d$. In practice the goniometer is scanned for values of 2θ varying from a few degrees to about 150°. However, for higher values of 2θ an angular dispersion widens peak profiles. On the contrary, in the low-2θ range only a small fraction of x-rays emitted by the sample is intercepted unless the analyzer crystal is very long. The surface of the crystal projected on a plane perpendicular to the x-ray beam may be smaller than the width of the beam emerging from the collimator. Thus 2θ values as well as macroscopic dimension determine the useful spectral range of a crystal.

For x-rays of low energies the interplanar distances of real crystals become too small. Some substances, namely salts of heavy metals with organic acids with long chains (otherwise the soaps of heavy metals), may take over the role of dispersive structures if their organic chains are orderly arranged. There are special techniques for making such quasi-crystals. The individual layers are called Langmuir-Blodgett films. The ends of hydrophobic chains of adjacent layers join each other, and the heavy metal ends are connected to the next heavy atom ends from the next molecules. Many features of such films depend on the length of the chain. A great number of similar structures with a wide variety of interplanar distances $2d$ (up to 156 Å in the case of lead melissate) have been synthesized. Langmuir-Blodgett structures have different negative properties: for example, they are soft and not very stable, they need very careful service, they are strongly hydrophobic, and they have low reflectivity. Some are strongly poisonous substances, such as thallium adipate.

e. Curved Crystals

Not only flat crystals are used in the goniometer. Some focusing systems, such as the Johann, Johansson, and Cauchois arrangements, may be used (Fig. 18). Then the crystal must be curved with the curvature radius equal to the diameter of the Rowland circle and at the same time to the Bragg curvature, being half that of the Rowland circle. Three-dimensional crystals can sometimes meet both these conditions. The focusing arrangements are extremely useful in all cases in which the sample area is small, losses of intensity are prohibitive, and scanning over large angles is not necessary. In these arrangements x-rays are collected in one point by focusing on the detector. An obvious virtue of such an arrangement is found in the analysis of small sample areas and by using focused exciting beam as in electron or proton microprobes. In such situations curved crystal geometry allows the collection of x-rays from a relatively large solid angle of diverging x-rays to a small spot (Fig. 19). A consequence is that the collimator normally placed between sample and crystal and crystal and detector must be replaced by slits and pinholes. The focusing geometry is a very economical arrangement for saving intensity (the brightness of such a device is estimated to be up to two orders of magnitude greater than that of conventional flat geometry). The curved crystal geometry is applied in simultaneous instruments in which scanning is not used and an optimized arrangement is chosen for each wavelength (element) implemented.

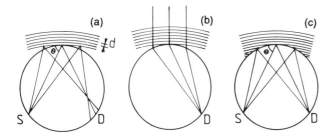

Figure 18 Curved crystal x-ray arrangements: (a) Johann, (b) Cauchois (in transmission), (c) Johansson. S, entrance slit of radiation; D, exit slit of diffracted radiation; d, crystal lattice spacing.

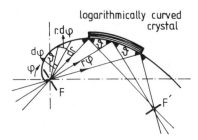

Figure 19   Focusing of x-rays by a logarithmically curved crystal. (Courtesy of Siemens AG.)

2. Multilayers

The limitation of XRF analysis in the region of soft x-rays is one of the most serious limitations of crystal monochromators, although from the point of view of principal assumptions constituting the background of XRF method there is no reason for such a limitation. Moseley's law determines the rules for qualitative analysis of any element, and the Shiraiwa and Fujino equations govern quantitative analysis. Irresolvable problems occurred until a few years ago because of troublesome absorption and bad detection while passing to longer wavelengths. The lack of an adequate dispersive device was one of the most serious problems. The concept of multilayers or, strictly speaking, layered synthetic microstructures (LSM) or multilayer interference mirrors (MIM), solved at least part of the question.

Two kinds of approaches to the idea of a suitable analyzing device in the soft x-ray domain have been applied from the very beginning. The first resulted from consideration of Bragg's equation (2). This indicates that if any crystal of a longer interplanar distance is found, it will be possible to disperse longer wavelengths in the soft x-ray range. The potential candidates for long-period crystals or quasi-crystals, such as clay minerals of the chlorite (14 Å) or illite-montmorillonite (25–30 Å) type, are not sufficiently ordered in this respect [51]. Unfortunately, natural crystals, irrespective of their different crystallographic planes, offer only limited possibilities in this spectral range: specifically, their reflectivity is very low. The class of compounds called Langmuir-Blodgett films was more promising, but the reflectivity and thermal, mechanical, and chemical stability was rather lower than the analytical demands.

The correct solution resulted from generalization of Bragg's law, namely that diffraction also occurs in media consisting of layers of different refraction coefficients—in other words, in sites of different periodically changing electron densities. Such structures may be formed artificially. They may be treated as crystallike substances ordered along the c axis, that is, spacing d or $2d$, but with uncontrolled arrangement within the ab plane. The structure in the ab plane can be amorphous, provided that even in these directions the optical and mechanical features are uniform.

The first trial of synthesis was made in 1940, but the Au-Cu multilayer of Du Mond and Youtz [52] did not survive more than a few days because of deterioration by interdiffusion. The problems of thermal, diffusional, and chemical instability of multilayers has recently been solved. Significant progress in this field has taken place in the last two decades.

The second approach was a result of an increasing interest in x-ray optics. As a branch of physics, x-ray optics is nearly as old as the discovery by Röntgen, but because of the specific properties of x-rays its progress has been slow. All existing substances exhibit complex refractive indices:

$$\mu = 1 - \delta - i\beta \tag{8}$$

where μ is the refractive index, δ is the so-called unit decrement, a real number, and the imaginary term β is related to absorption. The refractive indices in the x-ray region are all below but very close to 1. This makes all trials of x-ray focusing by lenses an impractical task. Franks [53] gives the example of a lens with a curvature radius of 10 mm built of a material with $\delta = 5 \times 10^{-5}$. The focal length of this lens is 100 m. In such a situation only glancing, grazing, and diffracting features of materials are of practical interest. Otherwise the grazing optics conditions for x-rays are extremely severe. However, the need for mirrors, monochromators, and focusing devices for the x-ray region is increasing rapidly in the fields of synchrotron radiation, plasma research and diagnostics, x-ray microscopy, x-ray fluorescence analysis, diffraction, thermonuclear fusion, x-ray laser, x-ray astronomy, and even x-ray waveguides. The invention of multilayers treated this time as a complex set of particular layers consisting of materials of periodically different refractive indices gave great impetus to progress in this branch of optics. Some important features of materials used to produce the multilayers are discussed later.

a. Nature of the Materials

In principle, three kinds of materials are necessary for the synthesis of multilayers. First, there must be a support (sometimes called a substrate), such as a piece of flat or curved silica layer, with the surface very precisely polished. This condition is very important because the roughness of the surface is translated to the deposited layers, correctable almost exclusively when one of the layers consists of amorphous materials, such as carbon. Two sources of relatively cheap and precisely polished supports were found: silicon wafers as used in microelectronics (curved) and glass or fused silica optical elements for low-scatter mirrors (flat or curved). An even more complicated role for the substrate is described by Nicolosi et al. [54] in whole coal analysis. The authors determined the content of C, N, O, and F using microstructure PX2 and the contents of P, S, K, and Ca by taking advantage of the supporting wafer of silicon (111) on which the multilayer was settled.

Heavy metals (Hf, Pb, W, Mo, Pt, Ir, Rh, and Os) and light elements or compounds (Mg, Si, SiO_2, and C) may be used as materials for alternating high and low optical densities. This list is not limited to these materials. The Ovonic Synthetic Materials Company [55] recommends its new multilayer structure, MoB_4C, for the analysis of B and Be. As a rule the low electron density compounds are chosen from elements with K absorption edges above the energy of spectral lines to be measured as boron and carbon for multilayers of spacing above 50 Å and beryllium or carbon for spacings above 120 Å.

b. Smoothness

This parameter is of great importance on each level, starting from the supporting substrate to the last upper layer. All rough places, uneven layers, disturbances in interfaces, or defects act in the manner analogous to the thermal motions in classical diffraction analysis. During synthesis, the extent of roughness can vary in a different manner: in some cases it tends to smoothness, in others to further growth. The use of amorphous materials to form some layers favors smoothing.

c. Reflectivity and Thickness of a Multilayer

Classical dispersive crystals exhibit low reflectivity in the soft x-ray range. The situation was greatly improved in multilayer reflectors in which a single layer at normal incidence angles exhibits an intensity reflection coefficient of the order of 10^{-4} rad. This coefficient is, however, proportional to the square of the electrical field amplitude, that is, to a value of the order of 10^{-2} for the amplitude reflection coefficient. Thus a multilayer consisting of 100 single layers may provide nearly the maximum reflectivity if the absorption can be neglected [56,57]. As a rule, commercial structures made by the Ovonic Synthetic Materials Company consist of 200 layers. These results are in accordance with the calculations by Rosenbluth concerning numbers of layers necessary for "saturation" of reflectivity (data cited by Barbee [58]). Peak reflectivities of some commercially available multilayers are often compared to classical pseudocrystal dispersive structures, including Philips materials PX1, 2, and 3 [54,59] and the OVONYX line of multilayers [55] with lead octodecanate and thallium acid phthalate. The predominance of the multilayer structures over classical structures is striking, and spectral resolution is the only parameter in which the classic crystals are better [59,60].

d. Quality of Layers and Boundaries

The structure of the multilayers ought to exhibit certain geometric features, and this condition is very severe. Thus, flatly layered or figured structures must be uniform with controlled curvature, without defects but with clearly defined optical parameters. All deviations from an ideal model, easy to estimate by the reflectivity characteristics, are as a rule coupled by the use of the Debye-Waller factor into the effective roughness parameter σ_{eff}, containing both thickness errors and imperfect boundaries [61]. The Debye-Waller factor is the Gaussian-type term describing the reduction of reflected amplitude at a boundary:

$$DW = \exp\left[-2\left(\frac{2\pi\sigma \cos \alpha}{\lambda}\right)^2\right] \tag{9}$$

where σ is the width of a smooth transition layer and α is the propagation angle of electromagnetic (EM) radiation, measured from the normal on transition from a medium with one refractive index to another.

The question arises, however, whether the errors result from a severe or from a mild change of the optical features on the boundaries. Any roughness of a given layer diminishes its reflectivity, but a random distribution of roughness may bear higher reflectivities of the higher order peaks. Moreover, many other physical factors can influence the uniformity and assumed layer characteristics: both the support and the particular monolayers can have different slopes, densities are subject to fluctuations, or some chemical elements can react mutually, for example W with C. It should be emphasized that despite so many obstacles the magnitude of the roughness can be kept in the range of an atom radius [58].

The numerical description of multilayers is based on the matrix method (a matrix expresses the features of a layer or interlayer region, and for the characteristics of a multilayer the product of matrices is written) or by the recursive use of layer reflection-transmission characteristics. Deviations of empirical results (mainly reflectivity measurements) from model results are associated with roughness, in which case σ_{eff}, the roughness parameter, results from the best fit of empirical data to the model.

The selection of an adequate method of synthesis and subsequent control of production and quality is the key question. Many methods were checked, and the aim of all these processes was the deposition in which one atom follows another. The physical vapor condensation, chemical vapor deposition, electrochemical covering, and sequential adsorption of films are the most promising methods. However, for the time being the first type prevails. Two basic kinds of physical vapor deposition are used—with thermal and sputter sources. In the thermal source technique the covering materials are evaporated by electron beam or laser evaporation. The vapor source and samples are in a fixed position. The time of evaporation is a parameter controlled by shutters or by pulsed laser irradiation. An in situ monitoring system was used as a feedback system for controlling the shutter. Spiller [62] invented a system for thermal deposition. He used a soft x-ray reflectometer (a source plus a detector), placed in an evaporation chamber to register the reflectivity of successive layers. The open loop is set according to the indications of maximum reflectivity for a given layer.

It is possible by this method to obtain a multilayer with an area up to 25 cm^2. The rates of deposition vary in the range of 2–100 Å/s but must be carefully controlled because the successive layers must be commensurate. Sometimes, especially when multilayers of large surface are produced, a special motion of the substrate in the evaporation chamber may be necessary to guarantee the same evaporation rate for all points of the surface. Other difficulties arise from this type of deposition: the high vacuum required in the evaporation chamber ($<10^{-4}$ torr), radiation heating of the deposited film, and troubles with thermal sources (alloys may segregate during melting).

The sputter source systems [63] belong to another processing method. The glow discharge, magnetron, or triode devices may serve as sputter sources. The plasma discharge is always the first step of the process. The plasma flux is directed by the electrical potential change to the cathode surface, where the ions and atoms dislodge the maternal atoms. The secondary atoms and ions establish the proper flux, which passes to the substrate to deposit a layer on it. Samples are placed on a special rotatory table, appearing at fixed intervals under the secondary flux courses. The rotation speed, precisely controlled, is the regulating factor for a designated thickness of layer. Purity of secondary fluxes and their energies also determine the quality of the multilayer. The main errors in this method result from ion and atom inclusion into layers (there are high energetic tails in the fluxes) and from heating caused by flux energy deposition in thin layers. However, substrate rotation may overcome many of these defects. Sputter rates range from 1 to 20 Å/s; the covered surfaces may be quite large.

After presenting two basic methods of layer formation, some attention must be paid to methods of control. As mentioned earlier, reflectivity measurements are of fundamental importance in multilayer quality estimation, both experimentally and theoretically [62]. Comparison of reflectivity predicted and obtained in an experiment is instructive [60,61]. Peak, integrated, and absolute reflectivities were measured with the use of both laboratory x-ray sources and synchrotron facilities. These results confirmed the layered structures of the synthesized materials; reflectivity was as a rule smaller than predicted, but the results show that the optical parameters estimated until now are reasonably correct, with the shorter the wavelength concerned the better [64].

The methods just discussed only allow the deduction of an indirect picture of the multilayer structure. Other methods give a direct insight into the structure. Thus, the diffraction data may be very useful but they require large sample volumes. Much smaller samples are needed for obtaining the structural information by electron microscopy, especially in combination with a special technique of sample preparation, the so-called microcleavage [65]. Wedge-shaped strips of materials are placed in the electron beam in such a manner that the planes of particular layers are parallel to the beam direction. Both the transmission image and the diffraction pattern can be obtained. The in-depth and lateral variations, the roughness, and the visual image of the structure are in the field of observation. High-resolution electron microscopy can give even better results, and electron energy loss spectroscopy (EELS) may provide the complementary chemical information on the subsequent layers.

Requirements for the application of multilayers in XRF may be summarized as follows: they should satisfy Bragg's law for the assumed wavelength domain within permissible goniometer angles, have a high reflectivity and a good spectral resolution, and suppress diffraction peaks of higher order. Flat or curved multilayers can be used. Dedicated designs are described in the literature for the application of curved multilayer optics [59,66]. The geometric requirements imposed on the curvature of multilayers for XRF are modest with respect to their application in other fields of x-ray optics. The increasing spectral distance between K lines of neighboring elements when going to lower Z elements may require the limited use of a dedicated multilayer for each element. In this case a simultaneous instrument could be the better solution, with a optimized channel for each element. Multilayers can also be used in electron microprobes, and these are as a rule curved [55].

From what has just been described, it may be concluded that any spectroscopist wanting to analyze a useful range of soft x-rays must have at least three different multilayers. Such combinations are not only synthesized on a laboratory scale but are also available commercially, for example from Philips [54,59]:

PX1	$2d = 50$ Å	fluorine to silicon
PX2	$2d = 120$ Å	carbon to fluorine
PX3	$2d = 200$ Å	boron and possibly for beryllium [67]

and from the Ovonic Synthetic Materials Company [55]:

OV-040A $2d = 40$ Å   fluorine to silicon

by multilayers changing $2d$ spacing every 20 Å to

OV-140B $2d = 140$ Å   carbon to oxygen

and also the MoB_4C structure:

OV-H series $2d = 244$ Å   beryllium to boron.

These materials are described in the current literature as completely thermally stable, and thus they do not need crystal stabilization when used in x-ray spectrometers [54,58,59]. Some sensitivity to the damaging action of intense radiation (above 0.15 J cm^{-2} [68]) may be a limitation of multilayer application in monochromatization of a dense flux of x-rays, but it is more concerned with the situation in synchrotron research. These data may be compared with the report by Barbee [58], in which during the process of multilayer synthesis by sputtering very high temporary increments of energy can be deposited in the layers (power $\simeq 10^5$ W cm^{-3} s^{-1}), which involves a significant increase in local temperatures (50–180°C).

The relative spectral resolution of dispersive devices is often expressed in the form of the important parameter $\Delta E/E$ or $\Delta\lambda/\lambda$, which is the basis for their choice in XRF application. Unfortunately, this is the weakest point of the multilayer. In a commercially available specimen the relative spectral resolution changes from about 2.5% to over 10% while passing from the structures with $2d = 40$ Å to those with $2d = 244$ Å, respectively, for different measured lines [55], and for the structure with $2d = 75$ Å it changes from 2.5 to 4% in the wavelength range of 7–75 Å [60], which is about three times worse [55] or more [60] than for a good classical pseudocrystal. The comparison of spectra of the same sample by the use of multilayers with different spacing, as in Fig. 20, is very instructive [59]. Relatively broad internal bandpasses of multilayers diminish the resolution of the whole spectrographic device. This is of special importance to the analysis of L or M series.

The suppression of higher order diffraction peaks gives better spectral purity and may at least partially compensate for the poorer resolution of multilayers. The isolation of the weak Mg $K\alpha$ line from the dominant Ca $K\alpha$ [3] lines is possible by the use of the PX1 structure for the analysis of cement (Fig. 21). This is an extreme example of the superiority of multilayers over classical crystal in particular cases [67]. Another great advantage of multilayers is their great reflectivity.

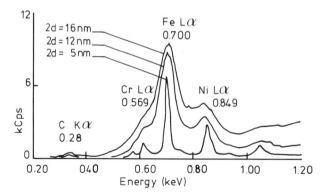

Figure 20 Comparison of sensitivity and resolution for layered synthetic microstructures with 2*d* values of 5, 12, and 16 nm from bottom to top, respectively. Increased sensitivity together with decreased resolution is evident when 2*d* increases. (From Ref. 59. Reprinted by permission of the author and Plenum Publ. Corp.)

It should be added that the peak reflectivity is 3–10 times greater than that of corresponding classical crystals.

Multilayers were used even in routine analyses of light elements. Many examples can be cited, including carbon analysis [54,59,67], boron analysis [59,67], and beryllium detection [67]. Analyses were made using both flat optics sequential spectrometers and curved optics in simultaneous instruments [59]. After overcoming the technical difficulties, the results obtained from such analyses may seem quite trivial [67]. One should note some special features of XRF spectrometry in the soft x-ray range: the type of analysis changes from the bulk to surface measurements; the influence of chemical state and the subsequent wavelength shifts on the analysis can be significant; the wavelength of soft x-rays is com-

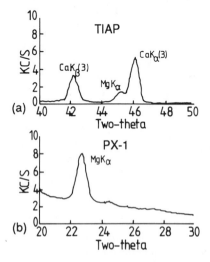

Figure 21 2θ scans for Mg in cement using TlAP and PX1. (From Ref. 67. Reprinted by permission of the author and Plenum Publ. Corp.)

parable to the surface roughness, even on the atomic level; and the problem of surface quality becomes extremely important.

Considering the so-called total information depth of x-rays, it can be seen that in the case of iron determination in a steel sample, the effective energy of exciting photons being 20 keV, the secondary x-rays pass to the detector from depths to 62 μm. A carbon signal emitted by a coal sample excited in the same manner passes to the detector from depths to 8.3 μm. The problem looks even more drastic if a carbon analysis is made of a steel sample, because in this case signals emerge from depths as small as 0.45 μm! This means that the analytical volume for iron is about 140 times greater than that for carbon.

The influence of the chemical state can be seen during analysis of the same elements using soft x-rays. It should be noted that $L\alpha_{1,2}$ lines of chromium and manganese are equal to 0.573 and 0.637 Å and must be analyzed in the region otherwise reserved for the K series of oxygen and fluorine, hence by the use of the PX2 or OV 120B structure. The question is whether the multilayer resolution is sufficient to show such subtle effects as chemical shifts.

The wavelengths of characteristic x-rays in the K series are longer than atomic radii starting from Ga (Ga K_α = 1.337 Å; atomic radius of Ga = 1.26 Å) and passing to the lighter elements. Wavelengths comparable to the dimensions of some natural obstacles always result in surface problems with the analysis. It should be taken into account that even the most perfectly polished sample surfaces have the roughness of the order of at least one atomic radius and that the best layers in multilayer structures have a roughness measured as effective roughness dimension σ_{eff} of the order of one to several atomic radii [58].

These last remarks are not intended to weaken interest in multilayer optics and its application in XRF, but only to evoke a special awareness of the problems encountered.

3. Spectral Resolution

Figure 22 shows the features of commonly used dispersive devices [69–80]. The width of the bandpass ΔE (expressed as FWHM whenever possible) is plotted versus the energy of radiation E; the parameter is relative to the spectral resolution $\Delta E/E$. A line of given value of $\Delta E/E$ is called the isoresolving line. These lines divide the whole figure into two zones of different resolutions (or bandpasses). Note that typical dispersing devices and typical detectors are collected in Fig. 22. Only a few detecting devices with no resolving power exist (Geiger-Müller counter), and no detector exists with infinite resolution. Real dispersive devices exhibit very contrasting features, with the parameter $\Delta E/E$ varying from almost 1 (scintillation counter or semiconductor photocathode) to systems with $\Delta E/E$ close to 10^{-4} [Ge (111) crystals and two- or three-crystal spectrometers]. Even better dispersive devices are needed, for example, for x-ray inelastic scattering spectroscopy in which energy resolution near 10^{-7} is wanted. XRF spectrometry has rather modest demands for resolution.

Attention must be paid to two groups of curves in Fig. 22. The first group concerns spectral distances between analogous lines of adjacent elements in the K and L series. These curves lie in a bandpass of 10^{-1}–10^{-2}, a medium spectral resolution range, and many single dispersive devices shown in Fig. 22 exhibit

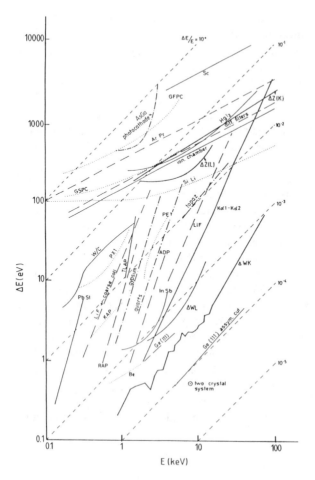

Figure 22 Spectral resolution of different devices: ΔZ = spectral difference between analogous lines of adjacent elements; ΔW = width of spectral line; $K\alpha_1 - K\alpha_2$ = difference between $K\alpha_1$ and $K\alpha_2$ lines.

better features. This range of resolution may be insufficient if many adjacent lines in densely populated areas of the spectrum must be separated (e.g., L lines of lead, gold, thallium, mercury, bismuth, platinum, and selenium in the energy range of 10–14 keV). Another group of curves shows the spectral widths of $K\alpha_1$ and $L\alpha_1$ lines and the spectral distances between $K\alpha_1$ and $K\alpha_2$ lines. Separation of these lines requires high-performance spectrometers. According to what one wants to measure, Fig. 22 is instructive for deciding which resolution is needed to do the job. It is clear that all research concerning isomer or chemical shifts in XRF demands high resolution [74]. The choice is limited.

One more interesting conclusion can be drawn from Fig. 22. Nearly all dispersive structures exhibit variable spectral resolution, but only KAP, topaz, and GE (111) crystals can be considered the isoresolving systems and, to some extent, also multilayer structures. As we can see from the shape of the ΔK and ΔL curves, nature does not demand isoresolving behavior from dispersive structures. Rather, these structures, which are parallel to the curves mentioned, would be optimal for the spectrographic aims; the Si-Li detector seems to have advantageous features. Generally speaking, a bandpass range of 10^{-1}–10^{-3} can be considered analytically useful. Systems with bandpasses below 10^{-3} are used only in spectrophysical work.

Spectral resolution is only one parameter determining the choice of a spectrometer. Other parameters, however, such as speed of measurement, intensity, an ease of operation, may affect the choice of the most suitable instrument.

E. The Goniometer

The goniometer is basically a very simple construction and is the analog of a diffractometer as used in x-ray diffraction. The essential features are shown in Fig. 6. The sample is positioned in front of the tube window. To account for possible heterogeneities, the sample is spun about its axis at 30 rotations per minute. The analyzer crystal is positioned in the center of the diffractometer circle and rotates about an axis perpendicular to the plane of the drawing and passing through the macroscopic plane of the crystal. The detector moves along the diffractometer circle supported by the goniometer arm. As imposed by the Bragg condition, the crystal rotates over an angle θ and the detector moves over an angle 2θ, realizing equal angles of incidence and diffraction with respect to the crystal plane. The crystal is diffracting the incident fluorescent beam of opening angle determined by the slit width of the collimator. The rotation of crystal and detector are coupled through a mechanism with gear ratio 1:2. In most setups both rotations are mechanically coupled, but in Applied Research Laboratories instruments both rotations are brought about by separate computer-controlled stepping motors (see *Sec. VI*).

The goniometer arm carries two detectors mounted in tandem: a flow proportional counter for the low-energy range (to about 5 keV), with two windows allowing the high-energy photons to pass to the scintillation detector. In the ARL instrument both detectors are moved on independent arms, each controlled by stepping motors. In this instrument the position is determined very accurately by an optical encoder using the moiré fringe principle (see *Sec. VI*).

Low-energy radiation is strongly attenuated by air. The $K\alpha$ radiation of a pure copper sample (radius of 2 inches) or of iron and chromium (with aluminum filter) from a stainless steel sample (measured on a Philips 1410 sequential spectrometer excited by a chromium anode tube at 45 kV and 6 mA) gives respective count rates of 67,800, 33,200, and 2660 cps in vacuum and 48,000, 16,800, and 890 cps in air. To decrease the absorption of photons, all wide-range spectrometers can operate in a controlled atmosphere (vacuum or helium). The vacuum cabinet contains the entire spectrometer except the scintillation detector. Attenuation of the high-energy radiation is low, and to reduce the size of the vacuum chamber the scintillation detector is placed outside the chamber but rotated about a common shaft with the flow counter. A long rectangular window of low-density material (Mylar) along the diffractometer circle allows the high-energy photons to reach the detector. Here, too, the ARL construction is different and both detectors (and eventually two goniometers) are fitted to the vacuum chamber. The samples are introduced by a sample lock that can adapt to only one sample (Philips and

Rigaku) or to the entire internal sample changer (Siemens and ARL), necessitating a secondary vacuum pump. The source is either a SWT or an EWT.

High angular precision is required for setting the goniometer arm, and thus severe mechanical constraints are imposed on its construction. Typical is an angular reproducibility of 0.001° and a mechanical resolution of a few thousandths of a degree. Special attention must be paid to the reduction of rotation from 2θ to θ with gears or with a coupled double-worm gear (for all manufacturers except ARL, which has a gearless construction). When scanning, the rotation speed is only a relatively small part of the total analysis time. Typical angular scan speeds (2θ) are in the range from 0.1 to 1000°/minute. Because of ARL's special design, the scan speeds are in the range of 0.25–128°/minute but with slewing speeds up to 4800°/min. The slewing speed is the maximum rotation speed of the goniometer arm for passing from one position to another without recording. The 2θ range covered is from 4 to 152°, with slight differences between manufacturers.

Jenkins et al. [81] and Croke and Nicolosi [82] announced a dual-channel sequential spectrometer. A twin primary collimator was installed consisting of an upper fine collimator and a lower coarse collimator. The beam is "divided" into two parts, each striking its own crystal. Two crystals are mounted close together but with a 15° inclination difference. The diffracted beams are detected by two detectors 30° apart. This fixed angle between both detectors, however, limits the benefits of a dual-channel instrument.

Two alternative constructions should be discussed here. Simultaneous or multichannel instruments are equipped with monochromators set by the manufacturer (or by the customer) at a fixed position for the detection of a single element. Usually curved crystal optics is used, and the whole setup is optimized for that specific element with appropriate curved crystal, entrance and exit slits, and detector. The crystals are logarithmically curved as explained in *Sec. III.D.1.* To give some more flexibility to multichannel instruments, one or two channels can be equipped with a scanner. The optics are either focusing (Siemens) or nonfocusing (Philips). The flat crystal arrangement is principally a reduced size of the normal monochromator. The focusing arrangement is complicated by the fact that crystal and detector must be moved to focus the diffracted beam on the detector at any position. The scan range of 2θ for the Siemens scanner is 30–120° and 10–100° for the Philips scanner. Full details of typical construction differences between manufacturers are given in *Sec. VI* on instrumentation.

F. Detectors

The objective of a detector is the transformation of photon energy into an electrical pulse. Pulses are counted over a period of time, and the count rate, expressed in counts per second or in any other unit of time, is a measure of the intensity of the detected x-ray beam. From the theoretical point of view, the interaction of photons with matter is explained by the particle character of electromagnetic radiation, not by its wave character. Three main classes of detectors may be distinguished for use in x-ray spectrometry according to the medium responsible for the energy transformation: gas detectors, scintillation detectors, and semiconductor detectors. They differ in the efficiency of detecting photons of a given energy, that is, the range of energy or wavelength they are suited for, and in the distribution of pulse amplitudes as a result of impact of photons of the same energy, that is, the spectral resolution. The best performance in energy discrimination is obtained with semiconductor detectors and is for the x-ray domain in the range of 130 eV. This allows pulse-height selection discriminating the energy of the impacting photons (at the expense of count rate, a maximum of 30 kcps). Detectors of this type are not used in WD spectrometers but constitute the heart of energy-dispersive spectrometers and are discussed extensively in *Chap. 3*. The other two detectors are discussed here.

1. Gas-Filled Detectors

In this type of detector the energy exchange process occurs between photons and gas atoms or molecules. When a photon strikes an atom, there is a given probability that the quantum of energy $h\nu$ of the photon is imparted completely to an orbital electron of the atom. As a result of this gain in energy, the electron emerges from the atom with a kinetic energy of $E_{kin} = h\nu - W$ (Einstein's photoelectric equation). W is the work function of the electron for leaving its orbital and is a specific constant of the atom. The electron imparts its kinetic energy to other atoms, creating a series of electron-cation pairs on its path through the gas in a number equal to E_{kin} divided by the energy needed to expel an electron. Values for different gases are listed in Table 4. Subsequently, the cations and electrons are accelerated toward the cathode and anode of the detector by the electrical field gradient between both electrodes, forming on their way to the electrodes an avalanche of new ion pairs. The discharge at the electrodes gives rise in the external electrical circuit to a current pulse. For a range of electrode potentials the current pulse is proportional to the energy of the photon, and in spectrometers they are exclusively used in this range. This is why these detectors are called proportional counters. Above this potential, in the Geiger-Müller plateau (Fig. 23), any photon entering the detector gives rise to a pulse of maximum amplitude without discriminating the photon energy. The ratio of the number of ion pairs discharged at the electrodes to the ion pairs originally formed is called the amplification factor, generally denoted A.

The basic parts of a gas-filled detector are a metal rectangular housing as cathode, a thin wire passing through the central axis as anode isolated from the cathode, a filling gas, and a window either on one side or on two opposite sides of the housing. This gas flows through or is enclosed in the housing (Fig. 24). The diameter of the anode wire should be small. Realistic diameters are in the range of 40–80 μm.

Table 4 Effective Ionization Potentials (eV)

He	27.8
Ne	27.4
Ar	26.4
Kr	22.8
Xe	20.8
NaI (Tl)	50
Si (Li)	3.8
Ge	3

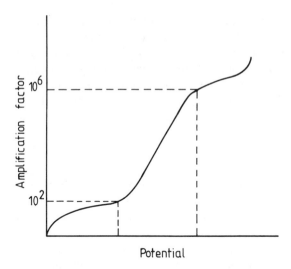

Figure 23 Region of proportionality for gas detectors.

a. Flow-Proportional Counters

These detectors cover the widest wavelength range and are used for the long wavelengths (>2 Å) in sequential spectrometers (exclusively) and also in simultaneous instruments. The windows are made of thin foils of Mylar or polyester of 0.6, 1.5, 2, or 6 μm or polypropylene 1, 2, or 6 μm thick. On the inner side the windows are coated with gold or aluminum to create a homogeneous electrical

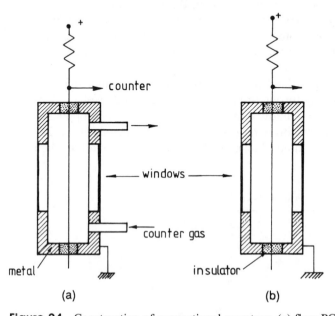

Figure 24 Construction of proportional counters: (a) flow PC; (b) sealed PC.

field inside the detector. Because of these very thin windows there is a leak of the filling gas that is compensated by a constant flow of gas through the counter (Fig. 24a). This is called a flow proportional counter. The gas is Ar with CH_4 at a concentration of 10%, and the flow is about 0.5 L/h. For detectors in simultaneous instruments helium is also used, or 88% He + 12% CO_2. For very specific applications other combinations of gases may be useful. The gas is stabilized for pressure, flow, and temperature. The amplification factor in the proportional range is between 10^2 and 10^6. The current pulse is conducted over a resistor by which it is converted into a voltage pulse that is amplified, shaped, possibly discriminated in height, and counted. The count rate is a maximum of 2×10^6 cps. The operating voltages range from 1000 to 3000 V.

b. Sealed Proportional Detector

Basically the construction is the same as in the flow proportional detector but the windows are thicker and no constant flow compensation for leak is needed (Fig. 24b). The gases used are Ne, Kr, and Xe with Al, Be, mica, Mylar, or polypropylene windows. According to the element of which the radiation is to be detected, the most suitable combination crystal, gas and window is chosen, for example, a PET crystal with a Ne detector with Al (for the determination of magnesium!) or a Be window, LiF with an Ar or Kr detector. If the filament is 80 μm thick, it is stiff and easy to install, has a long lifetime, and is easily exchanged.

Because the proportional counter is often used in tandem with a scintillation counter, two windows are present on both sides of the main body. The absorption probability of photons decreases with photon energy; the high-energy photons pass to the scintillation counter through the proportional counter without being absorbed. The sealed detectors are applied in simultaneous instruments and are not used in tandem with a scintillation counter. There is no need for a tandem arrangement because each channel is optimized for one element with the best suited combination of detector and crystal.

c. Pulse-Height Distribution

In the detector a photon generates a number of ion pairs proportional to the pulse amplitude. The formation of ion pairs is subject to statistical fluctuations, which means that the number of pairs formed oscillates around the most probable value. A record of pulses and their amplitudes as they emerge from the detector as a function of time is represented in Fig. 25. When these pulses are according to amplitude a distribution emerges as shown on the right side of Fig. 25. The width of the distribution is measured at half-maximum and is designated FWHM. This determines the energy resolution of the detector. It is small for semiconductor detectors, such as Si(Li) (about 150 eV for the 5.9 keV Mn $K\alpha$ lines from a standard ^{55}Fe source; see *Chap. 3*), about 900 eV for gas proportional counters, and 3500 eV for a scintillation counter, all measured for the same photon energy (see later). Note that this distribution differs from a distribution in intensity for a normal spectral line (intensity versus wavelength) because of natural and instrumental line broadening.

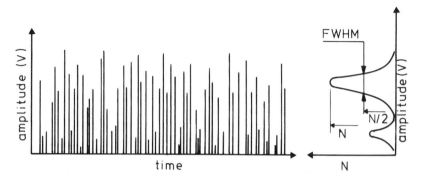

Figure 25 Amplitude-time record of impulses from the detector (left) and pulse-height distribution (right).

d. Escape Peak

As mentioned in the beginning of this section, a very probable way of energy exchange between a photon and an atom of the counter gas is the removal of an outer shell electron. The probability of removal of inner shell electrons is not zero, however, and this phenomenon happens in a number of interactions, in which case the specific K or L radiation is emitted. Because the wavelength of this radiation is on the long-wavelength side of the absorption edge, the mass attenuation coefficient is low and this characteristic photon has a good chance of escaping from the detector. The remainder of the photon energy is transferred to the electron of which the kinetic energy is generating a number of ion pairs but necessarily in a smaller number as an electron having received the whole photon energy. Thus, a number of pulses is created distributed around an amplitude proportional to the energy of this electron. The absorption edge for Ar K is 3203 eV; the energy of the Fe $K\alpha$ line is 6403 eV. Pulses centered around 6403 − 3023 = 3200 eV are generated, and about 3200/26 = 123 ion pairs are formed, 26 eV being the energy needed per electron-ion pair. These pulses are present in the pulse amplitude distribution of iron as a second maximum on the low-energy side of iron. For Ne, Kr, or Xe, photon energies greater than 0.87, 14.32, or 35.58 keV, respectively, are needed to excite the K radiation. For lower energies only L radiation is excited, resulting in an escape peak closer to the main pulse distribution.

e. Pulse-Height Discrimination

The energy resolution of a proportional counter, a few hundreds of eV at least, is not sufficient to be used in energy-dispersive spectrometers. The spectral line separation in wavelength-dispersive spectrometers is done by the crystal monochromator. As is obvious from Bragg's law, however, higher order lines of other elements and the continuum radiation of the tube may be present in the same spectral window of the crystal monochromator. In spectrometry in the ultraviolet or visible region, the higher order diffractions of the gratings are removed by filters. This is not easily done with x-rays. For the same angle 2θ, a second-order line has a substantially different wavelength or energy and is far enough apart to give two distinct pulse amplitude distributions at the exit of the counter. For x-rays electronic pulse height selection or discrimination fulfills the role of filters in the optical region. Within an interval of 0.1° around $2\theta = 37°$ for LiF200 as analyzer crystal, we find Au $L\alpha_1$ in first order, $E = 1.2764$ keV, Th $L\gamma_8$ in second order, $E = 0.6390$ keV, Sn $K\beta_2$ in third order, $E = 0.4259$ keV, and La $K\beta_2$ in fourth order, $E = 0.3201$ keV. The energy difference with the second-order line is 0.6374 keV, and the energy resolution of a proportional counter is about 0.3 keV. The electronic circuits of all commercial instruments are able to discriminate the pulses of both lines and consequently also the other higher order lines.

2. Scintillation Detectors

The energy exchange occurs in this type of detector in a medium of higher density and with high-Z elements in the matrix, namely a thallium-doped sodium iodide [NaI(Tl)], and as can be expected, this detector is only efficient for high-energy photons (<2 Å or >6 keV). The outer orbital electron for an iodide ion requires about 30 eV to be knocked out. Hence, the originally ejected electron is imparted with almost all the initial energy of the photon. The ejected electron dissipates its energy by promoting valence band electrons to an excited state 3 eV above ground level, an energy emitted on deexcitation as a photon of 3 eV or $\lambda = 410$ nm. The intensity of the emitted light pulse is proportional to the number of electrons excited by the x-ray photon ($\approx h\nu/3\text{eV}$). The pulses are detected by a photomultiplier in which the light pulse produces a few photoelectrons from the cathode material (e.g., indium antimonide). These electrons are accelerated inside the vacuum tube between the cathode and the first intermediary anode, called a dynode (Fig. 26). They gain kinetic energy and generate in turn a higher number of electrons from the dynode. This process is most often repeated 10 times, resulting in a substantial multiplication of the original number of electrons. The multiplication depends upon the potential differences between the successive dynodes (100–150 V), and the order of magnitude of this multiplication is 10^6, which is called the amplification factor A. The photomultiplier is generally a front-end tube, and the iodide crystal is in close contact with the tube window. The crystal itself has on its outer faces a reflective coating (but transparent for the photons above 6 keV) and a coating against moisture. In the external electronic circuit

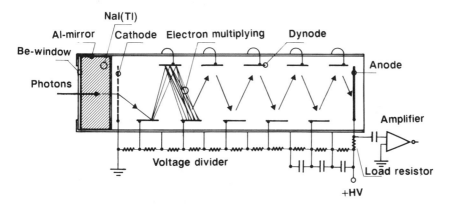

Figure 26 Layout of a scintillation counter.

the electrons provoke a current pulse of which the amplitude is ultimately proportional to the energy of the photon captured in the crystal. The conversion of the x-ray photons into a current pulse occurs over different steps: the initial formation of a photoelectron ($I^0 \rightarrow I^+ + e^-$), the quantified dissipation of the kinetic energy of this electron in 3 eV steps, emission of 410 nm light photons, the production of photoelectrons from the photocathode of the photomultiplier, and the multiplication of electrons. Although the production of the number of light photons in the crystal ($h\nu/3$ eV) is higher than the number of ion pairs created in a flow counter ($h\nu/30$ eV), the subsequent conversion into electrons is inefficient and roughly 1 electron is produced for every 10 light photons! The statistical errors of each process, particularly in the photomultiplier tube, propagate into the error of the output current pulse, thus substantially widening the distribution of pulse amplitudes and consequently reducing the resolution. The resolution ΔE is approximately two to three times larger than for a flow proportional counter (see Fig. 22), making a scintillation counter unsuitable for ED spectrometry. However, the resolution is still sufficient for a discrimination of higher order spectra.

Hardware for pulse amplifying, shaping, and counting are much the same as for the proportional counters. Also an escape peak is present, with energy the initial photon energy minus the energy at the iodine K absorption edge.

As already discussed, in many sequential spectrometers flow proportional and scintillation detectors are used in tandem. In another arrangement both detectors are shifted over 30°. A secondary collimator is placed in front of the scintillation detector but inside the vacuum chamber; the scintillation counter is outside the chamber, but close to the exit window. According to the manufacturer (Rigaku), this allows a gain in count rate of 15%.

a. Alternative Configurations

Some interesting modifications of the detector have been proposed in the paper of Ebel et al. [83]. These authors considered the possibility of reducing count losses between analyzer crystal and detector (see Fig. 27) by removing the secondary collimator and allowing the detector to collect a divergent beam and at the same time by using position-sensitive wire detector (with an energy resolution of 100 eV for detected energies of about 10 keV). This construction has been described only once but may lead to new solutions for WDXRF.

Another interesting effort is the use of a scintillation gas instead of the scintillation crystal [8,84]. This is the so-called gas-scintillation proportional counter. As is clear from Fig. 22, the resolution is substantially better than that of its "parents." The production of avalanches of ion pairs by the expelled electrons is the same as described previously. The discharge currents at electrodes is not measured, however, but rather the scintillation pulses in the ultraviolet part of the optical spectrum are detected by a photomultiplier as in other scintillation counters. Their amplitude is also proportional to the energy of the incident photon.

3. Dead Time and Shift of Maximum Pulse Amplitude

At high count rates (from a few thousands of counts per second on!), the recorded count rate deviates proportionally from the real count rate. The reason for this discrepancy can be found in the slow return of the detector gas to the ground

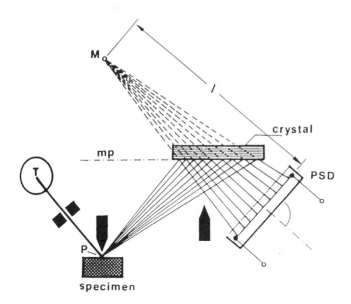

Figure 27 Geometry with position-sensitive detector: T = x-ray tube; specimen with point P irradiated by primary x-rays; crystal; mp = mirror plane; M = mirror point of P; PSD = position-sensitive detector; l = distance between mirror point and detector wire. (From Ref. 83. Reprinted by permission of John Wiley and Sons, Ltd.)

state. Cation sheaths screen the cathode. This reduces the effective voltage between the electrodes, resulting in a lower acceleration of the electrons and consequently in a smaller number of ion pairs formed. Fouling of the anode wire by impurities in the counter gas has the same effect. This phenomenon shifts the maximum of the pulse amplitude distribution to pulses from apparently lower energy photons. Moreover, when a photon is detected, the detector is for a short time unable to detect a next incident photon. This period of time is called dead time. The value of dead time is about 1–2 μs for scintillation counters, about 0.2 μs for proportional counters, and $\approx$200 μs for a Geiger-Müller detector. To overcome this problem one can either choose another detector (but the choice is very limited), lower the intensity of the exciting beam by masks or absorbers, change the geometry (as in done in the case of synchrotron radiation), or use electronic anticoincidence circuits or appropriate mathematical corrections. The relation between the real and recorded count rate is

$$N_r = \frac{N_m}{1 - N_m t_d} \tag{10}$$

where N_r = the real count rate (counts s^{-1} or cps), N_m = the measured count rate (cps), and t_d = the dead time (s).

If we allow a given level of discrepancy, say $a\%$, then the following equation is valid:

$$\frac{a}{100 - a} = N_r t_d \tag{11}$$

The real count rate and dead time are thus connected by a hyperbolic equation. If the allowed value of discrepancy must be kept constant, the left side of the hyperbola is constant.

However, this is only part of the problem. As already stated, a photon, arriving just at the end of the dead time, is registered with reduced amplitude. It is only after the so-called recovery time that a new photon is registered with normal amplitude.

An improved method for calculation of dead time was proposed by Bonetto and Riveros [85]. These authors used the well-known fact that second-order peaks are significantly less intensive than first-order peaks but the ratio of the intensities

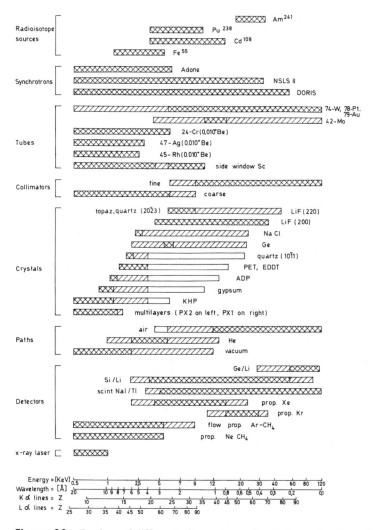

Figure 28 Review of different elements serving the construction of a spectrometer: optimal range (crossed area), feasible range (hatched area), and possible range (open area). (Figure adapted from Bertin, Ref. 13. Reprinted by permission of Plenum Publ. Corp.)

should be constant. If it is not constant, the deviation results fully from dead-time losses in the first-order peak. They derived the relation

$$\frac{N_{m1}}{N_{m2}} = B - AN_{m1} \qquad (12)$$

where $B = N_{r1}/N_{r2}$ and $A = Bt_d$. The indices 1 and 2 refer to the order of the spectrum.

A summary of constituting parts of spectrometers and the wavelength ranges in which they are used is given in Fig. 28. Note that different excitation sources are used in different branches of XRF, namely radioisotopes for EDXRF and tubes for WDXRF; there are also examples of synchrotron radiation. The x-ray laser is mentioned but does not yet exist.

IV. QUALITATIVE AND QUANTITATIVE ANALYSIS

In this section we review a number of items important to both qualitative and quantitative analysis. The qualitative application of x-ray spectrometry is obvious. In simple cases quite unequivocal assignments of spectral lines can be made based on a library of $K\alpha$ or $L\alpha$ lines, or in case of doubt, when overlapping exists, $K\beta$ or other L or M lines are checked to confirm or reject a line assignment. A few more complex situations are discussed. Quantitative analysis may also be simple. In the low concentration range, quasi-linear relationships are found between intensity and concentration. The same holds for determinations of higher concentrations but in a very narrow range for samples with quite comparable matrices. Interpolation on standard curves, obtained on suitable standard samples, allows quick and sufficiently accurate quantitative determination. In these boundary cases, matrix effects are negligible but *always* exist. In normal practice they compromise the conversion of intensity into concentration in such a way that more or less complicated mathematical methods must be used for obtaining any accurate quantitative determination. The approach by fundamental parameter programs or correction algorithms is dealt with in Chap. 5. Between simple qualitative work and rigorous quantitative work, however, an area of application is situated in which knowledge of the exact concentration or the complete qualitative composition is less important than the overall aspect of a spectrum for recognizing classes of samples, that is, pattern recognition. An introductory discussion of this aspect is given in this section. Another point of interest to qualitative as well as to quantitative work is measurement of background. A good estimation of background is necessary for obtaining net count rates, but decreasing the background allows lowering detection limits. The origin and magnitude of background is bound to the method of acquisition of a spectrum and thus to the type of spectrometer. This aspect is also treated in this section.

A. Background

The background in WDXRF has three main sources: (1) the coherent and incoherent scattering of source radiation, (2) the presence of characteristic radiation of other than sample origin (materials of sample cup masks, sample cups, collimators, spectrometer housing, for example), and (3) the detector noise. The background from characteristic radiation from construction materials of the spectrometer can be minimized by careful construction and geometric arrangement of the spectrometer. Scattering from sample cup masks can efficiently be reduced by masks in front of the primary collimator. This contribution is especially great when small samples are analyzed. The beam mask reduces the irradiated area and may totally eliminate scatter from sample cup masks. Higher order scatter increases with increasing wavelength. Pulse-height selection, as explained earlier, suppresses counts from higher order diffraction, but this technique is most efficient in the long-wavelength region and is decreasing efficient toward the short-wavelength end. The short-wavelengths contribution to the background can be reduced by installing adequate filters. A 100 μm titanium foil removes the characteristic lines of chromium from a chromium anode tube, significantly improving the conditions for the detection of Cr, Mn, Fe, Co, and Cu [15].

For Cr tube radiation and LiF (200) analyzer, crystal removal of scattered chromium radiation may be necessary when detecting the neighboring first-order lines. As Table 5 shows, not very many common elements are interfering. For steel analysis really only manganese represents a problem. Because chromium is present in prevailing concentrations, manganese determination is also difficult with an aluminum filter. The simplest solution in such a case is the use of LiF (220) crystal with better resolution in this region (but also with lower intensity). If tubes with other targets are used, tables are similar to Table 5 but with even more complex interferences if targets like Mo and Rh are used. The other elements listed refer mostly to less common elements and less intense lines. Other higher order lines may interfere as well if their concentration is high. For special cases a convenient combination of filters, tube targets, and analyzer crystals can almost always be constructed. Let us look at an example for the rhodium target tube. In the determination of palladium, Pd $K\alpha_1$ overlaps with the Compton peak Rh $K\beta$ (Fig. 29). A molybdenum filter in the primary beam reduces this peak to a very low level, allowing a perfect estimation of the location of the maximum and the determination of the intensity.

Another efficient solution for background suppression is the use of linearly polarized x-rays. An excellent example was given by Knöchel et al. [86] and is reproduced in Fig. 30. The analysis of nitrogen with a small concentration of xenon was made in a synchrotron excited XRF. Some special geometries allowing the use of polarized x-rays have been proposed by Wobrauschek and Aiginger

Table 5 Interfering Lines for Cr $K\alpha$ and $K\beta$ with LiF 200 Crystal

			ΔE (eV)	Δ (°2θ)					ΔE (eV)	Δ (°2θ)
Cr	$K\alpha$	Pm	$L\alpha_2$	−10	0.05	Cr	$K\beta$	Ho	Ll	< −10	0.04
		La	$L\beta_2$	−30	0.31			Pr	$L\beta_7$	−20	0.24
		Ba	$L\gamma_5$	−40	0.60			Mn	$K\alpha$	−50	0.61
		La	$L\beta_7$	30	−0.23			Pm	$L\beta_1$	10	−0.17
		Ce	$L\beta_6$	20	−0.31			Ba	$L\gamma_4$	30	−0.30
		Pm	$L\alpha_1$	20	−0.31						
		V	$K\beta$	10	−0.56						

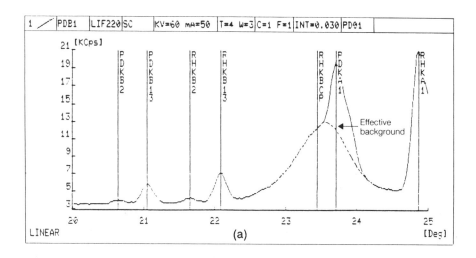

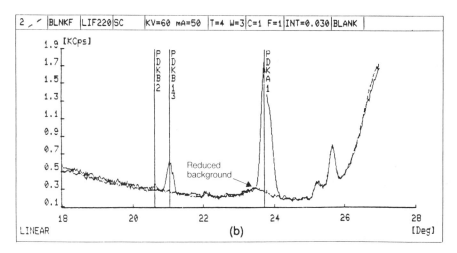

Figure 29 Spectra of Pd excited by a Rh tube and recorded with and without a primary beam molybdenum filter (0.1 mm Mo). The peaks are labeled as Pd $K\alpha$, Pd $K\beta$, Rh $K\alpha$, and Rh $K\beta$ with their respective indices. (Courtesy of ARL.)

[87–90] for the more conventional XRF spectrometers. The subject is discussed in detail in *Chap. 9*.

A careful estimation of background is apparently the simplest way to take background into account! A first approach, but with limited application, is measurement on blanks—simple at first glance, but the adequate blank is often a rarity. Moreover, the blank ought to have the same scattering properties. Some trials of mathematical background modeling were performed but for other purposes.

The most common method of background evaluation is the linear (or nonlinear) interpolation from background countings on a suitable angle above and below the peak position [*suitable* means not too far away for an easy interpolation

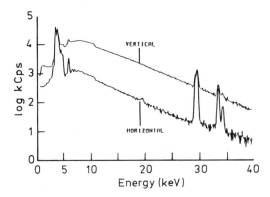

Figure 30 Synchrotron excited spectra of gaseous nitrogen with a small amount of xenon taken in the direction of polarization and perpendicular to the orbit plane. Counts normalized to stored electron current integrated over the lifetime of the detection system. DORIS is operated at 3.3 GeV and about 55 mA. (From Ref. 86. Reprinted by permission of the author and Elsevier Science Publishers.)

(linear), outside the tailings of the peak, and not at the position of an adjacent peak].

Finally, we refer to a method based on the relationship of fluorescent and scattered radiation intensities versus the value of attenuation coefficients. For this no measurement of background is necessary for the estimation of net line intensities [91].

B. Qualitative Spectrometry

Although the number of spectral lines is very limited compared to atomic emission spectra in the ultraviolet or visible region, overlapping of lines in x-ray spectra does exist. A classic example can be found in the papers of Gentry et al. [92], Sparks et al. [93], and Sparks [76], and an example is reproduced in Fig. 31. At energies at which lines of palladium and cesium (K series; about 22 and up to 28 keV) can be present, tails of L lines of uranium and thorium are found, and also the M lines of the (hypothetic) superheavy elements of $Z = 110$ and 126 should be found in this region. The erroneous attribution of some lines to these hypothetical elements was the subject of extended research and long discussions [76,93]. In the software of modern spectrometers, special searching blocks exist for the assignment of lines: the detected line is compared with data from the memory. After the initial assignment other lines of the element are identified according to data from the library. The initial assignment is eventually rejected if not confirmed by other lines (Fig. 32). Multichannel instruments are not suited for this type of work, except for elements implemented on the spectrometer (maximum 28 elements). Sequential instruments can cover all elements (from beryllium on). The spectra can be plotted on paper or displayed on the monitor screen of all modern spectrometers. Particular parts of the spectrum can be scanned with highest resolution and sensitivity (step scanning) according to the time available for the analysis and the information required. Parts of particular interest can be

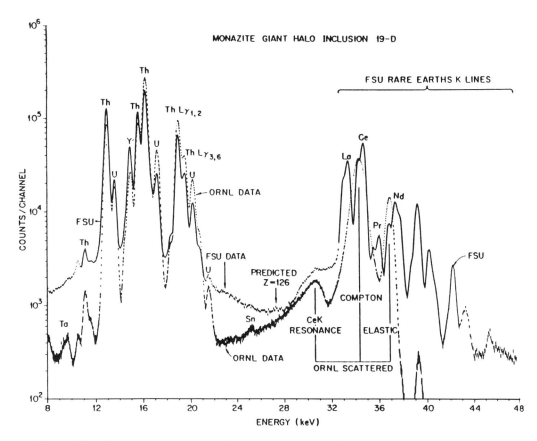

Figure 31 Fluorescence spectrum from a monazite particle excited with 37 keV synchrotron radiation (ORNL DATA) shows an improved signal-to-background ratio over that excited with 5.7 MeV protons (FSU DATA). Data points every 20 eV in ORNL spectrum and every 61.8 eV in FSU spectrum. (From Ref. 76. Reprinted by permission of the author and Plenum Publ. Corp.)

displayed separately with linear intensity axis or rescaled to square root or logarithmic scales. The peaks may be identified on the display by the angle 2θ, by element name, or by wavelength.

The spectrum can be scanned linearly with respect to 2θ or in a more optimal way. If Bragg's and Moseley's laws are combined, the following relationship between Z, the atomic number, and θ is found:

$$Z = \sqrt{\frac{m}{k2d \sin \theta}} + \sigma \tag{13}$$

Z is proportional to the inverse square root of $\sin \theta$. Jenkins et al. [94] proposed computer monitoring of the scan speed according to this equation. It optimizes scan speed, which represents a high gain in time especially in the low-Z range where line interdistances are long.

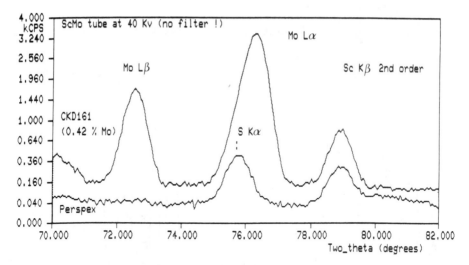

Figure 32 Example of qualitative analysis by rejection: suspected Mo $L\alpha$ peak in the lower spectrum was rejected in comparison with the standard containing molydenum (upper spectrum). (Courtesy of Philips Analytical.)

C. Special Quantitative Applications

Some special and nonconventional applications were revealed to have potential practical meaning. As stated earlier, monochromatic excitation is simplifying considerably the quantitative treatment of spectral data, but this case is only randomly met in WDXRF. The commonly used excitation source is the x-ray tube, which delivers a spectrum of characteristic lines superimposed on a white spectrum. For this reason no rigorous and at the same time simplified correction algorithms have been found. On the contrary, how the correction terms can be derived in a rigorous manner for the case of monochromatic excitation was shown in a paper by Kuczumow [95]. In its original formulation [96] the Lachance-Traill equation is

$$W_i = R_i(1 + \sum_{j \neq i} \alpha_{ij} W_j) \tag{14}$$

with α_{ij} defined as a constant written as

$$\alpha_{ij} = \frac{\mu_{j\Psi 1}(E) + \mu_{j\Psi 2}(E_i) - \mu_{i\Psi 1}(E) - \mu_{i\Psi 2}(E_i)}{\mu_{i\Psi 1}(E) + \mu_{i\Psi 2}(E_i)} \tag{14a}$$

where μ denote the mass absorption coefficients measured for primary or secondary photons of energy E and E_i, respectively, corrected for angles of incidence Ψ_1 or takeoff Ψ_2. It has been proved [97] that this expression is valid only for boundary conditions: monochromatic excitation and absence of enhancement. In all other conditions α are not constant as experimentally shown [15,98]. It is impossible to derive the more flexible algorithm directly from Shiraiwa and Fujino's equation [95,99]. Other algorithms of the type of Tertian [100] and Lachance COLA [98] are not recognized as fundamental. An algorithm that can be considered partially fundamental is that of Claisse-Quintin [97], but it suffers from inaccuracies in the approximations introduced. In recent years Rousseau [101] and Kuczumow and Holland [102] gave examples of algorithms derived from fundamental assumptions. Kuczumow and Holland describe the change in α when enhancement is included (monochromatic excitation), as follows:

$$W_i = R_i(1 + \sum_{j \neq i,k} \alpha_{ij} W_{ij} + \Delta\alpha_{ik} W_k) \tag{15}$$

where all α are defined as previously but $\Delta\alpha$ as

$$\Delta\alpha_{ik} = \frac{\mu'_{k\Psi 2}(E_i) - \mu_{r\Psi 2}(E_i)}{\mu_{i\Psi 1}(E) - \mu_{i\Psi 2}(E_i)} \tag{15a}$$

The term $\mu_{k\Psi 2}(E_i)$ was substituted by $\mu'_{k\Psi 2}(E_i)$ in Eq. (15). Although dependent on excitation condition E and composition, this new term has the very great advantage of being dependent in a simple way on concentration. The possibility of deriving the Lachance-Traill equation directly from the Shiraiwa and Fujino equation on conditions given in Eq. (15) is also well established [102]. Even third-order effects can be included, resulting in the expression

$$W_i = R_i[(1 + \sum_{j \neq i,k} \alpha_{ij} W_j) + \sum_{j=j,k} \alpha_{ij} W_j + R_j\beta_{jk} W_k] \tag{16}$$

The next interesting case concerns scattered radiation as a source of analytical algorithms. Similar to the derivation made in Ref. 95, an expression on the basis of coherently scattered radiation is obtained [103]:

$$W_i = R_i^{\text{coh}} + \sum_{j \neq i} (R_i^{\text{coh}} \alpha_{ij}^{\text{coh}} - \beta_{ij}^{\text{coh}})W_j \tag{17}$$

where

$$\alpha_{ij}^{\text{coh}} = \frac{\mu_{j\Psi 1}(E_0) + \mu_{j\Psi 2}(E_0) - \mu_{i\Psi 1}(E_0) - \mu_{i\Psi 2}(E_0)}{\mu_{i\Psi 1}(E_0) + \mu_{i\Psi 2}(E_0)} \tag{17a}$$

$$\beta_{ij}^{\text{coh}} = \frac{\sigma_j^{\text{coh}}(E_0)}{\sigma_i^{\text{coh}}(E_0)} \tag{17b}$$

and R_i^{coh} results from the comparison of a coherently scattered line from the sample and from the pure element i, and σ^{coh} is the mass coherent scatter coefficient. For Compton scattered radiation analogous expressions are obtained:

$$W_i = R_i^{\text{com}} + \sum_{j \neq i} (R_i^{\text{com}} \alpha_{ij}^{\text{com}} - \beta_{ij}^{\text{com}})W_j \tag{18}$$

with

$$\alpha_{ij}^{\text{com}} = \frac{\mu_{j\Psi 1}(E_0) + \mu_{j\Psi 2}(E_{\text{com}}) - \mu_{i\Psi 1}(E_0) - \mu_{i\Psi 1}(E_{\text{com}})}{\mu_{i\Psi 1}(E_0) + \mu_{i\Psi 2}(E_{\text{com}})} \tag{18a}$$

$$\beta_{ij}^{\text{com}} = \frac{\sigma_j^{\text{com}}(E_0)}{\sigma_i^{\text{com}}(E_0)} \tag{18b}$$

R_i^{com} is the ratio of the Compton peak of the sample to that of the pure element. Equations (17) and (18) may be called Lachance-Traill equations for scattered radiation by their general appearance as well as by the way they have been derived. These equations are less sensitive to the type of components and changes in composition than their analogs for fluorescent radiation, which otherwise is an advantage in analyses of samples widely varying in composition. In Table 6 the coefficients α and β from Eqs. (14), (15), (17), and (18) are collected, calculated for the K lines of La and Sm, the main components of samples consisting of La_2O_3 + Sm_2O_3 excited by 60 keV photons and 80 and 70° as incidence and emergence angles (the example is taken from EDXRF, but it also applies here).

When applied to nonhomogeneous samples, the α are also proved to be dependent on particle size for dispersions or layer thickness for piles of layers of different composition [104,105]. When applied to a dispersion of particles of radius r in a matrix of another element, the Lachance-Traill equation is written as

$$W_i = R_i[1 + \sum_{j \neq i} \alpha_{ij}(r)W_j] \tag{19}$$

where the α change according to the equation

$$\alpha_{ij}(r) = \frac{\mu_{j\Psi 1}(E_0) + \mu_{j\Psi 2}(E_{i,r}) - \mu_{i\Psi 1}(E_0) - \mu_{i\Psi 2}(E_i)}{\mu_{i\Psi 1}(E_0) + \mu_{i\Psi 1}(E_i)} \tag{19a}$$

Moreover, when the variation of α is transformed on more fundamental grounds by the procedure similar to that of Ref. 102, the relation is

$$b = [\mu_j(E_{i,r}) - \mu_j(E_i)_{\lim}](r - r_{\lim}) \tag{20}$$

which enables the presentation of $\alpha(r)$ in the form

$$a = [\alpha_{ij}(r) - \alpha_{ij\lim}](r - r_{\lim}) \tag{21}$$

These equations do not have a universal application because they are only derived for constant composition of the two phases and spherical particles. Equation (20) widens the possibilities of using this procedure in the case of polychromatic excitation.

D. Pattern Recognition

The spectrum of a sample of complex composition is characterized by a set of line intensities. If we imagine an n-dimensional space with the intensities on the coordinate axes, the set of intensities determines a point in this space. Such an intensity space is convertible into an n-dimensional space of concentrations, and the point in the intensity space has its equivalent in the concentration space. The

Table 6

		La_2O_3		Sm_2O_3
α^f	$(La_2O_3Sm_2O_3)$	0.292	$(Sm_2O_3La_2O_3)$	1.009
$\Delta\alpha^f$		−0.895		—
α^{coh}		0.285		−0.222
β^{coh}		1.153		0.867
α^{com}		0.281		−0.219
β^{com}		1.000		1.000

aim of pattern recognition is to attribute a sample to a given class. A class is in a discrete way a set of points of similar features in the intensity or concentration space or, in a continuous way, a subspace inside a complete n-dimensional space. This subspace is restrained by a closed hyperplane, which may be approximated by a number of hyperplanes of smaller dimensionality. Inherent to all experimental determinations, the determination of intensity is also bound to a certain degree of uncertainty. This implies that a point cannot be determined, only a small subspace of limited thickness, limited by assumed confidence levels. Equally ill-defined for the same reasons is the boundary hyperplane surrounding the subspace occupied by a class of samples. The "thickness" of this "space ring" surrounding the subspace is determined in relation to each coordinate axis by the appropriate multiplicity of standard deviations.

1. Delimitation of Classes

If we consider classes as separate subspaces in an n-dimensional configuration space, it is obvious that each of these classes has its own boundary hyperplane or closed space ring. If these hyperplanes can be determined in an analytical way, we are able to know the boundaries of classes. In simple cases a class can be delimited by construction of a low-dimensional hyperplane that closely approximates the real boundary. If classes are separated by empty spaces, the differentiation of classes is also easy.

2. Method of Nearest Neighbors

Here the classes are considered small subsets of points. The classification of each new point relies on comparison with the nearest point neighbors. Knowledge of the distance between sample and the nearest neighbors is essential. The accuracy of the method depends on the way the distance is determined and the degree of complexity of the boundaries between classes, although it can be simplified by imposing proper statistical weights on learning samples.

3. Reduction in Number of Dimensions

All points and classes are included in the multidimensional space generated by the number of elements present in the sample; however, some variables may be less dependent on the related coordinates, and then the number of variables necessary to differentiate the sample may be reduced. The number of dimensions may also be reduced if some variables are correlated.

4. Examples

Samples may be classified into groups according to their origin. A classic reference here is the paper by Kowalski et al. [106], which was followed by many other papers on archeology. Kowalski recognized the origin of obsidian artifacts by studying a set of intensities of 10 trace elements and projected the image of this set from a 10-dimensional configuration space into a 2-dimensional space (nonlinear mapping). Similar applications in geology are obvious.

If only a limited amount of data is available, samples can only be subdivided in groups. Discrimination of the alloys 663, B-1900, and 1455 can be done on the

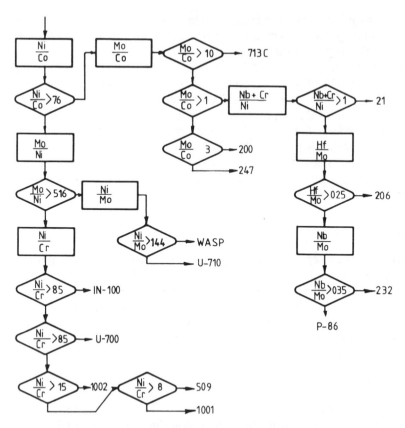

Figure 33 Selective action of KEVEX Material Analyzer Model 6500 by successive analysis of element ratios. (Courtesy of KEVEX Corp.)

basis of their zirconium content (Kevex Corporation, prospectus describing the Kevex Material Analyzer Mode 6500) [107]. Nickel-cobalt alloys were subdivided on the basis of a logical analysis of spectral response: is $A/B > X$ (Fig. 33)? Complete analysis of the alloy was not necessary. The same is often done when samples can be compared to standards, for example, the "framing action" by software (description of Siemens software [108]).

An important application of pattern recognition may be found in "two-stage analysis." For complex samples with nonlinear dependence of intensity on concentration, the analysis may be broken down in steps with smaller number of dimensions, or the samples may first be subdivided in classes before applying correction algorithms. The appropriate set of differential equations exists, establishing the relation between the absolute value of relative intensities (Claisse-Thinh systems [109]) or between increments of intensities (Ebel-Mantler [110] and Kuczumow [111] systems) and increments of concentrations. These increments are calculated with respect to a central standard that represents the origin of a local system of coordinates. In this local system a certain set of differential equations is valid and is of linear character (if not, it is not profitable to use it in such a restricted space). After attributing the samples to given classes, the right set of differential equations may be used for accurate determination of composition. The origin of the local system of coordinates may be at the same time the central point of a given class.

When the delimitation of classes becomes problematical, a point in the configuration space may be designated and a set of differential equations, valid about this point, may be determined. From this the boundaries of classes can be found and determined by some leading vector introduced in the central points of classes.

Yap and Tang [112–115] posed the reciprocal question in relation to "pattern recognition": how to extract some number of differentiating features from the known classes, differing by their time and place of origin. This was used to analyze pottery samples.

E. Detection Limits

The importance of correct determination of detection limits has been acknowledged by many chemists, especially those involved in environmental regulation. An accepted definition is as follows: "The limit of detection is the lowest concentration level that can be determined to be statistically significant from an analytical blank" [115]. The limit of detection is expressed as a concentration c_L or an amount q_L and is derived from the smallest measure x_L, x being the instrument reading. How then do we quantify "statistically significant"?

The simplest model to start from is the calibration curve, reading x versus concentration c, obtained by linear regression on the data:

$$x = mc + i \tag{22}$$

where m is the slope or analytical sensitivity and i is the intercept. The more points available, the better is the slope determined and also the intercept. The background x_B is obtained for $c = 0$. All measurements are subject to error, and x_B represents only a mean value in the center of a symmetrical distribution curve. When this distribution is normal or Gaussian, it is characterized by its standard deviation σ, or s if the number of observations is finite (and small). The standard deviation within this distribution is calculated as the square root of the variance:

$$s_B^2 = \frac{\sum_{j=1}^{n}(X_{B_j} - X_B)^2}{n-1} \tag{23}$$

If we are concerned with background determination, the distance of a measurement with value x may be expressed with respect to x_B in the number of standard deviations ks_B. The International Union of Pure and Applied Chemistry (IUPAC) determines the limit of detection in the following way:

$$X_L = \overline{X_B} + ks_B \tag{24}$$

The attributed concentration [through a calibration curve of the type of Eq. (22)] is

$$C_L = \frac{(X_L - \overline{X_B})}{m} \tag{25}$$

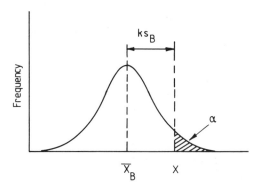

Figure 34 Probability curve: the probability that results has a positive deviation greater than ks_B from the most probable value $\overline{x_B}$ and is equal to α (dashed area under the curve).

or by the substitution of (24) in (25) and rearrangement,

$$C_L = \frac{ks_B}{m} \tag{26}$$

where k is a factor depending on the confidence level desired. For $k = 3$, the value recommended by IUPAC and others for x_L or c_L, a confidence level is allowed of 99.86%. This means that the shaded area α in Fig. 34 is 0.0013 (total area under the curve being 1), or for errors following a normal distribution, the signal value $x_L \geq x_B + 3s_B$ has a probability of 0.13% to belong to the population of values determining the mean value x_B. c_L is recommended to be reported as $c_{L(k=3)}$ to avoid any confusion about the choice of k.

Different values for c_L are encountered when the standard deviation of mean s_B is used, or the pooled standard deviation s_p, or the relative standard deviation (RSD):

$$s_{\overline{B}} = \frac{s_B}{\sqrt{n_B}} \tag{27a}$$

$$s_p = \sqrt{\frac{s_B^2}{n_S} + \frac{s_B^2}{n_B}} \tag{27b}$$

$$\text{RSD} = \frac{s_B}{X_B} \tag{27c}$$

where n_s is the number of determinations on the sample and n_B the number of background determinations. In these cases k is replaced by another k value calculated for a t distribution (Student's distribution). For example, when RSD is applied, c_L becomes

$$C_L = \frac{k\text{RSD}X_B}{m} \tag{28}$$

It is also obvious that the values of c_L change considerably when the distribution of errors is not normal. For a thorough discussion we may refer to the paper by Long and Winefordner [116]. This paper also contains interesting references to papers of Kaiser, Boumans, and others, which are therefore not repeated here.

Let us apply the IUPAC recommendation to XRF. The standard deviation s for counting of photons emitted at completely random intervals of time t obeys for an average number of accumulated counts N the equation

$$s_N = \sqrt{N} \tag{29}$$

For the derivation of this relation we may refer to a small, almost forgotten, but instructive book of Beers [117]. If we express this as relative standard deviation, Eq. (29) becomes

$$\text{RSD} = \frac{\sqrt{N}}{N} = \frac{\text{RSD}}{\sqrt{N}} = \frac{100}{\sqrt{N}} \tag{29a}$$

The detection limit is then calculated taking $k = 3$ and expressed as a function of intensity $R = N/t$ with t as the counting time for relative concentration units and with $m = \text{cps}/\%$:

$$c_L = \frac{3\sqrt{R_B/t_B}}{m} \tag{29b}$$

A standard format used by most manufacturers is to give c_L for $t_B = 100$ s. In Table 7 a series of values for c_L are given for randomly chosen elements collected

Table 7 Detection Limits[a]

Matrix	Element	c_L	Reference
Terephthalic acid	Fe	0.15	118
	Co	0.18	
Low-alloy steel	C	80, 240	118, 119
	Al	4, 4.5	
	Si	2.1, 4.1	
	Cr	2, 1.7	
Al-Mg alloy	Mg	7	118
	Si	5	
	Ti	3	
	Mn	2	
	Cu	1	
Aluminum	Mg	10.5	119
	P	1.3	
Glass powder	B	1%	120
Phosphosilicate glass	B_2O_3	0.4%	118
Cement	Na_2O	36	119[b]
	MgO	27	
	Al_2O_3	22	
	SiO_2	50	
	SO_3	24	
	P_2O_5	32	
Copper alloys	Be	0.2%	118

[a] ppm (unless otherwise specified) for 100 s counting time and $k = 3$.
[b] With use of multilayer OVO 55.

from manufacturer's documents. No general lists can be included because detection limits as well as sensitivities depend heavily on matrix composition. Instead this is intended as an indication of order of magnitude.

The detection limits previously discussed provided methods for deciding whether a value x_U for an unknown was greater than x_B with a given level of confidence. Another important approach is advocated and applied by Clayton and collaborators [121]. They state that "Traditional techniques for determining detection limits have been concerned only with providing protection against type I errors, or false positive conclusions (i.e., reporting an analyte as present when it is not). They have not considered type II errors, or false negative assertions (i.e., reporting an analyte as not present when it is)." They discuss in their paper a test of the null hypothesis $X = 0$ versus alternatives that $X > 0$, where X denotes the true but unknown concentration of an analyte in a medium and for two models of calibration curves with known or with unknown parameters. The test is applied to chromatographic data. Their theory can certainly be applied to XRF data but has not yet been reported.

Finally, we can recommend the excellent discussion of the physical origin of background sources and detection limits in PIXE and synchrotron excited XRF given by Sparks [76], which is also instructive for WDXRF.

F. Limits of XRF

Compared to ultraviolet (UV)-visible emission spectroscopy, x-ray spectra are "oligoline" spectra with respect to the resolution of the spectrometers. The emission spectrum of iron contains some 2000 lines between 200 and 800 nm; its complete x-ray counterpart has a dozen important lines between 1.757 Å ($K\beta_1$) and 0.615 Å (L_l)! For everyday work, however, only two lines 1.936 and 1.940 Å ($K\alpha_{1,2}$) and 1.757 Å ($K\beta$) play a role. A densely populated L series spectrum such as that of gold has a set of about 20 lines between 0.9 and 1.5 Å, a spectrum that can be managed by a spectrometer with a relative spectral resolution of 10^{-2}. The prospects for this spectrometric range must be bright, but what are the limits to this method?

1. Application Limits

All elements from boron (or beryllium?) on can be detected and/or quantitatively determined. On the high-Z element or low-λ side, the spectral resolution may be a problem for pure spectroscopic studies but seldom or never for qualitative or quantitative applications. Because the emitted energy is high the attenuation coefficients are low, so the critical depth is great and consequently the analyzed volume is very representative of the composition of the bulk. Critical depth is the thickness d_{crit} of a layer parallel to the surface from which 99% of the intensity sensed by the detector is recruited. It is a function of takeoff angle Ψ_2 (if the penetration depth of the exciting beam is much higher), the absorption coefficient of the matrix, and the density:

$$d_{\text{crit}} = \frac{\ln\left(1 - \dfrac{I_t}{I_\infty}\right)}{-\overline{\mu}\rho} \tag{30}$$

where I_t is the intensity emitted by a the thin foil and I_∞ is the intensity emitted by an infinitely thick sample.

For long wavelengths, $\mu\rho$ becomes high and d_{crit} and analyzed volume become low. The sampled volume is low, and the analysis may be called a near surface, not representative for the bulk. It is a disadvantage or true limit by force of nature, although it may also be exploited for near surface analysis of light elements to 1 or 2 μm. From the concept of critical depth it is obvious that coating analysis can be done up to thicknesses depending on the absorption (and enhancement!) characteristics of the coating, the overall thickness range going from 0.01 μm to a few hundreds μm.

Apart from the analysis of macrosamples, WD spectrometers may also be fitted to electron microprobes, where they supplement the ED spectrometer for better resolution of light elements. Curved crystal optics is generally used because of optimum efficiency (compare the quantum efficiency of a classical spectrometer $\approx 10^{-8}$, curved crystal spectrometer $\approx 10^{-6}$, and ED spectrometer $\approx 10^{-4}$). WD spectrometers are also needed for heavy particle-induced emission; because of the presence of many satellite lines, good resolution is required (Ref. 122 and *Chap. 11*).

The dynamic range of analysis covers about five decades from 10^{-3} to 100%. Below $10^{-3}\%$ preconcentration procedures must be used.

The intensity of exciting beams was not sufficient to allow spot analysis. Recently, Boehme et al. [34,35] published the results of two-dimensional mapping executed with small focused beams from high-power sources (rotating anode tubes) allowing lateral resolution of about 30 μm. This offsets the boundaries to a new domain: high spatial resolution x-ray spectrometry.

2. Limits of Precision and Accuracy

An analyst is always threatened by errors in the analysis: outliers and random and systematic errors. When a given operation or measurement provokes an offset between the true and measured values, a systematic error is said to be introduced. This offset may be due to a wrong calibration of any kind or to a wrong standard. Such errors can only be traced by analyzing a sample from the same batch with other techniques. Random errors and outliers can be dealt with by statistical methods. One example of such a method is described by Plesch [123] and implemented in the software package Spectra 310 (Siemens). A calibration with n standards is assumed to give a standard deviation. The standard deviation s_s is calculated according to

$$s_s = \sqrt{\sum \frac{(C_i' - C_i)^2}{n - p}} \tag{31}$$

where C_i is the true weight fraction and C_i' the result obtained; p is the number of parameters used in the fit: $p = 2$ for a linear and $p = 3$ for a parabolic fit. If the concentration C_a of one specimen differs too much from C_a', the presence of an outlier is suspected. The suspected value is eliminated, and a new standard deviation is calculated. C_a is an outlier with a probability of $P\%$ according to the inequality (F test) $(s_1/s_2)^2 > F_p$. The threshold values for F_p are listed in the current statistical literature. Application of the F test in fact tests whether a statistically significant difference exists between the standard deviation calculated with and without the suspected result. Possible causes of outliers are numerous, including all kinds of human mistakes, temporary instrument failures, and heterogeneity of specimens.

A measurement of fluorescent radiation is, as in any other experimental determination, subject to minimal error. The standard deviation of a radiation measurement is equal to the square root of the number of accumulated counts N. The real standard deviation is greater than $\sqrt{N}$ because of errors of instrumental origin. Another source of error is the conversion of intensity into concentration and, last but not least, inherent to the sample, namely microheterogeneity, which is difficult to foresee and to quantify and present even when all normal precautions for careful sample preparation are taken.

a. Microheterogeneity

Heterogeneities caused by sample preparation are dealt with in a subsequent section. These artifacts should be absent here. Segregation, however, which is the origin of what we call microheterogeneity, is a bulk (or surface) property of physical origin. Solid solutions of compounds or elements that are not perfectly soluble in each other result in multiphase components of which the "degree" of heterogeneity in composition of the segregated phase as well as in the size of the phases depends upon the history of the solution (thermal history, mechanical treatment, and so on). It is not easily mastered and difficult to quantify. Its importance for XRF analysis has been acknowledged by several authors: examples are tin and lead in solders [124], silicates in fused beads [125], and silicon in Si-Al alloys [126]. The first report we are aware of is a paper by Claisse in 1957 on the determination of FeS in a sulfur matrix [127]. A systematic study has been devoted to this phenomenon by Helsen and Vrebos [104,105,128–131], both by Monte Carlo simulation on hypothetical mixtures and by measurements on samples of known segregation. It was found by simulation on a (hypothetical) dispersion of spherical particles of iron in a chromium matrix that even for particle diameter of 1 μm an error is introduced in the relative intensity of Fe $K\alpha$ of 6% (on the level of $c_{Fe} = 0.5$) with respect to the perfectly homogeneous solution. The effect was proved experimentally on alloys of Al-Si, Al-Mo, and Al-Zn. The extent of the effect changes with time as in fused beads. This was demonstrated by the determinations of Novosel-Radovic et al. [125] on fluxed silicates and also confirmed by Monte Carlo simulation by Vrebos and Helsen [129]. The effect can theoretically be accounted for but is difficult to surmount in practice. Thus, microsegregation imposes a true physical limit to the accuracy obtainable by XRF spectrometry on quite a substantial number of samples, including minerals and alloys.

b. Correction Algorithms

The application of correction algorithms is limited to perfectly homogeneous samples, but even then, they do not allow a correct conversion of intensity into concentration over a wide concentration range. To have a quantitative idea about these errors, Rousseau [132] and Vrebos and Helsen [133] compared different algorithms both on experimental values as well as starting from relative intensities

Table 8 Analysis of a Cr-Co-Mo-Fe-Ni Alloy

Analyte		Concentration (Weight %)			
		1	2	3	4
Cr	a	12.70	19.90	19.70	19.44
	b	12.79	20.06	18.92	19.19
Co	a	0.70	42.00	11.55	13.50
	b	0.73	41.96	11.53	13.45
Mo	a	6.08	4.00	10.30	4.34
	b	6.10	3.75	10.50	4.35
Fe	a	36.20	3.19	0.46	2.09
	b	36.20	3.17	0.45	2.07
Ni	a	40.10	20.00	53.30	55.70
	b	40.19	20.20	53.39	55.51

[a] Chemical analysis.
[b] Application of Lachance's algorithm.

calculated from fundamental parameter programs (no counting errors and no errors on the composition of standards). As an illustration we cite a few results obtained by Lachance's three-coefficient algorithm on a rather complex but not uncommon sample.

The results are shown in Table 8. They are taken from a preprint of a paper of Pella et al. [134]. The errors are somewhat greater than expected on the basis of calculations starting from relative intensities calculated from fundamental parameters. Moreover, the errors are hard to predict (compare Cr for specimens 1 and 3) but unavoidable for the time being. The errors here are also greater than the counting errors.

3. High-Z Limit

We already insisted on the low-Z limit in Sec. III.D.1 and 2, but what about the high-energy end of the spectrum?

Commercially available generators and side-window tubes allow work at high voltages up to 100 kV, bringing even the K line of bismuth into the potential analytical spectrometric range. If a tungsten target is used, however, characteristic lines represent only a minority of the integrated emitted intensity and thus excitation is bound mostly to continuous radiation, the maximum of which lies at 50 kV. Thus, only elements with K absorption edges below this value can be efficiently excited (Eu, $Z = 63$, and all elements below). This is a first limitation.

If we return to Fig. 22 and compare the spectral resolution of the LiF spectrometer with spectral distances between analogous lines in the K series, it is obvious that spectral resolution is worse above 35 kV (for K lines for Pr, $Z = 59$). This is a second limitation.

From the description of goniometers we have seen that each of these devices has its optimal 2θ range between 15 and 70°: at the higher 2θ end the angular dispersion becomes insufficient; at the lower 2θ end the intensity becomes very low. If we choose LiF (210) with the shortest interplanar distance, $2d = 1.802$

Å, useful measurements can be made at 10.0° corresponding to photons of 39.6 keV (Sm, $Z = 62$ in the first-order spectrum). This is a third limitation.

Studying Fig. 22 further, we note that the spectral resolution of the LiF spectrometer is worse than the spectral resolution of the semiconductor Si-Li detector for photons of energy exceeding 17 keV, which corresponds to Mo $K\alpha$, $Z = 42$. This is the fourth and most serious limitation in the K series. For analysis in the L series there is no such limitation because the whole energy range lies within the capabilities of wavelength-dispersive devices.

The wide voltage range of generators should not be used without some awareness of inherent drawbacks. For example, the K absorption edge of Mo is 20 keV. Thus, each x-ray tube with characteristic lines above 20 keV and working under a potential above twice this 20 kV is quite good for the excitation of Mo. The potential, however, should not exceed this value by very much because the maximum of Bremsstrahlung would vary more and more from the molybdenum absorption edge and even the spectral characteristics do not improve. If we divide the mass coefficient of photoelectron excitation by the sum of mass coefficients for Compton and coherent scatters (this is a kind of measure of peak to background ratio), we obtain the values 76 and 18 for tube excitation potentials of 40 and 100 kV, respectively! This is a substantial loss in peak-to-background ratio.

In the preceding we tried to show that, although the spectrometers are almost perfect, there are limits to precision and accuracy due to the nature of the sample and to the mathematical conversion of intensity to concentration. These uncertainties are generally greater than those introduced by instrumental parameters.

V. CHEMICAL SHIFT AND SPECIATION

The photoelectric excitation of atoms and subsequent specific deexcitation by emission of x-rays is the basis of all qualitative and quantitative use of XRF to which we have made reference. The deexcitation process was considered to take place between unique excited and ground quantum states. If this was true, x-ray emission lines were to be found at strictly determined wavelengths or energies. However, electron density and quantum states are influenced by the environment of the atom with consequent wavelength shifts, or so-called chemical shifts, of the emitted lines. Such techniques as nuclear magnetic resonance or Mössbauer spectroscopy collect their spectroscopic information essentially from measurement of chemical shifts. As a matter of fact, all kinds of spectral lines of electromagnetic radiation are subject to chemical shifts. All these shifts are small and detectable only if the spectral line widths are sufficiently small and if an instrument is available with sufficient spectral resolution. For x-rays the line shift is at most a few electronvolts. An angular error $\Delta\theta = 0.02°$ on the determination of a peak position corresponds at $\lambda = 1$ Å to 5 eV. Moreover, small temperature changes of the analyzing crystal already shifts the positions over similar values.

The sensitivity of peak position to the chemical environment of the atom was observed at the very beginning of x-ray spectrometry by Lindh and Lundquist [135]. Three characteristics of lines are sensitive to change: line position (chemical shift) [136–138], line shape [139], and mutual ratios of line intensities [140]. A detailed summary of the effect can be found in the book of B. K. Agarwal [141], for example. In spectrometers with limited resolution, such as ED systems or amplitude spectrometry, most spectral shifts are hardly detectable.

Chemical shift research has been done on high-resolution spectrometers. Gohshi et al. [142] constructed an excellent two flat crystal instrument for studying chemical shifts. Asada et al. [140] and Haycock and Urch [143] improved the use of commercially available instruments for this purpose: by Fourier transform and iterative procedures for peak position detection or by using finer collimators. When analyzing crystals with large thermal expansion coefficients, such as PET, are used, a temperature variation of 1°C gives rise to peak shifts greater than the chemical shifts. Helsen and Wijnhoven [144] solved this problem by inserting standards at regular intervals and by using statistical techniques for determination of peak position. Sumbaev [136–138] made experiments with an instrument in a Cauchois arrangement. In the paper by Kataria et al. [145], it was proved that even using common solid-state Si-Li detectors with moderate energy resolution it was possible to detect chemical effects.

Theoretically, chemical shifts can be calculated by the self-consistent field approach for the four quantum states involved [146]. Unfortunately, the calculations are largely dependent on initial assumptions in which the choice is difficult and somewhat random and the results are far from unequivocal. The experimental results of Sumbaev seem to be more spectacular. He connected the chemical shift ΔE_{KL} to Pauling's ionicity i, valency m and p, which was determined by lifetime ratios of electrons on the levels K or L:

$$\Delta E_{KL} = im[p_K \, \Delta E_{K,(Z:Z-1)} - p_L \, \Delta E_{L,(Z:Z-1)}] \tag{32}$$

He achieved an even more surprising result by presenting chemical shifts on unit ionicity versus valence (Fig. 35). This dependence is linear and may serve as proof of both the importance of chemical shift and the accuracy of Pauling's scale. The shift can be related to the character of the bond in the molecule or the crystal. During the formation of a chemical bond, a rearrangement of electrons occurs, and thus the formation of ions or bonds of varying degree of ionicity, the influence of which is reflected on all electron energy levels, sometimes more on the K level

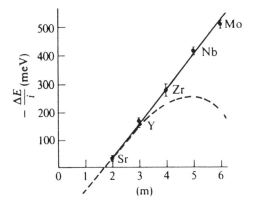

Figure 35 Chemical shift as a function of valence. The energy shift divided by the ionicity is plotted in a solid line. The dotted curve illustrates the uncorrected values of ΔE. (From Ref. 135. Reprinted by permission of Prof. J. B. Adashko.)

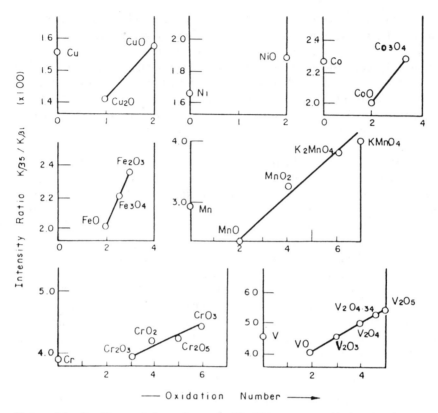

Figure 36   Satellite peak intensity ratio $K\beta_5/K\beta_1$ versus oxidation number. (From Ref. 139. Reprinted by permission of John Wiley and Sons, Ltd.)

than on others. An example may be the shift between the $K\alpha$ position of metallic aluminum and aluminum in different coordination states (IV and VI) [144].

Asada et al. [140,147,148] also made measurements in which the resulting chemical shift was attributed to the ionic bond character. They also investigated the changes in the ratio of some spectral lines, such as those for the $K\beta_5$ and $K\beta_1$ satellite line, and connected them to the oxidation number of the analyte element. The results are again surprisingly elegant (Fig. 36). Clear proof for the dependence between the shape of lines and the oxidation number was given for manganese [139] and also between the chemical shift and the coordination number for different elements [149]. The shift of Al $K\alpha$ for coordination states 0, IV, and VI was given by Helsen and Wijnhoven [144].

The application of the determination of chemical shift to chemical speciation is obvious. In geological samples sulfur can be distinguished according to its different oxidation states (sulfur, sulfide, sulfite, sulfate, and hyposulfate) [118]. Birks and Gilfrich [150] made measurements for sulfur using a portable low-power WDXRF instrument. In environmental and biological samples the oxidation state of chromium is important: Cr(III) does not pass the cell wall, but Cr(VI) does and is suspected to be carcinogenic. Iwatsuki and Fukasawa [151] reported the determination of chemical states of arsenic, selenium, and bromine. The deter-

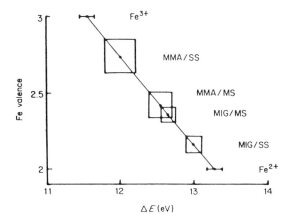

Figure 37 Relative chemical shift ΔE and the average valence of iron in standard samples and four welding fumes. The squares indicate the statistical uncertainties: SS = stainless steel; MS = mild steel; MMA = manual metal arc technique; MIG = metal inert gas technique. (From Ref. 150. Reprinted by permission of John Wiley and Sons, Ltd.)

mination of the valence of iron in welding fumes by measuring the energy shift and the intensity ratio of $K\beta'$ to $K\beta_{1,3}$ has yielded very fine results (Fig. 37) [152]. Kataria et al. [145] managed to measure the varying ratio between the total line intensities of $K\beta$ and $K\alpha$ as due to different oxidation states of manganese using an instrument of very moderate resolving power.

This discussion of other than elemental or quantitative information from x-ray spectra has been intentionally restricted to x-ray spectrometry. A counterpart exists in spectrometry of the ejected electrons, but that discussion belongs to the domain of spectrometry of Auger electrons or primary ejected electrons, for which higher resolutions are available and better speciation is possible. The other and sometimes severe drawback of these techniques is the required high vacuum and surface charge for nonconducting specimens. It should also be mentioned that Raman spectrometry is possible in the x-ray domain, but this field of research is only really accessible when synchrotron radiation is available [153–155], although performance on conventional instruments is feasible [156]. In analytical determinations by synchrotron radiation this field is often viewed as a drawback [157].

Not many researchers are working in the field of speciation by WDXRF; nevertheless, it is a field of increasing importance, with theoretical as well as practical interest. It gives more than simple elemental information. Moreover, the shift of lines affects the intensity measurements for quantitative analysis and stresses for specific cases the use of the right standards. In extreme cases it may even be compulsory for quantitative determinations to scan over the peak maximum for each determination and locate in each case the intensity at the maximum for good precision and accuracy. If ultimate accuracy should be required, correction for peak heights may be involved. The few examples given point out that for the purpose of chemical shift, shape, and intensity measurements of spectral lines, the use of dedicated instruments is indicated, although a certain amount of useful work can be performed on commercially available instruments.

VI. INSTRUMENTATION

The physical principles of a WD spectrometer do not allow dramatic differences in the general layout of the spectrometer. A practical instrument unites a number of construction compromises for speed, versatility, resolution, intensity, cost, and other factors, and all combinations have their own merits and their own drawbacks. Buying an instrument is a decision that in itself is a compromise because all benefits wanted by a customer are seldom or never united by a single instrument. In this section the main features of the most important commercially available instruments are juxtaposed without formulating a global or final judgment.

A. Electrical and Electronic Features

The functions controlled externally by push buttons or computer are summarized in Table 9. The philosophy of an integrated system is that groups of functions may be controlled by their own microcontroller of lower hierarchy; at the top of the hierarchy a master microprocessor organizes the collaborative operation. In Table 9 it is shown that a number of functions may be grouped around a slave microprocessor (as is done by Philips sequential spectrometer PW 1480). All slave microprocessors communicate with a master microprocessor. A function "remote diagnostics" is implemented on the lowest, which is linked by modem to a centralized diagnostic station (e.g., for one country) to service a customer's instrument remotely, giving advice or detecting the faulty parts. This considerably re-

Table 9   Summary of Functions to be Interfaced to Controller

X-ray path	Master microprocessor
Sample-related controls	
Beam attenuation	
Calibration	
Simple R to C conversion	
Stability tests	
Local diagnostics	
Generator control (kV, mA) displays	Generator microprocessor
Goniometer functions	Goniometer microprocessor
Scan mode	
Angle settings	
Sin θ	
Signal processing	Signal processing microprocessor
Intensity measurement	
Dead time	
Pulse-height discrimination	
Attenuation	
Output of data	Output microprocessor
Digital (to computer)	
Analogue (recorder)	
Printer	
Remote diagnostic computer	Modem diagnostic center

duces service time and is certainly something that will become a general feature of all computerized instruments. The philosophy described is implemented in the various commercial instruments to a greatly varying degree of complexity and is a substantial part of the cost of an instrument, although there is a steady evolution. On the master microprocessor, which is eventually the only one in the instrument, simple algorithms for intensity to concentration conversion are implemented so that the instrument can function as a stand-alone unit independent of a computer.

B. Spectrometer Layout

Spectrometer designs are discussed in the following order:

- Sequential instruments with EWT, sample down or sample up
- Sequential instruments with SWT, sample down or sample up
- Simultaneous instruments all with EWT and sample up
- Simultaneous instruments with EWT and sample down

For each type one commercial instrument is taken as an example, not necessarily implying that it is the only one or the best one available.

1. Sequential Spectrometers with EWT

A typical example of geometry with sample down, that is, the x-ray tube positioned under the sample, is represented in Fig. 38a (layout of Siemens SRS-300). The positions of the primary beam filter and spectrometer optics are also shown. The front end of the x-ray tube, beam filter, and specimens are enclosed in a vacuum chamber in which a 10-sample changer is housed. Alternatively, the sample chamber may be connected to an automatic 72-position sample changer. The optics from collimator to flow counter are isolated from the sample area during sample loading. The sample enclosure and spectrometer are each evacuated by their own vacuum pump. The scintillation counter is the only spectrometer part outside the vacuum chamber. Philips introduced recently a spectrometer with the same geometry (PW2400) with a front end x-ray tube, called Super Sharp Tube (SST). The duplex detector (replacing the classical tandem construction) is a combination of a flow counter and a sealed xenon detector (improved performance for $K\alpha$ of Ti to Cu and for $L\alpha$ to W). The scintillation counter is offset with respect to the gas counters and is enclosed in the vacuum chamber.

Sample-up position is used by Applied Research Laboratories in their Model 8420 (Fig. 38b). In this setup the sample is introduced in a separate compartment, which is evacuated before moving the sample to the measurement position. A sample changer may be also connected, enabling the automatic processing of 8 or 52 samples. The vacuum housing contains the whole spectrometer, including the scintillation counter.

2. Sequential Spectrometer with SWT

A typical sample-down geometry is used by Philips PW 1404 (Fig. 38c). The four-sample stage is loaded through a small lock. The lock is evacuated before transport of the sample into the vacuum chamber. Only one vacuum pump is used, but the

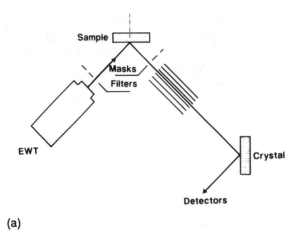

(a)

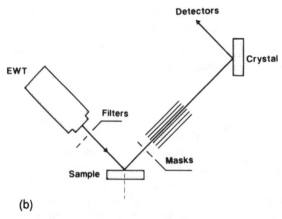

(b)

Figure 38 Some possible configurations of sequential instruments: (a) with EWT, sample down; (b) with EWT, sample up; (c) with SWT, sample down; (d) with SWT, sample up.

vacuum inside the spectrometer hardly changes during loading as a result of the small lock volume. Automatic sample changers with up to 102 positions and with bar code identification of the holders are available.

Figure 38d shows a spectrometer with sample-up position as manufactured by Rigaku (System 3030 or 3070). The main feature distinguishing this instrument from others is the possibility of offsetting the flow counter and introducing a second fine collimator between the crystal and the scintillation counter. This increases the intensity for heavy elements with respect to the usually applied tandem position.

The very large scale integration in the electronic industry requires quick and accurate analysis of boron and phosphorus on whole 6 inch silicon wafers. Rigaku and Philips offer a sequential spectrometer with a modified sample stage accepting wafers up to 8 inches. Spinning about any point on the wafer can be done with the Philips instrument only.

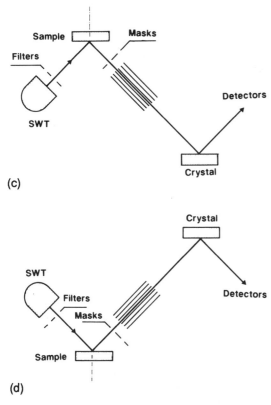

(c)

(d)

Figure 38 *Continued*

3. Simultaneous Instruments

All simultaneous instruments are equipped with end-window tubes with very similar construction features. Up to 28 monochromators may be arranged radially around the tube head, with alternating high and low takeoff angles. One or more monochromators may be replaced by programmable goniometers of the flat (a

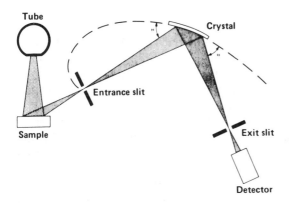

Figure 39 Layout of a simultaneous spectrometer. (Courtesy of Philips Analytical.)

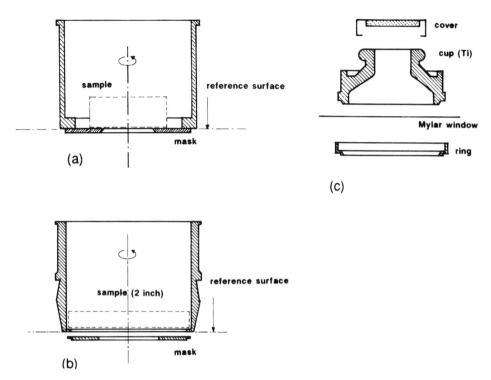

Figure 40 Design of sample cups: (a) a cup with masks fitted to the reference surface; (b) a cup with sample masks fitting to the inside; (c) a cup for analyzing liquids or powders with a disposable Mylar window; this sample holder is now being replaced by disposable all-plastic cups.

normal flat crystal goniometer of reduced size) or curved crystal type. The latter is small but limited in angular range; the former has a large angular range but is large and occupies at least two positions otherwise occupied by fixed monochromators. As already mentioned, Philips offers a small, low-power target transmission tube (TTT tube) used in the spectrometer PW 1660 capable of the simultaneous determination of only 14 fixed elements or 12 elements when fitted with a programmable goniometer. Most instruments irradiate the sample in the up position. Some instruments (e.g. Oxtuzol) irradiate in sample down position. The monochromators are either fitted to the outside of the vacuum cabinet, in which case the angle setting is facilitated but they must be vacuum tight, or are completely enclosed in the vacuum cabinet, in which case they need not be vacuum tight but tuning is more time consuming. Figure 39 shows a generally used geometry.

In Philips' sequential spectrometer (PW2400) an optional two fixed channels may be installed, making this instrument partly simultaneous. This is made feasible

Table 10 Sequential Spectrometers[a]

Manufacturer	1	1'	2	3	4	4'	5	6
Generator								
kV	100	60	60	75	60	60	60	60
mA	75	125	60	100	80	80	100	75
Switch time, s	<4	<2	≥30	—				
X-ray tubes	SW[b]	EW[c]	EW[c]	EW[d]	SW	EW[c]	SW[e]	SW[f]
Primary beam collimator	—	—	—	—	—	x[g]	—	—
Filters								
Optional or not	—	—	—	—	—	x	x	
Positions[h]	1	6	3	—	2	5	1	
Mask positions	2	4	4	1		5		
Primary collimator positions	2	3	2	2	1	2	2	
Crystal changer positions	8	8	6	6	3	8	6	4
Goniometer								
Gear	x	x[i]	x		x	x	x	
Gearless			x				x	
Angular setting								
Reproducibility	<0.001	<0.0001	0.005	0.003	0.001	0.001	0.001	0.002
Scan speed, °2θ minute^{-1}	0.06–75	0.06–120	0.2–500	0.25–128		0–200	0.25–4	0.1–10
Slew speed °2θ minute^{-1}	1000	2400	500	4800	1000	1000	350	600
Range °2θ								
Flow proportional counter	0–148	13–148	0–148	43–155	8–147	8–147	0–148	0–146
Sealed proportional counter	—	13–148	—	0–125			—	—
Scintillation counter	1–115	0–109	4–93	0–98	5–117	5–117	?	?
Analyzing range								
Standard	B–U	Be–U	B–U	B–U	O–U	O–U	O–U	B–U
Option				C	B, C	C		

[a] Manufacturers are listed in Table 12.
[b] Dual-anode tubes: Cr/Au, Sc/Mo.
[c] OEG 75 Machlett or AG66 AEG.
[d] OEG 76 Machlett.
[e] Toshiba.
[f] EA-75 dual-target tube; W/Mo 2600 W or Cr/Pt 2100 W.
[g] An attenuator with a 10% transmission may be inserted.
[h] The choice may be Al, Cu, messing, Ti, Ni, or Zr in thicknesses of 0.15–0.80 mm.
[i] With direct optical position sensor (DOPS)

by the use of the x-ray front end tube, model SST. In the ARL sequential instruments also fixed channels (up to 8) can be added.

Sample cups of different designs are offered by manufacturers. Figure 40 shows three examples of sample-down cups.'' Sample-up cups are basically the same but the sample is press fit to the mask by a spring.

C. Intercomparison of Instruments

In Tables 10 through 12 some important features of the main commercial instruments are summarized. Some manufacturers, such as Microspec Corporation, specialize in the production of WD attachments to electron microprobes (for details see Chap. 12). Dapple Systems does not figure in this list. It is a company offering hardware and software for automation and data processing of existing

Table 11 Simultaneous Instruments[a]

	1	1′	2	3	4	5	7
Generator							
kV	50	60	60	65	60	60	50
mA	4	75	80	100	80	100	4
Switch time		[b]					
Stab. %/% Δmains	0.0005	0.0005	0.002	0.001	0.001	0.001	0.006
X-ray tubes							
Remarks	[c]	[d]	[e]	[e]	[e]	[e]	[f]
Anodes[g]	Sc	Rh, Mo	Mo	Mo, W, Pt	Ag, W	Ag, W	W, no
Cr							
Beam attenuators	—	10	12	—	15	—	
Beam filter	—	—	6	—			
Monochromator							
Maximum number	14	28	28	31	24	24	10
Cryst. curved							
log.	x	x	x	—	x	—	
cyl.				x		x	
Goniometer							
Maximum number	1	2	2	4	2	1	1
Type							
Flat crystal	x	x	—	x	x	x	x
Curved crystal			x				
Angular range	10–120	5–120	30–120	0–152	10–87	0–110	x[h]
Crystal changer	—	—	—	6	—	—	
Element range	Ti–U	Ca–U	F–U	C–U	Ca–U	Ti–U	K–U
Mounting							
Outside vacuum chamber	x	x	—	—	x		x
Inside vacuum chamber			x	x			
Temperature stability (°C)	0.05	0.05	0.5	0.5	0.5	1.0	0.0075(?)
Room temperature range (°C)	5–40	5–40	14–31		17–28	18–28	15–30

[a] Manufacturers are listed in Table 12.
[b] Specified only by Philips: 2 s for switching from 20 to 60 kV. Switching occurs along the isowatt curve.
[c] Transmission target tube 200 W, no cooling required.
[d] Philips 60 kV EWT.
[e] Same as for sequential instruments.
[f] 200 W tube with Rh of W target; cooling by internal oil circulation.
[g] All offer Rh or Cr targets; only additional options are given.
[h] Kα lines from K to Mo and Lα lines from Sn to U.

instruments, which is a cost-effective alternative to be considered before buying a new instrument when budgets are small. Intercompared are the following:

	Sequential	Simultaneous	Referred to as
Philips	PW1404	PW1660	1
	PW2400	PW1606	1′
Siemens	SRS303	MRS404	2
ARL[a]	8400 (84XX)	8600 (86XX)	3
Rigaku[a]	3030	3530	4
	3070 (XY7Z)		4′
Shimadzu	VF320	VXQ150	5
Diano	2000	—	6
Oxford	—	QX	7

[a] For ARL and Rigaku a series number is given. XX or XYZ specifies further the instrument according to the options implemented on the instrument. These details are not further included.

Table 12 Manufacturers

Name	Address
1. ARL, Applied Research Laboratories	En Vallaire, CH-1024 Ecublens, Switzerland
2. Diano Corporation	30 Commerce Way, P.O. Box 1005, Woburn, MA 01801
3. Oxford Analytical Instruments	20 Nuffield Way, Abingdon, Oxon OX14 1 TX, England
4. N. V. Philips' Gloeilampenfabrieken	I & E Export Department, Building TQ 111-3, 5600 MD Eindhoven, The Netherlands
5. Rigaku Industrial Corporation	14-8, Akaoji-cho, Takatsuki, Osaka, Japan
6. Simadzu Corporation	Shinjuku Mitsui Bldg., 1-1, Nishi-Shinjuku 2-chome, Shinjuku-ku, Tokyo 160, Japan
7. Siemens Aktiengesellshaft Mess- und Prozesstechnik Vertrieb Analysentechnik, E 68 A	Postfach 211262, D 7500, Karlsruhe 21, Germany

In Table 10 the stability of the generators is not indicated because the manufacturers do not use the same units. It may be advocated to express the stability in the percentage variation of the voltage with respect to percentage variation of the mains. The same problem arises with precision and accuracy of angular settings of the goniometer, where again no uniform format is used. Vacuum operation or presence of helium atmosphere facility is not given in Table 10 because all instruments allow both possibilities in sequential as well as in simultaneous instruments.

ACKNOWLEDGMENTS

All manufacturers of instruments are gratefully acknowledged for supplying documents about their instruments. Dr. B. Volbert and Dr. ir. B. A. R. Vrebos from Philips Analytical and Dr. K. E. Mauser and N. Broll from Siemens AG deserve our gratitude for invaluable advice and critical reading of the manuscript. Dr. A. Kuczumow thanks the Catholic University of Leuven for a fellowship that enabled him to participate in this work.

REFERENCES

1. W. C. Röntgen, *Nature 53*:274 (1896).
2. M. E. Weebs and H. M. Leicester, *Discovery of the Elements*, 7th Ed., J. Chem. Educ., Easton, PA, 1967.
3. A. A. Markowicz and R. E. Van Grieken, *Anal. Chem. 58*:279R (1986).

4. R. Van Grieken, A. Markowicz, and S. Török, *Fresenius Z. Anal. Chem. 324*:825 (1986).
5. R. Jenkins, *Anal. Chem. 56*:1099A (1984).
6. J. L. de Vries, *Fresenius Z. Anal. Chem. 324*:492 (1986).
7. R. Klockenkämper, *Spectrochim. Acta 42B*:423 (1987).
8. J. N. Kikkert, *Spectrochim. Acta 38B*:1497 (1983).
9. A. H. Compton and S. K. Allison, *X-Rays in Theory and Experiment*, Van Nostrand, New York, 1935.
10. L. S. Birks, *X-Ray Spectrochemical Analysis*, 2nd Ed., Interscience, New York, 1969.
11. N. A. Dyson, *X-Rays in Atomic and Nuclear Physics*, Longman, London, 1973.
12. R. Jenkins, *An Introduction to X-Ray Spectrometry*, Heyden, London, 1974.
13. E. P. Bertin, *Principles and Practice of X-Ray Spectrometric Analysis*, 2nd Ed., Plenum, New York, 1975.
14. R. Jenkins and J. L. De Vries, *Practical X-Ray Spectrometry*, 2nd Ed., Springer-Verlag, New York, 1967.
15. R. Tertian and F. Claisse, *Principles of Quantitative X-Ray Fluorescence Analysis*, Heyden, London, 1982.
16. H. G. J. Moseley, *Philos. Mag. 26*:1024 (1913).
17. H. G. J. Moseley, *Philos. Mag. 27*:703 (1914).
18. E. Gillam and H. T. Heal, *Br. J. Appl. Phys. 3*:353 (1952).
19. J. Sherman, *Am. Soc. Test. Mater. Spec. Tech. Publ. 157*:27 (1954).
20. J. Sherman, *Spectrochim. Acta 7*:283 (1955).
21. T. Shiraiwa and N. Fujino, *Jpn. J. Appl. Phys. 5*:886 (1966).
22. T. Shiraiwa and N. Fujino, *Bull. Chem. Soc. Jpn. 40*:2289 (1967).
23. C. J. Sparks, Jr., *Adv. X-Ray Analysis 19*:19 (1976).
24. Z. Li-Xing, *X-Ray Spectrom. 13*:52 (1984).
25. R. P. Gardner and A. P. Hawthorne, *X-Ray Spectrom. 4*:138 (1975).
26. W. H. Bragg and W. L. Bragg, *Proc. Roy. Soc. 88A*:428 (1913).
27. H. Freiser and G. H. Nancollas, *IUPAC Compendium of Analytical Nomenclature: Definitive Rules 1987*, 2nd Ed., Blackwell Scientific, Oxford, 1987.
28. T. Arai, *Adv. X-Ray Spectrom. 30*:213 (1987).
29. Y. Cauchois and C. Senemaud, *International Tables of Selected Constants. 18. Wavelengths of X-Ray Emission Lines and Absorption Edges*, Pergamon Press, Oxford, 1978.
30. D. J. Nagel, in *Advances in X-Ray Spectroscopy* (C. Bonelle and C. Mande, Eds.), Pergamon Press, Oxford, 1982.
31. W. D. Coolidge, *Phys. Rev. 2*:409 (1913).
32. B. Vrebos and J. A. Helsen, *X-Ray Spectrom. 14*:27 (1985).
33. N. Nordfors, *Ark. Fys. 10*:279 (1956).
34. D. R. Boehme, *Adv. X-Ray Spectrom. 30*:39 (1987).
35. M. C. Nichols, D. R. Boehme, R. W. Ryon, D. Wherry, B. Cross, and G. Aden, *Adv. X-Ray Spectrom. 30*:45 (1987).
36. S. Thordarson, *Ann. Phys. 35*:135 (1939).
37. P. J. M. Botden, B. Combee, and J. Houtman, *Philips Tech. Rev. 14*:165 (1952).
38. Philips Materials, *Simultaneous X-Ray Spectrometer SYSTEM PW1660*.
39. H. Kuhlenkampff, *Ann. Phys. 69*:548 (1922).
40. H. A. Kramers, *Philos. Mag. 46*:836 (1923).
41. M. G. Brunetto and J. A. Riveros, *X-Ray Spectrom. 13*:60 (1984).
42. R. Tertian and N. Broll, *X-Ray Spectrom. 13*:134 (1984).
43. P. A. Pella, L. Feng, and J. A. Small, *X-Ray Spectrom. 14*:125 (1985).
44. A. A. Markowicz and R. E. Van Grieken, *Anal. Chem. 56*:2049 (1984).

45. M. Green and V. E. Cosslett, *Proc. Phys. Soc.* 78:1206 (1961).

46. M. Murata and H. Shibahara, *X-Ray Spectrom.* 10:41 (1981).

47. J. V. Gilfrich and L. S. Birks, *Anal. Chem.* 40:1077 (1968).

48. J. V. Gilfrich, P. G. Burkhalter, R. R. Whitlock, E. S. Warden, and L. S. Birks, *Anal. Chem.* 43:943 (1971).

49. J. V. Gilfrich, in *Handbook of Specroscopy*, Vol. I (J. W. Robinson, Ed.), CRC Press, Cleveland, 1974, p. 232.

50. Philips Materials, *Sequential X-Ray Spectrometer System PW 1404*, p. 7.

51. C. E. Weaver and L. D. Pollard, *The Chemistry of Clay Minerals*, Elsevier, Amsterdam, 1973, p. 91.

52. J. Du Mond and J. P. Youtz, *J. Appl. Phys.* 11:357 (1940).

53. A. Franks, *Sci. Prog. Oxf.* 64:371 (1977).

54. J. A. Nicolosi, J. P. Groven, D. Merlo, and R. Jenkins, *Opt. Eng.* 25:964 (1986).

55. Ovonic Synthetic Materials Co., advertising matter and letter to customers, March 30, 1987, Troy, Michigan.

56. J. H. Underwood and D. T. A. Atwood, *Phys. Today 37*:44 (1984).

57. J. H. Underwood, *Optics News 12* (March 1986).

58. T. W. Barbee, Jr., *Opt. Eng. 25*:898 (1986).

59. A. Van Eenbergen and B. Volbert, *Adv. X-Ray Anal. 30*:201 (1987).

60. B. L. Henke, J. Y. Uejio, H. T. Yamada, and R. E. Tackaberry, *Opt. Eng. 25*:937 (1986).

61. E. Spiller and A. E. Rosenbluth, *Opt. Eng. 25*:954 (1986).

62. E. Spiller, in *Low Energy X-Ray Diagnostics* (D. T. Atwood and B. L. Henke, Eds.), AIP Conf. Proc., Vol. 75, AIP, New York, 1981, p. 124.

63. T. W. Barbee, Jr., in *Low-Energy X-Ray Diagnostics* (D. T. Atwood and B. L. Henke, Eds.), AIP Conf. Proc., Vol. 75, AIP, New York, 1981, p. 124.

64. B. L. Henke, P. Lee, T. J. Tanaka, R. L. Shimabukuro, and B. K. Fujikawa, *Atomic and Nuclear Data Table*, Vol. 27, Academic Press, New York, 1982.

65. Y. Lepêtre, J. K. Schiller, G. Rasigni, R. Rivoira, R. Philip, and P. Dhez, *Opt. Eng. 25*:948 (1986).

66. J. V. Gilfrich, D. J. Nagel, N. G. Loter, and T. W. Barbee, Jr., *Adv. X-Ray Anal. 25*:355 (1982).

67. J. A. Nicolosi, J. P. Groven, and D. Merlo, *Adv. X-Ray Anal. 30*:183 (1987).

68. D. Kohler, J. L. Guttman, and B. A. Watson, *Rev. Sci. Instrum. 56*:812 (1985).

69. K. F. J. Heinrich, in *Energy Dispersive X-Ray Spectrometry*, NBS Spec. Publ. 604 (K. F. J. Heinrich, D. E. Newbury, R. L. Myklebust, and C. E. Fiori, Eds.), National Bureau of Standards, Washington, D.C., 1981.

70. R. Caciuffo, S. Melone, F. Rustichelli, and A. Boeuf, *Physics Reports* (Review Section of Physics Letters), Vol. 152, North Holland, Amsterdam, 1987, p. 1.

71. R. J. Plotnikow and G. A. Pszenicznyj, *Fluorescentnyj Rentgenoradiometriczeskij analiz*, Atomizdat, Moscow, 1973.

72. R. Fitzgerald and P. Gantzel, *Energy Dispersion X-Ray Analysis: X-Ray and Electron Probe Analysis*, ASTM Spec. Techn. Pub. 485, 1971.

73. P. G. Burkhalter and W. J. Campbell, *Oak Ridge National Laboratory Report ORNL-IIC-10*, Vol. 1, 1967.

74. Y. Gohshi, H. Kamada, K. Kohra, T. Utaka, and T. Arai, *Appl. Spectrosc. 36*:171 (1982).

75. P. J. Potts, P. C. Webb, and J. S. Watson, *Analyst 110*:597 (1985).

76. C. J. Sparks, Jr., in *Synchrotron Radiation Research* (H. Winick and S. Doniach, Eds.), Plenum, New York, 1980, p. 459.

77. S. I. Salem and P. L. Lee, *Atomic Data and Nuclear Data Tables*, Vol. 18, 1976, p. 233.

78. H. Bent, *Spectrochim. Acta 25B*:613 (1970).
79. D. Bandas, E. Kellogg, S. Murray, and R. Enck, Jr., *Rev. Sci. Instrum. 49*:1273 (1978).
80. J. V. Gilfrich, *Adv. X-Ray Anal. 30*:35 (1987).
81. R. Jenkins, B. Hammell, A. Cruz, and J. Nicolosi, Norelco Rep., Vol. 31, 1984.
82. J. E. Croke and J. Nicolosi, *Adv. X-Ray Anal. 30*:225 (1987).
83. H. Ebel, M. Mantler, N. Gurker, and J. Wernisch, *X-Ray Spectrom. 12*:47 (1983).
84. A. J. P. L. Palicarpo, *Space Sci. Instrum. 3*:77 (1978).
85. R. D. Bonetto and J. A. Riveros, *X-Ray Spectrom. 13*:44 (1984).
86. A. Knöchel, W. Petersen, and G. Tolkiehn, *Nucl. Instrum. Methods 208*:659 (1983).
87. P. Wobrauschek and H. Aiginger, *X-Ray Spectrom. 9*:57 (1980).
88. P. Wobrauschek and H. Aiginger, *X-Ray Spectrom. 12*:72 (1983).
89. P. Wobrauschek and H. Aiginger, *Adv. X-Ray Anal. 28*:64 (1985).
90. P. Wobrauschek and H. Aiginger, *Fresenius Z. Anal. Chem. 324*:865 (1986).
91. H. Bougault, P. Cambon, and H. Toulhoat, *X-Ray Spectrom. 6*:66 (1977).
92. R. V. Gentry, T. A. Cahill, N. R. Fletcher, H. C. Kaufman, L. R. Medsker, J. W. Nelson, and R. G. Flocchini, *Phys. Rev. Lett. 37*:11 (1976).
93. C. J. Sparks, Jr., S. Raman, E. Ricci, R. V. Gentry, and M. O. Krauze, *Phys. Rev. Lett. 40*:507 (1978).
94. R. Jenkins, Y. H. Hahm, and S. Pearlman, *Norelco Rep. 26*:27 (1979).
95. A. Kuczumow, *X-Ray Spectrom. 11*:112 (1982).
96. G. R. Lachance and R. J. Traill, *Can. Spectrosc. 11*:43 (1966).
97. F. Claisse and M. Quintin, *Can. Spectrom. 12*:129 (1967).
98. G. R. Lachance, paper presented at the International Conference on Industrial Inorganic Analysis, Metz, France, June 1981.
99. R. M. Rousseau and F. Claisse, *X-Ray Spectrom. 3*:31 (1974).
100. R. M. Rousseau, *X-Ray Spectrom. 3*:102 (1974).
101. R. M. Rousseau, *X-Ray Spectrom. 13*:115, 121 (1984).
102. A. Kuczumow and G. Holland, *X-Ray Spectrom. 18*:5 (1989).
103. A. Kuczumow, *Spectrochim. Acta 43B*:737 (1988).
104. J. A. Helsen and B. A. R. Vrebos, *Spectrochim. Acta 39B*:751 (1984).
105. J. A. Helsen and B. A. R. Vrebos, *Analyst 109*:295 (1984).
106. B. R. Kowalski, F. T. Schatzki, and F. H. Stross, *Anal. Chem. 44*:2176 (1972).
107. Kevex Corp., Kevex Material Analyzer Model 6500 brochure, p. 2.
108. Siemens, *Factors Affecting X-Ray Intensities (Sensitivity) in Sequential X-Ray Spectrometers*.
109. F. Claisse and T. P. Thinh, *Anal. Chem. 51*:954 (1979).
110. H. Kloyber, H. Ebel, M. Mantler, and S. Koitz, *X-Ray Spectrom. 9*:170 (1980).
111. A. Kuczumow, *X-Ray Spectrom. 13*:16 (1984).
112. C. T. Yap and S. M. Tang, *Appl. Spectrom. 38*:527 (1984).
113. C. T. Yap and S. M. Tang, *X-Ray Spectrom. 14*:157 (1985).
114. C. T. Yap, *X-Ray Spectrom. 16*:229 (1987).
115. Nomenclature, symbols, units and their usage in spectrochemical analysis-II, *Spectrochim. Acta 33B*:242 (1978).
116. G. L. Long and J. D. Winefordner, *Anal. Chem. 55*:712A (1983).
117. Y. Beers, *Theory of Error*, Wesley, Reading, MA, 1958.
118. Philips Analytical, *X-Ray Spectrometry Application Notes*, nb. 504, 622, 913, 623, 737, 808, 916.
119. Siemens SRS Anwendung, 84/8, 86/1, 86/5, 86/6.
120. N. Broll, *C. R. Coll. Rayons X*, Siemens, Grenoble, 1985, pp. 284–304.
121. C. A. Clayton, J. W. Hines, and P. D. Elkins, *Anal. Chem. 59*:2506 (1987).

122. R. L. Watson, A. K. Leeper, and B. I. Sonobe, *Nucl. Instrum. Methods 142*:311 (1977).
123. R. Plesch, *X-Ray Spectrom. 10*:8 (1981).
124. G. H. Glade and H. R. Post, *Appl. Spectrom. 24*:193 (1970).
125. V. Novosel-Radovic, D. Maljkovic, and N. Nenadic, *X-Ray Spectrom. 13*:148 (1984).
126. R. E. Michaelis and B. A. Kilday, *Adv. X-Ray Anal. 5*:405 (1962).
127. I. Claisse, *Norelco Rep. 4*:95 (1957).
128. B. Vrebos and J. A. Helsen, *Spectrochim. Acta 38B*:835 (1983).
129. B. A. R. Vrebos and J. A. Helsen, *Adv. X-Ray Anal. 28*:37 (1985).
130. J. A. Helsen and B. A. R. Vrebos, *X-Ray Spectrom. 15*:173 (1986).
131. J. A. Helsen and B. A. R. Vrebos, *Int. Lab. 16*(10):66 (1986).
132. R. M. Rousseau, *X-Ray Spectrom. 16*:103 (1987).
133. B. A. R. Vrebos and J. A. Helsen, *X-Ray Spectrom. 15*:167 (1986).
134. P. A. Pella, G. Y. Tao, and G. Lachance, *X-Ray Spectrom. 15*:251 (1986).
135. A. E. Lindh and O. Lundquist, *Ark. Mat. Astro Fys. 18*(14, 34, 35) (1924).
136. O. I. Sumbaev, E. V. Petrovic, Y. P. Smirnov, A. I. Ergorov, U. S. Zykov, and A. I. Grushko, *Soviet Phys.-JETP* (Eng. Trans.) *26*:891 (1968).
137. O. I. Sumbaev, *Sov. Phys.-JETP* (Eng. Trans.) *30*:927 (1970).
138. O. I. Sumbaev, in *Modern Physics in Chemistry* (E. Fluck and V. I. Goldanskii, Eds.), Academic Press, London, Vol. I, 1976, p. 31.
139. D. S. Urch and P. R. Wood, *X-Ray Spectrom. 7*:9 (1978).
140. E. Asada, T. Takiguchi, and Y. Suzuki, *X-Ray Spectrom. 4*:186 (1975).
141. B. K. Agarwal, *X-Ray Spectroscopy*, Springer-Verlag, Berlin, 1979, ISBN 3-540-09268-4.
142. Y. Gohshi, Y. Hukao, and K. Hori, *Spectrochim. Acta 27B*:135 (1982).
143. D. E. Haycock and D. S. Urch, *X-Ray Spectrom. 7*:206 (1978).
144. J Helsen and J. Wijnhoven, *Bull. Groupe Fr. Argiles 24*:15 (1972).
145. S. K. Kataria, R. Govil, A. Saxena, and H. N. Bajpei, *X-Ray Spectrom. 15*:49 (1986).
146. R. Manne, in *Inner-Shell and X-Ray Physics of Atoms and Solids* (D. J. Fabian, H. Kleinpoppen, and L. M. Watson, Eds.), Plenum, New York, 1981, p. 699.
147. E. Asada, *Jpn. J. Appl. Phys. 12*:1946 (1973).
148. E. Asada, *Jpn. J. Appl. Phys. 15*:1417 (1976).
149. Y. Gohshi, *Spectrochim. Acta 36B*:763 (1981).
150. L. S. Birks and J. V. Gilfrich, *Spectrochim. Acta 33B*:305 (1978).
151. M. Iwatsuki and T. Fukasawa, *X-Ray Spectrom. 16*:73 (1987).
152. V. P. Tanninen, E. Mikkola, H. K. Hyvärinen, A. Grekula, and P. L. Kalliomäki, *X-Ray Spectrom. 14*:188 (1985).
153. C. J. Sparks, Jr., *Phys. Rev. Lett. 33*:262 (1974).
154. P. Eisenberger, P. M. Platzman, and H. Winick, *Phys. Rev. Lett. 36*:623 (1976).
155. P. Eisenberger, P. M. Platzman, and H. Winick, *Phys. Rev. B., Solid State 13*:2377 (1976).
156. T. Suzuki, T. Kishimoto, and T. Kaji, *J. Phys. Soc. Jpn. 29*:730 (1970).
157. J. M. Jaklevic, R. D. Giauque, and A. C. Thompson, *Anal. Chem. 60*:482 (1988).

3

Energy-Dispersive X-ray Fluorescence Analysis Using X-ray Tube Excitation

Joseph M. Jaklevic and Robert D. Giauque
Lawrence Berkeley Laboratory, University of California, Berkeley, California

I. INTRODUCTION

The term "x-ray fluorescence analysis" (XRF) refers to the measurement of characteristic fluorescent emissions resulting from the deexcitation of inner shell vacancies produced in the sample by means of a suitable source of radiation. Numerous variants of the basic process have been studied. They differ both in the type and source of ionizing radiation and in the method employed to measure the fluorescent emission [1–3]. For routine XRF analysis, two major approaches are distinguishable based on the type of detector used to measure the characteristic x-ray emission spectra. Wavelength-dispersive x-ray fluorescence (WDXRF) analysis depends upon the use of a diffracting crystal to determine the characteristic wavelength of the emitted x-rays and is described in the previous chapter. Energy-dispersive x-ray fluorescence analysis (EDXRF) employs detectors that directly measure the energy of the x-rays by collecting the ionization produced in a suitable detecting medium.

Early approaches to EDXRF used gas proportional counters or scintillation detectors to determine the energy of the x-rays but were limited in application because of the inherently poor energy resolution that precluded the separation of characteristic x-rays of adjacent elements in the periodic table. In the early 1970s the technology of solid-state semiconductor diode detectors and associated pulse-processing circuits was developed to the point at which practical x-ray spectroscopy with an energy resolution of 200 eV or less became possible [4–6]. Although the energy resolution capabilities of semiconductor detectors remained inferior

151

to that achieved by wavelength-dispersive systems, the increased efficiency inherent in the energy-dispersive method compensated in many analytical applications and permitted the use of a multiplicity of experimental geometries not practical with WDXRF. A wide variety of EDXRF analytical systems based on radioisotope sources, x-ray tubes, charged particle accelerators, microprobe electron beams, and synchrotron light sources have been developed in recent years and are the subject of this and succeeding chapters.

The introductory material in this chapter discusses the instrumental aspects of the energy-dispersive method with particular emphasis on the use of semiconductor spectrometers as the detecting element. Most of this material is applicable to the discussions of related energy-dispersive methods in later chapters. In addition, a detailed discussion of EDXRF analysis using laboratory x-ray tubes is presented, including applications. For a more complete discussion of the instrumental aspects of semiconductor detectors and associated electronics, the reader is referred to any one of a number of reviews of the topic [7–9].

II. SEMICONDUCTOR DETECTORS FOR X-RAY SPECTROMETRY

Figure 1 illustrates the basic components involved in an EDXRF spectrometry measurement. The excitation radiation consists of a suitable flux of ionizing radiation, that is, photons, electrons, or heavy charged particles provided from one of many sources, such as a radioisotope, x-ray tube, nuclear particle accelerator, or relativistic synchrotron beam. The energy of the ionizing radiation must be sufficient to efficiently induce inner shell vacancies in the atoms within the sample. For x-ray excitation, the incident energy must be above the binding energy of the ejected photoelectron. For electrons or heavy charged particles, the energy of the incident particle must be adequate to transfer the energy and momentum required to eject the atomic electrons from the bound state. This is typically 10–100 keV for electrons and several MeV for protons or heavier charged particles.

The sample is placed in the excitation flux, and the characteristic x-rays resulting from the deexcitation of inner shell vacancies are detected by the energy-dispersive semiconductor spectrometer. In this type of detector, the total ionization produced by each x-ray striking the detector is converted to a voltage signal with amplitude proportional to the incident energy. Specially designed processing electronics are employed to maintain the linearity of the voltage signal with respect to the original charge pulse. A multichannel analyzer accumulates an energy spectrum of the sequential events in a histogram memory. Since the energy analysis does not depend in any way upon the diffraction or focusing of the x-rays, the geometry of the system is relatively insensitive to the placement of the detector with respect to the sample. This provides for a large solid angle and increased detection efficiency. Also, the mechanism by which the ionization signal is measured is not restricted to a narrow energy region, thus allowing the simultaneous detection of x-rays over a wide dynamic range of the emission spectrum. The principal advantages of EDXRF derive from the capability for simultaneous detection of characteristic x-rays from multiple elements with high geometric efficiency.

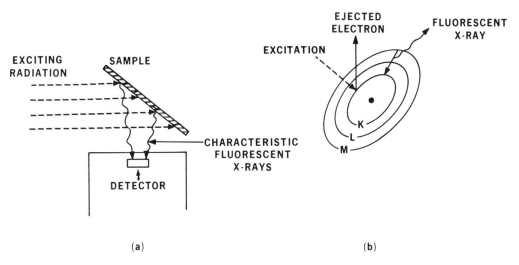

Figure 1 Basic elements in EDXRF measurements. (a) The exciting radiation creates vacancies in the inner shells of the sample atoms. The characteristic x-rays that are emitted are observed by the energy-dispersive detector. (b) The fluorescence process.

A. Detector Fabrication

The energy-dispersive detecting element is based on the simple semiconductor diode structure shown in Fig. 2. The example shown is a structure typical for lithium-drifted silicon detectors, Si(Li); however, the basic elements are similar for high-purity germanium detectors, HPGe. The device is fabricated on a cylindrical wafer of high-quality semiconductor material with rectifying p^+ or n^+ contacts on opposing surfaces. The bulk of the material is characterized by a very low concentration of free charge carriers. This reduced free carrier concentration is achieved either with the use of extremely pure material, in the case of HPGe detectors, or through the charge compensation of p-type silicon with lithium donor atoms in the case of Si(Li) detectors. A typical dimension for such a diode is 1–2 cm in diameter with a total thickness of 3–5 mm for the case of Si(Li) and 5–10 mm for Ge devices.

In the geometry shown, the lithium-diffused region acts as an n^+ contact; the metal surface barrier (typically an evaporated Au film) serves as the p^+ rectifying contact. When the diode is reverse biased, any remaining free carriers are swept out of the bulk by the applied field and an active depletion region is created. In this condition, the only current that flows between electrodes on the respective contacts is due to thermally generated charge carriers, which are excited above the narrow band gap of the semiconducting material.

When a photon is incident upon the active volume or depletion region of the diode, it normally interacts by photoelectric absorption to create an inner shell vacancy in the semiconductor material together with an energetic photoelectron. This photoelectron interacts with the atoms in the semiconductor crystal lattice to produce multiple low-energy ionization events. This process continues until

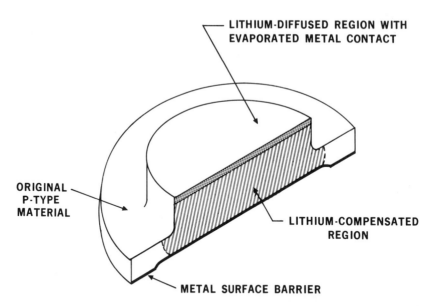

Figure 2 Cross section of a typical Si(Li) detector showing evaporated electrode contacts and active volume. HPGe differs in that the bulk is formed by high-purity Ge, which does not require the compensation of electrically active centers by lithium drifting.

the electron comes to rest at the end of a total range, which is short compared with the dimensions of the crystal. The energy associated with the inner shell vacancy is also absorbed in the crystal, in most cases following the emission of Auger electrons or multiple low-energy x-rays and subsequent reabsorption. The result of these multiple ionization processes is the production of a large number of free electron-hole pairs in the sensitive volume of the diode structure. These free charge carriers are then collected by the applied field as a current pulse. The number of these carriers is directly proportional to the energy of the x-ray quantum incident on the detector.

The types of energy-dispersive semiconductor x-ray spectrometers differ primarily in the properties of the material used to make the device. Although the most common material for semiconductor detectors is silicon or germanium, devices based on the use of compound semiconductors have also been fabricated and tested for specific applications [10,11]. Table 1 lists properties relevant to radiation detector applications for the case of the elements Ge and Si, which are the materials most commonly employed in commercial semiconductor spectrometer applications. A variety of other compound semiconductor materials, such as GaAs, CdTe, and HgI_2, have been used in research applications in which specific properties are emphasized. These properties include higher atomic number for increased photoelectric cross section, larger band gap for lower thermal leakage at room temperature, and a lower band gap corresponding to a smaller average energy for creation of hole-electron pairs [7]. In the following sections the details of detector operation are discussed, with particular emphasis on those aspects that impact on EDXRF most directly. These include the detector energy resolution, detection efficiency, spectral response, and system throughput.

Table 1 Detector Properties of Silicon and Germanium

	Si	Ge
Atomic number	14	32
Atomic weight	28.09	72.60
Density (300 K), g/cm^3	2.33	5.33
Band gap (300 K), eV	1.115	0.665
Average energy ϵ per hole-electron pair (77 K), eV	3.76	2.96
Fano factor (77 K)	0.12	0.08

Source: Adapted from Ref. 7.

B. Energy Resolution

The energy resolution of the semiconductor spectrometer system determines the ability of a given system to resolve characteristic x-rays from multiple-element samples and is normally defined as the full width at half-maximum (FWHM) of the pulse-height distribution measured for a monoenergetic x-ray at a particular energy. A convenient choice of x-ray energy at which resolution is quoted is the Mn $K\alpha$ line at 5.932 keV since this emission is readily available from [55]Fe radio-isotope sources. These x-rays are of sufficiently low energy that the contribution to the line width from the $K\alpha$ doublet structure and the intrinsic radiation width of the emission lines can be neglected. This is not true for higher energy characteristic x-rays, for which the width of the emission line must considered, particularly when precision spectral deconvolution is attempted. Figure 3 shows a typical pulse-height spectrum of Mn x-rays taken simultaneously with a calibrated pulser input. The purpose of the pulser measurement is to monitor the resolution of the electronic system independent of any peak broadening due to the detector itself. Typical state-of-the-art Si(Li) or HPGe detectors achieve a FWHM of 175 eV at 5.9 keV, although values as low as 160 eV have been reported. However, this number is but one indicator of the quality of an EDXRF detector system, and other factors, such as maximum count rate capability or the presence of background artifacts, may be more important in many analytical applications.

If one neglects the intrinsic emission width of the x-ray lines, the instrumental energy resolution of a semiconductor detector x-ray spectrometer is a function of two independent factors. One is determined by the properties of the detector itself; the other is dependent upon the details of the electronic pulse processing employed. The FWHM of the x-ray line (ΔE_{total}) is described as the convolution of a contribution due solely to detector processes (ΔE_{det}) together with a component associated with limitations in the electronic pulse processing (ΔE_{elec}):

$$(\Delta E_{\text{total}})^2 = (\Delta E_{\text{det}})^2 + (\Delta E_{\text{elec}})^2 \tag{1}$$

The detector contribution to the resolution (ΔE_{det}) is determined by the statistics of the free charge production process occurring in the depleted volume of the diode. The average number of electron-hole pairs produced by an incident photon can be calculated as the total photon energy divided by the mean energy required to produce a single electron-hole pair. If the fluctuation in this average were governed by Poisson statistics, then the variance would be the square root of that

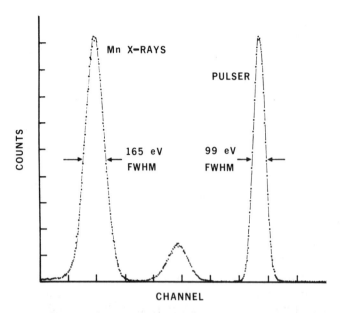

Figure 3 Pulse-height spectrum of Mn K x-rays obtained with a Si(Li) energy-dispersive detector. The pulser peak is acquired simultaneously as an indicator of the electronic component to energy resolution.

number. In semiconductor devices, the details of the energy loss process are such that the individual events are not strictly independent and a departure from Poisson behavior is observed. This departure is taken into account by the addition of a Fano factor in the expression for the detector contribution to the FWHM:

$$(\Delta E_{det})^2 = (2.35)^2 E \epsilon F \tag{2}$$

where ϵ is the average energy required to produce a free electron-hole pair, E is the energy of the photon, and F is the Fano factor. E/ϵ is the total number of electron-hole pairs; the factor 2.35 converts the root mean square deviation to FWHM.

Examination of the values of ϵ and F listed in Table 1 shows that for an equivalent energy, the detector contribution to the resolution is 28% less for the case of Ge compared to Si. This could be an important consideration in the choice of detectors for certain experiments, although this advantage of Ge over Si is mitigated in many applications because of the predominance of electronic noise at lower x-ray energies.

The contribution to resolution associated with electronic noise (ΔE_{elec}) is the result of random fluctuations in thermally generated leakage currents within the detector itself and in the early stages of the amplifier components. Although these processes are intrinsic to the overall measurement process, there are methods for limiting the impact on the final system resolution. Figure 4 is a schematic diagram of a typical pulse-processing system employed in a semiconductor detector x-ray spectrometer. The pulse processing can be divided between the charge integration, which takes place in the preamplifier, and the voltage amplification and pulse

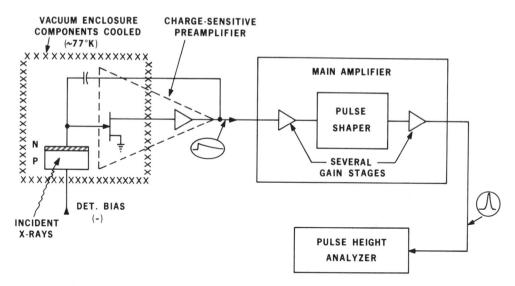

Figure 4 Typical pulse-processing system used for energy-dispersive detectors.

shaping, which occur in the main amplifier. To achieve the noise performance necessary for x-ray fluorescence applications, the contribution of thermally generated charge to the electronic noise is reduced by cooling the detector and the first stage of the preamplifier to cryogenic temperatures. This is normally performed using a liquid nitrogen reservoir in contact with a cold finger on which the detector and associated components are mounted. It was recently demonstrated that noise performance adequate for many applications can likewise be achieved using thermoelectric coolers [12,13]. However, because of the limitations on the maximum temperature differential that can be maintained using the Peltier effect, the best performance is still achieved with liquid nitrogen cooling.

The function of the charge-sensitive preamplifier and subsequent amplification stages is to convert the integrated charge pulse produced by collection of the photoelectrically induced ionization to a voltage pulse that can be measured and stored in the multichannel pulse-height analyzer (MCA). However, the pulse processing must also achieve a very important goal of amplifying this weak charge signal to a measurable level while suppressing random fluctuations in the signal amplitude produced by thermal noise. This is achieved by generating a carefully defined pulse shape in the main amplifier, which restricts the Fourier frequency components in the final signal in such a way that the signal contributions are emphasized relative to the noise fluctuations. The most common pulse shapers employed in modern semiconductor spectrometers generate output pulses that are Gaussian, triangular, or cusp shaped [14–16]. Each is capable of achieving adequate energy resolution for EDXRF analysis; the differences are due mainly to the effective time interval required to process a pulse. Since the relative amplitude of the noise contribution is a strong function of the characteristic time constant associated with the pulse shaper, the difference between pulse shapes becomes important for applications in which high counting rates are important.

A detailed analysis of the effects of various pulse-shaping options on spectrometer performance is presented in Ref. 17.

The frequency distribution of the noise components in a typical x-ray spectrometer is illustrated in Fig. 5, in which the square of the deviation in the average number of thermally generated ion pairs is plotted as a function of the characteristic peaking time of a simple pulse shaper. To calculate the noise in electron-volts, the equivalent noise in electrons must be multiplied by the energy per ion pair given in Table 1. The solid lines show the contributions arising from two independent processes in the detector and in the first-stage amplification elements. The dashed line is the sum of the two. The plots demonstrate that in general there exists a shaping time constant at which the contribution of electronic noise to the total resolution is a minimum. As pointed out later in the section on throughput, however, it is not always desirable to operate at this minimum noise shaping time since an unacceptably large dead time may result. Although modern commercial x-ray spectrometers are designed with a suitable compromise between energy resolution and count rate capabilities, an understanding of these tradeoffs on the part of the experimenter is important for optimum usage of the instrument.

Figure 6 summarizes the results of the energy resolution discussion by presenting plots of Eq. (1) as a function of energy for realistic detectors. Data for both Si and Ge have been calculated based on parameters given in Table 1. It is clear from the plots that the energy resolution at low energies is influenced primarily by the electronic noise, whereas for energies above 10 keV, detector properties dominate. Also, for this optimistic choice of detector properties, the Ge

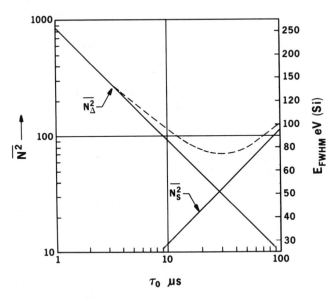

Figure 5 Relative contribution of electronic noise sources to the total electronic resolution as a function of the characteristic time constant, τ_0, of the pulse shaper. $\overline{N_s^2}$ is the component associated with fluctuations in the detector leakage current; $\overline{N_\Delta^2}$ is due to leakage currents in subsequent amplification stages. The dashed curve is the sum of the two.

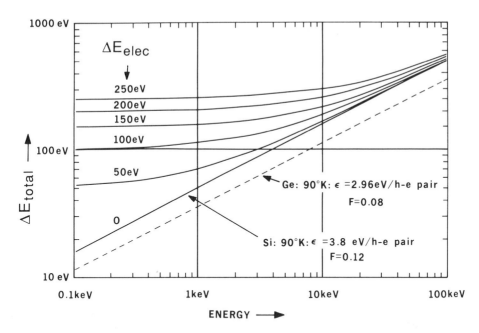

Figure 6 Calculated total energy resolution ΔE_{total} (FWHM) versus incident photon energy for both Si and Ge detectors. Asymptotic values show resolution limits assuming no contribution due to electronics processing.

devices exhibit significantly better energy resolution as a result of the favorable average energy loss per ion pair and the smaller Fano factor. Although energy resolution is one of the primary factors that determine the performance of an x-ray fluorescence analysis system, it should be emphasized that there are other properties, such as spectral background and counting rate capabilities, that are equally critical when evaluating a given system for analytical applications. The spectral resolution of semiconductor spectrometers has been compared with that of other detection methods in Fig. 22 of Chap. 2.

C. Detector Efficiency

One of the more important advantages of semiconductor spectrometers is the absolute efficiency with which fluorescent x-rays are detected and their energies measured. This is the result of the intrinsically high photoelectric absorption efficiencies for semiconductor materials in the x-ray energy range and from the large solid angles achieved in typical fluorescence geometries. If one assumes a point source of x-rays and neglects edge effects around the detector periphery, then the total efficiency E_{tot} of a given geometry can be expressed as the product of the solid angle of the detector with respect to the sample and the intrinsic efficiency of the diode itself:

$$E_{\text{tot}} = \frac{\Omega}{4\pi} \, \epsilon(E) \tag{3}$$

where Ω is the total solid angle and $\epsilon(E)$ is the energy-dependent intrinsic efficiency of the detector. The solid angle is determined by the area of the detector and the sample-detector distance and varies with the design of the system. Typical areas are 30–80 cm^2 for Si(Li) detectors and 80–320 cm^2 for HPGe detectors. Although the additional solid angle is advantageous for many applications, one must realize that the added capacitance associated with this area results in an increased contribution in the electronic noise of the system.

The intrinsic efficiency of the semiconductor device can be approximated by a simple model in which the probability of detecting an x-ray incident on the detector is assumed to be the probability of photoelectric absorption within the sensitive volume. This can be expressed as

$$\epsilon(E) = e^{-\mu t}(1 - e^{-\sigma d}) \qquad (4)$$

where t is the thickness of any absorptive layer between the sample and the detector, $\mu = \mu(E)$ is the mass absorption coefficient of the absorptive layer, d is the detector thickness, and $\sigma = \sigma(E)$ is the photoelectric mass absorption coefficient for the detector material. Figure 7 shows plots of Eq. (4) for the case of a 3 mm thick Si(Li) and a 5 mm thick HPGe detector. The poor efficiency at low energies is assumed to be determined by the combined absorption of the 25 μm Be cryostat window and a nominal 2 cm air path between sample and detector. The energies of the characteristic K emission lines for several elements are also shown. The plots show the near unity intrinsic efficiencies for both detectors over a wide range of useful x-ray energies. The Ge detector is efficient at much higher energies than Si because of the higher atomic number and subsequent larger photoelectric cross section.

This simple model gives a semiquantitative picture of the efficiency behavior of semiconductor spectrometers. However, several other factors must be considered for quantitative calibration of a fluorescence spectrometer. The concept of a thin entry window that either absorbs or transmits an incident x-ray does not describe cases in which the secondary electrons, either photoelectric or Auger, are emitted into the active volume from the window layer. Similarly, detailed studies of the low-energy efficiency have indicated the presence of an absorbing layer on the surface of both Si and Ge devices associated with the surface layer of the semiconductor material itself. Detailed experiments have indicated that the charge collection efficiency for events near the surface depends in a complex way upon low-energy x-ray and charge transport properties [18,19]. As a consequence of these entry window-related effects, the efficiency for low-energy photons can be reduced. The events lost from the photopeak can then appear in a continuum background below the full energy peak, where they reduce detectability and interfere with spectral analysis.

Similarly, secondary electrons or photons originating from the photoelectric absorption events that occur initially in the active volume can escape. This results in a collected charge pulse of reduced amplitude. An easily observed manifestation involves the observation of discrete escape peaks in the spectrum. These are associated with the loss of characteristic Si or Ge x-rays from the active volume, which produces a distinct peak in the spectrum at an energy a corresponding amount below the full energy peak. Continuous loss processes involving electron escape can also occur, although the probability is small. These and related mech-

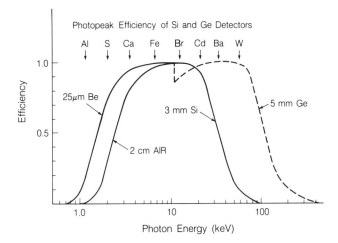

Figure 7 Calculated intrinsic photopeak efficiency for Si and Ge detectors as a function of incident photon energy. The high-energy limits are established by the photoelectric cross section of the detector material and the diode thickness. The low-energy cutoff is determined either by absorption in the thin Be window or in a nominal 2 cm air path from sample to detector or fluorescence source.

anisms that reduce the amplitude for a given event can lower the photopeak efficiency relative to the simple model described. There have been studies in which the efficiencies of Si(Li) and Ge have been carefully measured using calibrated sources of x-rays spanning the energy region of interest [20–22]. These indicate that the maximum intrinsic photopeak efficiency is reduced by a few percent relative to the curves shown in Fig. 7 and is slightly higher at the upper energy cutoff than calculated. These results substantiate the overall validity of the simple photoelectric absorption model but demonstrate the limitations if precise results are required. For most quantitative applications of EDXRF, it is not necessary to explicitly determine the photoelectric efficiency function since it is included in the overall calibration factor of the instrument.

D. Detector Background

Although artifacts associated with the partial collection of the photoelectric signal have a small effect upon the efficiency of the spectrometer, these processes can have far more serious consequences on analytical sensitivity through their effect upon spectral background. When photon excitation is employed to induce fluorescence, there exist competing physical processes in which the incident photons are scattered from the sample into the detector with very little loss in energy. These effects are particularly severe in trace analysis applications in which the number of matrix atoms that produce scattering is large compared to the number of atoms that fluoresce. In these cases, the major feature in the spectrum is a composite scatter peak consisting typically of coherent and incoherent contributions. The effect of incomplete charge collection of these events is to produce a continuum of events at lower energies in the spectrum. These background events can then interfere with the measurement of lower energy fluorescent x-rays, as shown in the schematic spectrum of Fig. 8. Here the fluorescent x-rays are shown superimposed on a continuum background whose amplitude is proportional to the area of the composite coherent-incoherent scattering peak at approximately 17 keV. This background is the limiting factor in establishing statistical accuracy in the determination of trace quantities.

Studies designed to reduce the source of this continuum background have indicated that in addition to fundamental x-ray and electron energy loss processes, a more significant background resulting from incomplete charge collection from the detector active volume is normally the dominant contribution [23,24]. This process is an artifact of the detector operation in which the collection of the free charge from the depleted volume is inhibited as a result of nonuniformities in the applied electrical field associated with edge effects at the periphery of the cylindrical detector. The impact of this process on the quality of the spectrum can be reduced either by external collimators that prevent the incident radiation from

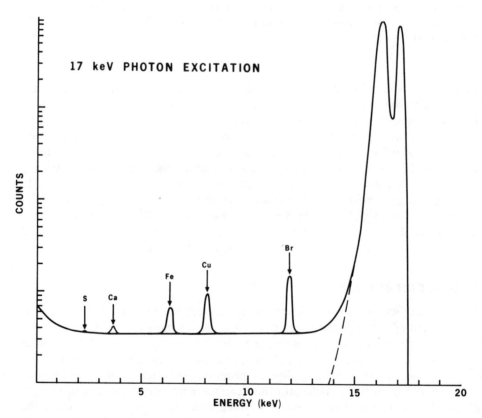

Figure 8   Schematic spectrum calculated assuming 17 keV excitation on a carbon matrix with a typical contribution to the background primarily due to incomplete charge collection. Peak heights of elements are calculated assuming equal mass concentrations for each, neglecting differences in absorption of the characteristic x-rays. Variation in intensity is due to a decrease in the excitation efficiency of monoenergetic excitation for lighter elements.

interacting in the periphery of the device or by internal electronic collimation brought about by the use of a guard-ring structure [23].

E. Count Rate Performance

In many applications involving x-ray tubes or synchrotron sources, the available excitation intensity is sufficient to induce pulse-counting rates that exceed the processing capabilities of the semiconductor spectrometer system. It then becomes necessary to reduce the excitation intensity or decrease the solid angle viewed by the detection system. For most analytical applications, optimum performance is achieved when maximum throughput of events is maintained, although there are specialized applications in which this is undesirable.

The count rate limitations associated with a semiconductor spectrometer are an inherent property associated with the finite pulse-processing time required by the electronic shaping network. When a random sequence of pulses is incident on the detection system, some of the events cannot be processed without ambiguity. To appreciate the fundamental nature of this limitation and its relationship to system performance, some elementary concepts of electronic pulse processing must be considered.

Figure 9 illustrates the time sequence of pulses that occur at various stages in the pulse-processing chain. Trace A shows the output of the charge-intergrating preamplifier. The steps at times 1, 2, and 3 represent the charge integrals of discrete events. Traces B and C are the outputs of the fast (short shaping time) amplifier and an associated discriminator that serves as a timing marker for the individual events. The main shaping amplifier output is shown as the Gaussian pulse shapes in trace D. For each event, a total pulse-processing time (neglecting amplitude-to-digital conversion in the pulse-height analyzer) of τ_d is required after the arrival of the pulse before the system is ready to accept the next event.

Although the average counting rate detected by the system can be well below the frequency defined by the reciprocal of the pulse width τ_d, that the events are statistically uncorrelated implies that the events are not uniformly distributed in time. There then exists a probability that two pulses will occur within the same processing time interval. This is illustrated by the overlap of pulses 2 and 3 in the trace.

At a low average counting rate, this overlap is not a limiting factor. As the average counting rate increases, however, a point is reached at which there is a significant probability that a second event will occur before the first event has been fully analyzed. If the two events occur within a time less than the shaping time of the amplifier, the charge signals are indistinguishable and an erroneous "pileup" energy signal results.

Modern spectrometers employ some form of rejection circuitry to eliminate these pileup events from the pulse-height spectrum. Typical systems rely on the inspection of the fast discriminator output to determine if two pulses have occurred in rapid succession. Appropriate logic is then employed to gate the output of the processor to eliminate the resultant ambiguous energy signals. This is shown in traces E and F, in which the logic signal causes the output to be inhibited when the pulses overlapped to produce an ambiguous pileup energy output. Since the fast shaper that generates the discriminator output has inherently poorer energy

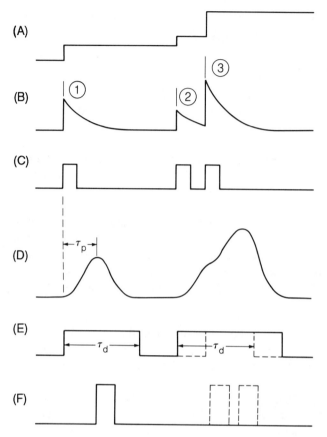

Figure 9 Time sequence of pulses during the processing of a series of x-ray events. (A) The output of the preamplifier represented as a series of voltage steps reflecting the integrated charge from the individual events. (B) The output from a fast shaping amplifier that operates with a shorter time constant than the main shaping system. (C) A fast discriminator timing pulse derived from B. (D) The output of the main shaper, with the pileup of pulses 2 and 3 indicated. (E) The dead-time gates and (F) the final output with the ambiguous 2 and 3 pileup pulses rejected.

resolution relative to the slow channel, the pileup rejection circuit is limited in the minimum energy pulses it can detect. However, in typical photon-excited EDXRF, the major fraction of pulses occur from scattered high-energy radiation and a pileup rejector system is an important feature in the system design.

The pileup probability is obviously a function of the characteristic shaping time τ_p, which in turn establishes the effective dead time τ_d of the system. This probability is independent of the details of the specific type of pileup rejection circuit used to eliminate the ambiguous events. The number of events that experience pileup and are eliminated from the spectrum as a consequence can be estimated for the case of a totally random arrival distribution. For a series of events that are randomly distributed in time with an average frequency N_0, the probability P that no pileup events occur within a characteristic time τ_d after a given event can be expressed as

$$P = N_0 e^{-N_0 \tau_d} \tag{5}$$

From this expression, we can calculate the fraction of events transmitted through the system. Figure 10 is a plot of nonpileup output rate versus input rate expressed in terms of a characteristic pulse-processing time. The input rate for which the output rate is a maximum is seen to be the reciprocal of the shaping time. The output rate at this point is reduced by a factor of $1/e$. It should be emphasized that this behavior is a fundamental consequence of random arrival statistics and a finite measurement time. However, one should be aware that it is always possible to reduce the characteristic processing time to achieve an increase in counting rate at some sacrifice in energy resolution. For example, a typical pulse peaking time of 10 μs would imply a dead time of 20 μs and a maximum output rate of 18.4 kHz. If one were willing to sacrifice electronic resolution by reducing the shaping time to 2 μs peaking time, a maximum output rate of 92 kHz would be possible. According to Fig. 5, this tradeoff would result in an increase in electronic noise from 100 eV to approximately 160 eV. If the application involved the detection of high-energy x-rays, this compromise might be advantageous.

Although different algorithms can be used to correct for events lost to pileup, there is no way to eliminate the effect through passive pulse processing [25–27]. Since different manufacturers of x-ray equipment vary in their approaches to pileup rejection, it is important to evaluate the throughput of the system using a variable intensity source of radiation. The input rate can be determined by scaling the fast discriminator output. The output rate can be simultaneously measured in the pulse-height analyzer. A plot of nonpileup output rate versus input rate can then be generated and compared to the ideal case shown in Fig. 10.

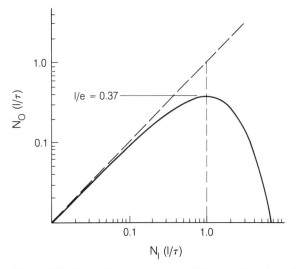

Figure 10 Nonpileup output rate (N_0) as a function of input rate (N_1) scaled as a function of a characteristic shaping time. Maximum output is 0.37 of the input rate that is equal to the reciprocal of the time constant.

When the excitation source can be switched off in a time interval that is short compared to the characteristic pulse-processing time, it is possible to increase the average output counting rate by eliminating the effects of pileup. Such pulsed excitation systems rely on the ability to detect an event in the fast channel and to shut off the excitation before a second pileup event can occur. In this mode of operation, the output rate can equal the input rate up to the point at which the system is continuously busy; that is, $N = 1/\tau$. This method of pileup control has been implemented in several laboratories using pulsed x-ray tubes, particle accelerators, and electron probe beams [28–31]. In addition to increasing the maximum output counting rate, this method has the secondary advantage of reducing the effects of heating in either the x-ray tube anode or the ion beam target since the excitation is turned off during the pulse-processing interval.

Finally, it should be pointed out that there is a very important class of measurements in which the optimum counting rate is not achieved with maximum throughput. These are experiments, such as fluorescence extended x-ray absorption fine structure (EXAFS), in which modulation of the x-ray yield is the quantity of interest [32]. Referring to Fig. 10, it is apparent that, at the maximum output rate, the slope is zero, and consequently, any modulation in the input rate will not affect the output rate. For EXAFS and similar applications, a lower input rate corresponding to a slope nearer unity is recommended.

F. Dead-Time Correction

The presence of pulse pileup must be considered in any system designed for quantitative measurements since it causes the efficiency with which pulses are processed to be rate dependent. EDXRF analytical systems are designed to correct for this discrepancy through a variety of approaches. The most straightforward involve the use of a live-time clock consisting of a gated oscillator and scaler. The oscillator clock is turned off when the system is busy processing pulses so that the duration of the measurement in terms of a live-time interval is corrected for those periods when the system is incapable of processing a pulse. Alternative methods rely on the direct measurement of the fast discriminator pulses to keep track of those events that were missed. An additional correction is added to compensate for those intervals in which pileup is occurring [33,34].

A simple empirical method used to correct for dead-time losses is a measurement of the ratio of input to output counts as a function of input rate over the range of values normally encountered. Subsequent analyses require that the input rate be measured for each experiment. The live-time correction can then be applied using the previously determined response function.

Since all live-time correction methods have some limitation on the range of counting rates over which they can be used, it is important that methods be devised to evaluate the precision with which such corrections are generated. A carefully prepared series of standards of varying concentrations represents a direct approach. A potentially more precise method involves the use of a single thin-film standard. Variable mass targets are placed behind the standard to vary the total counting rate over the range of interest. If the variable mass target is chosen so that the variable intensity of scattered or fluorescence radiation does not induce fluorescence in the thin-film standard, the measured intensity of fluorescence from the standard should be independent of total counting rate. The ability of the dead-time correction system to compensate for pileup can then be empirically evaluated.

G. Low-Energy X-Ray Detection

The efficiency curve shown in Fig. 7 illustrates that for conventional EDXRF measurements, the absorption of fluorescent x-rays in air and in the Be entry window limits the accessible energy range to photons greater than approximately 2 keV. The absorption losses in the air path can be significantly reduced by employing a He atmosphere. However, it is normally advisable to maintain a thin layer of air near the thin Be window on the cryostat since diffusion of He through the window and into the vacuum enclosure can affect long-term spectrometer operation.

The necessity of a Be window to maintain vacuum integrity between the cryostat enclosure and atmospheric pressure does not constitute a serious limitation for most analyses. However, there are applications in which the detection of x-rays with energies below 1 keV becomes necessary, particularly in microbeam analysis. It is therefore of interest to explore the fundamental limitations of semiconductor detector spectrometers in terms of subkilovolt x-ray detection.

If one neglects external sources of x-ray absorption, then the low-energy efficiency of a semiconductor detector is determined by the effective thickness of the entry window on the diode structure. This window is comprised of an evaporated contact used to form the rectifying Schottky barrier and a thin surface layer of inactive semiconductor material from which charge cannot be collected. The thickness of evaporated material, normally Au, can be determined by direct measurement during the manufacturing process and is typically in the range of 100 Å. The effective "dead layer" of semiconductor material is a more complicated parameter to determine. Empirical studies that have attempted to measure the absorption inherent in the semiconductor surface layer have established that the effective thickness is determined to a large extent by the absorption length of the low-energy photons and the charge transport characteristics of the associated ionization products. In the simplest model, there is competition between the rate of diffusion of the electron-hole distribution against the gradient of the applied field. For distances near the entry contact, a part of the charge can diffuse into the contact and be lost to the signal before the electrical field can sweep it to the opposing electrode. This loss of charge can be interpreted in terms of an effective "window" thickness of typically 0.2 μm silicon equivalent. In addition, thin evaporated contacts are deposited on the entry surface that can absorb incident low-energy photons. Figure 11 shows calculated efficiencies for the subkilovolt region assuming the typical values discussed earlier [35]. Absorption of both Au and Al contacts are illustrated as possible alternative choices for evaporated contact material. The results show that it is possible to measure x-ray energies to the limits imposed by the noise of the counting electronics. Experiments have demonstrated the detection of x-rays as low as the $K\alpha$ line of boron at 190 eV [36,37].

To perform experiments at these low energies, it is necessary to eliminate the Be foil interposed between the sample and detector, which serves as an entry

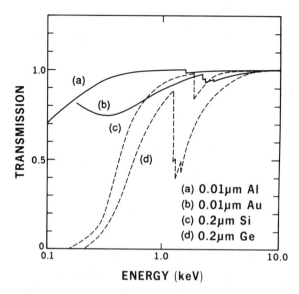

Figure 11 Calculated efficiencies for the very low energy region showing the limits to efficiency determined by the entry window of the detector itself. Curves a and b show the absorption due to an evaporated Al or Au layer of typical thicknesses required for contacts. Curves c and d show the effect of an "intrinsic" Si or Ge window.

window to the vacuum enclosure. This has typically been done by including the sample and excitation source in the same vacuum enclosure as the detector. This introduces additional problems, such as potential contamination on the detector window and the effect of low-energy photons, that is, optical fluorescence, on the detector output. Several methods have been devised to address these problems, involving the insertion of thin barriers between detector and sample within the same vacuum system. Alternative approaches that recently became available involve the use of high-strength thin windows made from low atomic number elements that are capable of withstanding a full 1 atm pressure differential. These include self-supporting 0.5 μm diamond polycrystalline films [38] and 0.25 μm windows composed of a vapor-deposited amorphous material consisting of 90% boron by weight with nitrogen and oxygen for the remainder [39]. These windows are reported to exhibit significant x-ray transmission for photons well below 1 keV. These developments will facilitate the detection of light elements primarily in electron probe applications in which a vacuum path is an inherent component of the measurement system and the excitation process favors low atomic numbers.

III. TYPICAL X-RAY TUBE EXCITATION SYSTEMS FOR EDXRF

X-ray fluorescence using photons as an excitation source for inducing inner shell ionizations in the sample atoms is perhaps the most widely used method for EDXRF. Photons have the advantage of being readily available from either radioisotope sources or x-ray tubes. The mean free path of x-rays in the energy range of interest for chemical analysis is sufficient that operation of the spectrometer in air is possible. Finally, the interaction of x-rays with the sample material is well understood and relatively insensitive to the chemical state of the elements. It is thus possible to perform many analyses with a minimum of sample preparation and with little concern for unknown matrix artifacts. A major disadvantage of photon excitation is the limited flux available from conventional sources and the difficulty in forming focused beams. With the increasing availability of synchrotron sources, however, these objections are becoming less serious, at least in large research laboratories where access to such facilities is possible.

Photon excitation with x-ray tubes continues to be the most practical method for routine EDXRF analysis. Compared with radioisotopes, the flux available with a typical low-power x-ray tube is sufficient to provide for maximum counting rates for most applications. It is also possible to fluoresce secondary targets to generate high-intensity fluxes of characteristic x-rays for monochromatic excitation. In contrast to a synchrotron source, the cost and size of a tube-excited system is consistent with operation in a typical analytical laboratory. In this section the general features of x-ray tube excitation for EDXRF are discussed and the relevant parameters determining the choice of excitation options are described. A more general discussion of x-ray tubes is given in Chap. 2.

The basic elements of an EDXRF x-ray tube excitation system are shown in Fig. 12. The x-ray tube consists of a vacuum enclosure containing an electron source and target anode. A thin window is mounted in the vacuum wall to allow the x-rays to exit the tube enclosure. Commercial tubes are typically manufactured from glass, although metal tubes with glass or ceramic electrical insulators are common [40]. The exit window is typically a thin mica or Be foil. The electron beam is generated using a thermionic cathode and is accelerated toward the anode by an electrostatic potential of typically 30–100 keV. In most applications the accelerating voltage is achieved with the anode assembly at ground potential and with the insulated cathode assembly at the appropriate negative high voltage. This configuration is generally preferable since the thermal power generated in the anode by the high-energy electron beam can be more easily dissipated at ground potential.

The tube structure shown in Fig. 12 also has a control grid that can be used to modulate or pulse the emission current. This can be used to modulate the electron beam current to ensure constant output of radiation from the tube in dc operation, or it can be used to rapidly switch the tube on and off for pulsed operation. In tubes without control grids, the output of the tube is maintained constant by regulating the temperature of the thermionic cathode. Reproducibility of analytical results is achieved for tube excitation by operating for a measured amount of time at a fixed emission current (dc operation) or for a fixed amount of integrated charge (pulsed operation). This approach depends upon the reasonable assumption that the total output x-ray flux is proportional to the number of electrons striking the anode.

The spectrum of x-rays produced in the anode is a continuum of energies upon which is superimposed the characteristic emission lines of the target material. The continuous x-ray "bremsstrahlung" spectrum is the result of slowing down the electrons through interactions with the nuclear charges of the target

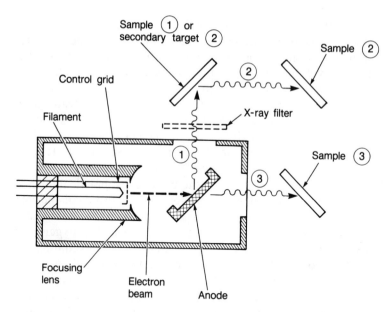

Figure 12   An x-ray tube and typical excitation geometries. The high-energy electron beam strikes the anode to produce x-rays. The flux (1) is used directly to strike the sample, or in the second case, the primary flux (1) strikes a secondary target, which generates a flux of x-rays (2), which strikes the sample. In the third case, the direct beam (3) is directed through the anode in a transmission geometry to the sample.

atoms. The x-ray spectrum ranges from the lowest energies that can escape the anode target to a maximum energy equal to the electron beam voltage. The characteristic x-rays are generated either through direct ionization of the target atoms by the incident electrons or through secondary excitation by photons in the high-energy continuum. Figure 13 shows a schematic spectrum obtained from an x-ray tube with a Mo anode (K absorption edge at 20 keV) operated at an accelerating voltage of 60 kV. The step in the continuum distribution indicates the effect of self-absorption by the target material for x-ray energies above the absorption edge.

The x-ray spectrum emitted by the anode can be used in any number of excitation and detection geometries. In Fig. 12 the beam (1) represents the configuration used for direct excitation of the sample by the x-rays emitted from the anode. A thin x-ray filter (typically the same element as the anode material) is used to preferentially transmit the characteristic x-rays generated in the anode. This configuration has the advantage of providing the most efficient use of the photons generated at the anode but has the disadvantage of lack of flexibility in choice of excitation energies.

In the second option shown, the sample is replaced by a secondary target whose characteristic x-ray spectrum is excited by the direct tube output. These x-rays are then incident on the sample. This geometry has the advantage of flexibility in the choice of excitation energies but does not make efficient use of the output flux. The total system efficiency is proportional to the product of two solid angles. They are the solid angle of the secondary target with respect to the anode

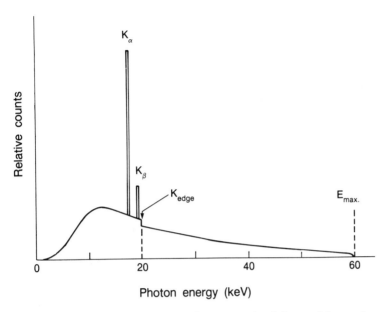

Figure 13 Schematic spectrum of x-rays emitted from a Mo anode x-ray tube operated at 60 keV.

and the solid angle of the sample with respect to the secondary target. Excitation geometries that make use of secondary targets must be carefully designed to achieve close coupling between the respective components and at the same time minimize scattering of direct, unfiltered radiation into the detector.

The third option is a variant of direct excitation in which the transmission of the x-rays through the anode itself is used to filter the continuum radiation. This method has certain advantages in the design of compact geometries, particularly if the thin-film anode material is incorporated into the exit window. However, this design is limited in the amount of anode power dissipation that can be achieved in the anode structure.

In any of the excitation options chosen, the maximum flux available is limited by the amount of power that can be dissipated in the anode structure. X-ray tubes can be roughly classified according to the type of anode cooling employed. Low-power x-ray tubes are typically air cooled and operate at a maximum of 100 W at the anode. Tubes with water-cooled anodes can operate at 1000 W. Rotating anode tubes can achieve 10 kW dissipation. For a 50 kV anode potential, these ratings correspond to anode currents of 2, 20, and 200 mA, respectively.

The yield of x-rays per incident electron is small since the majority of energy loss processes result in ionization of the outer shell electrons from which little useful x-ray flux is derived. Calculations of the yield of characteristic and continuum photons for the case of typical anode materials is a difficult task since the details of the electron energy loss process and x-ray transport properties can be extremely complicated. For the purpose of acquiring order of magnitude estimates of total yields, however, approximate calculations can be made [41–43]. These are summarized in Table 2 for a number of anode materials. A convenient number

Table 2 X-ray Tube Anode Characteristics

Anode material	Characteristic radiation	Energy (keV)	Anode[a] voltage (kV)	Approximate[b] X-ray yield
Cr	K	5.4	18	2.0×10^{-4}
Cu	K	8.0	27	2.2
Mo	K	17.4	60	2.2
Rh	K	20.1	66	2.3
Ag	K	22.1	75	2.4
W	L	8.4, 9.7	30[c]	1.5
Au	L	9.7, 11.4	40[c]	1.5

[a] Anode voltage is assumed to be three times the respective K or L absorption edge energy.
[b] Expressed as total K quanta per electron per steradian (Adapted from Ref. 41.)
[c] The high-Z transits are often chosen because of the intense continuum radiation available. In these cases higher anode voltages may be desirable.

to remember when estimating fluorescent intensities of x-ray tube sources is that the conversion efficiency for electron flux to either characteristic or continuum x-rays is approximately 1×10^{-5} photons per electron when the electron energy is three times the K shell binding energy of the anode material.

The choice of three times the K shell binding energy is a recommended operating point for most applications in which the characteristic x-rays are the dominant source of excitation. For lower beam energies, the intensity of characteristic x-rays resulting from direct electron excitation has been calculated to vary as the $\frac{3}{2}$ power of the applied anode voltage [41]. For beam energies much above the recommended value, the depth of penetration of the electron beam is sufficient that excessive x-ray absorption can occur. For secondary target systems, the voltage dependence is more complicated since it involves the properties of both the anode material and the secondary targets. Measurements performed using a tungsten anode tube with a Mo secondary target have indicated the x-ray yield varies approximately as the square of the applied tube voltage [44]. In either case, it is important to realize that the excitation efficiency is affected by the beam voltage and that appropriate precaution must be taken to stabilize the applied voltage, particularly under variable current loads.

The choice of x-ray tube or excitation geometry is dictated largely by the analytical applications. For thick samples with heavy element matrices, maximum count rates can easily be achieved even with low incident excitation fluxes such as those obtained from a low-power tube or an inefficient secondary target arrangement. For thin films or trace element studies, more intense excitation fluxes are required. Finally, in specialized applications, such as microbeam analysis, low background secondary target systems with inefficient geometries, or polarized scattering excitation [45–48], the elaboration of a rotating anode tube may be warranted.

For the majority of EDXRF applications, monochromatic x-rays (or nearly monochromatic characteristic x-rays) provide the most sensitive source for fluorescence excitation. This is because the maximum efficiency for photoabsorption in any given element is realized for incident energies immediately above the absorption edge of the particular element and decreases rapidly for higher energies. Furthermore, the flux of photons with energies below the edge produces no excitation but can interfere with the measurement of the fluorescent x-rays in the energy-dispersive semiconductor detector. As a practical matter, the energy of the monoenergetic excitation is chosen sufficiently far above the absorption edge such that radiation elastically and inelastically scattered from the matrix does not overlap the photopeaks of the fluorescent x-rays.

Figure 14 shows the calculated efficiency for the production of inner shell vacancies for incident photon energies corresponding to three typical secondary target materials. The steep slope exhibited for atomic numbers below the maximum for each energy and the correspondingly narrow range of elements accessible for any one excitation energy demonstrate the desirability of multiple secondary targets for EDXRF.

Figures 15 and 16 illustrate the use of multiple excitation energies for the analysis of an ambient air pollution sample consisting of a thin deposit of fine

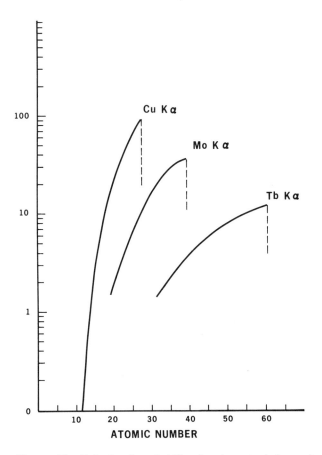

Figure 14 Calculated probability for characteristic production of x-rays as a function of target atomic number for three choices of incident characteristic x-rays: Cu $K\alpha$ at 8.0 keV, Mo $K\alpha$ at 17.4 keV, and Tb $K\alpha$ at 44.2 keV.

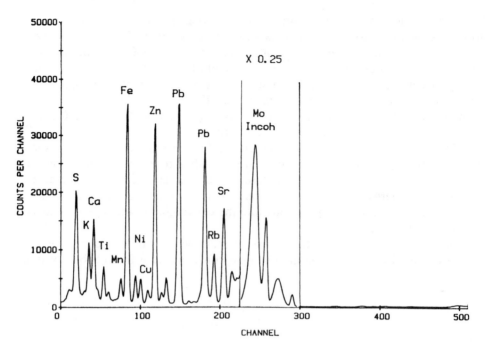

Figure 15   Plot of x-ray fluorescence spectrum obtained using radiation from a Mo secondary target for excitation of an atmospheric aerosol sample.

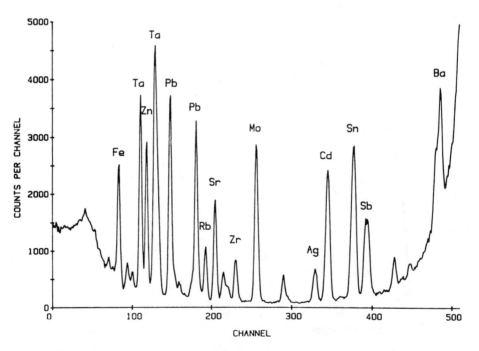

Figure 16   Spectrum of aerosol sample shown in Fig. 15 but acquired using excitation radiation from a Sm secondary target. The Ta lines in the low-energy region are from the Ta collimator. Note the order of magnitude difference in the vertical scale.

particles (diameters less than 2 μm) collected on a polytetrafluoroethylene filter. The spectra were acquired using Mo K (17.4 and 19.6 keV) and Sm K (39.9 and 45.6 keV) secondary targets for excitation. The data are normalized to represent equal number of incident photons in each case. Note the order of magnitude difference in the vertical scale between the two spectra. Typical elemental concentrations (ng/cm^2) measured on this particular thin sample are S = 15,000, Fe = 2900, Cu = 60, Sr = 370, Sb = 230, and Ba = 500. The Ta lines observed in the Sm excited spectrum are the result of collimator fluorescence.

When optimum sensitivity for the detection of trace quantities is not necessary, the option of employing continuum excitation for uniform analytical sensitivity can be exploited. In this case, the fluorescence intensity does not depend strongly upon the energy of the absorption edge since a more or less continuous spectrum of excitation energies is available. However, since there also exists a continuum of energies available for scattering from the matrix, the minimum detectability for any element is reduced. An exception to this rule is the analysis of very light elements ($Z < 20$), for which the additional flux together with the improved ratio of photoabsorption to scattering cross sections combine to provide improved sensitivity for analysis.

IV. APPLICATIONS OF TUBE-EXCITED EDXRF

Among the variety of implementations of EDXRF, tube excitation is probably the most versatile and widely applied. The size and cost of a facility are appropriate for use in typical analytical laboratories, and relatively high excitation fluxes can easily be achieved. For example, a 1 mA current in a typical x-ray tube produces the equivalent flux of a 1.85×10^{13} Bq (500 Ci) radioactive source. This excitation intensity is adequate to provide for analysis times of the order of 15 minutes duration even when secondary target geometries are employed. Although the stability of an x-ray tube cannot approach that of radioisotope sources, analytical precisions of 1% or less can typically be achieved for most tube-excited EDXRF applications.

Although radioisotope sources are limited in many applications as a result of the lack of intensity, there are situations in which they can be used to advantage. They can be designed for use in very compact geometries, which compensates for the lower intensity. They are often used for portable EDXRF equipment in which stringent power requirements must be met. Also, the range of energies available with radioisotope sources is not constrained by high-voltage engineering considerations. Radioisotopes with characteristic emissions of 50–150 keV can be employed in conjunction with HPGe detectors to study the fluorescence from higher atomic number elements.

Recently, the use of synchrotron sources has become more extensively studied in EDXRF applications. Synchrotron emissions have the advantage that they are extremely intense, highly collimated, and polarized perpendicular to the beam direction. The ability to select a tunable band of wavelengths can be exploited as a near ideal excitation source that can provide excitation for small samples either directly or in a focusing geometry. The use of radioisotope and synchrotron sources is discussed in Chaps. 7 and 8, respectively.

In the following section we discuss several analytical applications in which the use of x-ray tube-excited EDXRF has been prominent. In most cases the analysis could have been performed using alternative EDXRF procedures with appropriate consideration for differences in source intensity and counting times. In particular, when discussing sensitivities for tube-excited EDXRF, we choose as a typical analysis time interval 100–300 s. Most photon-excited EDXRF analyses have in common many of the sample preparation, data acquisition, calibration, and spectral analysis procedures. These are also the subjects of subsequent chapters.

A. Analysis of Liquids

Samples in the liquid form are homogeneous for XRF elemental determinations. For some industrial processes, the analysis of solutions is achieved using flow cells through which the concentrations of specific elements are closely monitored. These analyses can include the determination of trace and minor constituents of atomic number elements ($Z > 14$) in oils, gasolines, and aqueous solutions. If one or more characteristic x-ray lines from major or minor constituents do not completely dominate the spectrum, detection limits in the range of 1–10 μg/g can often easily be realized.

Sufficient sensitivities or selectivities often are not attainable to permit the direct determination of the elements in liquids. In these cases, the use of preconcentration procedures are required to reduce the matrix scattering. Some approaches that have been undertaken for the preconcentration of trace elements in waters and solutions include (1) evaporation, (2) precipitation of the desired elements followed by collection on a suitable thin substrate, (3) the application of cation- and anion-exchange resins and ion-collecting filters, (4) adsorption of elements onto activated carbon, and (5) chelation with subsequent sorption or solvent extraction. Most of these approaches are capable of forming thin-film or intermediate-thickness samples that are well-suited for multielement XRF determinations, as discussed in Chap. 6.

B. Analysis of Solids

X-ray tube systems are used for the determination of major, minor, and some trace elements in metals, alloys, steels, catalysts, and chemical compounds. In many cases, count rate limitations are easily imposed on the semiconductor detection system by high x-ray intensities from major or minor element constituents. This, in turn, can dictate sensitivities that are attainable. In some cases, the use of x-ray absorbing filters, placed between the sample and the detector, can be used to preferentially reduce high-intensity signals from major elements. This condition exists when major elements give rise to x-rays lower in energy than the desired weak higher energy signals. The use of photons provided by selected secondary targets whose fluorescence x-rays are not of sufficient energy to excite characteristic x-rays from some of the major constituents is often a good approach to improve analytical sensitivities. In the direct excitation mode, this same effect can also be achieved to some degree by optimization of the x-ray tube voltage chosen. For the determination of lower atomic number elements in alloys, highly polished sample surfaces are required. For many elemental determinations, high precisions are easily achieved.

A number of papers have been published regarding the use of x-ray tube excitation for broad-range multielement determinations for sediments, clays, soils, minerals, cements, ores, and fly ash. In some cases, pulverization and pelletization is the only sample preparation technique used. For the accurate determination of major elements (Na–Fe), the pulverized samples are usually fused with lithium metaborate-tetraborate mixtures or some similar type of mixtures. The resulting glass samples are homogeneous for the determination of the lower atomic number elements. Optimum conditions for the determination of both major and trace elements in silicate rocks have been developed [49]. Excitation radiation is provided directly by a pulsed low-power Ag anode x-ray tube. Measurements are carried out in vacuum. The x-ray tube is operated at 10 kV for the determination of the elements Na through Fe in fused glass disks. Pellets prepared from pulverized rock material are used for the determination of the trace elements. For these later measurements, an external Ag filter is employed and the x-ray tube is operated at 45 kV. The use of the Ag filter substantially enhances the signal-to-background ratios realized. A high degree of accuracy and precision is achieved for a wide range of major and trace elements. Limits of determination for many trace elements are in the range of 4–10 μg/g.

Tissues, plant materials, and dried biological fluids are generally pulverized

Table 3 Elemental Concentrations in NBS
SRM 1571 Orchard Leaves (μg/g ± 2σ)

Element	XRF	NBS
Ti	18.0 ± 8.5	—
V	<8	—
Cr	<5	2.6 ± 0.3
Mn	86.5 ± 4.9	91 ± 4
Fe	274 ± 19	300 ± 20
Co	<6	(0.2)
Ni	1.2 ± 0.5	2.3 ± 0.2
Cu	11.5 ± 1.0	12 ± 1
Zn	25.3 ± 2.1	25 ± 3
Ga	<0.5	—
Ge	<0.4	—
As	10.1 ± 0.8	10 ± 2
Se	<0.3	0.08 ± 0.01
Br	9.0 ± 0.5	(10)
Rb	11.5 ± 0.6	12 ± 1
Sr	36.3 ± 1.3	(37)
Y	<1	—
Zr	<3	—
Hg	<1	0.155 ± 0.015
Pb	40.7 ± 3.0	45 ± 3
Th	<1	—
U	<2	0.029 ± 0.005

before analysis. In some cases, small quantities of the samples are spread on a thin substrate and the inelastic Compton scattered radiation intensity is used as a measure of the sample mass in the beam path. For the accurate determination of a broad range of trace elements, the pulverized samples are pressed into thin pellets. Sensitivities in the range of 0.5–10 ppm are often realized for higher atomic number elements ($Z > 20$) using count intervals of 10 minutes or less. Using a low-power Mo x-ray tube, a total of 22 elements were measured in light element matrices [50]. Results ascertained for NBS SRM 1571 Orchard Leaves are listed in Table 3. When x-ray counting statistics are not the limiting factor, results determined for many biological samples are typically accurate to within ±10%.

Multilayer thin films, coatings, and air particulates collected on thin substrates are analyzed directly. Generally, two to four excitation conditions are chosen to achieve high sensitivities for a broad range of elements. Maximum signal-to-background ratios are realized if the materials are collected on very thin low atomic number substrates ($Z < 10$) and if the samples are analyzed in vacuum or within a helium atmosphere. Using an x-ray spectrometer that employed a W anode pulsed tube and three secondary targets, high sensitivities have been achieved for the analysis of aerosols collected on filters of mass thickness <1 mg/cm^2 [42]. Samples are analyzed in a helium atmosphere. Table 4 lists the detection limits for 34 elements determined with this system.

Table 4 Minimum Detection Limits (ng/cm^2) for Teflon Filters in Helium Atmosphere

Element	MDL	Element	MDL	Element	MDL
Al	130	Ti	30	Zr	8
Si	45	V	20	Mo	5
S	15	Cr	16	Ag	5
Cl	13	Mn	12	Cd	6
K	6	Fe	12	In	6
Ca	5	Ni	5	Sn	8
		Cu	6	Sb	8
		Zn	5	Te	10
		Ga	4	I	13
		Ge	3	Cs	24
		As	4	Ba	40
		Se	2		
		Br	2		
		Rb	3		
		Sr	3		
		Hg	7		
		Pb	8		

X-ray tube voltage, kV: 50, 60, 70.
Secondary target: Ti, Mo, Sm.
Time interval, minutes: 2, 3, 4.
All three runs were normalized to an equivalent amount of integrated current in the pulsed x-ray tube.

V. SUMMARY

The technology that makes chemical analysis with EDXRF practical is based on the use of semiconductor x-ray detectors and associated pulse-processing and data acquisition electronic components. The present chapter attempted to explain to the analyst the basic concepts behind the operation of these components and the manner in which they influence overall system performance. The tradeoffs one must make between such parameters as detector resolution, count rate, excitation efficiency, and other design variables determine how effectively one can tailor a given instrument or experimental apparatus to a specific application. Furthermore, a thorough understanding of the factors that limit performance should enable one to implement experimental tests to determine the effectiveness of a particular approach and evaluate various commercial options.

REFERENCES

1. R. Jenkins, *An Introduction to X-Ray Spectrometry*, Heyden and Son, New York, 1974.
2. H. A. Leibhatsky, H. G. Pfeiffer, E. H. Winslow, and P. D. Zemany, *X-Rays, Electrons, and Analytical Chemistry*, Wiley-Interscience, New York, 1972.
3. E. P. Bertin, *Principles and Practice of X-Ray Spectrometric Analysis*, 2nd Ed., Plenum Press, New York, 1975.
4. E. Elad and M. Nakamura, *Nucl. Instr. Methods, 41*:161 (1966).
5. H. R. Bowman, *Science 151*:562 (1966).
6. D. A. Landis, F. S. Goulding, R. H. Pehl, and J. T. Walton, *IEEE Trans. Nucl. Sci. 18*:115 (1971).
7. G. F. Knoll, *Radiation Detection and Measurement*, John Wiley and Sons, New York, 1979.
8. F. S. Goulding and D. A. Landis, Semiconductor detector spectrometer electronics, in *Nuclear Spectroscopy and Reactions*, Part A, Academic Press, New York, 1974.
9. E. E. Haller and F. S. Goulding, Nuclear radiation detectors, in *Handbook on Semiconductors*, Vol. 4 (C. Hilsson, Ed.), North Holland, Amsterdam, 1981.
10. D. E. Leyden, A. R. Harding, and K. Goldbach, *Adv. X-Ray Anal. 27*:527 (1984).
11. W. K. Warburton and J. S. Iwanczyk, *Nucl. Instr. Methods Phys. Res. A254*:123 (1987).
12. N. W. Madden, J. M. Jaklevic, J. T. Walton, and C. E. Wiegand, *Nucl. Instr. Methods 159*:337 (1979).
13. N. W. Madden, G. Hanepen, and B. C. Clark, *IEEE Trans. Nucl. Sci. 33*:303 (1986).
14. E. Fairstein and J. Hahn, *Nucleonics 23*:50 (1965).
15. K. Kandiah, A. J. Smith, and G. White, *IEEE Trans. Nucl. Sci. 22*:2058 (1972).
16. D. A. Landis, C. P. Cork, N. W. Madden, and F. S. Goulding, *IEEE Trans. Nucl. Sci. 29*:619 (1982).
17. F. S. Goulding and D. A. Landis, *IEEE Trans. Nucl Sci. 29*:1125 (1982).
18. J. Llacer, E. E. Haller, and R. C. Cordi, *IEEE Trans. Nucl. Sci. 24*:53 (1977).
19. F. S. Goulding, *Nucl. Instr. Methods 142*:213 (1977).
20. D. D. Cohen, *Nucl. Instr. Methods 178*:481 (1980).
21. I. M. Szoghy, J. Simon, and L. Kish, *X-Ray Spectrom. 10*:168 (1981).
22. J. L. Campbell and P. L. McGhee, *Nucl. Instr. Methods Phys. Res. A248*:393 (1986).

23. F. S. Goulding, J. M. Jaklevic, B. V. Jarrett, and D. A. Landis, *Adv. X-Ray Anal.* *15*:470 (1972).

24. J. M. Jaklevic and F. S. Goulding, *IEEE Trans. Nucl. Sci. 19*:384 (1972).

25. D. A. Gedcke, *X-Ray Spectrom. 2*:129 (1972).

26. P. J. Statham, *X-Ray Spectrom. 6*:95 (1977).

27. J. M. Hayes, D. E. Matthews, and D. A. Schoeller, *Anal. Chem. 50*:25 (1978).

28. J. M. Jaklevic, D. A. Landis, and F. S. Goulding, *Adv. X-Ray Anal. 19*:253 (1976).

29. J. E. Stewart, H. R. Zulliger, and W. E. Drummond, *Adv. X-Ray Anal. 19*:153 (1976).

30. H. Thiebeau, J. Stadel, W. Cline, and T. A. Cahill, *Nucl. Instrum. Methods 111*:615 (1973).

31. P. J. Statham, G. White, J. V. P. Long, and K. Kandiah, *X-Ray Spectrom. 3*:153 (1974).

32. H. J. Winick and S. Doniach (Eds.), *Synchrotron Radiation Research*, Plenum Press, New York, 1980.

33. D. J. Bloomfield, G. Love, and V. D. Scott, *X-Ray Spectrom. 12*:2 (1983).

34. J. M. Hayes, D. E. Matthews, and D. A. Schoeller, *Anal. Chem. 50*:25 (1978).

35. J. M. Jaklevic, J. T. Walton, R. E. McMurray, Jr., N. W. Madden, and F. S. Goulding, *Nucl. Instrum. Methods A266*:598 (1988).

36. R. G. Musket, *Nucl. Instrum. Methods 117*:385 (1974).

37. C. E. Cox, B. G. Low, and R. A. Sarren, Proceedings of the 1987 IEEE Nucl. Sci. Symposium, to be published in *IEEE Trans. Nucl. Sci.* (February 1988).

38. Product information available from Crytallume Corp., 125 Constitution Drive, Menlo Park, CA 94025.

39. Product information available from Kevex Corp., 355 Shoreway Road, P.O. Box 3008, San Carlos, CA 94070.

40. B. Skillicorn, *Adv. X-Ray Anal. 25*:49 (1982).

41. M. Green and V. E. Cosslett, *Proc. Phys. Soc. (Lond.) 78*:1206 (1961).

42. R. Tertian and N. Broll, *X-Ray Spectrom. 13*:135 (1984).

43. R. C. Placious, *J. Appl. Phys. 38*:2030 (1967).

44. J. M. Jaklevic, R. C. Gatti, F. S. Goulding, B. W. Loo, and A. C. Thompson, Environmental Protection Agency Technical Report No. EPA-60/4-78-034 (May 1980).

45. M. C. Nichols, D. R. Boehme, R. W. Ryon, D. Wherry, B. Cross, and G. Aden, *Adv. X-Ray Anal. 30*:45 (1987).

46. W. Michaelis, J. Knoth, A. Prange, and H. Schwenke, *Adv. X-Ray Anal. 28*:75 (1985).

47. H. Aiginger and P. Wobrauschek, *Adv. X-Ray Anal. 28*:1 (1985).

48. R. W. Ryon, J. D. Zahrt, P. Wobrauschek, and H. Aiginger, *Adv. X-Ray Anal. 25*:633 (1982).

49. P. J. Potts, P. C. Webb, and J. S. Watson, *X-Ray Spectrom. 13*:2 (1984).

50. R. D. Giauque, R. B. Garrett, and L. Y. Goda, *Anal. Chem. 51*:511 (1979).

4

Spectrum Evaluation

Pierre J.M. Van Espen and Koen H.A. Janssens
University of Antwerp, Antwerp, Belgium

I. INTRODUCTION

Spectrum evaluation is a crucial step in x-ray analysis, as much as sample preparation and quantification. As with any analytical procedure, the final performance of x-ray analysis is determined by the weakest step in the process. Spectrum evaluation in energy-dispersive (ED) analysis is certainly more critical than in wavelength-dispersive (WD) spectrometry because of the relatively low resolution of the solid-state detectors employed.

As a result, most of the research done in the field of spectrum evaluation was and still is situated in ED spectrometry. Although in WD spectrometry for a long time rate meters and/or strip-chart recorders have been employed, the processing of ED spectra by means of computers has always been more evident because of their inherent digital nature. For some of the techniques discussed here, the foundations have been laid in γ-ray spectrometry; for others (notably spectrum fitting procedures), ED x-ray analysis has developed its own specialized data-processing methodology. The availability of relatively cheap and fast microcomputers together with the implementation of mature spectrum evaluation packages on these machines has brought sophisticated spectrum evaluation within the reach of each x-ray spectrometry laboratory.

In this chapter various methods for spectrum evaluation are discussed, with emphasis on energy-dispersive x-ray spectra. Most of the methods are relevant for x-ray fluorescence, particle-induced x-ray emission, and analytical electron microscopy x-ray spectra. The principles of the methods and their practical use are discussed. Least-squares fitting, which is of importance not only for spectrum evaluation but also for quantification procedures, is discussed in detail in Section IX. Computer implementations of the main algorithms are presented in Section X.

II. FUNDAMENTAL ASPECTS

The aim of spectrum evaluation is to extract analytically relevant information from experimental spectra. This information can be quantitative in nature, involving, for example, the localization and identification of peaks, but is most often quantitative, in the form of the net number of counts under a peak. Obtaining this information is not straightforward, as the spectral data is contaminated with noise.

A. Amplitude and Energy Noise

In x-ray spectra, we can distinguish between amplitude and energy noise. Amplitude noise is due to the statistical nature of the counting process, in which random events (the arrival of x-ray photons at the detector) are observed during a finite time interval. For such a process the probability of observing N counts when the "true" number of counts is N_0 is given by the Poisson distribution [1]

$$P(N, N_0) = \frac{N_0^N}{N!} e^{-N_0} \tag{1}$$

An interesting property of a Poisson random variable is that the population standard deviation equals the square root of the true number of counts:

$$\sigma_{N_0} = \sqrt{N_0} \tag{2}$$

The sample standard deviation, which is an estimate of the true standard deviation, therefore can be calculated as the square root of the observed number of counts:

$$s_N = \sqrt{N} \approx \sigma_{N_0} \tag{3}$$

The Poisson statistics, also known as counting statistics, cause the typical channel-to-channel fluctuations observed in x-ray spectra. That the uncertainty of the data can be calculated from the data itself is of great importance for spectrum evaluation methods.

Energy noise, on the other hand, causes the characteristic x-ray lines to appear much wider than their natural line width of about 5–10 eV. Part of this line broadening is due to the nature of the photon-to-charge conversion process in the detector, and part of it is associated with the electronic noise in the pulse amplification and processing equipment, as discussed in Chapter 3. As a result, an x-ray with energy E, which on average corresponds to a pulse height stored in channel i, from time to time gives rise to slightly higher or lower pulses, causing the x-ray event to be stored in channels above and below i, respectively. Accordingly, characteristic x-ray lines appear as relatively broad (140–250 eV), nearly Gaussian peaks.

B. Information Content of a Spectrum

In the absence of these two noise contributions, spectrum evaluation would be trivial. A spectrum would consist of a well-defined background on which sharp characteristic lines were superimposed, and the intensity of the background and thus the net x-ray line intensities could be obtained without error. Any remaining peak overlap (e.g., between As $K\alpha$ and PB $L\alpha$, where the separation of 8 eV is less than the natural line width of As $K\alpha$) could be dealt with in an exact manner.

Unfortunately, we cannot eliminate the noise in the measurements completely. It is possible, however, to reduce the noise in various ways. Amplitude noise (i.e., counting statistics) can be diminished by acquiring the spectrum for a longer period of time or by using a more intense primary beam. Energy noise can be diminished by using a detector and associated electronics of good quality and by eliminating all kinds of electronic interference. Although these suggestions may appear naive, straightforward, or not appropriate to the context of spectrum evaluation, it is important to realize that once a spectrum has been acquired, its information content is constant. Indeed, no spectrum-processing procedure, no matter how sophisticated, can produce more information than that implicitly present. It is therefore much more efficient to employ optimal experimental conditions when acquiring the data rather than to rely on mathematical techniques in an attempt to obtain information that is not present in the first place [2].

From this point of view, spectrum processing can be seen as any (mathematical) procedure that transforms the information content of a measured spectrum into a form that is more useful for our purposes (i.e., more accessible). As is indicated in Figure 1, most of the procedures that calculate this "useful" information require some form of additional input. Sometimes this extra information is intuitive and not clearly defined; in other cases additional information is used in the form of a mathematical model of the spectral data. In this respect, not the complexity but rather the ability of the model to accurately describe the physical reality is important. When we use a procedure to estimate the net peak area of a characteristic line by summing the appropriate channels and interpolating the

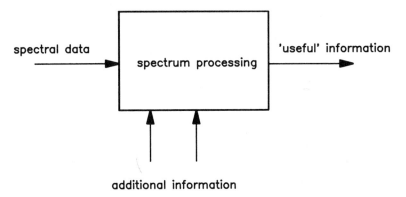

Figure 1   Spectrum evaluation seen as an information process: spectral data require additional information to produce useful information.

measured background left and right of the peak, explicitly some "additional information" is given to this "spectrum-processing" method in the form of peak and background boundaries. At the same time (implicitly), a certain mathematical model is assumed regarding the shape of the peak and the background. Provided this model is correct, even this simple procedure returns a correct estimate of the net peak area, comparable to that of a complex peak-fitting procedure.

The important distinction between simple and more sophisticated spectrum evaluation methods lies in the flexibility that is incorporated. Whereas the preceding simple area determination procedure is not generally valid in real-world situations (in which peak overlap, curved backgrounds, and tailing phenomena occur, for example), procedures employing more complex models are adaptable to each specific situation, yielding reliable peak estimates. Spectrum evaluation procedures should therefore be compared on the basis of the explicit and implicit assumptions that are made in the model(s) they employ.

C. Components of an X-ray Spectrum

To evaluate an x-ray spectrum correctly it is necessary to understand all the phenomena that contribute to the final appearance of the spectrum. This includes not only the two main features, characteristic lines and background, but also a number of spectral artifacts, which become especially important for trace analysis work [3].

1. Characteristic Lines

The characteristic radiation of a particular x-ray line has a Lorentz distribution. Peak profiles observed with a semiconductor detector are the convolution of this Lorentz distribution with the nearly Gaussian detector response function, giving rise to what is known as a Voight profile [4]. Since the Lorentz width is of the order of only 10 eV for elements with atomic number below 50, whereas the width of the detector response function is of the order of 160 eV, a Gaussian function is an adequate first-order approximation of the line profile. Only for K lines of such elements as U and Th does the Lorentz contribution become significant and must be taken into account [5].

A more close inspection of the peak shape reveals some distinct tailing a the low-energy side of the peak and a shelf extending to zero energy. This is mainly due to incomplete charge collection caused by detector imperfections (dead layer and regions of low electrical field), as discussed in Chapter 3. The effect is most pronounced for low-energy x-rays. For photons above 15 keV Compton scatter in the detector also contributes to deviation from the Gaussian shape. The distortion caused by incomplete charge collection has been described theoretically [6,7]. Various functions have been proposed to model the real peak shape more accurately [8].

The observed emission spectrum of an element is the result of many transitions, as explained in Chapter 1. The resulting x-ray lines, including possible satellite lines, must be considered in the analysis of an x-ray spectrum. A detailed discussion of the representation of the K and L spectra and the peak shape is given in Section VII.

2. Background

Background in x-ray spectra results from a variety of processes. The background in electron-induced x-ray spectra is almost completely due to the retardation of the electrons (*bremsstrahlung*). The intensity distribution of the continuum radiation emitted by the sample is in first approximation given by Kramer's formula (Chap. 1). Absorption in the detector windows and in the sample causes this continuous decreasing function to fall off at low energies, giving rise to the typical background shape observed. This attenuation, especially by major elements in the sample, further causes absorption edges to appear. Background modeling for electron-induced x-ray spectra has been studied in detail by a number of authors [9–13].

For particle-induced x-ray emission a similar background is observed, also mainly due to secondary electron bremsstrahlung. Other nuclear processes contribute to this background, making physical description virtually impossible. Special absorbers placed between sample and detector further alter the shape of the background.

In x-ray fluorescence the background is mainly due to coherent and incoherent scattering of the excitation radiation by the sample. The shape can therefore be very complex and depends both on the initial shape of the excitation spectrum and on the sample composition. When white radiation is used for excitation, the background is mainly radiative and absorption edges can also be observed. With quasi-monoenergetic excitation (secondary target), the incomplete charge collection of the intense coherently and incoherently scattered peaks is responsible for most of the background (see Chap. 3). Based on this, attempts were made to describe empirically the background as a function of the intensity of these peaks [14,15]; no realistic physical model for the background is in use, however.

The incomplete charge collection of intense fluorescence lines in the spectrum complicates the background in all three types of excitation. The cumulative effect of the incomplete charge collection of all lines causes the apparent background at lower energies to be significantly higher than expected on the basis of the primary background processes.

3. Escape Peaks

Escape peaks result from the escape of Si K photons from the detector after photoelectric absorption of the impinging x-ray photon near the edge regions of the detector. Because of this process, the energy deposited in the detector by the incoming x-ray is diminished with the energy of the Si K photon. Typical examples of the interference of escape peaks are the interference of Ti $K\alpha$ (4.51 keV) by the Fe $K\alpha$ escape at 4.65 keV and the interference of Fe $K\alpha$ by the Cu $K\alpha$ escape.

The position of the escape peak is thus expected 1.742 keV (Si $K\alpha$) below the parent peak. Experimentally it is observed that the energy difference is slightly but significantly higher, 1.750 keV [3]. The width of the escape peak is smaller than the width of the parent peak and corresponds to the spectrometer resolution at the energy of the escape peak.

The escape fraction f is defined as the number of counts in the escape peak N_e divided by the number of detected counts (escape + parent). Assuming normal incidence to the detector and escape only from the front surface, the following formula can be derived for the escape fraction [16]:

$$f = \frac{N_e}{N_p + N_e} = \frac{1}{2}\omega_K\left(1 - \frac{1}{r}\right)\left[1 - \frac{\mu_K}{\mu_I}\ln\left(1 + \frac{\mu_I}{\mu_K}\right)\right]$$ (4)

where μ_I and μ_K are the mass attenuation coefficient of silicon for the impinging and the Si K radiation, respectively, ω_K is the K fluorescence yield of silicon, and r the K jump ratio of silicon. Using 0.047 for the fluorescence yield and 10.8 for the jump ratio, the calculated escape fraction is in very good agreement with the experimentally determined values for impinging photons up to 15 keV [3].

Knowing the energy, width, and intensity of the escape peak, corrections for its presence can be made in a straightfoward manner.

4. Pileup and Sum Peaks

With modern pulse-processing electronics pileup effects are suppressed to a large extent. Within a pulse pair resolution time of a few microseconds or less, only true sum peaks are observed. The sum peaks are located very close to their expected position, with a deviation below a few electronvolts, and they are slightly wider (5%) than normal peaks located at the same energy in the spectrum [3]. The count rate of a sum peak is given by

$$\dot{N}_{11} = \tau\dot{N}_1\dot{N}_1$$ (5)

and

$$\dot{N}_{12} = 2\tau\dot{N}_1\dot{N}_2$$ (6)

with $\dot{N}_{11}$ the count rate (counts/s) in a sum peak due to the coincidence of two pulses with the same energy, $\dot{N}_{12}$ the count rate of a sum peak resulting from two pulses with different energies, and τ the pulse pair resolution time. Sum peaks are expected when a few large peaks at lower energy dominate the spectrum. It is important to note that they are count rate dependent. Sum peaks are often encountered in proton induced x-ray emission (PIXE) spectra of biological and geological material. The high count rate of the K and Ca K lines produces sum peaks that are easily observed in the high-energy region of the spectrum, where the background becomes low. A method to correct for the contribution of sum peaks in least-squares analysis has been proposed by Johansson [17] and is discussed further in Section VII.

5. Other Artifacts

A number of other features might appear in an x-ray spectrum and cause problems during spectrum evaluation.

In the K x-ray spectra of elements with atomic number between 20 and 40, one can detect a peaklike structure with a rather poorly defined maximum and a slowly declining tail [18]. This structure is due to the K-LL radiative Auger transition, which is an alternative decay mode of the K vacancy. The maximum is observed at the energy of the K-LL Auger electron transition energy. The intensity of the radiative Auger band varies from approximately 1% of the $K\alpha$ line for elements below Ca to 0.1% for elements above Zn. For chlorine and lower atomic number elements the radiative Auger band overlaps with the $K\alpha$ peak. In most analytical applications this effect does not cause serious problems. The structure can be considered part of the non-Gaussian peak tail.

The scattering of the excitation radiation in x-ray fluorescence is responsible for most of the background observed in the spectrum. When characteristic lines are present in the excitation spectrum, two peaks can be observed. The Rayleigh (coherently) scattered peak has a position and width as expected for a normal fluorescence line. The Compton (incoherently) scattered peak is shifted to lower energies according to the well-known Comptom formula and is often much broader than the normal characteristic line at the energy. This broader structure, resulting from scattering over a wider range of angles, is difficult to model. The structure can be further complicated by multiple scattering.

Apart from these commonly encountered scattering processes, it is possible to detect x-ray Raman scattering [19]. Again a bandlike structure is obtained with an energy maximum given by the incident photon energy minus the binding energy of the electron of the material analyzed. The Raman effect is expected in the analysis of elements having atomic number $Z - 2$ to $Z + 7$ when the K radiation of element Z is used for excitation. In this case Raman scattering occurs on L electrons. For x-ray excitation energies between 15 and 25 keV, the Raman scattering on the K electrons of Al through Cl can also be observed. Because of its high-energy edge, the effect may appear as a peak in the spectrum, with possible erroneous identification as a fluorescence line. The intensity of the Raman band increases as the incident photon energy comes closer to the binding energy of the electron. The observed intensity can amount to as much as 10% of the L fluorescence intensity for the elements Rh through Cs when excitation with Mo K x-rays is used.

III. SPECTRUM PROCESSING METHODS

Spectrum processing refers to a number of mathematical techniques that alter the general outlook of a spectrum. This implies the use of some filtering technique, for example to reduce the noise or the continuum background. Spectrum processing is most often implemented using digital filters. The effect of the filter, however, is best understood in the Fourier space. In this section various methods of filtering used in smoothing and peak search are discussed. Because of its relation to the frequency domain, the concept of Fourier transformation is introduced first.

A. Fourier Transformation, Convolution, and Deconvolution

One can think of a x-ray spectrum as consisting of a number of components with different frequencies. In the spectrum shown in Figure 2, one recognizes a nearly constant component (the background) as well as a component that fluctuates from channel to channel (fast). The latter is obviously due to counting statistics. The frequency characteristics of a spectrum can be studied in the Fourier space.

For any discrete function $f(x)$, $x = 0, \ldots, n - 1$, for example a pulse-height

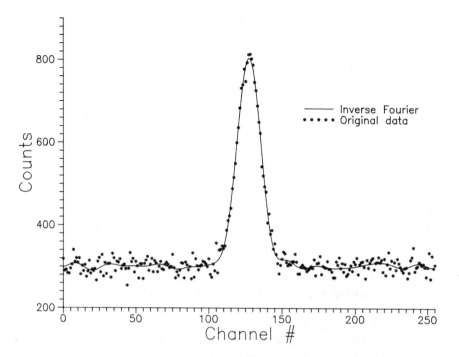

Figure 2 A 256-channel pulse-height spectrum (single Gaussian on a constant background) and the Fourier-filtered spectrum.

spectrum, the discrete Fourier transform is defined as

$$F(u) = \frac{1}{n} \sum_{x=0}^{n-1} f(x) \, e^{-j2\pi ux/n} \tag{7}$$

$$= \frac{1}{n} \sum_{x=0}^{n-1} f(x) \left(\cos 2\pi \frac{ux}{n} - j \sin 2\pi \frac{ux}{n} \right)$$

with $j = \sqrt{-1}$ and $u = 0, \ldots, n-1$. $F(u)$ is a complex number of which the real part $R(u)$ and the imaginary part $I(u)$ represent the amplitude of the cosine and the sine functions, respectively, that are necessary to describe the original data. The square of $F(u)$ is called the power spectrum:

$$|F(u)|^2 = R^2(u) + I^2(u) \tag{8}$$

and gives an idea which frequencies dominate in the original spectrum. Since there are n different nonzero real and imaginary coefficients, no information is lost by the Fourier transform and the inverse transformation is always possible:

$$f(x) = \sum_{u=0}^{n-1} F(u) \, e^{j2\pi ux/n} \tag{9}$$

For the pulse-height distribution given in Figure 2 (a single Gaussian on a constant background), Figure 3 shows the power spectrum. The frequency (inverse channel number) is defined as u/n, with $n = 256$ and $u = 0, \ldots, n/2$. The amplitude of the zero frequency $|F(0)|^2$, which is equal to the average of the spectrum, is not shown. The dominating low frequencies originate from the background and from the Gaussian peak, whereas the higher frequencies are caused mainly by the counting statistics. It is clear that if we eliminate those high frequencies we reduce this noise. This can be done by multiplying the Fourier transform with a certain function:

$$G(u) = F(u) \cdot H(u) \tag{10}$$

An example of such a function is a high-frequency cutoff filter:

$$H(u) = \begin{cases} 1 & u \le u_{\text{crit}} \\ 0 & u > u_{\text{crit}} \end{cases} \tag{11}$$

If we apply this filter to the Fourier transform of Figure 3 using $u_{\text{crit}} = 0.05$, thus effectively setting all real and imaginary coefficients above a frequency of 0.05–0 and we do the inverse Fourier transformation [Eq. (9)], the result as shown by the solid line in Figure 2 is obtained. The peak shape is preserved, but most of the statistical fluctuations are eliminated. If we cut off at even lower frequencies, peak distortion at the top and the base of the peak would become more pronounced.

This Fourier filtering can also be done directly in the original data space. Indeed, the convolution theorem says that multiplication in the Fourier space is equivalent to convolution in the original space:

$$G(u) = F(u) \cdot H(u) \Leftrightarrow f(x) * h(x) = g(x) \tag{12}$$

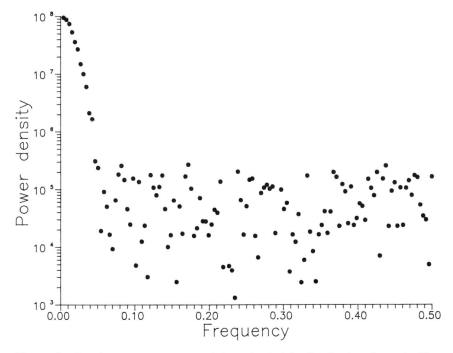

Figure 3 Fourier power spectrum of the pulse-height distribution shown in Figure 2.

The convolute at data point x is defined as the sum of the products of the original data and the filter centered around point x:

$$g(x) = f(x)*h(x) = \sum_{x'} f(x - x')h(x') \tag{13}$$

$h(x)$ is called a digital filter and is the inverse Fourier transform of $H(u)$. In general, the convolution or filtering of a spectrum y_i with some weighing function is expressed as

$$y_i^* = \frac{1}{N} \sum_{j=-m}^{j=m} h_j y_{i+j} \tag{14}$$

where h_j are the convolution integers and N a suitable normalization factor. The filter width is given by $2m + 1$. Fourier filtering with the purpose of reducing or eliminating some (high or low) frequency components in the spectrum can thus be implemented as a convolution of the original data with a digital filter. This convolution also alters the variance of the original data. Applying the concept of error propagation can show that the variance of the convoluted data is given by

$$s_{y_i^*}^2 = \frac{1}{N^2} \sum_{j=-m}^{m} h_i^2 y_{i+j} \tag{15}$$

when the original data follow a Poisson distribution ($s_y^2 = y$).

Since the measured spectrum $y(x)$ is itself a convolution of the original (x-ray emission) signal $f(x)$, with the instrument (or detector) response function $h(x)$, it is in principle possible to restore the measured signal if this response function is known. This can be accomplished by dividing the Fourier transform (FT) of the measured spectrum by the Fourier transform of the (nearly Gaussian) response function, followed by the inverse Fourier transform (IFT) of the resulting quotient:

$$\left. \begin{array}{l} y(x) \xrightarrow{\text{FT}} Y(u) \\ h(x) \xrightarrow{\text{FT}} H(u) \end{array} \right\} \frac{Y(u)}{H(u)} = F(u) \xrightarrow{\text{IFT}} f(x) \tag{16}$$

The detector response function changes with energy (becomes broader) and, more importantly, the presence of noise prohibits the straightforward use of this Fourier deconvolution technique. Indeed in the presence of noise the measured signal must be presented by

$$y(x) = f(x)*h(x) + n(x) \tag{17}$$

and its Fourier transform,

$$Y(u) = F(u) \cdot H(u) + N(u) \tag{18}$$

or

$$\frac{Y(u)}{H(u)} = F(u) + \frac{N(u)}{H(u)} \tag{19}$$

At high frequencies the response $H(u)$ goes to zero while $N(u)$ is still significant, so the noise is emphasized in the inverse transformation. This once more clearly shows that the noise (counting statistics) is the ultimate limitation in any spectrum-processing and analysis method.

A clear discussion of Fourier transformations related to signal processing can be found in Massart et al. [20]. Algorithms for Fourier transformation and related topics are given in Press et al. [21]. Detailed discussion of Fourier deconvolution can be found in many textbooks [22,23]. Fourier deconvolution in x-ray spectrometry based on maximum a posteriori or maximum entropy principles is discussed by several authors [24–26]. Gertner [26] implemented this method for the analysis of real x-ray spectra and compared the results with those obtained by simple peak fitting. The major problem—that the deconvolution algorithms are limited to systems exhibiting translational invariance—was overcome by a transformation of the spectrum so that the resolution becomes independent of the energy.

B. Smoothing

Because of the uncertainty $\sqrt{y}$ on each channel content y_i, fictitious maxima can occur both on the background and on the slope of the characteristic peaks. Removal or suppression of these fluctuations is often of considerable aid during the visual inspection of spectra (e.g., for locating small peaks on a noisy background) and is also used in most automatic peak search and background estimation procedures. Although smoothing can be useful in qualitative analysis, its use is not recommended before quantitative spectrum evaluation. Smoothing, although attempting to reduce the local uncertainty in the channel content, redistributes the original channel content over the neighboring channel, thus introducing a distortion in the spectrum. Accordingly, smoothing can provide a (small) improvement in the statistical precision obtainable with simple peak integration but is of no advantage when used with least-squares fitting procedures in which assumptions are made about the peak shapes.

1. Moving Average

The most straightforward way of smoothing (any) fluctuating signal is to employ the "boxcar" or moving average technique. Starting from a measured spectrum y, a smoothed spectrum y^* can be obtained by calculating the mean channel content around channel i:

$$y_i^* = \overline{y}_i = \frac{1}{2m + 1} \sum_{j=-m}^{+m} y_{i+j} \tag{20}$$

This can be seen as a convolution [Eq. (14)] with all coefficients $h_j = 1$. The smoothing effect obviously depends on the width of the filter $2m + 1$. The operation being a simple averaging, the standard deviation of the smoothed data is simply reduced by a factor $\sqrt{2m + 1}$ in regions where y_i is nearly constant. On the other hand, such a filter introduces a considerable amount of peak distortion. This distortion depends on the ratio of the width of the filter to the width of the peak. Figure 4 shows the peak distortion when a moving average filter of width 9, 17, and 25 is applied to a peak with full width at half maximum (FWHM) equal to 9 channels. Being a unit area filter, $\sum h_j/N = 1$ with $N = 2m + 1$, the total counts in the peak is not affected in an appreciable way other than by rounding

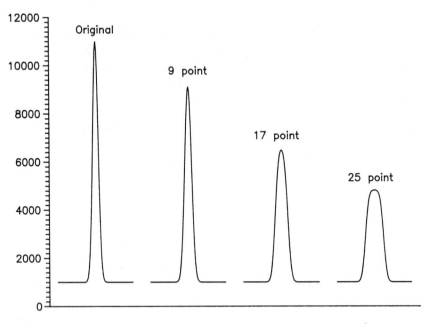

Figure 4 Effect of smoothing a peak with a moving average filter. The FWHM of the original peak is 9 channels.

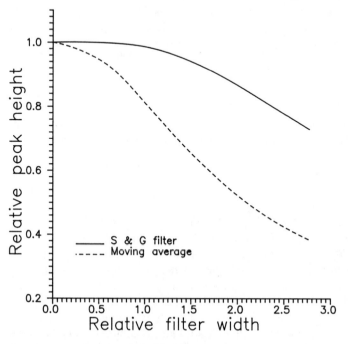

Figure 5 Reduction in height of a smoothed peak as a function of the ratio of the width of the filter to the width of the original peak.

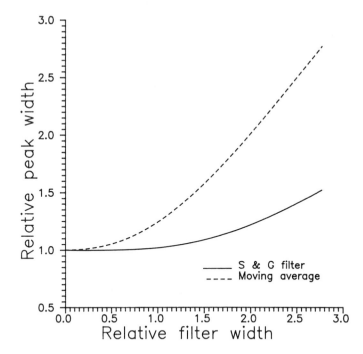

Figure 6 Increase in the width of a smoothed peak as a function of the ratio of the width of the filter to the width of the original peak.

errors. Figures 5 and 6 show the peak-height reduction and width increase caused by this filter as a function of the filter width to peak width ratio, $(2m + 1)/\text{FWHM}$.

The peak distortion effect is caused by the fact that in the calculation of y_i^*, the content of all neighboring channels is used with equal weight. Consequently, by employing a nonuniform filter h, which places more weight on the central channels and less on the channels near the edge of the filter, smoothing can be achieved with less broadening effects.

2. Savitsky and Golay (Polynomial) Filters

Another way of dealing with statistical fluctuations in experimental data is by drawing a best-fitting curve through the data points. This idea resulted in the development by Savitsky and Golay [27] of a general type of smoothing filter with very interesting features. The method is based on the fact that nearly all experimental data can be modeled by a polynomial of some order r, that is, $a_0 + a_1 x + a_2 x^2 + \cdots + a_r x^r$, when the data is confined to sufficiently small interval Δx. If we consider a number of data points around a central channel i_0, such as y_{i_0-2}, y_{i_0-1}, y_{i_0}, y_{i_0+1}, and y_{i_0+2}, a least-squares fit with the function

$$y(i) = a_0 + a_1(i - i_0) + a_2(i - i_0)^2 \tag{21}$$

can be made. Once we have determined the coefficients a_j, the value of the polynomial at the central channel i_0 can then be used as the smoothed value:

$$y_i^* = y(i_0) = a_0 \tag{22}$$

as is schematically illustrated in Figure 7. By moving the central channel to the right (from i_0 to i_{0+1}), the next smoothed channel content can be calculated by repeating the entire procedure.

Although at first sight this smoothing method would require a least-squares fit around each channel in the spectrum, implying a considerable computational effort, a careful consideration of the mathematics involved, especially that the x values are equidistant, reveals that the fitting coefficients can be expressed as fixed linear combinations of only the y_i values:

$$a_k = \frac{1}{N_k} \sum_{j=-m}^{j=m} C_{k,j} y_{i+j} \tag{23}$$

This means that it is possible to implement the least-squares fitting procedure more efficiently as a convolution of the spectrum with a filter having appropriate weights. For this second-order polynomial, one obtains the coefficients

$$\frac{C_{0,j}}{N_0} = \frac{3(3m^2 + 3m - 1 - 5j^2)}{(2m - 1)(2m + 1)(2m + 3)} \tag{24}$$

so that, for example, for second-degree polynomial smooth with 5 points ($2m + 1 = 5$), one gets

$$y_i^* = a_0 = \frac{1}{35}(-3y_{i-2} + 12y_{i-1} + 17y_i + 12y_{i+1} - 3y_{i+2}) \tag{25}$$

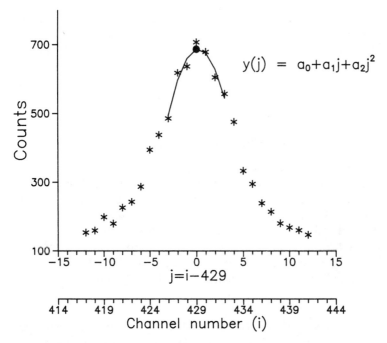

Figure 7 Concept of polynomial smoothing. A parabola is fitted through the points $i_0 - 3$ to $i_0 + 3$. The value of the parabola at i_0 is the smoothed value (large dot).

More in general for the case of a polynomial of degree r fitted to $2m + 1$ points, this can be written as

$$y_i^* = \frac{1}{N_{r,m}} \sum_{j=-m}^{j=m} C_{r,m,j} y_{i+j} \tag{26}$$

where the convolution integers $C_{r,m,j}$ and their normalization factors $N_{r,m}$ do not depend on the data to be smoothed but are a function only of the polynomial degree r and the filter half-width m. Table 1 lists the coefficients for polynomial smoothing filters with width between 5 and 25 points. The coefficients for a second- and a third-order polynomial are identical. In comparison with the moving average filter of the same width, polynomial smoothing filters are less effective in removing noise but have the advantage of causing less peak distortion. The distortion effect as a function of the ratio of filter width to peak width is given in Figures 5 and 6. When the filter width becomes larger than the peak, however, the smoothed spectrum features oscillations near the peak boundaries, as illustrated in Figure 8.

Another interesting feature of this type of filters is that it can produce not only a smoothed spectrum but also a smoothed first and second derivative of the spectrum. If we differentiate Eq. (21) and take the value at the center position,

$$y^{*\prime} = \left. \frac{dy(i)}{di} \right|_{i=i_0} = a_1 \tag{27}$$

$$y^{*\prime\prime} = \left. \frac{d^2y(i)}{di^2} \right|_{i=i_0} = 2a_2 \tag{28}$$

it follows from Equation (23) that the first and second derivatives of the smoothed spectrum can also be calculated by means of suitable convolution coefficients.

Table 1 Savitsky and Golay Coefficients for Second and Third-degree ($r = 2$ and $r = 3$) Polynomial Smoothing[a]

| | | j ($C_{r,m,j} = C_{r,m,-j}$) | | | | | | | | | | | |
m	$N_{r,m}$	0	1	2	3	4	5	6	7	8	9	10	11	12
2	35	17	12	−3										
3	21	7	6	3	−2									
4	231	59	54	39	14	−21								
5	429	89	84	69	44	9	−36							
6	143	25	24	21	16	9	0	−11						
7	1105	167	162	147	122	87	42	−13	−78					
8	323	43	42	39	34	27	18	7	−6	−21				
9	2261	269	264	249	224	189	144	89	24	−51	−136			
10	3059	329	324	309	284	249	204	149	84	9	−76	−171		
11	805	79	78	75	70	63	54	43	30	15	−2	−21	−42	
12	5175	467	462	447	422	387	342	287	222	147	62	−33	−138	−253

[a] $y_i^* = \dfrac{1}{N_{r,m}} \sum_{j=-m}^{j=m} C_{r,m,j} y_{i+j}$ and filter width $= 2m + 1$.

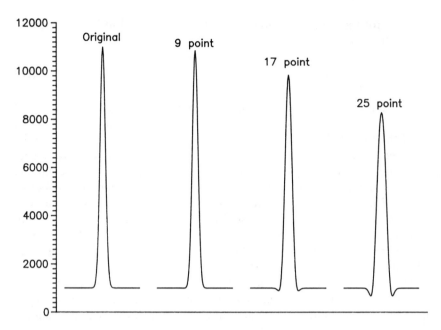

Figure 8 Effect of the smoothing of a peak with a Savitsky and Golay polynomial filter. The FWHM of the original peak is 9 channels.

For instance, for the first derivative of a second-order polynomial, using 5 data points this becomes

$$y_i^{*\prime} = \frac{1}{10}(-2y_{i-2} - y_{i-1} + y_{i+1} + 2y_{i+2}) \qquad (29)$$

The corresponding convolution integers for the calculation of the first and second smoothed derivatives are listed in Tables 2 through 4. The use of the derivative spectra is illustrated in next section dealing with peak search methods.

Variations on these smoothing strategies, such as the use of variable-width filters, are reviewed by Yule [28]. The effect of repeated smoothing on the accuracy and precision of peak area determination is discussed by Nielson [29]. A more comprehensive treatment of polynomial smoothing can be found in Enke and Nieman [30] and its references.

3. Low Statistics Digital Filter

Most smoothing techniques originate from signal processing and were initially introduced in the field of γ-ray spectroscopy and consequently have been in existence for about 10–20 years, but more recently in the PIXE community, considerable attention has been devoted to a number of aspects of spectrum processing and evaluation. Within the framework of background estimation (see later), a smoothing algorithm was developed that removes noise from a spectrum on a selective basis [31]. The method provides a n-point mean smoothing in regions of low statistics (low counts) while avoiding spreading of the base of peaks and degradation of minima between peaks.

Table 2 Savitsky and Golay Coefficients for Second-degree ($r = 2$) Polynomial, First Derivative[a]

| m | $N_{r,m}$ | \multicolumn{13}{c}{$j\ (C_{r,m,j} = C_{r,m,-j})$} |
		0	1	2	3	4	5	6	7	8	9	10	11	12
2	10	0	1	2										
3	28	0	1	2	3									
4	60	0	1	2	3	4								
5	110	0	1	2	3	4	5							
6	182	0	1	2	3	4	5	6						
7	280	0	1	2	3	4	5	6	7					
8	408	0	1	2	3	4	5	6	7	8				
9	570	0	1	2	3	4	5	6	7	8	9			
10	770	0	1	2	3	4	5	6	7	8	9	10		
11	1012	0	1	2	3	4	5	6	7	8	9	10	11	
12	1300	0	1	2	3	4	5	6	7	8	9	10	11	12

[a] $y_i^{*'} = \dfrac{1}{N_{r,m}} \sum\limits_{j=-m}^{j=m} C_{r,m,j} y_{i+j}$ and filter width $= 2m + 1$.

For each channel i in the spectrum, two windows, one on each side of the channel of width $f \times \text{FWHM}(E_i)$ channels are considered. In both windows, the channel contents are summed, yielding a left sum L and a right sum R. Both windows are subsequently reduced in width until either the total sum $S = L + y_i + R$ falls below some constant minimum M or until two conditions are met:

1. S is less than a cutoff value $N = A\sqrt{y_i}$, with A a constant.
2. The slope $(R + 1)/(L + 1)$ lies between $1/r$ and r, with r a constant.

The minimum constant M sets the base degree of smoothing in a region of vanishing counts. The first condition ensures that smoothing is confined to the low-statistics region of the spectrum; the second condition avoids the incorporation of the edges of the peaks in the averaging.

When the preceding conditions are satisfied, the average $S/(2f \times \text{FWHM} + 1)$ is adopted as a smoothed channel count. The following parameters are found to yield good results when treating PIXE spectra: $f = 1.5$, $A = 75$, $M = 10$, and $r = 1.3$. The method is illustrated in Figure 9, where it is compared with the other smoothing methods discussed here.

A FORTRAN implementation of the Savitsky and Golay polynomial smoothing and the low-statistics digital filter is given in Section X.

C. Peak Search Methods

Several methods have been developed for the automatic localization of peaks in a spectrum. Nearly all methods follow a strategy in which the original spectrum is transformed into a form that emphasizes the peaklike structures and reduces the background, followed by a decision about whether this peaklike structures

Table 3 Savitsky and Golay Coefficients for Third-degree ($r = 3$) Polynomial, First Derivative[a]

m	$N_{r,m}$	\(j\ (C_{r,m,j} = C_{r,m,-j})\)												
		0	1	2	3	4	5	6	7	8	9	10	11	12
2	12	0	8	−1										
3	252	0	58	67	−22									
4	1,188	0	126	193	142	−86								
5	5,148	0	296	503	532	294	−300							
6	24,024	0	832	1,489	1,796	1,578	660	−1,133						
7	334,152	0	7,506	13,843	17,842	18,334	14,150	4,121	−12,922					
8	23,256	0	358	673	902	1,002	930	643	98	−748				
9	255,816	0	2,816	5,363	7,372	8,574	8,700	7,481	4,648	−68	−6,936			
10	3,634,092	0	29,592	56,881	79,564	95,338	101,900	96,947	78,176	43,284	−10,032	−84,075		
11	197,340	0	1,222	2,365	3,350	4,098	4,530	4,567	4,130	3,140	1,518	−815	−3,938	
12	1,776,060	0	8,558	16,649	23,806	29,562	33,450	35,003	33,754	29,236	20,982	8,525	−8,602	−30,866

[a] $y_i^{*'} = \dfrac{1}{N_{r,m}} \displaystyle\sum_{j=-m}^{j=m} C_{r,m,j} y_{i+j}$ and filter width $= 2m + 1$.

Table 4 Savitsky and Golay Coefficients for Second and Third-degree ($r = 2$ and $r = 3$) Polynomial, Second Derivative[a]

m	$N_{r,m}$	\multicolumn{13}{c}{$j\ (C_{r,m,j} = C_{r,m,-j})$}
		0	1	2	3	4	5	6	7	8	9	10	11	12
2	7	−2	−1	2										
3	42	−4	−3	0	5									
4	462	−20	−17	−8	7	28								
5	429	−10	−9	−6	−1	6	15							
6	1,001	−14	−13	−10	−5	2	11	22						
7	6,188	−56	−53	−44	−29	−8	19	52	91					
8	3,876	−24	−23	−20	−15	−8	1	12	25	40				
9	6,783	−30	−29	−26	−21	−14	−5	6	19	34	51			
10	33,649	−110	−107	−98	−83	−62	−35	−2	37	82	133	190		
11	17,710	−44	−43	−40	−35	−28	−19	−8	5	20	37	56	77	
12	26,910	−52	−51	−48	−43	−36	−27	−16	−3	12	29	48	69	92

[a] $y_i^{*''} = \dfrac{1}{N_{r,m}} \displaystyle\sum_{j=-m}^{j=m} C_{r,m,j} y_{i+j}$ and filter width $= 2m + 1$.

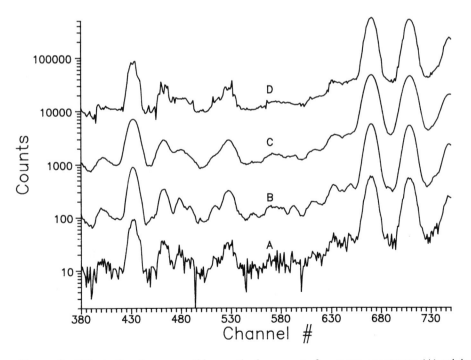

Figure 9 Effect of various smoothing methods on part of an x-ray spectrum: (A) original spectrum, (B) 9 point Savitsky and Golay filter, (C) 9 point moving average, (D) low-noise statistical filter.

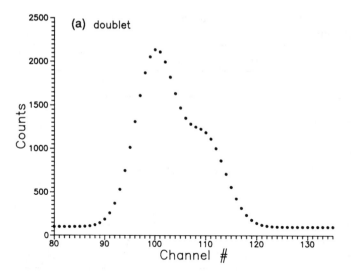

Figure 10 Peak search using first and second derivatives: doublet of two peaks, $\sigma = 4$ channels, separation 8 channels, on a constant background (a), first (b), and second (c) smoothed derivative using a 5 point Savitsky and Golay filter.

are statistically significant. The latter involves some adjustable parameter(s) ultimately controlling the sensitivity of the peak search.

Although visual inspection of the spectrum still appears to be the best method, peak search methods, which are heavily used in γ-ray spectrometry, may have some value in x-ray analysis. Their use in energy-dispersive x-ray analysis as part of an automated qualitative analysis procedure is hampered by the extreme peak overlap in these spectra, however. More elaborate procedures involving artificial intelligence techniques are more appropriate for this [32,33].

Peak search procedures usually involve three steps: (1) transformation of the original spectrum so that background contributions are eliminated, peaks are readily locatable, and overlapping peaks are (partially) resolved; (2) significance test and approximate location of the peak maximum; and (3) more accurate peak position estimate in the original spectrum.

The various peak search algorithms mainly differ in the choice of the transformation. Some methods use the first and second smoothed derivative of the spectrum. This method is illustrated in Figure 10. The sign change (crossing of the x axis) of the first derivative and, even more, the minimum of the second derivative are quite suitable to detect the peaks in the original spectrum.

Other methods employ some form of correlator technique, which is basically the convolution of the original spectrum with a filter that approximates the shape of the peak and therefore emphasizes the peak. If a zero area correlator (filter) is used, the background is at the same time effectively suppressed. The simplest and most effective correlators belong to the group of zero area rectangular filters. These filters have a central window with constant and positive coefficients and two side lobes with constant and negative coefficients. Convolution of an x-ray spectrum with this kind of filter yields spectra in which the background is removed

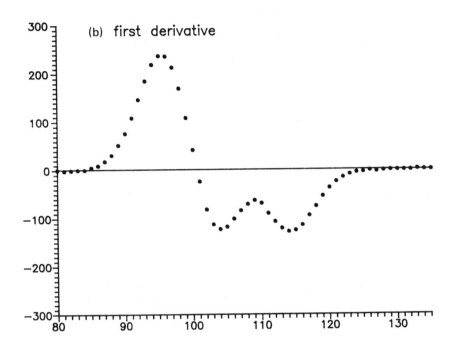

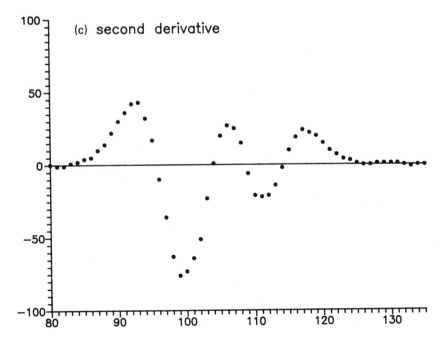

Figure 10 *Continued*

and peaks are easily locatable. They are similar to inverted second derivative spectra. Compared to the latter, however, the smoothing effect of these filters is somewhat bigger. An important representative of this group of filters is the "top hat" filter, which has a central window with an odd number of channels w and two side windows each v channels wide. The value of the filter coefficients follows from the zero area constraint:

$$
h_k = \begin{cases}
-\dfrac{1}{2v} & -v - \dfrac{w}{2} \le k < -\dfrac{w}{2} \\[2ex]
1/w & -\dfrac{w}{2} \le k \le +\dfrac{w}{2} \\[2ex]
-\dfrac{1}{2v} & +\dfrac{w}{2} < k \le \dfrac{w}{2} + v
\end{cases}
\tag{30}
$$

The filtered spectrum is obtained by the convolution of this filter with the original spectrum:

$$
y_i^* = \sum_{k=-v-w/2}^{k=+v+w/2} h_k y_{i+k}
\tag{31}
$$

The effect of this filter on a typical spectrum is shown in Figure 11. The variance of the filtered spectrum is obtained by simple error propagation:

$$
s_{y_i^*}^2 + \sum_{k=-v-w/2}^{k=+v+w/2} h_k^2 y_{i+k}
\tag{32}
$$

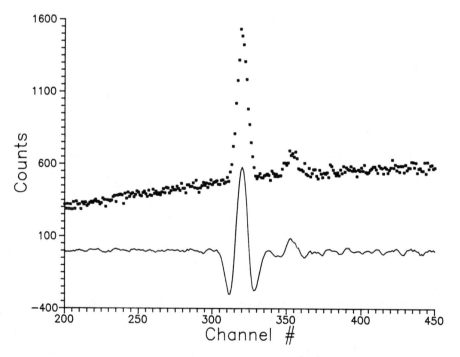

Figure 11   Result of the application of a top hat filter: (dots) typical x-ray spectrum; (solid line) filtered spectrum.

If y_i^* is significantly different from zero, a peak structure is found and the top of the peak can approximately be located by searching for the maximum. Thus i is a peak maximum in the original spectrum if

$$y_i^* > rs_{y_i} \tag{33}$$

and

$$y_{i-1}^* \leq y_i^* > y_{i+1}^* \tag{34}$$

In Figure 12 the positive part of the filtered spectrum ($w = 9$ and $v = 5$) and the associated decision line (rs_y^*) for $r = 1$ and 4 are displayed.

If required, other peak features can be obtained from the filtered spectrum: the distance between the two local minima is a measure of the width of the peak, and the height at the maximum is related to the net peak area.

Since the width and heights of the peaks in the filtered spectrum strongly depend on the dimensions of the filter, it is important that its dimensions are matched to the peak widths in the original spectrum. From considerations of peak detectability (signal-to-noise ratio) and resolution (peak broadening), it follows that the optimum width of the positive window w is equal to the FWHM of the peak(s) to be filtered [34]. The width of the negative side windows should be chosen as large as the curvature of the background allows. A reasonable compromise between sensitivity to peak shapes and rejection of continuum is reached when v equals FWHM/2–FWHM/3. Typical values for the sensitivity factor r are

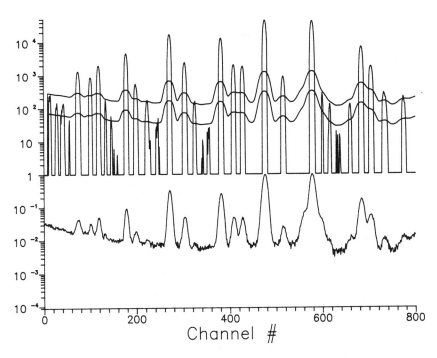

Figure 12 Peak search using the positive part of the top hat filtered spectrum and the associated decision level for one and four times the standard deviation. The original spectrum is shown at the bottom.

between 2 and 4. Higher values result in the loss of small peaks; lower values cause background noise to be interpreted as peaks.

Other zero area rectangular filters, variations on the top hat filter, are also in use, such as the square-wave filter with typical coefficient sequence -1, -1, $2, 2, -1, -1$ [35,36] and the symmetrical square-wave filter with coefficients -1, $1, 1, -1$ [37]. A detailed account of the performance of this filter is given in Reference 37. A method using a Gaussian correlator function is discussed by Black [38].

Once the peak top is approximately located in the filtered spectrum, a more precise maximum can be found by fitting a parabola over a few channels around the peak. For a well-defined peak on a low background (or after background subtraction), the channel content near the top of the peak can be approximated by a Gaussian:

$$y_i \approx h e^{[-(x-\mu)^2/2\sigma^2]} \tag{35}$$

The logarithm of the data is then a simple polynomial,

$$\ln y_i = \left(\ln h - \frac{\mu^2}{2\sigma^2} \right) + \frac{\mu}{\sigma^2} x - \frac{\mu^2}{2\sigma^2} x^2 \tag{36}$$

If we fit $\ln y_i$ with a polynomial $a_0 + a_1 x + a_2 x^2$, where x represents the channel number, the position of the peak μ is obtained from

$$\mu = -\frac{a_1}{2a_2} \tag{37}$$

with an accuracy of 0.1 channel or better if the peak is not interfered with background or other peaks [that is, if the model of Eq. (35) accurately describes reality]. An estimate of the peak's width and height is obtained at the same time:

$$\text{FWHM} = 2\sqrt{2 \ln 2}\, \sigma = 2.3548 \sqrt{-\frac{1}{2a_2}} \tag{38}$$

$$h = e^{a_0 + a_1^2/4a_2} \tag{39}$$

To obtain a reliable estimate of the parameters, it is recommended that only those channels in the FWHM or at most within the FWTM region of the peak be included in the fit.

As a somewhat simpler and faster alternative, one can find an estimate of the peak maximum by fitting the parabola over the three top channels of the peak. If i is the peak maximum found in the filtered spectrum, a better estimate of the maximum in the original spectrum is found by

$$\mu = i + \frac{1}{2} \frac{y_{i-1} - y_{i+1}}{y_{i-1} + y_{i+1} - 2y_i} \tag{40}$$

This method might be preferred for small peaks when the background cannot be disregarded.

A FORTRAN implementation of a peak location algorithm is given in Section X.

IV. BACKGROUND ESTIMATION METHODS

Except for some particular analysis procedures (e.g., the peak-to-background method in electron microscopy), the relevant analytical information is found in the net peak areas and background is considered a nuisance. There are in principle three ways to deal with the background: (1) the background can be suppressed or eliminated by a suitable filter; (2) the background can be estimated simultaneously with the other features in the spectrum; and (3) the background can be estimated and subtracted from the spectrum before further evaluation of the peaks. Approach 1 is discussed in Section VI, where the background is removed from spectra by applying a top hat filter followed by linear least-squares fit of the spectrum with a number of (also filtered) reference spectra. Least-squares fit (linear or nonlinear) with analytical functions (Sec. VII) allows the simultaneous estimation of background and peaks, provided a suitable mathematical function can be found for the background. In this section we discuss a number of procedures that aim to estimate the background independently of the rest of the information present in the spectrum. Once estimated, this background can be subtracted from the original spectrum and all methods for further processing, ranging from simple peak integration to least-squares fitting (but without a model for the background), can be applied.

Any background estimation procedure must fulfill two important requirements. First, the method must be able to reliably estimate the background in all kinds of situations, for example small isolated peaks on a high background as well as in the proximity of a matrix line. Second, to permit processing of a large number of spectra, the method must be nearly free of user-adjustable parameters: that is, it must be sufficiently robust to handle a wide variety of background shapes without modification.

Although recently a number of useful background estimation procedures have been developed, it must be realized that their accuracy in estimating the background is not optimal. In one way or another they rely on the difference in frequency response of the background compared to other structures, such as peaks, the former mainly consisting of low frequencies (slowly varying). Since the peaks also exhibit low frequencies at the peak boundaries, it is difficult to control the method in such a way that it correctly discriminates between peaks and background. This results in either a small under- or overestimation of the background, introducing potentially large relative errors for small peaks. In this respect, fitting the background with analytical functions may provide more optimal results. A considerable advantage of the methods discussed here is that they do not assume a mathematical model of the background. Constructing a detailed and accurate physical model for the background is nearly impossible except for some simple geometries and for particular excitation conditions, so that most often some polynomial type of function must be chosen when fitting a portion of the spectrum with analytical functions.

A. Peak Stripping

These methods are based on the removal of rapidly varying structures in a spectrum by comparing the channel content y_i with the channel content of its neighbors.

Clayton et al. [39] proposed a method that compares the content of channel

i with the mean of its two direct neighbors:

$$m_i = \frac{y_{i-1} + y_{i+1}}{2} \tag{41}$$

If m_i is smaller than y_i, the content of channel i is replaced by the mean m_i. If this transformation is executed once for all channels, one can observe a slight reduction in the peak height and the rest of the spectrum virtually remains the same. Repetition of the entire procedure gradually causes the peak to be "stripped" away. Since the method tends to connect local minima, it is very sensitive to local fluctuations in the background due to counting statistics. This makes smoothing of the spectrum before the stripping process, as discussed in the previous section, mandatory. Depending on the width of the peaks after typically 1000 cycles, the stripping converges and a more or less smooth background remains. To reduce the number of iterations it might be advantageous to perform a logarithmic or square-root transformation to the data before the stripping; $y_i' = \log(y_i + 1)$ or $y_i' = \sqrt{y_i}$. After stripping, the background shape is obtained by applying the reverse transformation. A major disadvantage of this method is that after a number of cycles the bases of partially overlapping peaks are transformed into broad "humps," which take much longer to be removed than isolated peaks. The method was originally applied to PIXE spectra but proves to be generally applicable for pulse-height spectra.

In Figure 13 this method is applied to estimate the background of an x-ray spectrum in the region between 1.6 and 13.0 keV. The spectrum results from a 200 mg/cm^2 pellet of National Institute of Science and Technology standard reference material (NIST SRM) bovine liver sample excited with the white spectrum of an Rh anode x-ray tube filtered through a thin Rh filter (Tracor Spectrace 5000). As a result of the white tube spectrum, a considerable background intensity is observed. Also, the background increases quit steeply in the region above 10 keV. To obtain the background, the following algorithm was used: (1) the square root of the original spectrum was taken, (2) these data were smoothed with a 10 point Savitsky and Golay filter, (3) a number of iterations were performed applying Equation (41) over the region of interest, and (4) the square of each data point was taken (backtransformation) to obtain the final background shape. In Figure 13 the background after 10, 100, and finally 500 iterations is shown.

As a generalization of this method the average of two channels a distance w away from i can be used:

$$m_i = \frac{1}{2}(y_{i-w} + y_{i+w}) \tag{42}$$

Ryan et al. [31] proposed to use two times the FWHM of the spectrometer at channel i as value for w. They reported that only 24 passes are required to produce acceptable background shapes in PIXE spectra. During the last 8 cycles, w is progressively reduced by $\sqrt{2}$ to obtain a smooth background. To compress the dynamic range of the spectrum a double-logarithmic transformation of the spectrum, $\log[\log(y_i + 1) + 1]$, was proposed before the iterative stripping. In combination with the low-statistics digital filter, this procedure is called the SNIP algorithm (statistical nonlinear iterative peak clipping).

A variant of this procedure is implemented in the subroutine SNIPBG given in Section X. Instead of the double logarithm we employed a square-root transfor-

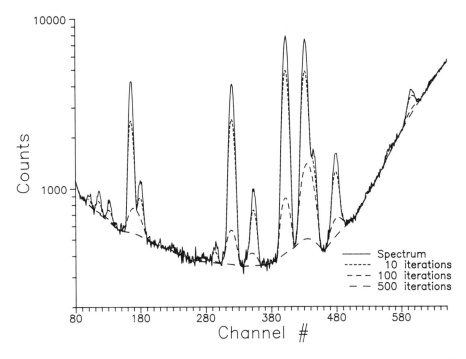

Figure 13 Background estimate of NIST bovine liver x-ray spectrum using a simple iterative stripping method. The background estimate after 1, 100, and 500 iterations is shown.

mation, and a Savitsky and Golay smoothing is performed on the square-root data. In addition the width w is kept constant over the entire spectrum. The value of w is also used as the width of the smoothing filter.

Using this implementation, the background of the spectrum just discussed is calculated and represented in Figure 14. The width was set to 11 channels, approximately corresponding to the FWHM of the peaks in the center of the spectrum, and 24 iterations were done. Apart from delivering a smoother background with fewer humps, a considerable gain in execution speed is also obtained. The first method requires several minutes for a 1024 channel spectrum, but only a few seconds is needed for the second method.

B. Parabolic Envelope Method

In the algorithm proposed by Kajfosz and Kwiatek [40], the background is modeled as an envelope of a family of parabolas. Since the background is a slowly varying structure, one can consider it to be composed of objects that are several times broader than the x-ray peaks. Moreover, as the amplitudes of x-ray peaks are always positive, the value of the background can nowhere exceed the value of the spectrum itself. Thus, in each channel i of the spectrum, a downward concave parabola:

$$D(j) = H(i) \left[1 - \frac{(i - j)^2}{x^2(u + v*i)} \right] \tag{43}$$

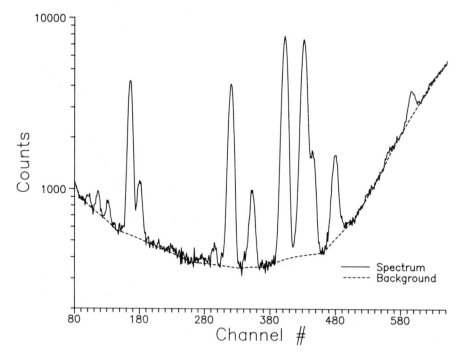

Figure 14   Background estimate using a variable-width stripping method (SNIP algorithm).

just touching the spectrum from below can be considered. $H(i)$ represents the amplitude of the parabola (its height), and x, u, and v are parameters by which the width of the parabola can be adjusted. Using the condition that for each channel j the parabola's $D(j)$ can nowhere exceed the spectrum $S(j)$, their amplitudes $H(i)$ can be determined via

$$H(i) = \min_{j} \left\{ \frac{S(j)}{1 - [(i - j)^2/x^2(u + v*i)]} \right\} \tag{44}$$

($S(j)$ represents the number of counts in channel j of the original spectrum, after elimination of statistical fluctuations), and a first approximation to the background $B(i)$ can be obtained as the envelope of these parabolas:

$$B(i) = \max_{j} \left[H(j) \left(1 - \frac{(i - j)^2}{x^2(u + v*i)} \right) \right] \tag{45}$$

If calculated in this way, the background $B(i)$ never exceeds the spectrum. Under the peaks it sometimes appears to be overestimated, however, especially if these peaks are situated on a slope. To avoid this problem, a second envelope is defined:

$$U(i) = \min_{j} \left[S(j) \left(1 + \frac{3(i - j)^2}{y^2(u + v*i)} \right) \right] \tag{46}$$

(y is also an adjustable parameter), this time consisting of upward concave parabolas, which touch the spectrum from above. When in Equation (44) the smoothed spectrum $S(i)$ is replaced by $S'(i)$, defined as

$$S'(i) = \min(S(i), U(i)) \tag{47}$$

satisfactory results can be obtained.

Since the number of computations to calculate the background according to this algorithm is proportional to the square of the number of channels in the spectrum, the background estimation of a 1024 channel spectrum would require considerable computation time. Therefore, a reduction in the number of channels by a factor of 5, for example, is performed, resulting in a decrease in computing time by a factor of 25. In the last step of the calculation, $B(i)$ is transformed back to the original spectrum length through linear interpolation.

Although Equations (44) through (46) employ the general expression for a parabola, in practice the first-order coefficient v is always set to zero. Consequently, the parameter u can be set to a fixed value (typically 100), leaving the two parameters x and y to determine the widths of the lower and upper family of parabolas, respectively. The x parameter determines the width of the family of downward concave parabolas that touch the spectrum from below. Using $u = 100$, typical values for x are in the range 3–20 for x-ray spectra taken with a resolution 10–40 eV/channel. When too large a value for x is chosen, the parabolas become too narrow, that is, more narrow than the peaks superimposed on the continuum, resulting in an overestimation of the latter. If on the other hand the value of x is too large, the enveloping parabolas become wider than broad variations in the background structure, resulting in an underestimate of the continuum. Similar to x, the parameter y determines the width of the upward concave parabolas. When in peak overlap the width of the valley between two partially overlapping peaks is of the same order of magnitude as the peak widths, a similar but reversed behavior than with the x parameter can be observed: that is, when too large a value of y is selected, the estimated background level tends to be higher than the bottom level of such valleys. From our own experience and also as reported by Ryan et al. [31], relatively small y values (typically 3–5) are required to obtain background shapes that coincide with the continuum in curved and peak-free regions of the spectrum. In Figure 15 the background estimate using this parabolic envelope method is shown using as parameters $x = 10$ and $y = 3$.

C. Automatic Background Estimation Using Orthogonal Polynomials

Another interesting background estimation procedure was introduced by Steenstrup [41], who applied the method to energy-dispersive x-ray diffraction spectra. The spectrum is fitted using orthogonal polynomials, and the weights of the least-squares fit are iteratively adjusted so that only background channels are included in the fit. The method is generally applicable to all pulse-height spectra and can be implemented as a computer algorithm that needs little or no supervision from the user in terms of input and control parameters.

The background is described by a set of polynomials up to degree m:

$$y(i) = \sum_{j=0}^{m} c_j P_j(x_i) \tag{48}$$

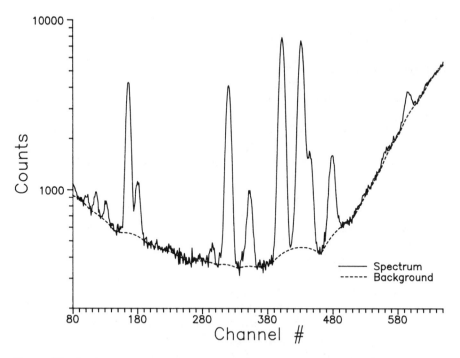

Figure 15 Background estimate using a parabolic envelope method.

where $P_j(x_i)$ is an orthogonal polynomial of degree j. As an example of $m = 3$ the fitting function becomes

$$y(i) = c_0 + c_1(x_i - a_0) + c_2[(x_i - a_1)(x_i - a_0) - b_1] \tag{49}$$

where a_0, a_1, and b_1 are coefficients independent of the dependent variable y_i. The least-squares estimate of the parameters c_j are given by

$$c_j = \sum_{i=1}^{n} \frac{w_i y_i P_j(x_i)}{\gamma_j} \tag{50}$$

where w_i are the weights of the fit. A detailed discussion on orthogonal polynomial fitting can be found in Section IX. Because the polynomial terms $P_j(x_i)$ are orthogonal, no matrix inversion is required to obtain the results. Moreover, polynomials with a much higher degree can be fitted to the experimental data with this method without running into problems with ill-conditioned normal equations and oscillating terms.

The goal of the method is to fit the background by the described orthogonal polynomial of degree m and to interpolate under the peaks. This can be achieved by careful manual selection of only those data pairs (x_i, y_i) that belong to the background. The more elegant approach proposed by Steenstrup consists of using all channels and choosing the weights in such a way that only the background contributions are emphasized. If $y(i)$ is a polynomial approximation of the background of degree m, then $y_i \approx y(i)$ if i is a background channel; otherwise $y_i > y(i)$. A better approximation to the background can then be found by choosing small weights for the data points where $y_i > y(i)$ and repeating the fit. The following weighting scheme is proposed by Steenstrup:

$$w_i = \frac{1}{y(i)} \qquad \text{if} \qquad y_i \leq y(i) + r\sqrt{y(i)} \tag{51}$$

$$w_i = \frac{1}{[y(i) - y_i]^2} \qquad \text{if} \qquad y_i > y(i) + r\sqrt{y(i)} \tag{52}$$

where r is an adjustable parameter. Typically $r = 2$ is used. If y_i is normally distributed (which is approximately the case for $y_i > 30$ counts), $y_i \leq y(i) + 2\sqrt{y(i)}$ holds for 97.7% of the channels containing only background. Too high a value of r includes the tails of the peaks in the fit. With the new weights a new polynomial fit of degree m is performed. The process is stopped when the new polynomial coefficients c_j are within 1 standard deviation from the previous coefficients. We also obtained good results by setting the weights effectively to zero when the channel content is statistically above the fitted background [Eq. (52)].

The method can be made even more unsupervised by including a procedure to automatically select the best degree of the polynomial. This can be done by fitting (as described earlier) successive polynomials with higher degrees and testing the significance of each additional polynomial coefficient. If the coefficient C_{m+1} is statistically not significant different from zero:

$$|C_{m+1}| < 2s_{C_{m+1}} \tag{53}$$

a polynomial of degree m is retained.

The FORTRAN subroutine OPOLBAC in Section X contains the complete pro-

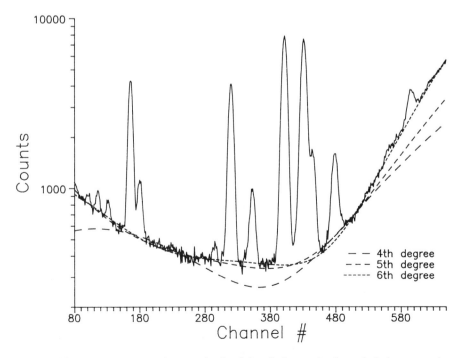

Figure 16 Background estimate obtained by fitting a 4, 5, and 6 degree orthogonal polynomial.

Table 5 Fit Background of the Spectrum in Fig. 16 Using Orthogonal Polynomials

Degree of polynomial	Number of channels used	Number of weight adjustments	Highest degree coefficient and SD
4	126	7	$(3.98 \pm 0.20)10^{-6}$
5	199	16	$(1.88 \pm 0.09)10^{-7}$
6	267	6	$(7.66 \pm 0.28)10^{-10}$
7	266	6	$(3.66 \pm 2.03)10^{-13}$

cedure to estimate the background of a given spectrum by an mth degree orthogonal polynomial. The routines were used to estimate the background of the bovine liver x-ray spectrum. A value of 1.5 was chosen for r [Eqs. (51) and (52)], and the weights were set to zero for the channels not belonging to the background [$w_i = 0$ in Eq. (52)].

Figure 16 shows the spectrum and the fitted background using a 4, 5, and 6 degree orthogonal polynomial. Table 5 gives the data of the fits. The final number of background channels retained, the number of weight adjustments performed, and the value and standard deviation of the highest polynomial coefficient are shown. The total number of channels in the fitting region was 571. On the basis

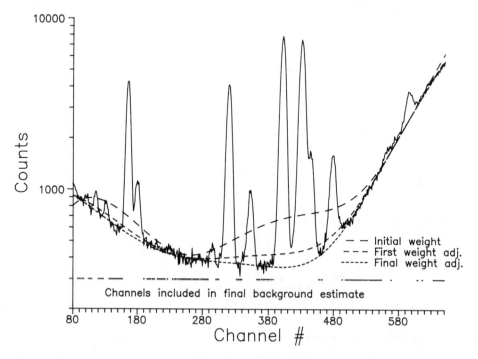

Figure 17 Iterative adjustment of the weights when fitting the background with a 6 degree orthogonal polynomial. The channels included in the final background estimate are indicated by the horizontal line.

of these data it must be concluded that a 6 degree orthogonal polynomial is needed to model the background.

In Figure 17 the effect of the weight adjustments is illustrated. The initial background with all weights $w_i = 1/y_i$ and the background after the first adjustment and after the sixth (last) adjustment are shown. The channels retained for the final background fitting are marked.

V. SIMPLE NET PEAK AREA DETERMINATION

In both WDXRF and EDXRF the number of counts under the characteristic x-ray line corrected for the background is proportional to the concentration of the analyte. At constant resolution this proportionality also exists for the net peak height. In EDXRF most often preference is given to the net peak area, mainly because this results in a lower statistical uncertainty for the small peaks normally encountered with this method. In WDXRF the acquisition of the entire peak profile is often too time consuming and the count rate is measured only at the peak maximum.

A. Peak Area Determination in EDXRF

The most straightforward method to obtain the net area of an isolated peak in a spectrum consists of interpolating the background under the peak and summing the background corrected channel contents in a window over the peak.

The net peak area N_p of an isolated peak on a background is given by

$$N_P = \sum_{i_{PR}}^{i_{PL}} (y_i - y_B(i)) = \sum_i y_i - \sum_i y_B(i) = N_T - N_B \tag{54}$$

where N_T and N_B are the total number of counts and the background counts in the integration window $i_{PL} - i_{PR}$. The uncertainty due to counting statistics in this estimate is

$$s_{N_p} = \sqrt{N_T + N_B} \tag{55}$$

With reference to Figure 18, the background in channel i can be obtained by interpolation, assuming a straight-line background:

$$y_B(i) = Y_{BL} + (Y_{BR} - Y_{BL}) \frac{i - i_{BL}}{i_{BR} - i_{BL}} \tag{56}$$

Y_{BL} and Y_{BR} are the values of the background at the channels i_{BL} and i_{BR}, left and right from the peaks. These values are best estimated by averaging over a number of channels:

$$Y_{BL} = \frac{1}{n_{BL}} \sum_{i=i_{BL1}}^{i_{BL2}} y_i = \frac{N_{BL}}{n_{BL}} \tag{57}$$

$$Y_{BR} = \frac{1}{n_{BR}} \sum_{i=i_{BR1}}^{i_{BR2}} y_i = \frac{N_{BR}}{n_{BR}} \tag{58}$$

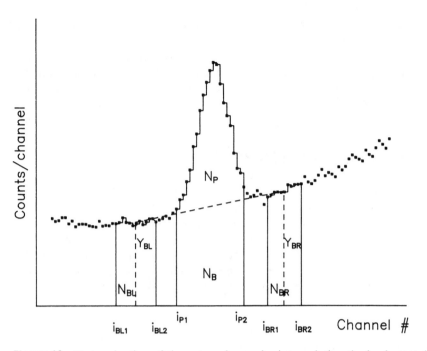

Figure 18   Determination of the net peak area by interpolating the background.

The number of channels in the background windows are $n_{BL} = i_{BL2} - i_{BL1} + 1$ and $n_{BR} = i_{BR2} - i_{BR1} + 1$. The center position of the background windows (not necessarily an integer number!) used in Equation (56) is $i_{BL} = (i_{BL1} + i_{BL2})/2$ and $i_{BR} = (i_{BR1} + i_{BR2})/2$.

If both background windows have equal width, $n_{BL} = n_{BR} = n_B/2$, and are positioned symmetrically with respect to the peak window ($i_P - i_{BL} = I_{BR} - i_P$), Equations (54) and (56) reduce to a much simpler expression:

$$N_P = N_T - \frac{n_P}{n_B}(N_{BL} + N_{BR}) \tag{59}$$

where N_{BL} and N_{BR} are the total counts in the left and right background windows, each $n_B/2$ channels wide, and n_P equals to the number of channels in the peak window. Applying the principle of error propagation, the uncertainty in the net peak area is then given by

$$s_{NP} = \sqrt{N_T + \left(\frac{n_P}{n_B}\right)^2(N_{BL} + N_{BR})} \tag{60}$$

From this equation, it can be seen that in principle the background should be estimated using as many channels as possible ($n_B \gg$) to minimize random errors due to counting statistics. In practice, the width is limited by curvature in the background and by the presence of other peaks. Most often the total width of the background ($n_B = n_{BL} + n_{BR}$) is taken to be equal to or slightly larger than the width of the peak window.

The optimum width of the peak window to minimize counting statistics depends on the peak-to-background ratio [42]. For low peak-to-background ratios, the peak height being only a fraction of the background height, the optimum width of the integation window is near 1.17 times the FWHM of the peak, assuming a Gaussian peak shape. In theory, for a peak height-to-background height ratio greater than 1, a slightly larger window should be used, although in practice the improvement in precision is negligible.

This method does not deliver the total net peak area but only a fraction of it. Integrating over $1.17 \times$ FWHM (from -1.378σ to 1.378σ) covers 83% of the peak. To cover 99% of the peak the window should be $2.19 \times$ FWHM ($\pm 2.579\sigma$).

If position or width changes are observed from one spectrum to the other, care should be taken to define the windows in such a way that they cover the same part of the spectrum. The wider the peak window the less sensitive is the method for peak shift.

Although such a peak area determination method seems naively simple, if correctly used it can provide results that are as accurate and as precise as the most sophisticated procedures. The premises for use are that the peak window should be known to be free from interferences, that there should be no peaks in the background windows, and that the background should be linear over the extent of the windows. The peak search procedures can in principle be used to set up the windows automatically. Its practical use is limited, however, by the complexity of the EDXRF spectra. Moreover, the use of such automated procedures is hazardous since no measure can be given for the presence or absence of systematic errors.

Because of these restrictions, a simple peak integration method cannot be used as a general tool for spectrum processing; in a limited number of applications, however, such as some types of routine analysis, good analytical results can be obtained.

There are a number of variants on these peak integation methods. An evaluation of their performance is given by Hertogen et al. [43].

Another important aspect of spectral data that should be mentioned is that the pulse-height spectrum is a discrete histogram representing a continuous function. Figure 19 shows the effect of digitization of a Gaussian peak into discrete channels. If the FWHM is less than 2.5 channels, the peak area obtained by summing the channel contents is largely overestimated. This corresponds to a spectrometer resolution of 60 eV/channel for a peak with FWHM of 150 eV. In practice, 40 eV/channel is the highest useful value because otherwise peak position and width determinations and the results of spectrum fitting become unreliable.

B. Net Count Rate Determination in WDXRF

In WDXRF, most often the count rate at the 2Θ angle of the peak maximum corrected for the background is used as analytical signal. The background is estimated at a 2Θ position on the left or right side of the peak. Background interpolation as described in the previous section is also possible.

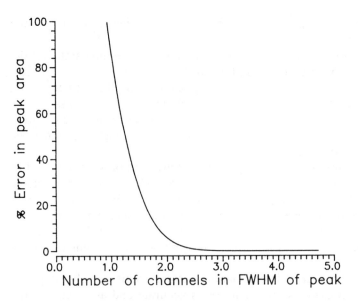

Figure 19 Systematic error in the peak area due to digitization in discrete channels.

If N_P is the number of counts accumulated during a time interval t_P at the top of the peak and N_B is the number of the background counts accumulated during time t_B, then the net count rate I is given by

$$I = I_P - I_B = \frac{N_P}{t_P} - \frac{N_B}{t_B} \tag{61}$$

and the uncertainty in the net count rate due to counting statistics is given by

$$s_I = \sqrt{\frac{N_P}{t_P^2} + \frac{N_B}{t_B^2}} \tag{62}$$

In WDXRF various counting strategies can be considered and the effect on the precision can be estimated using Equation (62) [44]. In a "optimum fixed time" strategy, the minimum uncertainty is obtained when, for a total measurement time $t = t_P + t_B$, t_P and t_B are chosen in such a way that their ratio is equal to the square root of the peak-to-background ratio:

$$\frac{t_P}{t_B} = \sqrt{\frac{I_P}{I_B}} \tag{63}$$

Under these conditions the uncertainty in the net intensity is given by

$$s_I = \frac{\sqrt{I_P} + \sqrt{I_B}}{\sqrt{t_P + t_B}} \tag{64}$$

VI. LEAST-SQUARES FITTING USING REFERENCE SPECTRA

Assuming that a measured spectrum of an unknown sample can be described as a linear combination of spectra of pure elements constituting the sample

$$y(i) = \sum_{j=1}^{m} a_j x_{ji} \tag{65}$$

with $y(i)$ the content of channel i of the model spectrum and x_{ji} the content of channel i of the jth reference spectrum, then the coefficients a_j are a measure of the contribution of the reference spectra to the unknown spectrum and can be used in subsequent quantitative analysis. The coefficients a_j can be obtained by multiple linear least-squares fitting, minimizing χ^2 with respect to all the parameters a_j:

$$\chi^2 = \sum_{i=n_1}^{n_2} \frac{1}{\sigma_i^2} (y_i - y(i))^2 = \sum_{i=n_1}^{n_2} \frac{1}{\sigma_i^2} \left(y_i - \sum_{j=1}^{m} a_j x_{ji} \right)^2 \tag{66}$$

where y_i and σ_i are the channel content and the associated uncertainty of the unknown spectrum and n_1 and n_2 are the limits of the fitting region. A detailed discussion of the least-squares fitting method is given in Section IX.

The assumption of linear additivity holds very often for the characteristic lines in the spectrum, but not for the continuum background. To apply this principle, the background can be removed from the unknown spectrum and from the reference spectra using one of the background estimation procedures described in Section IV before the least-squares fitting. Another, quite interesting, approach is to apply a digital filter to both unknown and reference spectra before performing the least-squares procedures. This method is known as the filter-fit method [45–47], which is discussed in some detail.

1. Filter-Fit Method

By the discrete convolution of a spectrum y_i with the top hat filter [Eq. (30)], the low-frequency component, that is, the slowly varying background, can effectively be suppressed as discussed in Section III. Apart from removing the slowly varying background, a rather severe distortion of the peaks is introduced. If we apply this filter to both the unknown spectrum and the reference spectra, the nonadditive continuum is removed and the same type of peak distortion is introduced in all spectra, allowing us to apply the method of multiple linear least-squares fitting to the filtered spectra. Equation (66) then becomes

$$\chi^2 = \sum_{i=n_1}^{n_2} \frac{1}{\sigma_i'^2} \left(y_i' - \sum_{j=1}^{m} a_j x_{ji}' \right)^2 \tag{67}$$

where y_i' and x_{ji}' are the filtered unknown and reference spectra; $\sigma_i'^2$ is the variance of y_i' given by

$$\sigma_i'^2 = \sum_k h_k^2 y_{i+k} \tag{68}$$

The least-squares estimates of the contribution of each reference spectrum are obtained by minimizing χ^2 with respect to these parameters. As explained in Section IX, this leads to the solution

$$a_j = \sum_{k=1}^{m} \alpha_{jk}^{-1} \beta_k \qquad j = 1, \ldots, m \tag{69}$$

with

$$\beta_j = \sum_{i=n_1}^{n_2} \frac{1}{\sigma_i'^2} y_i' x_{ji}' \tag{70}$$

$$\alpha_{jk} = \sum_{i=n_1}^{n_2} \frac{1}{\sigma_i'^2} x_{ki}' x_{ji}' \tag{71}$$

The uncertainty in each coefficient a_j can directly be estimated from the error matrix

$$s_{a_j}^2 + \alpha_{jj}^{-1} \tag{72}$$

Schamber [45] suggested the following equation for the uncertainties, taking into account the effect of the filter:

$$s_{a_j}^2 = \frac{v \cdot w}{v + w} \alpha_{jj}^{-1} \tag{73}$$

where w is the width of the central positive portion of the filter and v the width of the negative wings. A measure of the goodness of the fit is available through the reduced χ^2 value:

$$\chi_v^2 = \frac{1}{(n_2 - n_1 + 1) - m} \chi^2 \tag{74}$$

which is the χ^2 value according to Equation (67) divided by the number of points in the fit minus the number of reference spectra. A value close to 1 indicates a good fit; that is, the reference spectra are capable of adequately describing the unknown spectrum.

Most of the merits and the disadvantages of the filter-fit method can be deduced directly from the mathematical derivation given in the preceding discussion. The most interesting aspect of the filter-fit method is that it does not require any mathematical model for the background and that at least in principle the shapes of the peaks in the unknown spectrum are exactly represented by the reference spectra. Reference spectra should be acquired with good counting statistics, at least better than the unknown spectrum, as the least-squares method assumes that there is no error in the independent variable x_{ji}'. Reference spectra can be obtained from single-element standards. Only the portion of the spectrum that contains peaks must be retained as reference in the fit. This also allows multielement standards to be used, provided the groups of peaks of each element are well separated.

The reference spectra must provide an accurate model of the peak structure present in the unknown spectrum. This requires that reference and unknown spectra are acquired under strictly identical spectrometer conditions. Resolution changes, especially energy calibration changes, can produce large systematic errors. The magnitude of this error depends on the degree of peak overlap. Peak shifts of more than a few electronvolts should be avoided, which is readily possible with good modern detector electronics. If shifts are observed over long periods of operation of the spectrometer, careful recalibration of the spectrometer is required, or better, the reference spectra should be acquired again. Also, peak shift due to differences in count rate between standards and unknown should be avoided. Differential absorption is another problem that might influence the accuracy of the model. Because of the difference in x-ray attenuation in the reference sample and the unknown sample, $K\beta/K\alpha$ ratios might be different in the two spectra. This becomes especially problematic if the $K\beta$ line is above and the $K\alpha$ line below an absorption edge of a major element of the unknown sample. The magnitude of the error depends on the peak overlap. Careful selection of the samples to produce the reference spectra is therefore required.

The procedure requires that a reference spectrum is included for each element present in the unknown. The method is more sensitive to unreferenced elements than a least-squares procedure with analytical functions. Positive as well as negative systematic errors in the contributions a_j can be observed when references are missing because of the negative peak structure in the filtered spectra. The method allows no mechanism to deal with sum peaks. Apart from removing the continuum, the filter also has some smoothing effect on the spectrum and causes the peak structure to be spread out over more channels. This is equivalent to fitting a spectrum with a somewhat lower resolution than that originally acquired. Therefore, the precision and detection limits attainable with the filter-fit method are slightly higher than ideally obtainable. The width of the filter is important in this respect. Schamber [45] suggests taking the width of the top of the filter equal to the FWHM resolution of the spectrometer $u = $ FWHM. The width of the wings can be taken as $v = u/2$.

2. Application

The calculation procedure is thus quite simple and requires the following steps. The top hat filter is applied to the unknown spectrum and the m reference spectra [Eq. (30)], and the modified uncertainties are calculated using Equation (68). Next the vector β of length m and the $m \times m$ square matrix α are formed using Equations (70) and (71), summing over the part of the spectrum one wants to analyze $(n_1 - n_2)$. After calculating the inverse matrix α^{-1}, the contribution of each reference to the unknown and its uncertainty can be calculated using Equations (69) and (72) or (73). Only the relevant part, such as $K\alpha$ or $K\alpha$ plus $K\beta$, of the reference spectra must be retained; the rest of the filtered spectrum can be set to zero.

A test implementation of the filter-fit method is given in Section X with the C program FILFIT. This program was used to fit part of an x-ray spectrum of a brass sample (Fig. 20a). The measurements were carried out using a Mo x-ray tube and a Zr secondary target and filter assembly. The spectrum to be evaluated was obtained from a polished NIST SRM 1103 brass sample. Spectra of pure metals (Fe, Ni, Cu, Zn, and Pb) were measured under identical experimental conditions. A top hat filter of width $w = 5$ was used. Table 6 shows how the spectral data were divided in regions of interest to produce the reference spectra. Since considerable x-ray attenuation is present in brass, separate references were

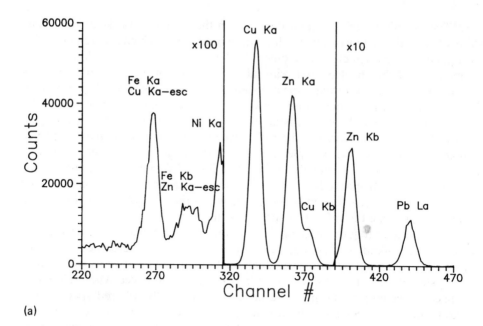

(a)

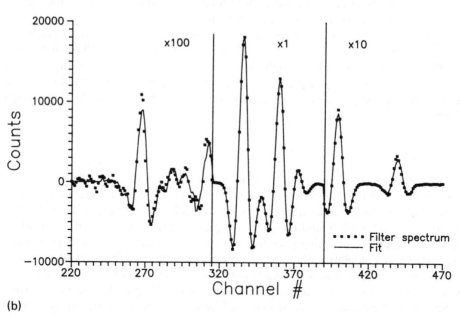

(b)

Figure 20 (a) X-ray spectrum of NIST SRM 1103 brass sample. (b) Top hat filtered spectrum and result of fit using reference spectra.

TABLE 6 Relation Between Spectra and References Used for Fitting the Unknown Spectrum Using the Filter-Fit Method

Spectrum	Region of interest (keV)	Used as
Pure Fe	4.25–8.12	Fe $K\alpha$ + $K\beta$ reference
Pure Ni	4.73–9.57	Ni $K\alpha$ + $K\beta$ reference
Pure Cu	5.70–8.46	Cu $K\alpha$ reference
	8.46–9.81	Cu $K\beta$ reference
Pure Zn	6.59–9.09	Zn $K\alpha$ reference
	9.09–10.30	Zn $K\beta$ reference
Pure Pb	8.36–13.68	Pb $L\alpha$ reference
SRM 1103	5.22–11.26	Unknown

created for the $K\alpha$ and $K\beta$ of Cu and Zn. This was not done for Fe, Ni, and Pb, which are present only as minor constituents in the brass sample. Figure 20b shows the filtered brass spectrum and the resulting fit using the seven (filtered) reference spectra. The region below Cu $K\alpha$ is expanded 100 times, and the region above the Zn $K\alpha$ is expanded 10 times. As can be seen, the agreement between the filtered brass spectrum and the fit is very good. The reduced χ^2 value is 8.5. This high value is probably due to small peak shifts in the reference spectra compared to brass spectrum. For this experiment standard nuclear instrumentation electronics were used rather than highly stabilized modern x-ray amplifiers.

Table 7 compares the results of the filter fit with the results of the evaluation of the spectrum by nonlinear least-squares fitting using analytical functions (see Sec. VII). Although the χ^2 value of the nonlinear fit was slightly better (2.7), one observes an excellent agreement between the two methods for the analytically important data (intensity ratios). Further, a slightly higher uncertainty is observed for the filter-fit results of the low-intensity peaks, as explained previously.

The filter-fit method is fast and relatively easy to implement. It can produce reliable results when the spectrometer calibration can be kept constant within a few electronvolts and suitable standards for each element present in the sample are available. The method performs well when one must deal with a background that is complex and difficult to model. If information on trace elements and major elements is required (very large peaks next to very small peaks), the method may not be optimal. For these reasons this filter-fit method is most often used to

TABLE 7 Comparison of Spectrum Evaluation Results Using Filter-Fit Method and Nonlinear Least-Square Fitting[a]

Element	Filter fit	Nonlinear fit	% difference
Fe	0.0053 ± 0.0002	0.0047 ± 0.0001	11
Ni	0.0019 ± 0.0002	0.00201 ± 0.00008	−6
Cu	0.546 ± 0.001	0.551 ± 0.001	−0.9
Zn	0.390 ± 0.001	0.3912 ± 0.0008	−0.31
Pb	0.0310 ± 0.0006	0.0308 ± 0.0003	0.65

[a] The intensity ratio in the SRM 1103 standard to the pure element intensity is given.

process spectral data in analytical electron microscopy, often in combination with a ZAF correction procedure.

VII. LEAST-SQUARES FITTING USING ANALYTICAL FUNCTIONS

A widely used and certainly the most flexible method to obtain net peak areas from a spectrum is least-squares fitting of the spectral data with an analytical function. The method is conceptually simple but not trivial to implement.

A. Concept

In this method an algebraic function, including analytically important parameters, such as the net areas of the fluorescence lines, is used as a model for the measured spectrum. The χ^2 is defined as the weighted sum of squares over a region of the spectrum of the differences between this model and the measured spectrum y_i:

$$\chi^2 = \sum_{i=n_1}^{n_2} \frac{1}{\sigma_i^2} [y_i - y(i, a_1, \ldots, a_m)]^2 \tag{75}$$

where σ_i^2 is the variance of data point i, normally taken as $\sigma_i^2 = y_i$, and a_j are the parameters of the model. The optimum values of the parameters are those for which χ^2 is minimal. They can be found by setting the partial derivative of χ^2 to the parameters to zero:

$$\frac{\partial \chi^2}{\partial a_j} = 0 \quad j = 1, \ldots, m \tag{76}$$

If the model is linear in all the parameters a_j, these equations result in a set of m linear equations in m unknowns a_j, which can be solved algebraically. This is known as linear least-squares fitting. If the model has one or more nonlinear parameters, no direct solution is possible and the optimum value of the parameters must be found iteratively. An initial value is given to the nonlinear parameters, and all the parameters are varied in some way until a minimum for χ^2 is obtained. The latter is equivalent to searching for a minimum in the m-dimensional χ^2 hypersurface. This is known as nonlinear least-squares fitting. The selection of a suitable minimization algorithm is very important because it determines to a large extent the performance of the method. A detailed discussion of linear and nonlinear least-squares fitting is given in Section IX.

The most difficult problem to solve when applying the least-squares procedure is the construction of an analytical function that accurately describes the observed spectrum. Over the region one wants to fit, the model must be capable of accurately describing the spectral data. This requires an appropriate model for the background, the characteristic lines of the elements, and all other feature present in the spectrum such as absorption edges and escape and sum peaks. Although the response function of the x-ray detection system is to a very good approximation Gaussian, deviation from the Gaussian shape must be taken into account. Failure to construct an accurate model results in systematic errors, which under certain conditions may lead to gross positive or negative errors in the estimated peak areas. On the other hand, the fitting function should be as simple as possible. Especially for nonlinear fitting, a large number of parameters can cause problems in the minimization of χ^2.

In general, the fitting model consists of two parts:

$$y(i) = y_B(i) + \sum_P y_P(i) \tag{77}$$

where $y(i)$ is the calculated content of channel i and the first part describes the background contribution and the second part the contributions from all peaks.

Since the fitting functions for both linear and nonlinear least-squares fitting have many features in common, we treat the detailed description of the fitting function for the most general case of nonlinear least-squares fitting. Moreover, if a nonlinear least-squares fitting program uses the Marquardt algorithm, the linear least-squares fit is also computational, a particular case of the nonlinear least-squares fitting. Such programs can perform linear as well as nonlinear fitting using the same computer code. A large part of the discussion given here is based on the computer code AXIL, developed by the authors for spectrum fitting of photon-, electron-, and particle-induced x-ray spectra [48–50].

B. Description of the Background

For the background model, various analytical functions are in use, depending on the excitation conditions and on the width of the fitting region. Except for electron microscopy it is extremely difficult, not to say impossible, to construct an acceptable physical model that describes the background, mainly because of the large number of processes that contribute to the background. For this reason very often some type of polynomial expression is used.

1. Linear Polynomial

A linear polynomial of the type

$$y_B(i) = a_0 + a_1(E_i - E_0) + a_2(E_i - E_0)^2 + \cdots + a_k(E_i - E_0)^k \tag{78}$$

is useful to describe the background over a region of 2–3 keV wide. Wider regions often exhibit too much curvature to be described by this type of polynomial. In Equation (78) E_i is the energy in kiloelectronvolts of channel i (see later) and E_0 is a suitable reference energy, often the middle of the fitting region. Expressing the polynomial as a function of $E_i - E_0$ rather than as a function of the channel number is done for computational reasons [$(E_i - E_0)^3$ is at most of the order of 10^3, whereas i^3 can be as high as 10^9]. The degree of the polynomial k can often be selected by the user: $k = 0$ produces a constant background, $k = 1$ a straight line, and $k = 2$ a parabolic background. Values of k in excess of 4 are rarely useful because physical nonrealistic oscillations occur. Equation (78) is linear in the fitting parameters $a_0, \ldots, a_k$; this function can thus be used in linear as well as in nonlinear least-squares fitting.

2. Exponential Polynomial

A linear polynomial cannot be used to fit the background over the entire spectrum or to fit regions of high positive or negative curvature. Higher curvature can be modeled by functions of the type

$$y_B(i) = a_0 e^{[a_1(E_i - E_0) + a_1(E_i - E_0)^2 + \cdots + a_k(E_i - E_0)^k]} \qquad (79)$$

where k is the degree of the exponential polynomial. A value of k as high as 6 might be required for an accurate description of a background between 2 and 16 keV. This function is nonlinear in the fitting parameters $a_1, \ldots, a_k$, requiring a nonlinear least-squares procedure and some initial guess of these parameters. Initial values for these nonlinear parameters cannot be determined on physical grounds. One possibility provided in the AXIL program, other than trial and error, is to estimate the background using one of the procedures described in Section IV and to perform a linear fit of the logarithm of the calculated background:

$$\ln y_B(i) = \ln a_0 + a_1(E_i - E_0) + a_2(E_i - E_0) + \cdots + a_k(E_i - E_0)^k \qquad (80)$$

to obtain an initial value for the parameters $a_1, \ldots, a_k$, which are then further optimized in the nonlinear fitting procedure.

The high degree of nonlinearity of this background parameters imposes high requirements on the nonlinear fitting algorithm. The computation time is increased considerably, and a large number of iterations is required to obtain stable values. When fitting a large number of similar spectra it might be advantageous to obtain a good estimate of all the background parameters using one representative spectrum and to treat the higher order terms as constant for the remaining spectra, optimizing only a_0, a_1, and a_2.

3. Bremsstrahlung Background

The exponential polynomial is often not suitable to describe the background shape observed in electron- and particle-induced x-ray spectra, mainly because such backgrounds have a large positive curvature in the low-energy region and negative curvature at higher energies. The reason for this is that, in contrast to secondary target XRF, for example, the background is radiative in nature. It consists of photons emitted from the sample because of the retardation of fast electrons. The slope of the emitted continuum is essentially an exponentially decreasing function according to Kramer's formulas. At low energies the emitted photons are strongly absorbed in the detector windows and in the sample. For this reason, a suitable function to describe such radiative backgrounds is an exponential polynomial, as already described, multiplied by the absorption characteristics of the spectrometer:

$$y_B(i) = a_0 e^{[a_1(E_i - E_0) + \cdots + a_k(E_i - E_0)^k]} \cdot T(E_i) \qquad (81)$$

A detailed discussion of the function $T(E)$ is given subsequently. To be physically correct, the absorption should be convoluted with the detector response function, as any sharp adsorption edge from window elements (Si or Au) or from elements present in the sample is smeared by the finite resolution of the detector.

4. Background Removal

An alternative to an algebraic function for the background is to calculate the background first, using one of the procedures outlined in Section IV, and to strip this background from the measured spectrum before least-squares fitting. This requires that the weights $1/\sigma_i^2$ are also changed. If y_i' is the measured spectral data after subtraction of the background,

$$y_i' = y_i - y_B(i) \tag{82}$$

χ^2 becomes

$$\chi^2 = \sum_i \frac{1}{\sigma_i'^2} [y_i' - y_P(i, a_1, \ldots, a_m)]^2 \tag{83}$$

and the variance of y_i' is given by

$$\sigma_i'^2 = \sigma_i^2 + \sigma_{y_{B(i)}}^2 \tag{84}$$

None of the background estimation procedures gives a value for the uncertainty in the background estimate, but a reasonable approximation is to take the variance equal to the background value so that Equation (84) becomes

$$\sigma_i'^2 = y_i + y_B(i) \tag{85}$$

If this adjustment of the weights is not made, the uncertainties in the net peak area are underestimated, especially for small peaks on a high background.

It is rather difficult for an inexperienced user to select the appropriate background model for a given spectrum. The following might serve as a general guideline. For fitting regions 2–3 keV wide, a linear polynomial background is often adequate. To fit large regions of XRF spectra, the exponential polynomial provides the most accurate results, with k typically equal to 4–6. The same holds for the bremsstrahlung background for electron microscopic (EM) and PIXE spectra. The simplest procedure from a users' point of view is to strip the background before fitting, but this method does not provide optimum results. A slight under- or overestimate might occur, resulting in large relative systematic errors for small peaks.

C. Description of Fluorescence Lines

Since the response function of Si(Li) detectors is so nearly Gaussian, it is obvious that all mathematical forms to describe the x-ray spectrum involve this function. Only when describing K lines of high atomic number elements, such as Pb or U, does the influence of the natural line shape become appreciable, requiring the use of a Voight profile [4,5].

1. Single Gaussian

A Gaussian peak is characterized by three parameters: the position, width, and height or area. It is desirable to describe the peak in terms of area rather than height because the area is the physically meaningful parameter but the height depends on the spectrometer resolution. The first approximation to the profile of a single peak is then given by

$$\frac{A}{\sqrt{2\pi}\sigma}\, e^{-[(x_i-\mu)^2/2\sigma^2]} \tag{86}$$

where A is the peak area (counts), σ the width of the Gaussian expressed in channels, and μ the location of the peak maximum. σ, the distance between the peak center and the inflection point, is related to the more (to the spectroscopist) appealing FWHM by the factor $2\sqrt{2\ln 2}$ or FWHM $= 2.35\sigma$. Only the peak area is a linear parameter; both width and position are nonlinear parameters. This implies that linear least-squares procedures assume knowing the position and the width very accurately from calibration.

To describe a part of a spectrum, the fitting function must contain a number of such functions, one of each peak. With 10 elements and 2 peaks per element we need to fit 60 parameters. It is highly unlikely that such a nonlinear least-squares fit will terminate successful at the global minimum. To improve the performance of the fitting procedure it is recommended that the fitting function be altered.

2. Energy and Resolution Calibration Function

The first obvious step is to drop the idea of fitting the position and width of each peak individually. In x-ray spectrometry the energies of fluorescence lines are very accurately known, often better than within 1 eV. The observed pattern of peaks in a spectrum can be translated to elements present in the sample. Once we know which elements are present, we can predict all x-ray lines that constitute the spectrum and their energies. Thus we can rewrite our peak model in terms of energy rather than channel number. Defining ZERO as the energy of channel 0 and expressing the spectrum GAIN in eV/channel, the energy in eV of channel i is given by

$$E_i = \text{ZERO} + \text{GAIN} \times i \tag{87}$$

and the Gaussian peak can be written as

$$G(i, En_j) = \frac{\text{GAIN}}{S\sqrt{2\pi}}\, e^{[-(En_j-E_i)^2/2S^2]} \tag{88}$$

with En_j the energy of the x-ray line in eV and S the peak width given by

$$S^2 = \left(\frac{\text{NOISE}}{2.3548}\right)^2 + 3.85 \times \text{FANO} \times En_j \tag{89}$$

NOISE is the electronic contribution to the peak width (typical 80–100 eV, expressed in FWHM units, with the factor 2.3548 to convert to σ), FANO is the Fano factor (typically around 0.114), and 3.85 the energy required to produce an electron-hole pair in silicon. The term GAIN/$S\sqrt{2\pi}$ in Equation (88) is required to normalize the Gaussian so that the sum over all channels is unity.

For the case of linear least-squares fitting, ZERO, GAIN, NOISE, and FANO are physically meaningful constants. In the case of nonlinear least squares they might be considered constants but also as parameters to be optimized during the fitting. The great advantage of fitting the energy and resolution calibration rather than the position and width of each peak is a vast reduction in the number of parameters. The nonlinear fit of 10 peaks now requires 14 parameters compared to 30. Even more importantly, all available information in the fitted spectrum is now used to estimate ZERO, GAIN, NOISE, and FANO rather than the position and the width of each peak individually. Imagine a severe overlapping doublet with low counting statistics with a well-defined peak at both sides of this doublet. These two peaks contribute the most to the determination of the four calibration parameters, virtually fixing the position and the width of the two peaks in the strong overlapping doublet. As a consequence their areas can be determined much more accurately.

Returning to our discussion in Section II on information content, we did not achieve this extra performance for nothing: we supplied extra information under the form of the energy of the peaks and the relations between energy and channel number and between peak width and energy. Such an approach [fitting with Eq. (88)] requires that the extra information we supply is indeed correct.

With modern electronics the linearity of the energy calibration [Eq. (87)] holds very well in regions above 2 keV. Including the low-energy region in the fit of an entire spectrum might require more complex energy calibration functions. To fit PIXE spectra from 1 to 30 keV, Maenhaut and Vandenhaute [51] suggested the following function:

$$i = C_1 + C_2 E + C_3 e^{-C_4 E} \qquad (90)$$

The relation between the square of the peak width and the energy [Eq. (89)] is based on theoretical considerations and is obeyed over the energy region under consideration (1–30 keV), provided one accounts for the doublet splitting of the x-ray lines of the various elements. Indeed, the $K\alpha_1 - K\alpha_2$ separation increases from a negligible value for Ca (3.5 eV) to nearly 100 eV for higher atomic number elements, such as Mo. The observed peak shape of the K lines are actually envelopes of two peaks. This envelope can be represented rather well by a single Gaussian, but failing to take this doublet splitting into account (i.e., fitting with single Gaussians where doublets are required) results in peak widths that do not obey Equation (89). To illustrate this, the observed width of a number of $K\alpha$ lines as a function of x-ray energy is presented in Figure 21. The dotted line represents the width of the $K\alpha$ doublet fitted as one peak; the solid (straight) line shows the width of the individual lines in the doublet.

3. Response Function for an Element

This brings us to another aspect of the fitting model. We actually want to model not so much a single peak but rather an entire element. We can consider a number of lines as logically belonging together, such as the $K\alpha_1$ and $K\alpha_2$ of the preceding doublets or all the K lines of an element, and fit them as one group, with one area parameter A representing the total number of counts in all the lines of the group.

The spectrum of an element can then be represented by

$$y_P(i) = A \sum_{j=1}^{N_P} R_j G(i, En_j) \qquad (91)$$

where G are the Gaussians representing the various lines and R_j the relative tran-

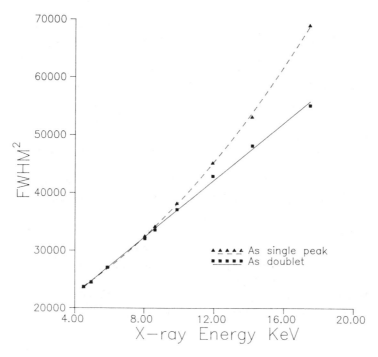

Figure 21   FWHM of various $K\alpha$ lines, fitted as a single peak and as a $K\alpha_1$-$K\alpha_2$ doublet.

sition probabilities of the lines. The summation runs over all lines in the group N_P, with $\sum R_j = 1$.

In principle, the transition probabilities of all lines originating from a vacancy in the same (sub)shell (K, L_1, L_{II}, and so on) are constant independent of the excitation. The relative intensities may vary, however, because of absorption in the sample and in the detector windows. To take this into account, the x-ray attenuation must be included in Equation (91). The relative intensity ratios are obtained by multiplying the transition probabilities with the absorption correction term

$$R'_j = \frac{R_j T(En_j)}{\sum\limits_j R_j T(En_j)} \tag{92}$$

The transition probabilities between various subgroups (i.e., between K and L or between L_I and L_{II}) depend on the type of excitation (photons, electrons, or protons) and on the excitation energy. General values can therefore not be given and must be determined for the particular excitation if one wants to combine lines of different subgroups. Transition probabilities of lines in various groups, determined experimentally and calculated from first principles, can be found in the literature [52–57].

Which lines should be included to describe the fluorescence of an element? As already mentioned, doublet splitting needs to be taken into account. Not only the $K\alpha$ but most other x-ray lines are actually doublets ($K\beta = K - M_2 M_3$; $L\alpha =$

$L_3 - M_4M_5$). Considering both the energy difference between the two peaks in the doublet and the spectrometer resolution, it was found [3] that if the quantity $T = (E_1 - E_2)/\text{FWHM}$, with E_1 and E_2 the energy of the two lines in the doublet, is greater or equal to 0.1, the doublet should be described with two Gaussians. Based on this criterion and also considering their relative importance, the various lines to describe the K and L fluorescence of the elements are given in Table 8. Some transitions, such as $K\beta_2$, only become significant from a certain element on. In Figure 22 the fit of a tungsten L line spectrum using 24 transitions to form the 3 L subshells is shown. The relative intensities from the L lines within each subshell were taken from the literature [57]. The fit was thus done with one peak area parameter for the L_1, the L_2, and the L_3 sublevels.

Fitting elements rather than individual peaks enhances the capability of the method to resolve overlapping peaks. The area of the Cr $K\alpha$ peak, interfered by a V $K\beta$ peak, can be obtained with higher precision (lower standard deviation) because the area of the V $K\beta$ peak is related to the area of the (uninterfered with) V $K\alpha$ peak. Again we introduce more a priori knowledge into our model. If this information (the relative intensity ratio) is not correct, we introduce systematic errors; that is, the Cr $K\alpha$ peak area, although having a low standard deviation, is not correct. In practice there is a tradeoff between the gain in accuracy and the gain in precision. Errors in the values of the transition probabilities and in the value of the absorption correction term are sufficiently small for small to

TABLE 8 K(Na-Ba) and L(Fe-U) Line Transitions Required for Complete Description of the Fluorescence of the Element as Single Peak or Doublet

Transition	Single Peak	Doublet
$K-L_2,L_3$	Na-Co	Ni-Ba
$K-M_2,M_3$	S-Ru	Rh-Ba
$K-N_2,N_3$	Se-Ba	
$L_1-M_{3,2}$	Fe-Ru	Rh-U
$L_1-N_{3,2}$	Rb-La	Ce-U
$L_1-O_{3,2}$	In-Pb	Bi-U
$L_1-M_{5,4}$	Ag-Ba	La-U
$L_1-N_{5,4}$	Lu-Pt	Au-U
L_2-M_4	Fe-U	
L_2-N_4	Rh-U	
L_2-M_1	Fe-U	
L_2-N_1	Rb-U	
L_2-O_4	Ta-U	
L_2-O_1	Eu-U	
L_2-M_3	Hf-U	
$L_3-M_{5,4}$	Fe-Te	I-U
$L_3-N_{5,4}$	Rh-Re	Os-U
$L_3-O_{5,4}$	Hf-U	
L_3-M_1	Fe-U	
L_3-N_1	Rb-U	
L_3-O_1	In-U	

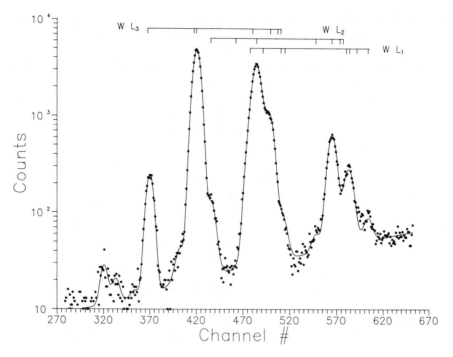

Figure 22 Fit of a complex L line spectrum of tungsten. In total 24 transitions, divided over the three L subshells, are required for the description of the spectrum.

moderately high peaks (up to 10^5 counts peak area) to be fitted as one group. For much higher peaks systematic errors become significant, requiring fitting as individual peaks.

4. Modified Gaussians

When fitting Gaussians to very large peaks, the deviation from the pure Gaussian shape becomes evident. This inaccuracy is always present but only becomes significant when the peak has very good statistics. In Figure 23 a Mn K spectrum with 10^7 counts in the Mn $K\alpha$ peak is shown. One observes a tailing on the low-energy side of the peaks and the background is also higher before the $K\alpha$ peak than behind the $K\beta$. These observations are explained by a number of effects. For lower energetic x-rays (<10 keV), incomplete charge collection and other artifacts dominate. At higher energies Compton scattering may also contribute. Apart from these detector-related effects, x-ray satellite transitions, such as KLM radiative Auger transitions on the low-energy side of the $K\alpha$ peak and *KMM* transitions on the low-energy side of the $K\beta$, contribute to the overall shape of the spectrum. In the literature considerable attention has been given to the explanation and to the accurate mathematical description of the observed peak shape [8,58–61].

Failure to account for the deviation from the Gaussian peak shape causes a number of problems when fitting x-ray spectra. Small peaks riding on the tail of large peaks (e.g., Mn $K\alpha$ in front of Fe $K\alpha$) cannot be fitted accurately, resulting in large systematic errors for the small peaks. Since the least-squares method seeks to minimize the difference between the observed spectrum and the fitted function, the tail might become filled with peaks of elements that are not really present. Also, the background over the entire range of the spectrum becomes difficult to describe.

A number of analytical functions have been proposed to account for the true line shape. Nearly all include a flat shelf and an exponential tail, both convoluted with the Gaussian response function. Similar functions are also used to describe the peak shape in γ spectra. A typical representative of the group of modified Gaussians is the "Hypermet" function [62]:

$$F(i) = G(i) + S(i) + D(i) \tag{93}$$

with $G(i)$ the Gaussian of height H_G:

$$G(i) = H_G e^{[-(i-i_0)^2/2\sigma^2]} \tag{94}$$

$S(i)$ a step function of height H_S convoluted by the Gaussian to account for the shelf:

$$S(i) = \frac{H_S}{2} \operatorname{erfc} \frac{i - i_0}{\sigma\sqrt{2}} \tag{95}$$

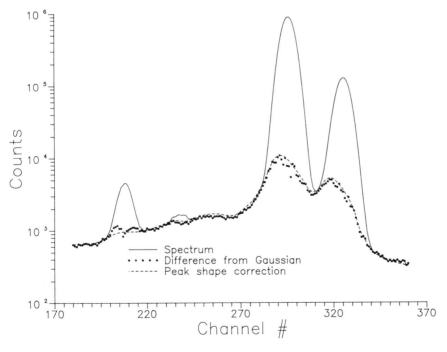

Figure 23 Mn K line spectrum with very good counting statistics. The difference from the Gaussian response is obtained by subtracting all Gaussian peaks. From this the peak shape correction is calculated.

and $D(i)$ an exponential tail of height H_D, convoluted by the Gaussian:

$$D(i) = \frac{H_D}{2} e^{[(i-i_0)/\beta]} \, \text{erfc} \left(\frac{i - i_0}{\sigma\sqrt{2}} + \frac{\sigma}{\beta\sqrt{2}} \right) \tag{96}$$

In these equations i represents the channel number and i_0 the center of the peak, σ the width of the peak, and β the width of the exponential. The complement of the error function, used to convolute the step and the tail, is defined as

$$\text{erfc}(u) = 1 - \text{erf}(u) = 1 - \frac{2}{\sqrt{\pi}} \int_0^u e^{-t^2} \, dt \tag{97}$$

and can be calculated via series expansion [21]. The shape of these functions is depicted in Figure 24.

Campbell et al. [8] used a Gaussian, a short tail, and a long tail exponential to fit the peak in energy-dispersive spectra obtained from $K\alpha_1$ and $L\alpha_1$ lines selected by Bragg reflection from a curved crystal. In this way the influence of the doublet structure and the satellite lines is eliminated. Excellent fits with reduced χ^2 values between 1.02 and 1.16 were obtained for peaks having 10^6 counts.

Fitting real x-ray spectra with modified Gaussians as given by Equation (93) would dramatically increase the number of parameters to be optimized (H_S, H_D, and β) for each peak. This shows, however, that all phenomena contributing to the deviation from the Gaussian shape depend on the energy of the x-ray line, so

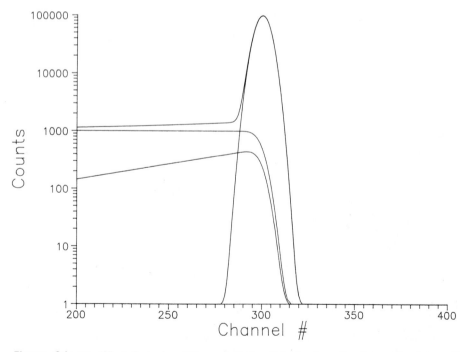

Figure 24 Modified Gaussian (Hypermet) function to describe the peak shape of an x-ray line. The step function, the exponential tail, and the Gaussian are shown, together with the envelope.

that the parameters of the function can be related to the parameters of the Gaussian (e.g., H_S/H_G and H_D/H_G) and that these ratios can be expressed as smooth functions of energy. In a calibration step the parameters of these functions are determined using pure element spectra with very good counting statistics. An additional problem is that the tailing of the $K\beta$ peaks tends to be larger because of the more intense KMM radiative transition, thus requiring a separate set of parameters for $K\alpha$ and $K\beta$.

An alternative procedure to describe the peak shape more accurately is used in the AXIL program [58]. The deviation from the Gaussian shape is stored as a table of numerical values, representing the difference between the observed shape and the pure Gaussian. The table extends from zero energy to the high-energy side of the $K\beta$ peak and is normalized to the area of the $K\alpha$ peak. The deviation is obtained from pure element spectra having very good counting statistics ($K\alpha \simeq 10^7$ counts). Preferably thin films are used to keep the background as low as possible and to avoid absorption effects. The area, position, and width of all peaks in the spectrum are determined by fitting Gaussians on a constant background over the FWTM of the peaks. The Gaussian contributions are then stripped from the spectrum. The resulting non-Gaussian part as shown in Figure 23 is further smoothed to reduce the channel-to-channel fluctuations and subsequently used as a numerical peak shape correction. The fitting function for an element is then given by

$$y_P(i) = A\left\{R'_{K\alpha}[G(i, E_{K\alpha}) + S(i)] + \sum_{j=2}^{N_P} R'_j G(i, En_j)\right\} \qquad (98)$$

Values in the table are interpolated to account for the difference between the energy scale of the correction and the actual energy calibration of the spectrum. Similar to the parameters of the non-Gaussian analytical functions, the shape of the correction seems to vary slowly from one element to another. This allows us to interpolate the peak shape correction for all elements from a limited set of corrections for the more important elements (e.g., V, Cr, and Mn corrections can be calculated from measured Ti and Fe peak shape corrections).

A major disadvantage of this method is that it is quite difficult and laborious to obtain good experimental peak shape corrections. Although they are in principle detector dependent, experience has proven that the same set of corrections can be used for different detectors with reasonable success, once again proving the fundamental nature of the observed non-Gaussian shape. Another disadvantage is that the peak shape correction for the $K\beta$ becomes underestimated if strong differential absorption takes place since the correction is related to the $K\alpha$ peak area only. Also, it is nearly impossible to apply this method to the description of L-line spectra. A mayor advantage, however, is the computational simplicity of the method and that no extra parameters are required.

5. Absorption Correction

The absorption correction term $T(E)$ used in Equations (81) and (92) includes x-ray attenuation in all layers and windows between the sample surface and the active area of the detector. For high energetic photons the transparency of the Si crystal must also be taken into account. In x-ray fluorescence, the attenuation in the sample, causing additional changes in the relative intensities, can also be considered relatively easily. The total correction term is thus composed of a number of contributions:

$$T(E) = T_{\text{det}} + T_{\text{path}} + T_{\text{sample}} \tag{99}$$

The detector contribution is given by

$$T_{\text{det}}(E) = e^{-\mu_{\text{Be}}(\rho d)_{\text{Be}}} e^{-\mu_{\text{Au}}(\rho d)_{\text{Au}}} e^{-\mu_{\text{Si}}(\rho d)_{\text{Si}}} (1 - e^{-\mu_{\text{Si}}(\rho D)_{\text{Si}}}) \tag{100}$$

where μ, ρ, and d are the mass attenuation coefficient, the density, and the thickness of the Be window, the gold contact layer, and the silicon dead layer. In the last term D is the thickness of the detector crystal.

Any absorption in the path between the sample and the detector can be modeled in a similar way. For an air path the absorption is given by

$$T_{\text{path}}(E) = e^{-\mu_{\text{air}}(\rho d)_{\text{air}}} \tag{101}$$

The mass attenuation coefficient of air can be calculated assuming a composition of 79.8% N_2, 19.9% O_2, and 0.3% Ar. The density of air under standard conditions is 1.2×10^{-3} g/cm^2. In a PIXE setup it is common practice to use additional absorbers between the sample and the detector, with one absorber sometimes having a hole (funny filter). The absorption behavior of such a structure can be modeled by

$$T_{\text{path}}(E) = e^{-\mu_f(\rho d)_f}[h + (1 - h)e^{-\mu_{ff}(\rho d)_{ff}}] \tag{102}$$

where the first term account for the absorption in (all) solid filters and the second term accounts for the absorption of the filter having a hole. The fraction of the detector solid angle subtended by the hole is denoted by h.

The sample absorption correction in the case of x-ray fluorescence, assuming excitation with an (equivalent) energy E_0, is given by

$$T_{\text{samp}} = \frac{1 - e^{-\chi_s(\rho d)_s}}{\chi_s(\rho d)_s} \tag{103}$$

and the sample attenuation coefficient is given by

$$\chi_s = \frac{\mu_s(E)}{\sin \theta_1} + \frac{\mu_s(E_0)}{\sin \theta_2} \tag{104}$$

This sample attenuation coefficient can only be calculated provided the weight fraction of all elements constituting the sample are known. The sample composition is often not known since the aim of the spectrum evaluation is to obtain the net peak areas from which the concentrations are to be calculated. Although x-ray attenuation in solid samples might become very large, it is important to realize that not the absolute value of the absorption correction but the ratio is important. The latter changes less dramatically especially since the energy difference of the lines involved is small. For an infinitely thin Fe sample the $K\beta/K\alpha$ intensity ratio is 0.134. In an infinitely thick Fe matrix the absorption correction terms $T(E_{K\beta})$ and $T(E_{K\alpha})$ are 161 and 134, respectively, assuming Mo $K\alpha$ excitation, causing the $K\beta/K\alpha$ ratio to change to 0.161. Therefore a rough estimate of the sample composition is often sufficient. Van Dyck and Van Grieken [63] demonstrated the integration of spectrum analysis software with quantitative analysis software. The sample absorption term is recalculated based on the estimated sample composition and spectrum fitting is repeated.

6. Sum and Escape Peaks

As indicated in Section II, escape peaks can be modeled by a Gaussian with an energy 1.750 keV below that of the parent peak. The area relative to the area of the parent peak can be calculated from the escape fraction [Eq. (4)]

$$\eta = \frac{N_e}{N_P} = \frac{f}{1 - f} \tag{105}$$

Including the escape peaks the description of the fluorescence of one element then becomes

$$y_P(i) = A \sum_{J=1}^{N_P} R'_J[G(i, E_J) + \eta G(i, E_J - 1.75)] \tag{106}$$

Various polynomial functions expressing the escape ratio as a function of the energy of the parent peak are also in use. The coefficients of the function can be determined by least-squares fitting from experimental escape ratios. In Figure 25 an example of a fit of the Cu $K\alpha$ peak with a Cu $K\alpha$ escape in the region of Fe $K\alpha$ is given.

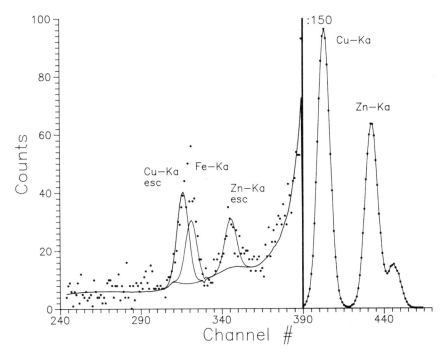

Figure 25 Fit of Cu and Zn escape peaks. The interference of these escape peaks with the Fe $K\alpha$ line can cause serious overestimates of Fe in brass samples.

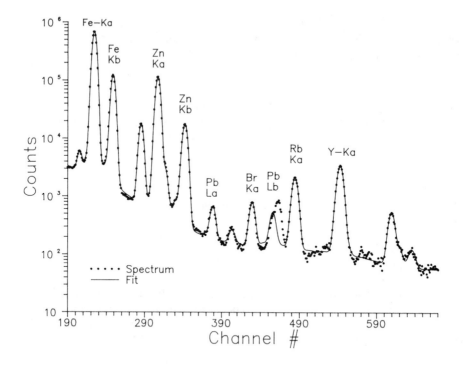

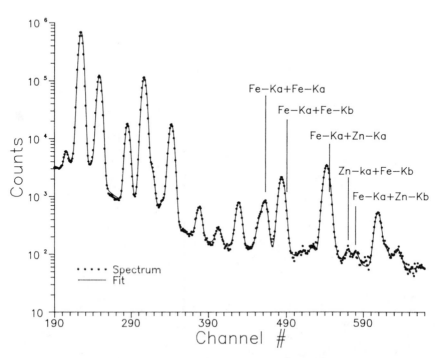

Figure 26 Fit of part of a PIXE spectrum having very high count rates for Fe and Zn: (top) without considering sum peaks; (bottom) sum peaks included in the fitting model.

Incorporation of sum peaks in the fitting model is more complex. The method discussed here was first implemented by Johansson in the HEX program [17]. First the spectrum is fitted without considering pileup peaks. The peaks are then sorted according to their height and the n largest peaks are retained. Peaks that differ less than 50 eV are combined in one peak. Using Equations (5) and (6), the relative intensities of all possible $n(n + 1)/2$ pileup peaks and their energies can be calculated and the m most intense are retained. Knowing the relative intensities and the energies of these m pileup peaks, they can be included in the fitting model as one "pileup element." In the next iteration the total peak area A of this pileup element is obtained. The construction of the pileup element can be repeated during the next iterations as more reliable peak areas become available. In Figure 26 part of an PIXE spectrum is shown fitted with and without the inclusion of sum peaks.

D. Special Aspects of Least-Squares Fitting

1. Constraints

When using nonlinear least-squares fitting, it can be advantageous to impose some kind of limit upon the fitting parameters to eliminate physically meaningless results. Some examples of what can happen if too much freedom is given to the fitting parameters are provided by Statham [46].

The incorporation of the energy and resolution calibration functions [Eqs. (87) and (89)] in the fitting model already places a severe constraint on the fit, avoiding the situation in which two peaks that are very close together swap their positions or become one broader peak.

One way to really constrain the fitting parameter is by defining the real physically meaningful parameter P_j [e.g., the GAIN in Eq. (87) or the peak position when Eq. (86) is used to fit] as an arctangent function of the fitting parameter a_j [47,64]:

$$P_j = P_j^0 + \frac{2L_j}{\pi} \arctan a_j \qquad (107)$$

where P_j^0 is the expected value (initial guess) of the parameter P_j and L_j defines the range. As a result of this transformation the parameter P_j is always in the range $P_j^0 \pm L_j$. Apart from significantly adding to the mathematical complexity, such a transformation has the disadvantage that the χ^2 minimum, although lying within the preselected interval, cannot be reached because the path toward the minimum passes a "forbidden" region. Also this constraint makes no distinction between the more probable values of the parameter lying near the center of the interval and the unlikely values at the limits.

An alternative approach, proposed by Nullens et al. [64], relies on modification of the curvature of the χ^2 surface. The fitting parameters, such as ZERO, GAIN, NOISE, and FANO, are random variables with an expected value equal to the initial guess a_j^0 and having some uncertainty Δa_j, and they can be included in the definition of χ^2 just as the observed data points $y_i \pm \sigma_i$:

$$\chi^2 = \sum_i \frac{1}{\sigma_i^2} (y_i - y(i))^2 + \sum_j \frac{1}{(\Delta a_j)^2} (a_j^0 - a_j)^2 \qquad (108)$$

Applying this definition to the Marquardt nonlinear least-squares fitting algorithm results in modified equations for the diagonal terms of the α matrix and for the β vector (see Sec. IX):

$$\alpha_{jj} = \sum_i \frac{1}{\sigma_i^2} \left(\frac{\partial y_0(i)}{\partial a_j} \right)^2 + \frac{1}{(\Delta a_j)^2} \tag{109}$$

and

$$\beta_j = \sum_i \frac{1}{\sigma_i^2} \left(y_i - y_0(i) \frac{\partial y_0(i)}{\partial a_j} \right) + \frac{a_j^0 - a_j}{(\Delta a_j)^2} \tag{110}$$

If no significant peaks are present in the fitting interval, the derivatives of the fitting function with respect to the position and width of the peaks are zero. Therefore, the second term in β_j dominates, causing the parameter estimate in the next iteration to be such that a_j tends toward a_j^0 (the initial value). If well-defined peaks are present, however, the second term in β is negligibly small compared to the first term and the method acts as if no constraints were present.

Another way to look at this is by considering the curvature of the χ^2 surface near the minimum. Figure 27 shows a cross section through the χ^2 surface along the peak position parameter. The χ^2 function is shown with and without constraints for a small and a large peak on a constant background (peak area 100 and 10,000 counts; background 100 counts/channel). The true peak position is at channel 100, and the FWHM of the peak is 8 channels; the constraint Δa is 1 channel. From this figure it is evident that the minimum in the modified χ^2 surface is much better defined for a small peak, whereas the modification has no influence on the shape of χ^2 for a large peak. This method has been implemented in the AXIL program to constrain the energy and resolution calibration parameters.

2. Weighting the Fit

The weights w_i used in least-squares methods [Eq. (75)] are defined in terms of the "true" (population) standard deviation of the data y_i, which can be obtained directly from the data itself (channel content):

$$w_i = \frac{1}{\sigma_i^2} \approx \frac{1}{y_i} \tag{111}$$

The approximation is valid for moderate to good statistics. When regions of the spectrum with very bad statistics (small channel contents) are fitted, estimating the weight from the measured channel content causes a systematic bias and leads to an underestimate of the peak areas. To overcome this problem, Phillips suggested that the weights be estimated on a 3 channel average [65]. In the AXIL program initially the weights are based on the measured channel content, but when the overall χ^2 value of the fit falls below 3 the weights are calculated as the inverse of the fitted channel content. This is based on the idea that when the calculated spectrum approaches the measured spectrum, the calculated channel contents are a better estimate of the true channel content than the measured values. The effect of the weighting on the fitting results are considerable, as can be seen in Figure 28.

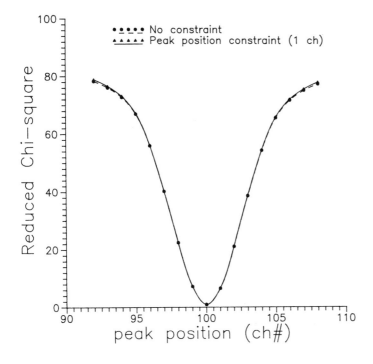

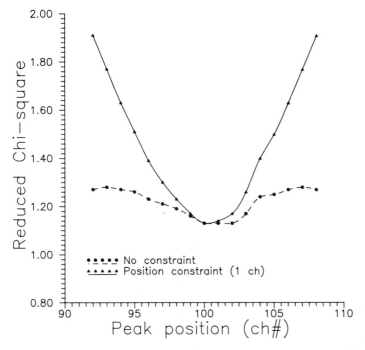

Figure 27 Effect of the use of a constraint on the shape of the χ^2 response surface: (top) marginal effect for a large peak; (bottom) important contribution for a small peak.

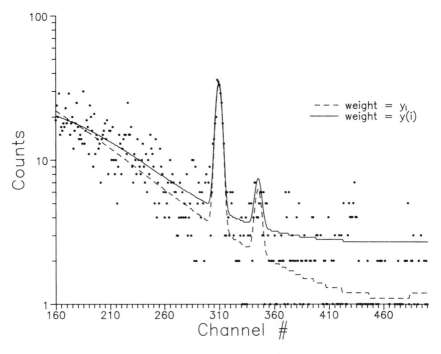

Figure 28 Effect of weighing the least-squares fit is shown on part of an PIXE spectrum with very small number of counts per channel.

E. Examples

To illustrate the working of the nonlinear least-squares fitting method, an artificial spectrum with Cu and Zn K lines is fitted with four Gaussian peaks on a constant background. Using the Marquardt algorithm, the position, width, and area of each peak are determined. The fitting function thus becomes

$$y(i) = a_0 i + \sum_{k=1}^{4} a_k e^{-[(i-a_{k+1})^2/2a_{k+2}^2]} \tag{112}$$

with the i the channel number (independent variable) and a_k the parameters to be determined, 13 in total. In Table 9 the true parameters of the spectrum, the initial guesses for the nonlinear parameters, and the final fitted values are given. The initial guesses were deliberately taken rather far from the optimum values. Figure 29a shows the fitted spectrum after the first and second iterations. During the second iteration the Marquardt algorithm evolved into a gradient search, drastically changing the position and width of the peaks. Even after 5 iterations, the calculated spectrum deviates considerably from the "measured" spectrum, as can be seen in Figure 29b. Finally, after 10 iterations a perfect match between them is obtained with a reduced χ^2 value of 0.96.

By observing how the iteration procedure changes the peak width and position parameters, one can imagine that something might go wrong. Especially if the spectrum is more complex, the changes are so great that the iteration stops at a

TABLE 9 Artificial Spectrum in Figure 29 Generated by Nonlinear Least-Squares Fitting[a]

Parameter	True value	Initial guess	Fitted value
Area, counts			
Cu $K\alpha$	100,000	0	$100,134 \pm 321$
Cu $K\beta$	13,400	0	$13,163 \pm 169$
Zn $K\alpha$	30,000	0	$30,092 \pm 213$
Zn $K\beta$	4,106	0	$4,138 \pm 83$
Position (channel number)			
Cu $K\alpha$	402.05	395	402.03 ± 0.01
Cu $K\beta$	445.25	450	445.35 ± 0.06
Zn $K\alpha$	431.55	435	431.59 ± 0.04
Zn $K\beta$	478.60	485	478.68 ± 0.09
Width (channel)			
Cu $K\alpha$	3.913	3	3.91 ± 0.01
Cu $K\beta$	4.033	3	3.99 ± 0.05
Zn $K\alpha$	3.995	3	4.02 ± 0.03
Zn $K\beta$	4.123	3	4.06 ± 0.08

[a] True value of spectrum parameters, initial guesses, and fitted values of the peak area, position, and width.

false minimum or even drifts away completely. In both cases physically incorrect parameter estimates are obtained. In practice it is off course possible to give much better initial estimates for the peak position and width parameters than used in this example. From Table 9 it is evident that the fit was quite successful, with all peak area, position, and width parameters estimated correctly, within the calculated standard deviation. One also observes that the uncertainties in the peak area parameters are approximately equal to the square root of the peak area and that the position and the width of the peaks are estimated very precisely (within 0.001 channel, or 0.2 eV). The uncertainty in these parameters is higher for smaller peaks, as expected.

The other extreme case of least-squares fitting is illustrated by the next example. Rather than fitting the area, position, and width parameters of each peak, linear least squares is used, thus assuming that the position and width of each peak are known (or equivalently, the spectrum energy and resolution calibration and the energy of the lines). The spectrum consists of simulated Fe and Co K lines on a constant background. The linear fitting function includes the background and the four Gaussians:

$$y(i) = a_0 E_i + \sum_{k=1}^{4} a_k G_k(i, En_k) \tag{113}$$

where E_i is the energy of channel i given by $E_i = \text{ZERO} + \text{GAIN} \times i$ and the Gaussian peaks G_k are given by Equation (88). Using (multiple) linear least-squares fitting, the background height a_0 and the peak area parameters $a_1, \ldots,$ a_4 are determined. In Figure 30 the spectrum and the different components are shown, but the fitting function (sum of the components) is not shown for clarity.

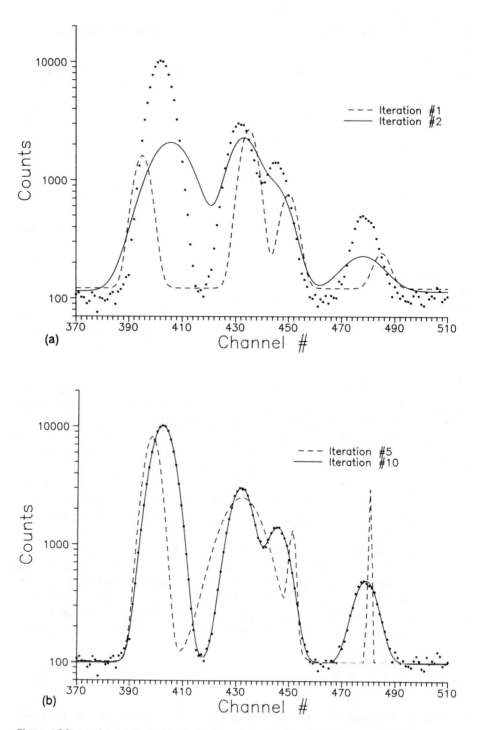

Figure 29 Artificial Cu and Zn *K* line spectrum fitted with a nonlinear least-squares procedure to optimize peak area, position, and width. (Top) Fitted spectrum after first and second iteration; (bottom) after fifth and final iteration.

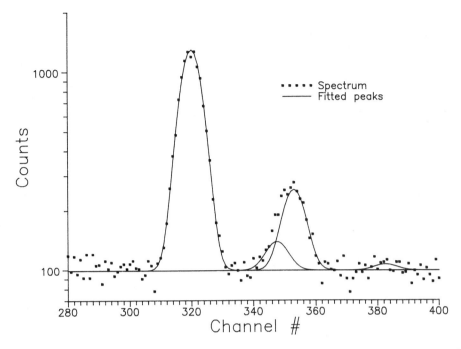

Figure 30 Artificial Fe and Co spectrum, background, and peak components obtained by linear least-squares fitting.

The reduced χ^2 value is 0.92. In Table 10 the true values are compared with the fitting results. The fitted values agree with the true values, taking into account the uncertainties of the estimates.

As a third example the nonlinear least-squares fitting, taking into account the energy and resolution calibration of the spectrum, is considered. This represents an intermediate case between the two previous examples. The artificial spectrum used has the same Fe and Co peaks as in the previous example, but a large Cu $K\alpha$ line is added. The fitting function is the same as given in Equation (113), except that ZERO, GAIN, NOISE, and FANO are now considered parameters of the fit. The iteration procedure of the nonlinear least-squares method is illustrated in

TABLE 10 True Values and Linear Least-Squares Estimates for the Fe Co Spectrum in Figure 30

Parameter	True value	Fitted value
Peak area, counts		
Fe $K\alpha$	10,000	10,144 ± 110
Fe $K\beta$	1,343	1,358 ± 63
Co $K\alpha$	400	348 ± 55
Co $K\beta$	54	67 ± 40
Background, counts/channel	100	99 ± 1

Figure 31, where the fitted function after the first, the second, and the last iteration are shown. Again the initial guesses for the nonlinear parameters (ZERO, GAIN, NOISE, and FANO) were considerably removed from the optimum values. As a consequence, 14 iterations were required to reach the χ^2 minimum at 1.09. In practical situations with better initial values, typically between 3 and 5 iterations are required for much more complex spectra. The evolution of χ^2 and the calibration parameters during the iterations are given in Figure 32. During the first iteration, an increase in χ^2 is observed. The Marquardt algorithm (see Sec. IX) therefore evolves into a gradient search, increasing the parameter λ until a lower χ^2 value is found. The same happens during the second iteration. From iteration 3 on, the algorithm proceeds in the normal way using linearization of the fitting function. In Table 11 the fitted values are compared with the true spectrum parameters. One also observes that the nonlinear fitting method using energy and resolution calibration produces net peak areas and uncertainties that are nearly identical to the results obtained by linear least squares (Table 10).

As a final example, a complex spectrum (geological reference material JG1, excited with Mo K x-rays from a secondary target system) is evaluated using the nonlinear least-squares fitting program AXIL [48–50]. In Figure 33 the spectrum and the fit are shown together with the residuals of the fit (see further). For the description of the spectrum from 1 to 16 keV, the fluorescence lines of 17 elements were used. Including escape and sum peaks, this amounts to as many as 84 Gaussians. The least-squares fit thus performed required the optimization of 32 pa-

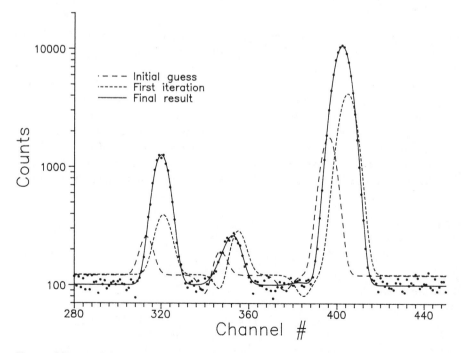

Figure 31 Artificial spectrum containing Fe, Co, and Cu K lines, fitted using nonlinear least squares and energy and resolution calibration. The fit after the first, the second, and the last iterations is shown.

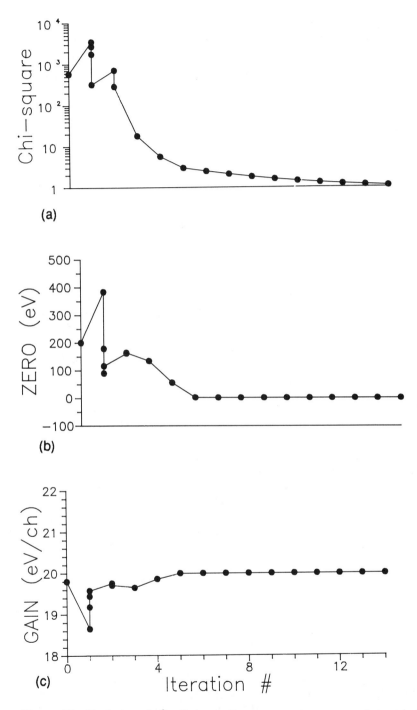

Figure 32 Evolution of χ^2 and the nonlinear parameters ZERO and GAIN during iterations when fitting the x-ray spectrum shown in Figure 31.

TABLE 11 True Values of the Spectrum in Figure 31, Initial Guesses, and Fitted Values by Nonlinear Least-Squares Fitting and Energy and Resolution Calibration Functions

Parameter	True value	Initial value	Fitted value
ZERO, eV	0	200	2.2 ± 4.1
GAIN, eV/channel	20	19.8	19.994 ± 0.010
NOISE, eV	100	60	103.2 ± 1.6
FANO	0.114	0.114	0.1146 ± 0.0020
Fe $K\alpha$, counts	10,000	0	10135 ± 112
Fe $K\beta$, counts	1,343	0	1349 ± 64
Co $K\alpha$, counts	400	0	335 ± 55
Co $K\beta$, counts	54	0	52 ± 39
Cu $K\alpha$, counts	100,000	0	99720 ± 321
Background, counts	100	0	100.4 ± 1.0

rameters (4 calibration parameters, 21 peak areas, and 7 background parameters). The output of the fitting program is shown in Figure 34. Starting from a fairly good initial guess of the calibration parameters, the χ^2 minimum of 1.5 is obtained in three iterations. For all the elements the energy, the relative intensity, the estimated net peak area and uncertainty, and the χ^2 value of the peak are indicated. For most elements, the lines are treated as one group with fixed intensity ratios. Only K, Ca, and Fe are fitted with individual $K\alpha$ and $K\beta$ peaks. Since the peaks of these elements are very high, escape peaks and a peak shape correction are included. The background was described by a sixth degree exponential polynomial. Because of the incomplete charge collection, the fitted background is lower than the observed background below 6 keV (Fe $K\alpha$) and the difference is taken up by the peak shape correction. The residuals (Fig. 33) indicate an overall good fit, with most of the residuals in the − 3 to 3 interval, although still some systematic patterns can be observed around channels 240, 360, 650, and 790, possibly due to the presence of lines from other elements (e.g. Ba L lines). In practice, however, it is not recommended that all possible elements be included, but including some elements whose presence is uncertain is possible, as illustrated by the inclusion of As in the fitting model. The net As $K\alpha$ intensity is estimated as 14 ± 35, indicating that this element is present below the detection limit.

F. Evaluation of Fitting Results

To understand and appreciate the capabilities and limitations of least-squares fitting using library spectra or linear or nonlinear analytical functions, it is useful to study in some detail the effect of random and systematic errors. Random errors are associated with the uncertainty σ_i of the channel content y_i. As we see subsequently, these uncertainties influence the precision of the net peak areas and determine the ultimate resolving power of the least-squares method. Systematic errors, on the other hand, are caused by discrepancies between the fitting model and the observed data and cause inaccuracies in the net peak areas. The mag-

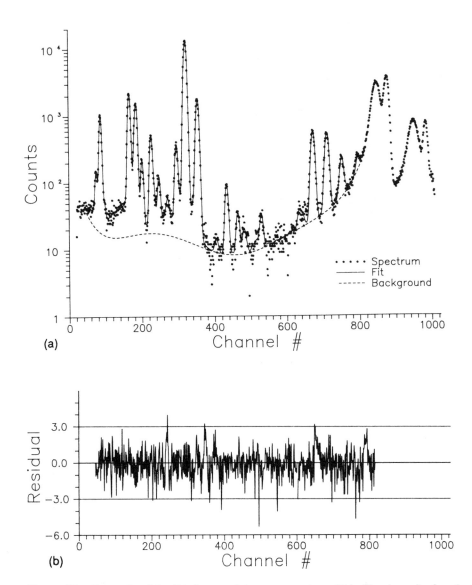

Figure 33 Example of the fit of a complex x-ray spectrum (thin-film deposit of geological reference material JG-1 excited with a Mo secondary target system). The residuals are plotted as an indicator of the quality of the fit.

nitudes of both types of errors depend in a rather complex way on the spectral data and on the fitting model used.

1. Error Estimate

Section IX explains how the least-squares fitting method (linear as well as non-linear) allows the estimation of the uncertainties in the fitted parameters. These uncertainties are the result of statistical fluctuations in the spectral data that are

```
┌─────────────────────────────────────────────────────────────────────┐
│ AXIL IBM-PC V3.00                        02-10-1991    19:37:28       │
│ Spectrum: JG1.SPE                                       3000.s        │
└─────────────────────────────────────────────────────────────────────┘
```

Fitting Region: channels 47 - 813	3 iterations done
ChiSquare = 1.5 last change = 0.31%	lambda= 1.E-05

CALIBRATION   DATA

	Initial estimate	Final estimate
ZERO (eV)	0.0 ± 100.0	3.7 ± .5
GAIN (eV/ch)	20.000 ± 2.000	19.936 ± .002
det NOISE (eV)	120.0 ± 40.0	116.0 ± 2.0
FANO factor	.114 ± .050	.109 ± .003

PEAK   DATA

# Line	E(KeV)	rel. int.	peak area	st. dev	chi-sq
1 Al-K			690. ±	37.	
KA1	1.487	1.00000	690. ±	37.	1.2
2 Si-K *			7080. ±	89.	
KA1	1.740	1.00000	7080. ±	89.	2.8
3 K -Ka *			16791. ±	135.	
KA1	3.313	.98890	16605. ±	134.	1.1
KA1 -esc	1.571	.01110	186. ±	1.	1.0
4 K -Kb			2128. ±	93.	
KB1	3.590	1.00000	2128. ±	93.	1.2
5 Ca-Ka *			11546. ±	133.	
KA1	3.691	.99040	11436. ±	131.	1.5
KA1 -esc	1.949	.00960	111. ±	1.	1.4
6 Ca-Kb			1519. ±	46.	
KB1	4.013	1.00000	1519. ±	46.	1.6
7 Ti-K			4631. ±	79.	
KA1	4.509	.88357	4092. ±	70.	1.3
KB1	4.932	.11643	539. ±	9.	4.4
8 V -K			332. ±	42.	
KA1	4.950	.88235	293. ±	37.	4.6
KB1	5.427	.11765	39. ±	5.	.9
9 Cr-K			193. ±	31.	
KA1	5.412	.88509	171. ±	28.	1.0
KB1	5.947	.11491	22. ±	4.	1.6

(a)

Figure 34 (a) Output of the spectrum analysis program AXIL, showing the results of the evaluation of the spectrum shown in Figure 33; and (b) results of the spectrum evaluation.

propagated into the parameters. A careful inspection of the uncertainties is necessary to evaluate the fitting results. To illustrate this, the previous example of the nonlinear least-squares fit of the (artificial) Fe-Co-Cu spectrum (Fig. 31 and Table 11) is used.

Quite a number of interesting observations can be made by studying Table 11. One might conclude that the calculated standard deviation of the peak area

# Line	E(KeV)	rel. int.	peak area	st. dev	chi-sq
10 Mn-K			3354. ±	69.	
KA1	5.895	.88113	2955. ±	61.	1.3
KB1	6.491	.11887	399. ±	8.	1.3
11 Fe-Ka *			119065. ±	350.	
KA1	6.399	.99660	118660. ±	349.	1.7
KA1 -esc	4.657	.00340	405. ±	1.	2.0
12 Fe-Kb			16595. ±	133.	
KB1	7.059	.99730	16550. ±	133.	4.7
KB1 -esc	5.317	.00270	45. ±	0.	.8
13 Cu-K			55. ±	17.	
KA1	8.048	.58259	32. ±	10.	.8
KA2	8.028	.29906	16. ±	5.	.8
KB1	8.905	.11834	7. ±	2.	1.2
14 Zn-K			968. ±	36.	
KA1	8.639	.58072	562. ±	21.	.8
KA2	8.616	.29858	289. ±	11.	.5
KB1	9.572	.12069	117. ±	4.	1.1
15 Ga-K			305. ±	24.	
KA1	9.252	.57895	176. ±	14.	1.0
KA2	9.225	.29790	91. ±	7.	.9
KB1	10.263	.12315	38. ±	3.	.7
16 As-K			16. ±	41.	
KA1	10.544	.57559	9. ±	23.	.7
KA2	10.508	.29653	5. ±	12.	.7
KB1	11.724	.12788	2. ±	5.	1.5
17 Rb-K			7751. ±	102.	
KA1	13.395	.56078	4347. ±	57.	.8
KA2	13.336	.29137	2258. ±	30.	.9
KB1	14.958	.13265	1028. ±	14.	.8
KB2	15.185	.01520	118. ±	2.	2.6
18 Sr-K			7402. ±	102.	
KA1	14.165	.55816	4132. ±	57.	.9
KA2	14.098	.29051	2150. ±	29.	.9
KB1	15.832	.13414	993. ±	14.	2.5
KB2	16.085	.01718	127. ±	2.	36.6
19 Y -Ka			1076. ±	64.	
KA1	14.958	.65720	707. ±	42.	.8
KA2	14.883	.34280	369. ±	22.	1.2
20 Pb-L			511. ±	66.	
L2M4	12.614	.26208	134. ±	17.	1.1
L3M5	10.552	.42017	215. ±	28.	.7
	...				
21 Pile_up			255. ±	50.	
	12.799	.44120	112. ±	22.	2.4
	9.712	.12303	31. ±	6.	1.5
	13.458	.12262	31. ±	6.	.7
	...				

(b)

Figure 34 *Continued*

should be close to square root of the peak area. This is indeed almost true for the Fe $K\alpha$ peak ($\sqrt{10{,}000} = 100$ versus 112) but not for the Fe $K\beta$ ($\sqrt{1343} = 37$ versus 64) and certainly not for the Co $K\alpha$ peak ($\sqrt{400} = 20$ versus 55). The reason for this is that the least-squares method correctly takes into account the fact that the peaks are sitting on a background and are overlapping, both phenomena causing higher uncertainty in the parameter estimate.

On the other hand, the value of the (constant) background is estimated very accurately ($1 \ll \sqrt{100}$). The reason is simply that this background was estimated using 121 channels of which many are actually only background and indeed $\sqrt{121} = 11$. Knowing that the background has been estimated almost without error, we can now prove that the uncertainties in the peak areas were indeed correctly estimated. In Section V we have seen that for simple peak integration $s_N^2 = s_T^2 + s_B^2$. Since $s_B^2 \approx 0$ in this case, the uncertainty in the net peak area must be close to the square root of the total number of counts in the peak $s_N = s_T = \sqrt{T} = \sqrt{(N + B)}$. The background under the peaks extends over about 16 channels, making the background of the order of 1600 counts, thus $\sqrt{(10{,}000 + 1600)} = 108$ versus 112 and $\sqrt{(54 + 1600)} = 41$ versus 39. Using this method of reasoning, one still observes that the estimated uncertainties of the Fe $K\beta$ and Co $K\alpha$ peaks are higher: $\sqrt{(1343 + 1600)} = 54$ versus 64 and $\sqrt{(400 + 1600)} = 44$ versus 55. This is because these two peaks overlap to a certain degree. The more closer are the peaks, the higher the uncertainties in the two peak areas will become. (Theoretically in the limit of complete overlap the uncertainty is infinite and the two peak areas will have complete erratic values, but their sum will still represent correctly the total net peak area of the two peaks; in practice, curvature matrix α is singular so that the matrix inversion fails.) It can thus be concluded that a least-squares method not only correctly estimates the net peak areas but also their uncertainty, taking into account the height of the background and the degree of peak overlap, providing of course that the fitting model is capable of describing the measured spectrum.

As a consequence the uncertainties in the net peak areas can also be used to decide whether a peak is indeed present in the spectrum and to calculate detection limits. The uncertainty in the Co $K\beta$ peak (net peak area of 54 counts on a background of 1600 counts) is 40 counts; the estimated peak area is 67 counts. On the basis of a 3σ criterion, this peak area is statistically not significantly different from zero as the confidence interval 67 ± 120 contains zero. One must thus conclude that in the spectrum the presence of the Co $K\beta$ peak cannot be ascertained. Actually, any value of the peak area between -120 and 120 is statistically equivalent to a value of zero. Any value above 120 gives us clear evidence that the peak is present; any value below -120 indicates that there is something wrong with the model since truly negative peak areas are physically meaningless. Because the uncertainty in the net peak area includes the effect of the background and the peak overlap, they can be used to calculate correctly the a posteriori detection limits of the elements (peaks) present in the spectrum. Three situations can occur:

1. Estimated peak area > 3 standard deviation.
   → Report: area ± standard deviation.

2. -3 standard deviation $\leq$ estimated area ≤ 3 standard deviation.
  $\rightarrow$ Report: detection limit equal to 3 standard deviation.
3. Area < -3 standard deviation.
  $\rightarrow$ Revise the fitting model.

2. Criteria for Quality of Fit

By definition of the least-squares method, the χ^2 value [Eq. (75)] estimates how well the model describes the data. The reduced χ^2 value, obtained by dividing χ^2 by the number of degrees of freedom:

$$\chi_v^2 = \frac{1}{v} \chi^2 = \frac{1}{n - m} \chi^2 \qquad (114)$$

has an expectation value of 1 for a "perfect" fit. The number of degrees of freedom equals the number of data points minus the number of parameters estimated during the fit. Since χ^2 is also a random variable, the observed value is often slightly larger or smaller than 1. Actually, χ^2 follows (approximately) a χ^2 distribution and the 90% confidence interval can be determined:

$$\chi_{v,0.95}^2 \leq \chi_v^2 \leq \chi_{v,0.05}^2 \qquad (115)$$

where $\chi_{v,0.95}^2$ and $\chi_{v,0.05}^2$ are the tabulated χ^2 values at the 95 and 5% confidence intervals. For $v = 20$ the confidence interval is 0.543 to 1.571 and for $v = 200$, 0.841 to 1.170. An observed value of χ_v^2 in this interval indicates that the model describes the experimental data within the statistical uncertainty of the data. In other words, all remaining differences between the data and the fit can be attributed to the noise fluctuations in the channel contents and are statistically not significant. When fitting complex spectra with good statistics, much higher χ^2 values are obtained due to small imperfections of the background or peak model. This does not mean that the estimated peak areas are no longer useful, but values in excess of 3 may indicate that the fitting model needs improvement.

The reduced χ^2 value as defined in Equation (114) gives an overall estimate of the fit quality over the entire fitting region. Locally, in some part of the spectrum, the fit may actually be worse than indicated by this value. It is therefore useful to define χ^2 for each peak separately:

$$\chi_P^2 = \frac{1}{n_2 - n_1} \sum_{i=n_1}^{n_2} \frac{1}{\sigma_i^2} (y_i - y(i))^2 \qquad (116)$$

where n_1 and n_2 are the boundaries of the peak at FWTM and $n_2 - n_1$ approximates the number of degrees of freedom. High values of χ_P^2 indicate that the peak is fitted badly and the resulting peak area should be used with caution. In this case ($\chi_P^2 > 1$) it is advisable to give a conservative estimate for the uncertainty in the net peak area by multiplying the calculated uncertainty with the square root of the χ^2 value:

$$s_A' = s_A \sqrt{\chi_P^2} \qquad (117)$$

Although the χ^2 values give an indication of the goodness of fit, visual inspection of the fit is highly recommended. Because of the large dynamic range of the data, a plot of the spectrum and the fit on a linear scale nearly always give the impression of a perfect fit. A plot of the logarithm or the square root of the data is more appropriate. The best method is to plot the residuals of the fit, defined as

$$r_i = \frac{y_i - y(i)}{\sigma_i} \tag{118}$$

Residuals in excess of 3 or -3 then indicate regions of bad fit, as can be seen in Figure 33. It is the sum of the squares of these residuals that were minimized by the least-squares method.

G. Available Computer Codes

A number of computer programs for spectrum evaluation based on the least-squares method are reported in the literature. Without attempting to be complete, the main characteristics of a number of programs are summarized here. Wätjen made a compilation of the characteristics of eight computer packages for PIXE analysis [66].

An intercomparison of five computer programs for the analysis of PIXE spectra revealed a very good internal consistency among the five programs [67]. PIXE spectra of biological, environmental, and geological samples were used, and their complexity placed high demands on the spectrum analysis procedures. The programs that tested were AXIL, University of Gent, Belgium; HEX, University of Lund, Sweden; SESAM-X (Marburg, Germany), the Guelph program; and PIXAN, Lucas Heights. It was concluded that the most serious disagreement occurs for small peaks on the low-energy tails of very large peaks, pointing to a need for a more accurate description of the tail functions. Also very good agreement between the linear (SESAM-X) and the nonlinear least-squares approach was observed.

The Los Alamos PIXE data reduction software [68] contains three components. The K and L relative x-ray intensities of the elements making up the sample are computed, taking into account the detector and sample absorption. Using Gaussian peak shape the energy and resolution calibration of the spectrum are calculated. With the relative peak areas and the calibration functions obtained in this way, the spectrum is fitted using a Gaussian peak shape and a polynomial background. Escape and pileup peaks can be included. A linear least-squares fit is done, with the relative elemental concentrations and the polynomial background coefficient as unknowns. The background and the relative concentrations are constrained to nonnegative values, and all elements having x-ray lines in the spectrum interval considered are included.

The PIXASE computer package [69] performs spectrum analysis using nonlinear least-squares fitting. Elements are represented by groups of lines with fixed (absorption-corrected) relative intensities and including escape peaks. Each peak is modeled by a Gaussian with the addition of an exponential tail and an error function. The square of the peak width is a first-order polynomial of the peak energy and the position a second-order polynomial of the peak energy. The background is described as the sum of an exponential polynomial and two simple exponentials. Pileup effects are treated as one pileup element. The nonlinear least-squares fitting is done using a simple grid search technique. The search space of each parameter is limited by user-supplied minimum and maximum values. To fit large series of similar spectra, linear least-squares fitting using a library of calculated spectrum components can be done.

Bombelka et al. [70] described an PIXE analysis program based on linear least squares. The peak shape includes a Gaussian, a low-energy tail function to account for the incomplete charge collection and the escape peak. The position and the square of the width of the peaks are given by a first-order linear function of the energy. The fluorescence lines of an element are modeled as a sum of those peak shapes with relative intensity ratios corrected for absorption in the detector windows and absorbers. The background is composed of an fourth-order exponential polynomial multiplied by the x-ray attenuation term (bremsstrahlung background) and a second-order linear polynomial. Pileup is taken into account as a pileup element. The energy and resolution calibration parameters are obtained from selected peaks in the spectrum. The parameters of the exponential background are calculated from background spectra. The parameters obtained by the linear least-squares fit are the amplitude parameters of each element and of one pileup element, the amplitude parameter of the bremsstrahlung background, and the linear polynomial background parameters. The computer implementation for PIXE is called SESAM-X and is highly interactive with graphical representation of the spectral data. Other computer programs for tube-excited [72] and synchrotron radiation-excited [72] XRF were developed based on this code.

SAMPO-X [73] is intended for the analysis of electron-induced x-ray spectra and is based on the well-known SAMPO code originally developed for γ-ray spectroscopy [74]. A Gaussian with two exponential tails as in the original SAMPO program is used to represent the peaks. The height and the position parameter are obtained by the nonlinear least-squares fit. The peak width and the tail parameters are obtained from shape calibration tables by interpolation. The background is modeled by the semiempirical electron bremsstrahlung intensity function proposed by Pella et al. [75]. The thickness of the detector beryllium window and the atomic number of the sample, which occur in this formula, are adjusted by least-squares fit. The program also includes an element identification based on the energy and intensity of the fitted peaks and standard ZAF matrix correction.

Jensen and Pind [76] describe a program for the analysis of energy-dispersive x-ray spectra. The program uses a sum of Gaussians, one for each fluorescence line. The background is subtracted first using a linear, parabolic, or exponential function fitted from peak-free regions in the spectrum. The peak width is obtained from a calibration function that expresses the logarithm of the peak width as a linear function of the peak position. The width calibration is done using nonoverlapping peaks in a calibration spectrum or in the spectrum to be analyzed. Peak positions are determined using a peak search method or entered by the operator with the aid of a graphical display of the spectrum. The peak heights are then determined using linear least-squares fitting.

The computer code developed at the Technical University of Graz [77] is primarily intended for the evaluation of x-ray fluorescence spectra. A Gaussian response function with a low-energy tail is used to describe the peaks. The square of the peak width is a linear function of the peak energy. A straight-line equation relates the peak position to the energy of the peaks. A parabola is used to describe the background, and absorption edges are modeled by a complementary error function. The fitting parameters are the peak heights, the three background parameters, the height of the absorption edges, and the two energy calibration parameters. Nonlinear least-squares fitting is done with the Marquardt algorithm using a tangent transformation to constrain the fitting parameters to physical meaningful values [64]. Provision for escape peaks and Auger peaks is also made [78].

The AXIL program [48] was originally developed for the analysis of x-ray fluorescence spectra and later modified to allow the evaluation of electron- and particle-induced x-ray spectra [49,50]. It uses the Marquardt algorithm for nonlinear least-squares fitting with a modified (constrained) χ^2 function. Linear, exponential, and bremsstrahlung polynomials can be used to model the background as well as background stripping. X-ray lines are described by Gaussian functions with an optional numerical peak shape correction. Escape and sum peaks can be included in the model. The peak position and the square of the peak width are related to the x-ray energy by linear functions. Provision is made to correct for the absorption in detector windows, in filters, and in the sample.

VIII. METHODS BASED ON THE MONTE CARLO TECHNIQUE

The Monte Carlo technique can be used to simulate x-ray spectra. These simulated spectra are useful to study the behavior and performance of various spectrum-processing methods. The Monte Carlo technique has also been used to evaluate x-ray spectra in combination with quantitative analysis, as discussed later in this section.

A. Simulation of X-ray Spectra

During the development and test phase of a spectrum-processing method, it is often advantageous to use computer-simulated spectra. For these spectra, such features as the position, width, and area of peaks are exactly known in advance. They can be generated to any desired complexity. To make any real use of them, the simulated spectra must possess the same channel-to-channel variation according to a Poisson distribution as experimental spectra.

A simple and adequate procedure consists of first calculating, over the channel range of interest, the ideal spectrum y^0 using, for example, a polynomial background and a series of Gaussians as given by Equation (112). More complex functions, including a physically realistic model for the background and tailed peaks, can be used if desired. The next step is to add or subtract some number of counts from each channel content to obtain the desired counting statistical noise. In other words, the "true" content y_i^0 must be converted into a random variable y_i, so that it obeys a Poisson distribution [Eq. (1)] with $N_0 = y^0$.

Poisson-distributed random variables can be generated by various computer algorithms. An example [21] is given in Section X. For $y^0 > 30$ the Poisson-distributed random variable can be approximated by the much easier to calculate normally distributed random variable. The probability of observing y counts in a channel, assuming a normal distribution, is given by

$$P(y) = \frac{1}{\sigma\sqrt{2\pi}} e^{[-(y-\mu)^2/2\sigma^2]}$$

(119)

with $\mu = \sigma^2 = y^0$. For large y^0 the normal distribution is a very good approximation of the Poisson distribution. Even for small channel contents this approximation is quite satisfactory. The probability of observing 6 counts, assuming the true value is 4, is 0.10 according to a Poisson distribution and 0.12 according to a normal distribution.

The problem of adding counting statistics to the calculated spectrum is thus reduced to calculating a normally distributed random variable y with mean value $\mu = y^0$ and variance $\sigma^2 = y^0$. Starting from a uniformly distributed random variable U in the interval $[0,1]$, which can be generated by a pseudo-random number generator, normally distributed random variables with zero mean and unit variance can be obtained with the Box-Muller method [21].

$$v_1 = 2U_i - 1$$

$$v_2 = 2U_{i+1} - 1$$

$$r = v_1^2 + v_2^2$$

if $r < 1$ \hfill (120)

$$z_1 = v_1 \sqrt{\frac{-2 \ln r}{r}}$$

$$z_2 = v_2 \sqrt{\frac{-2 \ln r}{r}}$$

v_1, v_2, and r are calculated from two uniformly distributed random numbers U_i and U_{i+1}. If r is less than 1, two normally distributed random numbers z_1 and z_2 can be calculated. If $r \geq 1$, v_1, v_2, and r are recalculated using a new set of uniform random numbers. The normally distributed random number y, with mean $\mu = y^0$ and variance $\sigma^2 = y^0$, is then obtained by simple scaling:

$$y = \mu + z\sigma = y^0 + z\sqrt{y^0} \hfill (121)$$

Applying this to all channels produces the desired counting statistics. Since z is normally distributed with mean 0 and unit variance, z can be negative as well as positive. The count rate in each channel is thus increased or decreased randomly and proportionally to $\sqrt{y^0}$. The final step in the computer simulation is to convert the real numbers that were used during the calculation to integers.

Another interesting procedure is to generate artificial spectra from parent spectra. The parent spectrum is a spectrum acquired for a very long time so that it exhibits very good counting statistics (high channel content). A large number of child spectra can be generated with lower and varying counting statistics by the procedure explained subsequently. This method is useful to study the effect of counting statistics on spectrum-processing algorithms [31].

From the parent spectrum y_i, which might first be smoothed to reduce the noise even further provided not too much distortion is introduced, the normalized cumulative distribution function Y_j is calculated:

$$Y_j = \frac{\sum\limits_{i=0}^{j} y_i}{\sum\limits_{i=0}^{n} y_i} \hfill (122)$$

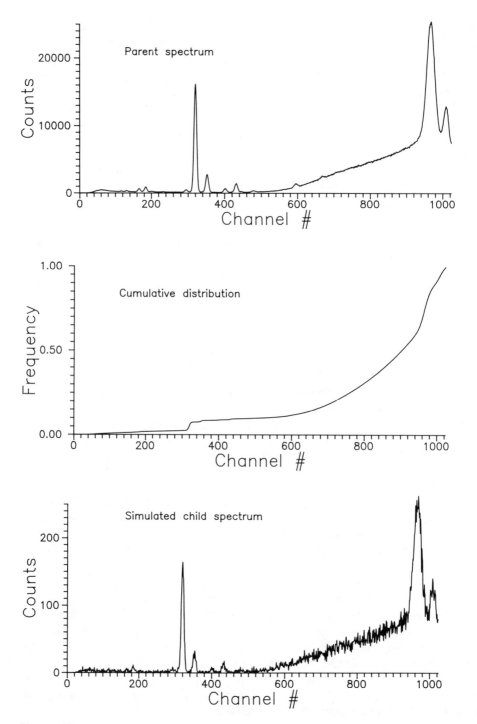

Figure 35 Simulation of a ''child'' spectrum from a ''parent'': (top) original spectrum; (middle) the cumulative distribution; (bottom) generated child.

n is the number of channels in the spectrum and Y_j is in the interval [0–1]. To generate the child spectrum we select N times a channel i according to the equation

$$i = Y^{-1}(U) \tag{123}$$

where Y^{-1} is the inverse cumulative distribution and U is a uniformly distributed random number. Each time channel i is selected, one count is added to that channel. Since N is the total number of counts in the child spectrum, the counting statistics can be controlled by varying N. In practice Y^{-1} cannot be calculated, so for each random number U we select the channel i for which $Y_i \leq U < Y_{i+1}$. Figure 35 shows an EDXRF spectrum from a 0.187 mg/cm^3 pellet of International Atomic Energy Agency (IAEA) animal blood reference material acquired for 1000 s with a Tracor Spectrace 5000 instrument, which is used as a parent spectrum. The cumulative distribution function and a child spectrum simulated with $N = 3 \times 10^4$ are also shown. The total number of counts in the original spectrum is 3×10^6. The child spectrum is equivalent to a spectrum that would have been acquired for 10 s. Some computer routines useful for simulation experiments are given in Section X.

B. Spectrum Evaluation Using Monte Carlo Techniques

Based on a detailed description of the detector response function and on Monte Carlo simulation of the interactions of primary excitation photons with the sample, Doster and Gardner developed a method to simulate the complete spectral response of an EDXRF system [79–81]. With Monte Carlo simulation the intensity of the characteristic lines and the scattered excitation spectrum are estimated, taking primary and secondary effects (absorption, enhancement, and single and multiple scattering) into account. The simulated pulse-height spectrum is then obtained by convolution with the detector response function. Quantitative analysis using the complete spectral response approach is based on finding the sample composition such that the difference in the least-squares sense between the simulated and the measured spectrum is minimal. The analysis involves the following steps: (1) simulation of the x-ray intensities over the expected composition range of the unknowns, (2) convolution with the detector response function to obtain simulated spectra, (3) construction of a χ^2 map (weighted sum of squares of differences between simulated spectra and experimental spectrum) as a function of the sample composition, and (4) interpolation of χ^2 for the composition corresponding to the minimum.

An interesting aspect of this method is that all the information present in the spectrum (characteristic lines, scattered radiation, and background) is considered in the analysis. Spectrum evaluation forms here an integrated part of the entire analytical procedure. The simulations require considerable computer power, especially for complex samples, and a very accurate detector response function, including a priori known energy, and resolution calibration of the spectrometer is required. Doster and Gardner demonstrated analytical accuracies of the order of 2% absolute for the analysis of Cr-Fe-Ni alloys with a ^{109}Cd radioisotope system [79].

Based on this work, two other approaches were developed. Yacout and Dunn [82] demonstrated the use of the inverse Monte Carlo method, which requires in principle only one simulation to analyze a set of similar samples. The second method is called Monte Carlo library least-squares analysis [83]. Starting from an initial guess of the composition of an unknown sample, a spectrum is simulated, taking into account all the interactions in the sample. During the simulation one keeps track of the response of each element to construct library spectra. After the simulation these library spectra are used to obtain the elemental concentrations by linear least-squares fitting (see Sec. VI). If the concentration in the unknown samples differs too much from the initial assumed concentration, the simulation is repeated. In contrast to the normal library least-squares method, this method has the advantage that the library spectra are simulated for a composition close to the composition of the spectrum to be analyzed, rather than measured from standards. This eliminates the necessity of applying the top hat filter and problems related to changes in $K\beta/K\alpha$ ratios, and again the method combines spectrum evaluation with quantitative analysis. Only the required computation time and accurate knowledge of the spectrometer response are limiting factors for the general applicability of this method.

IX. LEAST-SQUARES FITTING METHOD

The aim of the least-squares method is to obtain "optimum" values for the parameters of a function that models the dependence of the experimental data. The method has its roots in statistics but is also considered part of numerical analysis. Least-squares parameter estimate, also known as curve fitting, plays an important role in experimental science. In x-ray fluorescence it is used in many calibration procedures, and it forms the basis of a series of spectrum analysis techniques. In this section an overview of the least-squares method with emphasis on spectrum analysis is given.

Based on the type of fitting function, one makes a distinction between linear and nonlinear least-squares techniques because one requires numerical techniques of different complexities to solve these two problems. The linear least-squares method deals with the fitting of functions that are linear in the parameters to be estimated. For this problem a direct algebraic solution exists. If the fitting function is not linear in one or more of the parameters, one uses nonlinear least-squares techniques, of which the solution can only be found iteratively. A group of linear functions of general interest are the polynomials, the straight line being the simplest case. The special case of orthogonal polynomials is also considered. If more than one independent variable $(x_{1i}, x_{2i}, \ldots, x_{mi})$ is associated with each measurement of the dependent variable y_i, one speaks of multivariate regression. Spectrum analysis using a library function (e.g., filter-fit method) belongs to linear multivariate least squares. If analytical functions, for example, Gaussians, are fitted to a spectrum, the method of linear or nonlinear least squares is used depending on whether nonlinear parameters, such as the peak position and width, are optimized in the fit.

A. Linear Least Squares

Consider the problem of fitting experimental data with the following linear function:

$$y = a_1X_1 + a_2X_2 + \cdots + a_mX_m \tag{124}$$

This function covers all linear least-squares problems. If $m = 2$, $X_1 = 1$, and $X_2 = x$, the straight-line equation $y = a_1 + a_2x$ is obtained. For $m > 2$ and $X_k = x^{k-1}$ Equation (124) is a polynomial $y = a_1 + a_2x + a_3x^2 + \cdots + a_mx^{m-1}$ to be fitted to the experimental data points $\{x_i, y_i, \sigma_i\}i = 1, \ldots, n$. If $X_1, \ldots, X_m$ represent different independent variables, the case of multiple linear regression is dealt with. Because of this generality we discuss the linear least-squares method based on Equation (124) in detail.

Assume a set of n experimental data points:

$$\{x_{1i}, x_{2i}, \ldots, x_{mi}, y_i, \sigma_i\} \quad i = 1, \ldots, n \tag{125}$$

with x_{ki} the value of the kth independent variable X_k in measurement i, assumed to be known without error, and y_i the value of the dependent variable measured with uncertainty σ_i. The optimum set of parameters $a_1, \ldots, a_m$ that gives a least-squares fit of Equation (124) to these experimental data are those values of $a_1, \ldots, a_m$ that minimize the χ^2 function:

$$\chi^2 = \sum_{i=1}^{n} \frac{1}{\sigma_i^2} (y_i - a_1X_1 - a_2X_2 - \cdots - a_mX_m)^2 \tag{126}$$

The minimum is found by setting the partial derivatives of χ^2 with respect to the parameters to zero:

$$\frac{\partial \chi^2}{\partial a_k} = -2 \sum_{i=1}^{n} \frac{1}{\sigma_i^2} (y_i - a_1X_1 - a_2X_2 - \cdots - a_mX_m)$$

$$X_k = 0 \quad k = 1, \ldots, m \tag{127}$$

Dropping the weights $1/\sigma_i^2$ temporarily for clarity, we obtain a set of m simultaneous equations in the m unknown a_k:

$$\sum y_iX_1 = a_1 \sum X_1X_1 + a_2 \sum X_2X_1 + \cdots + a_m \sum X_mX_1$$
$$\sum y_iX_2 = a_1 \sum X_1X_2 + a_2 \sum X_2X_2 + \cdots + a_m \sum X_mX_2$$
$$\vdots$$
$$\sum y_iX_m = a_1 \sum X_1X_m + a_2 \sum X_2X_m + \cdots + a_m \sum X_mX_m \tag{128}$$

where the summations run over all experimental data points i. These equations are known as normal equations. The solution—the values of a_k—can easily be found using matrix algebra. Since two (column) matrices are equal if their corresponding elements are equal, the set of equations can be written in matrix form as

$$\begin{bmatrix} \sum y_iX_1 \\ \sum y_iX_2 \\ \vdots \\ \sum y_iX_m \end{bmatrix} = \begin{bmatrix} a_1 \sum X_1X_1 + a_2 \sum X_2X_1 + \cdots + a_m \sum X_mX_1 \\ a_1 \sum X_1X_2 + a_2 \sum X_2X_2 + \cdots + a_m \sum X_mX_2 \\ \vdots \\ a_1 \sum X_1X_m + a_2 \sum X_2X_m + \cdots + a_m \sum X_mX_m \end{bmatrix} \tag{129}$$

The right-hand column matrix can be written as the product of a square matrix α and a column matrix **a**:

$$
\begin{bmatrix} \sum y_i X_1 \\ \sum y_i X_2 \\ \vdots \\ \sum y_i X_m \end{bmatrix} = \begin{bmatrix} \sum X_1 X_1 & \sum X_2 X_1 & \cdots & \sum X_m X_1 \\ \sum X_1 X_2 & \sum X_2 X_2 & \cdots & \sum X_m X_2 \\ \vdots & & & \\ \sum X_1 X_m & \sum X_2 X_m & \cdots & \sum X_m X_m \end{bmatrix} \begin{bmatrix} \mathbf{a}_1 \\ \mathbf{a}_2 \\ \vdots \\ \mathbf{a}_m \end{bmatrix}
\tag{130}
$$

or

$$
\beta = \alpha \cdot \mathbf{a}
\tag{131}
$$

This equation can be solved for **a** by premultiplying both sides of the equation with the inverse matrix α^{-1}:

$$
\alpha^{-1}\beta = \alpha^{-1}\alpha \mathbf{a} = \mathbf{I}\mathbf{a}
\tag{132}
$$

or, **I** being the identity matrix,

$$
\mathbf{a} = \alpha^{-1}\beta
\tag{133}
$$

Introducing the weights again, the elements of the matrices are given by

$$
\beta_j = \sum_{i=1}^{n} \frac{1}{\sigma_i^2} y_i X_j \qquad j = 1, \ldots, m
\tag{134}
$$

$$
\alpha_{jk} = \sum_{i=1}^{n} \frac{1}{\sigma_i^2} X_k X_j \qquad j = 1, \ldots, m, \, k = 1, \ldots, m
\tag{135}
$$

and

$$
a_j = \sum_{k=1}^{m} \alpha_{jk}^{-1} \beta_k \qquad j = 1, \ldots, m
\tag{136}
$$

where α_{jk}^{-1} are the elements of the inverse of matrix α^{-1}.

The uncertainty in the estimate of a_j is due to the uncertainty of each measurement multiplied by the effect of measurement on a_j:

$$
s_{a_j}^2 = \sum_{i=1}^{n} \sigma_i^2 \left(\frac{\partial a_j}{\partial y_i}\right)^2
\tag{137}
$$

Since α_{jk}^{-1} is independent of y_i, the partial derivative is simply

$$
\frac{\partial a_j}{\partial y_i} = \frac{1}{\sigma_i^2} \sum_{k=1}^{m} \alpha_{jk}^{-1} X_k(i)
\tag{138}
$$

After calculation it can be shown that

$$
s_{a_j}^2 = \sum_{k=1}^{m} \sum_{l=1}^{m} \alpha_{jk}^{-1} \alpha_{jl}^{-1} \left[\sum_{i=1}^{n} \frac{1}{\sigma_i^2} X_k X_l\right]
\tag{139}
$$

the term between brackets being α_{kl}, and

$$
s_{a_j}^2 = \sum_{k} \sum_{l} \alpha_{jk}^{-1} \alpha_{jl}^{-1} \alpha_{kl} = \alpha_{jj}^{-1}
\tag{140}
$$

This results in the simple statement that the variance (square of uncertainty) of a fitted parameter a_j is given by the diagonal element j of the inverse matrix α^{-1}. The off-diagonal elements are the covariances. For this reason α^{-1} is often called the error matrix. Similarly, α is called the curvature matrix since its elements are a measure of the curvature of the χ^2 hypersurface in the m-dimensional parameter space. It can easily be shown that

$$\frac{1}{2} \frac{\partial^2 \chi^2}{\partial a_j \, \partial a_k} = \sum_i \frac{1}{\sigma_i^2} X_k X_j = \alpha_{jk} \tag{141}$$

If the uncertainties in the data points σ_i are unknown and the same for all data points $\sigma_i = \sigma$, these equations can still be used by setting the weights $w_i = 1/\sigma_i^2$ to 1. Assuming the fitting model is correct, σ can be estimated from

$$\sigma_i^2 = \sigma^2 \simeq s^2 = \frac{1}{n-m} \sum_i (y_i - a_1 X_1 - a_2 X_2 - \cdots - a_m X_m)^2 \tag{142}$$

The uncertainties in the parameters are then given by

$$s_{a_j}^2 = s^2 \alpha_{jj}^{-1} \tag{143}$$

If the uncertainties in the data points are known, the reduced χ^2 value can be calculated as a measure of the goodness of fit:

$$\chi_v^2 = \frac{1}{n-m} \sum_i \frac{1}{\sigma_i^2} (y_i - a_1 X_1 - a_2 X_2 - \cdots - a_m X_m)^2 = \frac{1}{n-m} \chi^2 \tag{144}$$

The expected value of χ_v^2 is 1.0, but because of the random nature of the experimental data values slightly smaller or larger than 1 are observed even for a "perfect" fit. χ_v^2 follows a χ^2 distribution with $n - m$ degrees of freedom, and a 90% confidence interval can be defined:

$$\chi^2(v, P = 0.95) \leq \chi_v^2 \leq \chi^2(v, P = 0.05) \tag{145}$$

where $\chi^2(v, P)$ is the (tabulated) critical value of the χ^2 distribution for v degrees of freedom at a confidence level P. Observed χ_v^2 values outside this interval indicate a deviation between the fit and the data that cannot be attributed to random statistical fluctuations.

B. Least-Squares Fitting Using Orthogonal Polynomials

A special group of linear functions are orthogonal polynomials. Orthogonality means that the polynomials are uncorrelated, and this has some distinct advantages. Let $P_j(x_i)$ be an orthogonal polynomial of degree j; a fitting function can then be constructed as a sum of these orthogonal polynomials of successive higher degree:

$$y(i) = \sum_{j=0}^{m} c_j P_j(x_i) \tag{146}$$

The least-squares estimates of the coefficients c_j are determined by minimizing the weighted sum of squares:

$$\chi^2 = \sum_{i=1}^{n} w_i(y_i - y(i))^2 \tag{147}$$

which results in a set of $m + 1$ normal equations in the $m + 1$ unknown. Since $P_j(x_i)$ are a set of orthogonal polynomials, they have the property that

$$\sum_{i=1}^{n} w_i P_j(x_i) P_k(x_i) = \gamma_k \delta_{jk} \tag{148}$$

with γ_k a normalization constant and $\delta_{jk} = 0$ for $j \neq k$.

Because of this property the matrix of the normal equations is diagonal and the polynomial coefficients are directly obtained from

$$C_j = \sum_{i=1}^{n} \frac{w_i y_i P_j(x_i)}{\gamma_j} \tag{149}$$

The variance of the coefficients is given by

$$\sigma_{cj}^2 = \frac{1}{\gamma_j} \tag{150}$$

Another advantage of the use of orthogonal polynomials is that the addition of one extra term $C_{m+1}P_{m+1}$ does not change the values of the already determined coefficients $c_0, \ldots, c_m$. Further, if the y_i are independent then also the c_j are independent; that is, the variance-covariance matrix is also diagonal. As a result much higher degree orthogonal polynomials can be fitted compared to ordinary polynomials without running into problems with ill-conditioned normal equations and oscillating terms.

Orthogonal polynomials can be constructed by recurrence relation

$$P_{j+1}(x_i) = (x_i - a_j)P_j(x_i) - b_j P_{j-1}(x_i) \qquad j = 0, \ldots, m - 1 \tag{151}$$

a_j and b_j are constants independent of y_i given by

$$a_j = \frac{\displaystyle\sum_{i=1}^{n} w_i x_i (P_j(x_i))^2}{\gamma_j} \qquad j = 0, \ldots, m \tag{152}$$

$$b_j = \frac{\displaystyle\sum_{i=1}^{n} w_i x_i P_j(x_i) P_{j-1}(x_i)}{\gamma_{j-1}} \qquad j = 0, \ldots, m \tag{153}$$

Further, the normalization factor is given by

$$\gamma_j = \sum_{i=1}^{n} w_i (P_j(x_i))^2 \tag{154}$$

and

$$b_0 = 0 \qquad \text{and} \qquad P_0(x_i) = 1 \tag{155}$$

Thus an example of a first-order orthogonal polynomial is

$$c_0 P_0 + c_1 P_1 = c_0 + c_1(x_i - a_0) \tag{156}$$

with

$$a_0 = \frac{\sum\limits_{i=1}^{n} w_i x_i}{\sum w_i} \tag{157}$$

C. Nonlinear Least Squares

In this section we consider the fitting of a function that is nonlinear in one or more fitting parameters. Examples of such functions are a decay curve,

$$y(x) = a_1 e^{a_2 x} \tag{158}$$

or a Gaussian on a linear background,

$$y(x) = a_1 + a_2 x + a_3 e^{-(x - a_4)^2 / 2 a_5^2} \tag{159}$$

Equation (158) is nonlinear in a_2. Similarly, Equation (159) is nonlinear in the parameters a_4 and a_5. Equation (158) of representative for a group of functions on which linear least squares can be applied after suitable transformation. Fitting with Equation (159) implies the application of truly nonlinear least-squares fitting with iterative optimization of the fitting parameters.

1. Transformation to Linear Functions

Taking the logarithm of Equation (158),

$$\ln y = \ln a_1 + a_2 x \tag{160}$$

and defining $y' = \ln y$, $a_1' = \ln a_1$, a linear (straight-line) fitting function is obtained:

$$y' = a_1' + a_2 x \tag{161}$$

and the method discussed earlier can be applied, but not without making the following important remark. We have transformed our original data y_i to $y_i' = \ln y_i$. Consequently also the variance $\sigma_i'^2$ has been changed according to the general error propagation formula:

$$\sigma_i'^2 = \sigma_i^2 \left(\frac{\partial y'}{\partial y} \right)^2 \tag{162}$$

or in this particular case,

$$\sigma_i'^2 = \frac{\sigma_i^2}{y_i} \tag{163}$$

Thus even if all original data points had the same uncertainty ($\sigma_i = \sigma$) and unweighed linear least-squares fitting could have been used, after the transformation a weighted linear least-squares fit is required. The results of the fit are the parameters with their associated uncertainties of the transformed equation, and to obtain the original parameter we must perform a backtransformation with the appropriate error propagation,

$$a_1 = e_1^{a'}$$ (164)

$$S_{a'}\frac{2}{1} = S_{a'}\frac{2}{1}\left(\frac{\partial a_1}{\partial a_1'}\right)^2 = a_1 S_{a'}\frac{2}{1}$$ (165)

3. General Nonlinear Least Squares

For the general case of least-squares fitting with a function that is nonlinear in one or more of its fitting parameters, no direct solution exists. Still, we can define the object function χ^2:

$$\chi^2 = \sum_i \frac{1}{\sigma_i^2}(y_i - y(x_i, a))^2$$ (166)

whose minimum will be reached when the partial derivative with respect to the parameters are zero, only this results in a set of m nonlinear equations for which no general solution exists. The other approach to the problem is then to consider χ^2 as a continuous function of the parameters a_j (i.e., χ^2 takes a certain value of each set of values of the parameters a_j for a given data set $\{x_i, y_i, \sigma_i^2\}$). χ^2 thus forms a hypersurface in the m-dimensional space formed by the fitting parameter a_j. This surface must be searched to locate the minimum of χ^2. Once found, the corresponding coordinate values of the axes are the optimum values of the fitting parameters. Figure 36 illustrates how such a χ^2 hypersurface might look for a fitting function with two parameters.

The problem of nonlinear least-squares fitting is thus reduced to the problem of finding the minimum of a function in an m-dimensional space. Any algorithm that performs this task should operate according to the following:

1. Given some initial set of values for the parameters a_{ini}, evaluate χ^2:

$$\chi^2_{old} = \chi^2(a_{ini})$$

2. Find a new set of values a_{new} such that $\chi^2_{new} < \chi^2_{old}$
3. Test the minimum of the χ^2 value:
    if χ^2_{new} is the (true) minimum
            accept a_{new} as the optimum values of the fit
    else
            $\chi^2_{old} = \chi^2_{new}$ and repeat step 2.

From this scheme the iterative nature of nonlinear least-squares fitting methods becomes evident. Moreover, it shows some other important aspects of these methods: initial values are required to start the search, we need a procedure to obtain a new set of parameters, which preferably are such that χ^2 is decreasing, and we need to be sure that the true minimum, not some local minimum, is finally reached.

A variety of algorithms has been proposed, ranging from brute-force mapping procedures, dividing the m-dimensional parameter space into small cells and evaluating χ^2 in each point, to more subtle simplex search procedures [84]. The most important group of algorithms is nevertheless based on evaluation of the curvature

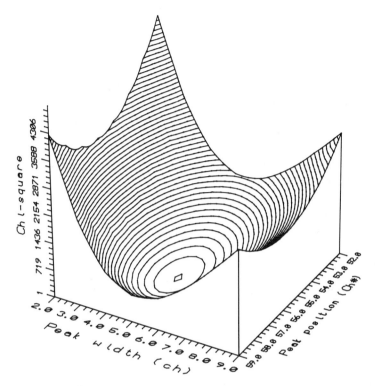

Figure 36 χ^2 surface showing the variation in χ^2 as a function of the peak position and the peak width parameter (peak 10,000 counts high, peak maximum at channel 56, and width five channels).

matrix. The gradient method and first-order expansion are discussed briefly since they form the basis of the most widely used Leverberg-Marquardt algorithm [1,21,85].

a. The Gradient Method

Having a fitting function $y = y(x, a)$ and χ^2 defined as a function of the m parameters a_j:

$$\chi^2 = \chi^2(a) = \sum_{i=1}^{n} \frac{1}{\sigma_i^2} (y_i - y(x_i, a))^2 \tag{167}$$

the gradient of χ^2 in the m-dimensional parameter space is given by

$$\nabla\chi^2 = \sum_j \frac{\partial\chi^2}{\partial a_j} \mathbf{j} \tag{168}$$

where $\mathbf{j}$ is the unit vector along the axis j and the components of the gradient are given by

$$\frac{\partial\chi^2}{\partial a_j} = -2 \sum_i \frac{1}{\sigma_i^2} (y_i - y(x_i, a)) \frac{\partial y}{\partial a_j} \tag{169}$$

It is convenient to define

$$\beta_j = -\frac{1}{2}\frac{\partial \chi^2}{\partial a_j} \tag{170}$$

The gradient gives the direction in which χ^2 increases most rapidly. A minimum search method can thus be developed on this basis. Given the current set of parameters a_j, a new set of parameters a_j' is calculated (for all j simultaneously):

$$a_j' = a_j + \Delta a_j \beta_j \tag{171}$$

which follows the direction of steepened descent and guarantees a decrease in χ^2 (at least if the appropriate step size Δa_j are taken).

The gradient method works quite well away from the minimum, but near the minimum the gradient becomes very small (at the minimum even zero). Fortunately, the method discussed next behaves in the opposite way.

b. First-Order Expansion

If we write the fitting function $y(x_i, a)$ as a first-order Taylor expansion of the parameters a_j around y_0,

$$y(x, a) = y_0(x, a) + \sum_j \frac{\partial y_0(x, a)}{\partial a_j} \delta a_j \tag{172}$$

we obtain an (approximation) to the fitting function that is linear in the parameter increments δa_j. $y_0(x, a)$ is the value of the fitting function for some initial set of parameter a. Using this function we can now express χ^2 as

$$\chi^2 = \sum_i \frac{1}{\sigma_i^2}\left(y_i - y_0(x_i, a) - \sum_j \frac{\partial y_0(x_i, a)}{\partial a_j}\delta a_j\right)^2 \tag{173}$$

and we can use the method of linear least squares to find the parameters δa_j so that χ^2 is minimal. We are thus fitting the difference $y_i' = y_i - y_0(x_i, a)$ with the derivatives as variables and the increments δa_j as unknowns. With reference to the section on linear least-squares fitting [Eq. (126)],

$$X_j = \frac{\partial y_0(x_i)}{\partial a_j} \tag{174}$$

and [Eqs. (134) and (135)],

$$\beta_j = \sum_{i=1}^{n} \frac{1}{\sigma_i^2}(y_i - y_0(x_i))\frac{\partial y_0(x_i)}{\partial a_j} \tag{175}$$

$$\alpha_{jk} = \sum_{i=1}^{n} \frac{1}{\sigma_i^2}\frac{\partial y_0(x_i)}{\partial a_j}\frac{\partial y_0(x_i)}{\partial a_k} \tag{176}$$

defining a set of m normal equations in the unknowns δa_j

$$\beta = \alpha \cdot \delta a \tag{177}$$

with solution

$$\delta a_j = \sum_{k=1}^{m} \alpha_{jk}^{-1}\beta_k \tag{178}$$

It is not very difficult to prove that

$$\beta_j = -\frac{1}{2}\frac{\partial \chi_0^2}{\partial a_k} \tag{179}$$

that is, the component of the gradient of χ^2 at the point of expansion and

$$\alpha_{jk} \simeq \frac{1}{2}\frac{\partial^2 \chi_0^2}{\partial a_j\, \partial a_k} \tag{180}$$

Thus α_{jk} in Equation (176) is the first-order approximation to the curvature matrix whose inverse is the error matrix.

The first-order expansion of the fitting function is closely related to the first-order Taylor expansion of the χ^2 hypersurface itself:

$$\chi^2 = \chi_0^2 + \sum_j \frac{\partial \chi_0^2}{\partial a_j}\, \delta a_j \tag{181}$$

where χ_0^2 is the χ^2 function at the point of expansion:

$$\chi_0^2 = \sum_{i=1}^{n} \frac{1}{\sigma_i^2} (y_i - y_0(x_i, a))^2 \tag{182}$$

At the minimum the partial derivation of χ^2 with respect to the parameter a_k is zero:

$$\frac{\partial \chi^2}{\partial a_k} + \frac{\partial \chi_0^2}{\partial a_k} + \sum_j \frac{\partial^2 \chi_0^2}{\partial a_j\, \partial a_k}\, \delta a_k = 0 \tag{183}$$

This results in a set of equations in the parameters δa_k:

$$\frac{\partial \chi_0^2}{\partial a_k} = -\sum_{j=1}^{m} \frac{\partial \chi_0^2}{\partial a_j\, \partial a_k}\, \delta a_k \tag{184}$$

$$\beta_k = \sum \alpha_{jk}\, \delta a_k \tag{185}$$

which is the same set of equations, except that in the expansion of the fitting function only a first-order approximation of the curvature matrix is used.

Since near the minimum the first-order expansion of the χ^2 surface is a good approximation, we can also conclude that the first-order expansion of the fitting function (which is computationally more elegant since only derivatives of functions, not of χ^2 are required) will yield parameter increments δa_j, which will direct us toward the minimum. For each linear parameter in the fitting function the first-order expansion of the function in this parameter is exact and the calculated increment δa_j is such that the new value $a_j + \delta a_j$ is optimum (for the given set of nonlinear parameters that may not yet be at their optimum value).

c. The Marquardt Algorithm

Based on the observation that away from the minimum the gradient method is effective and near the minimum the first-order expansion is useful, Marquardt developed an algorithm that combines both methods using a scaling factor λ that moves the algorithm either in the direction of the gradient search or into the direction of first-order expansion [85].

The diagonal terms of the curvature matrix are modified in the following way:

$$\alpha'_{jk} = \begin{cases} \alpha_{jk}(1 + \lambda) & j = k \\ \alpha_{jk} & j \neq k \end{cases} \tag{186}$$

where α_{jk} is given by Equation (176) and the matrix equation to be solved for the increments δa_j is

$$\beta_j = \sum_k \alpha'_{jk} \, \delta a_k \tag{187}$$

When λ is very large ($\lambda \gg 1$), the diagonal elements of α dominate and Equation (187) reduces to

$$\beta_j \approx \alpha'_{jj} \, \delta a_k \tag{188}$$

$$\delta a_k \simeq \frac{1}{\alpha'_{jj}} \beta_j \simeq \frac{1}{\alpha'_{jj}} \frac{\partial \chi^2}{\partial a_k} \tag{189}$$

which is the gradient, scaled by a factor α'_{jj}. On the other hand, for small values of λ ($\lambda \ll 1$) the solution is very close to first-order expansion.

The algorithm proceeds as follows:

1. Given some initial values of the parameters a_j evaluate $\chi^2 = \chi^2(\mathbf{a})$ and initialize $\lambda = 0.0001$.
2. Compute β and α matrices using Equations (175) and (176).
3. Modify the diagonal elements $\alpha'_{jj} = \alpha_{jj} + \lambda$, and compute δa.
4. If $\chi^2(\mathbf{a} + \delta\mathbf{a}) \geq \chi^2(\mathbf{a})$
   increase λ by a factor of 10 and repeat step 3.
   If $\chi^2(\mathbf{a} + \delta\mathbf{a}) < \chi^2(\mathbf{a})$
   decrease λ by a factor of 10.
   accept new parameters estimates $\mathbf{a} \leftarrow \mathbf{a} + \delta\mathbf{a}$ and repeat step 2.

The algorithm thus performs two loops, the inner loop incrementing λ until χ^2 starts to decrease and the outer loop calculating successive better approximations to the optimum values of the parameters. The outer loop can be stopped when χ^2 decreases by a negligible absolute or relative amount.

Once the minimum is reached the diagonal elements are an estimate of the uncertainty in the fitting parameters just as in linear least squares:

$$s^2_{aj} = \alpha_{jj}^{-1} \tag{190}$$

which is equal to α'^{-1}_{jj}, provided the scaling factor λ is much smaller than 1.

In Section X a number of computer programs for linear and nonlinear least-squares fitting are given. Further information can be found in many textbooks [21]. The book by Bevington [1] contains a very clear and practical discussion of the least-squares method.

X. COMPUTER IMPLEMENTATION OF VARIOUS ALGORITHMS

In this section a number of computer routines related to spectrum evaluation are listed. The calculation routines are written in FORTRAN. Some example programs, using the FORTRAN routines, are written in C. The programs were tested using

MicroSoft FORTRAN Version 4.0A and C Version 5.1. Most of the routines are written for clarity rather than optimized for speed or minimum space requirement.

A. Smoothing

1. Savitsky and Golay Polynomial Smoothing

Subroutine SGSMTH calculates smoothed spectrum using a second-degree polynomial filter (see Sec. III.B.2).

Input: Y   original spectrum
         NCHAN   number of channels in the spectrum
         ICH1, ICH2   first and last channel number to be smoothed
         IWID   width of the filter ($2m + 1$), IWID < 42
Output: S   smoothed spectrum, defined only between ICH1 and ICH2

```
      SUBROUTINE SGSMTH (Y, S, NCHAN, ICH1, ICH2, IWID)
      INTEGER*2 NCHAN, ICH1, ICH2, IWID
      REAL*4 Y(0:NCHAN-1), S(0:NCHAN-1)
      REAL C(-20:20)
C --  Calculate filter coefficients
      IW = MIN(IWID, 41)
      M = IW/2
      SUM = FLOAT ((2*M-1) * (2*M+1) * (2*M+3))
      DO 10 J = -M, M
        C(J) = FLOAT (3*(3*M*M+3*M-1-5*J*J))
10    CONTINUE
C --  Convolute spectrum with filter
      JCH1 = MAX(ICH1, M)
      JCH2 = MIN(ICH2, NCHAN-1-M)
      DO 30 I = JCH1, JCH2
        S(I) = 0.
        DO 20 J = -M, M
          S(I) = S(I) + C(J) * Y(I+J)
20      CONTINUE
        S(I) = S(I)/SUM
30    CONTINUE
      RETURN
      END
```

2. Low Statistics Digital Filter

Subroutine LOWSFIL smooths spectrum using low statistics digital filter algorithm (see Sec. III.B.3).

Input: Y   original spectrum
         NCHAN   number of channels
         ICH1, ICH2   first and last channel to be smoothed
         IFWHM   FWHM in channels of a peak in the middle of the smoothing region

Output: S   smoothed spectrum, defined only between ICH1 and
                              ICH2

```
      SUBROUTINE LOWSFIL (Y,S,NCHAN,ICH1,ICH2,IFWHM)
      INTEGER*2 NCHAN, ICH1, ICH2, IFWHM
      REAL Y(0:NCHAN-1), S(0:NCHAN-1)
      LOGICAL NOKSLOPE, STOOHIGH
      REAL AFACT, FFACT, MFACT, RFACT
      PARAMETER (AFACT=75., FFACT=1.5, MFACT=10.,
     *RFACT=1.3)
C -- Adjust smoothing region
      IW = NINT(FFACT*IFWHM)
      JCH1 = MAX(ICH1-IW, IW)
      JCH2 = MIN(ICH2+IW, NCHAN-1-IW)
      DO 100 I = JCH1, JCH2
        IW = NINT (FFACT*IFWHM)
        SUML = 0.
        SUMR = 0.
        DO 20 J = I-IW, I-1
          SUML = SUML + Y(J)
20      CONTINUE
        DO 30 J = I+1, I+IW
          SUMR = SUMR + Y(J)
30      CONTINUE
C -- Adjust window
50      CONTINUE
        SUMT = SUML + Y(I) + SUMR
        IF (SUMT .GT. MFACT) THEN
          SLOPE = (SUMR+1.)/(SUML+1.)
          NOKSLOPE = SLOPE.GT.RFACT .OR. SLOPE .LT.
     *    1./RFACT
          STOOHIGH = SUMT .GT. AFACT*SQRT (Y(I))
          IF (NOKSLOPE .OR. STOOHIGH .AND. IW .GT. 1)
     *    THEN
            SUML = SUML - Y(I-IW)
            SUMR = SUMR - Y(I+IW)
            IW = IW - 1
            GOTO 50
          ENDIF
        ENDIF
C -- Smoother value
        S(I) = SUMT/FLOAT(2*IW+1)
100   CONTINUE
C -- Copy data points that could not be smoothed
      DO 110 I = ICH1, JCH1-1
        S(I) = Y(I)
110   CONTINUE
      DO 120 I = JCH2+1, ICH2
        S(I) = Y(I)
120   CONTINUE
      RETURN
      END
```

B. Peak Search

Subroutine LOCPEAKS locates peaks in a spectrum using positive part of top hat filter (see Sec. III.C).

Input:	Y	spectrum
	NCHAN	number of channels in the spectrum
	R	peak search sensitivity factor, typically 2–4
	IWID	width of the filter, approximately equal to the FWHM of the peaks
	MAXP	maximum number of peaks to locate (size of array IPOS)
Output:	NPEAK	number of peaks found
	IPOS	peak positions (channel number)

The routine is optimized for speed and requires no other arrays than the spectrum and a table to store the peak maxima found. This is achieved by using a variant of the top hat filter; that is, for a filter width of 5 the coefficients are -1, -1, $+1$, $+1$, $+1$, $+1$, $+1$, -1, -1, -1. The next point in the filtered spectrum can thus be calculated from the current by subtracting and adding, and only the value of the previous, the current, and the next point in the filtered spectrum are retained. This makes the routine quite cryptic but it works very fast and reliably.

```
      SUBROUTINE LOCPEAKS (Y, NCHAN, IWID, R, IPOS,
     *NPEAKS, MAXP)
      INTEGER*2 NCHAN, IWID, NPEAKS, MAXP
      INTEGER*2 IPOS (MAXP)
      REAL Y (NCHAN), R
C --  Number of channels in the top of the filter must
C     be odd, and at least 3
      NP = MAX ((IWID/2) *2+1, 3)
      NPEAKS = 0
C --  Calculate half-width and start and stop channel
      N = NP/2
      I1 = NP
      I2 = NCHAN-NP
C --  Initialize running sums
      I = I1
      TOTAL = 0.
      TOP = 0.
      DO 20 K = -N*2, NP
        TOTAL = TOTAL + Y(I1+K)
20    CONTINUE
      DO 22 K = -N, N
        TOP = TOP + Y(I1+K)
22    CONTINUE
C --  Loop over all channels
      LASTPOS = 0
      SENS = R*R
      FI = 0.
      FNEXT = 0.
```

```
          SNEXT = 0.
          DO 100 I = I1+1, I2
            TOP = TOP - Y(I-N-1) + Y(I+N)
            TOTAL = TOTAL - Y(I-NP) + Y(I+NP)
            FPREV = FI
            FI = FNEXT
            SI = SNEXT
            FNEXT = TOP + TOP - TOTAL
            SNEXT = TOTAL
C     Significant?
            IF(FI.GT.0. .AND. (FI*FI.GT.SENS*SI)) THEN
C     Find maximum
              IF (FI .GT. FPREV .AND. FI .GT. FNEXT) THEN
                IF (FPREV.GT.0. .AND. FNEXT.GT.0.) THEN
C     and store (channel number is array index - 1 and
C     FI refers to I-1)
                  NEWPOS = I-2
                  IF (NEWPOS.GT.LASTPOS+2) THEN
                    NPEAKS = NPEAKS+1
                    IPOS (NPEAKS) = NEWPOS
                    LASTPOS = NEWPOS
                    IF (NPEAKS .EQ. MAXP) RETURN
                  ENDIF
                ENDIF
              ENDIF
            ENDIF
100       CONTINUE
          RETURN
          END
```

C. Background Estimation

1. Peak Stripping

Subroutine SNIPBG, variant of SNIP algorithm, calculates background via peak stripping (see Sec. IV.A).

Input:	Y	spectrum
	NCHAN	number of channels in the spectrum
	ICH1, ICH2	first and last channel of region to calculate the background
	FWHM	width parameter for smoothing and stripping algorithm, set to the average FWHM of peaks in the spectrum; typical value 8.0 (floating point!)
	NITER	number of iterations of SNIP algorithm, typically 24
Output:	YBACK	calculated background in the region ICH1–ICH2

Comment: Uses subroutine SGSMTH

```
          SUBROUTINE SNIPBG (Y, YBACK, NCHAN, ICH1, ICH2,
     *FWHM, NITER)
          INTEGER*2 NCHAN, ICH1, ICH2, NITER
```

```
      REAL*4 Y (0:NCHAN-1), YBACK (0:NCHAN-1), FWHM
      PARAMETER (SQRT2=1.4142, NREDUC=8)
C-- Smooth spectrum
      IW = NINT (FWHM)
      I1 = MAX (ICH1-IW, 0)
      I2 = MIN (ICH2+IW, NCHAN-1)
      CALL SGSMTH (Y, YBACK, NCHAN, I1, I2, IW)
C-- Square-root transformation over required spectrum
C   region
      DO 10 I = I1, I2
         YBACK (I) = SQRT (MAX(YBACK(I), 0.))
10    CONTINUE
C-- Peak stripping
      REDFAC = 1.
      DO 30 N = 1, NITER
C.. Set width, reduce width for last NREDUC iterations
         IF (N .GT. NITER-NREDUC)REDFAC = REDFAC/SQRT2
         IW = NINT (REDFAC*FWHM)
         DO 20 I = ICH1, ICH2
            I1 = MAX (I-IW, 0)
            I2 = MIN (I+IW, NCHAN-1)
    *       YBACK (I) = MIN (YBACK (I), 0.5*(YBACK (I1)
    *       +YBACK (I2)))
20    CONTINUE
30    CONTINUE
C-- Backtransformation
      DO 40 I = ICH1, ICH2
         YBACK (I) = YBACK (I) *YBACK (I)
40    CONTINUE
      RETURN
      END
```

2. Orthogonal Polynomial Background

Subroutine OPOLBAC fits the background of a pulse-height spectrum using an orthogonal polynomial. Background channels are selected by adjusting the weight of the fit (see Sec. IV.C).

Input:	NPTS	number of data points (channels)
	X	array of channel numbers
	Y	array of spectrum
	R	adjustable parameter [Eqs. (51) and (52)], typical value 2
Output:	YBACK	array of fitted background
	W	array of weights
	A, B	coefficients of the orthogonal polynomial
	C	fitted parameters of the orthogonal polynomial
	SC	uncertainties of C
	FAILED	logical variable TRUE if no convergence after MAX-ADJ weight adjustments
	RCHISQ	reduced χ^2 value of the fitted background
Workspace:	WORK1, WORK2 of size NPTS	

The routine calls ADJWEIG to adjust the weights according to Equations (51) and (52). Further, the subroutine ORTPOL is used to fit the polynomial. The iteration (adjustment of weights) stops when all coefficients c_j change less than 1 standard deviation or when the maximum number of iterations is reached.

```
          SUBROUTINE OPOLBAC (NPTS,X,Y,W,YBACK,WORK1,WORK2,
         *NDEGR,A,B,C,COLD,SC,FAILED,RCHISQ,R)
          INTEGER NPTS, NDEGER
          REAL*4 X(NPTS),Y(NPTS),W(NPTS),YBACK(NPTS),
         *WORK1(NPTS),WORK2(NPTS)
          REAL*4 A(NDEGR),B(NDEGR),C(NDEGR),COLD(NDEGR),
         *SC(NDEGR)
          REAL*4 RCHISQ, R
          LOGICAL*2 FAILED
          PARAMETER (MAXADJ=20)
          LOGICAL*2 NEXT
C -- Initialize
          DO 10 J = 1, NDEGR
             COLD(J) = 0.
10        CONTINUE
          DO 20 I = 1, NPTS
             YBACK(I) = 0.
20        CONTINUE
C -- Main iteration loop
          DO 100 K = 1, MAXADJ
C .. Calculate weights
          CALL ADJWEIG (NPTS,Y,W,YBACK R,NBPNTS)
C .. Fit orthogonal polynomial
          CALL ORTPOL (NPTS,X,Y,W,YBACK,WORK1,WORK2,
         *   NDEGR,A,B,C,SC SUMSQ)
          RCHISQ = SUMSQ/FLOAT (NBPNTS - NDEGR)
          S = SQRT (RCHISQ)
C .. Test if further adjustments of weights is required
          NEXT = .FALSE.
          DO 30 J = 1, NDEGR
            SC(J) = S * SC(J)
            IF (ABS (COLD(J) -C(J)) .GT. SC(J)) NEXT =
         *.TRUE.
            COLD(J) = C(J)
30        CONTINUE
C .. convergence
          IF(.NOT.NEXT) THEN
             FAILED = .FALSE.
             RETURN
          ENDIF
100       CONTINUE
C -- No convergence after MAXADJ iterations
          FAILED = .TRUE.
          RETURN
          END

          SUBROUTINE ADJWEIG (NPTS,Y,W,YFIT,R,NBPNTS)
C ** Adjust weights to emphasize the background
          INTEGER NPTS, NBPNTS
```

```
          REAL*4 Y(NPTS), W(NPTS), YFIT(NPTS), R
          NBPNTS = 0
C -- Loop over all data points
          DO 10 I = 1, NPTS
            IF (YFIT (I) .GT. 0.) THEN
              IF (Y(I) .LE. YFIT(I) + R*SQRT(YFIT(I)))
          *THEN
C .. Point is considered as background
                W(I) 1./YFIT(I)
                NBPNTS = NBPNTS+1
            ELSE
C .. Point is NOT considered as background.
                W(I) = 1. / (Y(I) - YFIT(I)) **2
              ENDIF
          ELSE
C .. Background <= 0, weight based original data
C    (initial condition)
                W(I) = 1./MAX(Y(I),1.)
                NBPNTS = NBPNTS+1
            ENDIF
10        CONTINUE
          RETURN
          END
```

D. Filter-Fit Method

The C program FILFIT is a test implementation of the filter-fit method (see Sec. VI). This program simply coordinates all input and output, allocates the required memory space, and calls two FORTRAN routines that do the actual work. The subroutine TOPHAT returns the convolute of a spectrum with the top hat filter or the weights (the inverse of the variance of the filtered spectrum). The general-purpose subroutine LINREG is called to perform the multiple linear least-squares fit. The output includes the reduced χ^2 value, the parameters of the fit a_j (which is an estimate of the ratio of the intensity in the analyzed spectrum to the intensity in the standard for the considered x-ray lines), and their standard deviation. The program spends most of the time on input; calculation times are of the order of seconds for a few hundred channels and 5–10 references. The routine GETSPEC reads the spectral data and must be supplied by the user.

```
/* Program FILFIT */
#include <stdio.h>
#include <malloc.h>
#include <float.h>
#include <math.h>

void fortran TOPHAT ();
void fortran LINREG ();
float spec [2048];

main ()
  {
    int nchan, first_ch_fit, last_ch_fit, width, ierr;
    int i, first_ch_ref, last_ch_ref, ref, num_ref;
```

```
int num_points;
int filter_mode = 0, weight_mode = 1, ioff;
float meas_time, ref_meas_time, *scale_fac;

float *x, *xp, *y, *w, *yfit, *a, *sa, chi;
double *beta, *alpha;
char filename [64];

/* input width of tophat filter */
 scanf ("%hd", &width);

/* input spectrum to fit and fitting region */
 scanf ("%s", filename);
 scanf ("%hd %hd", &first_ch_fit, &last_ch_fit);
 nchan = Getspec (spec, filename, &meas_time);
 num_points = last_ch_fit - first_ch_fit + 1;

/* filter spectrum and store in y[] */
 y = (float *)calloc(num_points, sizeof(float));
 TOPHAT (spec, y, &nchan, &first_ch_fit,
 &last_ch_fit, &width, &filter_mode);

/* calculate weights of fit and save in w[] */
 w = (float *)calloc(num_points, sizeof(float));
 TOPHAT (spec, w, &nchan, &first_ch_fit,
 &last_ch_fit, &width, &weight_mode);

/* read reference spectra, filter and store in x[] */
 scanf ("%hd", &num_ref);
 scale_fact = (float *)calloc(num_ref,
 sizeof(float));
 x = (float *)calloc(num_points*num_ref,
 sizeof(float));
 for( ref = 0; ref < num_ref; ref++ )
  {
    scanf ("%s", filename);
    nchan = Getspec(spec, filename, &ref_meas_time);
    scale_fac [ref] = ref_meas_time / meas_time;
    scanf ("%hd %hd", &first_ch_ref, &last_ch_ref);
    if( first_ch_ref < first_ch_fit )
    first_ch_ref = first_ch_fit;
    if( last_ch_ref > last_ch_fit )
    last_ch_ref = last_ch_fit;
    ioff = ref * num_points + first_ch_ref -
    first_ch_fit;
    xp = x + ioff;
    TOPHAT(spec,xp,&nchan,&first_ch_ref,&last_ch_
  *ref,&width,&filter_mode);
  }

/* perform least squares fit */
 yfit = (float *)calloc(num_points, sizeof(float));
 a = (float *)calloc(num_ref, sizeof(float));
 sa = (float *)calloc(num_ref, sizeof(float));
 beta = (double *)calloc(num_ref, sizeof(double));
 alpha = (double *)calloc(num_ref* (num_ref+1)/2,
```

```
 *sizeof(double)));
  LINREG(y, w, x, &num_points, &num_ref, &num_points,
*&num_ref, yfit, a, sa, &chi, &ierr, beta, alpha);

  if(ierr == 0)
    {
     printf("Filter fit:       Chi-square=%f\n", chi);
     printf("Standard    Int. in analyse spectrum/Int.
*in standard\n");
     for(i=0; i<num_ref; i++)
      printf("%hd         %f    %f\n", i+1, a[i]*scale_
*fac[i], sa[i]*scale_fac[i]);

     for(i=0; i<num_points; i++)
       {
        printf("%4hd %7.0f %9.2f", first_ch_fit+i,
       *y[i], yfit[i];
        for(ref=0;ref<num_ref;ref++)
          printf(" %7.0f", x[ref * num_points+i]);
        printf("\n");
       }
    }
}


      SUBROUTINE TOPHAT(IN,OUT,NCHAN,IFIRST,ILAST,IWIDTH,
     *MODE)
      INTEGER*2 NCHAN,IFRST,ILAST,IWIDTH,MODE
      REAL*4 IN(NCHAN), OUT(1)
C ** Tophat filter of width IWIDTH, MODE = 0 calculate
C    filtered spectrum,
C    MODE !=0 calculate weights (1/variance of filtered
C    spectrum)
C -- Calculate filter constants.
      IW = IWIDTH
      IF(MOD(IW,2) .EQ.0) IW = IW+1
      FPOS = 1./FLOAT (IW)
      KPOS = IW/2
      IV = IW/2
      FNEG = -1./FLOAT(2*IV)
      KNEG1 = IW/2 + 1
      KNEG2 = IW/2 + IV
      N = 0
C -- Loop over all requested channels.
      DO 30 I = IFRST+1, ILAST+1
C .. Central positive part,
        YPOS = 0.
        DO 10 K = -KPOS, KPOS
          IK = MIN(MAX(I+K,1),NCHAN)
          YPOS = YPOS + IN(IK)
10      CONTINUE
C .. Left and right negative part,
        YNEG = 0.
        DO 20 K = KNEG1, KNEG2
          IK = MIN(MAX(I-K,1),NCHAN)
```

```
                  YNEG = YNEG + IN(IK)
                  IK = MIN(MAX(I+K,1),NCHAN)
                  YNEG = YNEG + IN(IK)
      20      CONTINUE
            N = N + 1
            IF( MODE.EQ.0 ) THEN
              OUT(N) = FPOS * YPOS + FNEG * YNEG
            ELSE
              VAR = FPOS*FPOS*YPOS + FNEG*FNEG*YNEG
              OUT(N) = 1. / MAX(VAR, 1.)
            ENDIF
      30      CONTINUE
            RETURN
            END
```

E. Fitting Using Analytical Functions

The C program NLRFIT is an example implementation of nonlinear spectrum fitting using an analytical function (see Sec. VII). The program only coordinates input and output. The actual fitting is done using the Marquardt algorithm with the FORTRAN subroutine MARQFIT (see later). The fitting function consists of a poly-nomial background with NB terms and NP Gaussians. The background parameters and the area, position, and width of each Gaussian are optimized during the fit. The fitting function is calculated using the routine FITFUNC. The derivatives of the fitting function with respect to the parameters are calculated by the routine DERFUNC.

```
/* Program NLRFIT */
#include <stdio.h>
#include <malloc.h>
#include <float.h>
#include <math.h>
#define MAX_PEAKS 10
#define MAX_CHAN 1024

void fortran MARQFIT();
float spec[MAX_CHAN];
/* Fortran common block structure COMMON /FITFUN/ NB, NP */
struct common_block { short NB, NP ;};
extern struct common_block fortran FITFUN;

main ()
  {
  char specfile [64];
  int nchan, first_ch_fit, last_ch_fit, nb, np;
  int i, j, n, num_points, num_param, ierr, max_iter;
  float ini_pos[MAX_PEAKS], ini_wid[MAX_PEAKS];
  float *x, *xp, *y, *w, *yfit, *a, *sa, chi, lamda;
  float crit_dif;
  float *b, *beta, *deriv, *alpha;
  double *work;
```

```
/* Input of parameters and spectral data */
   scanf ("%s", specfile);
   scanf ("%hd %hd %hd %f",&first_ch_fit,&last_ch_fit,
   &max_iter,&crit_dif);
   scanf ("%hd %hd", &np, &nb);
   for (i=0; i<np; i++)
     scanf ("%f %f",&ini_pos [i], &ini_wid[i]);
   nchan = GetSpec(spec, specfile);

   num_points = last_ch_fit - first_ch_fit + 1;
   num_param = nb + 3*np;

  /* Allocate memory for y[], w[], x[] */
  y = (float *)calloc(num_points, sizeof(float));
  w = (float *)calloc(num_points, sizeof(float));
  x = (float *)calloc(num_points, sizeof(float));

  /* Store independent var. (spectrum), weights and
  dep. var. (channel #) */
   for (i=first_ch_fit, n=0; i<=last_ch_fit; i++, n++)
     {
       y[n] = spec[i];
       w[n] = (spec[i] > 0.)? 1./spec[i] : 1.;
       x[n] = (float)i;
     }

  /* allocate memory for other arrays required */
  yfit = (float *)calloc(num_points, sizeof(float));
  a = (float *)calloc(num_param, sizeof(float));
  sa = (float *)calloc(num_param, sizeof(float));
  b = (float *)calloc(num_param, sizeof(float));
  beta = (float *)calloc(num_param, sizeof(float));
  deriv = (float *)calloc(num_param, sizeof(float));
  alpha = (float *)calloc(num_param*(num_param+1)/2,
  sizeof(float));
  work = (double *)calloc(num_param*(num_param+1)/2,
  sizeof(double));

  /* initialize, all linear parameters to zero, peak
  position and width to their initial guesses */
  lamda = 0.001;
  for(i=0; i<np; i++)
    {
      a[nb+np+i] = ini_pos[i];
      a[nb+2*np+i] = ini_wid[i];
    }

 /* perform least squares fit */
 FITFUN.NP = np;
 FITFUN.NB = nb;
 MARQFIT(&ierr, &chi, &lamda, &crit_dif, &max_iter,
 x, y, w, yfit, &num_points, a, sa, &num_param,
 b,  beta, deriv, alpha, work);

 if(ierr == 0)
   {
```

```c
      printf("\nNon-linear fit:  Chi-square=%f\n", chi);
      printf("Polynomial background parameters\n");
      for(i=0; i<nb; i++)
        printf("%hd     %f   %f\n", i+1, a[i], sa[i]);
      printf("Peak parameters Area Position Width\n");
      for(i=0; i<np; i++)
        {
          printf("%hd %10.0f   %-10.0f", i+1, a[nb+i],
        *sa[nb+i]);
          printf("%10.3f   %-10.3f", a[nb+np+i],
        *sa[nb+np+i]);
          printf("%10.3f    %-10.3f\n", a[nb+2*np+i],
        *sa[nb+2*np+i]);
        }
      for(i=0; i<num_points; i++)
        printf("%4hd %7.0f %9.2f\n", first_ch_fit+i,
        y[i], yfit[i]);
    }
}
```

```fortran
      SUBROUTINE FITFUNC(X, YFIT, NPTS, A, NTERMS)
      REAL*4 X(NPTS), YFIT(NPTS), A(NTERMS)
      COMMON/FITFUN/NB, NP
C ** Fitting function, polynomial background and NP
C    gaussians
C    with position, width, and area as parameters
      PARAMETER(SQR2PI=2.50663)
C -- Loop over all channels
      DO 100 I = 1, NPTS
C .. background
         YFIT(I) = A(1)
         DO 20 J = 2, NB
           YFIT(I) = YFIT(I) + A(J) * X(I)**(J-1)
20       CONTINUE
C .. Peaks
         DO 30 K = 1, NP
           AREA = A(NB+K)
           POS = A(NB+NP+K)
           SWID = A(NB+2*NP+K)
           Z = ((POS-X(I))/SWID)**2
           IF(Z.LT.50.) THEN
             G = EXP(-Z/2.)/SWID/SQR2PI
             YFIT(I) = YFIT(I) + AREA*G
           ENDIF
30       CONTINUE
100   CONTINUE
      RETURN
      END

      SUBROUTINE DERFUNC(X, NPTS, A, NTERMS, DERIV, I)
      REAL*4 X(NPTS), A(NTERMS), DERIV(NTERMS)
      COMMON/FITFUN/NB, NP
```

```
C ** Derivatives of fitting function: polynomial
C      background and NP gaussians
C      with position, width, and area as parameters
       PARAMETER(SQR2PI=2.50663)
C -- Derivatives of function with respect to the
C      background parameters
       DERIV(1) = 1.
       DO 10 J = 2, NB
       DERIV(J) = X(I)**(J-1)
10     CONTINUE
C -- Derivatives of function with respect to the peak
C      parameters
       DO 30 K = 1, NP
       AREA = A(NB+K)
       POS = A(NB+NP+K)
       SWID = A(NB+2*NP+K)
       Z = ((POS-X(I))/SWID)**2
       IF(Z.LT.50.) THEN
          G = EXP(-Z/2.)/SWID/SQR2PI
C .. Peak area
          DERIV(NB+K) = G
C .. Peak position
          DERIV(NB+NP+K) = -AREA*G*(POS-X(I))/SWID/SWID
C .. Peak width
          DERIV(NB+2*NP+K) = AREA*G*(Z-1.)/SWID
       ELSE
          DERIV(NB+K) = 0.
          DERIV(NB+NP+K) = 0.
          DERIV(NB+2*NP+K) = 0.
       ENDIF
30     CONTINUE
       RETURN
       END
```

F. Monte Carlo Methods

1. Uniform Random Number Generator

Function URAND is a FORTRAN function returning uniform distributed random numbers in the interval $0 \le U < 0$. The random number generator is based on Knuth's "subtractive" method [21] (see Sec. VIII.A)

Input:	ISEED	set to any negative number to initialize the random generator
Output:	URAND	uniform random number in the interval $0 \le$ URAND < 1

```
       REAL*4 FUNCTION URAND(ISEED)
       INTEGER*2 ISEED
       REAL*4 UTABLE(56)
       REAL*4 UBIG, USEED
```

```
      PARAMETER (UBIG=4000000., USEED=1618033.)
      SAVE I1, I2, UTABLE, INIT
C -- Initialize table
      IF( ISEED.LT.0 .OR. INIT.EQ.0 ) THEN
         U = USEED + FLOAT(ISEED)
         U = MOD( U, UBIG )
         UTABLE(55) = U
         UTMP = 1.
         DO 10 I = 1, 54
           II = MOD( I*21, 55 )
           UTABLE(II) = UTMP
           UTMP = U - UTMP
           IF( UTMP.LT.0. ) UTMP = UTMP + UBIG
           U = UTABLE(II)
10       CONTINUE
         DO 30 K = 1, 4
           DO 20 I = 1, 55
             UTABLE(I)=UTABLE(I)-UTABLE(1+MOD(I+30,55))
             IF(UTABLE(I).LT.0) UTABLE(I) = UTABLE(I) +
     *         UBIG
20           CONTINUE
30       CONTINUE
         I1 = 0
         I2 = 31
         ISEED = 1
         INIT = 1
      ENDIF
C -- Get next "random" number
      I1 = I1+1
      IF(I1.EQ.56) I1=1
      I2 = I2+1
      IF(I2.EQ.56) I2=1
      U = UTABLE(I1) - UTABLE(I2)
      IF(U.LT.0.) U = U + UBIG
      UTABLE(I1) = U
      URAND = U/UBIG
      RETURN
      END
```

2. Normal Distributed Random Deviate

Function NRAND returns a normally distributed random number with zero mean
and unit variance using the Box-Muller method (see Sec. VIII.B).

Input: ISEED   set to any negative number to initialize the random
                           sequence
Output: NRAND   normally distributed random number deviates with
                           zero mean and unit variance

```
      REAL*4 FUNCTION NRAND(ISEED)
      INTEGER*2 ISEED
      SAVE NEXT, FAC, V1, V2
      IF(NEXT.EQ.0 .OR. ISEED.LT.0) THEN
```

```
10      CONTINUE
        V1 = 2. * URAND(ISEED) - 1.
        V2 = 2. * URAND(ISEED) - 1.
        R = V1*V1 + V2*V2
        IF(R.GE.1. .OR. R.EQ.0.) GOTO 10
        FAC = SQRT( -2. * LOG(R)/R)
        NRAND = V1*FAC
        NEXT = 1
  ELSE
    NRAND = V2*FAC
    NEXT = 0
  ENDIF
  RETURN
  END
```

3. Poisson Distributed Random Deviate

Function PRAND can be used to produce approximately Poisson-distributed random deviates. For small numbers (<20) the direct method is used; for larger numbers the Poisson distribution is approximated by the normal distribution.

Input: Y   (population) mean of deviate
           ISEED   set to any negative number to initialize the random sequence

Output: PRAND   Poisson-distributed random deviate with mean Y.

```
        REAL*4 FUNCTION PRAND(Y,ISEED)
        INTEGER*2 ISEED
        REAL*4 Y
        REAL*4 NRAND, URAND
        IF(Y.LT.20.) THEN
C -- Use direct method
        G = EXP(-Y)
        PRAND = -1.
        T = 1.
10      CONTINUE
            PRAND = PRAND + 1.
          T = T * URAND(ISEED)
        IF(T.GT.G) GOTO 10
        ELSE
C -- Approximate by normal distribution
            PRAND = Y + SQRT(Y) * NRAND(ISEED)
        ENDIF
        RETURN
        END
```

G. Least-Squares Procedures

1. Linear Regression

Subroutine LINREG is a general-purpose (multiple) linear regression (see Sec. IX.A).

Input: Y   array of dependent variable
          W   array of weights $(1/\sigma_i^2)$
          X   matrix of independent variables
          N   number of data points
          M   number of independent variables (columns of X)
          NMAX,   size of X matrix
          MMAX
Output: YFIT   array of fitted Y values
          A   estimated least-squares parameters
          SA   standard deviation of A
          CHI χ^2 value
          IERR   error condition; -1 if fit failed (singular matrix)
Work-   BETA   of size M
space: ALPHA   of size $M*(M + 1)/2$

```
      SUBROUTINE LINREG(Y, W, X, N, M, NMAX, MMAX,
     *YFIT, A, SA, CHI, IERR, BETA, ALPHA)
      INTEGER*2 N, M, NMAX, MMAX, IERR
      REAL*4 Y(N), W(N), YFIT(N), A(M), SA(M), CHI
      REAL*4 X(NMAX,MMAX)
      REAL*8 BETA(M),ALPHA(1)
c     Accumulate BETA and ALPHA matrices
      JK = 0
      DO 10 J = 1, M
        BETA(J) = 0.0D0
        DO 2 I = 1, N
          BETA(J) = BETA(J) + W(I)*Y(I)*X(I,J)
    2   CONTINUE
        DO 6 K = 1, J
          JK = JK + 1
          ALPHA(JK) = 0.0D0
          DO 4 I = 1, N
            ALPHA(JK) = ALPHA(JK) + W(I)*X(I,K)*X(I,J)
    4     CONTINUE
    6   CONTINUE
   10 CONTINUE
c     Invert ALPHA matrix
      CALL LMINV(ALPHA, M, IERR)
      IF(IERR .EQ. -1) THEN
        RETURN
      ENDIF
c     Calculate fitting parameters A
      DO 20 J = 1, M
        A(J) = 0.
        JJ = J*(J-1)/2
        DO 12 K = 1, J
          JK = K + JJ
          A(J) = A(J) + ALPHA(JK)*BETA(K)
   12   CONTINUE
        DO 14 K = J+1, M
          JK = J + K*(K-1)/2
          A(J) = A(J) + ALPHA(JK)*BETA(K)
```

```
     14    CONTINUE
     20 CONTINUE
c     Calculate uncertainties in the parameters
         DO 30 J = 1, M
           JJ = J*(J+1)/2
           SA(J) = DSQRT(ALPHA(JJ))
     30 CONTINUE
c     Calculate fitted values and chi-square
         CHI = 0.
         DO 40 I = 1, N
           YFIT(I) = 0.
           DO 32 J = 1, M
             YFIT(I) = YFIT(I) + A(J)*X(I,J)
     32      CONTINUE
           CHI = CHI + W(I)*(YFIT(I)-Y(I))**2
     40 CONTINUE
         CHI = CHI/FLOAT(N-M)
         RETURN
         END
```

2. Orthogonal Polynomial Regression

Subroutine ORTPOL fits a degree orthogonal polynomial to a set data points $[x_i, y_i, w_i]$ (see Sec. IX.B)

Input:	NPTS	number of data points
	X	array of independent variables
	Y	array of dependent variables
	W	array of weights ($w_i + 1/\sigma_i^2$)
	NDEGR	degree of orthogonal polynomial to be fitted
Output:	A, B	parameters of the orthogonal polynomials
	C	fitted orthogonal polynomial coefficients
	SC	standard deviation of C
	SUMSQ	χ^2 value
Work-space:	PJ, PJMIN	of size NPTS

```
      SUBROUTINE ORTPOL(NPTS, X, Y, W, YFIT, PJ, PJMIN,
     *NDEGR, A, B, C, SC, SUMSQ)
      INTEGER NPTS, NDEGR
      REAL*4 X(NPTS), Y(NPTS), W(NPTS), YFIT(NPTS)
      REAL*4 PJ(NPTS), PJMIN(NPTS)
      REAL*4 A(NDEGR), B(NDEGR), C(NDEGR), SC(NDEGR),
     *SUMSQ
C --  Initialize
      DO 10 I = 1, NPTS
        PJ(I) = 1.
        PJMIN(I) = 0.
        YFIT(I) = 0.
 10     CONTINUE
      GAMJMIN = 1.
```

```
C -- Loop over all polynomial terms
      DO 100 J = 1, NDEGR
C .. Accumulate normalization factor, A and B constants
C    for term j
      GAMJ = 0.
      A(J) = 0.
      B(J) = 0.
      DO 20 I = 1, NPTS
        GAMJ = GAMJ + W(I)*PJ(I)*PJ(I)
        A(J) = A(J) + W(I)*X(I)*PJ(I)*PJ(I)
        B(J) = B(J) + W(I)*X(I)*PJ(I)*PJMIN(I)
20      CONTINUE
      A(J) = A(J)/GAMJ
      B(J) = B(J)/GAMJMIN
C .. Least squares estimate of coefficient C
      C(J) = 0.
      DO 30 I = 1, NPTS
        C(J) = C(J) + W(I)*Y(I)*PJ(I)
30      CONTINUE
      C(J) = C(J)/GAMJ
      SC(J) = SQRT(1./GAMJ)
C .. Contribution of this term to the fit
      DO 40 I = 1, NPTS
        YFIT(I) = YFIT(I) + C(J)*PJ(I)
40      CONTINUE
C .. Next polynomial term
      IF(J .LT. NDEGR) THEN
        DO 50 I = 1, NPTS
          PJPLUS = (X(I)-A(J))*PJ(I) - B(J)*PJMIN(I)
          PJMIN(I) = PJ(I)
          PJ(I) = PJPLUS
50        CONTINUE
        GAMJMIN = GAMJ
      ENDIF
100     CONTINUE
C -- Weighted sum of squares value
      SUMSQ = 0.
      DO 110 I = 1, NPTS
        SUMSQ = SUMSQ + W(I)*(Y(I)-YFIT(I))**2
110     CONTINUE
      RETURN
      END
```

3. Nonlinear Regression

Subroutine MARQFIT performs nonlinear least-squares fitting according to the Marquardt algorithm (see Sec. IX.C).

Input: MAXITER   maximum number of iterations
        X   array of independent variables
        Y   array of dependent variables
        W   array of weights ($w_i = 1/\sigma_i^2$)

	NPTS	number of data points
	NTERMS	number of parameters
	A	array of initial values of the parameters
Output:	IERR	error status; -1 indicates failure of fit
	CHISQR	reduced χ^2 value
	FLAMDA	Marquardt control parameter
	YFIT	array of fitted data points
	A	least-squares estimate of the fitting parameters
	SA	standard deviation of A
	CRIDIF	minimum percentage difference in two χ^2 values to stop the iteration
Work-space:	B, BETA, DERIV	of size NTERMS
	ALFA, ARR	of size NTERMS*(NTERMS + 1)/1

The routine requires two user-supplied subroutines: FITFUNC to evaluate the fitting function $y(i)$ with the current set of parameters a and DERFUNC to calculate the derivatives of the fitting function with respect to the parameters.

```
      SUBROUTINE MARQFIT(IERR, CHISQR, FLAMDA, CRIDIF,
     *MAXITER, X, Y, W, YFIT, NPTS, A, SA, NTERMS,
     *B, BETA, DERIV, ALFA, ARR)
      INTEGER*2 IERR, NPTS, NTERMS
      REAL*4 CHISQR, FLAMDA, CRIDIF
      REAL*4 X(NPTS), Y(NPTS), W(NPTS), YFIT(NPTS)
      REAL*4 A(NTERMS), SA(NTERMS)
      REAL*4 B(1), BETA(1), DERIV(1), ALFA(1)
      REAL*8 ARR(1)
C ** Marquardt algorithm for nonlinear least-squares
C    fitting
      PARAMETER (FLAMMAX=1E4, FLAMMIN=1E-6)
C -- Evaluate the fitting function YFIT for the current
C    parameters and save the chi-square value
      NITER = 0
      CALL FITFUNC(X, YFIT, NPTS, A, NTERMS)
      CHISQR = CHIFIT(Y, YFIT, W, NPTS, NTERMS)
      FLAMDA = 0.
C -- Set ALFA and BETA to zero, save the current value
C    of the parameters A
100   CONTINUE
      NITER = NITER+1
      CHISAV = CHISQR
      DO 110 J = 1, NTERMS
        B(J) = A(J)
        BETA(J) = 0.
110   CONTINUE
      DO 112 J = 1, NTERMS*(NTERMS+1)/2
        ALFA(J) = 0.
112   CONTINUE
```

```
C -- Accumulate Alpha and Beta matrices
      DO 120 I = 1, NPTS
        D = Y(I) - YFIT(I)
C .. Calculate derivatives at point i
        CALL DERFUNC(X, NPTS, A, NTERMS, DERIV, I)
        DO 120 J = 1, NTERMS
          BETA(J) = BETA(J) + W(I)*D*DERIV(J)
          JJ = J*(J-1)/2
          DO 120 K = 1,J
            JK = JJ + K
            ALFA(JK) = ALFA(JK) + W(I)*DERIV(J)*DERIV(K)
120   CONTINUE
C -- Test and scale ALFA matrix
      DO 140 J = 1, NTERMS
        JJ = J*(J-1)/2
        JJJ = JJ + J
        IF(ALFA(JJJ) .LT. 1.E-20) THEN
          DO 130 K = 1, J
            JK = JJ + K
            ALFA(JK) = 0.
130       CONTINUE
          ALFA(JJJ) = 1.
          BETA(J) = 0.
        ENDIF
        SA(J) = SQRT(ALFA(JJJ))
140   CONTINUE
      DO 160 J = 1, NTERMS
        JJ = J*(J-1)/2
        DO 150 K = 1, J
          JK = JJ + K
          ALFA(JK) = ALFA(JK)/SA(J)/SA(K)
150     CONTINUE
160   CONTINUE
C -- Store ALFA in ARR, modify the diagonal elements
C    with FLAMDA
200   CONTINUE
      DO 210 J = 1, NTERMS
        JJ = J*(J-1)/2
        DO 205 K = 1, J
          JK = JJ + K
          ARR(JK) = DBLE(ALFA(JK))
205     CONTINUE
        JJJ=JJ+J
        ARR(JJJ) = DBLE(1.+FLAMDA)
210   CONTINUE
C -- Invert matrix ARR
      CALL LMINV(ARR, NTERMS, IERR)
      IF(IERR .NE. 0) RETURN
C -- Calculate new values of parameters A
      DO 220 J = 1, NTERMS
        DO 220 K = 1, NTERMS
          IF(K .GT. J) THEN
            JK = J + K*(K-1)/2
```

```
          ELSE
            JK = K + J*(J-1)/2
          ENDIF
          A(J) = A(J) + ARR(JK)/SA(J) * BETA(K)/SA(K)
220     CONTINUE
C -- Evaluate the fitting function YFIT for the new
C    parameters and chi-square
          CALL FITFUNC(X, YFIT, NPTS, A, NTERMS)
          CHISQR = CHIFIT(Y, YFIT, W, NPTS, NTERMS)
          IF(NITER.EQ.1) FLAMDA = 0.001
C -- Test new parameter set
          IF(CHISQR .GT. CHISAV) THEN
C .. Iteration NOT successful, increase flamda and try
C    again
            FLAMDA = MIN(FLAMDA * 10.,FLAMMAX)
            DO 300 J = 1,NTERMS
              A(J) = B(J)
300         CONTINUE
            GOTO 200
          ENDIF
C .. Iteration successful, decrease LAMDA
          FLAMDA = MAX(FLAMDA/10.,FLAMMIN)
C .. Get next better estimate if required
          PERDIF = 100.*(CHISAV-CHISQR)/CHISQR
          IF(NITER.LT.MAXITER .AND. PERDIF.GT.CRIDIF) GOTO
     *100
C -- Calculate standard deviations and return
          DO 320 J = 1, NTERMS
            JJ = J*(J+1)/2
            SDEV = DSQRT( ARR(JJ) ) / SA(J)
            SA(J) = SDEV
320       CONTINUE
          RETURN
          END

          FUNCTION CHIFIT(Y, YFIT, W, NPTS, NTERMS)
          REAL*4 Y(NPTS), YFIT(NPTS), W(NPTS)
C ** Evaluate chi-square
          CHI = 0
          DO 300 I = 1, NPTS
            CHI = CHI + W(I) * (Y(I) - YFIT(I))**2
300       CONTINUE
          CHI = CHI/FLOAT (NPTS - NTERMS)
          CHIFIT = CHI
          RETURN
          END
```

4. Matrix Inversion

Subroutine LMINV is a general-purpose routine to invert a symmetrical matrix.

Input:	ARR	upper triangle and diagonal of real symmetrical matrix stored in linear array, size $= N*(N + 1)/2$.
	N	order of the matrix (number of columns)

Output: IERR   error status; IERR = 0 inverse obtained, IERR = -1
                                singular matrix
           ARR   upper triangle and diagonal of inverted matrix

```
      SUBROUTINE LMINV (ARR, N, IERR)
      INTEGER*2 N, IERR
      REAL*8 ARR(1)
      REAL*8 DIN,WORK,DSUM,DPIV
      INTEGER*2 I,IND, IPIV, J, K, KEND, KPIV, L, LANF,
     *LEND, LHOR, LVER, MIN
      KPIV = 0
      DO 10 K = 1, N
        KPIV = KPIV + K
        IND = KPIV
        LEND = K - 1
        DO 4 I = K, N
          DSUM = 0.D0
          IF(LEND .GT. 0) THEN
            DO 2 L = 1,LEND
              DSUM = DSUM + ARR(KPIV-L) * ARR(IND-L)
2           CONTINUE
          ENDIF
          DSUM = ARR(IND) - DSUM
          IF(I .EQ. K) THEN
            IF(DSUM .LE. 0.D0) THEN
              IERR = -1
              RETURN
            ENDIF
            DPIV = DSQRT(DSUM)
            ARR(KPIV) = DPIV
            DPIV = 1.D0/DPIV
          ELSE
            ARR(IND) = DSUM * DPIV
          ENDIF
          IND = IND + I
4       CONTINUE
10    CONTINUE
      IERR = 0
      IPIV = N*(N+1)/2
      IND = IPIV
      DO  20  I = 1, N
        DIN = 1.D0/ARR(IPIV)
        ARR(IPIV) = DIN
        MIN = N
        KEND = I -1
        LANF = N - KEND
        IF(KEND .GT. 0) THEN
          J = IND
          DO  14  K = 1, KEND
            WORK = 0.D0
            MIN = MIN -1
            LHOR = IPIV
```

```
            LVER = J
            DO  12  L = LANF, MIN
               LVER = LVER + 1
               LHOR = LHOR + L
               WORK = WORK + ARR(LVER) * ARR(LHOR)
12          CONTINUE
            ARR(J) = -WORK * DIN
            J = J - MIN
14          CONTINUE
        ENDIF
        IPIV = IPIV - MIN
        IND = IND -1
20      CONTINUE
        DO  30  I = 1, N
          IPIV = IPIV + I
          J = IPIV
          DO  24  K = I, N
            WORK = 0.D0
            LHOR = J
            DO  22  L = K,N
               LVER = LHOR + K - I
               WORK = WORK + ARR(LVER) * ARR(LHOR)
               LHOR = LHOR + L
22          CONTINUE
            ARR(J) = WORK
            J = J + K
24        CONTINUE
30      CONTINUE
        IERR = 0
        RETURN
        END
```

REFERENCES

1. P. R. Bevington, *Data Reduction and Error Analysis for the Physical Sciences*, McGraw-Hill, New York (1969).
2. P. J. Statham and T. Nashashibi, in *Microbeam Analysis*, (D. E. Newbury, Ed.), San Francisco Press, San Francisco, (1988), p. 50.
3. P. Van Espen, H. Nullens, and F. Adams, *X-ray Spectrom.* 9:126 (1980).
4. D. H. Wilkinson, *Nucl. Instrum. Methods* 95:259 (1971).
5. R. Gunnink, *Nucl. Instrum. Methods* 143:145 (1977).
6. D. C. Joy, *Rev. Scient. Instrum.* 56:1772 (1985).
7. J. Heckel and W. Scholz, *X-ray Spectrom.* 16:181 (1987).
8. J. L. Campbell, A. Perujo, and B. M. Millman, *X-ray Spectrom.* 16:195 (1987).
9. P. J. Statham, *X-ray Spectrom.* 5:154 (1976).
10. P. J. Statham, *X-ray Spectrom.* 5:16 (1976).
11. D. G. W. Smith, C. M. Gold, and D. A. Tomlinson, *X-ray Spectrom.* 4:149 (1975).
12. W. M. Sherry and J. B. Vander Sande, *X-ray Spectrom.* 6:154 (1977).
13. D. J. Bloomfield and G. Love, *X-ray Spectrom.* 14:8 (1985).
14. P. Van Espen and F. Adams, *X-ray Spectrom.* 5:123 (1976).
15. R. Cirone, G. E. Gigante, and G. Gualtieri, *X-ray Spectrom.* 13:110 (1984).

16. S. J. B. Reed and N. G. Ware, *J. Phys. E 5*:582 (1972).
17. G. I. Johansson, *X-ray Spectrom. 11*:194 (1982).
18. P. Van Espen, H. Nullens, and F. Adams, *Anal. Chem. 51*:1325 (1979).
19. P. Van Espen, H. Nullens, and F. Adams, *Anal. Chem. 51*:1580 (1979).
20. D. L. Massart, B. G. M. Vandeginste, S. N. Deming, Y. Michotte, and L. Kaufman, *Chemometrics: A textbook*, Elsevier, Amsterdam, 1988.
21. W. H. Press, B. P. Flannery, S. A. Teukolsky, and W. T. Vetterling, *Numerical Recipes in C, The Art of Scientific Computing*, Cambridge University Press, Cambridge, 1988.
22. P. A. Jansson, *Deconvolution with Applications in Spectroscopy*, Academic Press, New York, 1984.
23. D. Brook and R. J. Wynne, *Signal Processing Principles and Applications*, Edward Arnold, London, 1988.
24. L. A. Schwalbe and H. J. Trussell, *X-ray Spectrom. 10*:187 (1981).
25. J. Nunez, L. E. Rebollo Neira, A. Plastino, R. D. Bonetto, D. M. A. Guérin, and A. G. Alvarez, *X-ray Spectrom. 17*:47 (1988).
26. I. Gertner, O. Heber, J. Zajfman, D. Zajfman, and B. Rosner, *Nucl. Instrum. Methods B36*:74 (1989).
27. A. Savitzky and M. J. E. Golay, *Anal. Chem. 36*:1627 (1964).
28. P. H. Yule, *Nucl. Instrum. Methods 54*:61 (1967).
29. K. K. Nielson, *X-ray Spectrom. 7*:15 (1978).
30. C. G. Enke and T. A. Nieman, *Anal. Chem. 48*:705A (1976).
31. C. G. Ryan, E. Clayton, W. L. Griffin, S. H. Sie, and D. R. Cousens, *Nucl. Instrum. Methods B34*:396 (1988).
32. K. Janssens and P. Van Espen, *Anal. Chim. Acta 184*:117 (1986).
33. K. Janssens, W. Dorriné, and P. Van Espen, *Chemometrics Intelligent Lab. Systems 4*:147 (1988).
34. A. Robertson, W. V. Prestwich, and T. J. Kennett, *Nucl. Instrum. Methods 100*:317 (1972).
35. Philips and Marlow, *Nucl. Instrum. Methods 137*:525 (1976).
36. McCullagh and Helmer, *Report EGG-PHYS-5890*, Idaho National Engineering Laboratory, Idaho Falls, ID, (1982).
37. J. P. Op De Beeck, J. Hoste, *Atomic Energy Rev. 13*:151 (1975).
38. W. W. Black, *Nucl. Instrum. Methods 71*:317 (1969).
39. E. Clayton, P. Duerden and D. D. Cohen, *Nucl. Instrum. Methods B22*:64 (1987).
40. J. Kajfosz and W. M. Kwiatek, *Nucl. Instrum. Methods B22*:78 (1987).
41. S. Steenstrup, *J. Appl. Crystalogr. 14*:226 (1981).
42. R. Jenkins, R. W. Gould, and D. Gedcke, *Quantitative X-ray Spectrometry*, Marcel Dekker, New York (1981).
43. J. Hertogen, J. De Donder, and R. Gijbels, *Nucl. Instrum. Methods 115*:197 (1974).
44. E. P. Bertin, *Principles and Practice of X-ray Spectrometric Analysis*, Plenum Press, New York, (1970).
45. F. H. Schamber, in *X-ray Fluorescence Analysis of Environmental Analysis* (T. Dzubay, Ed.), Ann Arbor Science Publishers, Ann Arbor, MI, 1977, p. 241.
46. P. J. Statham, *X-ray Spectrom. 7*:132 (1978).
47. J. J. McCarthy and F. H. Schamber, *NBS Special Publ. 604*:273 (1981).
48. P. Van Espen, H. Nullens, and F. Adams, *Nucl. Instrum. Methods 142*:243 (1977).
49. P. Van Espen, H. Nullens, and W. Maenhaut, in *Microbeam Analysis* (D. E. Newbury, Ed.), San Francisco Press, San Francisco, (1979), p. 265.
50. P. Van Espen, K. Janssens, and J. Nobels, *Chemometrics Intelligent Lab. Systems 1*:109 (1986).
51. W. Maenhaut and J. Vandenhaute, *Bull. Soc. Chim. Belg. 95*:407 (1986).

52. J. H. McCrary, L. V. Singman, L. H. Ziegler, L. D. Looney, C. M. Edmonds, and C. E. Harris, *Phys. Rev. A 4*:1745 (1971).
53. S. I. Salem, B. G. Saunders, and G. C. Melson, *Phys. Rev. A 1*:1563 (1970).
54. S. I. Salem and R. J. Wimmer, *Phys. Rev. A 2*:1121 (1970).
55. J. H. Scofield, *Phys. Rev. 179*:9 (1970).
56. J. H. Scofield, *Phys. Rev. A 9*:1041 (1974).
57. J. H. Scofield, *Phys. Rev. A 10*:1507 (1974).
58. P. Van Espen, H. Nullens, and F. Adams, *Nucl. Instrum. Methods 145*:579 (1977).
59. L. A. McNelles and J. L. Campbell, *Nucl. Instrum. Methods 127*:73 (1975).
60. J. L. Campbell, B. M. Millman, J. A. Maxwell, A. Perujo, and W. J. Teesdale, *Nucl. Instrum. Methods B9*:71 (1985).
61. R. P. Gardner, A. M. Yacout, J. Zhang, and K. Verghese, *Nucl. Instrum. Methods A242*:299 (1986).
62. G. W. Phillips and K. W. Marlow, *Nucl. Instrum. Methods 137*:525 (1976).
63. P. Van Dyck and R. Van Grieken, *X-ray Spectrom. 12*:111 (1983).
64. H. Nullens, P. Van Espen, and F. Adams, *X-ray Spectrom. 8*:104 (1979).
65. G. W. Phillips, *Nucl. Instrum. Methods 153*:449 (1978).
66. U. Wätjen, *Nucl. Instrum. Methods B22*:29 (1987).
67. J. L. Campbell, W. Maenhaut, E. Bombelka, E. Clayton, K. Malmqvist, J. A. Maxwell, J. Pallon, and J. Vandenhaute, *Nucl. Instrum. Methods B14*:204 (1986).
68. C. J. Duffy, P. S. Z. Rogers, and T. M. Benjamin, *Nucl. Instrum. Methods B22*:91 (1987).
69. L. Zolnai and Gy. Szabó, *Nucl. Instrum. Methods B34*:118 (1988).
70. E. Bombelka, W. Koenig, and F.-W. Richter, *Nucl. Instrum. Methods B22*:21 (1987).
71. R. Breschinsky, E. Krush, and R. Wehrse, *Diplomarbeit, Fachbereich Physik*, Universität Bremen, Germany, 1979.
72. W. Petersen, P. Ketelsen, and A. Knöchel, *Nucl. Instrum. Methods A245*:535 (1986).
73. P. A. Aarnio and H. Lauranto, *Nucl. Instrum. Methods A276*:608 (1989).
74. J. T. Routti and S. G. Prussin, *Nucl. Instrum. Methods 72*:125 (1969).
75. P. A. Pella, L. Feng, and J. A. Small, *X-ray Spectrom. 14*:125 (1985).
76. B. B. Jensen and N. Pind, *Anal. Chim. Acta 117*:101 (1985).
77. E. Marageter, W. Wegscheider, and K. Müller, *Nucl. Instrum. Methods B1*:137 (1984).
78. E. Marageter, W. Wegscheider, and K. Müller, *X-ray Spectrom. 13*:78 (1984).
79. J. M. Doster and R. P. Gardner, *X-ray Spectrom. 11*:173 (1982).
80. J. M. Doster and R. P. Gardner, *X-ray Spectrom. 11*:181 (1982).
81. A. M. Yacout, R. P. Gardner, and K. Verghese, *Adv. X-ray Anal. 30*:121 (1987).
82. A. M. Yacout and W. L. Dunn, *Adv. X-ray Anal. 30*:113 (1987).
83. K. Verghese, M. Mickael, T. He, and R. P. Gardner, *Adv. X-ray Anal. 31*:461 (1988).
84. C. E. Fiori, R. L. Myklebust, and K. Gorlen, *NBS Spec. Publ. 604*:233 (1981).
85. D. W. Marquardt, *J. Soc. Ind. Appl. Math. 11*:431 (1963).

5
Quantification by XRF Analysis of Infinitely Thick Samples

J. L. de Vries*   *Eindhoven, The Netherlands*

Bruno A. R. Vrebos   *Philips Analytical Research Laboratories, Almelo, The Netherlands*

I. CORRELATION BETWEEN INTENSITIES AND CONCENTRATION

In quantitative analysis the measured x-ray fluorescent (XRF) intensity of a given element is converted into its weight concentration in the sample. As a first approximation one expects a linear relationship. Each atom of the analyte element i has the same probability to be excited by the primary photons and to emit its characteristic photons λ_i. Indeed, if we are dealing with separate atoms or ions, as in a gas or in a very dilute solution, the relation holds:

$$I_i = MW_i \tag{1}$$

where W_i is the weight fraction of i.

The constant M consists of many physical and instrumental factors:

- Shape and intensity of the primary beam
- Probability that an atom i emits its characteristic radiation λ_i
- Probability that these photons λ_i pass through the measuring channel: collimators, diffracting crystal, and pulse-height window
- Probability that these photons are being detected and registered

For a given instrument, voltage, and power on the x-ray tube, these factors remain constant for the analyte i and can be determined by measuring the fluorescent

---

* Retired.

intensity of the pure element i. However, we are dealing with compact specimens in which the atoms are bound into chemical compounds.

Both the primary rays and the fluorescent x-rays are absorbed by the different atoms in the sample.

A. General Relationship Between Intensity and Concentration

1. Excitation by Monochromatic Radiation

Let us first consider excitation by monochromatic radiation of wavelength λ_p (Fig. 1). Let us assume that the number of primary photons of wavelength λ_p striking the sample surface at an angle ψ_1 is $I_{p,0}$ per unit of time. These photons are absorbed by the various atoms in the sample. The remaining intensity $I_{p,x}$ at a layer Δx a distance x within the sample is thus

$$I_{p,x} = I_{p,0}e^{-\mu_s(\lambda_p)\rho_s \csc\psi_1 x} \tag{2}$$

where $\mu_s(\lambda_p)$ is the mass attenuation coefficient of the sample for λ_p and ρ_s its density. $\mu_s(\lambda_p)$ can be calculated from the weight fractions of the elements present in the sample and the mass attenuation coefficients:

$$\mu_s(\lambda_p) = \sum_{i=1}^{n} \mu_i(\lambda_p)W_i \tag{3}$$

where n is the total number of elements in the sample. In the layer Δx a certain fraction of the primary photons is absorbed by element i:

$$W_i \frac{\mu_i(\lambda_p)}{\mu_s(\lambda_p)} \tag{4}$$

To yield characteristic radiation, the photon must be absorbed by electrons in the appropriate shell. For example, if $K\alpha$ radiation is considered, only the fraction of primary photons absorbed by K electrons can yield the desired radiation. Absorption in the K shell is

$$\frac{r_i - 1}{r_i} \tag{5}$$

where r_i is the absorption jump ratio. Only a fraction of the excited atoms i in the K shell give rise to fluorescent rays λ_i; this fraction is called the fluorescence yield ω_i. Furthermore, only a fraction g_i emits $K\alpha$ rays; other possibilities are $K\beta$ photons. Only a certain fraction of these fluorescent rays, emitted in the

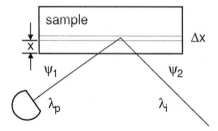

Figure 1 The sample is excited by radiation of wavelength λ_p.

direction of the detector, $d\Omega/4\pi$ ($d\Omega/4\pi$ is the solid angle, intercepted by the detector system), can pass through the collimator at an angle ψ_2 with respect to the sample surface. These secondary rays are absorbed from the layer Δx to the sample surface according to

$$e^{\mu_s(\lambda_i)\rho_s\csc\psi_2 x} \tag{6}$$

In total the contribution to the total emitted fluorescent radiation λ_i from the layer Δx is thus

$$I_i = I_{p,0}(\lambda_p)e^{-\mu_s(\lambda_p)\rho_s\csc\psi_1 x - \mu_s(\lambda_i)\rho_s\csc\psi_2 x}\ W_i\ \frac{r_i - 1}{r_i}\ \mu_i(\lambda_p)\omega_i g_i\ \Delta x\ \csc\ \psi_1\ \frac{d\Omega}{4\pi} \tag{7}$$

Integrating Equation (7) for x from zero to infinity gives

$$I_i = I_0(\lambda_p)P_i\ \frac{\mu_i(\lambda_i)}{\mu_s(\lambda_p)\ +\ G\mu_s(\lambda_i)}\ W_i \tag{8}$$

where P_i is a constant for a given element and a given spectrometer and G is equal to $\sin\psi_1/\sin\psi_2$. In this derivation we assumed that

1. The sample is completely homogeneous.
2. The primary rays are not scattered on their way to the layer Δx.
3. No enhancement effects occur.
4. The characteristic radiation is not scattered on its way to the sample surface.

2. Excitation by Continuous Spectra

When continuous radiation is used to excite the characteristic photons λ_i, all the primary wavelengths λ between the minimum wavelength λ_{min} and the absorption edge wavelength λ_{edge} of the analyte i can contribute to the excitation of i. Thus,

$$I(\lambda) = \int_{\lambda_{min}}^{\lambda_{edge}} J(\lambda)\ d\lambda \tag{9}$$

$J(\lambda)$ being the x-ray spectrum. By substituting this expression into Equation (8) and considering that the absorption of the primary and secondary rays by all elements must be taken into account, one obtains

$$I_i = P_iW_i \int_{\lambda_{min}}^{\lambda_{edge}} \frac{J(\lambda)\mu_i(\lambda)}{\sum_{j=1}^{n} W_j\mu_j(\lambda)\ +\ G\sum_{j=1}^{n} W_j\mu_j(\lambda_i)}\ d\lambda \tag{10}$$

The constant P contains, among other factors, the fluorescent yield and the weight in the series. $\mu_j(\lambda)$ is the absorption coefficient for element j for the primary photons and $\mu_j(\lambda_i)$ the absorption coefficient for element j for the characteristic wavelength λ_i of the analyte i.

B. Some Conclusions from this Equation

The integration over x [Eq. (8)] is taken from 0 to infinity. It is obvious that the first layers contribute more to the intensity of λ_i than the layers more inward. Theoretically, even at large values of x, still a very minor contribution to I_i is expected. Often infinite depth is defined arbitrarily as that thickness x of which the contribution of the layer Δx is 0.01% of the surface layer. The value of this infinite thickness depends on the value of the absorption coefficients and the density of the sample. In practice it may vary from a few micrometers for heavy matrices and long wavelengths to even centimeters for short wavelengths and light matrices, as in solutions.

For a given element and a fixed geometry, an efficiency factor $C(\lambda_p, \lambda_i)$ can be introduced:

$$C(\lambda_p, \lambda_i) = \frac{\mu_i(\lambda_p)}{\sum_{j=1}^{n} W_j \mu_j(\lambda_p) + G \sum_{j=1}^{n} W_j \mu_j(\lambda_i)} \tag{11}$$

It is obvious that the terms $\sum W_j \mu_j$ are the origin of nonlinear calibration lines, as variations in W_j influence the value of the denominator, the so-called matrix effect. For a pure metal this reduces to

$$C(\lambda_p, \lambda_i) = \frac{\mu_i(\lambda_p)}{\mu_i(\lambda_p) + G \mu_i(\lambda_i)} \tag{12}$$

The absorption jump ratio r varies for most common elements from about 4 to 10 for the light elements; G has for most equipment a value of 1.5–2. It follows that for a pure element and all values of λ_p, $\mu_i(\lambda_p)$ in the denominator dominates over the second term, which in the first approximation may thus be neglected. Thus $C(\lambda_p, \lambda_i)$ becomes almost independent of λ_p! It is thus not necessary to use a tube anode with characteristic rays close to the absorption edge when analyzing pure elements.

In the general case, however, the wavelengths in the primary spectrum close to the absorption edge are most effective in exciting the analyte i. The efficiency factor $C(\lambda_p, \lambda_i)$ is thus a combination of the absorption curve of analyte i as a function of λ and the spectral distribution. In the first approximation an "effective wavelength" λ_e can be introduced, which has the same effect for excitation of element i as the total primary spectrum. The exact value of this λ_e is influenced by the characteristic tube lines if they are active in exciting i. Otherwise λ_e in general has a value of approximately two-thirds of the absorption edge λ_{edge}. Its actual value, however, is dependent on the chemical composition of the specimen. For instance, for Fe, $\lambda_{edge} = 174$ pm: one expects a $\lambda_e = 116$ pm. In the ZnO-Fe_2O_3 system, λ_e was found to vary from 130 pm for 100% Fe_2O_3 to 119 pm for 10% Fe_2O_3 in ZnO.

Another interesting case is the analysis of a heavy element i in a light matrix. In the summation in the denominator of Equation (10), the term $W_i \mu_i(\lambda_p)$ and $W_j \mu_i(\lambda_i)$ is then the most important. The other terms can, in the first approximation, be neglected if W_i is not small. This means, however, that the terms W_i in the nominator and the denominator cancel and the measured intensity becomes independent of W_i; thus the analysis becomes impossible in this extreme case. A solution to this problem is found by making the influence of the terms $W_i \mu_i(\lambda)$ in the denominator less dominating by adding a large term $W_a \mu_a(\lambda)$. This can be done by making W_a large, for example by diluting, or $\mu_a(\lambda)$ large, by adding a heavy absorber. Equation (11) enables one to calculate beforehand how large this term should be to eliminate fluctuations in the concentration of the other elements.

In the derivation of the fluorescence of the analyte i in the preceding discussion, the following simplifications were made:

First, it was assumed that the primary rays follow a linear path to the layer Δx at height x. However, the primary rays may also be scattered. In general, the loss in intensity of the primary rays due to scattering may be neglected. The scattering effects become more important the more energetic the primary rays and the lighter the matrix. This scatter may give a higher background in the secondary spectrum, thus leading to poorer precision of the analysis. On the other hand, the excitation efficiency may be enhanced as the primary rays "dwell longer" in the active layers, thus having more probability to encounter atoms i. This effect may overrule the increase in background radiation. A case in hand is the analysis of Sn in oils, an analysis that gives better results using the Sn K lines at a high x-ray tube voltage than using the Sn L lines at moderate voltages. Incidentally, this scattering of the primary radiation makes it possible to check the voltage over the x-ray tube. According to Bragg's law, the primary spectrum is zero at angle

$$\theta_0 = \sin^{-1} \frac{n\lambda_{min}}{2d_{crystal}} \tag{13}$$

where $2d_{crystal}$ is the $2d$ spacing of the crystal used and n is an integer number. λ_{min} is given by

$$\lambda_{min} = \frac{1240 \text{ pm}}{kV} \tag{14}$$

In practice, a lower value of kilovoltage is found when Compton scattering dominates over Rayleigh scattering, and thus the λ_{min} found has a value too low by the Compton shift, which is about 24 pm for most spectrometers.

The integral in Equation (8) was taken from zero to infinity; further, it was assumed that the sample is completely homogeneous. This, of course, is never realized in practice because we deal with discrete atoms in chemical compounds. In powders, the different compounds may have a tendency to cluster. The particles in general have different sizes and shapes. Placing the sample into solution, either aqueous or solid (melt), may overcome this problem.

It was stated that infinite thickness may vary from 20 μm to a few centimeters, but the most effective layers are much thinner. Thus the number of discrete particles actually contributing to the fluorescent radiation may be rather small.

All atoms in the sample may be excited, emitting their characteristic radiation in all directions. These characteristic rays are absorbed by other atoms in the sample. If these x-rays are energetic enough, they may excite other atoms of the analyte i. Thus the intensity I_i measured is higher than expected from Equation (10). The intensity I_i is enhanced by the contributions of heavier elements in the sample. This enhancement is more pronounced if the rays of the enhancing elements are slightly more energetic than the energy of the absorption edge of the element i. The enhancement may contribute up to 40–50% of the total fluorescent radiation I_i, especially when the concentration of the enhancing elements is much greater than the concentration of the analyte. This applies even more for the light elements, for which the primary spectrum may not be very effective: the most intense wavelengths are far removed from the absorption edge of these light elements. Besides these first enhancement effects, a second enhancement may occur. The x-rays of element A, for example, may have sufficient energy to excite element B (where Z_A is higher than Z_B), which in turn emits its characteristic radiation λ_B, exciting the atoms of the analyte i. This phenomenon is referred to as secondary enhancement or tertiary fluorescence and may often be neglected. Even in favorable cases it rarely contributes more than a few percent to the intensity I_i of the analyte i.

II. FACTORS INFLUENCING THE ACCURACY OF THE INTENSITY MEASUREMENT

The total error of the analysis consists of many errors whose source may be

- Measurement of the intensity
- Reproducibility of the sample preparation
- Conversion of intensity into concentration

The error of a single determination may be found from n determinations of the same analysis, giving a mean result x_{mean} of all the separate results x. If n is sufficiently large the standard deviation s may be found from

$$s = \sqrt{\frac{\sum_{i=1}^{n} (x_i - x_{mean})^2}{n - 1}} \tag{15}$$

In this total deviation s random and systematic errors are combined. Random errors give an indication of the precision of an analysis, the scatter of results around a mean, whereas systematic errors are the reason for deviations of the mean value from the "true" value. An analysis may thus be precise but not very accurate if systematic errors are present, whereas accurate values could be found from the mean of widely scattered measurements if only large random errors were present. The total error of a measurement is comprised of all the separate errors. If only random errors are considered, the resulting standard deviation s is given by

$$s^2 = s_1^2 + s_2^2 + \cdots + s_n^2 \tag{16}$$

where $s_1, s_2, \ldots$ are the errors associated with, for example, intensity measurements, sample preparations, and instrumental settings. In practice it is often found that s is dependent on the concentration of the analyte W_i [1]. For example,

$$s = K\sqrt{W_i + W_b} \tag{17}$$

where W_b is a small concentration offset (typically 0.001). Thus K, rather than s, becomes an indication of the accuracy of the determination.

A. Random Errors

1. Counting Statistics

If an x-ray measurement consisting of the determination of a number of counts N is repeated n times, the results $N_1, N_2, N_3, \ldots, N_n$ would spread about the true value N_0. If n is large, the distribution of the measurements would follow a Gaussian distribution:

$$W(N) = \frac{1}{\sqrt{2\pi N}} e^{[-(N - N_0)^2/2N]} \tag{18}$$

provided N is also large. The standard deviation σ of the distribution is equal to $\sqrt{N_0}$, which approximates to $\sqrt{N_{\text{mean}}}$ if n and N are large, where N_{mean} is the mean of n determinations.

From the properties of the Gaussian distribution it follows that

- Of all values N, 68.3% are within $N_0 \pm \sigma$
- Of all values N, 95.4% are $N_0 \pm 2\sigma$
- Of all values N, 99.7% are $N_0 \pm 3\sigma$

Similarly, there is a certain probability that the true result N_0 will lie between $N - \sqrt{N}$ and $N + \sqrt{N}$, assuming the same distribution for N and N_0. The concentration analyzed for is dependent on the net count rate, which is the peak count rate R_p minus the background count rate R_b. The total measuring time T equals $t_p + t_b$, where t_p and t_b are the times measuring peak and background, respectively. In modern equipment there is no significant statistical error in the measurement of t. We can thus assume that R follows the same Gaussian distribution as N with the same relative standard deviation ϵ_N. ϵ_N is defined as

$$\epsilon_N = \frac{\sigma_N}{N} \tag{19}$$

Hence,

$$\epsilon_N = \frac{\sqrt{N}}{N} = \frac{1}{\sqrt{N}} = \frac{1}{\sqrt{R}\sqrt{t}} = \epsilon_R \tag{20}$$

and

$$\sigma_R = \epsilon_R R = \frac{\sqrt{R}}{\sqrt{t}} \tag{21}$$

It is obvious that the relative counting error decreases as t increases.

When a net count rate must be determined, we must measure the peak R_p and the background R_b: thus we have two independent variables. The standard deviation of the net intensity σ_d is given by

$$\sigma_d = \sqrt{\sigma_p^2 + \sigma_b^2} = \sqrt{\frac{R_p}{t_p} + \frac{R_b}{t_b}} \tag{22}$$

and the relative standard deviation ϵ_d by

$$\epsilon_d = \frac{\sqrt{R_p/t_p + R_b/t_b}}{R_p - R_b} \tag{23}$$

In general, there are three methods for dividing the total measuring time T available:

1. Fixed time: $t_p = t_b$ and $T = t_p + t_b$.
2. Same relative error; hence $N_p = N_b$ or

$$\frac{t_p}{t_b} = \frac{R_b}{R_p} \tag{24}$$

3. Optimum division of t_p and t_b, where σ_d reaches a minimum.

For a fixed time the resulting standard deviation is

$$\sigma_d = \sqrt{\frac{2}{T}} \sqrt{(R_p + R_b)} \tag{25}$$

For fixed counts,

$$\sigma_d = \sqrt{\frac{1}{T}} \sqrt{(R_p + R_b)} \sqrt{\frac{R_p}{R_b} + \frac{R_b}{R_p}} \tag{26}$$

for the optimum division,

$$\sigma_d = \sqrt{\frac{1}{T}} (\sqrt{R_p} + \sqrt{R_b}) \tag{27}$$

and

$$\epsilon_d = \frac{1}{\sqrt{T}} \frac{1}{\sqrt{R_p} - \sqrt{R_b}} \tag{28}$$

It can be demonstrated that for a given total measuring time T,

$$\sigma_{FTO} < \sigma_{FT} < \sigma_{FC} \tag{29}$$

where FTO refers to the method of fixed time optimized and FT and FC are for fixed time and fixed counts, respectively. When R_p is very large compared with R_b, σ_{FTO} approximates σ_{FT}; thus the fixed time method is commonly used. It follows that when aiming for optimum instrumental conditions the term $\sqrt{R_p} - \sqrt{R_b}$ is a good quality function. This parameter is often used as a figure of merit (FOM). Obviously, the highest value of $\sqrt{R_p} - \sqrt{R_b}$ gives the best result. When it can be assumed that $\sqrt{R_p} - \sqrt{R_b}$ approximates R_p (for low intensities and high background), optimizing this term equals optimizing

$$\frac{M}{\sqrt{R_b}} \tag{30}$$

where M is $R_p - R_b$, the slope of the calibration line in counts per second per percent.

If the ratio of two count rates R_1 and R_2 must be determined, the methods of fixed time and fixed count give the same result, whereas the method of optimum division, where

$$\frac{t_1}{t_2} = \sqrt{\frac{R_2}{R_1}} \tag{31}$$

always gives the best result.

The total counting error is the combination of instrumental error and counting statistics. When the count rate is very high the counting error is small, and it may be worthwhile to apply a ratio method to reduce possible instrumental errors. If the count rate is low, however, it is better to spend all the available time in analyzing the sample to reduce the counting error.

2. Instrumental Errors

If the instrumental and counting errors are random and independent variables, then

$$\epsilon_{tot} = \sqrt{\epsilon_{instr}^2 + \epsilon_{count}^2} \tag{32}$$

$$\epsilon_{instr} = \sqrt{\epsilon_{tot}^2 - \epsilon_{count}^2} \tag{33}$$

Although the counting error is dependent on the instrumental error, Equation (33) is still a good approximation. σ_{tot} can be found from a series of repeated results of one measurement with known σ_{count}. To check the instrumental instability, all the functions should be measured separately. A radioactive source ^{55}Fe, for example, can be used to check the detector and electronic circuitry. The x-ray tube can be checked by repeating the measurements with specimen and goniometer fixed. Recycling checks the goniometer, repositioning the sample checks the sample holder, and so on. Sometimes the error σ_{tot} found is smaller than expected, or even smaller than σ_{count}. This may indicate that an unexpected systematic error is involved, or there may be an uncorrected dead time in the equipment, that is, the true counting rate is higher than measured, which means that the relative error is smaller.

3. Detection Limit

A characteristic line intensity decreases with decreasing concentration of the analyte and finally disappears in the background noise. The true background intensity may be constant, but its measurements fluctuate around a mean value R_{bmean}. To be significantly different from the background, a signal R_p must, although it is larger than R_{bmean}, be distinguished from the spread in R_b. In other words, if we measure a signal R_p larger than R_b and we assume the analyte to be present, what, then, is the probability that our assumption is correct? If the measurements are random and follow a Gaussian distribution, then this probability is determined by σ_{R_b}. If the measurement R_p is higher than $R_b + 2\sigma_{R_b}$, then the probability that our assumption is correct is approximately 95%; if more certainty is required, say 99.7%, then R_p should be higher than $R_b + 3\sigma_{R_b}$. The net intensity is thus

$3\sigma_{R_b}$ and the detection limit (DL) is

$$DL = \frac{3\sigma_{R_b}}{M} \tag{34}$$

where M is the sensitivity in counts per second per percent. In x-ray spectrometry, however, the background signal is sample dependent and cannot be measured independently as in radioactivity measurements. Hence R_b must be measured in an off-peak location in the spectrum. The result R_b found in measuring time t seconds is assumed to be R_{bmean}, and σ_{R_b} is assumed to be $\sqrt{(R_b/t)}$. Thus two measurements must be made: R_p and R_b, each in time t seconds. The detection limit thus becomes equal to

$$DL = \frac{3\sqrt{2}}{M} \sqrt{\frac{R_b}{T}} \tag{35}$$

where $T = 2t$. If we are satisfied with a 95% probability that our assumption is correct, then

$$DL = \frac{2\sqrt{2}}{M} \sqrt{\frac{R_b}{T}} \tag{36}$$

which is roughly equal to

$$DL = \frac{3}{M} \sqrt{\frac{R_b}{T}} \tag{37}$$

It is obvious that the detection limit decreases if the counting time increases. However, the total error in R_b contains the instrumental error as well. There is thus no sense increasing the counting time when the instrumental error dominates.

4. Variation in X-ray Spectrum

The fluorescent intensity of the analyte I_i is in first approximation dependent on the primary spectrum by

$$I_i = Ki(V_0 - V_c)^p \tag{38}$$

where K is a constant, i is the current of the x-ray tube, V_0 the working voltage, V_c the excitation voltage, and p varies between 1 and 2, depending on the ratio of excitation by characteristic tube rays and white continuum.

With modern instruments the tube voltage is not dependent on the mains cycle, but they run on constant potential, which still may fluctuate. If the working voltage V_0 or the region or line of highest excitation probability is rather close to V_c, then small fluctuations in V_0 introduce a considerable error in R_i. For instance, if $V_0 = 1.5\ V_c$, then a 1% error in V_0 gives an error in I_i of 6% with $p = 2$. It is therefore better to run the tube at three to five times the excitation voltage of the analyte. Too high a voltage might introduce an unproportionly high background.

5. Other Instrument Errors

Other possible random instrumental errors include setting sample and goniometers. These errors must be checked by repeated measurements of one sample in a systematic way, for example,

- Repeated count with stationary sample and goniometer
- Repeated count with stationary sample, repositioning goniometer
- Repeated count with stationary goniometer, repositioning sample
- Repeated count with stationary goniometer, reloading sample holder

Some diffracting crystals have a rather high coefficient of expansion, and therefore their d value may fluctuate with fluctuation in temperature. This results in a setting of the goniometer, slightly off-peak, which introduces a change in measured intensity. Most modern equipment is therefore thermostabilized.

6. Particle Statistics

Only a limited volume of the sample can actually contribute to the fluorescent radiation. As long as this active volume is the same in standards and actual samples and the atomic distribution is completely homogeneous, this poses no problem. The atoms are bound into chemical compounds, however, forming finite particles with different chemical compositions. The analyte may occur only in particles with a certain chemical composition but not in other particles. Thus only these specific particles can contribute to the fluorescent radiation of i. The count rate R_i measured thus depends on the number of those particles present in the active volume, where, evidently, the first layers contribute most to the fluorescent radiation.

Table 1 gives an indication of the penetration depth of radiation of various wavelengths into matrices with varying absorption power. It is evident that for most solid samples most of the fluorescent radiation originates within 20 μm from the surface. Assuming an irradiated area of 1.5 cm² and an effective active layer of 10 μm, the active volume would be 15×10^8 μm³. If we were dealing with cubic particles of 100 μm for all the particles and the analyte i were present in only one kind of particle and its concentration was 10%, then only 150 particles could contribute to the fluorescent radiation. If we further assume a Gaussian distribution of all particles, equally shaped and of the same size, then the amount of the active particles N would have a standard deviation s of $\sqrt{N}$. In this case $N = 150$ and therefore s is approximately 12; thus there are 150 ± 12 "active" particles contributing to fluorescent radiation. These fluctuations would introduce

TABLE 1 "Infinite" Thickness, (μm) for Certain Analytical Lines as a Function of the Matrix[a]

Analytical line	Fe base	Mg base	H_2O solution	Borate	Borate/La_2O_3 10%
Sn $K\alpha$	300	10,000	100,000	70,000	10,000
Mo $K\alpha$	100	3,700	30,000	30,000	2,600
Ni $K\alpha$	12	340	2,400	2,000	300
Cr $K\alpha$	33	120	900	800	250
Al $K\alpha$	1.5	4	10	15	5
Na $K\alpha$	0.7	20	9	6	5
C $K\alpha$	0.3	0.3	4	1	0.4

[a] Both the incidence and exit angles are 45°, excitation by a Rh tube at 60 kV. (The influence of element-specific absorption can be seen from the values for Ni $K\alpha$ and Cr $K\alpha$ in the Fe base matrix, for example.)

an error into the measurements of $s = 8\%$ relative $(12 \div 150)$. The concentration found would for this reason alone vary as $10 \pm 0.8\%$. For a concentration of 1%, this error would be 25%. In practice these errors may be even larger: the irradiation of the sample is not homogeneous. The primary spectrum originates in a rather small anode and passes through a larger window and is thus cone shaped. Spinning the sample in its own plane during the analysis reduces this error. Furthermore, the first layers are the most effective, having only a small number of particles containing the analyte with corresponding larger relative errors.

It is evident that in extreme cases the effective layer is very thin, less than 1 μm. Care should thus be taken that the sample surface is as smooth as possible; surface irregularities, such as grooves and ridges, introduce a considerable error. Spinning the sample, again, reduces this error.

It is therefore vital that the sample be completely homogeneous. If this is not possible and powders must be analyzed, care should be taken that the particles are very small, less than a few micrometers in diameter. Sample preparation is outlined in Section III and discussed in more detail in Chapter 13.

B. Systematic Errors

1. Dead Time

It takes a certain time after an x-ray photon is detected in the counter and accompanying electronics before the counting circuit is ready to accept a following photon. Any photon entering the counter within this period, called the dead time of the counter circuit, is simply not registered and is thus lost. This dead time is of the order of a few microseconds. The losses are thus dependent on the actual count rate. The measured count rate R_m is always lower than the true count rate R_T. Their relation can be approximated by the expression

$$R_T = \frac{R_m}{1 - \tau_d R_m} \tag{39}$$

where τ_d is the dead time. For instance, if $\tau_d = 1$ μs and $R_m = 10^5$ cps, then the dead-time loss is approximately 10%. With modern equipment very high count rates can be handled, to reach sufficient precision in a short time. It is therefore necessary to reduce these losses. In most instruments an automatic dead-time correction circuit is included.

2. Matrix Effects

The fluorescent intensity of the analyte i is, as discussed before, not only dependent on its concentration but can also be strongly dependent on the composition of the sample itself. The primary rays are absorbed and scattered, and secondary fluorescence or enhancement may occur. All these effects depend on the chemical composition of the sample. The importance of these effects depends on the concentration of these matrix elements and their influence, for example, their absorption of primary and secondary rays. These matrix effects may introduce large systematic errors when they are not properly accounted for, as discussed in Section IV.

C. Choice of Optimal Conditions

Some consideration in choosing the analytical line are as follows:

1. High sensitivity, thus preferably the strongest line in the spectra, commonly the $K\alpha$
2. Low background radiation
3. Constant angle of Bragg diffraction
4. No coincidence with lines of other elements

Each of these items is discussed in more detail in the following sections.

1. For the light and medium Z elements, the $K\alpha$ line is by far the strongest line in the spectrum and is thus often chosen as the analytical line. For the elements excited over 40 keV, the L lines are preferred because the K lines in general cannot be used, the maximum voltage for most spectrometers being limited to 100 kV; an overvoltage of three times V_c or more is needed to get high characteristic intensity.
2. Another reason the L lines are preferred for the heavier elements is that in direct tube-excited XRF, the background due to scattering of the primary x-rays is much lower in the L region.
3. The wavelength of some analytical lines may shift slightly with the valence state of the elements, especially for the light elements, and thus the standard used in setting the goniometer to the analytical line should correspond to the sample in this respect. Another reason for an apparent shift in angle may be the change in d value of the analyzing crystal with temperature.
4. The analytical line should be, ideally, completely free of any disturbing lines. However, there are many sources of disturbing influences.

a. Spectral Overlap

Two or more characteristic lines may not be completely separated from the analytical line. This separation may be improved by using a crystal with better dispersion, that is, a lower d value. However, the choice must often be made between high intensity and high dispersion. If the disturbing line is due to a high-order crystal reflection, its influence may be strongly reduced by the proper setting of the pulse-height selector. In some cases, however, the escape peak of the analytical line originating in the detector may be able to pass the selector window. For instance, the third-order reflection using a penta erythritol (PE) crystal of the scattered tube lines of a Sc anode slightly interfere with the analysis for Al in a light matrix because their escape peak energy is very close to the energy of the Al $K\alpha$ line.

Often the overlap is due to a diagram line of an element of which another diagram line is free of overlap. Because, in general, two diagram lines of one element have a constant intensity ratio, the measured intensity of the nonoverlapped line of the disturbing element multiplied by a constant factor (experimentally determined) may be subtracted from the measured intensity of the analytical line to give the characteristic intensity.

b. Primary Radiation Scattered by the Specimen

The whole white spectrum is scattered by the specimen, including the characteristic lines, giving rise to a continuous background. If the sample consists of rather coarse grains, however, it may happen that a crystallite is in a favorable position for Bragg diffraction for a wavelength of the continuum; thus a sharp peak will be found in the spectral analysis. The influence of primary tube lines, coherently or incoherently scattered, may be eliminated by proper choice of the anode.

c. Spurious Reflections by the Analyzing Crystal

LiF (110) is a common analyzing crystal. The second order of reflection is used, as the first order is crystallographically forbidden. When a very, very high intensity is observed, however, a first-order reflection is still found, due to asymmetry of the electronic cloud. Similarly, a second order may be observed using a Ge crystal.

d. Satellite Lines

The common wavelength tables give only characteristic K, L, and M lines for most elements, and M lines of heavy elements may interfere with the K lines of light elements. Nondiagram satellite lines may also occur, often giving rise to an unexpected background level.

III. REMARKS ON SAMPLE PREPARATION
A. Metals

In general, metal samples need only surface treatment; a flat surface is required—no grooves or ridges. If the sample is spun during analysis, these requirements are less strict. Care should be taken that the softer metals in an alloy are not smeared over the surface during lapping or by abrasive paper, such as Pb in bronzes.

B. Powders

The characteristic particles should be very small, in particular if the grains have different chemical composition. In milling, differences in the mechanical hardness of the grains may have a disturbing effect. Care should be taken that even the hardest grain is sufficiently reduced in size to obtain reproducible intensities. For instance, in the cement industry the raw mix consists of calcite and quartz. It is well known that the intensity of Ca first increases considerably and Si decreases; later Ca decreases and Si increases with increased milling, as the softer grains

These powders, consisting of grains of different hardness, can be better analyzed in solid solutions.

C. Liquids

In most spectrometers liquids can be analyzed as such, even under reduced pressure. Care should be taken that the analyzed surface remains reproducible and that no gas bubbles are formed under influence of the primary x-rays, as in acid solutions. Liquid solutions are very well suited for reducing matrix effects and for adding various amounts of internal standards. However, they exhibit a high background level as a result of scattering.

D. Fusion

Placing the sample into solid solution by fusion is a general method to obtain homogeneous samples, but it is often rather time consuming (typically 5 to 15 minutes). The flux and fusion conditions should be chosen so that a homogeneous glass is formed without internal strain or residual grains. This may be checked by observing the surface in a scanning electron microscope or by looking at the glass bead through a polarizing microscope.

IV. CONVERTING INTENSITIES TO CONCENTRATION

The most simple equation relating intensity to concentration is

$$I_i = MW_i \tag{40}$$

where M is assumed to be a constant. This equation holds in general when the total effect of the matrix on the analyte i is constant, as in low-alloy steels or thin-film specimens. The intensity I_i in Equation (40) is a net intensity—measured intensity corrected for background, line overlap, and so on. In practice, the measured intensity is often used directly, without the subtraction of background, leading to a more general equation

$$I_i = B + MW_i \tag{41}$$

where B is the measured intensity when $W_i = 0$. If there is no line overlap, B is the background. This equation can be rearranged to

$$W_i = B' + M'I_i \tag{42}$$

The constant M is called the sensitivity and is expressed in counts per second per unit concentration (e.g., percent or micrograms per liter).

The most common way to determine the constants B and M (or B' and M') is by linear regression on a number of standard samples. Linear regression can be done by minimizing the sum of squared residuals of W or I; see Figure 2. The two lines obtained (one by minimizing ΔW, the other by minimizing ΔI) are not the same. Since, for analysis, the intensity I_i is measured, it is recommended to

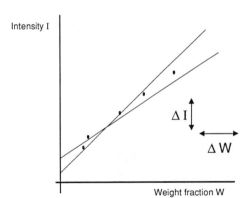

Figure 2 Straight lines through the data points can be determined by minimizing the sum of the squared residuals ΔI or ΔW.

minimize for W. In Equation (42), B' and M' can be expressed as

$$M' = \sum_{j=1}^{n} \frac{W_{ij}I_{ij} - \dfrac{\sum_{j=1}^{n} W_{ij} \sum_{j=1}^{n} I_{ij}}{n}}{\sum_{j=1}^{n} I_{ij}^2 - \dfrac{\sum_{j=1}^{n} I_{ij} \sum_{j=1}^{n} I_{ij}}{n}} \tag{43}$$

$$B' = \frac{\sum_{j=1}^{n} W_{ij} - M' \sum_{j=1}^{n} I_{ij}}{n} \tag{44}$$

where the sums are over the standard samples, $j = 1, 2, \ldots, n$, with n the number of standard samples. Equally important, however, are the variances on the parameters determined. Formulas to calculate estimates of these variances can be found in the literature [2].

Some important conclusions from the preceding section are as follows:

1. The concentration of the standard samples must cover the expected range of concentrations.
2. The calculated concentration is more accurate at the center of the line than at the extremities; the estimated variance on W_x increases with $(W_x - W_{\text{average}})$.
3. Since a calibration line is derived using data from several standards, the analysis of the unknown can be more accurate than the accuracy of the individual standard samples.

Ideally, all standards must be similar to the unknown in all aspects considered: matrix effect, homogeneity, and so on. This would lead to the use of standard samples with a very limited concentration range. Such a set of standards is in disagreement with the observation that the variance on the slope factor is smaller with increasing range. On the one hand, this advocates the use of a set of standards with a wide range of concentrations; on the other hand, the requirement of similarity in matrix effects tends to limit the range. Obviously, a compromise must be made.

As a first approximation, the degree of variation in matrix effect between two samples for a given analyte i can be estimated by calculating, for both compositions, the parameter

$$I_i = P_i \frac{\mu_i}{\mu_s(\lambda_0) + G\mu_s(\lambda_i)} \tag{45}$$

where $\mu_s(\lambda_0)$ and $\mu_s(\lambda_i)$ are the mass attenuation coefficients of the sample considered for wavelengths λ_0 and λ_i, respectively, and G is the geometrical factor. The relative difference between these expressions should not exceed a few percent; otherwise the matrix effects become too important to ignore.

With increasing range of concentrations, deviations from linearity are observed due to variations in matrix effects between samples and standards. The analyst must then resort to other methods to obtain accurate results.

Matrix effects are studied most easily by considering binary systems, that is, samples with only two elements or compounds.

In the case of absorption (both primary and secondary absorption must be considered), three cases can be distinguished:

1. A simple, linear relationship between relative intensity R and weight fraction W (line 1 in Fig. 3)
2. Curve 2 (Fig. 3) is obtained when the analyte's radiation is absorbed by the matrix element. This is usually called positive absorption.
3. Curve 3 (Fig. 3) is obtained when the matrix element absorbs less than the analyte itself. This is called negative absorption.

Enhancement generally leads to a calibration like curve 4 (Fig. 3).

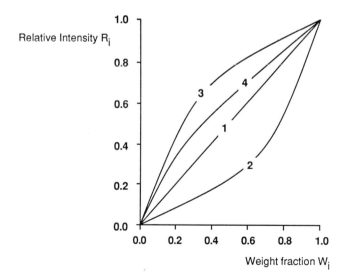

Figure 3 Calibration curves for binaries. Curve 1, no net matrix effect; curve 2, net absorption of the analyte by the matrix (positive absorption); curve 3, net absorption of the analyte's radiation by the analyte (negative absorption); curve 4, enhancement of the analyte by the matrix.

It can be shown that the behavior of the calibration curves can be explained in terms of attenuation coefficients only, if absorption is the only matrix effect. Furthermore, if monochromatic excitation is used, a single constant (calculated from attenuation coefficients) suffices to express the effect of one element on the intensity of another. In the following sections, various methods to deal with matrix effects are discussed.

A. Elimination or Evaluation of the Total Matrix Effect: Compensation Methods

1. Scattered Radiation

If the variation in matrix effect is mainly due to absorption, scattered x-rays can be used to obtain an estimate of the absorption coefficient of the sample at a certain wavelength λ_s. The intensity of the scattered radiation can be shown to be inversely proportional to the mass attenuation coefficient μ_s of the sample:

$$I_s(\lambda_s) \cong \frac{1}{\mu_s(\lambda_s)} \qquad (46)$$

where λ_s is the wavelength of the scattered radiation and $\mu_s(\lambda_s)$ is the mass attenuation coefficient of the sample for wavelength λ_s. The intensity of the fluorescent radiation is also inversely proportional to the absorption coefficient [Eq. (45)], but at a different wavelength:

$$I_i \cong \frac{W_i}{\mu_s(\lambda_0) + G\mu_s(\lambda_i)} = \frac{W_i}{\mu_s^*} \qquad (47)$$

where

$$\mu_s^* = \mu_s(\lambda_0) + G\mu_s(\lambda_i) \qquad (48)$$

Mass attenuation coefficients at two different wavelengths are virtually proportional, provided there are no significant absorption edges between the two wavelengths considered [3]. Hence, the ratio I_i/I_s is proportional to the concentration of the analyte.

Both coherently and incoherently scattered primary radiation, such as tube lines for tube-excited XRF, can be used, as well as the scattered continuous radiation. The method also corrects to some degree for surface finish, grain size effects, and variations in tube voltage and current, but it does not correct for enhancement, thus limiting its use to analytes that are mainly influenced by absorption only. Also, no absorption edges of major elements may be situated between the two wavelengths considered. This reduces the range of analytes that can be covered using the scattered tube lines. The use of scattered continuum radiation close to the analyte peak may be disadvantageous because of limited intensity, leading to either long measurement times or poor counting statistics.

2. Internal Standard

In this method an element is added to each sample in a fixed proportion to standard samples as well as the unknowns. The element added should be similar to the analyte in terms of absorption and enhancement properties. Such an element is called an internal standard. The intensity of the internal standard is affected in much the same way as the intensity of the analyte, provided there are no absorption edges (absorption) or characteristic lines, including scattered tube lines (enhancement) between the two wavelengths considered. Since

$$I_i = M_i W_i \tag{49}$$

for the analyte, and

$$I_s = M_s W_s \tag{50}$$

for the internal standard, dividing Equation (49) by Equation (50) gives

$$\frac{I_i}{I_s} = K W_i \tag{51}$$

where

$$K = \frac{M_i}{M_s W_s} \tag{52}$$

M_i and M_s are not constants (otherwise linear calibration would suffice) but depend on the matrix elements. Both M_i and M_s, however, vary in a similar manner with the matrix. Therefore, the ratio M_i/M_s is less sensitive to variation in the matrix effect and in practice can be considered a constant. In practice, the constant K is determined using linear regression. The main advantage of the internal standard method over the scattered radiation method is its ability to correct effectively for enhancement as well as for absorption. It also corrects, at least partially, for variations in density of pressed samples. Since there should be no absorption edges and no characteristic wavelengths between the lines of analyte (atomic number Z) and internal standard, often the element with atomic number $Z + 1$ or $Z - 1$ is used as internal standard, using K lines. The method has some important limitations, however:

- Sample preparation is made more complicated and is more susceptible to errors.
- Addition of reagents and the requirement of homogeneity of the sample limits the practical application of the method to the analysis of liquids and fused specimens.
- Although the rule $Z + 1$ or $Z - 1$ can serve as a rule of thumb, it is quite clear that for samples in which many elements are to be quantified, a suitable internal standard cannot be found for every analyte element.
- Although the internal standard method is easier to apply to liquids, this can generate some problems. Heavier elements (for example Mo) are more difficult to analyze using this method because liquid samples are generally not of infinite thickness for the K wavelengths of these heavier elements. The method provides some compensation in this case, however.

Theoretically, L lines of a given element can be used as internal standards for K lines, and vice versa, if these wavelengths are reasonably close to each other and no interfering lines nor edges occur between them.

Compound phases, such as oxides, can also be used as the internal standard. The composition of the additive must be constant, as it will affect the matrix effect.

The range of concentrations over which this method is suitable can be quite large, up to 10–20%, but the internal standard technique is most effective at low concentrations (or high dilutions).

3. Standard Addition Methods

Another method of analysis involves the addition of known quantities of the analyte to the specimen and is referred to as the standard addition method.

If the analyte element is present at low levels and no suitable standards are available (e.g., the matrix is unknown), standard addition and/or dilution may prove to be an alternative, especially if the analyst is interested in only one analyte element. The principle is the following: adding a known amount of the analyte i (ΔW_i) to the unknown sample gives an increased intensity $I_i + \Delta I_i$. Assuming a linear calibration, the following equations apply:

$$I_i = M_i W_i \tag{49}$$

for the original sample and

$$I_i + \Delta I_i = M_i(W_i + \Delta W_i) \tag{53}$$

for the sample with the addition. Thus, the method assumes that linear calibration is adequate throughout the range of addition because it assumes that an increase in the concentration of the analyte by an amount ΔW_i will increase the intensity by $M_i \Delta W_i$. These equations can be solved for W_i. To check the linearity of the calibration, the process can be repeated by adding different amounts of the analyte to the sample and plotting the intensity measured versus the concentrations added (Fig. 4). The intercept of the line on the concentration axis equals $-W_i$. The intensities used for calibration must be corrected for background and line overlap.

The method is suitable mainly for determination of trace and minor concentration levels because the amounts ΔW_i added to the sample must be in proportion to the amount W_i in the sample itself. The extrapolation error can be quite large if the slope of the line is not known accurately. Adding significant amounts of additives to the sample, however, may lead to nonlinearity, because it alters the matrix effect.

Compounds and solutions can be used for the standard addition. If the analyte

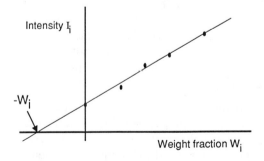

Figure 4   Standard addition method. Net intensity is plotted versus weight fraction added to the sample, and a best fit line is determined.

i in the original sample is in a different phase than in the additive, care must be taken in the calculation of the concentration sought. The relevant stoichiometric or gravimetric factors must be included. This also applies if the analyte is present under elemental or ionic form in, for example, the original sample and in a compound phase in the additive, or vice versa.

Another way to alter the concentration of the analyte is by diluting the liquid or solid solution of the sample. By diluting several times by known amounts, a line can be established. By repeating this procedure with a standard solution containing a known amount of i the unknown concentration can be found.

4. Dilution Methods

Dilution methods can also eliminate or reduce the variation of the matrix effect, rather than compensating for such variation. The dilution method can be explained using Equation (11), which can be rewritten as

$$C(\lambda_p, \lambda_i) = \frac{\mu_i(\lambda_p)}{\mu_s(\lambda_p) + G\mu_s(\lambda_i)} \tag{54}$$

where $\mu_s(\lambda_p)$ and $\mu_s(\lambda_i)$ are the mass attenuation coefficients of the sample for the primary wavelength λ_p and analyte wavelength λ_i, respectively. Apparently, deviations from linearity are due to variations in $\mu_s(\lambda_p)$ and/or $\mu_s(\lambda_i)$, ignoring enhancement. If one adds to the sample D grams of a diluent d for each gram of sample, the denominator of Equation (54) becomes

$$\frac{1}{1 + D} (\mu_s(\lambda_p) + G\mu_s(\lambda_i)) + \frac{D}{1 + D} (\mu_d(\lambda_p) + G\mu_d(\lambda_i)) \tag{55}$$

If the term $D/(1 + D)(\mu_d(\lambda_p) + G\mu_d(\lambda_i))$ is much larger than $1/(1 + D)(\mu_s(\lambda_p) + G\mu_s(\lambda_i))$, the factor $C(\lambda_p, \lambda_i)$ becomes essentially a constant and variations due to varying matrix effects between samples become negligible. This can be done in two different ways:

1. Making $D/(1 + D)$ large: diluting each sample by adding a large, known amount of a diluent.
2. Adding a smaller quantity of diluent than in the previous case, but with a much larger value for

   $$\mu_d(\lambda) + G\mu_d(\lambda_i)$$

   this is called the technique of the heavy absorber.

Both these procedures, however, require addition of reagents to the sample. This can easily be done for dissolved samples, in either liquids or fused samples, but is more difficult for powdered samples (homogeneity!).

These methods do not eliminate the matrix effects completely but reduce their influence. On the other hand, they reduce the line intensity of the analyte, and thus a compromise must be sought.

Dilution methods also have the advantage of reducing the enhancement effect if one uses a nonfluorescing diluent (for example, H_2O or $Li_2B_4O_7$). If the diluent can excite the analyte as well as some matrix elements, then the contribution of those unknown quantities of matrix elements to the total enhancement is reduced.

This method allows analysis of all measurable elements in the sample, as opposed to the standard addition method, in which an addition must be made for each element of interest.

B. Mathematical Methods

The term "mathematical methods" refers to those methods that *calculate* rather than *eliminate* or *measure* the matrix effect.

Mathematical methods are independent of sample preparation. The actual calculation method used to convert intensities to concentrations does not affect the choice of the sample preparation method. The aim of the sample preparation is limited to the presentation to the spectrometer of a sample that is homogeneous (with respect to the XRF technique) and that has a well-defined, flat surface, representative of the bulk of the sample. Mathematical methods usually require knowledge of all elements in the standard samples and allow determination of all elements in the unknowns. In practice, trace elements can be neglected: they are neither subject to an important (and variable) matrix effect, nor do they contribute significantly to the matrix effect of other elements. Their concentration is often found by linear regression.

1. Fundamental Parameter Method

The fundamental parameter method is based on the physical theory that enables one to calculate the intensity of fluorescent radiation originating from a sample of known composition. The equations usually consider both primary and secondary fluorescence (enhancement). Higher order effects and effects due to scattered radiation are usually neglected. Formulas to calculate intensities of fluorescent radiation were proposed very shortly after the introduction of the commercial XRF spectrometers in the early 1950s. The equation for the primary contribution was derived earlier [see Sec. I.A and Eq. (10)], and is repeated here

$$I_{i,1} = P_i W_i \int_{\lambda_{min}}^{\lambda_{edge,i}} \frac{J(\lambda)\mu_i(\lambda)}{\mu_s(\lambda) + G\mu_s(\lambda_i)} d\lambda \tag{56}$$

where $I_{i,1}$ is the contribution of the primary fluorescence to the total intensity and $\lambda_{edge,i}$ is the wavelength of the edge of the analyte i. Similarly, the contribution of the secondary fluorescence (enhancement) $I_{i,2}$ is

$$I_{i,2} = W_i \sum_{j,q} \int_{\lambda_{min}}^{\lambda_{edge,j}} \frac{P_i W_j P_j J(\lambda)\mu_j(\lambda)\mu_i(\lambda_j)}{\mu_s(\lambda) + G\mu_s(\lambda_i)} d\lambda$$

$$\times \left[\frac{\sin \psi_1}{\mu_s(\lambda)} \ln \left(1 + \frac{\mu_s(\lambda)}{\mu_s(\lambda_j) \sin \Psi_1} \right) + \frac{\sin \psi_2}{\mu_s(\lambda_i)} \ln \left(1 + \frac{\mu_s(\lambda_i)}{\mu_s(\lambda_j) \sin \Psi_2} \right) \right] \tag{57}$$

where

$$P_i = \omega_i g_i \frac{r_i - 1}{r_i} \tag{58}$$

The incident spectrum in Equation (57) is now limited from λ_{min} to $\lambda_{edge,j}$, the edge of element j. The summation in Equation (57) is over all elements j that have characteristic lines that can excite the analyte i. For each of these elements j, *all* characteristic lines q must be considered. This is quite simple if none of the L lines or M lines of element j can excite the analyte. In this case, only the $K\alpha$ and $K\beta$ lines need be considered. If the L lines are energetic enough for enhancement, the sheer number of L lines (for example, W has between 15 and 20 characteristic L lines that can be considered) would make the calculation very time consuming. Therefore, most programs consider only 3 to 5 L lines for each element. A similar reasoning holds for the M lines.

However, application of these formulas was limited to prediction of intensities for samples with given compositions. The application of those formulas for analysis was not pursued until the 1960s. The method of analyzing samples by fundamental parameter equations was developed independently by Criss and Birks [4] and by Shiraiwa and Fujino [5].

Application of a fundamental parameter method for analyzing samples consists of two steps: calibration and analysis. Both steps are discussed in more detail in the following sections.

a. Calibration

The fundamental parameter equation is used to predict the intensity of characteristic lines for a composition that is identical to the standard used. If the analyst uses more than one standard sample, the calculations are repeated for each of the standards.

The calculations are performed using the appropriate geometry: that is, incidence and takeoff angles are taken in agreement with those of the spectrometer used, and the parameters of the tube spectrum (the anode material, voltage, thickness of beryllium window, and so on) correspond to those used in the spectrometer for the measurements. These theoretically predicted intensities are then linked to the actually measured intensities. The intensities predicted are (almost always) net intensities, void of background, line overlap, crystal fluorescence, and so on. Hence, the measured intensities must be corrected for such spectral artifacts. For each characteristic line measured, net intensities can be plotted versus the calculated intensities and a straight line can be determined. The slope of such a line is the proportionality factor between predicted (calculated) and measured intensity. In general, this relationship is determined more accurately if more standards are used.

A special case of calibration ensues when one uses pure elements as standards. Dividing each of the measured (net) intensities of the unknown by the measured (net) intensity of the corresponding pure element gives the relative intensity. This relative intensity can be calculated by fundamental parameter programs. In fact, some programs use equations that express the intensity of characteristic radiation directly in terms of relative intensity. Relative intensity in this respect is thus defined as the intensity of the sample divided by the intensity of the corresponding pure element (or compound, if the concentration of the analyte is defined in compound concentration) under identical conditions for excitation and detection.

Basically, the calibration function accounts for instrument-related factors only. Matrix effects are accounted for using physical theory.

The instrument-related parameters for a wavelength-dispersive spectrometer are as follows:

- Collimation
- Crystal reflectivity
- Efficiency of detector(s)
- Fraction of emergent beam allowed into the detectors after Bragg reflection (also dependent on Bragg angle)

b. Analysis

Step 1. For every unknown specimen a first estimate of composition is made. There are many different ways to obtain such a first estimate. These vary from using a simple, fixed composition (e.g., equal to 100%, divided by the number of elements considered; the first estimate of the concentration of each of the elements is thus taken to be equal) to a composition derived from the measured intensities in combination with the calibration curves. Using the calibration data, it is possible to estimate, for each element, the intensity that should be obtained if the pure elements were measured. These numbers are then used to divide the intensity of the corresponding element, measured in the unknown. The resulting fractions are scaled to 100% and used as first estimate.

Step 2. For this estimate of composition, the theoretical intensities are calculated. These are converted to measured intensities using the calibration data.

Step 3. The next estimate of composition is then obtained based on the difference between measured and calculated intensities. Again, many different methods are available:

- The simplest method is based on *linear interpolation*. If for a given element the measured intensity is 10% higher than the calculated intensity, the concentration of that element is increased by 10%.
- Rather than a linear relationship, some authors use a *hyperbolic* relationship. This is done because the relationship between concentration and intensity is usually not linear over a wider range. If the sample is pseudobinary, hyperbolic relationships have proven to be better approximations. For more complex samples, it still works out quite well, because the concentrations of the other elements are considered fixed at this stage. The hyperbolic equation requires a minimum of three points for its parameters to be determined. The points requiring the least additional calculation time are (1) the origin (net intensity zero, at concentration zero); (2) the pure element $W = 100\%$; the corresponding intensity has already been calculated; and (3) the point $W =$ current estimate; the intensity has already been calculated. These three points allow a hyperbolic relation to be established around the current, estimated composition. From this curve and the measured intensity, the new concentration estimate is derived. This approach is repeated for every analyte element. This method usually provides a faster convergence than using the simple straight line.
- It is also possible to use *gradient methods* to determine the next composition estimate. The formulas for the first derivative, with respect to concentration, of the fundamental parameter equation have been published [6], but the calculation is cumbersome and time consuming. Also, the derivatives can be ob-

tained by a finite difference method when the effect of a small change in composition on the intensity is observed.

Step 4. The process, starting at step 2, is now repeated, until convergence is obtained. Different convergence criteria exist. Calculation can be terminated if one of the following criteria is satisfied for *all* the elements (compounds) concerned:

- The intensities calculated in step 2 do not change from one step to another by more than a preset level.
- The intensities calculated in step 2 agree to within a preset level with the measured intensities.
- The compositions calculated in step 3 do not change from one step to another by more than a preset level (e.g., 0.05 or 0.01%).

One or more of these criteria might be incorporated in the program. These criteria, however, are no guarantee that the final result is accurate to within the level specified in the convergence criteria.

c. Extensions to the Method

More complex scenarios are possible. The most common includes calculation of a set of influence coefficients (based on theoretical calculations) to obtain a composition quickly, close to the final result. From there on, the fundamental parameter method takes over. This should yield faster convergence in terms of computation time, because the calculation by influence factors of the preliminary composition is very fast. This method reduces the number of evaluations of the fundamental parameter equation.

Also, the different programs available differ quite markedly in their treatment of the intensities of the standards measured. Some programs use weighting of the standards, stressing the standard(s) closest (in terms of intensity) to the unknown. Such programs thus use a different calibration for each unknown sample. The unknown sample dictates which standards are given a high weighting factor and which standards are used with less weighting. Other programs use all standards with equal weighting.

d. Typical Results

The early results of the fundamental parameter method on stainless steels are given by Criss and Birks [4]. The average relative differences between x-ray results and certified values were about 3–4% relative. Later, Criss et al. [7] reported accuracies of about 1.5% relative for stainless steels using a more accurate fundamental parameter program. Typical results for tool steel alloys are given in Table 2.

e. Factors Affecting Accuracy

The accuracy of the final result is determined by the measurement; the sample preparation; the physical constants used in the fundamental parameter equation; the limited physical processes that are considered in the fundamental parameter

TABLE 2 Analysis of Tool Steels with a
Fundamental Parameter Program[a]

Element	Min (%)	Max (%)	SD (%)
W	1.8	20.4	0.52
Co	0.0	10.0	0.20
Mn	0.21	0.41	0.01
Cr	2.9	5.0	0.13
Mo	0.2	9.4	0.04
S	0.015	0.029	0.003
P	0.022	0.029	0.003
Si	0.14	0.27	0.03
C	0.65	1.02	0.16

[a] XRF11, from CRISS SOFTWARE, Largo, MD.
One standard was used. Min., max = minimum and
maximum concentrations in the set of analyzed samples; SD = standard deviation, calculated from the
difference between concentration values found and
certified.
Source: Data courtesy of Philips Analytical, Almelo,
The Netherlands.

equations; and the standards and the calibration. In the following discussion, the
effect of measurements and sample preparation is not considered.

The physical constants used in the fundamental parameter equations are incidence and exit angle; spectrum of incident beam; mass attenuation coefficients;
fluorescence yields; absorption jump ratios; ratios of intensity of different lines
within a given series (e.g., $K\alpha/K\beta$ ratio); and wavelengths (or energies) of absorption edges and emission lines.

The incidence angle in most wavelength- and energy-dispersive (WD and ED)
spectrometers is in fact a relatively wide cone with a different intensity at the
boundaries compared to the center. This incident cone is neglected, and the incident radiation is considered parallel, along a single, fixed direction. A similar
observation holds for the exit angle. This effect is far less pronounced if diffraction
from a plane crystal surface is used for dispersion, as is done in most sequential
wavelength-dispersive spectrometers, for example. This has been studied to some
extent by Muller [8]. To our knowledge none of the fundamental parameter programs available takes this into account. Its influence, however, is to some extent
compensated for by calibration with standard samples.

The incident tube spectrum requires more attention. Parts of the primary
spectrum might excite an element B that, in turn, excites element A very efficiently. In such a case, the intensity of A is enhanced by B. In some cases, this
enhancement may make the intensity of element A sensitive to small errors in the
tube spectrum representation that would not be compensated for if the pure A is
used for calibration. This can arise, for example, in the analysis of silica-zirconia
samples with a Rh tube (g). Pure silica is relatively insensitive to the intensity of
the characteristic K lines of Rh. In combination with Zr, however, the situation

is different. Indeed, the Rh K lines are strongly absorbed by Zr. Zr then emits K and L lines that enhance Si. As a result, the Si intensity is more sensitive to the Rh K lines in SiO_2-ZrO_2 mixtures than in pure SiO_2. Tube spectra have been calculated using the algorithm of Pella et al. [10], for example.

Several compilations of mass attenuation coefficients are published in the literature. A continuing effort to compile the most comprehensive table has been undertaken by the National Bureau of Standards (now called National Institute of Standards and Technology, Gaithersburg, MD).

When selecting a table of mass attenuation coefficients for use in a fundamental parameter program, the following question must be addressed: Does the table cover all the analytical needs? (In practice, does it cover the complete range of interest from the longest wavelength considered to the excitation potential of the tube?)

The analyst should be aware that the use of formulas to generate mass attenuation coefficients can lead to values that can be significantly different from the corresponding table values.

At present in XRF, the compilations of McMaster et al. [11], Heinrich [12], Leroux and Thinh [13], or Veigele [14] are most often used. A short discussion on agreement between some of these compilations has been presented by Vrebos and Pella [15].

A comprehensive reference to fluorescence yields, including Coster-Kronig transitions, can be found in the work of Bambynek et al. [16]; see also Chapter 1 and Appendix V of Chapter 1.

Absorption jump ratios can be derived from the tables of attenuation coefficients.

Data for the K spectra can be found in the work by Venugopalo Rao et al. [17]; see also Chapter 1.

A comprehensive table of wavelengths of absorption edges and emission lines was published by Bearden [18] and is also presented in Appendices I and II of Chapter 1. Because attenuation coefficients are wavelength-dependent, an error in a wavelength of any characteristic line, will automatically lead to a bias in the corresponding attenuation coefficient.

Although the formula for *tertiary fluorescence* was derived by Shiraiwa and Fujino [5], among others, it is not included in most fundamental parameter programs. Usually the tertiary fluorescence effect is considered small enough to be negligible. Shiraiwa and Fujino presented data showing a maximum contribution of tertiary fluorescence of about 3% relative to the total intensity of Cr in Fe-Cr-Ni samples [5]. Higher order enhancement is also possible, but it is even less pronounced than tertiary fluorescence.

Other processes not considered in the fundamental parameter methods are coherent and incoherent scatter of both the primary spectrum and the fluorescent lines. This is usually justified by pointing out that the photoelectric effect is by far the major contribution to the total absorption. It is believed that the contribution by scattered photons to the excitation of characteristic photons is negligible. In some cases, however, the scattered primary spectrum may have a considerable influence, as illustrated earlier in this chapter.

The use of good type standards (similar to the unknown) almost always leads to more accurate results, compared to a situation in which the standards used have a widely different composition from that of the unknown. This is because most of the errors caused by inaccuracies in the physical constants cancel out. The degree of similarity between standards and unknown has an important effect on the accuracy of the analysis. The user has very little interaction during calibration. Most programs do not provide, for example, an easy way to delete a standard for a certain analyte. The calibration step is intimately mixed with the analysis phase. The use of a sample with known composition as a quality assurance standard to assess accuracy is highly recommended. The sample(s) used should be typical of the unknowns and should not be used for calibration to avoid biases and overly optimistic estimates of accuracy.

2. Constant Influence Coefficient Algorithms

Another class of mathematical methods calculates the matrix effect by means of coefficients rather than evaluates the fundamental parameter equation for each unknown. It will be shown that these coefficients can also be calculated from theory using fundamental parameters. Some of these algorithms use only a single coefficient per interfering element; others use more than one. The distinction used here, however, depends on whether the influence coefficient is a constant for a given application or the value of the coefficient varies with composition. The latter methods are discussed in Section IV.B.3. Only two algorithms that use constant influence coefficients are discussed here: the Lachance-Traill and the de Jongh algorithms. The practical application of the resulting equations, that is, calibration and analysis, are treated separately in Section IV.B.4 because these have many elements in common with all the influence coefficient algorithms discussed here and in Section IV.B.3.

a. The Lachance-Traill Algorithm

i. Formulation. In 1966, Lachance and Traill proposed a correction algorithm based on influence coefficients [19]. The equations are, for a ternary,

$$W_A = R_A(1 + \alpha_{AB}W_B + \alpha_{AC}W_C) \tag{59a}$$

$$W_B = R_B(1 + \alpha_{BA}W_A + \alpha_{BC}W_C) \tag{59b}$$

$$W_C = R_C(1 + \alpha_{CA}W_A + \alpha_{CB}W_B) \tag{59c}$$

R_A, R_B, and R_C are the relative intensities of A, B, and C, respectively. The coefficients α_{ij} are called *influence coefficients*. The relative intensity is defined as the measured intensity divided by the intensity of the *corresponding pure element*, measured under identical conditions. In practice, the relative intensity is often derived from measurements on standards, and the pure element (or compound) is not required. A more general notation of the Lachance-Traill algorithm is, for analyte i,

$$W_i = R_i\left[1 + \sum_{j \neq i}^{n} \alpha_{ij}W_j\right] \tag{60}$$

where the summation covers all n elements (or compounds) in the sample, *except* the analyte itself. Hence, there are $n - 1$ terms in the summation. This is common to all currently used algorithms. Algorithms developed before 1966 had n terms, rather than $n - 1$, for samples with n elements.

Equations (59a), (59b), and (59c) are linear equations in the concentrations of the elements W_A, W_B, and W_C. Note that there are only two coefficients for each analyte element. Consider, for example, the first equation of the set (59):

$$W_A = R_A (1 + \alpha_{AB} W_B + \alpha_{AC} W_C) \tag{59a}$$

element A is the analyte, and its concentration is equal to the relative intensity R_A, multiplied by the matrix correction factor $[1 + \alpha_{AB} W_B + \alpha_{AC} W_C]$. This matrix correction factor has only two coefficients: one (α_{AB}) to describe the effect of element B on the intensity of A and, similarly, one to describe the effect of element C on the intensity of A. α_{AA}, which would correct for the effect of A on its own intensity (often referred to as self-absorption), is zero. Similarly, α_{BB} and α_{CC} are also zero. The effect of A on A, however, is taken into account, as is shown in the next section.

ii. Calculation of the Coefficients. Lachance and Traill also showed that the influence coefficients α_{ij} should be calculated for monochromatic excitation (wavelength λ_0), assuming absorption only from the expression

$$\alpha_{ij} = \frac{\mu_j(\lambda_0) \csc \psi_1 + \mu_j(\lambda_i) \csc \psi_2}{\mu_i(\lambda_0) \csc \psi_1 + \mu_i(\lambda_i) \csc \psi_2} - 1 \tag{61}$$

When secondary fluorescence (enhancement) is involved, the coefficients are calculated in the same way. Enhancement is thus treated as negative absorption. This assumption is not valid when enhancement is quite severe. Differences in primary absorption may easily be confused with enhancement (see Fig. 2). From Equation (61) it follows clearly that α_{ij} is a measure for the effect of j on i, *relative to* i, and that α_{ii} is always zero. Therefore, it is not included in the summation of Equation (60). Also from Equation (61), it follows that the value of the coefficients cannot be less than -1. It must be stressed that Equation (61) is, strictly speaking, valid only for monochromatic excitation and for those analytes that are subject to absorption only (no enhancement). In this case the influence coefficient is concentration independent: it is a constant. It does, however, depend on such parameters as the wavelength of the primary photons and the incidence and exit angle. In all other cases (polychromatic excitation and/or enhancement), Equation (61), *strictu sensu*, cannot be used. A polychromatic beam (from an x-ray tube, for example) can be "replaced" by a monochromatic beam by resorting to the effective wavelength. The effective wavelength, however, is composition dependent (see Sec. I.B). The value of the coefficients calculated using Equation (61) is thus also dependent on composition, although neither W_i nor W_j figures explicitly in Equation (61). If enhancement is dominant, another method must be applied to calculate the coefficients.

Influence coefficients for the algorithm of Lachance-Traill, for example, can also be calculated based on actual measurements. Rewriting,

$$W_i = R_i[1 + \alpha_{ij} W_j] \tag{62}$$

to

$$\alpha_{ij} = \frac{W_i/R_i - 1}{W_j} \tag{63}$$

yields an expression that can be used to obtain α_{ij} based on the composition of the binary and the relative intensity R_i. The drawbacks associated with this method are as follows:

1. Calculation of R_i requires the measurement of the intensity on the pure i (element or compound). This could lead to large errors if the intensity of i in the binary is much lower than that of the pure i, as a result, for example, of the nonlinearity of the detectors.
2. The pure elements (or compounds) are not always easy to come by or could be unsuitable to present to the spectrometer (e.g., pure Na or Tl).
3. Equation (63) is very prone to error propagation when W_i is close to 1. The nominator is then a difference between two quantities of similar magnitude, and the denominator is then close to zero, magnifying the errors.
4. Also, the binary samples required can pose problems: some alloys tend to segregate, and homogeneous samples are then difficult to obtain.

The coefficients α_{ij} can also be calculated from theory: calculate R_i for the binary with composition (W_i, W_j) and substitute in Equation (63). This method eliminates problems 1, 2, and 4.

The Lachance-Traill algorithm assumes that the influence coefficients can be treated as constants, independently of concentration; this limits the concentration range when the matrix effects change considerably with changes in composition; and the influence coefficients are invariant to the presence and nature of other matrix elements. So, α_{FeCr}, determined for use in the Fe-Cr-Ni ternary samples, is the same as α_{FeCr} in Fe-Cr-Mo-W-Ta or Fe-Cr samples.

b. The De Jongh Algorithm

i. Formulation. In 1973, de Jongh proposed an influence coefficient algorithm [20] based on fundamental parameter calculations. The general formulation of his equation is

$$W_i = E_i R_i \left(1 + \sum_{j \neq e}^{n} \alpha_{ij} W_j \right) \tag{64}$$

where E_i is a proportionality constant (determined during the calibration). The summation covers $n - 1$ elements (as is the case with the algorithm of Lachance and Traill), but the eliminated element e is the same for *all* equations. If for a ternary sample element C is eliminated, the following equations are obtained:

$$W_A = E_A R_A (1 + \alpha_{AA} W_A + \alpha_{AB} W_B) \tag{65a}$$

$$W_B = E_B R_B (1 + \alpha_{BA} W_A + \alpha_{BB} W_B) \tag{65b}$$

$$W_C = E_C R_C (1 + \alpha_{CA} W_A + \alpha_{CB} W_B) \tag{65c}$$

Note that, to obtain the concentration of elements A and B, the concentration of C, W_C, is not required. This is different from Lachance and Traill's algorithm: to calculate the concentration of A using Equation (59a), the concentrations of both B and C are required. If the user is not really interested in element C (for example, element C is iron in stainless steels), Equation (65c) need not be considered and the analysis of the ternary sample can be done measuring R_A and R_B and solving Equations (65a) and (65b).

ii. Calculation of the Coefficients. De Jongh also presented a method to calculate coefficients from theory. The basis is an approximation of W_i/R_i by a Taylor series around an "average composition":

$$\frac{W_i}{R_i} = E_i(1 + \delta_{i1} \Delta W_1 + \delta_{i2} \Delta W_2 + \cdots + \delta_{in} \Delta W_n) \tag{66}$$

where E_i is a constant and

$$\Delta W_i = W_i - W_{i,\text{average}} \tag{67}$$

δ_{ij} are the partial derivatives of W_i/R_i with respect to concentration:

$$\delta_{ij} = \frac{\partial \ W_i/R_i}{\partial \ W_j} \tag{68}$$

In practice, these derivatives are calculated as finite differences. W_i/R_i is calculated for a sample with the average composition $W_{1,\text{average}}$, $W_{2,\text{average}}$, . . . , $W_{n,\text{average}}$ (symbol $[W_i/R_i]_{\text{average}}$). Then, the concentration of each element j in turn is increased by a small amount, such as 0.1% (0.001 in weight fraction), and W_i/R_i is calculated for that composition (symbol $[W_i/R_i]_{Wj+0.001}$). Substituting in Equation (68) yields

$$\delta_{ij} = \frac{\partial \ W_i/R_i}{\partial \ W_j} = \frac{[W_i/R_i]_{Wj+0.001} - [W_i/R_i]_{\text{average}}}{0.001} \tag{69}$$

This process is repeated for each of the elements j to calculate all the coefficients for analyte i. This is also repeated for the other analyte elements. The coefficients calculated from Equation (69) can be used in Equation (66) for analysis. Equation (66), however, has n factors, rather than $n - 1$, and uses ΔW_j rather than W_j. Using

$$\sum_{j=1}^{n} \Delta W_j = 0 \tag{70}$$

or

$$\Delta W_e = -\Delta W_1 - \Delta W_2 - \cdots - \Delta W_n \tag{71}$$

one element e can be eliminated. The resulting equation is similar to Equation (66) but has only $n - 1$ terms:

$$\frac{W_i}{R_i} = E_i(1 + \beta_{i1} \Delta W_1 + \beta_{i2} \Delta W_2 + \cdots + \beta_{in} \Delta W_n) \tag{72}$$

with

$$\beta_{i1} = \delta_{i1} - \delta_{ie} \tag{73}$$

for all β_{ij} except β_{ie}, which is equal to zero. Equation (72) has $n - 1$ terms, but they are still in ΔW rather than W. Transformation of ΔW to W is done by substituting Equation (67) in Equation (72):

$$\frac{W_i}{R_i} = E_i \left(1 - \sum_{j \neq e}^{n} \beta_{ij} W_{j,\text{average}} + \sum_{j \neq i}^{n} \beta_{ij} W_j \right) \tag{74}$$

which can be rearranged to Equation (64) with

$$\alpha_{ij} = \frac{\beta_{ij}}{1 - \sum_{j \neq e}^{n} \beta_{ij} W_{j,\text{average}}} \tag{75}$$

iii. Typical Results. Tables with de Jongh's coefficients have been used for a wide variety of materials ranging from high-temperature alloys to cements. An example for stainless steels is given in Table 3.

Other elements, such as Si, P, S, and C, are also present in trace amounts. The coefficients are calculated at a given composition [see Eq. (69)]. The practical range of concentration over which these coefficients yield accurate results varies from 5 to 15% in alloys to the whole range from 0 to 100% in fused oxide samples.

iv. Comparison of the De Jongh and Lachance-Traill Algorithms. The following points can be noted. (1) The basis of the de Jongh algorithm is a Taylor series expansion around an average (or reference) composition. (2) De Jongh can eliminate any element; Lachance and Traill eliminate the analyte itself: α_{ii} is zero. De Jongh eliminates (i.e., fixes the coefficient to zero) the same element for all analytes. Eliminating the base material (for example, iron in steels) or the loss on ignition (for beads) generally leads to smaller numerical values for the coefficients and avoids the necessity to analyze for all elements. (3) De Jongh's coefficients are calculated at a given reference composition. They are composition dependent and take into account all elements present. A coefficient α_{ij} represents the effect of element j on element i in the presence of all other elements: they are *multielement coefficients* rather than binary coefficients. (4) Calculation of the coefficients is based on theory, treating both absorption and enhancement effects. Hence, the coefficients are susceptible to the errors described earlier (fundamental parameter methods). However, the calculation of the coefficients involves a division of the matrix correction terms for the "slightly affected" composition by the corresponding term of the reference composition [Eq. (75)]. This compensates for some of the biases introduced by the fundamental parameters. (5) For a sample containing n elements, there are n equations (one for each of the elements) if one

TABLE 3 Analysis of Stainless Steels with Theoretical Influence Coefficients (De Jongh)[a]

Element	Min. (%)	Max. (%)	SD (%)
Mn	0.64	1.47	0.015
Cr	12.40	25.83	0.06
Ni	6.16	20.70	0.06

[a] Min., max = minimum and maximum concentration in the set of analyzed samples; SD = standard deviation, calculated from the difference between concentration values found and certified.
Source: Data courtesy of Philips Analytical, Almelo, The Netherlands.

uses the Lachance-Traill algorithm. Each of these equations has $n - 1$ influence coefficients, and the coefficient of the analyte in each of the equations has been set to 0. For the same sample, de Jongh requires only $n - 1$ equations using $n - 1$ coefficients per equation. One element has been eliminated throughout. This element is usually one of no or little interest to the analyst, such as iron in steels or the loss on ignition in fused beads. Indeed, the de Jongh method allows us to eliminate the loss on ignition [21]. The nth equation (for the eliminated element) can also be written, and its form is identical to the others: it also has $n - 1$ terms, and the coefficients can be calculated following exactly the same procedure as for the other coefficients. (6) The coefficients in the Lachance-Traill equation have been calculated empirically using measured data from many more standards than used in de Jongh's algorithm, which always used theoretical coefficients. These coefficients could be obtained from Philips, eliminating the need for a large computer at each user's site. Recent developments in more powerful computers make it possible to calculate the coefficients for the Lachance-Traill algorithm from theory as well.

3. Algorithms with Variable Coefficients

Both algorithms discussed thus far use a single, constant coefficient for each (except one) interfering element. The expression from Lachance and Traill uses coefficients expressing the effect of one element on the characteristic intensity of the analyte, relative to the analyte, ignoring all other elements. Such coefficients are therefore referred to as binary coefficients. The algorithm of de Jongh effectively calculates multielement influence coefficients. Such coefficients predict the effect of one element on the intensity of another in a given matrix.

This distinction between binary and multielement coefficients can clearly be seen in a ternary sample. Algorithms based on binary coefficients add interelement effects from each of the constituent elements. Assume a Ni-Fe-Cr sample. The total matrix effect on Cr is accounted for using a coefficient expressing the influence of Ni on Cr (α_{CrNi}) and a similar coefficient for Fe on Cr (α_{CrFe}). Both these coefficients are calculated for the corresponding binaries (Ni-Cr and Fe-Cr). In a ternary sample (e.g., Ni-Fe-Cr), however, the effect of Ni on Cr is affected by the presence of Fe (and Ni similarly affects the effect of Fe on Cr). This effect is called the crossed effect, and is discussed in a later section.

The expression presented by Lachance and Traill to calculate the influence coefficients required the use of monochromatic excitation. An equivalent wavelength was used to calculate the coefficients when polychromatic excitation is applied. The equivalent wavelength, however, has been shown to vary with composition, as treated earlier in this chapter. The theoretical coefficients calculated according to de Jongh are also composition dependent, since the reference composition is used explicitly in the calculations.

This variation is due to the fact that the composition of the matrix varies considerably if analysis is required over a wide range of concentrations. This was recognized early in the development of influence coefficient algorithms, and many different algorithms with variable coefficients have been proposed. A variable influence coefficient in this respect is an influence coefficient that varies explicitly with concentration of one or more components in the sample. Some of these are discussed in subsequent sections.

a. The Claisse-Quintin Algorithm

i. Formulation. Claisse and Quintin [22] extended Lachance and Traill's algorithm by considering a polychromatic primary beam. The resulting equation for W_A can be expressed as

$$W_A = R_A\left(1 + \sum_{j\neq A}^{n} \alpha_{Aj}W_j + \sum_{j\neq A}^{n} \alpha_{Ajj}W_j^2 + \sum_{j\neq A}^{n}\sum_{k\neq A,k>j}^{n} \alpha_{Ajk}W_jW_k\right) \qquad (76)$$

where the summation over j has $n - 1$ terms (all n elements, except i), and the summation over k has $(n - 2)/2$ terms (all n elements, except the analyte i and element j; furthermore, if α_{ijk} is used, then α_{ikj} is not). For a binary sample, Equation (76) reduces to

$$W_A = R_A(1 + \alpha'_{AB}W_B) \qquad (77)$$

with

$$\alpha'_{AB} = \alpha_{AB} + \alpha_{ABB}W_B \qquad (78)$$

clearly showing that the influence coefficient α'_{AB} varies linearly with composition (i.e., W_B). For binaries, $W_A = 1 - W_B$; hence Equation (78) can also be rearranged to

$$\alpha'_{AB} = \alpha_{AB} + \alpha_{ABA}W_A \qquad (79)$$

Equations (78) and (79) are, theoretically at least, identical. It has been shown, however, that Equation (79) is preferable to Equation (78) if samples with more than two elements (or compounds) are analyzed [23]. Note that the value of α_{AB} in Equations (78) and (79) is different.

ii. Cross-product Coefficients. For a ternary sample, Claisse-Quintin's algorithm can be written as

$$W_A = R_A(1 + \alpha_{AB}W_B + \alpha_{ABB}W_B^2 + \alpha_{AC}W_C + \alpha_{ACC}W_C^2 + \alpha_{ABC}W_BW_C) \qquad (80)$$

The terms

$$\alpha_{AB}W_B + \alpha_{ABB}W_B^2$$

and

$$\alpha_{AC}W_C + \alpha_{ACC}W_C^2$$

are the matrix corrections due to B and C, respectively. The term $\alpha_{ABC}W_BW_C$ corrects for the simultaneous presence of both B and C and is referred to as a *cross-product coefficient.*

Tertian and Vie le Sage [24] assume that a multielement influence coefficient α_{ij}^M can be approximated as the sum of the binary coefficient α_{ij}^B, and a linear variation with the other elements:

$$\alpha_{ij}^M = \alpha_{ij}^B + \tau_{ijk}W_k \qquad (81)$$

where τ_{ijk} is a coefficient expressing the effect of element k on the influence coefficient α_{ij}^M. Similarly,

$$\alpha_{ik}^M = \alpha_{ik}^B + \tau_{ikj} W_j \tag{82}$$

Substituting Equations (81) and (82) in

$$1 + \alpha_{ij}^M W_j + \alpha_{ik}^M W_k \tag{83}$$

yields

$$1 + \alpha_{ij}^B W_j + \alpha_{ik}^B W_k + \alpha_{ijk} W_j W_k \tag{84}$$

with

$$\alpha_{ijk} = \tau_{ijk} + \tau_{ikj} \tag{85}$$

It must be realized that the crossed effect is introduced by the use of binary coefficients; the use of multielement coefficients would not lead to the crossed effect.

iii. Calculation of the Coefficients. Claisse and Quintin also published methods to calculate the coefficients [22] from measurements on binary and ternary mixtures or from theory. These methods, however, are now generally superseded by theoretical calculations, as discussed in Sections IV.B.3.c and d.

b. The Rasberry-Heinrich Algorithm

Following a systematic study of the Fe-Cr-Ni ternary system, Rasberry and Heinrich concluded that the two different phenomena, absorption and enhancement, should be described by two different equations. They introduced the algorithm [25]

$$W_A = R_A \left[1 + \sum_{j \neq A}^{n} A_{Aj} W_j + \sum_{k \neq A}^{n} \frac{B_{Ak}}{1 + W_A} W_k \right] \tag{86}$$

where only one coefficient is used for each interfering element. The coefficient A is used when absorption is the dominant effect. In this case, the coefficient B is taken equal to zero. If for a given analyte all B coefficients are zero, Equation (86) reduces to the Lachance-Traill expression. When enhancement by element k dominates, a B coefficient is used. The corresponding A coefficient is then taken equal to zero. Hence, the total number of terms in both summations is $n - 1$.

The correction factor for enhancement can be rewritten as

$$\alpha_{ij} = \frac{B_{ij}}{1 + W_i} \tag{87}$$

showing that α_{ij} varies with concentration in a nonlinear fashion.

Among the disadvantages of the Rasberry-Heinrich algorithm are the following:

1. It is not always clear which interfering elements should be assigned a B coefficient and which an A. In Pb-Sn alloys, the Sn $L\alpha$ line is fluoresced (enhanced) by both Sn K and Pb L lines, yet the calibration curve for Sn $L\alpha$ clearly shows that absorption is dominant (Fig. 5).

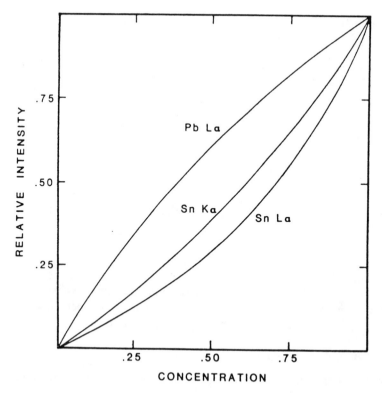

Figure 5 Calibration curves for Pb $L\alpha$, Sn $K\alpha$, and Sn $L\alpha$ in Pb-Sn binaries. The Sn $L\alpha$ is apparently dominated by absorption rather than by enhancement of Sn K and Pb L lines.

2. Furthermore, Equation (87) suggests that the value of B_{ij} at $W_i = 1$ is half the value of B_{ij} at $W_i = 0$. This is not generally valid. Mainardi and coworkers [26] therefore suggested replacing the 1 in the denominator by another coefficient.

3. Rasberry and Heinrich did not publish a method to calculate the coefficients from theory. Some of the disadvantages of calculating empirical coefficients have been discussed in Section IV.B.2.a.

For these reasons, the Rasberry and Heinrich algorithm is not generally applicable. However, the concept of a hyperbolically varying influence coefficient has been incorporated in the three-coefficient algorithm of Lachance.

c. The Three-Coefficient Algorithm of Lachance

i. Formulation. In 1981, Lachance [27] proposed a new approximation to the binary influence coefficient α_{ij}^B given by Equation (88):

$$\alpha_{ij}^B = \alpha_{ij1} + \frac{\alpha_{ij2} W_m}{1 + \alpha_{ij3}(1 - W_m)} \tag{88}$$

with

$$W_m = 1 - W_i \tag{89}$$

W_m is the concentration of all matrix elements. It has been shown by Lachance and Claisse [23], as well as by Tertian [28], that variable binary coefficients must be expressed in terms of W_m (or $1 - W_i$). For multielement samples, cross-product coefficients α_{ijk} are used to correct for the crossed effect, similar to Equation (76). The general equation for a multielement sample is

$$W_i = R_i \left\{ 1 + \sum_{j \neq i}^{n} \left[\alpha_{ij1} + \frac{\alpha_{ij2} W_m}{1 + \alpha_{ij3}(1 - W_m)} \right] W_j + \sum_{j \neq i}^{n} \sum_{k \neq i, k > j}^{n} \alpha_{ijk} W_j W_k \right\}$$

(90)

where the summation over j has $n - 1$ terms (all n elements, except i), and the summation over k has $(n - 2)/2$ terms (all n elements, except the analyte i and element j; furthermore, if α_{ijk} is used, then α_{ikj} is not).

Vrebos and Helsen [29] published some data on this algorithm clearly showing the accuracy of the algorithm using theoretically calculated intensities to avoid errors due to sample preparation and measurement.

ii. Calculation of the Coefficients. The coefficients α_{ij1}, α_{ij2}, and α_{ij3} are calculated using fundamental parameters at three binaries i-j. The cross-product coefficients are calculated from a ternary. The compositions are listed in Table 4. The samples referred to in Table 4 are hypothetical samples. The intensity is calculated from fundamental parameters and requires no actual measurements on real samples.

Step 1. Calculate the relative intensity R_i for the first composition in Table 4. If the analysis of interest has more than three elements, then divide that system in combinations of three elements i, j, k at a time. Element i is the analyte; j and k are two interfering elements.

Step 2. Using Equation (63), the corresponding influence coefficient α_{ij}^{B} can be calculated.

Step 3. For this composition, $W_m = 1 - W_i = W_j = 0.001$, which is small enough to be considered 0. Hence, Equation (88) reduces to

$$\alpha_{ij}^{B} = \alpha_{ij1}$$

(91)

α_{ij}^{B} has been calculated in step 2, so α_{ij1} can be computed.

TABLE 4 Composition of the Samples Used for Calculation of the Coefficients for Lachance's Three-Coefficient Algorithm (Weight Fraction)

Sample number	W_i	W_j	W_k
1	0.999	0.001	0.0
2	0.001	0.999	0.0
3	0.5	0.5	0.0
4	0.999	0.0	0.001
5	0.001	0.0	0.999
6	0.5	0.0	0.5
7	0.30	0.35	0.35

Step 4. Calculate the intensity for the second composition of Table 4, and use Equation (63) to calculate α_{ij}^B. In most cases, this value is different from that found in step 2 because the compositions involved are different.

Step 5. The $1 - W_m = W_i = 0.001$ is small enough to be considered zero; hence Equation (88) reduces to

$$\alpha_{ij}^B = \alpha_{ij1} + \alpha_{ij2} \tag{92}$$

α_{ij1} and α_{ij}^B are known, so α_{ij2} can be calculated.

Step 6. Calculate the intensity for the third composition of Table 4, and use Equation (63) to calculate α_{ij}^B. In most cases, this value is different from that found in step 2 or step 4 because the compositions involved are different.

Step 7. Using $W_m = 1 - W_i = 0.5 = W_i$, Equation (88) reduces to

$$\alpha_{ij}^B = \alpha_{ij1} + \frac{\alpha_{ij2}0.5}{1 + \alpha_{ij3}0.5} \tag{93}$$

which can be rearranged to

$$\alpha_{ij3} = \frac{\alpha_{ij2}}{\alpha_{ij}^B - \alpha_{ij1}} - 2 \tag{94}$$

All coefficients on the right-hand side are known, so α_{ij3} can be calculated.

Step 8. Repeat steps 1–7 for W_i and W_k to compute the coefficients α_{ik1}, α_{ik2}, and α_{ik3}.

Step 9. Calculate the intensity R_i for the ternary (composition 7 in Table 4). Calculate α_{ij}^B and α_{ik}^B, using Equation (88) and the coefficients determined earlier.

Step 10. Equation (90) for a ternary sample i-j-k reduces to

$$W_i = R_i(1 + \alpha_{ij}^B W_j + \alpha_{ik}^B W_k + \alpha_{ijk} W_j W_k) \tag{95}$$

which can be rearranged to solve for α_{ijk}:

$$\alpha_{ijk} = \frac{W_i/R_i - 1 - \alpha_{ij}^B W_j - \alpha_{ik}^B W_k}{W_j W_k} \tag{96}$$

all variables on the right-hand side of Equation (96) are known, so α_{ijk} can be calculated.

Step 11. Repeat for other interfering elements (j and k), and repeat for other analytes i.

Tao et al. [30] published a comprehensive computer program allowing the calculation of the coefficients and analysis of unknowns.

d. The Algorithm of Rousseau

i. Formulation. Rousseau and Claisse [31] used a linear relationship to approximate the binary coefficients and cross-product coefficients:

$$W_i = R_i\left[1 + \sum_{j \neq i}^{n} \left(\alpha_{ij} + \alpha_{ijm}W_m\right)W_j + \sum_{j \neq i}^{n} \sum_{k \neq i, k > j}^{n} \alpha_{ijk}W_j W_k\right] \tag{97}$$

The binary influence coefficients are thus approximated by

$$\alpha_{ij}^B = \alpha_{ij} + \alpha_{ijm}W_m \tag{98}$$

This model can be used as a stand-alone influence coefficient algorithm but has been proposed as the starting point for a fundamental parameter algorithm [32]. This equation was compared to the three-coefficient algorithm of Lachance by Vrebos and Helsen [29]. They show that the accuracy is somewhat less than that of Lachance but for most practical purposes should give equivalent results.

ii. Calculation of the Coefficients. Rousseau has shown that the fundamental parameter equation [Equations (56) and (57)] can be rearranged to

$$W_i = R_i\left(1 + \sum_{j \neq i}^{n} \alpha_{ij} W_j\right)$$

(99)

and he also proposed a method to calculate the α coefficients directly from fundamental parameters without first calculating the intensity [Equations (56) and (57)] [32]. As a matter of fact, Rousseau first calculates the coefficients for a given composition and then calculates the intensity using Equation (99). The coefficients in Equation (97) are calculated in a way very similar to the method described in Section IV.B.3.c. The compositions involved are given in Table 5. The samples referred in Table 5 are hypothetical samples. The intensity is calculated from fundamental parameters and requires no actual measurements on real samples. For the first two binaries of Table 5, the influence coefficient is calculated (symbol $\alpha_{ij}(0.20, 0.80)$ and $\alpha_{ij}(0.80, 0.20)$, respectively). Then the corresponding values are substituted in Equation (98):

$$\alpha_{ij}(0.20, 0.80) = \alpha_{ij} + \alpha_{ijm}0.80$$

(100a)

$$\alpha_{ij}(0.80, 0.20) = \alpha_{ij} + \alpha_{ijm}0.20$$

(100b)

These equations can be solved for α_{ij} and α_{ijm}. Similarly, using compositions 3 and 4 from Table 5, the corresponding coefficients for $i-k$ can be calculated. The cross-product coefficients α_{ijk} are calculated using Equation (96).

4. Application

In Sections IV.B.2 and IV.B.3 several influence coefficient algorithms have been discussed. Application of the resulting equations for calibration and analysis is discussed here. This discussion is equally valid for all the influence coefficient algorithms.

TABLE 5 Composition of the Samples Used for Calculations of the Coefficients for Rousseau's Algorithm (Weight Fraction)

Sample number	W_i	W_j	W_k
1	0.20	0.80	0.0
2	0.80	0.20	0.0
3	0.20	0.0	0.80
4	0.80	0.0	0.20
5	0.30	0.35	0.35

a. Calibration

Step 1. It is assumed that the coefficients have been calculated from theory [for example using Equation (63) or (75)].

Step 2. Calculate the matrix correction term [the brackets of Eqs. (60), (64), (76), and (86) and the braces of Eq. (90)] for all standard samples and for a given analyte. The coefficients are known (step 1) and for standard samples, all weight fractions W_i and W_j are known.

Step 3. Plot the measured intensity of the analyte multiplied by the corresponding matrix correction term against analyte weight fraction. Then, determine the "best" line:

$$W_i = B_i + M_i I_i [1 + \cdots] \tag{101}$$

by minimizing ΔW_i (see Sec. IV introduction). Note that Equation (101) is more general than Equation (42), which does not correct for matrix effects. This process is repeated for all analytes. Other methods are also feasible. The most common variant is

$$\frac{W_i}{[1 + \cdots]} = B_i + M_i I_i \tag{102}$$

This is nearly equivalent to Equation (101) but with parentheses:

$$W_i = (B_i + M_i I_i)[1 + \cdots] \tag{103}$$

The term $B_i + M_i I_i$ is related directly to the relative intensity R_i. Corrections for line overlap should affect only this term.

b. Analysis

For each of the analytes, a set of equations must be solved for the unknown W_i, W_j, and so on. If the matrix correction term used is the one according to Lachance and Traill [Eq. (60)] or de Jongh [Eq. (64)], then the set of equations can be solved algebraically (n linear equations in n unknowns for Lachance and Traill and $n - 1$ equations in $n - 1$ unknowns for de Jongh). Mostly, however, an iterative method is used. As a first estimate, one can simply take the matrix correction term equal to 1. This yields a first estimate of the composition W_i, W_j, ... This first estimate is used to calculate the matrix correction terms for all analytes. Subsequently, a new composition estimate can be obtained. This process is repeated until none of the concentrations change between subsequent iterations by more than a preset quantity.

If the matrix correction is done by algorithms that use more than one coefficient, for example Claisse and Quintin [Eq. (76)] or Rasberry and Heinrich [Eq. (86)], then the equations are not linear in the unknown concentrations and an algebraic solution is not possible. An iterative method, such as described here, can be used.

5. Empirical Coefficients

Empirical coefficients are coefficients that are not calculated from theory but from actually measured samples using regression analysis. They were the earliest correction methods but are now largely superseded by more theoretical methods. Such coefficients tend to mix the matrix correction with the sensitivity of the spectrometer. On the one hand, the matrix effect is determined by the composition of the sample and by "physical" parameters such as takeoff and incidence angles and tube anode and voltage. These are the same for spectrometers of similar design. The sensitivity of the spectrometer, on the other hand, depends on the reflectivity of the crystals, the efficiency of the detectors, and so on. These parameters are unique for each spectrometer. Also, if one of the analyte lines is overlapped by another x-ray line, some of this effect can also affect the value of the influence coefficients.

The coefficients thus determined are instrument specific and are not transferable to other instruments.

a. The Sherman Algorithm

Sherman was among the first to propose an algorithm for correction of matrix effects [33]. For a ternary system, the algorithm can be represented by the set of equations

$$(a_{AA} - t_A)W_A + a_{AB}W_B + a_{AC}W_C = 0$$

$$a_{BA}W_A + (a_{BB} - t_B)W_B + a_{BC}W_C = 0 \tag{104}$$

$$a_{CA}W_A + a_{CB}W_B + (a_{CC} - t_C)W_C = 0$$

where a_{ij} represents the influence coefficient of element j on the analyte i and t_i is the time, in seconds, required to accumulate a preset number of counts. The constants a_{ij} are determined from measurements on samples with known composition. Determination of the composition of an unknown involves solving the set of linear equations [Eq. (104)]. This set, however, is homogeneous: its constant terms are all equal to zero. Thus, only ratios among the unknown W_i can be obtained. To obtain the weight fractions W_i an extra equation is required. Sherman proposed using the sum of all the weight fractions of all the elements (or components) in the sample, which ideally should be equal to unity. For a ternary sample,

$$W_A + W_B + W_C = 1 \tag{105}$$

Using Equation (105), one of the equations in the set [Eq. (104)] can be eliminated. The solution obtained, however, is not unique: for a ternary, any two of the three equations in the set [Eq. (104)] can be combined with Equation (105). This yields three different combinations. Furthermore, any of the three elements can be eliminated in each of the combinations. Hence, a total of $3 \times 3 = 9$ different sets can be derived from Equation (104) and Equation (105), and each of these sets generates different results. In general, the algorithm yields n^2 different results for a system with n elements or compounds. This is clearly undesirable, since it is hard to determine which set will give the most accurate results. Another disadvantage is that the sum of the elements *determined* always equals unity, even if the most abundant element has been neglected. Furthermore, the numerical values of the coefficients depend, among other parameters such as geometry and excitation conditions, also on the number of counts accumulated. Nonquantifiable parameters, such as the reflectivity of the diffracting crystal used in wavelength-dispersive spectrometers or tube contamination, also affect the value of the coefficients. The coefficients determined on a given spectrometer cannot be used with another instrument: they are not transportable. The other algorithms discussed use some form of a ratio method: the Lachance and Traill algorithm, for example, uses relative intensities. The measurements are then made relative to a monitor; this reduces or eliminates the effect of such nonquantifiable parameters.

b. The Algorithm of Lucas-Tooth and Price

Lucas-Tooth and Price [34] developed a correction algorithm, in which the matrix effect was corrected for using the intensity (rather than concentration) of the interfering elements. The equation can be written as

$$W_i = B_i + I_i \left(k_0 + \sum_{j \neq i}^{n} k_{ij} I_j \right) \tag{106}$$

where B_i is a background term and k_0 and k_{ij} are the correction coefficients. A total of $n + 1$ coefficients must be determined, requiring at least $n + 1$ standards. Usually, however, a much larger number of standards is used. The coefficients are then determined by a least-squares method, for example. The corrections for the effect of the matrix on the analyte are done via the intensities of the interfering elements; their concentrations are not required. The method assumes that the calibration curves of the interfering elements themselves are all linear; the correction is done using intensities rather than concentrations. The algorithm therefore has a very limited range. Its use is limited to applications in which only one or two elements are to be analyzed (it still involves measurements of all interfering element intensities) and a computer of limited capabilities is used (although calculation of the coefficients involves more computer capabilities than the subsequent routine analysis of unknowns).

The advantages of the method are that it is very fast, because the calculation of composition of the unknowns requires no iteration; analysis of only one element is possible; this requires, however, determination of all relevant correction factors; and it is a very simple algorithm, requiring very little calculation.

c. Algorithms Based on Concentrations

Similar algorithms have been proposed using corrections based on concentrations rather than intensities. The main aim was to obtain correction factors that could be determined on one spectrometer and used, *without alteration*, on another. In practice, the coefficients still have to be adjusted because of the intimate and inseparable entanglement of spectrometer-dependent factors with matrix effects.

V. CONCLUSION

Among the advantages of XRF analysis are that the method is nondestructive and allows direct analysis involving little or no sample preparation. Major constituent analysis requires correction for matrix effects of variable (from one sample to another) magnitude. Several methods have been described. Each of these methods has its own advantages and disadvantages. These, by themselves, do not generally lead to the selection of the "best" method. The choice of the method to use is also determined by the particular application.

From the previous sections, it may appear that the mathematical methods are more powerful than the compensation methods. Yet if only one or two elements at trace level in liquids must be determined, compensation methods (either standard addition or the use of an internal standard) can turn out to be better suited than rigorous fundamental parameter calculations, for example. Compensation methods correct for the effect of an unknown, but constant, matrix. Also, they do not require the analysis of all constituents in the sample. Mathematical methods (fundamental parameters as well as methods based on theoretical influence coefficients), on the other hand, can handle cases in which the matrix effect is more variable from one sample to another. In this respect they appear more flexible than compensation methods but they require more knowledge of the *complete* matrix. All elements contributing significantly to the matrix effect must be quantified (either by x-ray measurement or by another technique), even if the determination of their levels is not required by the person who submits the sample to the analyst.

If complete analysis (covering all major elements) is required, the analyst has the choice between the fundamental parameter method and algorithms based on influence coefficients. Commonly, fundamental parameter methods are (or were) used in research environments rather than for routine analysis in industry. This choice is more often made on such considerations like availability of the programs and computers than on differences in analytical capabilities. Influence coefficient algorithms tend to be used in combination with more standards than fundamental parameter methods because their structure and simple mathematical representation make it easy to interpret the data (establishing a relationship between concentration and intensity, corrected for matrix effect). The final choice, however, must be made by the analyst.

REFERENCES

1. W. Johnson, International Report BISRA MG/D Conf. Proc., 610/67 (1967).
2. N. R. Draper and H. Smith, *Applied Regression Analysis*, John Wiley, New York, 1966, p. 7.
3. J. Hower, *Am. Miner., 44*:19 (1959).
4. J. W. Criss and L. S. Birks, *Anal. Chem., 40*:1080 (1968).
5. T. Shiraiwa and N. Fujino, *Japanese J. Appl. Physics, 5*:886 (1966).
6. T. Shiraiwa and N. Fujino, *Adv. X-ray Anal., 11*:63 (1968).
7. J. W. Criss, L. S. Birks, and J. V. Gilfrich, *Anal. Chem., 50*:33 (1978).
8. R. O. Muller, *Spectrochemical Analysis by X-ray Fluorescence*, Plenum Press, New York, 1972, Chap. 9.
9. J. W. Criss, *Adv. X-ray Anal., 23*:111 (1980).
10. P. A. Pella, L. Y. Feng, and J. A. Small, *X-ray Spectrom., 14*:125 (1985).
11. W. H. McMaster, N. K. Delgrande, J. H. Mallet, and J. H. Hubbel, *Compilation of X-ray Cross Sections*, UCRL 50174, Sec II, Rev 1 (1969).
12. K. F. J. Heinrich, in *The Electron Microprobe* (T. D. McKinley, K. F. J. Heinrich, and D. B. Wittry, Eds.), Wiley, New York, 1966, p. 296.

13. J. Leroux and T. P. Thinh, *Revised Tables of Mass Attenuation Coefficients*, Corporation Scientifique Claisse, Quebec, Canada, 1977.
14. W. J. Veigele, in *Handbook of Spectroscopy* (J. W. Robinson, Ed.), Vol. 1, CRC Press, Cleveland, OH, 1974, p. 28.
15. B. A. R. Vrebos and P. A. Pella, *X-ray Spectrom.*, *17*:3 (1988).
16. W. Bambynek, B. Crasemann, R. W. Fink, H. U. Freund, H. Mark, C. D. Swift, R. E. Price, and P. Venugopala Rao, *Rev. Mod. Phys.*, *44*:716 (1972).
17. P. Venugopala Rao, M. H. Chen, and B. Crasemann, *Phys. Rev. A*, *5*:997 (1972).
18. J. A. Bearden, *Rev. Mod. Phys.*, *39*:78 (1967).
19. G. R. Lachance and R. J. Traill, *Can. Spectrosc.*, *11*:43 (1966).
20. W. K. de Jongh, *X-ray Spectrom.*, *2*:151 (1973).
21. W. K. de Jongh, *X-ray Spectrom.*, *8*:52 (1979).
22. F. Claisse and M. Quintin, *Can. Spectrosc.*, *12*:129 (1967).
23. G. R. Lachance and F. Claisse, *Adv. X-ray Anal.*, *23*:87 (1980).
24. R. Tertian and R. Vie le Sage, *X-ray Spectrom.*, *6*:123 (1977).
25. S. D. Rasberry and K. F. J. Heinrich, *Anal. Chem.*, *46*:81 (1974).
26. R. T. Mainardi, J. E. Fernandez, and M. Nores, *X-ray Spectrom.*, *11*:70 (1982).
27. G. R. Lachance, paper presented at the International Conference on Industrial inorganic Elemental Analysis, Metz, France, June 1980.
28. R. Tertian, *Adv. X-ray Anal.*, *19*:85 (1976).
29. B. A. R. Vrebos and J. A. Helsen, *X-ray Spectrom.*, *15*:167 (1986).
30. G. Y. Tao, P. A. Pella, and R. M. Rousseau, *NBSGSC, a FORTRAN Program for Quantitative X-ray Fluorescence Analysis*, NBS Technical Note 1213, NBS, Gaithersburg, MD, 1985.
31. R. M. Rousseau and F. Claisse, *X-ray Spectrom.*, *3*:31 (1974).
32. R. M. Rousseau, *X-ray Spectrom.*, *13*:121 (1984).
33. J. Sherman, *The Correlation Between Fluorescent X-ray Intensity and Chemical Composition*, ASTM Special Publication 157, 1953, p. 27.
34. H. J. Lucas-Tooth and B. J. Price, *Metallurgia*, *64*:149 (1961).

SUGGESTIONS FOR FURTHER READING

R. Tertian and F. Claisse, *Principles of Quantitative X-ray Fluorescence Analysis*, Heyden and Son, London, 1982.
E. P. Bertin, *Principles and Practice of X-ray Spectrometric Analysis*, 2nd Ed., Plenum Press, New York, 1975.

6

Quantification in XRF Analysis of Intermediate-Thickness Samples

Andrzej A. Markowicz   *Academy of Mining and Metallurgy, Cracow, Poland*

René E. Van Grieken   *University of Antwerp, Antwerp, Belgium*

I. INTRODUCTION

In recent years, a number of approaches have been developed for quantitation in x-ray fluorescence (XRF) analysis of intermediate-thickness samples whose mass per unit area m fulfills the relation

$$m_{thin} < m < m_{thick} \tag{1}$$

where m_{thin} and m_{thick} are the values of mass per unit area for thin and thick samples [see for definition Chap. 1, Eqs. (85) and (87)]. Intermediate samples can be preferable to thick specimens because remaining uncertainties about mass attenuation coefficients have a smaller effect on the analysis results, less material is required, the sensitivity is more favorable for low-Z elements, and secondary enhancement effects are less important. Historically, the oldest correction method applied in quantitative XRF analysis of intermediate-thickness samples is the emission-transmission method in which the specific x-ray intensities from a sample are measured successively with and without a target positioned adjacent to the back of the sample in a fixed geometry. To avoid many additional measurements that are inevitable in the emission-transmission method, some alternative correction procedures based on the use of scattered primary radiation were recently developed.

The underlying principles as well as the ranges of applicability and the limitations of the correction procedures applied to XRF analysis of intermediate-thickness samples are outlined here.

II. EMISSION-TRANSMISSION METHOD

For homogeneous intermediate-thickness samples, the mass per unit area of the element i, m_i, can be calculated from the following equation for the monochromatic excitation [see Eq. (82) in Chap. 1]:

$$m_i = \frac{I_i(E_i)}{B_i} \, Ab_{corr} \tag{2}$$

where

$$B_i = \frac{GI_0(E_0)\epsilon(E_i)\tau_i'(E_0)\omega_i p_i(1 - 1/j_i)}{\sin \psi_1} \tag{3}$$

and Ab_{corr} is the absorption correction factor given by

$$Ab_{corr} = \frac{[\mu(E_0) \csc \psi_1 + \mu(E_i) \csc \psi_2]m}{1 - e^{-[\mu(E_0) \csc \psi_1 + \mu(E_i) \csc \psi_2]m}} \tag{4}$$

[The symbols used in Eqs. (2) through (4) are explained in Chap. 1, Eqs. (80) and (81).]

The value of the constant B_i (sometimes called the sensitivity factor) can be determined either experimentally as the slope of the straight calibration line for the ith element obtained with thin homogeneous samples or semiempirically based on both the experimentally determined $GI_0(E_0)$ value and the relevant fundamental parameters.

The absorption correction factor Ab_{corr} represents the combined attenuation of the primary and fluorescent radiations in the whole specimen and can be determined individually for each sample by transmission experiments [1,2]. These are done by measuring the x-ray intensities with and without the specimen from a thick multielement target located at a position adjacent to the back of the specimen, as shown in Figure 1. If $(I_i)_S$, $(I_i)_T$, and $(I_i)_0$ are the intensities after background correction from the sample alone, from the sample plus target, and from the target alone, respectively, then the combined fraction of the exciting and fluorescent radiations transmitted through the total sample thickness is expressed by

$$e^{-[\mu(E_0) \csc \psi_1 + \mu(E_i) \csc \psi_2]m} = \frac{(I_i)_T - (I_i)_S}{(I_i)_0} \equiv H \tag{5}$$

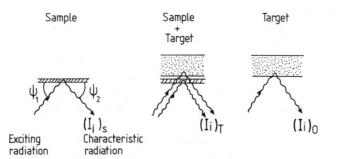

Sample   Sample + Target   Target

$(I_i)_S$ $(I_i)_T$ $(I_i)_0$

Exciting radiation Characteristic radiation

Figure 1 Experimental procedure used in the emission-transmission method for the correction of matrix absorption effects.

After a simple transformation, Equation (2) can be rewritten as

$$m_i = \frac{I_i(E_i)}{B_i} \frac{-\ln H}{1 - H} \tag{6}$$

The emission-transmission method can only be applied in the quantitative XRF analysis of homogeneous samples of which the mass per unit area is smaller than the critical value m_{crit}, defined by

$$m_{crit} = \frac{-\ln H_{crit}}{\mu(E_0) \csc \psi_1 + \mu(E_i) \csc \psi_2} \tag{7}$$

where H_{crit} is the critical value of the transmission factor defined by Equation (5); in practice, $H_{crit} = 0.1$ (or 0.05).

To minimize possible absorption correction errors resulting from enhancement of the specimen radiation by scattered target radiation, targets that yield a high ratio of scattered to fluorescent radiation should not be used.

Giauque et al. [3] developed a modified version of the emission-transmission method. Using data from the attenuation measurements and Equation (5), the values of $[\mu(E_0) \csc \psi_1 + \mu(E_i) \csc \psi_2]m$ are calculated for the energies of characteristic x-rays of all elements present in a thick multielement target. If these values are plotted versus the fluorescence x-ray energy on a log-log scale (Fig. 2), an approximate value for $m\mu(E_0) \csc \psi_1$ can be obtained by extrapolation of

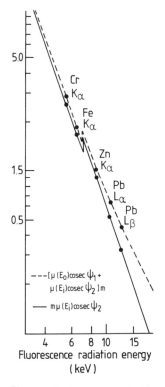

Figure 2 Curves of $[\mu(E_0) \csc \psi_1 + \mu(E_i) \csc \psi_2]m$ and $m\mu(E_i) \csc \psi_2$ values versus fluorescence x-ray energy for an NBS SRM 1632 coal specimen. (From Ref. 3. Reprinted with permission from *Anal. Chem.* Copyright American Chemical Society.)

the curve to the energy of the excitation radiation. In turn, values for $m\mu(E_i)$ csc ψ_2 can be calculated, a curve for these values drawn, and a new value for $m\mu(E_0)$ csc ψ_1 established. This last step is iterated several times. Using data from the latter curve, the absorption correction factors for all radiations of interest can be calculated from Equation (4). If some elements to be determined are major or minor constituents, a few separate curves for $m\mu(E_i)$ csc ψ_2 values should be plotted between the preselected x-ray energies corresponding to the relevant absorption edges. In the emission-transmission method of Giauque et al. [3], the incoherent scattered radiation corrected for matrix absorption is used as the internal standard to compensate for variations in sample mass, x-ray tube output, and sample geometry.

Van Dyck et al. [4] developed a correction method that allows calculations of the absorption coefficients (and absorption correction factors as well) at any energy for intermediate-thickness samples, without additional measurements, by using the ratio of the x-ray signals from a Zr wire positioned in front of the sample and from a Pd foil placed behind the sample, both in a fixed geometry as shown in Figure 3. The Zr wire provides an external reference signal (Zr $K\alpha$), which is applied for normalization of all measured fluorescent intensities to reduce considerably the effect of variations in exciting x-ray tube intensity and of dead-time losses. The coefficients for higher energies are calculated with an iterative program from the experimentally measured absorption coefficient at the Pd L energy (2.9 keV), $\mu(E_{Pd})$. In the first step, the total attenuation coefficient at the Pd L energy, caused exclusively by the low-Z elements (e.g., $Z < 17$) in the sample that show no characteristic peak above 3.0 keV in the spectrum, is calculated from the normalized measured intensities, $(I_{Pd})_T'$ and $(I_{Pd})_0'$ and the different characteristic peaks recorded in the spectrum. This attenuation coefficient due to low-Z elements, $\mu_{low-Z}(E_{Pd})$, is given by [4]

$$\mu_{low-Z}(E_{Pd}) \tag{8}$$

$$= \frac{\ln[(I_{Pd})_0'/(I_{Pd})_T']/m \; \text{csc} \; \psi_2 - \mu(E_0) \; \text{csc} \; \psi_1/\text{csc} \; \psi_2 - \sum_{j=1}^{n'} W_j\mu_j(E_{Pd})}{1 - \sum_{j=1}^{n'} W_j}$$

where n' is the number of characteristic peaks in the XRF spectrum and $\mu_j(E_{Pd})$ is the mass attenuation coefficient for the Pd L energy in the element j [5] giving rise to a characteristic peak in the XRF spectrum.

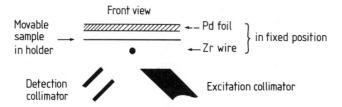

Figure 3 Measurement geometry applied in an automatic absorption correction method. (From Ref. 4. Reprinted by permission of John Wiley & Sons, Ltd.)

The weight fraction of the element j in the sample, W_j, is correlated with the recorded characteristic peak through the sensitivity factor, routinely obtained by measuring thin single- or multielement standards. As a first approximation, the characteristic peak intensities are not corrected for absorption. In a second step, based on the $\mu_{low-Z}(E_{Pd})$ value, the absorption coefficients of the low-Z matrix for other energies, $\mu_{low-Z}(E)$, are calculated quantitively by assuming nearly parallel properties of the logarithmic absorption curves $\ln(E)$ versus $\ln E$.

The total mass attenuation coefficient $\mu(E)$ for the characteristic x-rays of an element in the whole sample can now be calculated, taking into account the contributions $\mu_j(E)$ of the high-Z elements:

$$\mu(E) = \left(1 - \sum_{j=1}^{n'} W_j\right) \mu_{low-Z}(E) + \sum_{j=1}^{n'} W_j\mu_j(E) \tag{9}$$

The overall procedure is summarized and schematically represented in Figure 4.

Better $\mu(E)$ values [Eq. (9)] are obtained in the second and following loops by carrying out appropriate absorption corrections to the characteristic intensities from which the W_j values are derived using $\mu(E)$ and $\mu(E_0)$ values from the previous loop and by including in Equation (8) the $\mu(E_0)$ values, taken as zero in the first loop. The iteration is stopped when the difference in $\mu(E)$ between two loops is negligible.

Subroutine ENHANC is applied to evaluate the enhancement effect of the Pd L x-rays caused by all elements in the sample of which the x-ray energy is higher than the L_I, L_{II}, and L_{III} absorption edges of Pd. The intensity caused by this enhancement effect can be considered to result in an apparent increase in $(I_{Pd})'_0$. The enhancement contributions I_{enhPd} are added to the $(I_{Pd})'_0$ in Equation

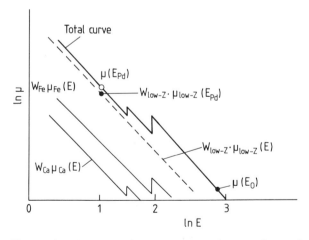

Figure 4 The calculation steps. From the experimentally measured total absorption coefficient at the Pd L energy, $\mu(E_{Pd})$, the calculated contribution from high-Z elements (giving a characteristic peak in the spectrum) is subtracted; through this low-Z matrix contribution at the Pd L energy, the total low-Z absorption curve (dashed line) is calculated; the contribution from high-Z elements (e.g., Ca and Fe) is then added, to yield the total absorption curve (thick line). (From Ref. 4. Reprinted by permission of John Wiley & Sons, Ltd.)

(8). This total $(I_{Pd})'_{0eff}$ value and the detected $(I_{Pd})'_T$ signal allow us to calculate the correct μ values via Equations (8) and (9).

A comprehensive discussion of the influence of secondary enhancement of the Pd L x-rays by the sample as well as of the influence of two other complicating factors, grain size effects and a heterogeneous sample load, is presented in Reference 4. The grain size and sample heterogeneity effects induce inaccuracies on the absorption coefficient determinations that may well reach 20% for particulate samples, such as intermediate-thickness deposits of geological materials. This approach thus has the same limitations as all emission-transmission methods applied to heterogeneous samples.

III. ABSORPTION CORRECTION METHODS VIA SCATTERED PRIMARY RADIATION

The use of scattered primary radiation in XRF analysis provides an alternative to the common problem of matching standards of similar composition to samples to be analyzed. The backscatter peaks are sometimes treated as fluorescent peaks from internal standards because they suffer matrix absorption similar to that of fluorescent peaks and behave similarly with instrumental variations. They also provide the only direct spectral measure of the total or average matrix of geological, biological, or other materials containing large quantities of light elements, such as carbon, nitrogen, and oxygen, usually not observed by their characteristic x-ray peaks.

XRF matrix correction methods based on the use of scattered radiation have mostly been applied in quantitative analysis of infinitely thick samples under a wide variety of experimental conditions, including discrete and continuum primary radiation sources and detection by both wavelength- and energy-dispersive systems [6–13]. The scattered radiation methods utilize the incoherent (Compton) or/and coherent (Rayleigh) scatter peaks from line excitation sources or the intense high-energy region from continuum sources.

A. Absorption Corrections Based on Incoherent Scattered Radiation

For specimens of less than infinite thickness, the intensity I_{Com} of Compton-scattered radiation can be expressed by [14]

$$I_{Com} = \frac{k' I_0 \sigma_{Com}(E_0)[1 - e^{-(\mu(E_0) \csc \psi_1 + \mu(E_{Com}) \csc \psi_2)m}]}{\mu(E_0) \csc \psi_1 + \mu(E_{Com}) \csc \psi_2} \tag{10}$$

where k' is a constant for a given measurement geometry and detection efficiency, $\sigma_{Com}(E_0)$ is the Compton mass-scattering coefficient of the sample material for the primary radiation of energy E_0 (cm^2 g^{-1}), $\mu(E_{Com})$ is the total mass attenuation coefficient of the sample material for Compton-scattered primary radiation of energy E_{Com} (cm^2 g^{-1}) [see Eq. (48) in Chap. 1)]. Equation (10) is valid for monochromatic excitation.

Assuming that the relation of the atomic number to the mass number is constant for every element to be found in the sample and that $\mu(E_0) \simeq \mu(E_{Com})$, the following simplified formula for the intensity of Compton-scattered radiation from a multielement sample can be obtained:

$$I_{Com} = \frac{k_1' I_0 [1 - e^{-\mu(E_0) \csc \psi_1 (1 + \csc \psi_2/\csc \psi_1) m}]}{\mu(E_0) \csc \psi_1 (1 + \csc \psi_2/\csc \psi_1)} \tag{11}$$

where k_1' is a constant for given geometry of measurement and the energy of the incident radiation. The value of the product $k_1' I_0$ is determined experimentally from a reference scatterer and is valid as long as the reference scatterer has a matrix that is not too different from that of the specimens.

Kieser and Mulligan [15] worked out a method based on the use of the incoherent scatter radiation, which gives accurate mass absorption coefficients for a limited average Z range. The mass absorption coefficient $\mu(E_0)$ for specimens of intermediate thickness is found from Equation (11) after a numerical solution. To obtain a value of the mass absorption coefficient $\mu(E)$ at any energy E, the authors [15] assumed that the slope of the curve log $\mu(E)$ versus log E is constant for all elements (approximately -2.7) over a range of x-ray energies. The proposed Compton-scattered method for the determination of an intermediate specimen's mass absorption coefficient at any energy can be applied as long as no absorption edge of a major or minor element intervenes. When the values of the $\mu(E)$ are determined, calculation of the absorption correction factor [Eq. (4)] is straightforward if, of course, the mass of the sample to be analyzed is known.

A modified fluorescent Compton correction method for quantitative XRF of intermediate specimens was developed by Holynska and Markowicz [16]. The method is based on the use of the measured x-ray fluorescent intensities of all determined elements and the intensity of Compton-backscattered radiation. The authors derived the following expression for the determination of the mass per unit area of the element i, m_i:

$$m_i = \frac{a_i I_i}{1 + \sum_{l=1}^{n'} a_{il} I_l + \frac{b_i I_{Com}}{m}} \tag{12}$$

where a_i, a_{il}, and b_i are constant coefficients obtained experimentally on the basis of standard samples and n' is the number of the elements to be determined, including the ith element.

As is seen from Equation (12), the absorption matrix correction is carried out via the intensities of characteristic x-rays of all elements determined and the intensity I_{Com}, reflecting for the most part the variations in the composition of light matrix. To apply the absorption matrix correction, Equation (12), the total mass per unit area of the specimen must be evaluated, for example by sample weighing. This fluorescent-Compton correction method can be used in the XRF analysis of homogeneous intermediate-thickness samples in a limited range of mass per unit area ($m < 10m_{thin}$).

B. Absorption Corrections Based on Both Coherent and Incoherent Scattered Radiations

In recent years several absorption correction methods based on both coherent and incoherent scattered radiations have been developed and applied in quantitative XRF analysis of intermediate samples. This group of correction methods is represented either by relatively simple approaches [17,18] or by very sophisticated fundamental parameter procedures [19–22] providing superior analytical flexibility.

For specimens of less than infinite thickness, the intensity I_{coh} of coherent scattered radiation can be calculated from

$$I_{coh} = \frac{k'' I_0 \sigma_{coh}(E_0)}{\mu(E_0)(\csc \psi_1 + \csc \psi_2)} [1 - e^{-\mu(E_0)(\csc \psi_1 + \csc \psi_2)m}] \qquad (13)$$

where k'' is a constant for a given measurement geometry and detection efficiency for the primary x-rays of energy E_0 and $\sigma_{coh}(E_0)$ is the coherent mass-scattering coefficient of the sample material for the primary radiation ($cm^2 g^{-1}$) [see Eq. (68) in Chap. 1].

Bazan and Bonner [17] showed, for the first time, a linear relation between the effective absorption coefficient (defined as the sum of the sample absorption coefficients for exciting and characteristic x-rays) and the ratio of incoherent to coherent scattering. However, the coefficients of the calibration line varied somewhat with the matrix, and this hampered practical applications of this simple approach.

Markowicz [18] found theoretically that the sensitivity of the absorption correction via the incoherent/coherent scattered x-ray intensities ratio is better than that of the absorption procedure involving each of the scattered radiations individually. For intermediate-thickness samples, in a limited range of rather small values of mass per unit area, the intensities of the Compton-scattered radiation I_{Com} and the coherent scattered radiation I_{coh} are different functions of the total mass attenuation coefficient of the incident radiation $\mu(E_0)$; the intensity I_{Com} is a linearly decreasing function, and the intensity I_{coh} appears to be a linearly increasing function of the $\mu(E_0)$. For a limited range of $\mu(E_0)$ values the following simple expression can be used to evaluate $\mu(E_0)$ [18]:

$$\mu(E_0) = C_1 + C_2 m + C_3 r + C_4 mr \qquad (14)$$

where

$$r = \frac{I_{Com}}{I_{coh}}$$

and C_1–C_4 are constants calculated by the least-squares fit on the basis of experimental results for standard samples.

The values of the total mass attenuation coefficient of the fluorescent radiation in a whole sample, $\mu(E_i)$, is obtained from the simple dependence of the $\mu(E_i)/\mu(E_0)$ ratio on the values $\mu(E_0)$, calculated separately for each element to be determined. Finally, sample weighing provides the value of mass per unit area m, and calculation of the absorption correction factor Ab_{corr} via Equation (4) can be simply performed if, of course, the values of the effective angles ψ_1 and ψ_2 are evaluated experimentally or theoretically. The applicability of the proposed matrix correction method [18] involving both incoherent and coherent scattered primary radiations is limited to XRF analysis of intermediate-thickness samples of mass per unit area smaller than about $10 m_{thin}$.

A backscattered fundamental parameters (BFP) method for quantitative XRF analysis of intermediate samples of variable composition and thickness was developed by Nielson [19]. The method utilizes thin-film multielement calibration of the spectrometer and mathematical matrix corrections in which the samples are modeled as a composite of heavy elements, which are quantified through their characteristic radiation, and of light elements, estimated through the coherent and incoherent x-ray scatter peaks. Figure 5 schematically illustrates the basis for analyzing the heavy elements ($Z > 13$) and the light elements (H, C, N, O, Na, and others), which must be estimated by the difference from the scattered x-ray peaks. The BFP method utilizes coherently and incoherently scattered x-rays to identify and estimate the quantities of two light elements representative of the light element portion of the sample matrix. The quantities of the two light elements a and b are estimated by solving for W_a and W_b in the simultaneous equations

$$gI_{coh} - \sum_{j=1}^{n'} W_j\sigma_{cohj}(E_0) = W_a\sigma_{coha}(E_0) + W_b\sigma_{cohb}(E_0) \tag{15}$$

and

$$hI_{Com} - \sum_{j=1}^{n'} W_j\sigma_{Comj}(E_0) = W_a\sigma_{Coma}(E_0) + W_b\sigma_{Comb}(E_0) \tag{16}$$

where g and h are the geometry-dependent calibration factors determined experimentally by using any standard of known total composition. Since several light element pairs may satisfy Equations (15) and (16), the pair is chosen whose incoherent/coherent scattering cross-sectional ratios lie immediately on either side of the ratio of the observed scatter attributable to light elements [19]:

$$\frac{\sigma_{Coma}(E_0)}{\sigma_{coha}(E_0)} < \frac{hI_{Com} - \sum_{j=1}^{n'} W_j\sigma_{Comj}(E_0)}{gI_{coh} - \sum_{j=1}^{n'} W_j\sigma_{cohj}(E_0)} < \frac{\sigma_{Comb}(E_0)}{\sigma_{cohb}(E_0)} \tag{17}$$

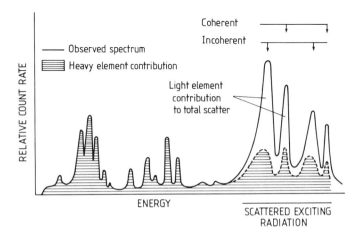

Figure 5 Light element contributions to x-ray scattering, from which absorption corrections are computed. (From Ref. 19. Reprinted with permission from *Anal. Chem.* Copyright American Chemical Society.)

The heavy and light element concentrations are used in computing the absorption correction factor Ab_{corr} [Eq. (4)] and the enhancement correction factor [Eq. (83) in Chap. 1]. Since the concentrations and corrections are interdependent, all calculations are carried out by iteration (for more details see Ref. 19).

To improve the sensitivity of the determination of low-Z elements, Sanders et al. [23] extended the previously described BFP method. The new method utilizes the coherent and incoherent backscatter intensities to compute matrix corrections [24] from the combined results of two separate energy-dispersive XRF (EDXRF) data from different (Ti $K\alpha$ and Zr $K\alpha$) excitation sources. The Ti-excited spectrum allows a more sensitive determination of elements in the Al–Ca range.

The coherent/incoherent scatter ratio is also applied in an absorption correction procedure developed by Van Dyck and Van Grieken [20] for monochromatic x-ray excitation. In this method, coherent and incoherent scattered radiations are used to calculate, first, the effective mass of the sample and, second, the absorption coefficients for x-rays of interest and hence the absorption correction factors. The effective thickness is the sample thickness weighted at every point for the excitation-detection efficiency, in the same way as the measured characteristic radiation is weighted. Assuming that the major elements of the sample do not differ too greatly in atomic number, the effective thickness m_{eff} can be calculated from [25]

$$m_{eff} = \frac{I_{Com}}{b_0 Ab_{corr}(E_{Com})(b_0 I_{coh} Ab_{corr}(E_{Com})/a_0 I_{Com} Ab_{corr}(E_{coh}))b_1/(a_1 - b_1)}$$

(18)

where a_0, a_1 and b_0, and b_1 are experimental constants obtained by fitting the results of measured standards (for mixtures or compounds the coherent and incoherent scatter factors S_{coh} and S_{Com}, in fact the relevant mass-scattering coefficients, are given by

$$S_{coh} = a_0 \sum_{j=1}^{n} W_j Z_j^{a_1}$$

(19)

and

$$S_{Com} = b_0 \sum_{j=1}^{n} W_j Z_j^{b_1}$$

respectively). $Ab_{corr}(E_{Com})$ and $Ab_{corr}(E_{coh})$ are the absorption correction factors for the incoherent and coherent scatter radiation, respectively, as defined in Equation (4).

A reasonably accurate effective mass is obtained by modeling the sample as a composite of high-Z elements, calculated from their characteristic peaks using Equation (2), and of a light matrix with mass per unit area m_{low-Z}, evaluated from the coherent and incoherent scatter peaks after subtraction of the high-Z element contribution. The method for the determination of the effective thickness allows the analysis of samples of heterogeneous thickness and irregular shape.

In the method proposed by Van Dyck and Van Grieken [20], calculation of the mass attenuation coefficient for x-rays of interest is preceded by an evaluation of the mass attenuation coefficient μ (2.956 keV) at the Ar $K\alpha$ energy (2.956 keV).

This energy is preferred because it is at the lower end of the energy range that can safely be used in conventional EDXRF analysis and because, when working under vacuum, it is situated in a peak-free part of an XRF spectrum. The value of μ (2.956 keV) is derived from the ratio coherent to incoherent scatter intensities R, based on the relationship (see Fig. 6) of the calculated mass attenuation coefficient at 2.956 keV versus the measured R ratio:

$$\mu \ (2.956 \ \text{keV}) = g_0 + g_1 R + g_2 R^2 \tag{20}$$

where g_0, g_1, and g_2 are constant coefficients derived by means of a least-squares fit based on the experimental results with standard samples. To improve the accuracy of the method, the mass absorption coefficient of the low-Z matrix at 2.956 keV, $\mu_{\text{low-}Z}$ (2.956 keV), must be calculated from the measured ratio of coherent to incoherent scatter intensities, corrected for the high-Z elements contribution using their characteristic x-ray intensities. Through this low-Z matrix contribution at 2.956 keV, the total low-Z absorption curve is calculated (in full analogy to low-Z matrix contribution at the Pd L energy on Figure 4). Finally the mass absorption coefficients for the different x-ray energies E are calculated by adding

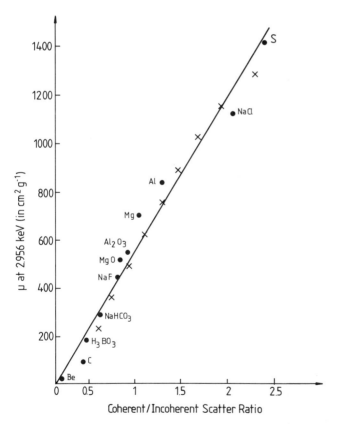

Figure 6   Calculated mass absorption coefficient at 2.956 KeV versus measured coherent/incoherent scattered intensity ratios for pure elements and compounds (circles) and a graphite-sulfur mixture (crosses). (From Ref. 20. Reprinted with permission from *Anal. Chem.* Copyright American Chemical Society.)

the low-Z and high-Z absorption contributions [see the total absorption curve, Eq. (9) and Fig. 4].

The proposed procedure for the automatic determination of the mass attenuation coefficient, based on the coherent/incoherent scatter ratio, has several obvious merits compared with the emission-transmission absorption correction method. First, there is no supplementary measurement or work needed [apart from experimentally obtaining the semiempirical dependence of the μ (2.956 keV) on the R ratio (Fig. 6) and sensitivity factors B_i, Eq. (2), on the basis of thin standard samples], since the additional information on a sample composition is present in the spectrum itself. Second, since the energies of coherently and incoherently scattered primary radiation, from which the information is extracted, are higher than that of the incident radiation used in transmission experiments, secondary effects, that is, grain size effects, inhomogeneous thickness of the sample, and irregular sample surface, are less important. A third positive point of the described procedure [20] is the independence, within certain limits, of the analytical results on the mass of the specimen. Also, it is worth emphasizing the capability of determining reasonably accurate mass absorption coefficients when the mean atomic number of the sample varies drastically.

Most of the existing correction procedures in XRF analysis of intermediate-thickness samples ignore the enhancement effect. It appears, however, that for some special cases the enhancement effect should be taken into account. Van Dyck et al. [26] derived theoretical formulas for secondary fluorescent x-ray intensities in medium-thickness samples based on the Sherman equations. Their computer routine for enhancement corrections was incorporated into an overall program for evaluation of x-ray spectra and calculation of x-ray absorption correction factors from scatter peaks.

IV. QUANTITATION FOR INTERMEDIATE-THICKNESS GRANULAR SPECIMENS

During the last few years much interest has gone to the quantitative XRF analysis of granular specimens. The fundamentals of the quantitative XRF analysis of granular specimens involve typical problems of radiation physics, that is, the interaction of photons with a specimen of finite size.

A. Particle Size Effects in XRF Analysis of Thin and Intermediate-Thickness Specimens

The analysis of granular materials by XRF, when traditional methods of fusing or grinding cannot be used requires careful consideration of the so-called particle size effects. These effects exist in XRF analysis of any granular materials, irrespective of the mass per unit area of the specimen, and may constitute a major source of error in quantitative analysis. The size of particles affects not only the intensity of characteristic x-rays but the intensity of both backscattered and transmitted x- and low energy γ-radiation as well [4,27].

Different models have been proposed to account for the influence of particle size on the characteristic x-ray intensity leaving the sample [27–33]. Many of these models involve relatively complex calculations, particularly when the particle size effects for thick granular samples must be evaluated. This section is limited to the particle size effects in the quantitative XRF analysis of thin and intermediate-thickness (monolayer) samples.

For granular intermediate-thickness samples, the mass per unit area of the element i, m_i, is given by [31]

$$m_i = \frac{I_i(E_i)}{B_i} \frac{1}{F_i} \tag{20}$$

where F_i is the heterogeneity factor which for a certain discrete particle size is defined by

$$F_i = \frac{1 - e^{-(\mu_f^*(E_0) \csc \psi_1 + \mu_f^*(E_i) \csc \psi_2)a_r}}{(\mu_f^*(E_0) \csc \psi_1 + \mu_f^*(E_i) \csc \psi_2)a_r} \tag{21}$$

where $\mu_f^*(E_0)$ and $\mu_f^*(E_i)$ are the linear attenuation coefficients (cm^{-1}) for primary and fluorescent radiation in fluorescent particles, respectively, and a_r is the radiometric particle diameter.

The radiometric diameter, introduced by Claisse and Samson [28], represents the mean geometrical path of x-rays through one particle. These and many other authors [27,29–31] have taken a_r as simply equal to the volume of the grain divided by the particle area presented to the radiation, averaged over all possible orientations of the grain. Hence, for spherical particles, the radiometric diameter is equal to $0.67a$ (with a = geometric diameter). One can easily visualize, however, that such an approach assumes equal weighting of the contribution from all possible radiative paths to the average. In view of the different absorption effects themselves, this is untrue. Recently, Markowicz et al. [34] introduced, instead of the commonly used radiometric diameter approach, the concept of an effective absorption-weighted radiometric diameter for fluorescent radiation, depending on both the geometry and absorption effects, and provided a comparison of these two approaches for single spherical particles for two excitation-detection geometries (π and $\pi/2$). This new approach allows quantitative evaluation of the discrepancies resulting from the concept a_r = particle volume/average area. It was concluded that for the π geometry both approaches give practically the same results (maximum relative differences amount only to 5%) but for $\pi/2$ geometry, the radiometric diameter approach can safely be applied only for very small particles and/or at relatively high energies of primary radiation [34].

For samples with a certain particle size distribution described by a function $f(a_r)$, the heterogeneity factor F_i can be calculated according to the formula [31]

$$F_i = \int_{a_{r\min}}^{a_{r\max}} \frac{f(a_r)(1 - e^{[-a_r(\mu_f^*(E_0) \csc \psi_1 + \mu_f^*(E_i) \csc \psi_2)]}) \, da_r}{[\mu_f^*(E_0) \csc \psi_1 + \mu_f^*(E_i) \csc \psi_2]a_r} \tag{22}$$

where $a_{r\min}$ and $a_{r\max}$ are the smallest and the largest radiometric particle diameters, respectively, and

$$f(a_r) = \frac{dV_f}{(V_f)_t \, da_r} \tag{23}$$

with dV_f = volume of fluorescent particles having a size between a_r and $(a_r + da_r)$, and $(V_f)_t$ = total volume of the fluorescent particles.

The theoretical predictions given by Equations (21) and (22) have been compared with the experimental results for the heterogeneity factors obtained for samples with some discrete particle sizes and various particle size distributions [35]. For granular samples of copper sulfide, a satisfactory agreement has been obtained; for samples of iron oxide some discrepancies due to agglomeration of the particles have been observed.

An in-depth study of the influence of sample thickness, excitation energy, and excitation-detection geometry on the particle size effects in XRF analysis of intermediate-thickness samples was recently carried out [36].

B. Correction Methods for the Particle Size Effect in XRF Analysis of Intermediate Specimens

As already mentioned, the heterogeneity factor F_i describing the magnitude of the particle size effect in XRF analysis of thin and monolayer samples can simply be calculated if the particle size or the function of the particle size distribution is known. This occurs in XRF analysis of air particulates (or aerosols), for example, when special sampling techniques, involving a cascade impactor, are applied. At different stages of the cascade impactor the particles of definite sizes are collected [37] and calculation of the heterogeneity factor F_i may be straightforward, if, of course, the kind of fluorescent particles is known.

A simple empirical particle size correction factor $(1 + ba)^2$ was proposed by Criss [38], in which a is the particle diameter and b is a coefficient that depends on particle composition and experimental conditions. The author has provided a table of the values for b for the determination of 48 different elements in 200 compounds using either a Cr- or W-target x-ray tube.

Another correction for the particle size effect based on the model of Berry et al. [27], also requiring evaluation of the particle size in a sample, is due to Nielson [19].

All the particle size corrections mentioned are of limited applicability since evaluation of a particle size or particle size distribution function must be done before XRF analysis. Moreover, in many cases it may be necessary to consider how the indirectly determined particle sizes relate to true sizes. However, even when there is some uncertainty in the sizes and compositions of the particles, which are input parameters in the particle size correction, it is better to make an appropriate correction than no correction at all.

1. Empirical Particle Size Correction Method Using Dual Measurements

To overcome the problems encountered when calculated particle size correction factors are applied in the quantitative XRF analysis of granular samples, a particle size correction method based on dual measurements of characteristic x-rays excited by x- or γ-radiation of two different energies was recently developed [39]. The method utilizes the difference in the particle size effect for two excitation energies and offers the possibility of experimental detection and correction of this effect. In general, these two excitation energies should be chosen so that the effect of particle size is small for one of them and large for the other. Thus, the measured intensities of the characteristic x-rays of the element to be determined are different functions of the particle size. The ratio of these two intensities is sensitive to the particle size, and it can be used for obtaining the particle size correction factor $K_i = 1/F_i$. First, a calibration curve giving the relationship between the correction factor K_{i2} and the ratio $(I_{i1}/I_{i2})_{rel}$ must be plotted. The ratio $(I_{i1}/I_{i2})_{rel}$ is given by

$$\left(\frac{I_{i1}}{I_{i2}}\right)_{rel} = \frac{I_{i1}/I_{i2}}{(I_{i1}/I_{i2})_{hom}} \tag{24}$$

where I_{i1}/I_{i2} is the ratio of the intensities of the characteristic x-rays of the ith element excited in a granular sample by primary x-rays of two different energies (indexes 1 and 2, respectively) and $(I_{i1}/I_{i2})_{hom}$ is the same for a thin homogeneous sample. Taking into account that in the thin-sample technique interelement effects may be neglected, single-element standard samples can be used for obtained the $(I_{i1}/I_{i2})_{hom}$ ratio. The calibration curve mentioned earlier can be obtained either theoretically or experimentally with the use of calibration samples of known discrete particle size fractions. It has been shown [39] that there is also a possibility of applying such a calibration curve for granular samples with various particle size distribution functions. For a π geometry the calibration curves for the determination of the correction factor K_{i2} are described by the equation [40]

$$K_{i2} = \left(\frac{I_{i1}}{I_{i2}}\right)_{rel} \frac{t'\chi a_r}{1 - e^{-t'\chi a_r}} \tag{25}$$

where

$$t' = \frac{\mu_f^*(E_{01}) + \mu_f^*(E_i)}{\mu_f^*(E_{02}) + \mu_f^*(E_i)} \tag{26}$$

$$\chi = \mu_f^*(E_{02}) + \mu^*(E_i) \tag{27}$$

$\mu_f^*(E_{01})$, $\mu_f^*(E_{02})$ are the linear absorption coefficients for primary radiation of two different energies (indexes 1 and 2, respectively) in fluorescent particles (cm^{-1}). The particle size correction factor K_{i2} is given by

$$K_{i2} = \frac{\chi a_r}{1 - e^{-\chi a_r}} \tag{28}$$

The proposed method of particle size correction was verified experimentally for the determination of copper, applying ^{238}Pu and ^{241}Am radioisotopes as sources of primary radiation, and a satisfactory agreement between theoretical predictions and experimental results was reported [39].

2. Applicability of the Particle Size Correction Method

Accurate determination of the correction factor K_{i2} is mainly affected by fluctuations resulting from counting statistics in all measured intensities of the characteristic radiation. The absolute error ΔK_{i2}, in determining the correction factor K_{i2} can be calculated from the formula

$$\Delta K_{i2} = \frac{t'\chi a_r}{1 - e^{-t'\chi a_r}} S_c \tag{29}$$

where S_c is standard deviation resulting from counting statistics for the ratio $(I_{i1}/I_{i2})_{rel}$.

The relative error in determining the particle size correction factor K_{i2} is given by

$$\frac{\Delta K_{i2}}{K_{i2}} = \frac{S_c}{(I_{i1}/I_{i2})_{rel}} = S_r \tag{30}$$

where S_r is relative standard deviation resulting from counting statistics for the ratio $(I_{i1}/I_{i2})_{rel}$.

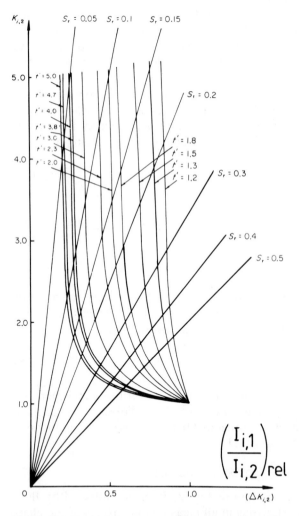

Figure 7 Theoretical relationship of the particle size correction factor K_{i2} with the ratio of the intensities $(I_{i1}/I_{i2})_{rel}$ for different values of the parameter t' (family of curves) and with the error $\Delta K_{i,2}$ for different values of the relative standard deviation S_r (family of straight lines). (From Ref. 40. Reprinted by permission of John Wiley & Sons, Ltd.)

Equations (25) and (30) enable us to estimate the applicability of the particle size correction method. This can be done with the aid of Figure 7. From two families of curves in Figure 7, one can estimate the maximum value of the particle size correction factor $(K_{i2})_{max}$ that can be determined with the particle size correction method for given values of S_r and t'. Figure 8 presents the theoretical relationship of the maximum value of the correction factor $(K_{i2})_{max}$ with the parameter t' for different values of the relative standard deviation S_r; Figure 9 shows the theoretical relationship of $(K_{i2})_{max}$ with S_r for different values of t'. The families of curves shown in Figures 8 and 9 allow us to determine the application limits of the particle size correction method, that is, to determine $(K_{i2})_{max}$ in various configurations. In practice, however, it is more interesting to know the maximum value of the radiometric particle diameter for which the particle size effect can still be corrected. This can be determined from Equation (25).

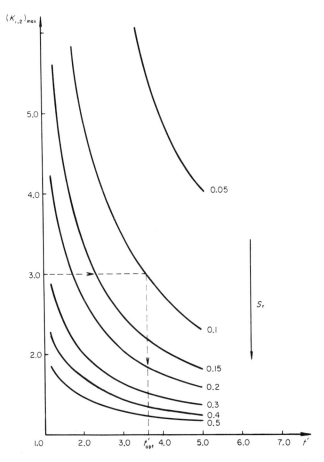

Figure 8 Theoretical relationship of the maximum value of the particle size correction factor $(K_{i2})_{max}$ with the parameter t' for different values of the relative standard deviation S_r. t'_{opt} = the optimum value of the parameter t' (see text for details). (From Ref. 40. Reprinted by permission of John Wiley & Sons, Ltd.)

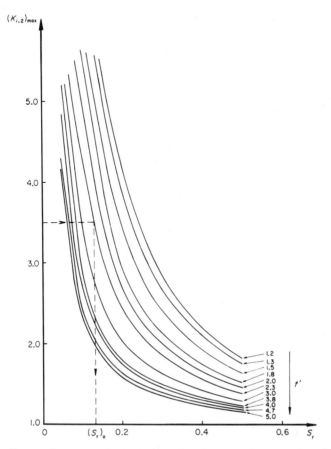

Figure 9 Theoretical relationship of the maximum value of the particle size correction factor $(K_{i2})_{max}$ with the relative standard deviation S_r for different values of the parameter t'. $(S_r)_0$ = the maximum acceptable value of the relative standard deviation S_r (see text for details). (From Ref. 40. Reprinted by permission of John Wiley & Sons, Ltd.)

3. Selection of the Optimum Measurement Conditions

When the ith element is present in a given chemical compound and the maximum values of the radiometric diameter of the particle in a sample are known, it is possible to define the optimum value t'_{opt} of the parameter t' for a given energy of exciting radiation E_{02}. The value of t'_{opt} can be determined from Figure 8 for a given value of the relative standard deviation S_r. Thus one can obtain the following inequality, which should be fulfilled by the parameter t':

$$t' \leq t'_{opt} \tag{31}$$

This means that the value of the energy of the exciting radiation E_{01} of the "correction" source should fulfill the inequality

$$E_{01} \geq (E_{01})_{opt} \tag{32}$$

where $(E_{01})_{opt}$ is the energy of the exciting radiation E_{01} for which $t' = t'_{opt}$.

Taking into account that the efficiency of photoexcitation depends on the energy of the primary radiation, the practical conclusion can be drawn that the values of both the parameter t' and the energy of the exciting radiation E_{01} should be as close as possible to the values of t'_{opt} and $(E_{01})_{opt}$, respectively.

On the other hand, for a given pair of sources of primary radiation, that is, for a given value of the parameter t', it is possible to estimate, using Figure 9, the maximum acceptable value $(S_r)_0$ of the relative standard deviation of the ratio $(I_{i1}/I_{i2})_{rel}$. In consequence, the appropriate measurement times and/or activities of the sources of primary radiation can be selected.

Although the correction method may look complicated, it is currently the only real correction procedure dealing with the particle size effect in XRF analysis of granular intermediate-thickness specimens. The idea of applying different excitation x-ray energies to estimate particle size corrections was recently exploited by Nielson and Rogers [41].

REFERENCES

1. J. Leroux and M. Mahmud. *Anal. Chem. 38*:76 (1966).
2. R. D. Giauque, F. S. Goulding, J. M. Jaklevic, and R. H. Pehl, *Anal. Chem. 45*:671 (1973).
3. R. D. Giauque, R. B. Garrett, and L. Y. Goda, *Anal. Chem. 51*:511 (1979).
4. P. Van Dyck, A. Markowicz, and R. Van Grieken, *X-ray Spectrom. 9*:70 (1980).
5. W. H. McMaster, M. Delgrande, J. H. Mallett, and J. M. Hubbell, University of California, Lawrence Radiation Laboratory Report, UCPL-50174 (1969).
6. G. Andermann and J. W. Kemp, *Anal. Chem. 30*:1306 (1958).
7. Z. H. Kalman and L. Heller, *Anal. Chem. 34*:946 (1962).
8. D. L. Taylor and G. Andermann, *Anal. Chem. 43*:712 (1971).
9. D. L. Taylor and G. Andermann, *Appl. Spectrosc. 27*:352 (1973).
10. L. Leonardo and M. Saitta, *X-ray Spectrom. 6*:181 (1977).
11. L. G. Livingstone, *X-ray Spectrom. 11*:89 (1982).
12. J. Kikkert, *Adv. X-ray Anal. 26*:401 (1983).
13. A. Markowicz, *X-ray Spectrom. 13*:166 (1984).
14. H. Meier and E. Unger, *J. Radioanal. Chem. 32*:413 (1976).
15. R. Kieser and T. J. Mulligan, *X-ray Spectrom. 8*:164 (1979).
16. B. Holynska and A. Markowicz, *X-ray Spectrom. 8*:2 (1979).
17. F. Bazan and N. A. Bonner, *Adv. X-ray Anal. 19*:381 (1976).
18. A. Markowicz, *X-ray Spectrom. 8*:14 (1979).
19. K. K. Nielson, *Anal. Chem. 49*:641 (1977).
20. P. M. Van Dyck and R. E. Van Grieken, *Anal. Chem. 52*:1859 (1980).
21. K. K. Nielson, R. W. Sanders, and J. C. Evans, *Anal. Chem. 54*:1782 (1982).
22. K. K. Nielson and V. C. Rogers, *Adv. X-ray Anal. 27*:449 (1984).
23. R. W. Sanders, K. B. Olsen, W. C. Weiner, and K. K. Nielson, *Anal. Chem. 55*:1911 (1983).
24. K. K. Nielson and R. W. Sanders, *The SAP3 Computer Program for Quantitative Multielement Analysis by Energy-Dispersive X-Ray Fluorescence*, U.S. DOE Report PNL-4173, 1982.
25. P. Van Espen, L. Van't dack, F. Adams, and R. Van Grieken, *Anal. Chem. 51*:961 (1979).
26. P. M. Van Dyck, Sz. B. Török, and R. E. Van Grieken, *Anal. Chem. 58*:1761 (1986).
27. P. F. Berry, T. Furuta, and J. R. Rhodes, *Adv. X-ray Anal. 12*:612 (1969).

28. F. Claisse and C. Samson, *Adv. X-ray Anal.* *5*:335 (1962).
29. A. Lubecki, B. Holynska, and M. Wasilewska, *Spectrochim. Acta B 23*:465 (1968).
30. C. B. Hunter and J. R. Rhodes, *X-ray Spectrom. 1*:107 (1972).
31. J. R. Rhodes and C. B. Hunter, *X-ray Spectrom. 1*:113 (1972).
32. A. R. Hawthorne and R. P. Gardner, *X-ray Spectrom. 7*:198 (1978).
33. N. N. Krasnopolskaya and V. F. Volkov, *X-ray Spectrom. 15*:3 (1986).
34. A. Markowicz, P. Van Dyck, and R. Van Grieken, *X-ray Spectrom. 9*:52 (1980).
35. B. Holynska and A. Markowicz, *X-ray Spectrom. 10*:61 (1981).
36. P. Van Dyck, A. Markowicz, and R. Van Grieken, *X-ray Spectrom. 14*:183 (1985).
37. M. Katz (Ed.), *Methods of Air Sampling and Analysis*, American Public Health Association, Washington, D.C. (1977), pp. 592–600.
38. J. W. Criss, *Anal. Chem. 48*:179 (1976).
39. B. Holynska and A. Markowicz, *X-ray Spectrom. 11*:117 (1982).
40. A. Markowicz, *X-ray Spectrom. 12*:134 (1983).
41. K. K. Nielson and V. C. Rogers, *Adv. X-ray Anal. 29*:587 (1986).

7
Radioisotope X-ray Analysis

John S. Watt   *Commonwealth Scientific and Industrial Research Organization, Sydney, New South Wales, Australia*

I. INTRODUCTION

Radioisotope x-ray fluorescence (XRF) and x-ray preferential absorption (XRA) techniques are used extensively for the analysis of materials, covering such diverse applications as analysis of alloys, coal, environmental samples, paper, waste materials, and metalliferous mineral ores and products [1–5]. Many of these analyses are undertaken in the harsh environment of industrial plants and in the field. Some are continuous on-line analyses of material being processed in industry, where instantaneous analysis information is required for the control of rapidly changing processes.

Radioisotope x-ray analysis systems are often tailored to a specific but limited range of applications. They are simpler and often considerably less expensive than analysis systems based on x-ray tubes, but these attributes are often gained at the expense of flexibility of use for a wide range of applications.

Operators making analyses in the field or in industrial plants are usually less skilled than those working in the laboratory with x-ray tube systems. Manufacturers of radioisotope x-ray analysis systems compensate for this by producing simple semiautomated or fully automated systems whose output, calibrated for the specific application, is given directly in terms of concentrations of elements required.

Radioisotope x-ray techniques are preferred to x-ray tube techniques when simplicity, ruggedness, reliability, and cost of equipment are important; when minimum size, weight, and power consumption are necessary; when a very con-

359

stant and predictable x-ray output is required; when the use of high-energy x-rays is advantageous; and when short x-ray path lengths are required to minimize the absorption of low-energy x-rays in air.

X-ray fluorescence techniques based on the x-ray tube–Bragg crystal spectrometer are usually considerably more sensitive than those based on radioisotope sources. This high sensitivity is due to the excellent x-ray resolving power of the crystal spectrometer, which is superior to that of the solid-state detector, the detector having the best x-ray resolution. Radioisotopes cannot be used with crystal spectrometers because of the low geometrical efficiency of this spectrometer coupled with the fact that the x-ray yield of radioisotope sources is very low, about six orders of magnitude less than that of x-ray tubes used with crystal spectrometers.

For some applications, x-ray preferential absorption and x-ray scattering (XRS) techniques are preferred to XRF techniques, particularly when coarse particulate material is to be analyzed. Radioisotopes are the only practical source of x-rays for these applications, because to penetrate deep into the material, high-energy x-rays are required. The most important applications of XRA and XRS techniques are the on-line analysis of particulate material on conveyors.

Some of the terminology used in this chapter is now briefly defined. The element whose concentration is to be determined is the *analyte*, and the other elements, the *matrix elements*. The common link between all techniques and applications discussed is the dependence of the analysis primarily on the photoelectric absorption of x- and γ-rays. Compton and coherent scattering are the other important interactions in the sample. The terms x-ray and γ-ray can often be used interchangeably. The term x-ray is always used when discussing fluorescent x-rays. Radioisotope sources emit either γ-rays directly from the nucleus of the unstable atom or fluorescent x-rays emitted following the ejection of an atomic electron. γ-rays emitted by radioisotopes usually have energies greater than 50 keV. X-ray fluorescence analysis depends on both x-ray and γ-ray excitation, but most XRA and XRS analyses are based on the use of γ-rays. The term "high-energy γ-ray" is used when the γ-ray interaction in the sample is essentially entirely due to Compton scattering (typically above 300 keV), this measurement being used to determine either the bulk density or mass per unit area of the sample. The term "low-energy γ-ray" is used when photoelectric interactions are important to the analysis.

This chapter reviews radioisotope x-ray fluorescence, preferential absorption, and scattering techniques. Some of the basic analysis equations are given. The characteristics of radioisotope sources and x-ray detectors are described, and then the x-ray analytical techniques are presented. The choice of radioisotope technique for a specific application is discussed. This is followed by a summary of applications of these techniques, with a more detailed account given of some of the applications, particularly those of considerable industrial importance.

II. BASIC EQUATIONS

The basic equations for x-ray analysis are given in Chapter 1. Some additional equations used for XRF, XRA, and XRS analyses are presented here. The typical

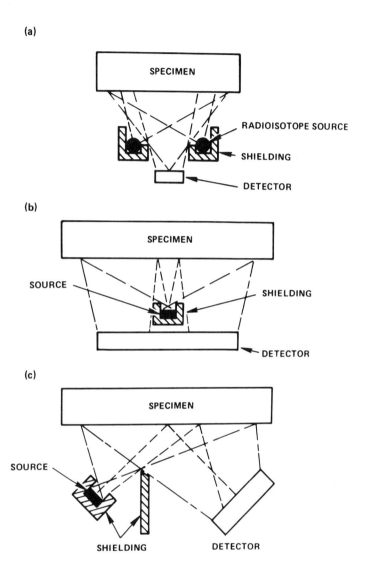

Figure 1 The three geometries [6] for radioisotope-excited x-ray fluorescence analysis: (a) annular source, (b) central source, and (c) side source.

geometries [5,6] of the radioisotope source, sample, and detector used are shown in Figures 1 and 2.

A. Absorption of X-rays

The intensity I of a narrow beam of monoenergetic x- or γ-rays transmitted through a sample [as already shown in Chap. 1, Eq. (40)] is given by

$$I = I_0 e^{-\mu \rho t} \tag{1}$$

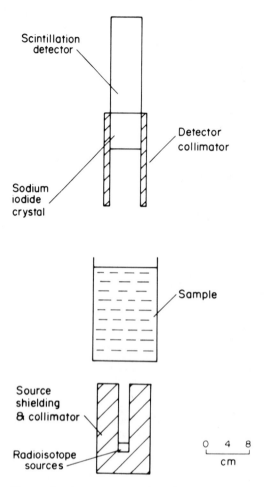

Figure 2 Arrangement of radioisotope γ-ray sources, collimators, and scintillation detector used in x-ray preferential absorption analysis [5].

where

$$\mu = \sum_{i=1}^{n} (W_i \mu_i) \tag{2}$$

$$\sum_{i=1}^{n} W_i = 1 \tag{3}$$

I_0 is the intensity of x-rays detected without the sample being present; μ and t are the mass absorption coefficient and path length of x-rays in the sample, respectively; ρ is the bulk density of the sample; and μ_i and W_i are the mass absorption coefficient and weight fraction of the ith element in the sample, respectively.

Equation (1) also holds for broad beams of x-rays when the cross section for photoelectric absorption is much greater than those for Compton and coherent scattering, that is, particularly for low energy x-rays and high atomic number (Z) elements. This assumption can be checked by reference to Appendices VI–VIII of Chapter 1.

B. Fluorescent X-ray Intensity

When a monoenergetic beam of x-rays excites the K x-rays of the analyte i in an infinite-thickness sample and both the incident and emitted x-rays are normal to the sample, the detected intensity I_i of the $K\alpha$ x-rays is given approximately by Equations (86) and (82) in Chapter 1, namely,

$$I_i = \frac{G\epsilon(E_i)a_i(E_0)I_0(E_0)}{\mu(E_0) + \mu(E_i)} \tag{4}$$

where

$$G = \text{geometrical constant}$$
$$\epsilon(E_i) = \text{intrinsic efficiency of the detector to the x-rays of the analyte}$$
$$a_i(E_0) = W_i\tau_i'(E_0)\omega_i p_i\left(1 - \frac{1}{j_i}\right)$$
$$I_0(E_0) = \text{the source emission, photons/s}$$
$$\mu(E_0), \mu(E_i) = \text{mass absorption coefficients for the exciting radiation with}$$
$$\text{energy } E_0 \text{ and the characteristic radiation with energy } E_i,$$
$$\text{respectively, in the sample, cm}^2/\text{g}$$
$$\tau_i'(E_0) = \text{total photoelectric mass absorption coefficient for the } i\text{th element at energy } E_0, \text{ cm}^2/\text{g}$$
$$\omega_i = \text{the } K\text{-shell fluorescent yield of the analyte}$$
$$p_i = \text{relative transition probability for } K\alpha \text{ lines}$$
$$j_i = \text{jump ratio}$$

Enhancement effects (Chap. Sec. C.3) have been assumed to be negligible.

The intensities of L- and M-shell fluorescent x-rays can be calculated from equations similar to Equation (4). For radioisotopes emitting x-rays of more than one energy, I_i can be separately calculated for each energy and the total fluorescent x-ray intensity determined by summing the products of I_i and the x-rays emitted per disintegration.

C. Scattered X-ray Intensities

X-rays are scattered from the sample to the detector by coherent and Compton-scattering interactions. There is no loss of energy with coherent scattering. The energy E of the Compton-scattered x-ray is given by [Eq. (48), Chap. 1]

$$E = \frac{E_0}{[1 + \gamma(1 - \cos\theta)]} \tag{5}$$

where E_0 is the energy (keV) of the incident x-ray; $\gamma = E_0/511$; and θ is the scattering angle. The scattering angle θ in most radioisotope XRF systems (Fig. 1) is in the range 90–150°. The loss in x-ray energy due to Compton scattering at 90, 120, and 150° is shown in Figure 3 and can seen to be very small at energies below 20 keV.

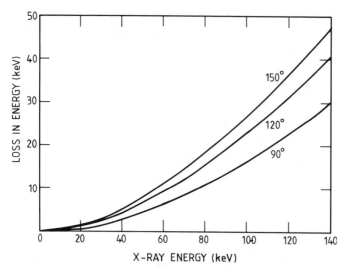

Figure 3 Loss in energy of x-rays on Compton scattering at angles 90, 120, and 150°.

The detected intensity I_s of x-rays scattered from an infinitely thick sample to the detector is given by

$$I_s = \frac{GI_0(E_0)T_s\epsilon_s \sum(\mu_{s_i}(\theta)W_i)}{\sum[(\mu_i + \mu_{s_i})W_i]} \tag{6}$$

where G, $I_0(E_0)$, and W_i are the same as in Equation (4); T_s is the transmission of the scattered x-rays through the filter and the detector window; ϵ_s is the efficiency of the detector for the scattered x-rays; $\mu_{s_i}(\theta)$ is the differential scattering cross section for the x-rays scattered by the ith element toward the detector; and μ_{s_i} is the mass absorption coefficient of the scattered x-rays in the ith element of the sample.

Equation (6) holds for both coherent and Compton scattering when the appropriate scattering cross section [Eqs. (68) and (61), Chap. 1] is used. It assumes that the photoelectric absorption cross section of the x-rays in the sample is much greater than that for scattering and that incident and emergent x-rays are normal to the sample.

The differential and total Compton-scattering cross sections per atom are proportional to Z/A, where A is the atomic weight of the atom, and, except for hydrogen, are almost independent of the atomic number of the atom. Hence, the scattered intensity is approximately inversely proportional to the sum of the mass absorption coefficients in the sample of the incident and emerging x-rays.

The cross section for coherent scattering is highest for small scattering angles, low-energy x-rays, and high atomic number atoms. For angles greater than 90°, the cross section is low and varies by only a factor of about 2.

D. X-ray Fluorescence Analysis

The concentration of the analyte is determined from measurement of the intensity of its fluorescent x-rays, often combined with measurements of the intensities of the fluorescent x-rays of matrix elements and/or the Compton-scattered x-rays. Often the denominator of Equation (4) is proportional to that of Equation (6), that is,

$$\mu(E_0) + \mu(E_i) \propto \sum[(\mu_i(E_0) + \mu_{si})W_i] \tag{7}$$

Hence Equations (4) and (6) can be combined to give

$$W_i \simeq k \frac{I_i}{I_{Com}} \tag{8}$$

where k is a constant. If a major matrix constituent has an absorption edge energy between that of the wanted element and the energy of the radioisotope x-rays, then

$$W_i \simeq kI_i\left(\frac{1}{I_{Com}} + k_1\mu_a W_a\right) \tag{9}$$

where k_1 is a constant and the subscript a refers to the major matrix constituent. Equations (8) and (9) can be checked for accuracy for any specific application by substituting mass absorption coefficients (Apps. VI–VIII of Chapter 1) and elemental concentrations into Equations (4) and (6).

E. X-ray Preferential Absorption Analysis

X-ray preferential absorption analysis is based on the measurement of the transmitted intensities of x-rays, at one or more energies, through the sample (Fig. 2). Sensitivity of analysis depends on the selective absorption of the x-rays by the analyte compared with absorption by the sample matrix.

From Equations (1) through (3), the concentration of the analyte is given by

$$W_i = \frac{\ln(I_0/I)/\rho t - \mu_M}{\mu_i - \mu_M} \tag{10}$$

where μ_i is the mass absorption coefficient of the x-rays in the analyte and μ_M is the mass absorption coefficient of the matrix, given by

$$\mu_M = \sum(\mu_j W_j) \tag{11}$$

and the subscript j refers to the matrix elements and $\sum W_j = 1 - W_i$. The concentration of the analyte can thus be determined if the product of the bulk density and thickness of the sample is known and the mass absorption coefficients of the matrix elements are approximately equal or the composition of the matrix does not vary.

In practice, XRA analysis usually involves measurements of transmission of narrow beams of x-rays, at two x-ray energies, through the sample [5]. This is called dual-energy (x- or γ-ray) transmission (DUET) analysis. The beams are usually coincident and thus not differentially affected by heterogeneity of the sample in the beam path (Fig. 2).

From Equation (10), the concentration of the wanted element W_i is

$$W_i = \frac{-(\mu'_M - R\mu''_M)}{[(\mu'_i - \mu'_M) - R(\mu''_i - \mu''_M)]} \tag{12}$$

where

$$R = \frac{\ln(I_0/I)'}{\ln(I_0/I)''} \tag{13}$$

$$R = \frac{(\mu_i' - \mu_M')W_i + \mu_M'}{(\mu_i'' - \mu_M'')W_i + \mu_M''} \tag{14}$$

and prime and double prime refer to the first and second x-ray energies. The concentration is thus determined independently of the density and thickness of the sample through which the coincident x-ray beams pass.

The sensitivity of analysis is high when $\mu_i' \gg \mu_M'$ and when ρt is large. The analysis is accurate when the ratio μ_M'/μ_M'' is constant, independently of variations in composition of the sample matrix. This ratio is approximately constant when the x-ray energies are just above and below the K-shell absorption edge energy of the analyte and at higher energies when, at each x-ray energy, the mass absorption coefficients of all matrix elements are about the same. In the latter case, the energy of the higher energy x-ray is usually chosen so that the mass absorption coefficients of the analyte and matrix elements are the same, the transmission measurement thus determining the mass per unit area of sample in the x-ray beam.

The error in determination of R [Eq. (13)] caused by counting statistics is

$$\delta R = \frac{R}{\rho t}\sqrt{\left[\frac{(\delta I/I)'}{\mu'}\right]^2 + \left[\frac{(\delta I/I)''}{\mu''}\right]^2} \tag{15}$$

where $\delta I/I$ is the relative counting statistical error and μ is the mass absorption coefficient of the x-ray in the sample. The corresponding error in determining the concentration of the analyte can be found by substituting $R + \delta R$ for R in Equation (12).

The error in determination of the concentration of the analyte, caused by an increase in the concentration (δC) of one matrix element k replacing another matrix element l, can be calculated by increasing the mass absorption coefficient of the sample matrix by

$$\mu_M[new] = \mu_M[old] + (\mu_k - \mu_l)\,\delta C \tag{16}$$

and substituting the new mass absorption coefficient into Equation (12).

These equations accurately predict all aspects of XRA analysis [5] except when the sample is so highly heterogeneous that, within the beam of x-rays, there are significant differences in absorption of the x-rays.

F. X-ray Scattering Analysis

Two x-ray scattering methods of analysis are comparison of the detected intensities of the Compton and coherent scattered x-rays [7] and determination of the intensity of the Compton-scattered x-rays [8]. The former method is essentially a measure of the ratio of the differential scattering cross sections of the two components, which is proportional to between Z and Z^2 [9]. The latter method depends on the absorption of x-rays in the sample, which in the photoelectric region is proportional to between Z^4/A and Z^5/A [9]. Thus, the method is very similar to XRA analysis and is considerably more sensitive than the Compton- to coherent-scattered x-ray method. Both methods are accurate only when the changes in detected x-ray intensities caused by changes in the concentration of the analyte are much greater than those caused by changes in the concentration of the matrix elements.

The sensitivity of both techniques and errors due to variations in concentrations of matrix constituents can be predicted using Equation (6), where the photoelectric absorption cross section in the sample is much greater than the scattering cross section.

Dual-energy scattering techniques [10], analogous to dual-energy preferential absorption techniques, are used to minimize the effects of sample heterogeneity. The x-ray scattering techniques are used in applications in which only one side of the sample is accessible and the thickness of the sample is too great to allow sufficient penetration of x-rays. Compared with DUET analysis, the main disadvantage is that narrow beams of x-rays cannot often be used because the lower geometrical efficiency of the source, sample, and detector. Hence multiple scattered x-rays are detected with a consequent loss in accuracy of analysis.

III. RADIOISOTOPE X-RAY SOURCES AND DETECTORS

The characteristics of radioisotope x-ray sources and detectors are described here. A full understanding of the different characteristics of scintillation, proportional, and solid-state detectors is essential because of the need to tailor their use to specific applications and to environmental conditions in the field and in industrial plants.

A. Radioisotope Sources

A few radioisotope sources are used frequently for x-ray analysis; these are listed with their most important characteristics in Table 1. Also listed are two radioisotopes that emit high-energy γ-rays and are used most frequently with the x-ray sources to correct for changes in sample mass per unit area, thickness, or bulk density.

The activity of radioisotopes is specified in terms of the rate of disintegration of the radioactive atoms, that is, decays per second or becquerels (Bq). The becquerel replaces the non-SI unit, the curie (Ci), which equals 3.7×10^{10} becquerels. The number of x- or γ-rays emitted per disintegration is given in Table 1 so that the essential parameter, x-rays emitted per second by the source, can be calculated. The emission rate decreases with time, the number of radioisotope atoms decaying from N_0 to N after an elapsed time t being given by

$$N = N_0 e^{-0.693t/T_{1/2}} \tag{17}$$

where $T_{1/2}$ is the half-life of the radioisotope. The source decays to half of its original emission rate during the time equal to its half-life. The radioisotope source is usually replaced after one to two half-lives.

Table 1 Properties of Radioisotope Sources Used for XRF, XRA, and XRS Analysis and Determination of Bulk Density ρ, Mass per Unit Area (ρt), and Thickness t in X-ray Analysis

Radio-isotope	Half-life (years)	X- or γ-ray energy (keV)	Photons per disintegration	Dose at 1 m from 1 GBq (μSv h^{-1})	XRF, XRA, XRS, or ρt
^{55}Fe	2.7	Mn K x-rays (5.9, 6.5)	0.28	—[a]	XRF
^{238}Pu	88	U L x-rays (13–20)	0.13	—[a]	XRF
^{244}Cm	17.8	Pu L x-rays (14–21)	0.08	—[a]	XRF
^{109}Cd	1.3	Ag K x-rays (22, 25)	1.07	—[a]	XRF
		88	0.04		
^{125}I	0.16	35	0.07	2.7	XRF
		Te K x-rays (27–32)	1.38		
^{241}Am	433	59.5	0.36	3.6	XRF, XRA, XRS
^{153}Gd	0.66	Eu K x-rays (41–48)	1.10	27	XRA
		97	0.30		
		103	0.23		
^{57}Co	0.74	122	0.86	24	XRF, XRA, XRS
		136	0.11		
^{133}Ba	10.3	81	0.34	65	XRA, XRS, ρt
		276	0.07		
		303	0.18		
		356	0.62		
		384	0.09		
^{137}Cs	30.2	662	0.85	83.7	ρt

[a] It is difficult to assign a radiological protection meaning to the dose of low-energy x-rays. The specific circumstances require individual consideration.

The physical size of radioisotope x-ray sources is small. Figure 4 shows the encapsulations of typical cylindrical and annular sources of ^{109}Cd [11].

There are international codes for the safe use of radioisotopes, and a simple introduction to radiation protection has been published [12]. Each organization using radioactive substances is required to hold a license, issued in most countries by a government health department or atomic energy authority. The International Commission on Radiological Protection [13] recommends that, for members of the public, it would be prudent to limit exposures to radiation on the basis of a lifetime average annual dose of one millisievert (mSv). Table 1 lists the dose rate at 1 m from each radioisotope source, assuming no absorption in the source or by air. The x-ray dose is inversely proportional to the square of the distance from the source. X-ray doses received during operation of x- and γ-ray instrumentation and gages are trivial compared with these maximum permitted doses because of the low x-ray output of radioisotope sources, careful design of operating techniques, and x-ray shielding.

The International Organization for Standardisation (ISO) has produced a system for classifying sealed radioisotope sources based on safety requirements for typical uses [11]. Prototype sealed radioisotope sources undergo temperature, external pressure, impact, vibration, and puncture tests (Table 2), which increase

Table 2 Classification of Performance Standards for Sealed Radioistope Sources[a]

Test	Class					
	1	2	3	4	5	6
Temperature	No test	−40°C (20 minutes) +80°C (1 h)	−40°C (20 minutes) +180°C (1 h)	−40°C (20 minutes) +400°C (1 h) and thermal shock 400° to 20°C	−40°C (20 minutes) +600°C (1 h) and thermal shock 600° to 20°C	−40°C (20 minutes) +800°C (1 h) and thermal shock 800° to 20°C
External pressure[b]	No test	25 kPa absolute to atmospheric pressure	25 kPa absolute to 2 MPa absolute	25 kPa absolute to 7 MPa absolute	25 kPa absolute to 70 MPa absolute	25 kPa absolute to 170 MPa absolute
Impact[c]	No test	50 g from 1 m	200 g from 1 m	2 kg from 1 m	5 kg from 1 m	20 kg from 1 m
Vibrations	No test	30 minutes 25–500 Hz at $5gn$ peak amplitude	30 minutes 25–50 Hz at $5gn$ peak amplitude; 50–90 Hz at 0.635 mm amplitude peak to peak; 90–500 Hz at $10gn$	90 minutes 25–80 Hz at 1.5 mm amplitude peak to peak; 80–2000 Hz at $20gn$		
Puncture[d]	No test	1 g from 1 m	10 g from 1 m	50 g from 1 m	300 g from 1 m	1 kg from 1 m

[a] Details of the testing procedures are given in ISO.2919 and BS.5288. A further class X can be used when a special test procedure has been adopted.

[b] External pressure 100 kPa = 1 atmo (approximate).

[c] The source, positioned on a steel anvil, is struck by a steel hammer of the required weight; the hammer has a flat striking surface, 25 mm in diameter, with the edges rounded.

[d] The source, positioned on a hardened steel anvil, is struck by a hardened pin, 6 mm long and 3 mm diameter, with a hemispherical end, fixed to a hammer of the required weight.

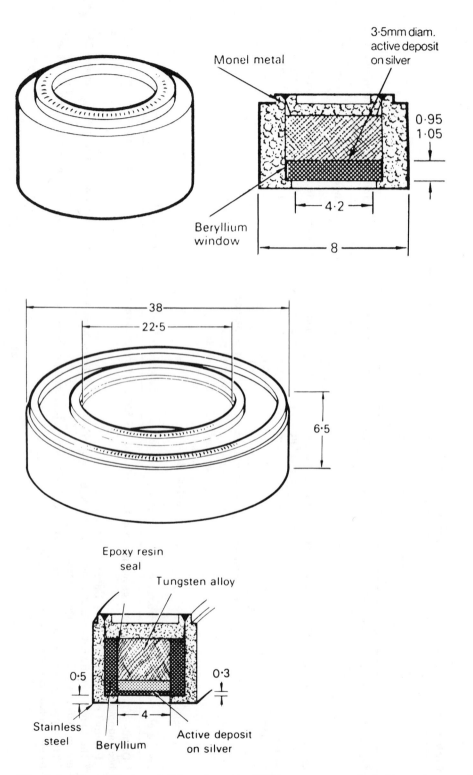

Figure 4   Encapsulation of disk and annular [109]Cd γ-ray sources [11]. Dimensions are in mm.

in severity as the classification (C) increases from 1 to 6. The ISO classifies the test requirements for specific types of applications of the sealed sources. The classification for low-energy γ-ray gages and XRF analysis instruments used in industry is C33222, that is, from Table 2, the first classification 3 is temperature (−40 and 180°C), the second classification 3 is 25 kPa to 2 MPa, and so on. Most radioisotope sources are designed and manufactured to have a greater integrity than required by this classification. For example, the ^{109}Cd sources (Fig. 4) are coded C64344 and C33344 according to the ISO classification, compared with C33222 required.

B. X-ray and γ-ray Detectors

Scintillation, proportional, and solid-state detectors are extensively used in radioisotope x-ray analysis. The important characteristics of these detectors are x-ray energy resolution, efficiency, the ratio of the full energy peak to total detection efficiency, the spectrum of x-rays not in the peak, the sensitive area and thickness of the detector, the complexity of the detector and associated electronics, the robustness of the overall system, and its cost.

The complexity and associated cost of equipment is greatest for solid-state detectors and least for scintillation detectors. The need to use liquid nitrogen with solid-state detector systems and their relative complexity and cost have proven to be a cost penalty but not a limiting factor, even for applications of on-line analysis in industry.

1. X-ray Energy Resolution

Energy resolution, expressed as the full width at half (peak) maximum (FWHM) and shown as continuous lines in Figure 5, was calculated from equations given by Jenkins et al. [14]. The energy resolution of solid-state detectors is much superior to that for proportional and scintillation detectors (see also Fig. 22 in Chap. 2). Figure 6 shows the calculated energy spectrum for the detection of 8 keV x-rays in each detector and also the energies of the $K\alpha$ x-rays in the 6–9 keV energy range. Figure 7 shows the difference in energy of $K\alpha$ x-rays between adjacent atomic number elements.

Table 3 compares the difference in $K\alpha$ x-ray energies with the energy resolution for the elements aluminium, iron, and tin based on data given in Figures 5 and 7. Solid-state detectors are the only detectors that can fully resolve the $K\alpha$ x-rays of adjacent Z elements. The factors affecting their resolution are discussed

Table 3 Difference in Energy of the $K\alpha$ X-rays of Adjacent Atomic Number Elements, and the Energy Resolution of Three Types of Detectors

Atomic number	Energy of $K\alpha$ x-rays (eV)	Difference in $K\alpha$ energies (eV)	Energy resolution of detector (eV)		
			Solid state	Proportional	Scintillation
13 (Al)	1,490	253	117	425	3,000
26 (Fe)	6,400	527	160	660	6,200
50 (Sn)	25,300	1087	275	1750	12,200

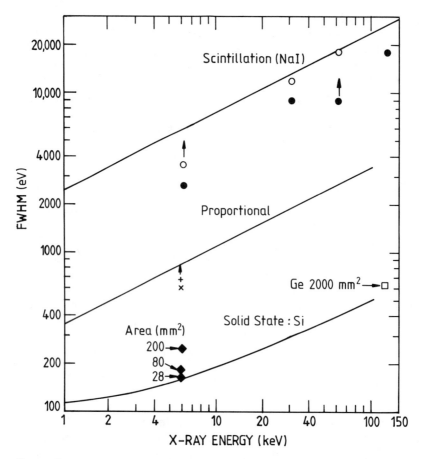

Figure 5 X-ray energy resolution (FWHM) of scintillation, proportional, and solid-state detectors. The continuous lines are calculated [14], the silicon detector results being based on a small detector (10 mm^2 × 3 mm): The (diamonds) resolutions for various area solid-state detectors [15]; (+) typical and (×) best for the high-resolution proportional detectors [16]; and (open circles) typical and (solid circles) best resolutions for specific NaI scintillation detectors [17].

in detail in Chapter 3, Section II.B. Proportional detectors have an energy resolution less than twice the energy difference in $K\alpha$ x-rays of adjacent Z elements. Hence their energy-resolving power is useful even if there are adjacent Z elements in the sample. Scintillation detectors have such limited resolving power that other techniques must be used to discriminate between adjacent Z elements, such as balanced filters (see Sec. IV.C). This is achieved, however, at the expense of some loss in sensitivity of analysis.

2. Efficiency of Detection of X-rays and Sensitive Area

Figure 8 shows the calculated efficiencies of scintillation and solid-state detectors over the energy range 1–150 keV. At low energies, the decrease in efficiency is due to the absorption of x-rays in the beryllium window at the front of the detector.

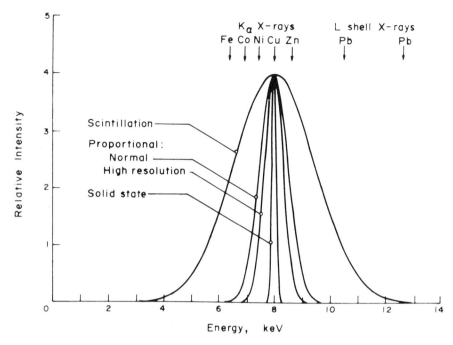

Figure 6 The calculated energy spectra for the detection of 8 keV x-rays in scintillation, proportional, and solid-state detectors [3].

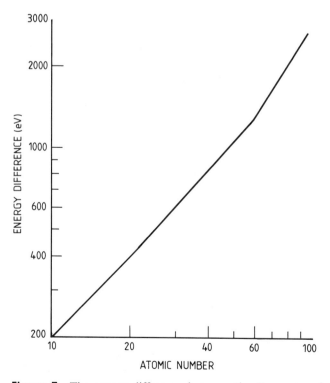

Figure 7 The energy difference between the $K\alpha$ x-rays of adjacent atomic number elements.

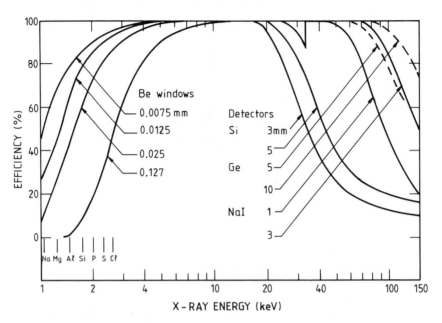

Figure 8 Calculated efficiencies of NaI scintillation detectors and silicon and germanium solid-state detectors used in XRF analysis.

The efficiency at high x-ray energies is determined by the probability that the x-ray interacts with atoms in the sensitive volume of the detector. The most efficient detectors are those with high atomic number and high mass per unit area. For x-ray energies above about 40 keV, solid-state detectors made of germanium are preferred to those made of silicon. The efficiency of proportional detectors depends on the type and pressure of the filling gas and the diameter of the detector (Fig. 9).

The sensitive area for scintillation detectors is usually from 1000 to 2000 mm²; for proportional detectors, it ranges from 500 to 1000 mm²; and for silicon solid-state detectors, it is 10–100 mm². Hence, in general, count rates are highest for the poorer resolution detectors. Central source geometry (Fig. 1b) is normally used with scintillation and proportional detectors and annular source geometry (Fig. 1a) with solid-state detectors. The efficiency of solid-state detectors is discussed in some depth in Chapter 3, Section II.D.

3. Full-Energy Peak to Total Detection Efficiency

The ratio of full-energy peak to total detection efficiency is critical to the sensitivity of XRF analysis. The spectrum outside the full-energy peak can be caused by three factors. The first is the "escape peak" [18], resulting from photoelectric absorption of the incident x-ray in the detector followed by escape of some of its fluorescent x-rays from the detector. The energy left in the detector equals the difference in energy of the x-ray entering the detector and that of the escaping fluorescent x-ray. The escape peak is greatest for the higher Z detectors. The ratio of x-rays in the escape and full-energy peaks is highest for proportional

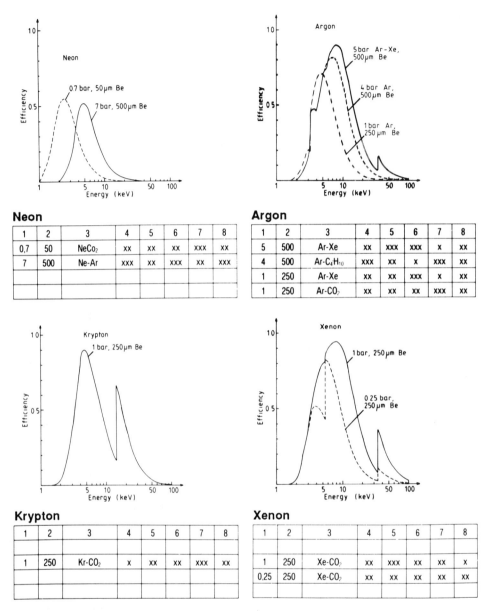

Neon

1	2	3	4	5	6	7	8
0,7	50	NeCo$_2$	xx	xx	xx	xxx	xx
7	500	Ne-Ar	xxx	xx	xxx	xx	xxx

Argon

1	2	3	4	5	6	7	8
5	500	Ar-Xe	xx	xxx	xxx	x	xx
4	500	Ar-C$_4$H$_{10}$	xxx	xx	x	xxx	xx
1	250	Ar-Xe	xx	xx	xxx	x	xx
1	250	Ar-CO$_2$	xx	xx	xx	xxx	xx

Krypton

1	2	3	4	5	6	7	8
1	250	Kr-CO$_2$	x	xx	xx	xxx	xx

Xenon

1	2	3	4	5	6	7	8
1	250	Xe-CO$_2$	xx	xxx	xx	xx	x
0,25	250	Xe-CO$_2$	xx	xx	xx	xx	xx

Figure 9 X-ray detection efficiences of Outokumpu [16] proportional detectors with different types of gas fillings. The number code below each graph is gas pressure, bar (1), window thickness in mm (2), gas mixture (3), low background (4), high efficiency (5), long lifetime (6), high count rate (7), and high resolution (8). The ratings shown are xxx, excellent; xx, good; and x, fair.

detectors with gas fillings of Xe ($K\alpha$ x-ray of 29.7 keV) and Kr (12.6 keV) and for NaI (iodine $K\alpha$ of 28.5 keV) scintillation detectors. However, even silicon (1.74 keV) has 2% of the detected counts in the escape peak when excited by 2 keV x-rays.

Other factors that lead to less than complete absorption of the energy of the x-ray in the detector are (1) Compton scattering of the incident x-ray in the detector, with the scattered x-ray, or Compton electron, escaping from the sensitive volume; or (2) the incident x-ray is photoelectrically absorbed in the detector, and the photoelectron escapes from the sensitive volume before losing all its energy. The full-energy peak to total detection efficiency is highest for the high-Z detectors and for low-energy x-rays. It is lowest for the low-Z gases used in some proportional detectors.

4. Comments on the Characteristics of Proportional Detectors

The characteristics of proportional detectors vary considerably with type of filling gas and pressure and are much more variable than the characteristics of scintillation and solid-state detectors. The best energy resolution is obtained using Penning mixtures [19]. Although the improvement in energy resolution is relatively small (Fig. 5), it is critically important for applications in the atomic number range 26–30 (iron to zinc). The low average ionization energy of Penning mixtures also leads to other important advantages [19]: the voltage required is lower, hence the gas pressure can be higher. This leads to higher efficiency of detection, fewer wall effects, and smaller escape peaks and hence to a higher ratio of peak to total efficiency. The life of the detector is also increased, because of the use of only noble gases, to more than 10^{13} counts.

The characteristics of proportional detectors supplied by Outokumpu Oy [16] are summarized in Figure 9. The recommended gas fillings for proportional detectors depend on the specific analysis application. The efficiency of detection of low-energy x-rays is limited by the absorption of the x-rays in the beryllium window. Proportional detectors with lower gas pressures are used in the detection of low-energy x-rays because the thinner windows do not withstand higher pressures.

5. Developments in Solid-State Detectors

The solid-state detector (SSD) is the best type of x-ray detector for XRF analysis, but its potential has not been fully realized, particularly in industrial and field use, because of the need for liquid nitrogen cooling. There has been much promising research into mercuric iodide, cadmium telluride, and gallium arsenide SSDs [20], which can operate at or near ambient temperatures. The field effect transistor (FET), the low-noise preamplifier associated with these SSDs, must be cooled, for example to $-20°C$. Mercuric iodide is considered the most promising of these detectors. Two U.S. companies recently reduced their extensive research and development programs on HgI_2 SSDs because of difficulties in achieving reliable detectors. The promise of these SSDs has not yet been realized in practical, commercially available, high-resolution x-ray detector systems. Even if further research leads to the successful development of these ambient temperature SSDs, one disadvantage in their use for XRF analysis remains: that is, being made of high-Z materials, the x-ray energy spectra are complicated by K and L x-ray escape peaks.

Recent developments have led to the production of silicon detectors that can be operated at temperatures much higher than that of liquid nitrogen ($-195°C$). Madden et al. [21] used these silicon detectors cooled in a Peltier cryostat. The front-end assembly, mounted in the cryostat, contains a silicon detector and a FET and is mounted on a four-stage Peltier cooling cell. With the assembly under high vacuum, a temperature of $-74°C$ is achieved with a cell power of 4.3 W. For a 16 mm^2 × 2 mm thick detector, an energy resolution of 190 eV at 5.9 keV was achieved. More recently, a commercial manufacturer [22] announced the development of a Peltier-cooled silicon detector system with x-ray energy resolutions of 155, 180, and 240 eV (at 6 keV), respectively, for 10, 30, and 80 mm^2 detectors, that is, as good as have been achieved for liquid nitrogen cooling. Other organizations have developed silicon detectors and Peltier cryostats that give good energy resolutions for x-rays [23,24].

These Peltier-cooled systems are likely to replace liquid nitrogen cooling systems, opening the way for the more widespread use of silicon detectors in industry.

C. Electronics

The electronics used with the various detectors are discussed in Chapters 2 and 3 and are covered in detail by Jenkins et al. [25]. The limits to accuracy and sensitivity of XRF analysis are usually determined by the limitations of the detector in energy resolution, and maximum count rate, for example, rather than of the electronics. With the excellent gain stabilization electronics now available, it is rare that the electronics system is a significant limiting factor even in the harsh environmental conditions of industrial plants.

IV. X-RAY AND γ-RAY TECHNIQUES

The range of radioisotope x- and γ-ray techniques used for analysis is far more extensive than the range based on x-ray tube techniques. Almost all x-ray tube systems are based on the high-energy resolution of wavelength-dispersive (the crystal spectrometer) or energy-dispersive (the solid-state detector) devices. With this high resolving power, there is less need to tailor a technique to the specific application. Radioisotope x-ray systems, especially those involving scintillation or proportional detectors, must be carefully matched to the specific application. This disadvantage is more than compensated for by such attributes as mechanical ruggedness, simplicity, and portability, which are so important in industrial and field applications.

The selection of the radioisotope source to analyze different elements depends on many factors, including whether the energy of the radioisotope x- or γ-rays is sufficient to excite the element, the energies of the x-rays scattered by the sample, and the energy resolution of the detector. Figure 10 is an approximate guide and, although prepared for proportional detectors [16], can be used for solid-state detectors and to a more limited extent for scintillation detectors. ^{238}Pu and ^{244}Cm, emitting x-rays similar in energy, can be used interchangeably. ^{57}Co can be used

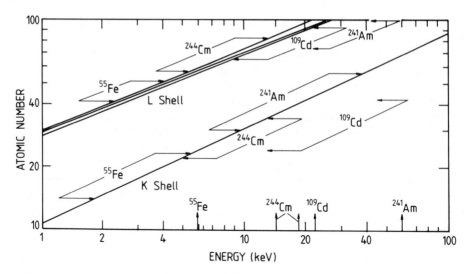

Figure 10 Appropriate radioisotope x-ray source for use with proportional or solid-state detectors to gain a high sensitivity of XRF analysis for elements in a specified atomic number range. The principal energies of the x-rays emitted by each source are indicated above the x axis. The diagonal lines are the K- and L-shell absorption edge energies. K-shell XRF analysis is preferred for most applications, except when a combination of K- and L-shell excitation is required to gain a wider coverage of atomic number elements using the same radioisotope source and, in some cases, for high atomic number elements.

with scintillation detectors or germanium solid-state detectors for the K-shell XRF analysis for high-Z elements, such as uranium and lead.

This section reviews the radioisotope XRF techniques used with solid-state, proportional, and scintillation detectors; x-ray preferential absorption techniques that are normally based on use of scintillation detectors; and x-ray scattering techniques that are often based on use of scintillation detectors. An example of the application of each technique is also given.

A. XRF Techniques Based on Solid-State Detectors

The analysis of samples of copper ores taken from the process streams of three different mineral concentrators is used to illustrate XRF analysis with a solid-state detector [26]. The ore samples were excited by x-rays from a 3.3 GBq ^{238}Pu source and the fluorescent x-rays detected by a 28 mm^2 × 3 mm thick silicon detector. The x-ray spectrum (Fig. 11) of one of the samples shows well-resolved peaks of $K\alpha$ x-rays from iron, copper and zinc, and $L\alpha$ x-rays from lead and the complex spectra of the Compton and coherently scattered x-rays. The count rates of the copper $K\alpha$ x-rays (Fig. 12a) lie within three bands, separated from each other owing to the large difference in absorption of x-rays in the sample matrix caused by the widely different iron concentrations (5, 20, and 50 wt%) of the different ores. The use of the scattered x-ray component to provide a correction for matrix absorption reduces the overall error to 0.15 wt% copper (Fig. 12b).

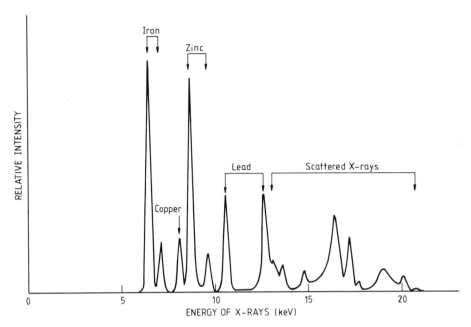

Figure 11 The spectrum of x-rays from a copper, lead, and zinc ore sample excited by ^{238}Pu x-rays and detected by a silicon solid-state detector.

Solid-state detectors are the only type of detector for which the x-ray energy resolution (Fig. 5) is sufficient to resolve the fluorescent x-rays of adjacent Z elements (Fig. 7). There are minor problems of overlap in some cases in which the energy of the $K\alpha$ x-ray of the wanted element overlaps the energy of the $K\beta$ x-ray of another element in a sample. These overlaps can easily be identified from the energies of fluorescent x-rays as a function of Z (Appendix II of Chapter 1). Fluorescent x-rays can also overlap slightly if the concentration of an adjacent Z element is very much higher than that of the wanted element. The extent of this overlap can be calculated using the fluorescent x-ray energies and the x-ray energy resolution (FWHM) of the detector.

The small sensitive area is the main limitation of solid-state compared with proportional and scintillation detectors. It takes longer to obtain the same counting statistics. It is not always possible to use higher activity sources to overcome this limitation because of self-absorption of x-rays in the source and, for some radio-isotopes, the cost of the source.

Figure 13 shows the 3σ minimum detection levels for low concentrations of various elements in a low-Z matrix [27,28]. The counting time was 600 s. The measurements with ^{109}Cd (185 MBq), ^{241}Am (370 MBq), and ^{57}Co (370 MBq) were made using a 30 mm^2 × 5 mm thick silicon solid-state detector (FWHM of 250 eV at 6.4 keV), and the sample matrix was silica gel. The measurements with ^{125}I (185 GBq) were made using a 50 mm^2 × 3 mm silicon detector (FWHM 250 eV) and a matrix of average atomic number of 10. The measurements with ^{133}Ba (370

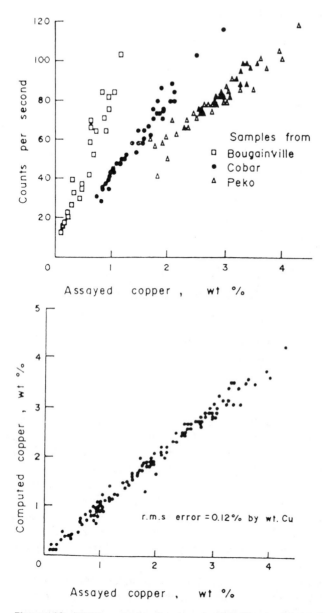

Figure 12 XRF analysis of copper in flotation feed samples from three mineral concentrators, based on the intensity of (a) copper $K\alpha$ x-rays and (b) the ratio of the intensities of copper $K\alpha$ and scattered x-rays. The analysis was based on a ^{238}Pu source and a silicon solid-state detector.

GBq) were made using a 800 mm^2 × 13 mm germanium detector (FWHM 590 eV at 122 keV) and a water matrix. The minimum detectable levels using one source vary greatly with atomic number; hence to maintain low minimum detectable levels over a wide atomic number range, several radioisotope sources, emitting x-rays of different energies, must be used. In this case, minimum de-

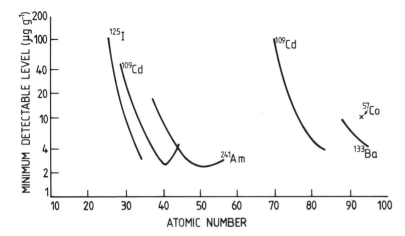

Figure 13   Minimum detectable levels (3σ) determined with various radioisotope x-ray sources and a solid-state detector [27,28].

tectable levels less than 10 μg g^{-1} are achieved for many elements. Iron 55 can be used to extend the range of sensitive detection down to $Z = 15$.

B. XRF Techniques Based on Proportional Detectors

The main advantages of proportional detectors over solid-state detectors are larger sensitive area, simpler equipment, and no need for cooling the detector to a very low temperature. The higher count rates possible with these detectors lead to shorter analysis times, except for those applications in which the energy resolution (Fig. 5) limits the sensitivity of analysis. Many important XRF applications are not limited by the poorer energy resolution, and many do not involve adjacent Z elements.

It is highly important to select the appropriate type of proportional detector for the specific XRF application. The best type is determined by optimizing the various characteristics summarized in Figure 9 for a specific application.

Figure 14 shows the spectra of x-rays, taken with 4 mm thick samples of pure water and water containing 100 μg g^{-1} of both iron and zinc, measured with a ^{244}Cm source and a proportional counter filled with a neon-argon Penning mixture gas to a pressure of 7 bar [19]. The minimum detectable levels are comparable with those obtained with a silicon solid-state detector with x-ray tube excitation [19]. If other elements in the atomic number range 26–30 had been present, however, there would have been incomplete resolution of the K x-rays emitted and hence poorer sensitivity of analysis.

Figure 15 shows the 3σ minimum detectable levels for low concentrations of elements in water [29,30]. The sensitivities are corrected to a common counting time of 100 s. The low minimum detectable levels were obtained by careful choice of filling gas and pressure in relation to the exciting x-ray energy used. For most except low atomic number elements, the minimum detectable levels were less than 10 μg g^{-1}, which is similar to the results for solid-state detectors (Fig. 13);

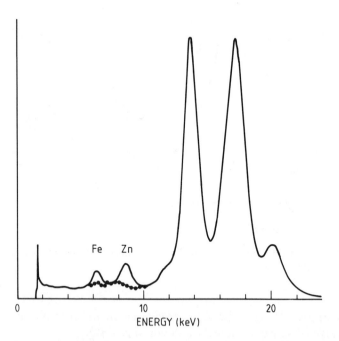

Figure 14 Spectrum of x-rays from water sample and from water containing 100 μg g^{-1} of iron and zinc as determined using a Ne + Ar-filled proportion detector and excited by x-rays from a ^{244}Cm source [19]. The spectra are the same except at the iron and zinc K x-ray peaks.

however, they were achieved in a time six times shorter than before. These low minimum detectable levels were achieved using different proportional counter gas fillings, and as a consequence the technique is less flexible than that for the solid-state detector. If adjacent Z elements had been present, the minimum detectable levels for the proportional counter would have been considerably worse but, for the solid-state detector, much less changed.

The relatively modest improvement in x-ray energy resolution of high-resolution compared with standard proportional detectors leads to considerable improvement in the accuracy of analysis in some applications. Hietala and Viitanen [31] indicate that an improvement in resolution from 16 to 10% at 8 keV results in the relative standard deviation for determination of zinc in copper-zinc tailings containing 0.1 wt% of both copper and zinc, to be reduced from 0.40 to 0.05.

C. XRF Techniques Based on Scintillation Detectors

The x-ray energy resolution of scintillation detectors is so poor (Fig. 5) that the detector cannot be used in most applications to resolve the K x-rays of the analyte and matrix elements. Selectivity to the analyte is obtained with filters and radiators and by the choice of the energy of the x-rays incident on the sample [3]. Scintillation detector XRF systems are used extensively in field work and in industrial plants because of their simplicity, high x-ray detection efficiency, portability,

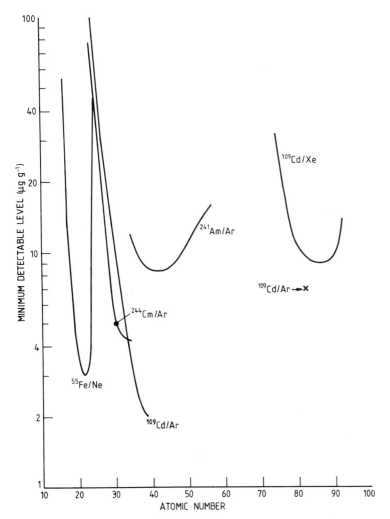

Figure 15 Minimum detectable levels (3σ) for low concentrations of elements in water using proportional detectors and 100 s counting time [29,30]. The measurements with the argon detector (gas pressure, 5 bar) were made with 9 mm thick water samples and 111 MBq of ^{109}Cd, 2.2 GBq of ^{244}Cm, and 1.67 GBq of ^{241}Am; with the neon detector (1 bar pressure, and a 0.05 mm thick beryllium window), with 20 mm thick water samples and 3.7 GBq of ^{55}Fe; and with the xenon detector (1 bar), with 110 mBq of ^{109}Cd.

ruggedness of the detector and electronics, and low cost. These systems are simpler than those based on proportional detectors and hence, if sufficiently sensitive and selective to the wanted element, are the preferred system. They are best used for applications requiring determination of the concentration of one or two elements only. Applications involving more elements are best undertaken with proportional and solid-state detector systems. The minimum detectable levels of scintillation techniques are at least a factor of 10 higher (i.e., worse) than those for solid-state detectors.

Three types of head units are used with scintillation detectors: direct excitation, γ-ray x-ray source excitation, and detector-radiator (Fig. 16). Filters can be used with all three assemblies.

1. Filters

Filters placed between the sample and detector (Fig. 16) increase the sensitivity of analysis by filtering out a higher proportion of fluorescent x-rays of matrix elements than those of the analyte. Zinc, for example, may be the analyte in samples also containing iron. Calculations based on Equation (1) and mass absorption coefficients [see Eq. (69) in Chap. 1] show that a 27 mg cm^{-2} aluminium filter transmits 27% of the zinc $K\alpha$ x-rays of 8.6 keV but only 4.5% of the iron $K\alpha$ x-rays of 6.4 keV (Fig. 17); that is, it reduces the iron K x-rays relative to the zinc K x-rays by a factor of 6. If the sample also contains lead, however, about 55% of the lead $L\alpha$ x-rays (10.5 keV) would be transmitted, twice that of the zinc $K\alpha$ x-rays. In this case (Fig. 17), an absorption edge type filter of copper (22.4 mg cm^{-2}) could be used to reduce the lead $L\alpha$ x-ray transmission to only 1% and also to reduce the iron K x-ray peak. This selective enhancement of the zinc compared with the iron and lead x-ray components partly compensates for the limitation of the poor resolution of the scintillation detector. Although some iron and lead fluorescent x-rays will still be detected within the pulse-height channel set about the zinc $K\alpha$ x-ray peak, in many applications this component will have been sufficiently reduced to make the analysis possible.

If the measurement with one absorption edge filter does not give sufficient selectivity to the wanted element's fluorescent x-rays, balanced filters are used (Fig. 18). The intensities of x-rays in the fluorescent x-ray channel are measured separately, first with one filter and then with the other. The atomic numbers of the two filters are chosen so that their K-shell absorption edge energies are just

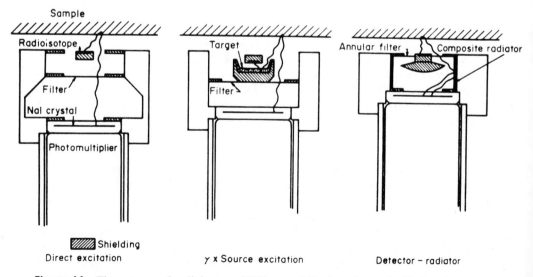

Figure 16 Three types of radioisotope XRF assemblies based on scintillation detectors.

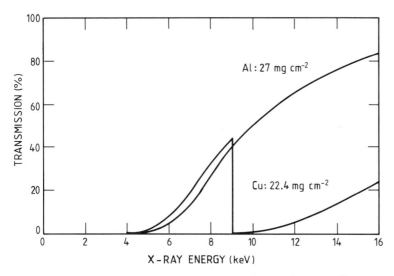

Figure 17 X-ray transmission through aluminium and copper filters.

above and below the energy of the $K\alpha$ x-rays of the analyte. The masses per unit area of the two filters are chosen so that the product $\mu\rho t$ [Eq. (1)] is the same for both filters, except within the energy window enclosing the $K\alpha$ x-ray energy of the wanted element. Hence the difference in the count rates using the two filters is proportional to the intensity of fluorescent x-rays of the analyte.

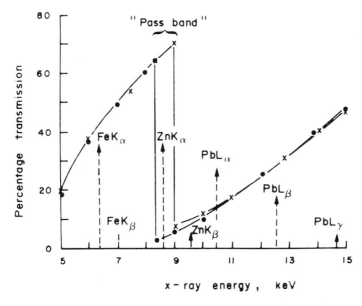

Figure 18 X-ray transmission through balanced filters of copper (x) and nickel (solid circle). Count rate measurements made first with one and then with the other filter are subtracted to give a count rate proportional to the intensity of zinc $K\alpha$ x-rays from a sample.

Figure 18 shows the transmission of x-rays by copper and nickel filters, which are chosen when the analyte is zinc [32]. Except in the energy window enclosing the zinc $K\alpha$ x-ray, the transmission is the same for the two filters. The balanced filter technique is thus highly selective to the zinc $K\alpha$ x-rays. The count rates, measured separately in the fluorescent x-ray channel with the two filters, are usually high and the difference in count rates can be quite small.

There are two disadvantages with filter techniques: the sensitivity of analysis is poor when the fluorescent x-rays of the main interfering matrix element have an energy just less than that of the fluorescent x-rays of the analyte (Fig. 17), and the sensitivity is considerably less than that obtained with detectors that have the inherent resolving power to isolate the fluorescent x-rays of wanted and matrix elements. These losses in sensitivity result from the only partial absorption of interfering x-rays in the filter and, for absorption edge filters, from the detection of filter K x-rays in the x-ray channel of the analyte. The latter is a direct consequence of the use of the broad-beam geometries of radioisotope XRF systems. The filter K x-rays are mainly excited by the x-rays backscattered by the sample.

2. Direct Excitation Assemblies

The direct excitation technique (Fig. 16) is the most widely used of the three scintillation detector assemblies. The energy of the radioisotope x-ray is usually chosen so that the fluorescent and backscattered x-rays are resolved by energy analysis (Fig. 19). The intensity of the backscattered x-rays is used to correct for

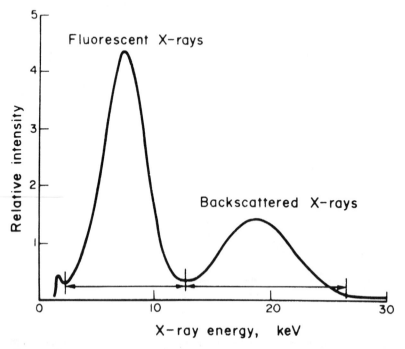

Figure 19 Typical spectrum of x-rays from a copper ore slurry excited by x-rays from a ^{238}Pu source and measured using a scintillation detector.

the absorption of the analyte's fluorescent x-rays by the sample matrix [Eq. (8) or (9)]. The filter enhances the sensitivity and selectivity of analysis.

Direct excitation assemblies are used extensively in industry, for example in laboratory and portable elemental analyzers [1] and in-stream analysis of mineral slurries [3].

3. Detector-Radiator Assemblies

The detector-radiator assembly (Fig. 16) discriminates well against interfering x-rays of energy just less than that of the fluorescent x-rays of the wanted element [33]. The basis of this discrimination is that the atomic number of the radiator element can be chosen so that, of the two x-ray components with nearly similar energies, only the higher of the two has sufficient energy to excite the K x-rays of the radiator element. The detector is shielded from the sample and hence sees only the x-rays emitted by the radiator. Balanced radiator techniques, analogous to balanced filter techniques, can also be used to improve selectivity to the analyte if there is another matrix element, in this case emitting fluorescent x-rays of energy higher than those of the analyte. The count rates obtained with detector-radiator assemblies are about 5% of those obtained with direct excitation assemblies, using a source of the same activity, because of the additional excitation stage of the radiator.

The intensity of higher energy x-rays scattered by the sample can be measured simultaneously in the one assembly by use of a second radiator element of atomic number considerably higher than that of the first radiator. The x-ray energies of the two components are well resolved and similar to that shown in Figure 19.

Detector-radiator systems are much less widely used than balanced filter techniques. Applications include the determination of lead in zinc concentrates, in which the zinc $K\alpha$ x-rays (8.6 keV) from the high concentration of zinc (e.g., 50 wt%) swamp the lead L x-rays (10.5 to 14.8 keV) from the low concentration of lead (e.g., 0.5 wt%). A radiator of zinc (absorption edge energy of 9.66 keV) is excited by the lead L but not by the zinc K x-rays. This radiator technique improves the sensitivity to lead to that of zinc by a factor of about 20 times.

4. γ-Ray-Excited x-ray Assemblies

A limited number of x- or γ-ray energies are emitted by radioisotope sources (Table 1). A secondary x-ray source, in which γ-rays from a radioisotope source excite the fluorescent x-rays of a target material, can be used to obtain essentially monoenergetic x-rays of energy determined by the atomic number of the target element (Fig. 16). Hence the energy of the x-rays incident on the sample can be chosen to suit the specific XRF application. The γ-ray-excited x-ray assembly [3] makes use of filters, including balanced filters, similarly to the direct excitation assembly. The count rates using the secondary excitation source assembly are about 5% of the count rate of a direct excitation assembly using the same activity source.

A balanced energy technique, in which separate measurements are made with two targets (in the secondary source) whose fluorescent x-rays straddle the K-shell absorption edge energy of the wanted element, can be used to obtain more selectivity to the wanted element.

The γ-ray-excited x-ray assembly is used as an alternative to direct excitation when no suitable energy x-ray is emitted by radioisotope source. One application is in the determination of the coating mass of tin on steel by detection of the tin *K* x-rays. If 60 keV x-rays are used to excite the tin, tin *K* x-rays from both sides of the steel are detected; by choosing the energy of the incident x-rays to be just greater than that of the tin *K*-shell absorption edge, tin *K* x-rays from only the one side are detected because of the high absorption of the lower energy x-rays in the steel.

D. X-ray Preferential Absorption Techniques

X-ray preferential absorption analysis is often based on the dual-energy γ-ray transmission technique (Fig. 2) because of important industrial applications involving the analysis of coarse and heterogeneous materials, such as coal [34,35] and metalliferous mineral ores. Low-energy γ-rays must be used in these applications to obtain sufficient transmission through the material, so that the only practical approach is to use radioisotope sources. Scintillation detectors are used to ensure efficient detection of the γ-rays, with pulse-height analysis to separate the two energies.

Figure 20 illustrates the results using the DUET technique to determine the lead content of zinc concentrate and residue samples [36]. The radioisotopes

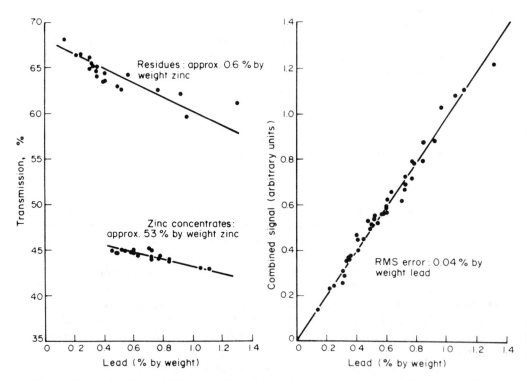

Figure 20   γ-ray preferential absorption analysis for lead, showing (left) the transmission of ^{153}Gd γ-rays (~100 keV), and (right) combination of the separate transmission measurements for ^{153}Gd and ^{241}Am (59.5 keV) γ-rays.

[241]Am and [153]Gd were used. Their γ-ray energies, 60 and 100 keV, respectively, straddle the K-shell absorption edge energy of lead (88 keV). A common calibration curve is obtained despite the great difference in absorption by the matrix of the concentrates (with about 50 wt% zinc) and of the residues (with about 0.4 wt% zinc).

This technique becomes more complicated when the γ-ray transmission measurements are made on material on fast-moving conveyors, as for the on-line determination of the ash content of coal [34,35]. Equations (12) and (13) hold only for time intervals, during which there is little change in mass per unit area, whereas there would be a linear summing of count rates in time although the correct response is logarithmic [Eq. (13)]. This problem can be overcome by counting for shorter intervals during which the mass per unit area changes little and summing the logarithms of the counts during many of these intervals [35].

E. X-ray Scattering Techniques

The most important applications of x-ray scattering techniques are for the continuous analysis of particulate material on conveyors and sample bylines in industrial plants. There is often no need to crush the material before analysis because of the penetration of the γ-rays in the material. Americium 241 γ-ray scattering has been used to determine the ash content of coal both in a sample byline [37] and, when sufficient thickness of coal is available, directly on a conveyor belt [38]. Dual-energy γ-ray scattering has been used in ore-sorting applications [10], in which the thickness of the ore lumps is too great for XRA techniques to be used.

F. Count Rates and Calibration

The count rates of fluorescent and backscattered x-rays from the sample are determined by many factors [Eqs. (4) and (6)]. Activities of the radioisotope sources used range from a hundred MBq to 10 GBq, the higher activities being used mainly with solid-state detectors, which have a smaller area than scintillation and proportional detectors. For x-ray preferential absorption analysis, higher activities of 1–10 GBq are used for industrial on-line applications in which high count rates are essential with the rapidly changing mass per unit area. The source activities used for x-ray scattering analysis are approximately the same as those for XRA analysis.

Radioisotope x-ray systems are usually calibrated by comparing the measured count rates with analyses of the same samples by more conventional techniques. The coefficients linking the count rates and chemical analysis are determined by linear multiple regression. This method of calibration is essential in most applications, especially when the materials to be analyzed have an unknown particle size. For the calibration to be valid, it is essential to calibrate with materials covering the full range of variations in elemental composition and particle size. In some industrial applications, these factors may change slowly with time and hence the calibration must be regularly updated.

V. CHOICE OF X-RAY TECHNIQUE FOR A SPECIFIC APPLICATION

The choice of radioisotope x-ray analysis technique for a specific application depends on several interacting factors: the overall accuracy of the sampling and analysis required; the time available to achieve this accuracy; techniques available to obtain a sufficiently representative sample of the material being analyzed; and the sample preparation requirements, such as grinding. Each of these factors is considered here.

A. Overall Accuracy and Time for Analysis

The overall accuracy of analysis depends on errors in sampling, sample preparation, and x-ray analysis. The maximum acceptable time must include not only the time for the x-ray analysis but also the time for sampling and sample preparation. Errors due to nonrepresentative sampling are too often underestimated in industrial and field applications. The accuracy of the overall analysis is thus often considerably worse when sampling errors are taken into account. Simpler analytical techniques could often have been used without significant loss of overall accuracy.

The time for analysis is critically important to the control of some industrial processes. In the coating industry, the time for determination of the coating mass of tin and zinc on steel must be less than a few seconds, and hence continuous analysis directly on the main coating line is essential. In mineral concentrators, rapid changes in the grade of ore entering the plant and the time taken for the ore to pass through the plant (about 15 minutes) make it essential for the process slurries to be analyzed within 5 minutes. This can be achieved by continuous analysis either directly in stream or on slurries in sample bylines. In scrap yards, alloys must be sorted into different types in short periods: otherwise the sorting operation is not economical. In this case, it is essential to have analysis equipment that is both portable and capable of producing rapid results.

B. Sampling and Sample Presentation

All analyses involve sampling, some more than others. In the continuous on-line measurement of material on conveyors or in process slurry streams, the analysis is averaged over large quantities of material but still has potential for sampling errors because not all the material on the belt or in the slurry stream is viewed. If material is continuously sampled from the main stream and fed through sample bylines past the x-ray analyzer, the errors in sampling the main stream and in viewing only part of the sample byline must both be considered.

A mathematical approach to sampling has been detailed by Gy [39]. The most readily accessible information on sampling is given in the various international standards for the sampling of process material from moving conveyors and from slurry streams [40]. These describe preferred sampling practices and estimate the accuracies of sampling, including the effects of different top sizes and the number and mass of sample increments taken. These standards are regularly amended.

Sampling errors are of critical importance to the exploration for and assessment of deposits and the mining of both metalliferous ores and coal. It is possible to analyze only a very small proportion of the ore. There are established practices for choosing where to sample (e.g., on a regular spaced grid), the number of samples to be taken, and the weight of each sample to be taken. This complex field of sampling geostatistics is covered in textbooks [41].

C. Choice of Radioisotope X-ray Technique

The choice of the most appropriate radioisotope x-ray technique for a specific application depends on the requirements of accuracy of sampling and x-ray analysis and the time available for the analysis. The simplest radioisotope x-ray technique that satisfies these requirements is usually chosen.

In the laboratory, where many different types of analyses may be required, XRF analysis with a solid-state detector is the most flexible method. For industrial and field applications, there is often a more restricted range of analyses coupled with a greater need for simple and reliable equipment. Cost may also be an important consideration. X-ray fluorescence analysis based on scintillation or proportional detectors is often, but not always, the best approach.

These considerations lead to the use of a much wider range of radioisotope x-ray techniques in industry than in the laboratory. The choice of technique is more complicated. Should an ore sample be ground before analysis and L-shell XRF be used for a high-Z element, such as uranium, or can K-shell XRF techniques determine the concentration with sufficient accuracy despite the heterogeneity of the sample? Should the analysis technique be chosen to make a direct measurement on-line, on a byline of the process stream, or on a sample taken to a laboratory? Should XRF techniques even be used in an on-line application given the inaccuracies introduced by the heterogeneity of the material to be analyzed? The practical alternative may be to use XRA or XRS techniques or a nuclear technique, depending on more penetrating radiation, such as high-energy γ-rays or neutrons. The applications discussed in the next section indicate the preferred solutions to some important analysis applications, particularly in industry.

VI. APPLICATIONS

Table 4 lists some important applications of radioisotope x-ray techniques based on XRF equipment, usually referred to as laboratory or portable elemental analyzers. The analysis techniques involve the use of scintillation or proportional detectors. These analyzers are in widespread use in many application areas and in many industries [1,2]. The total number used worldwide probably is about 2000–3000. Some suppliers are listed in Table 5. A photograph of a portable model of an element analyzer based on high-resolution proportional detectors is shown in Figure 21, and a laboratory model based on scintillation detectors is shown in Figure 22. Many solid-state detector systems are also in routine use in laboratories.

Table 6 lists types of on-line analysis systems in routine use. These include systems for the on-line analysis of mineral slurries, flowing powders, coal, coal slurries, paper, sulfur in oil and petroleum products, and coatings. Most of the analysis systems are based on scintillation and proportional detectors, but some

Table 4 Applications of Laboratory and Portable Elemental Analyzers

Metal sorting and identification: low-alloy steels; stainless steels; nickel alloys; cobalt in tungsten carbide

Mining and mineral: copper, lead, zinc, tin, arsenic, molybdenum, nickel, iron, chromium, bismuth, and uranium in commercial-grade ores, concentrates, and tailings; iron in silica sand; silicon, potassium, titanium, and iron in clays

Paper and pulp: silicone coatings on paper; calcium, titanium, filler in paper

Environmental: hazardous materials (e.g., lead, arsenic, chromium, or cadmium in waste sludge); trace elements in wastewater discharge

Fibers, films, and coatings: copper, zinc, tin, gold, silver, and chromium plating thicknesses; metals in plating solutions; silver in photographic film

Chemicals and process control: lead, titanium, and zinc in paint; sulfur and calcium in cement; vanadium in catalysts; palladium and gold coatings on silica spheres used as catalysts

Plastics: calcium, lead, tin, and chlorine in PVC; zinc in polystyrene; chlorine in urethane rubbers; bromine and chlorine in butyl rubbers; silicon in polythene; TiO_2 in nylon; bromine in Styrofoam

Agricultural: calcium in phosphates; copper, chromium, and arsenic in wood preservatives and treated wood; bromine in almonds; iron-zinc ratio in meat for grading

Energy: lead, calcium, sulfur, vanadium, and chlorine in gasoline or oil; sulfur in petroleum coke; sulfur and ash in coal

---

of the more recently developed systems also use solid-state detectors. Table 6 lists the number of plant installations of on-line analysis systems (when known) and some commercial suppliers of the equipment.

Examples of the application of these techniques are now discussed in detail, with emphasis given to applications of industrial importance. Some examples illustrate the interaction of sampling and sample presentation with the selection of the appropriate radioisotope technique.

Table 5 Manufacturers and Suppliers of Laboratory and Portable Elemental Analyzers

Manufacturer	Detector type
Amdel [42]	Scintillation
ASOMA [30]	Proportional or scintillation
BRGM [43]	Silicon solid state
Columbia Scientific [44]	High-resolution proportional
Data Measurement Corporation [45]	Scintillation
Mineral Control Instrumentation [46]	Scintillation
Outokumpu Oy [16]	High-resolution proportional
Oxford Instruments [47]	Proportional
Princeton Gamma-Tech [48]	Proportional
Texas Nuclear Corporation [49]	Scintillation

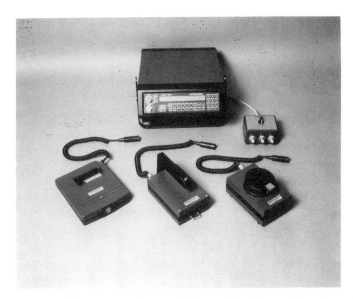

Figure 21 The Outokumpu [16] portable elemental analyzer based on high-resolution proportional detectors, with the probe units in front for use with liquid or solid samples of normal Z (left) and low-Z (center) elements and for direct analysis from the surface of materials (right).

Figure 22 A laboratory elemental analyzer based on use of a scintillation detector. (Reprinted courtesy of Amdel [42].)

Table 6 On-line Analysis Systems Based on Radioisotope X-ray Sources[a]

Determination of	Technique[b]	Manufacturer	Number of installations[c]
Metal content of mineral slurries	XRF/XRA, s and ss, in-stream	Amdel [42]	46
	XRF, p, by-line	Asoma [30]	—[d]
	XRF, ss, by-line	Outokumpu Oy [16]	15
	XRF, ss, in-stream	Texas Nuclear [49]	11
	XRF, ss, by-line	Ramsey [50]	8
Metal content of clay and mineral powders	XRF, p, by-line	Asoma	—[d]
	XRF, ss, by-line	Outokumpu Oy	5
Iron and chromium in ore on conveyors	Dual-energy XRS, s, on-line	Outokumpu Oy	4
Ore sorter	Dual energy XRS, s, on-line	Outokumpu Oy	1
Ash in coal on conveyor	Dual energy XRA, s, on-line	MCI [46]	57
		Harrison Cooper [51]	—[e]
		SAI [55]	—[e]
	XRS, s, on-line	EMAG [52]	20
	XRS, s, by-line	Humboldt-Wedag [53]	35
	XRS, p, by-line	Sortex [54]	30

Application	Technique	Manufacturer	Number of installations
Solids weight fraction and ash in coal in slurries	XRF, neutron, and γ transmission, s, on-line	Amdel	2
Tin content of galvanizing solutions	XRF, p	Rigaku [56]	7
Calcium in cement raw mix	XRF, s	Rigaku	6
		Outokumpu	4
Coating mass of Zn, Sn/Cr, Sn/Ni, Zn/Fe, Sn/Pb on steel, others	XRF, p, on-line	Data Measurement [45]	30
		FAG [57]	—[e]
		Rigaku	23
		YEW [58]	339
		Mitsubishi [59]	71
Sulfur in oil	XRA, ion ch, by-line	Sentrol [60]	25
	XRF, p, by-line	YEW	—[e]
	XRA, p, β-ray transmission, on-line	Paul Lippke [61]	—[e]
Ash content and/or mineral filler material in paper	XRF, p, on-line	Sentrol	10

[a] X-ray tube techniques can also be used in some of these applications.
[b] Detectors: s = scintillation; p = proportional; ss = solid state; ion ch = ion chamber.
[c] In most cases, these are underestimates. For example, the sulfur in oil gages is for Japan alone.
[d] No installations; product introduced mid-1987.
[e] Number of installations unknown.

A. Identification of Alloys

Rapid sorting of alloys is required in many areas of the metals industry, such as fabrication, inventory control, and the sorting of scrap [62]. Some common alloy groups include nickel alloys, copper alloys, stainless and high-temperature steels, and carbon and chromium-molybdenum steels. Although 40–50 elements are involved in the alloying process, in any given alloy there are only 10–20; of these only about 10 at the most are required for the identification of specific alloy.

The main requirements of analysis equipment for alloy identification are portability, speed and reliability of identification, and an ability to be used by unskilled operators. Balanced filter techniques have two main disadvantages. Concentrations of at least several elements must be determined, and hence separate measurements must be made with several sets of balanced filters. The sensitivity of analysis is insufficient for the lower concentrations of some specific elements in the alloys.

Piorek and Rhodes [62] showed that, using XRF analysis based on a 111 MBq ^{109}Cd source and a high-resolution proportional detector, many alloys can be identified in one measurement. Measurements are first made to identify the alloy by group. The spectrum of the unknown alloy is then compared with the key features of spectra of known alloys in the group, which are stored in a microprocessor in the equipment. Table 7 shows the identification results for different alloys by group. The probability of correct identification is greater than 80% for all alloys except the carbon steels. Some of the identification failures are for alloys very close in composition, such as stainless steels 409 and 410, for which the main elements differ in concentrations by less than 1%. The most difficult identification is for carbon steels in which the concentrations of alloying elements are very low in the presence of almost 100% iron, and the difference in concentrations of the same elements between two grades approaches the detection limit of the XRF technique. Improved identification can in many cases be achieved by using different radioisotope sources. Overall, the analyzer offers a much simpler ap-

Table 7 Identification Result of Alloys Using XRF Analysis with 111 MBq ^{109}Cd and a High-Resolution Proportional Detector

		Identification result	
Alloy group	measured elements	% Correct	% Incorrect
Nickel alloys	Ti, Cr, Fe, Co, Ni, Cu, Nb, Mo, W, BS[a]	100.0	0.0
Copper alloys	Mn, Fe, Ni, Cu, Zn, Pb, BS	87.3	12.7
Stainless and high-temperature steels	Ti, Cr, Mn, Fe, Co, Ni, Cu, Nb, Mo, BS	81.9	18.1
Cr/Mo steels	Cr, Fe, Ni, Mo, Ni, Mo	93.4	6.6
Carbon steels	Cr, Fe, Ni, Mo	61.9	38.1

[a] Backscattered radiation.
Source: From Reference 62.

proach to identification of alloys, with a reliability of identification as good as more complex techniques.

B. Determination of Uranium and Gold in Ore

During the exploration of mining of metalliferous ores, large numbers of ore samples must be analyzed to compensate for the inherent variability of expression of the ore. The ore may be analyzed in the laboratory or, for higher Z elements, at the mine face or in situ in boreholes.

K-shell XRF techniques are often preferred for the analysis for uranium; because of the penetration of the uranium K x-rays in the ore, little or no crushing of samples is required and the uranium concentration is averaged over much larger samples. Uranium can be determined to 20 μg g^{-1} (1σ) in 30 s using a 222 MBq ^{57}Co and a 28 mm^2 × 5 mm germanium detector [63]. This technique was used routinely to survey samples for uranium and other high-Z elements ($Z \geq 40$) in the laboratory by the Australian company Geopeko in the extensive exploration programs that found the large Ranger uranium deposit at Jabiru in the Northern Territory, Australia [64]. It is also used for borehole logging, with detection limits of 0.04 wt% (1σ) for uranium, tungsten, and lead and 6μg g^{-1} for gold, in 120 s counting time using a 518 MBq ^{57}Co source and a small silicon detector in a 32 mm diameter borehole probe [65].

An analysis system has been developed for the in situ determination of gold in ore at the mine face [66]. The hand-held probe consists of a 4.4 GBq ^{109}Cd (88 keV γ-rays) source, a 200 mm^2 × 7 mm thick germanium detector, and a small Dewar flask containing liquid nitrogen, which must be replenished after 6 h of operation. The sensitivity to gold has been optimized by careful choice of the incident γ-ray energy and by measuring gold $K\beta$ x-rays whose energy is greater than the energies of most of the Compton-scattered γ-rays. The precision for a 30 s scan time is 20 μg g^{-1} (1σ) and 2 μg g^{-1} for 100 × 30 s scans. The technique is suitable for use in high-grade gold mines but has insufficient sensitivity to be applied widely in gold mining, in which 0.3 μg g^{-1} (1σ) is normally required.

C. On-line Determination of Coating Mass

Tin on steel (tinplate), zinc on steel (galvanized iron), zinc and aluminium on steel (zincalum), and iron oxide on plastic (magnetic tape) are coated products manufactured in large quantities. The coatings are applied at high speed. Accurate control of the coating mass per unit area is essential to economize on the operation.

Two XRF techniques can be used to determine the mass of the coating. Fluorescent x-rays of the coating element can be excited and their intensity measured. The intensity increases with coating thickness. Alternatively, fluorescent x-rays of the base material can be excited. Their intensity decreases with increase in coating mass because of the absorption of the incident and excited x-rays in the coating. Both radioisotopes and x-ray tubes are used as the source of x-rays, with radioisotope sources preferred except for those applications requiring very fast response, such as 0.1 s.

Radioisotope XRF techniques for the on-line determination of coating mass are based on the use of proportional detectors. These are preferred to scintillation detectors because of their better energy resolution and because they can be used at the relatively high temperatures that occur above the hot tin and galvanized iron coating processes. In commercially available systems (Table 6), the analysis head unit continuously scans across the width of the strip so that coating mass can be controlled across the whole strip. These commercial systems are used worldwide in most high-throughput coating operations. They can also be used to determine the separate coating masses of multiple coatings.

Coated products are usually sold with a specified minimum coating mass. The accurate coating mass determination has led to coatings being controlled much closer to the minimum specification. The variations in coating mass obtained on the zinc galvanizing line of John Lysaght Pty. Ltd., Port Kembla, Australia, corresponding to no gage, gage with manual control, and gage with automatic control, are shown in Figure 23. In 1977, improved control of zinc coating mass led to savings of A$300,000 per year per line; similarly, at the nearby Australian Iron and Steel Pty. Ltd., savings of A$1 million per year [67] were made for tinplate.

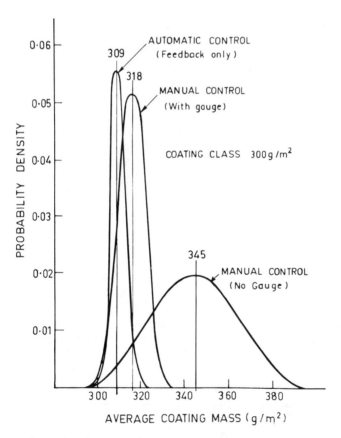

Figure 23 The frequency distribution for product zinc coating mass before and after installation of an on-line zinc coating mass gage at John Lysaght Pty. Ltd. [67]. Product minimum specification is 300 g m^{-2}.

D. On-stream Analysis of Metalliferous Mineral Slurries

Most metalliferous minerals are concentrated from their ores by froth flotation. The grade of ore fed to the concentrator can vary rapidly; hence to control the flotation process, the concentrations of valuable minerals in the plant process slurries should be determined continuously. The concentrations, in the slurry solids, of such base metal minerals as nickel, copper, zinc, and lead are usually in the range of 0.3–15 wt% for feed streams, tens of wt% in concentrate streams, and 0.03–0.3 wt% in residue streams. The solids weight fraction is in the range of 15–50 wt%. The time for analysis (less than 5 minutes) is too short for laboratory analysis of samples taken from the process streams.

A radioisotope x-ray system based on the use of scintillation detectors was introduced in 1973 [68]. Cesium 137 γ-ray transmission is used to determine the bulk density of the slurry and hence the solid weight fraction. X-ray fluorescence techniques based on direct excitation and detector-radiator assemblies (Fig. 16) are used to determine the concentration of all but some high-Z elements, such as lead, which are determined by ^{153}Gd γ-ray (100 keV) transmission. In each case, the x- or γ-ray measurements are combined with the solids determination to obtain the concentration of elements in the slurry solids. These techniques are sufficiently sensitive for all but a few residue streams containing very low concentrations of valuable mineral. This limitation was overcome in 1981 by the introduction of a solid-state detector probe.

The radioisotope x-ray system (Fig. 24) is based on probes, each containing a radioisotope source and a detector, which are immersed directly into the plant process slurries [68]. Electrical signals from the probes are fed to a signal analyzer unit and its output to a central computer. Thus, not only is there no need for sampling from the plant process streams but all streams are analyzed continuously rather than sequentially.

This radioisotope x-ray system is very different from x-ray tube and crystal spectrometer systems for on-stream analysis. These systems were developed in the 1960s to scan slurries sequentially in up to 14 simple bylines [69]. Analysis of the slurries involved sampling from the main process stream, running slurries through long lengths of sample bylines to the central analyzer, and subsampling from the bylines before presentation to the analyzer. This is a complex and expensive system.

X-ray tube analysis systems developed in the 1980s sequentially route the slurries from up to five process streams through a common flow cell viewed by an XRF analyzing unit [70]. The analyzer is mounted in the plant near the process streams and hence overcomes much of the mechanical complexity of the 14 stream system just discussed. Both the x-ray tube–crystal spectrometer system and the radioisotope XRF solid-state detector systems are routinely used with this new sample byline system. Both systems can also be used for the continuous analysis of fine powders. The radioisotope system is capable of determining elements of atomic number as low as 14 (silicon), because with the short x-ray path length, the absorption of the low-energy x-rays in air is minimized.

The development of different systems for the on-stream analysis of mineral slurries illustrates the strong influence of sampling and sample presentation of the type of XRF analysis system used. Various radioisotope (Table 6) and x-ray tube

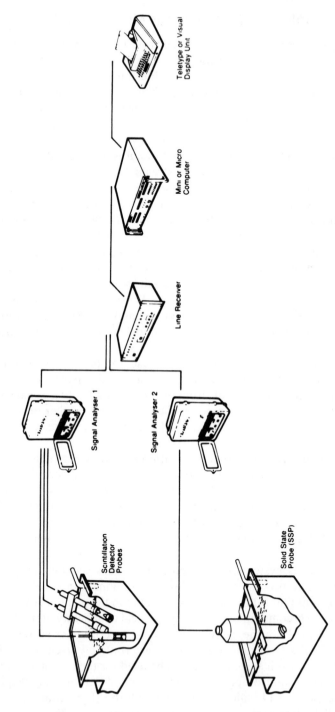

Figure 24 A system for the on-stream analysis of mineral slurries [42]. The microcomputer outputs solids weight fraction and concentrations of valuable minerals in the solids of each stream.

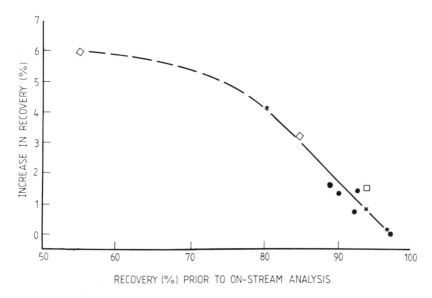

Figure 25 Improvements in recovery of mineral concentrators after installation of on-stream analysis systems. The symbols refer to tin (diamonds), lead (square), zinc (circles), and copper (asterisks).

systems are now in use, with about 200 plant systems being installed in concentrators throughout the world. The radioisotope systems, both in stream [68,71] and short sample byline, and the five-stream x-ray tube system, are preferred for new installations because they cost less and are mechanically less complex. Improvements in plant control based on this analysis information have led to an increase in the recovery of valuable minerals (Fig. 25), decreased reagent addition, and reduced need for assay and sampling staff [68]. Total savings per concentrator vary in the range from US$100,000 to several million dollars a year.

E. On-line Determination of the Ash Content of Coal

Continuous on-line determination of the ash content of coal on conveyors is required for the control of coal mining, blending, sorting, and preparation operations. The coal carried by the conveyors is usually in the range 100–600 t/h, the speed of the conveyor is about 3 m s^{-1}, and the coal particle top size in the range 10–150 mm, depending on the application. The process requirement for analysis time varies 1 or 2 s for the fast sorting of coal to 10 minutes for the steady-state control used in coal preparation plants.

Three x- or low-energy γ-ray techniques have been developed: backscatter of x-rays [72], backscatter of low-energy γ-rays [37,38], and dual-energy γ-ray transmission [34,35]. Each relies on the photoelectric effect, which depends on atomic number, and on the fact that ash (mainly SiO_2 and Al_2O_3, with smaller concentrations of Fe_2O_3) has an effective Z greater than that of the coal matter (carbon, hydrogen, nitrogen, and oxygen).

The x-ray technique [72] depends on the scatter of approximately 17 keV x-rays from a ^{238}Pu source in the coal and, at the same time, excitation of iron K x-rays in the coal to correct for the high absorption per unit mass by Fe_2O_3 compared with Al_2O_3 and SiO_2. Since the low-energy x-rays penetrate only thin layers of coal, the coal is sampled from the conveyor, subsampled, and ground to -5 mm top size particles, partially dried, and then presented in a moving stream of controlled geometry to the radioisotope x-rays analysis system. This system compensates for the effect of variations in Fe_2O_3 in the ash, a significant source of error in some applications. However, it involves complex sampling, sample handling and processing, and blockages occur when the coal is very wet. Before the gage manufacturer discontinued production in 1985 [72], 30 of these gages were installed in industry.

The low-energy γ-ray technique, using an ^{241}Am (60 keV) source, depends on measurement of the intensity of γ-rays scattered from thick layers (≥ 15 cm) of coal. It was first used on a high-throughput sample byline [37]. Although coal must be sampled, there is no need for the coal to be subsampled and crushed because of the high penetration of γ-rays in the coal. The technique has been adapted for use on-line [38], the analysis head unit riding on a raft that is spring loaded so that it is always touching the top of the coal on the conveyor. Its use is restricted to conveyor speeds of less than 2 m s^{-1} and to a minimum thickness of 15 cm of coal on the conveyor compared to the normal practice of 5–20 cm.

The dual-energy γ-ray transmission technique [34,35] measures coal directly on the conveyor belt (Fig. 26). There is no need for sampling the coal. The ash content is determined independently of vertical segregation of coal on the belt, and if segregation across the belt occurs, the narrow beam of γ-rays can be made to scan across the belt to obtain a representative sample. The coal mass per unit area in the γ-ray beam must be ≥ 3 or 4 g cm^{-2} to achieve sufficient sensitivity of analysis. Variations of iron in the ash limits the accuracy of ash determination in some applications.

The choice of a suitable x- or γ-ray analysis technique is highly influenced by the complexities in sampling of the coal on the conveyor, and the subsequent subsampling and grinding. Radioisotope techniques that measure directly on-line are preferred to those involving sampling and, if sampling is necessary, preference is given to those that minimize sampling and sample presentation.

Dual-energy γ-ray transmission is now the preferred technique for the on-line determination of ash content of coal, except for applications in which unacceptable errors in ash are caused by variations in iron in the ash. In this case, a high-energy γ-ray technique, which is based on the pair production interaction and is much less sensitive to variations in iron, is preferred. Its main disadvantages are cost and that it must operate on a sample byline [73]. Consultants have estimated that the use of these two types of on-line ash monitors in Australia has led, over a 5 year period, to a US$128 million increase in the value of product coal.

F. On-line Analysis of Paper

Continuous on-line analysis of paper is required for control of the production process. Paper consists of cellulose, water, and mineral matter. The characteristics and the quality of various types of paper are to a great extent dependent

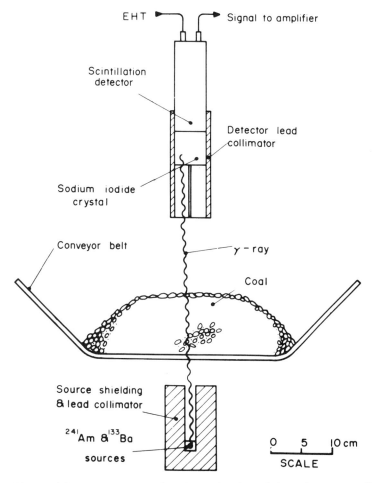

Figure 26 System for on-line determination of the ash content of coal based on dual-energy γ-ray transmission techniques.

on the quality, quantity, and distribution of filler materials. These filler materials occupy the spaces between fibers and improve the printing properties of the paper. Filler materials include $CaCO_3$, kaolin, talc, and TiO_2, and the concentrations of each may vary.

The analysis may be achieved by a combining XRF, XRA, and β-ray transmission techniques [74]. The mass absorption coefficients of x-rays in the 1–10 keV region are shown in Figure 27, with abrupt changes in the K-shell absorption edge energies of calcium (4.04 keV) and titanium (4.96 keV). X-ray preferential absorption measurements are made at x-ray energies of 5.9 (^{55}Fe), 4.51 (Ti $K\alpha$ x-rays), and 3.69 keV (Ca $K\alpha$ x-rays), the latter two energies being obtained by exciting a secondary target material with ^{55}Fe x-rays (Fig. 28). The total mass per unit area of the paper is determined by β-ray transmission. The distribution of the filler material through the paper is determined by making XRF measurements, using ^{55}Fe x-rays to excite calcium K x-rays, to give the difference in

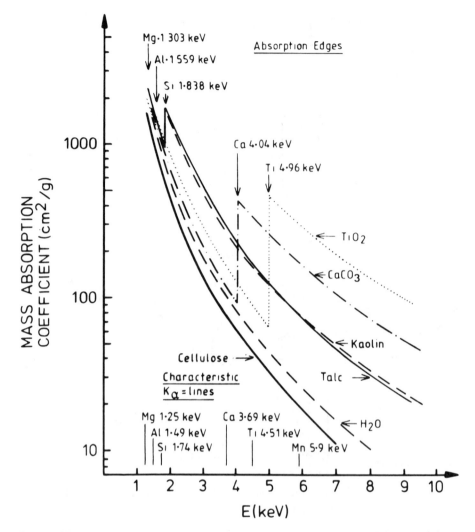

Figure 27 Mass absorption coefficients of the most frequently used filler materials, water and cellulose, relating to the manufacture of paper [74].

$CaCO_3$ concentrations near the surface on each side of the paper. These techniques can be used to determine the concentrations of cellulose, kaolin or talc, $CaCO_3$, and TiO_2. About 10 of these gages are now installed in plants worldwide.

G. Determination of Sulfur and Chlorine in Oil

Sulfur in oil is a source of pollution. Strict environmental controls are applied to limit sulfur release into the atmosphere. Oil in sample bylines from main pipelines is monitored routinely by one of two radioisotope x-ray techniques. One technique combines XRF (^{55}Fe and a proportional detector) and β-ray transmission. The other combines XRA, using 22 keV x-rays that are absorbed equally per unit

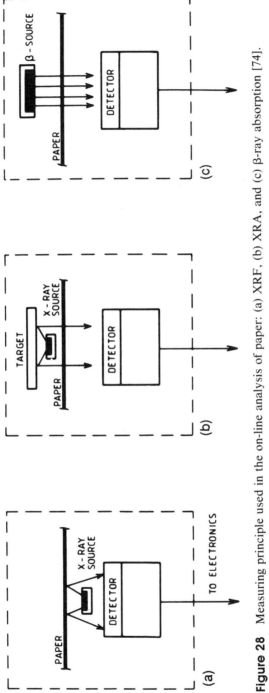

Figure 28 Measuring principle used in the on-line analysis of paper: (a) XRF, (b) XRA, and (c) β-ray absorption [74].

weight by both carbon and hydrogen, with a nonnuclear technique to determine the density of the oil. In Japan alone, 71 of the former gages and 339 of the latter are used (Table 6).

Spent engine oil is burned as fuel in industrial boilers. Since chlorinated solvents have previously been mixed with much of this oil, there is a danger that hazardous levels of these compounds or their derivatives will be released into the atmosphere. The U.S. Environmental Protection Agency has banned the sale of used oil for fuel if the total halogen level (interpreted as total chlorine level) exceeds 1000 μg g^{-1}. Piorek and Rhodes [75] and Gaskill et al. [76] showed that a portable XRF analyzer (Fig. 21), with a high-resolution proportional detector and a 740 GBq ^{55}Fe source, can be used to analyze for chlorine in oil. The used oil also contains phosphorus, sulfur, potassium, and calcium, and the intensities of the K x-rays from all these elements were incorporated into the calibration so that the measured chlorine K x-ray intensity could be properly corrected for spectral overlap. The analysis time was 30 s, and the overall accuracy achieved was quite adequate.

H. Analysis of Environmental Samples

The concentrations of particulates in air are determined to identify sources of air pollution. The air particulates are collected by drawing large volumes of air through a filter paper, which is then submitted to multielement analysis. Rhodes et al. [77] showed that energy-dispersive XRF analysis based on radioisotope sources and a silicon solid-state detector is a simple and cost-effective method for determining the elemental concentrations on the filter paper. The advantage of the solid-state detector is that simultaneous, multielement analysis is achieved with excellent sensitivity and short analysis time.

Analyses were made using 4.44 GBq ^{55}Fe, 14.8 GBq ^{238}Pu, and 444 MBq ^{109}Cd; a 80 mm^2 $\times$ 4 mm thick silicon detector (energy resolution of 180 eV at 5.9 keV); and counting for 10 minutes per filter paper with ^{238}Pu and ^{109}Cd and 5 minutes with ^{55}Fe. The minimum detectable level (3σ of the background under the relevant peak) for the 19 elements measured varied in the range 0.03 and 0.24 μg cm^{-2}. This is more than adequate for levels required in environmental pollution analysis. Rhodes and Rautala [2] have since shown that the same radioisotopes, with activities of 111–370 MBq, can be used with a high-resolution proportional detector to determine the same elements with detection limits between 0.4 and 12 μg cm^{-2} (3σ) in 4 minutes. The main competitive techniques are energy-dispersive techniques using x-ray tubes and solid-state detectors (Chap. 3) and accelerators using proton-induced x-ray emission (PIXE) techniques using particle accelerators (Chap. 11).

On-site analysis of metals in soils, sediments, and mine wastes is required as part of rehabilitation studies at hazardous waste sites. The distribution of the hazardous metals is heterogeneous owing to the exigencies of dumping and possible leaching by rainwater. Many samples must be taken over the surface area and at depth to define the zones of metal accumulation and metal depletion. Chappell et al. [78] and Piorek and Rhodes [79] demonstrated that the portable XRF analyzers based on a high-resolution proportional detector and a 3.7 GBq ^{244}Cm source can be used to determine concentrations of arsenic, copper, zinc, and lead in soils in the range of approximate 10–10,000 μg g^{-1}, with an accuracy that is quite adequate for this type of investigation. Rapid, on-site analyses can be incorporated immediately into a field investigation program, making it possible to change the density of sampling at any spot, depending on the results of previous analyses.

VIII. CONCLUSIONS

Radioisotope x- and γ-ray techniques are widely used for routine analysis in the laboratory, in industrial plants, and in the field. Because of their simplicity, they are preferred to x-ray tube analysis techniques in many applications, particularly the more routine analyses involving a limited range of sample compositions. The techniques are less sensitive, however, than those based on x-ray tube–crystal spectrometer systems and less flexible to widely different analysis applications.

Radioisotope x- and γ-ray techniques have had great impact on industrial and field applications. Their use of the continuous on-line analysis in the coating, mineral, coal, paper, and petroleum industries has led to better control of the industrial processes. This in turn has led to products that more closely meet specifications, with consequently large savings in production costs. The use of portable instruments in the field, for example for the sorting of alloys and for detection of hazardous waste materials, has led to much more rapid analysis of materials and hence to the much wider use of x-ray techniques in field analysis.

ACKNOWLEDGMENTS

The author thanks the many scientists, organizations, and companies who generously supplied information used in this chapter and to the following for permission to reprint figures used in this chapter: Pergamon Press (Figs. 2, 12, 16, 19, 20, and 26); EG&G Ortec (Fig. 1); The Radiochemical Centre, Amersham (Fig. 4); the International Atomic Energy Agency (Fig. 6); The Analyst (Fig. 18); Outokumpu Oy (Figs. 9 and 21); IEEE (Fig. 14); Automatica (Fig. 23); Australian Mineral Development Laboratories (Fig. 24); The Australasian Institute of Mining and Metallurgy (Fig. 25); and Acta Polytechnica Scandinavia (Figs. 27 and 28).

REFERENCES

1. J. R. Rhodes, *Energy Dispersion X-ray Analysis: X-ray and Electron Probe Analysis*, ASTM, Philadelphia, STP 485 (1971), p. 243.
2. J. R. Rhodes and P. Rautala, in *Nuclear Geophysics* (C. G. Clayton, Ed.), Pergamon Press, Oxford (1983), p. 333.
3. J. S. Watt, in *Nuclear Geophysics* (C. G. Clayton, Ed.), Pergamon Press, Oxford (1983), p. 309.
4. *Practical Aspects of Energy Dispersive X-ray Emission Spectrometry*, IAEA-216, IAEA, Vienna, 1979.
5. J. S. Watt and E. J. Steffner, *Int. J. Appl. Radiat. Isot.* 36:867 (1985).

6. R. Jenkins, R. W. Gould, and D. Gedcke, *Quantitative X-ray Spectrometry*, Marcel Dekker, New York (1981), p. 94.
7. H. P. Schatzler, *Int. J. App. Radiat. Isot. 30*:115 (1979).
8. R. A. Fookes, V. L. Gravitis, and J. S. Watt, *Anal. Chem. 47*:589 (1975).
9. K. Siegbahn (Ed.), *Alpha-, Beta- and Gamma-ray Spectroscopy*, Vol. 1, North Holland, Amsterdam (1965), p. 38.
10. Beltcon 100 GS: Technical Description, Outokumpu Oy, Finland.
11. Radiochemical Centre, Amersham, UK, *Industrial Gauging and Analytical Instrumentation Sources* (1986), pp. 15, 59–60.
12. A. Martin and S. A. Harbison, *An Introduction to Radiation Protection*, Chapman and Hall, London, 1986.
13. *Quantitative Bases for Developing a Unified Index of Harm*, Annals of the ICRP, ICRP Publication 45, Pergamon Press, Oxford, 1985.
14. R. Jenkins, R. W. Gould, and D. Gedcke, *Quantitative X-ray Spectrometry*, Marcel Dekker, New York (1981), pp. 120, 129, 192.
15. EG&G Ortec Catalogue: Nuclear Instruments and Systems (1986), pp. 3–16, 3–20.
16. Outokumpu Oy, PO Box 85, SF-02201, Espoo, Finland.
17. Harshaw Radiation Detectors, Harshaw/Filtrol Partnership, 6801 Cochran Road, Solon, OH 44139.
18. R. Jenkins, R. W. Gould, and D. Gedcke, *Quantitative X-ray Spectrometry*, Marcel Dekker, New York, (1981), p. 132.
19. M.-L. Jarvinen and H. Sipila, *IEEE Trans. Nucl. Sci. NS-31*:356 (1984).
20. M. Cuzin, *Nucl. Instr. Methods A253*:407 (1987).
21. N. W. Madden, G. Hanepen, and B. C. Clark, *IEEE Trans. Nucl. Sci. NS-33*:303 (1986).
22. Kevex Corporation, P.O. Box 4050, Foster City, CA 94404.
23. K. G. Gardner, CSIRO, Private Mail Bag 7, Menai, NSW 2234, Australia.
24. Tracor X-ray, Inc., 345 East Middlefield Rd., Mountain View, CA 94043.
25. R. Jenkins, R. W. Gould, and D. Gedcke, *Quantitative X-ray Spectrometry*, Marcel Dekker, New York, 1981.
26. V. L. Gravitis, R. A. Greig, and J. S. Watt, *Australas. Inst. Min. Metall. Proc. 249*:1 (1974).
27. P. Hoffmann, *Fresenius Z. Anal. Chem. 323*:801 (1986).
28. R. Spatz and K. H. Lieser, *Fresenius Z. Anal. Chem 288*:267 (1977).
29. M.-L. Jarvinen and H. Sipila, in *Advances in X-ray Analysis*, Vol. 27 (Cohen et. al, Eds.), Plenum Press, New York (1984), p. 539.
30. ASOMA Instruments, 12212-H Technology Boulevard, Austin, TX 78727.
31. M. Hietala and J. Viitanen, *Advances in X-ray Analysis*, Vol. 21, Plenum Press, New York (1978), p. 193.
32. J. R. Rhodes, *Analyst 91*:683 (1966).
33. J. S. Watt, *Int. J. Appl. Radiat. Isot. 23*:257 (1972).
34. R. A. Fookes et al., *Int. J. Appl. Radiat. Isot. 34*:63 (1983).
35. V. L. Gravitis, J. S. Watt, L. J. Muldoon, and E. M. Cochrane, *Nucl. Geophys. 1*:111 (1987).
36. W. K. Ellis, R. A. Fookes, V. L. Gravitis, and J. S. Watt, *Int. J. Appl. Radiat. Isot. 20*:691 (1969).
37. G. Fauth, D. Leininger, and H. Ludke, *Gamma, X-ray and Neutron Techniques for the Coal Industry*, IAEA, Vienna (1986), p. 165.
38. S. Cierpicz, *Gamma, X-ray and Neutron Techniques for the Coal Industry*, IAEA, Vienna (1986), p. 149.
39. P. M. Gy, Sampling of Particulate Materials: Theory and Practice, in *Developments in Geomathematics*, Vol. 4, Elsevier, New York, 1982.

40. International Standards Organisation, Central Secretariat, base Postale 56, CH-1211 Geneve 20, Switzerland.

41. M. David, *Geostatistical Ore Reserve Estimation*, Elsevier, Amsterdam, 1977.

42. Australian Mineral Development Laboratories, P. O. Box 114, Eastwood, SA 5063, Australia.

43. BRGM, BP 6009-45060 Orleans, Cedex 02, France.

44. Columbia Scientific Industries, PO Box 203190, Austin, TX 78720.

45. Data Measurement Corporation, PO Box 490, Gaithersburg, MD 20877.

46. Mineral Control Instrumentation Pty. Ltd., P.O. Box 64, Unley, SA 5061, Australia.

47. Oxford Instruments, 20 Nuffield Way, Abingdon, Oxon OX14 1TX, UK.

48. Princeton Gamma-Tech, Box 641, Princeton, NJ 08540.

49. Texas Nuclear Corporation, Box 9267, Austin, TX 78766.

50. Ramsey Ltd., 385 Enford Rd., Richmond Hill, Ontario, L4C 3G2, Canada.

51. Harrison Cooper Systems, Inc., AMF Box 22014, Salt Lake City, UT 84122.

52. Electrical Engineering and Automation (EMAG), Katowice, Poland.

53. Humboldt Wedag, P.O. Box 2729, 4630 Bochum, Germany.

54. Sortex Ltd., Pudding Mill Lane, London E15 2PJ, UK.

55. Science Applications Inc., 1257 Tasman Drive, Sunnyvale, CA 94089.

56. Rigaku Denki, 14-8 Akaoji, Takatsuki-shi, Osaka, Japan.

57. FAG Kugelfischer Georg Schafer KG Auf Aktien, Tennenloher Strasse 41, Erlangen, Germany.

58. Yokogawa Hokushin Electric Corporation, 9-32 Nakacho 2-chome, Musashino-shi Tokyo, 180, Japan.

59. Mitsubishi Corporation, 6-3, Marunouchi 2-chone, Chiyoda-Ku, Tokyo 100, Japan.

60. Sentrol Systems Ltd., 4401 Steeles Avenue West, North York, Ontario, Canada M3N2S4.

61. Paul Lippke GmbH & Co. KG, Postfach 1760, 5450 Neuwied 1, Germany.

62. S. Piorek and J. R. Rhodes, Application of a microprocessor based portable analyser to rapid non-destructive alloy identification, in Proceedings of ISA-1986, Houston, (1986), pp. 1355–1368.

63. EG&G Ortec, portable assay instruments for detection and/or measurement of ore values, undated brochure.

64. G. Sherrington, Geopeko Mineral Exploration Pty. Ltd., 25 Merriwa Gardens, Sydney, Australia, private communication, 1987.

65. MAP portable assayers: Specifications and technical information, Scitech Corporation, Kennewick, WA (undated).

66. R. F. Hill and W. Garber, *IEEE Trans. Nucl. Sci. NS-25*:790 (1978).

67. Watt, J. S. *Practical Aspects of Energy Dispersive X-ray Emission Spectrometry*, IAEA, Vienna (1978), p. 135.

68. J. S. Watt, *Proc. Aust. IMM 290*:57 (1985).

69. A. Leppala, J. Koskinen, T. Leskinen, and P. Vanninen, *Trans. Soc. Mining Eng. AIME 250*:261 (1971).

70. K. Saarhilo, Experiences of a new on-stream x-ray analyzer in a metal refinery, in Proceedings of IFAC Automation in Mining, Mineral and Metal Processing, Helsinki, Finland (1983), pp. 357–367.

71. P. Berry, W. Garber, and K. Blake, Application of an in-stream elemental analysis system, in Proceedings of 11th Annual Mining and Metallurgy Industries Symposium, Tucson, AZ (1983), pp. 31–42.

72. H. Fraenkel, Sortex Ltd., Pudding Mill Lane, London E15 2PJ, UK, private communication, 1987.

73. B. D. Sowerby, *Gamma, X-ray and Neutron Techniques for the Coal Industry*, IAEA, Vienna (1986), p. 131.

74. V. Kelha, M. Luukkala, and T. Tuomi, *Acta Polytech. Scand. Appl. Phys. Ser. 138*:90 (1983).
75. S. Piorek and J. R. Rhodes, Hazardous waste screening using a portable x-ray analyzer, in Proceedings of 15th Environmental Symposium, Long Beach, CA (1987), pp. 292–297.
76. A. Gaskill, E. D. Estes, and D. L. Hardison, *Evaluation of Techniques for Determining Chlorine in Used Oils*, Vol. 1, RTI Project Number 472U-3255-05, 1987.
77. J. R. Rhodes, A. H. Pradzynski, and R. D. Sieberg, *ISA Trans. 11*(4):337 (1972).
78. R. W. Chappell, A. O. Davis, and R. L. Olsen, Portable x-ray fluorescence as a screening tool for analysis of heavy metals in soils and mine wastes, in Proc. National Conf. on Management of Uncontrolled Hazardous Waste Sites, Washington, D.C. (1986), pp. 115–119.
79. S. Piorek and J. R. Rhodes, Hazardous waste screening using a portable x-ray analyzer, paper presented at the Symposium on Waste Minimization and Environmental Programs Within DOD, Long Beach, CA, April 1987.

8

Synchrotron Radiation-Induced X-ray Emission

Keith W. Jones   *Brookhaven National Laboratory, Upton, New York*

I. INTRODUCTION

Elemental analysis using emission of characteristic x-rays is a well-established scientific method. Its success is highly dependent on the properties of the excitation source. X-ray tubes have long existed as a principal excitation source. Electrons and protons have also had an impact as excitation sources. The rapid development of the synchrotron radiation x-ray source during the past 25 years is starting to have a major impact on the general field of x-ray analysis. Improvements to the synchrotron source being made on facilities now being designed and constructed promise to accelerate the development of the general scientific use of the synchrotron radiation-induced x-ray emission (SRIXE) method.

A short summary of the present state of SRIXE is presented here. It is hoped that it serves as a succinct introduction to the basic ideas for those not working in the field and possibly help to stimulate new types of work by those starting in the field as well as by experienced practitioners of the art.

The topics covered include short descriptions of (1) the properties of synchrotron radiation, (2) a description of facilities used for its production, (3) collimated microprobes, (4) focused microprobes, (5) continuum and monoenergetic excitation, (6) detection limits, (7) quantitation, (8) applications of SRIXE, (9) computed microtomography (CMT), and (10) chemical speciation using x-ray absorption near edge structure (XANES) and extended x-ray absorption fine structure (EXAFS). An effort has been made to cite a wide variety of work from different laboratories to show the vital nature of the field.

Many review articles and books cover all aspects of the production and use of synchrotron radiation [1–4]. The early article on SRIXE by Sparks [5] is a useful source of information on various aspects of the use of SRIXE as part of a high-resolution x-ray microscope (XRM) system. Several more recent review articles have covered new developments [6–10].

Microscopy using x-rays can be carried out in several ways. A large effort is aimed at producing extremely high resolution maps of the linear attenuation coefficient μ for low-energy x-rays in biological materials [11]. Another approach is based on detection of electrons emitted from the specimen [12]. Elemental detection can be accomplished by mapping μ above and below the absorption edge for the element. Rarback et al. [13] showed that this may be a preferable method for elements lighter than about calcium because of the small values of the fluorescence yield. This approach was used by Kennedy et al. [14] to study the calcium distribution in a bone specimen using the Ca L absorption edge for image formation.

Image formation using fluorescent x-rays is advantageous for the high-Z elements in terms of attaining the best possible values for minimum detection limits (MDLs) combined with the best possible spatial resolution. It is not possible to use x-ray detection in some situations in which the size of the specimen is much greater than the absorption depth of the fluorescent x-rays. In this case, also, above and below edge imaging using the technique of computed microtomography is possible. The use of these methods in common on a single specimen can be helpful. In the discussion that follows the use of all these methods is discussed, although the major emphasis is placed on SRIXE.

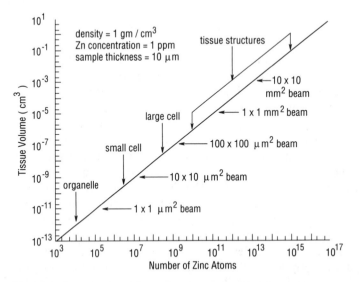

Figure 1 The number of zinc atoms contained in a given volume element is plotted as a function of volume. A constant zinc concentration of 1 ppm contained in an organic matrix with a density of 1 g/cm³ is assumed. This shows that the detection of trace amounts of an element at a given concentration level becomes increasingly difficult as the volume probed decreases. [From K. W. Jones and J. G. Pounds, *Biol. Trace Elem. Res., 12*:3–16 (1987).]

Images may also be formed using other types of information obtained through probing the sample. X-ray diffraction can be used to determine crystal structure. Information on the chemical state of the elements and position in the lattice can be obtained by use of XANES or EXAFS. These are, of course, methods widely applied with both conventional and synchrotron x-ray sources. A merging of these methods with technologies for producing micrometer-sized beams is a natural direction for development of the instrumentation to follow. This is beginning to happen now, and the rate of development is bound to increase in the future.

No matter what mode is used for image formation or to characterize a single volume element in a specimen, the work is dominated by the fact that only limited numbers of atoms are present to produce a signal. For example, consider the number of zinc atoms present in an organic material as a function of the specimen volume probed for a constant weight fraction of 1 ppm. Values are shown in Figure 1, and it can be seen that only 10^4 atoms are present in a 1 μm^3 volume. Detection of such a small number is a technical challenge no matter what mode of detection is used.

II. PROPERTIES OF SYNCHROTRON RADIATION

It has been known for almost 100 years that the acceleration of a charged particle results in the radiation of electromagnetic energy. The development of the betatron and synchrotron electron accelerators 50 years later led to the experimental observation of radiation from electrons circulating in a closed orbit [15] and to naming it specifically synchrotron radiation after the accelerator used to produce it. The first recognition of the unique properties of synchrotron radiation by Tombulian and Harman [16] then brought about an explosive development of activity in constructing improved sources for production of the radiation and to using the radiation in experimental science.

The original synchrotrons were designed for use in nuclear physics research. Later facilities were designed to optimize the conditions for production of x-rays. The components of the new facilities include a source of electrons or positrons and an accelerator to produce high-energy beams. This might be done by use of a linear accelerator to produce energies of around 100 MeV. These beams are then injected into a synchrotron and boosted to energies in the GeV range. Finally, the beam is stored in the same accelerator used to attain the final energy or in a separate storage ring. An acceleration field in the ring is used to supply energy to the beam to compensate for the radiation energy loss. The lifetime of the stored beam is many hours, so that in practice the synchrotron source is almost starting to become similar to a standard x-ray tube in use.

The synchrotron-produced x-ray beams have unique properties that make them desirable for use. They have a continuous energy distribution so that monoenergetic beams can be produced over a wide range of energies. The photons are highly polarized in the plane of the electron beam orbit, which is extremely important for background reduction in SRIXE experiments in particular. The x-rays are emitted in a continuous band in the horizontal direction but are highly collimated in the vertical direction. It is therefore possible to produce intense beams with little angular divergence. The source size is small, and as a result, the production of intense beams of small area is feasible. The synchrotron source is a pulsed source because of the nature of synchrotron accelerators. The x-rays are produced in narrow bursts, less than 1 ns in width, and have a time between pulses of around 20 ns or more.

The main parameters of interest in defining the synchrotron source are then:

1. Magnitude of the stored electron-positron current. Typically, currents are in the range from 100 to 1000 mA. Lifetimes of the stored beam are many hours. The lifetimes for stored electrons at the National Synchrotron Light Source (NSLS) at Brookhaven National Laboratory (BNL, Upton, NY) have typical values of around 24 h. At the Laboratoire pour L'Utilisation de Radiation Elecromagnetique (LURE; Orsay, France), where positron beams are used, the lifetimes are even longer since the positive beam does not trap positively charged heavy ions produced from the residual gases in the vacuum chamber.

2. Assuming that the size and angular divergence of the electron beam are not important, the source brightness is defined as [units of photons^{-1} mr^{-2} (0.1% bandwidth)]:

$$\frac{d^2I}{d\theta \, d\psi} = 1.327 \times 10^{13} E^2(\text{GeV})i(A)H_2 \frac{\omega}{\omega_c} \tag{1}$$

where E is the energy of the electron beam in GeV, i the current in amperes, $H_2(\omega/\omega_2)$ is a function tabulated in Reference 4, ω is the photon angular frequency, and ω_c is the critical frequency that splits the emitted power into halves and is given by the expression $3\gamma^3 c/2\rho$ where γ is the electron energy in units of the electron rest mass, c is the velocity of light, and ρ is the radius of curvature of the electron path. The angles θ and ψ are the angles of emission in the plane of the electron orbit and perpendicular to that plane, respectively.

3. The total photon emission is found by integrating over ψ and is given by (in units of photons s^{-1} mr^{-1} (0.1% bandwidth)$^{-1}$

$$\frac{dI}{d\theta} = 2.457 \times 10^{13} E(\text{GeV})i(A)G_1 \frac{\omega}{\omega_c} \tag{2}$$

where $G_1(\omega/\omega_c)$ is tabulated in Reference 4.

For electron beams with nonzero emittance (finite area and angular divergence), it is necessary to define another quantity, the brilliance, which is the number of photons emitted into angular intervals $d\theta$ and $d\psi$ at angles θ and ψ from an infinitesimal source area [in units of photons·s^{-1} mr^{-2} mm^{-2} (0.1% bandwidth)$^{-1}$]. The values of brilliance and brightness are important in evaluating the performance expected for focused and collimated x-ray microscopes. Kim [4] provides a detailed discussion of the emittance effects, polarization, and performance of wiggler and undulator insertion devices.

Values attained for these quantities at the 2.5 GeV x-ray ring of the NSLS are shown in Figures 2 through 4. This ring is typical of the present state of the art in synchrotron x-ray production. The ring energy is high enough to produce x-rays over an energy range sufficient to produce K x-rays from elements to about $Z = 40$ with good efficiency and L x-rays throughout the periodic table. Thus it

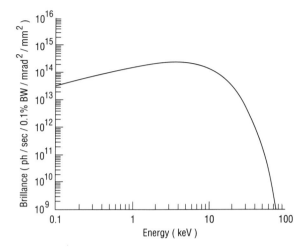

Figure 2   The brilliance of the NSLS x-ray ring is plotted as a function of the x-ray energy produced. The brilliance is important in defining limitations on production of x-ray images using focusing optics. [From K. W. Jones et al., *Ultramicroscopy 24*:313–328 (1988).]

is highly suitable for use as the basis of a system for x-ray microscopy-based SRIXE.

The high degree of linear polarization of the x-rays from the synchrotron source is a major factor in making the synchrotron XRM a sensitive instrument. The physics describing the interaction of polarized x-rays with matter is therefore an important topic. Hanson [17] carried out an extensive examination of the scattering problem and gives methods for assessing particular geometries used in the XRM.

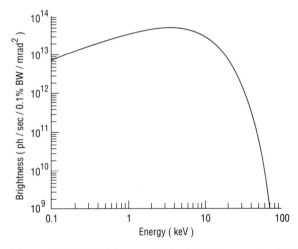

Figure 3   The brightness of the NSLS x-ray ring is defined in Equation (1) and plotted as a function of the x-ray energy produced. The brightness is defined in Equation (1). The brightness is of importance in defining the usefulness of x-ray beams produced by use of a collimator. [From K. W. Jones et al., *Ultramicroscopy 24*:313–328 (1988).]

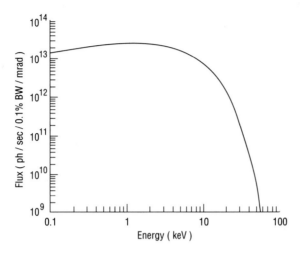

Figure 4 The x-ray flux produced by the NSLS x-ray ring is plotted as a function of the x-ray energy produced. The values are given after integration over the vertical opening angle of the beam. The definition of the flux is given in Equation (2). [From K. W. Jones et al., *Ultramicroscopy 24*:313–328 (1988).]

III. DESCRIPTION OF SYNCHROTRON FACILITIES

Many synchrotron facilities are now located around the world that are suitable for use in various types of x-ray spectrometry measurements. They fall roughly into three classes.

First-generation synchrotrons were often built primarily as high-energy physics machines and were used secondarily for synchrotron radiation production. The Cornell high-energy synchrotron source (CHESS) at Ithaca, New York is an example.

Second-generation synchrotrons were optimized as radiation sources and, as

Table 1 First-Generation Synchrotron Light Sources[a]

Storage ring (laboratory)	Energy (GeV)	Location	Project phase
ADONE (LNF)	1.5	Frascati, Italy	Dedicated
DCI (LURE)	1.8	Orsay, France	Partly dedicated
VEPP-3 (INP)	2.2	Novosibirsk, USSR	Partly dedicated
BEPC (IHEP)	2.2–2.8	Beijing, China	Partly dedicated
SPEAR (SSRL)	3.0–3.5	Stanford, CA	Partly dedicated
ELSA (Bonn University)	3.5	Bonn, Germany	Partly dedicated
DORIS II (HASYLAB)	3.5–5.5	Hamburg, Germany	Partly dedicated
VEPP-4 (INP)	5.0–7.0	Novosibirsk, USSR	Partly dedicated
CESR (CHESS)	5.5–8.0	Ithaca, NY	Partly dedicated
Acc. Ring (KEK)	6.0–8.0	Tsukuba, Japan	Partly dedicated
Tristan (KEK)	25–30	Tsubuka, Japan	Partly dedicated

[a] Most of these early facilities were built for research for the primary electron or positron beams. Synchrotron radiation research was parasitic. Today, all are at least partly dedicated as synchrotron light sources. See Refs. 18, 19.

Table 2 Second-Generation Synchrotron Light Sources[a]

Storage ring (laboratory)	Energy (GeV)	Location	Project phase
CAMD (LSU)	1.2	Baton Rouge, USA	Proposed
SOR (KU	1.5	Kyushu, Japan	Operating
DELTA (DU)	1.5	Dortmund, Germany	Operating
NSLS x-ray (BNL)	2.5	Upton, NY	Operating
Photon facility (KEK)	2.5	Tsukuba, Japan	Operating
Synchrotron radiation source (SRS)	?	Daresbury, UK	Operating

[a] This is the first generation of machines built as dedicated synchrotron radiation facilities. The use of bending magnet ports is emphasized, although some straight sections have insertion devices (e.g., undulators and wigglers). See Refs. 18, 19.

a result, produce x-ray beams with superior brilliance and brightness characteristics. The two rings of the Brookhaven NSLS fall in this category.

Finally, the third generation of synchrotrons is now being designed. They will be the first sources intended to incorporate insertion devices, wigglers, and undulators in the design phase. In some cases the ring energy is increased to give better performance with the insertion devices. Synchrotrons designed with these features will begin operation in the latter part of the 1990s. The European Synchrotron Radiation Facility (ESRF) at Grenoble, France, the Super Photon Ring-8 GeV (Spring-8) in Kansai, Japan, and the Advanced Photon Source (APS) at Argonne National Laboratory (Argonne, IL) illustrate this case.

A listing of synchrotron laboratories producing high-energy x-ray beams suitable for use in SRIXE is given in Tables 1 through 3 [18,19]. The number of XRM

Table 3 Third-Generation Synchrotron Light Sources[a]

Light source	Energy (GeV)	Location	Project phase
ALS	1.0–1.9	Berkeley, CA	Construction
SRRC	1.3	Hsinchu, ROC	Construction
INDUS II	1.4	Indore, India	Authorized
MAX II	1.5	Lund, Sweden	Design
BESSY II	1.5–2.0	Berlin, Germany	Design
ELETTRA	1.5–2.0	Trieste, Italy	Construction
LNLS	2.0	Campinas, Brazil	Authorized
PLS	2.0	Pohang, Korea	Authorized
SIBERIA II	2.5	Moscow, USSR	Construction
ESRF	6.0	Grenoble, France	Construction
APS	7.0	Argonne, IL	Construction
SPring-8	8.0	Kansai, Japan	Design
DAPS	???	Daresbury, England	Design

[a] This is the newest generation of dedicated synchrotron radiation facilities. The use of insertion devices (e.g., undulators and wigglers) is emphasized, but bending magnet ports will also be available. See Refs. 18, 19.

beam lines is growing rapidly, and their employment for research as a result is becoming widespread.

IV. APPARATUS FOR X-RAY MICROSCOPY

The apparatus used for x-ray microscopy measurements varies substantially from laboratory to laboratory. A schematic diagram of the components of a very comprehensive system is shown in Figure 5. All the components are not necessarily used in a specific instrument. The most important differences between systems lies in the treatment of the incident beam. The simplest approach is to use the white beam (full-energy spectrum) and a collimator. The more complex systems use focusing mirrors to collect more photons and demagnify the beam and monochromators to produce monoenergetic beams. At present, the performance of the various systems is quite comparable in terms of spatial resolution and minimum detection limits. Thus, a versatile and flexible approach to the choice and design of the components is advisable. A number of the different instruments are described briefly in the following sections.

A. Collimated X-ray Microscopes

A highly effective XRM system can be made by simply collimating the white beam (continuous energy) of x-rays produced by the synchrotron. This approach has

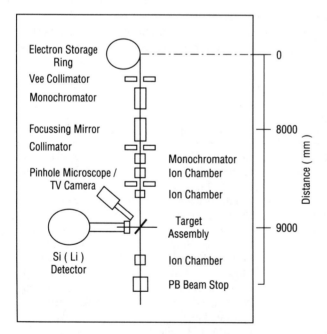

Figure 5   A comprehensive synchrotron beam line designed for use as a x-ray microscope. Not all the components would be utilized at a given time in practice. The rather varied uses of the system require a flexible approach to the design of the equipment.

been followed mainly by groups at the Hamburg storage ring [20], Hamburg Synchrotron Laboratory (HASYLAB), and at the Brookhaven NSLS [21]. The use of white radiation is feasible because the high brightness gives a high flux of photons in a small area and the high polarization of the synchrotron beams minimizes scattering from the sample into the detector. White radiation also makes possible efficient multielement detection over a very broad range of atomic number.

The collimators for the instrument can be made from a set of four polished tantalum strips. The strips can be spaced with thin plastic or metal foils to produce apertures usable to a beam size of 1 μm. Alternatively, the slits can be attached to individual stepper motors for producing variably sized beams of size greater than about 10 μm. The collimation approach is ultimately limited by beam spreading related to finite source and pinhole dimensions and to diffraction.

The collimated XRM (CXRM) was first operated at the now defunct Cambridge Electron Accelerator by Horowitz and Howell [22]. They used a pinhole made by evaporation of a thick gold layer around a 2 μm quartz fiber followed by subsequent etching away of the quartz. The x-ray energy was about 2.2 keV, and image contrast was achieved using determination of the linear attenuation coefficient. The specimen was moved past the incident beam, and fluorescent x-rays were detected with a proportional counter. A spatial resolution of 2 μm was measured with this pioneering early apparatus.

Later versions of the CXRM have been placed in operation at Hamburg and Brookhaven. Some of the details of the operations are given to illustrate the more important operational details.

It is often best to maximize the photon flux on the sample by employing a white beam of x-rays. The flux available at a point 9 m from a NSLS bending magnet source is shown in Figure 6. The flux from this source, integrated over the entire energy spectrum, is about 3×10^8 photons/s/μm² with a 100 mA stored electron current.

For best operation of the CXRM, care must be taken in shaping the incident x-ray spectrum using filters. The influence of the beam filters was studied by the

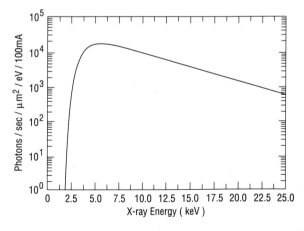

Figure 6   Energy distribution of photon flux produced by a bending magnet on the NSLS x-ray ring at a stored electron current of 100 mA. (From Ref. 24.)

Hamburg group [20]. Their results are shown in Figure 7. By varying the effective energy, it is possible to tailor the beam to give the best possible MDL for a given atomic number. Filters on the detector can be used in some cases to reduce the effect of a major element. In experiments that examine the distribution of trace elements in bone, a filter of polyimide can reduce the high rates caused by the calcium in the bone but do not have a large effect on the x-rays from iron, copper, or zinc.

An equally critical task is the alignment of the energy-dispersive x-ray detector. Kwiatek et al. [23] reported on this phase of the optimization procedure. The importance of aligning the detector in the horizontal plane can be seen by reference to Figure 8. In addition, the energy spectrum and degree of polarization change as a function of the vertical distance from the plane of the electron orbit in the storage ring. Figure 9 shows the relative photon flux for the two polarization states. Examination of the curves shows that the alignment needs to be made to an accuracy of better than a few hundred micrometers to obtain the best reduction of scattered background.

Results of the experimental background to peak ratio determination as a function of vertical displacement are shown in Figure 10 for elements from calcium to zinc in a gelatin matrix and for palladium in a pyrrhotite matrix. The dependence on scattering angle in the horizontal plane is shown in Figure 11. The details of the methods used to make the alignments are given by Kwiatek et al. [23].

The Brookhaven work has shown that it is possible to achieve spatial resolutions below the 10 μm range using the CXRM. The spatial resolution of the instrument was demonstrated by scanning a thin evaporated gold straight edge through the beam and recording the intensity of the L x-rays as a function of

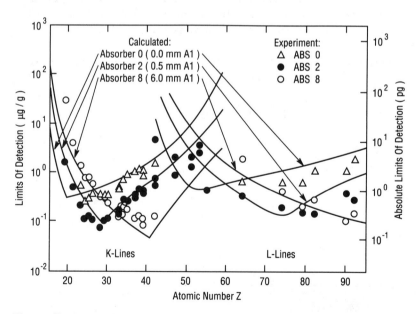

Figure 7 Values for MDLs obtained for a white-light XRM are shown as a function of atomic number. The change in the MDLs as a function of filtering of the incident beam using aluminum filters is an important feature of this type of arrangement. (From Ref. 20.)

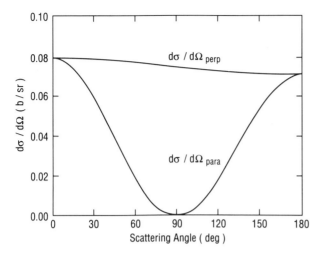

Figure 8 Dependence of incoherent scattering cross sections for x-rays polarized parallel and perpendicular to the plane of the stored electron orbit on the scattering angle. Observation at a scattering angle of 90° gives optimum signal-to-background conditions. [From K. W. Jones et al., *Ultramicroscopy* 24:313–328 (1988).]

distance. The resolution in this case was 3.5 μm. Some control over the resolution for a given collimator is obtained by inclining the collimator with respect to the incident beam.

The Brookhaven device is routinely used for trace element measurements with a spatial resolution of less than 10 μm. The sensitivity and MDL obtained under these conditions have been reported by Jones et al. [24], and their results are displayed in Figures 12 and 13.

It is interesting to realize that an absolute determination of the elemental concentrations can be made based either on the theoretical estimates of photon flux or from a direct determination using an ion chamber. Kwiatek et al. [23] point

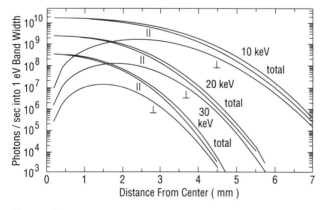

Figure 9 Polarization of NSLS x-ray beams is given as a function of distance from the plane of the electron orbit for x-ray energies of 10, 20, and 30 keV. [From A. L. Hanson, et al., *Nucl. Instrum. and Methods in Phys. Res. B24/25*: 400–404 (1987).]

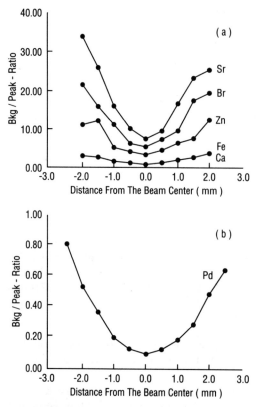

Figure 10 The background-to-signal ratio is shown as a function of displacement of the detector normal to the plane of the orbit of the stored electrons. The alignment becomes increasingly critical at the higher x-ray energies. The curves shown in Figure 10a were obtained for Ca, Fe, Zn, Br, and Sr contained in a gelatin matrix. The curve shown in Figure 10b was obtained for Pd contained in a pyrrhotite matrix. (From Ref. 23.)

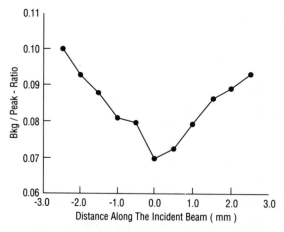

Figure 11 The background-to-signal ratio is plotted as a function of scattering angle in the horizontal plane. The curve was obtained by displacement of the detector and the equivalent angular range spanned was roughly ±4°. (From Ref. 23.)

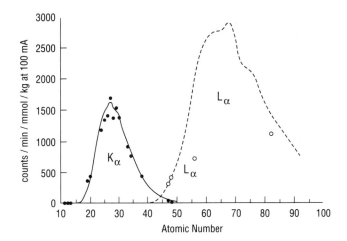

Figure 12 The sensitivity of the BNL-collimated XRM is given as a function of atomic number for K and L x-ray detection. Values are given in units of counts/(ppm-s) at a stored electron current of 100 mA with an acquisition time of 300 s and a beam size of 8 × 8 μm. A 3 mm diameter Si(Li) x-ray detector placed 40 mm from the beam was used as a detector. The curves were calculated from basic principles. (From Ref. 24.)

out that the expected counting rate for a given element is given by

$$Y(E_Z) = N_A \epsilon_k \frac{d\Omega}{4\pi} e^{-\mu(E_z)\chi d} \int_{E_{ab}}^{\infty} N(E) e^{[-\mu(E)\chi]\sigma_{fi}(E)} \, dE \tag{3}$$

where $Y(E_Z)$ is the count rate for an element with a characteristic x-ray energy E_Z, N_A is the number of target atoms within a beam spot, ϵ_k is the detector efficiency, $d\Omega/4\pi$ the solid angle, $N(E)$ the photon flux [number of photons/(s mm² 0.1 keV bandwidth)], $\mu(E)$ the linear attenuation coefficient for Al and $\mu(E_Z)$

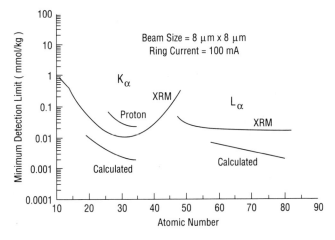

Figure 13 MDLs measured for the BNL collimated XRM under the conditions described in the text and for Figure 12. (From Ref. 24.)

Table 4 Experimental and Predicted Count Rates for Different Beam Filters[a]

Beam filter thickness (μm Al)	Fe (cps)		Zn (cps)	
	M[b]	T[c]	M	T
50	86.5	75.6	77.7	53.7
100	54.8	48.5	59.1	43.2
200	24.7	23.6	37.0	28.8

[a] Experimental errors are 10–15%.
[b] Measured values.
[c] Theoretical values.

that for polyimide, χ the Al thickness, χ_d the polyimide thickness, $\sigma_{fl}(E)$ the fluorescent cross section, E_{ab} the energy of the absorption edge, and E the photon beam energy.

Results of a comparison of measured and calculated rates for a section of gelatin with known amounts of iron and zinc are shown in Table 4 for three different incident beam spectra. The results of the comparison are excellent for iron, less so for zinc. It is clear, however, that it is feasible to make determinations of the abundances of the elements without reference to standards.

B. Focused X-ray Microscopes

Focused x-ray microprobes (FXRM) have been the subject of great interest over the years. The first version appears to have been developed by Sparks [5] and was used at Stanford Synchrotron Radiation Laboratory (SSRL). In an interesting application, it was used in a search for superheavy elements [25].

A variety of different schemes have been used in the intervening years. A summary of recent approaches at different laboratories is given here to illustrate the directions now being taken by this approach to production of intense x-ray beams. Early estimates of MDLs for specific configurations have been given by Gordon [26], Gordon and Jones [27], and Grodzins [28].

1. LURE

The equipment used for SRIXE at LURE was described by Chevallier et al. [29] and Brissaud et al. [30]. The white beam from the storage ring passes through a beryllium window and is incident on a curved crystal of mosaic pyrolytic graphite. The size of the incident beam is 2×1 cm before the monochromator, which produces a final focused beam of 14 keV photons with a size of 1×1 mm. The fluorescent x-rays are detected with a Si(Li) detector collimated using a 2.8 mm aperture. The MDLs achieved are around 1 ppm for thick geological specimens and much less for organic matrices. The probe has only been used for measurements that do not require high spatial resolution.

2. Photon Factory

The group working at the Photon Factory in Japan have made seminal contributions to the development of SRIXE and of synchrotron x-ray microscopy

(SXRM). An early instrument used Wolter focusing optics and achieved spatial resolutions of around 10 × 30 μm [31]. A later version was developed to provide higher photon flux and give improved values for the MDLs [32].

The design chosen was similar to one described by Jones et al. [33], which has been partially implemented at the NSLS. A schematic diagram of the Photon Factory apparatus is shown in Figure 14. The platinum-coated ellipsoidal mirror was designed to produce a demagnification of the beam of 1:9.5. The mirror accepted 0.6 mrad of beam in the horizontal direction and 0.05 mrad in the vertical direction and was run at a grazing angle of 7 mrad. A gain in intensity of a factor of 170 was found for a monoenergetic beam of 8 keV. The system produced a photon flux of about 1.4×10^3 photons/(s·μm²·mA). The ability to locate the system in close proximity to the electron beam is a key factor in maximizing the photon flux.

3. SRS

The instrumentation at the Daresbury synchrotron radiation source (SRS) was developed by a group of collaborators from the Free University in Amsterdam and Warwick University at Coventry. The latest device uses a silicon crystal as both a monochromator and focusing device [34–36]. A diagram of the instrument is shown in Figure 15. The crystal is a 100 μm thick silicon crystal bent with a radius of 100 mm in the sagittal plane and 5740 mm in the meridional plane. The beam is focused to a spot size of about 15 × 20 μm at an energy of 15 keV. The demagnification is 1000 in the horizontal plane and 15 in the horizontal plane with a photon flux increase greater than 10^4. The flux of 15 keV photons at the specimen of 10^4 photons/(s·μm²·mA) is obtained. This is an impressive result when it is remembered that the device is located about 80 m from the electron beam.

4. Lawrence Berkeley Laboratory (LBL)

The group at the Lawrence Berkeley Laboratory developed a focusing system based on the Kirkpatrick-Baez geometry [37–42]. A schematic diagram of their system is shown in Figure 16. The system uses a parallel beam of photons to produce an image that is demagnified by about a factor of 100 to produce final images of a few micrometers. The use of multilayer coatings of tungsten carbide

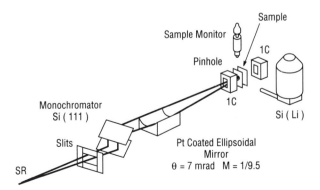

Figure 14 The focusing XRM used at the Photon Factory. (From Ref. 32.)

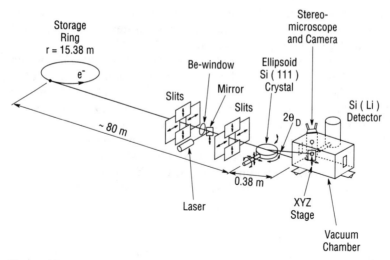

Figure 15   The focusing XRM in operation at the Daresbury XRM. (From Ref. 36.)

on the mirrors gives a quasi-monoenergetic beam with a bandwidth of about 10% at 10 keV and a high throughput. Much of the experimental work was carried out at the NSLS in collaboration with the BNL group. It is thus possible to make comparisons in performance between the CXRM and FXRM on the same storage ring.

The LBL FXRM has produced photon fluxes of about 3×10^5 photons/ $(s \cdot mA \cdot \mu m^2)$ at the 10 keV energy. It is important to note that the use of a final collimator is not required in this apparatus. Improvements in the device will make it possible to reach higher x-ray energies and provide easy tunability of the x-ray beam energy.

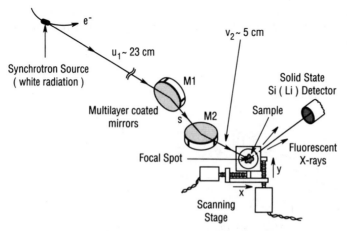

Figure 16   The focusing XRM of the Berkeley group used at the BNL X26 beam line. (From Ref. 40.)

5. Novosibirsk

Baryshev et al. [43] working at the Novosibirsk VEPP-3 synchrotron used both monoenergetic and white beams of x-rays. A single-crystal pyrolytic graphite monochromator was used to produce monoenergetic beams with energies between 8 and 35 keV. The spatial resolution was 60 μm for the monoenergetic beam and 30 μm for the white beam. The MDL for the monoenergetic beam was 10 ppm for elements from iron to strontium for a 1–3 s run.

V. CONTINUUM AND MONOCHROMATIC EXCITATION

Successful XRMs have been put into operation using both continuum and monoenergetic synchrotron-produced x-rays. The continuum radiation is extremely convenient to use since it is easy to construct a CXRM with a minimum of equipment and achieve excellent performance. Further, the broad-band excitation means that measurements can be made for essentially all elements in the periodic table in a single exposure. Monoenergetic radiation can be used at an energy optimized for production of a given element, thus reducing radiation damage in organic materials. Counting rate limitations in energy-dispersive detectors are reduced because of the elimination of scattered x-ray events. The energy can be tuned to eliminate interferences, such as lead arsenic, and to eliminate excitation of elements with Z higher than that of the element of interest. Maps can also be constructed by subtraction of images obtained above and below absorption edges.

Successful XRMs have also been produced in collimated and focused modes employing either continuum or monoenergetic radiation. The brute-force collimated continuum radiation microprobe employed at Brookhaven and Hamburg is comparable to the other types of probes in terms of spatial resolution and MDL. The performance of the Brookhaven instrument was compared with the performance of the LBL Kirkpatrick-Baez XRM operated on the same NSLS beam line [20]. A comparison of results obtained with the CXRM positioned at 10 m from the source with the FXRM at 20 m from the source is given by Rivers et al. [44]. Figure 17 shows the results of the comparison. The MDL obtained with the FXRM is somewhat better than for the CXRM, but the wider energy range of the CXRM is a substantial advantage in some cases. The quality of the Kirkpatrick-Baez optics makes it feasible to dispense with the use of a collimator placed close to the specimen, which is an advantage in the design of the experiments in some cases.

For ultimate performance in terms of the spatial resolution, it very obvious that the FXRM is the required choice. The construction of good collimators in the 1 μm^2 size region is difficult and could be impossible below 1 μm^2. Heat loading of the collimator by the beam changes the size, and resolution is limited by diffraction. The challenge is thus to

1. obtain better optics to improve the spatial resolution
2. improve the efficiency of the optics to obtain higher photon fluxes

These goals can be addressed most effectively by the use of focusing optics.

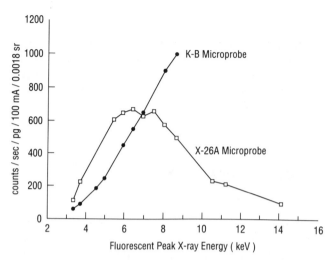

Figure 17 Sensitivity of the BNL-collimated white light XRM compared with the sensitivity obtained with the Berkeley focused XRM. (From Ref. 44.)

VI. QUANTITATION

Methods for making quantitative elemental determinations using x-ray fluorescence have been developed over many years. These approaches are discussed for conventional x-ray sources in Chapters 5 and 6. Some of the approaches used at synchrotron sources are given here to show how the methods developed for use with conventional tubes have been used with the new radiation source.

Giauque et al. [45] measured several U.S. National Bureau of Standards (NBS), now National Institute of Standards and Technology (NIST), and the Japan National Institute of Environmental Science (NIES) standard reference materials (SRM) at the 54-pole wiggler beam line at the Stanford Synchrotron Radiation Laboratory. They referred their measurements to a copper standard prepared by evaporation in which the weight was determined from gravimetric measurements. Multielement standards were prepared by dissolving known weights of an element in an acid solution. Monoenergetic radiation was used for the work at energies of 10 and 18 keV. The beam size was defined by an aperture 3 × 3 mm in size. The differences in sample mass were accounted for by normalizing to the intensity of the scattered radiation [46]. The results obtained for three NBS materials are shown in Table 5. The table also includes the ratio between the NBS value and the value obtained with SRIXE to demonstrate that there is excellent agreement.

The work of Giauque et al. [46] addressed quantitation in thin biological specimens in which matrix effects were negligible and Compton scattering could be used for normalization of masses. It is necessary to extend this approach if thick specimens are to be investigated. There are many geological experiments for which this situation holds. Methods for quantitation are discussed by Brissaud et al. [30] and by Lu et al. [47]. Corrections are made for attenuation of incoming and fluorescent x-rays by the sample matrix and by any filters employed as well as for secondary fluorescence. The composition of the major elements is generally

Table 5 Results Determined for Three NBS Standard Reference Materials (μg/g)

Element	Nonfat Milk Powder SRM 1549 Disk[a] 51 mg/cm² Weight[b] 37 mg Weight[c] 500 mg		Wheat Flour SRM 1567 Disk[a] 60 mg/cm² Weight[b] 43 mg Weight[c] 400 mg		Rice Flour SRM 1568 Disk[a] 60 mg/cm² Weight[b] 43 mg Weight[c] 400 mg	
	NBS	XRF	NBS	XRF	NBS	XRF
K	16,900 ± 3000	17,800 ± 2000	1360 ± 40	1220 ± 130	1120 ± 20	1360 ± 160
Ca	13,000 ± 500	12,000 ± 800	190 ± 10	174 ± 10	140 ± 20	158 ± 14
Cr	0.0026 ± 0.0007	<0.6		<0.3		<0.4
Mn	0.26 ± 0.06	0.33 ± 0.12	8.5 ± 0.5	8.2 ± 1.8	20.1 ± 0.4	22.1 ± 2.8
Fe	(2.1)	2.30 ± 0.16	18.3 ± 1.0	17.1 ± 4.8	8.7 ± 0.6	9.1 ± 1.2
Ni		0.24 ± 0.06	(0.18)	0.11 ± 0.06	(0.16)	0.18 ± 0.06
Cu	0.7 ± 0.1	0.65 ± 0.04	2.0 ± 0.3	1.88 ± 0.12	2.2 ± 0.3	2.21 ± 0.22
Zn	46.1 ± 2.2	46.9 ± 0.9	10.6 ± 1.0	10.3 ± 0.4	19.4 ± 1.0	21.9 ± 1.8
As	(0.0019)	<0.05	(0.006)	<0.03	0.41 ± 0.05	0.42 ± 0.09
Se	0.11 ± 0.01	0.09 ± 0.04	1.1 ± 0.2	0.92 ± 0.06	0.4 ± 0.1	0.38 ± 0.04
Br	(12)	12.1 ± 0.2	(9)	8.5 ± 1.4	(1)	1.19 ± 0.17
Rb	(11)	13.1 ± 0.2	(1)	0.94 ± 0.06	(7)	8.4 ± 0.9
Sr		3.69 ± 0.10		0.82 ± 0.04		0.19 ± 0.04
Hg	0.0008 ± 0.0002	<0.1	0.001 ± 0.0008	<0.06	0.0060 ± 0.0007	<0.08
Pb	0.019 ± 0.003	<0.1	0.020 ± 0.010	<0.1	0.045 ± 0.010	0.10 ± 0.09

[a] Mass thickness of disks.
[b] Weight of area scanned.
[c] Recommended sample weight.
Source: From Reference 45.

known for geological materials; hence concentrations can be referred to that of one of the major elements.

Brissaud et al. [30] compared the SRIXE results with several standards. The results are shown in Table 6. Table 6 gives the recommended concentration, the SRIXE value, and the ratio of the two. It can be seen that agreement is quite good. In this case, as noted before, a comparatively large 1 mm beam was used. Lu et al. [47] used a microprobe with a beam size of 30×60 μm to analyze different specimens of feldspars. They compared their SRIXE results with values obtained using an electron probe and atomic absorption spectroscopy. The agreement with the electron probe was good, but the atomic absorption values tended to be systematically lower than the SRIXE values. Results for the comparison with the electron probe are shown for iron and strontium in Figure 18.

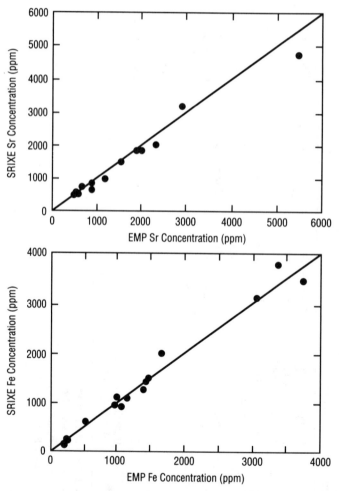

Figure 18 Comparison of concentration for Sr and Fe in feldspar obtained using the NSLS XRM with those obtained using an electron microprobe. The solid lines show the values expected for exact agreement between the two methods. The good agreement validates the use of the XRM method for geological analyses. (From Ref. 47.)

Table 6 16.6 and 21.7 keV SRXRF Analysis of Three International Geostandards[a]

Element	GSN			BEN			MICA-Fer		
		SRIXE			SRIXE			SRIXE	
	GST	16.6	21.7	GST	16.6	21.7	GST	16.6	21.7
K, %	3.84	3.54	3.34	1.15	1.05	0.86	7.26	6.10	5.89
Ca, %	1.78	1.68	1.66	9.85	8.98	8.74	0.31	0.15	0.07
Ti, %	0.41	0.41	0.38	1.57	1.54	1.49	1.50	1.35	1.47
V	65	125	188	235	185	360	135	42	257
Cr	55	45	117	360	346	407	90	125	134
Mn	433	479	551	1540	1662	1679	2695	3006	3200
Fe, %	2.63	3.13	3.12	8.99	9.66	9.20	17.96	18.95	19.15
Ni	34	36	35	267	320	278	35	39	—
Cu	20	16	25	72	62	85	4	—	14
Zn	48	60	66	120	117	129	1300	1320	1501
Ga	22	21	27	17	10	21	95	79	118
Rb	185	115	189	47	27	61	2200	1013	2696
Sr	570	469	905	1370	1550	1759	5	1	5
Y	19	—	22	30	—	34	25	—	50
Zr	235	—	296	265	—	343	800	—	1058
Nb	23	—	32	100	—	136	270	—	375
Ba	1400	—	1103	—	—	—	—	—	—
W	470	500	610	30	—	30	—	—	—
Pb	53	51	82	4	2	5	13	23	17
Th	44	11	13	11	2	3	150	51	50
U	8	—	4	2.4	—	2	60	—	33

[a] GST are the admitted values. Units are in ppm or %. The abbreviations, GSN, BEN, and MICA-Fer are the names of standards as given in K. Gavindaraju, *Geostandards Newletters* 3:3 (1979); 4:49 (1980); 8:173 (1984).
Source: From Reference 30.

For thin biological specimens, concentrations can be established by use of a sensitivity curve, such as that displayed in Figure 12. Corrections for differences in thickness can be made by normalizing to the intensity of the scattered incoherent peak if a monoenergetic beam is used or to the continuum comprising both incoherent and coherent scattering if a white beam is used [46].

Quantitation of SRIXE draws on many years of experience gained using tube-excited x-ray fluorescence. The main differences are when SRIXE is used with microbeams with dimensions on the micrometer scale. In this case strict attention must be given to the uniformity of the standards used and to the experience gained in calibration of the electron microprobe [48] and proton microprobe [49].

VII. SENSITIVITIES AND MINIMUM DETECTION LIMITS

The related questions of sensitivities and minimum detection limits have been addressed by calculations based on the known physical parameters of the XRM systems and by empirical determinations. The detailed understanding of the x-ray production process using synchrotron radiation is helpful in assessing the sensitivities and MDLs that can be achieved in SXRM. The results cited for the sensitivities and MDLs should be taken as representative of the current situation. The actual values depend on the particular experimental conditions, synchrotron ring currents, spatial resolutions, and so on, so that exact comparisons are not terribly meaningful.

Sparks [5], Gordon [26], Gordon and Jones [27], and Grodzins [28] presented calculations of the minimum detection limits expected using the second-generation synchrotron radiation sources, such as the NSLS, to produce x-ray microbeams. Many experimental determinations have been made for the two quantities. The results obtained by different groups using different types of XRMs are given here.

Figures 11 and 12 show the values obtained by Jones et al. [24] using the collimated microprobe at the NSLS. Their values are relevant to thin specimens with an organic matrix. Values for the calculated MDLs extrapolated from the work of Gordon and Jones [27] are included in Figure 12. It can be seen that the experimental values are in the same range as the calculations, although the experimental system was not the same as that considered theoretically. The sensitivities shown in Figure 12 show the agreement between predicted (solid curve) and measured (data points) counting rates. The curves were calculated from knowledge of the experimental arrangement, synchrotron energy spectrum, and specimen parameters. The experimental points show the measured values assuming the same specimen parameters.

Figure 19 shows the results obtained for the MDLs by Giauque et al. [41] using the Kirkpatrick-Baez XRM at the NSLS. The sensitivity curve was previously discussed in comparing sensitivities for the collimated and focused instruments (see Fig. 17). Jones et al. [20] compared the MDL values for the collimated white beam and monoenergetic focused beam approaches and showed that the MDL values were very similar. Ketelsen et al. [50] also compared MDL values for white beams and monoenergetic beams. The results are shown in Figure 20 and also demonstrate that the two approaches give comparable results. The

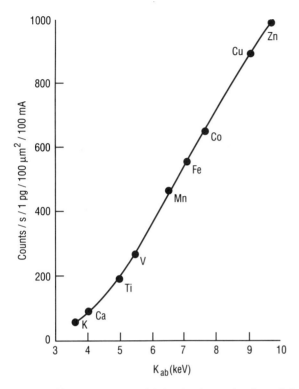

Figure 19 Relative sensitivity for determination of elements from K to Zn obtained using the LBL Kirkpatrick-Baez XRM at the NSLS X26 beam line. NIST (NBS) thin glass Standard Reference Materials 1832 and 1833 were used.

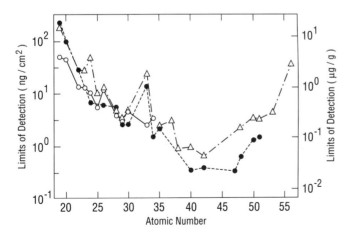

Figure 20 Detection limits observed for aerosol samples using the Hamburg synchrotron with a 70 mA stored beam of 3.7 GeV electrons for 300 s: (open triangles) white beam filtered with 8 mm of Al; (filled circles) a 31 keV monoenergetic beam, and (open circles) a 12.5 keV monoenergetic beam. (From Ref. 50.)

absolute values of the two experiments cannot be directly compared because of differences in the beam size and ring operating conditions.

VIII. BEAM-INDUCED DAMAGE

Passage of a photon beam through a specimen results in energy deposition through the photoelectric effect and Compton-scattering processes. This energy deposition results in a breaking of molecular bonds as the secondary electrons produced lose their energy by further ionization or scattering processes with other atoms. These effects have been examined in great detail for electron beams used in electron microscopy. Much less has been done for the XRM. This is not surprising since the field is much younger and less extensively developed than the field of electron microscopy. Another important reason is that the photon beam fluences employed thus far have been substantially smaller than those employed in the electron microscope, and as a result the magnitude of beam-induced damage has not been as important.

Biological and other organic materials are generally more susceptible to beam-induced damage than other materials. For this reason the discussion is limited to these materials. In the future, when more intense synchrotron x-ray sources are available, it will be necessary to expand the list of materials considered.

A qualitative calculation can be done to illustrate the energy deposition for various energy photon beams. Assume that the photon flux used for a typical fluorescence experiment is 10^6 photons/(s·μm^2). Typical run times are 10 minutes or less. The maximum photon fluence for current XRM instruments is thus about 6×10^8 photons/μm^2. It is interesting to note that these fluences are now starting to approach the range of the fluences found in use of the electron microprobe.

The x-ray dose needed to kill living biological systems has been examined in detail over the years [51–55]. A dose that exceeds 1 Gy is likely to cause serious damage to a biological cell or system.

Thus, there are limitations to the use of x-ray beams for the examination of living systems, as there are, of course, for all other beams. The limitations depend on the absorbed dose. One way that this examination is done is to measure the linear attenuation coefficients in CMT or in projection radiography experiments. Spanne [56] calculated the dose given to an object in obtaining an image with a signal-to-noise ratio of 5 for a water phantom with a contrasting detail with a diameter 0.005 of the phantom diameter. Spanne's results for objects with different diameters and compositions are shown in Figure 21. It can be seen that the dose is strongly energy dependent and that for relatively small objects (1 mm) the observed dose at the optimum energy is about 10^2 Gy. Examination of living systems using CMT with resolutions on a level of 5–10 μm thus may not be practicable. However, relaxation of the resolution criterion would reduce the dose to a point at which *in vivo* examination of living systems is feasible. It can also be seen from Figure 21 that the optimum energy for examining very thin specimens (cells) becomes very low. This represents a separate field of research and is not considered here. The use of projection radiography methods should also be useful. The dose is much less since fewer photons are needed to form an image. In this case the examination of living systems will be easier.

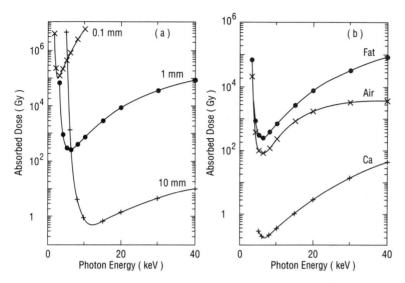

Figure 21　(a) Absorbed dose at the center of a circular water phantom for detection of an element of fat with a signal-to-noise ratio of 5 as a function of photon energy. The cylinder diameters are shown on the figure. (b) Absorbed dose at the center of 1 mm diameter water phantoms for detection of elements of fat, air, and calcium with a signal-to-noise ratio of 5. The element diameter is 0.005 of the phantom diameter of 5 μm. (From Ref. 56.)

If x-ray beams are used to make fluorescence measurements on nonliving systems, a loss of mass and possibly of trace elements can occur. Slatkin et al. [54] investigated changes in the morphology of human leukocytes and showed that severe damage resulted for fluences of 15 keV photons of about 10^{17}–10^{19} photons/cm². Figure 22 is a photomicrograph of the leukocytes after bombardment by fluences from 0.4 to 2.4×10^9 photons/cm². Damage to thin sections of kidney (10 μm) was much less. Mass loss in other types of organic materials was measured by Themner et al. [55]. They determined mass loss by measuring the change in scattered radiation counting rate as a function of total dose. The results they obtained for irradiation of a skin sample are shown in Figure 23. It can be seen that changes can be observed at dose values comparable to those used in many fluorescence experiments. Mass loss should therefore be carefully measured if the scattered radiation is used as a measure of the specimen areal density for quantitation purposes. The effects discussed can lead to the loss of trace elements and to errors in the assignment of concentrations, as is well known from the case of electron or proton microprobes. As in those cases, measurements need to be made of the yield of characteristic x-rays as a function of photon fluence for fluorescence measurements.

In summary, although beam-induced damage can be observed, it does not seem to have been a problem in experiments conducted to date. However, since much higher fluences are expected as focusing methods are improved and as the synchrotron source itself becomes more powerful, the beam damage effects will become of more central importance.

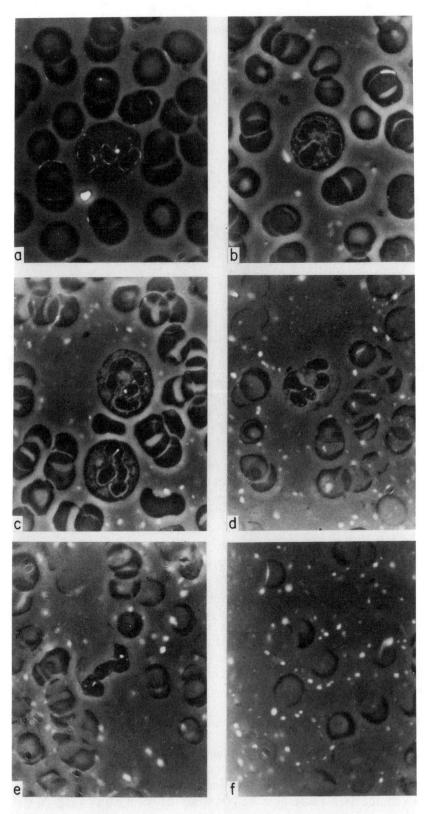

10 μm

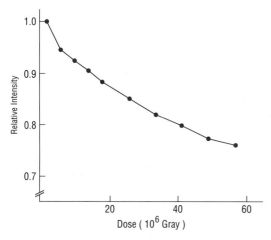

Figure 23 Mass loss observed for skin sample irradiated with a white synchrotron beam with energies from 4–17 keV is shown as a function of the dose. The mass loss was determined by observation of the scattered radiation intensity.

IX. APPLICATIONS OF SRIXE

There are now examples of the uses of the XRM in many different fields. A brief description of the diverse applications is given here to illustrate the rapid development of the field and to give an idea of the ways the XRM may be used in the future.

Brissaud et al. [30] made low-resolution measurements on a number of different materials. One case involved examination of three Gallic coins using synchrotron radiation and a comparison with the results obtained using proton-induced x-ray emission and neutron activation analysis. The x-ray and proton beams probe material close to the surface of the coin; the activation approach gives the bulk concentration. The results given in Figure 24 show differences in the concentrations found with these methods and also show that the activation approach is not applicable to all elements. This straightforward example shows that the synchrotron can be used to good effect in studying archeological and other materials with spatial resolutions of the order of 1 mm. It is a type of experiment, however, that should be viewed as a bridge between the use of conventional x-ray tube sources with high brilliance.

The distribution of trace elements in bone and other calcified tissues is generally of great interest since the concentrations of the essential trace elements are relevant to bone growth and disease. Therapeutic agents used to treat disease states may modify the trace element concentrations and deposit in particular patterns in the tissue themselves; finally, such toxic elements as lead are stored in the bone in localized patterns. Several experiments can be cited to illustrate these points.

Figure 22 Morphology of red blood cells is shown as a function of the fluence of incident 15 keV photons. The fluences for the different exposures are as follows: 0, 0.4, 1.1, 1.6, 2.0, and 2.4 × 10^{19} photons/cm^2 for frames a–f, respectively. (From Ref. 54.)

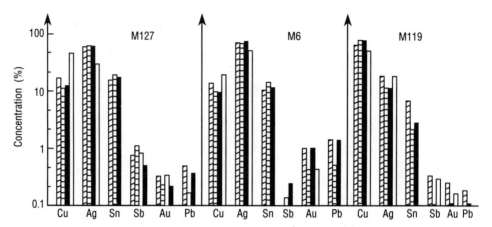

Figure 24 Chemical compositions of coins determined using SRIXE, PIXE (solid bar), and neutron activation analysis (open bar). The SRIXE work was done at 17 keV (diagonal crosshatch) and 35 keV (horizontal crosshatch). (From Ref. 30.)

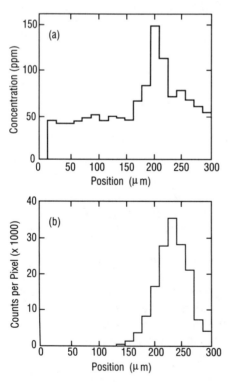

Figure 25 Scan over a 2 μm section of a neonatal hamster tooth germ showing the variation of zinc (a) and calcium (b). The zinc is located on the outside edge of the calcium distribution, but the reasons for difference remain to be determined. (From Ref. 57.)

Tros et al. [57] examined neonatal hamster tooth germs using both micro-PIXE and the XRM approach. In this work neonatal hamsters were treated with a fluoride compound commonly used to inhibit tooth decay. A tooth was obtained from the hamster after a 1 day interval, and thin sections were measured using the two types of beams. The fluorine concentrations were found using the proton microprobe at Amsterdam and the zinc distribution by XRM at Daresbury. The XRM was chosen for the determination of the essential trace elements because of its high sensitivity and low beam-induced damage. The proton beam was suited to the determination of the fluorine since the cross sections for inelastic proton scattering are high. The combined use of the two methods shows that ion beams can be used effectively in combination with the photon beams to cover all elements in the periodic table. The results of the XRM measurement of Zn and Ca are shown in Figure 25.

Gallium nitrate is another therapeutic element used to treat the accelerated bone resorption found in cancer patients. The mechanisms by which gallium interacts with the bone are as yet poorly understood. Bockman et al. used the NSLS X26 XRM to study sections of rat tibia obtained from animals treated with gallium nitrate [58]. A scan across the tibia from periosteum to endosteum giving the distribution of Ca, Ga, and Zn in a 12 μm thick section of bone is shown in Figure 26a. A two-dimensional map of the gallium and calcium distributions in a fetal rat ulna bone that was exposed to gallium (25 μM) in culture medium for 48 h is shown in Figure 26b. The bone structure is shown in an electron micrograph in the center section of Figure 26b. The noncalcified portion accumulates little gallium compared to the calcified portion, where the gallium accumulates in the metabolically active regions where new bone matrix is being formed. The method has been used to study the kinetics of gallium absorption in the bone for different doses and for different states of the bone metabolism. Changes in the concentrations of iron and zinc as a function of the gallium treatment indicate that it may be possible to infer particular enzymes that are targets for the gallium.

Lead can be used to show the uses of the XRM in the study of the effects of toxic metals. Lead is a major public health problem in many countries. Most of the lead in the body is stored in the skeleton and can be released to cause serious health effects under certain conditions. It is known to cause neurological and other problems in children and is associated with kidney and cardiovascular disease in adults. Understanding the deposition patterns and kinetics of lead in bone are therefore of great importance from a practical point of view. Jones et al. [59,60] reported measurements on lead distributions in the human tibia and in sections of deciduous teeth. Figure 27 shows the distribution of lead in such a tooth. The objective in this case is to attempt to correlate the distributions with the blood lead concentrations at birth from examination of the enamel (formed by the time of birth) and the dentine (later exposures). A knowledge of the time-integrated lead exposure can then be related to neurological deficits and other effects.

The XRM has been used for many different types of geological experiments. Many of the experiments used thin sections of rocks for studies of specific types of minerals in a heterogeneous matrix. Examples of this type of application are the study of zoned carbonate gangue cements found in Tennessee [61]. The XRM measurements were used in an attempt to interpret the effects of trace amounts of Mn and Fe on cathodoluminescence in carbonates. Further, the variations in

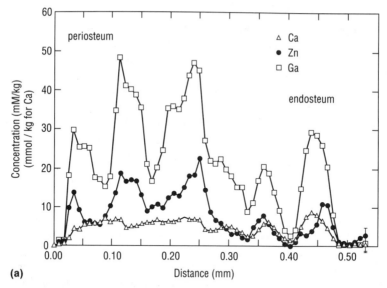

Figure 26 (a) Scan over a thin section of the tibia diaphysis of a rat treated with gallium nitrate, showing the distribution of the Ca, Zn, and Ga. Some of the variations come from irregularities in specimen thickness. The spatial distribution is a function of the gallium nitrate treatment of the rat. The step size was 9 μm. (b) Map of the distribution of gallium (left) and calcium (right) in a fetal rat ulna after exposure to gallium (25 μM) in a culture medium. The light regions have the highest elemental concentrations. A scanning electron micrograph (center) shows the bone structure in the same region. The dark portions at the top of the figure are the noncalcified cartilaginous portions of the bone, which accumulates little gallium. Most of the gallium is distributed in metabolically active regions of the metaphysis and diaphysis.

the trace element concentrations should aid in deducing the source and direction of flow of fluids responsible for the formation of dolomites.

Small particles can also be effectively examined using XRM. Sutton and Flynn [62,63] carried out a series of experiments dealing with analyses of extraterrestrial particles. These analyses were carried out on particles of sizes less than 100 μm. Minimum detection limits less than 10 ppm were obtained for particles less than 20 μm in size using the BNL X26 XRM. Tuniz et al. [64] also used this equipment for examination of fly ash taken from different types of power plants and incinerators. They measured elemental composition in individual fly ash particles with sizes down to a few micrometers and also made two-dimensional maps of the distributions of the elements in 10 μm thick sections of particles produced by a lapping technique. The results may be useful in verifying models for the production of toxic compounds in incinerators based on the presence of specific metals in the ash [65–67].

Fluorescence can also be used to advantage in the material sciences. For example, Isaacs et al. [68] studied the concentration gradients produced in a solution during the localized corrosion of stainless steels. The combination of high

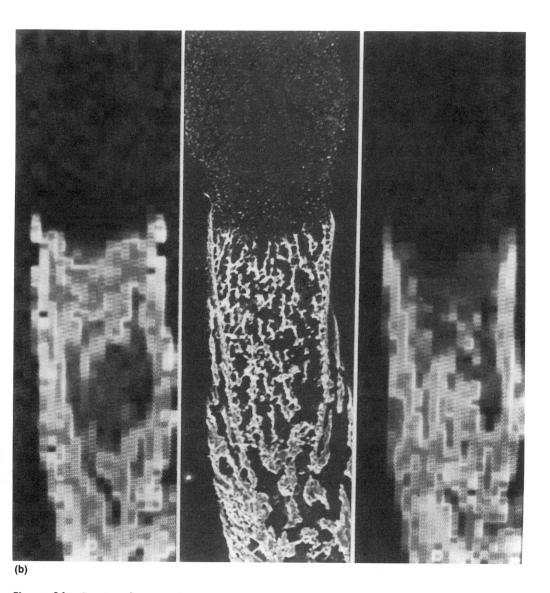

(b)

Figure 26 *Continued*

spatial resolution and excellent detection sensitivity enabled them to study the variation in the nickel concentration above a stainless steel surface immersed in a bulk chloride electrode. Figure 28 shows the electrochemical cell used in the work and the observed variation of nickel concentration above the stainless steel surface. From study of the concentration gradients, Isaacs et al. [68] were able to identify effects arising from silicon in the steel. In situ studies of kinetic effects should be of increasing interest, not only in corrosion measurements, but also for other types of chemical reactions.

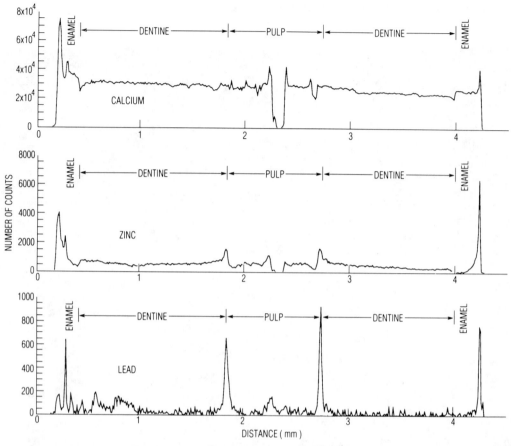

Figure 27 The relative concentrations of Ca, Zn, and Pb are for a scan across a section of a child's deciduous tooth that was thick compared to the absorption of the characteristic x-rays. The variation of the lead in the scan may be useful in the future in understanding the time dependence of lead exposure and uptake by the child. The spatial resolution and step size were both about 10 μm. (From Ref. 60.)

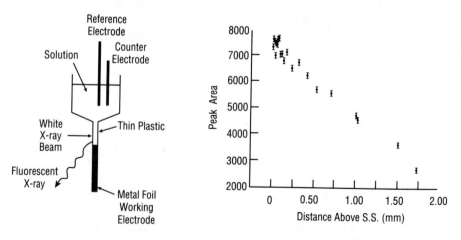

Figure 28 Apparatus used for studying the corrosion of stainless steel in a chloride environment (left). The spatial distribution of nickel observed above the stainless steel is also shown as a function of distance above the metal/liquid interface. (From Ref. 68.)

X. TOMOGRAPHY

Computed microtomography is an important approach to nondestructive analysis and has been extensively developed using conventional x-ray sources. The synchrotron source gives substantial advantages because of its high brilliance and continuous x-ray spectrum. The superior properties of the synchrotron source have led to CMT instrumentation capable of superior spatial resolution and shorter data acquisition times.

Almost all the work that has been done has concentrated on the use of CMT in the attenuation mode, in which determinations are made of the linear attenuation coefficients. The absorption mode is a simple and effective approach to CMT imaging. It can be used on samples with a variety of sizes and attenuation coefficients by choosing the appropriate x-ray energy. The use of SRIXE is more restricted since the specimen must be small enough to allow the escape of the characteristic x-ray of interest. As is the case with EXAFS and XANES, however, the absorption and emission (SRIXE) approaches are complementary, and both need to be available as part of the basic XRM.

Refinements to the method are necessary to obtain better information on the elemental composition of the materials. The absorption approach can be refined by producing tomograms above and below an elemental x-ray absorption edge. Subtraction of the tomograms gives the concentration of that element. The difference technique is valuable for study of major and minor elements to the 0.1% level.

The imaging of trace elements often necessitates the use of SRIXE so that specific elements are selected through the detection of their characteristic x-rays. The detection system must be chosen for high efficiency and high counting rate capability. The type of specimen that can be investigated is limited by the attenuation of the fluorescent x-rays in the sample being investigated. The sample dimension is strongly constrained because of this.

Several groups have done CMT work with synchrotrons. Flannery et al. [69] developed a third-generation system at the NSLS. They used a x-ray magnification system with a scintillation detector coupled to an image intensifier to produce images with a claimed resolution down to 1 μm. A similar approach was used by Bonse et al. [70] at SSRL. Spanne and Rivers [71] demonstrated a first-generation system at the NSLS X26 beam line. Later work with the apparatus produced images with a spatial resolution of 1×1 μm and a slice thickness of 5 μm, quite comparable to the results from third-generation devices. The first-generation approach takes longer to produce images but has the advantage that beam-scattering effects in the sample are eliminated and SRIXE measurements can be performed to produce elemental maps. In the third-generation systems, elemental maps are made by subtracting images taken above and below the absorption edge of interest. The MDL for such an approach is about 0.1%.

Spanne [72] carried out a pilot study with the aim of evaluating the potential for mapping of light elements at the cellular level in the rat sciatic nerve using fluorescence CMT. A comparison of the mean free path for characteristic x-rays from potassium and typical sciatic nerve sizes shows that it is feasible to make corrections for the attenuation of the potassium K x-rays in the nerve. The computed emission tomogram of the distribution of potassium in the epineurium of a

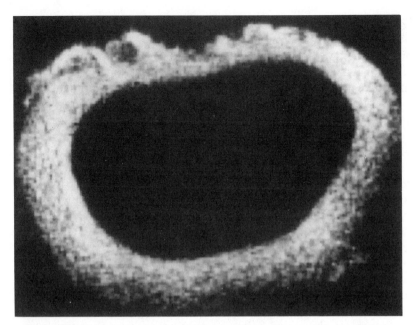

Figure 29 Computed-emission tomogram showing the distribution in the epineurium of a freeze-dried rat sciatic nerve. The pixel size was 3 × 3 μm and a slice thickness of 5 μm. The matrix size was 175 × 175 pixels. (From Ref. 72.)

rat sciatic nerve given in Figure 29 illustrates this point. Note that the short escape depth for the potassium x-rays that are observed necessitates special measures during the reconstruction of the image. Saubermann [73] points out that fluorescence CMT makes possible studies of elemental distributions in unsectioned samples. Examination of unsectioned samples also makes feasible *in vitro* analysis of sections of nerves several millimeters long. Longitudinal distributions of elements can then be conveniently studied by scanning at different heights, and it is even possible to return to a previously mapped region for a more detailed examination if steep concentration gradients are discovered. Longitudinal concentration gradients have been demonstrated in the rat sciatic nerve, although with a very poor longitudinal resolution, and may be of significance in nerve injury [74,75].

The active use of CMT is just beginning. Even now, it is apparent that both absorption and emission approaches are needed and that both approaches are required at any synchrotron XRM facility.

XI. EXAFS AND XANES

EXAFS and the related XANES have been widely applied to give information on the chemical state of elements in many different materials [2,76]. Many of the experiments that have been carried out have used relatively large x-ray beams and thick specimens to make possible absorption measurements on a time scale of minutes. This approach is not useful for elemental concentrations less than about 0.1–1%. The use of EXAFS and XANES at lower concentrations can be achieved by fluorescent x-ray detection. Cramer et al. [77] developed a 13-element Si(Li) x-ray detector that gives a high effective detection efficiency for this type of application. Other workers developed the means of acquiring EXAFS spectra on a millisecond time scale [78] and made measurements of chemical state on a finer spatial scale [79]. The future development of x-ray microscopy can thus be seen to include the use of SRIXE with EXAFS and XANES and the development of new techniques to make it possible to work with improved MDLs and beam sizes at the micrometer level.

The NSLS X26 group employed a simple channel-cut silicon monochromator with an energy resolution of about 1.1 eV for several demonstration XANES experiments. The beam was first defined with a four-jawed aperture whose size could be adjusted using computer-controlled stepping motor drivers. This was followed by the monochromator placed about 10 cm upstream of the target. The beam moved vertically on the target by about 60 μm during the scan. This was not important for the resolutions used but could be easily compensated for by a correlated motion of the target to keep the same spot under the beam. The first test used a thick specimen of NIST SRM 1570 Spinach Leaves, which contained

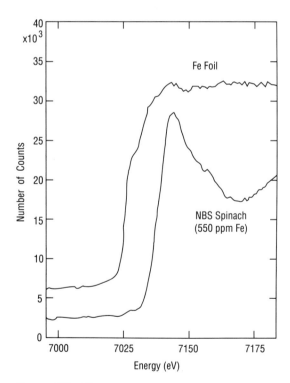

Figure 30 Fluorescent XANES spectrum obtained for Fe at the 550 pm level in a thick NIST SRM 1570 (spinach) sample compared to the spectrum obtained under the same conditions for an iron foil.

550 ppm iron. The iron x-rays were detected with the XRM equipment just described. The work was done with a beam size of 2 mm². The results of a scan of a pure iron specimen (beam size 200 × 200 μm) and the spinach leaves are shown in Figure 30. The spectra agree well with the results of a similar scan done on the NSLS X19 beam line using the 13-element Si(Li) detector. Extrapolation from these initial values showed that work with beam sizes to 100 μm and perhaps lower was feasible with the existing equipment for this particular target.

A more stringent demonstration is the use of thinner specimens. For this purpose measurements were made on the chromium contained in olivine and pyroxene components in a 30 μm section of lunar mare basalt 15555 from Apollo 15 and in a 10 μm section of a rat kidney. The lunar basalt study was undertaken since the oxidation state of the chromium could shed light on conditions existing at the time of the formation of the mineral studied [80]. The rat kidney measurement was needed to cast light on nephrotoxic effects resulting from environmental exposures. It is hypothesized that the oxidation state of the chromium changed during its passage from lungs to kidneys with related implications for health effects.

The basalt specimen contained chromium at a level of about 1000 ppm and the kidney at a level of about 50 ppm. Figure 31 shows the XANES spectrum for chromium in pyroxene and olivine contained in the lunar basalt taken with a beam resolution of 200 × 200 μm. Figure 32 shows the spectrum obtained for the rat kidney, but in this case with a 1 × 1 mm beam size. The beam was positioned over the medullar portion of the kidney. The XANES spectra show that chromium

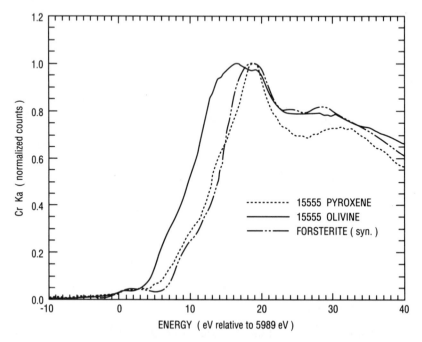

Figure 31 Fluorescent XANES spectrum obtained for Cr at the 1000 pm level in a 30 μm thick section of pyroxene and olivine contained in lunar basalt 15555. The spatial resolution was 200 × 200 μm. (From Ref. 80.)

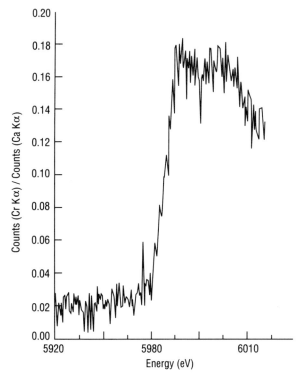

Figure 32 Fluorescent XANES spectrum obtained for Cr at the 50 ppm level contained in a thin section of rat kidney. The spatial resolution was 1×1 mm. The spectrum shows that little Cr (VI) is present in this portion of the kidney.

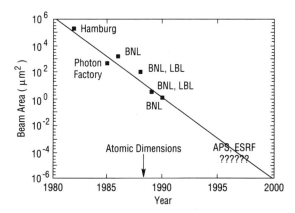

Figure 33 The spatial resolution (expressed as the beam area) obtained with high-energy x-ray microscopes is plotted as a function of time. The best spatial resolutions in 1990 are around 1–4 μm^2. If improvements were to continue at the same rate in the future, the resolution will approach atomic dimensions around the year 2000.

exists primarily in the 2^+ and 3^+ states in the lunar olivines but in the 3^+ state in the kidney.

The combination of SRIXE with EXAFS and XANES has barely begun to be applied to experimental investigations, but it has already been shown that measurements can be carried out with moderate spatial resolution and fairly good MDL values. The new combination of SRIXE and microbeam trace element detection with EXAFS and XANES will be invaluable in the study of heterogeneous materials. Several groups are developing the methodologies used in time- and spatially-resolved EXAFS and XANES which will include application of the SRIXE technique and result in a major addition to the XRM field.

XII. FUTURE DIRECTIONS

What does the future hold for synchrotron XRM? Several different developments can be predicted for the next decade (1992–2001), with some confidence.

First, continuation of the present types of experiments will go on at an expanded rate as the virtues of the method become better known. The applications will benefit from enhancements in the facilities, which will bring spatial resolutions to 1 μm^2 for two-dimensional maps or to 1 μm^3 for CMT. Relatively minor improvements to the existing systems are needed to achieve this level of resolution. MDL values should also improve somewhat with the introduction of improvements in the x-ray detector systems, again using known techniques. There will be a further merging of XRM techniques with those developed for EXAFS and XANES.

Second, development of new focusing methods will begin and this will lead to the achievement of spatial resolutions of the order of 0.1 μm by 1996–2001. An example is the production of zone plates suitable for focusing of 8 keV x-rays by Bionta et al. [81]. The use of SRIXE in this case might be restricted by the typical brilliance values found in second-generation synchrotrons.

Third, a new era will be signaled by the commissioning of third-generation synchrotron sources, such as the Advanced Photon Source at Argonne National Laboratory and the European Synchrotron Radiation Facility at Grenoble. These systems will produce beams with improvements of three orders of magnitude when using undulator insertion devices and with higher energies when using either bending magnets or wigglers. A general discussion of the impact of these new rings on x-ray microscopy has been given by Sparks and Ice [82]. They find detection limits for fluorescence detection for a 1 μm^2 beam and a data acquisition time of 1 s for 10^8 atoms in a solid and for 10^3 free atoms. A discussion of the impact of the APS on the geosciences was given by Sutton et al. [83]. Results from the first tests of an undulator designed for the APS have been reported by Rivers et al. [84].

Figure 33 summarizes the time dependence for the synchrotron XRM spatial resolution. The area resolution has decreased exponentially over the past 10 years. Extrapolating into the future, assuming the same rate of improvement, leads to an estimate of an area resolution corresponding to a beam size of roughly 10 nm when the third-generation rings are in operation and 1 nm by the year 2000. It will be interesting to see how far this extrapolation deviates from the reality of the year 2001! Whatever the exact details, it is obvious that the field of synchrotron XRM will be active and evolve very rapidly.

ACKNOWLEDGMENTS

I am particularly indebted to my colleagues for stimulating discussions and interactions that have influenced my views of x-ray microscopy. Among them are R. S. Bockman, R. D. Giauque, Y. Gohshi, L. Grodzins, A. L. Hanson, J. B. Hastings, J. G. Pounds, M. L. Rivers, A. J. Saubermann, J. V. Smith, P. Spanne, S. R. Sutton, A. C. Thompson, J. H. Underwood, and R. D. Vis. This work was supported in part by Processes and Techniques Branch, Division of Chemical Sciences, Office of Basic Energy Sciences, U.S. Department of Energy, Contract No. DE-AC02-76CH00016 for development of analytical techniques, and by the National Institutes of Health Biotechnology Research Resource Grant No. P41RR01838.

REFERENCES

1. C. Kunz, *Synchrotron Radiation: Techniques and Applications*, Springer-Verlag, Berlin, 1979.
2. H. Winick and S. Doniach (Eds.), *Synchrotron Radiation Research*, Plenum Press, New York, 1980.
3. G. Margaritondo, *Introduction to Synchrotron Radiation*, Oxford University Press, New York, 1988.
4. Kwang-Je Kim, *X-Ray Data Booklet* (D. Vaughan, Ed.), Lawrence Berkeley Laboratory Center for X-Ray Optics, Berkeley (1986), Section 4.
5. C. J. Sparks, Jr., *Synchrotron Radiation Research* (H. Winick and S. Doniach, Eds.), Plenum Press, New York, (1980), p. 459.
6. J. R. Chen, E. C. T. Chao, J. A. Minkin, J. M. Back, K. W. Jones, M. L. Rivers, and S. R. Sutton, *Nucl. Instr. Methods Phys. Res.* B49:533 (1990).
7. R. D. Vis and F. van Langevelde, *Nucl. Instr. Methods Phys. Res.* B54:417 (1991).
8. A. Knöchel, *Fresenius Z. Anal. Chem.* 337:614 (1990).
9. R. D. Vis, *Fresenius Z. Anal. Chem.* 337:622 (1990).
10. K. W. Jones and B. M. Gordon, *Anal. Chem.* 61(5):341A (1989).
11. D. Sayre, M. Howells, J. Kirz, and H. Rarback (Eds.), *X-ray Microscopy*, Vol. II, Springer-Verlag, Berlin, 1988.
12. H. Ade, J. Kirz, S. L. Hulbert, E. D. Johnson, E. Anderson, and D. Kern, *Appl. Phys. Lett.* 56:1841 (1990).
13. H. Rarback, F. Cinott, C. Jacobsen, J. M. Kenney, J. Kirz, and R. Rosser, *Biol. Trace Element Res.* 13:103 (1987).
14. J. M. Kennedy, C. Jacobsen, J. Kirz, H. Rarback, F. Cinotti, W. Thomlinson, and G. Schidlovsky, *J. Microsc.* 138(3):321 (1985).
15. F. R. Elder, A. M. Gurenwitsch, R. V. Langmuir, and H. C. Pollock, *Physiol. Rev.* 71:829 (1947).
16. D. H. Tombulian and P. L. Harman, *Physiol. Rev.* 102:1423 (1956).
17. A. L. Hanson, *Nucl. Instr. Methods Phys. Res.* A290:167 (1990).
18. H. Winick, *Nucl. Instr. Methods* 291:401 (1990); H. Winick, *Synchr. Rad. News* 2(2):25 (1989).

19. A. Jackson, *Synchr. Rad. News 3*(3):13 (1990); J. Fuggle, *Synchr. Rad. News 3*(1):24 (1990).
20. A. Knöchel, W. Petersen, and G. Tolkien, *Nucl. Instr. Methods 208*:659 (1983).
21. K. W. Jones, W. M. Kwiatek, B. M. Gordon, A. L. Hanson, J. G. Pounds, M. L. Rivers, S. R. Sutton, A. C. Thompson, J. H. Underwood, R. D. Giauque, and Y. Wu, *Advances in X-ray Analysis 31* (C. S. Barrett, J. V. Gilfrich, R. Jenkins, J. C. Russ, J. W. Richardson, Jr., and P. K. Predicki, Eds.), Plenum Press, New York (1988), p. 59.
22. P. Horowitz and J. Howell, *Science 178*:608 (1972).
23. W. M. Kwiatek, A. L. Hanson, and K. W. Jones, *Nucl. Instr. Methods Phys. Res. B50*:347 (1990).
24. K. W. Jones, B. M. Gordon, G. Schidlovsky, P. Spanne, Xue Dejun, R. S. Bockman, and A. J. Saubermann, in *Microbeam Analysis 1990* (D. B. Williams, P. Ingram, and J. R. Michael, Eds.), San Francisco Press, San Francisco, 1990, pp. 401–404.
25. C. J. Sparks, S. Raman, H. L. Yakel, R. V. Gentry, and M. O. Krause, *Phys. Rev. Lett. 38*:205 (1977); C. J. Sparks, S. Raman, E. Ricci, R. V. Gentry, and M. O. Krause, *Phys. Rev. Lett. 40*:507 (1978).
26. B. M. Gordon, *Nucl. Instr. Methods 204*:223 (1982).
27. B. M. Gordon and K. W. Jones, *Nucl. Instr. Methods Phys. Res. B10/11*:293 (1985).
28. L. Grodzins, *Neurotoxicology 4*:23 (1983).
29. P. Chevallier, C. Jehanno, M. Maurette, S. R. Sutton, and J. Wang, *J. Geophys. Res. 192*(B4):E649 (1987).
30. I. Brissaud, J. X. Wang, and P. Chevallier, *J. Radioanal. Nucl. Chem. 131*:399 (1989).
31. S. Hayakawa, A. Iida, S. Aoki, and Y. Gohshi, *Rev. Sci. Instrum. 60*:2452 (1989).
32. S. Hayakawa, Y. Gohshi, A. Iida, S. Aoki, and M. Ishikawa, *Nucl. Instr. Methods Phys. Res. B49*:555 (1990).
33. K. W. Jones, B. M. Gordon, A. L. Hanson, J. B. Hastings, M. R. Howells, and H. W. Kraner, *Nucl. Instr. Methods Phys. Res. B3*:225 (1984).
34. F. Van Langvelde, G. H. J. Tros, D. K. Bowen, and R. D. Vis, *Nucl. Instr. Methods Phys. Res. B49*:544 (1990).
35. F. Van Langevelde, D. K. Bowen, G. H. J. Tros, R. D. Vis, A. Huizing, and D. K. G. de Boer, *Nucl. Instr. Methods Phys. Res. A292*:719 (1990).
36. F. Van Langevelde, G. H. J. Tros, D. K. Bowen, and R. D. Vis, Non-imaging optics for photon probe microanalysis at the SRS, Daresbury (UK), Proceedings of XIIth IXCOM, Crakow, Poland, in press.
37. P. Kirkpatrick and A. V. Baez, *J. Opt. Soc. Am 39*:766 (1948).
38. A. C. Thompson, Y. Wu, J. H. Underwood, and T. W. Barbee, Jr., *Nucl. Instr. Methods Phys. Res. A255*:603 (1987).
39. J. H. Underwood, A. S. Thompson, Y. Wu, and R. D. Giauque, *Nucl. Instr. Methods Phys. Res. A266*:296 (1988).
40. A. C. Thompson, J. H. Underwood, Y. Wu, R. D. Giauque, K. W. Jones, and M. L. Rivers, *Nucl. Instr. Methods Phys. Res. A266*;318 (1988).
41. R. D. Giauque, A. C. Thompson, J. H. Underwood, Y. Wu, K. W. Jones, and M. L. Rivers, *Anal. Chem. 60*:855 (1988).
42. Y. Wu, A. C. Thompson, J. H. Underwood, R. D. Giauque, K. Chapman, M. L. Rivers, and K. W. Jones, *Nucl. Instr. Methods Phys. Res. A291*:146 (1990).
43. V. B. Baryshev, N. G. Gavrilov, A. V. Daryin, K. V. Zolotarev, G. N. Kulipanov, N. A. Mezentsev, and Y. V. Terekhov, *Rev. Sci. Instrum. 60*:2456 (1989).
44. M. L. Rivers, S. R. Sutton, and K. W. Jones, *X-ray Microscopy*, III, Springer-Verlag, (1992).
45. R. D. Giauque, J. M. Jaklevic, and A. C. Thompson, *Anal. Chem. 58*:940 (1986).
46. R. D. Giauque, R. B. Garrett, and L. Y. Goda, *Anal. Chem. 51*:511 (1979).

47. F. Q. Lu, J. V. Smith, S. R. Sutton, M. L. Rivers, and A. M. Davis, *Chem. Geol.* *75*:123 (1989).
48. J. J. Hren, J. I. Goldstein, and D. C. Joy (Eds.), *Introduction to Analytical Electron Microscopy*, Plenum Press, New York, 1979.
49. S. E. A. Johansson and J. L. Campbell (Eds.), *PIXE: A Novel Technique for Elemental Analysis*, John Wiley & Sons, New York, 1988.
50. P. Ketelsen, A. Knöchel, and W. Peterson, *Fresenius Z. Anal Chem. 323*:807 (1986).
51. J. Kirz and D. Sayre, in *Synchrotron Radiation Research* (H. Winick and S. Doniach, Eds.), Plenum Press, New York (1980), p. 277.
52. D. Sayre, J. Kirz, R. Feder, D. M. Kim, and E. Spiller, *Ann. N. Y. Acad. Sci. 306*:286 (1978).
53. D. Sayre, J. Kirz, R. Feder, D. M. Kim, and E. Spiller, *Ultramicroscopy 2*:337 (1977).
54. D. N. Slatkin, A. L. Hanson, K. W. Jones, H. W. Kraner, J. B. Warren, and G. C. Finkel, *Nucl. Instr. Methods 227*:378 (1984).
55. K. Themner, P. Spanne, and K. W. Jones, *Nucl. Instr. Methods Phys. Res. B49*:52 (1990).
56. P. Spanne, *Phys. Med. Biol. 34*(6):679 (1989).
57. G. H. J. Tros, F. Van Langevelde, and R. D. Vis, *Nucl. Instr. Methods Phys. Res. B50*:343 (1990).
58. R. S. Bockman, M. A. Repo, R. P. Warrell, Jr., J. G. Pounds, G. Schidlovdky, B. M. Gordon, and K. W. Jones, *Proc. Natl. Acad. Sci. U S A 87*:4149 (1990).
59. K. W. Jones, G. Schidlovsky, D. E. Burger, F. L. Milder, and H. Hu, in *In Vivo Body Composition Studies, Recent Advances* (S. Yasumura, J. E. Harrison, K. G. McNeill, A. D. Woodhead, and F. A. Dilmanian, Eds.), Plenum Press, New York (1990), p. 281.
60. K. W. Jones, G. Schidlovsky, P. Spanne, Xue Dejun, R. S. Bockman, M. B. Rabinowitz, P. B. Hammond, R. L. Bornschein, and D. A. Hoeltzel, *X-Ray Microscopy*, III, Springer-Verlag, (1992).
61. O. C. Kopp, D. K. Reeves, M. L. Rivers, and J. V. Smith, *Chem. Geol. 81*:337 (1990).
62. S. R. Sutton and G. J. Flynn, Stratospheric particles: Synchrotron x-ray fluorescence determination of trace element contents, in Proceedings of the 18th Lunar and Planetary Science Conference, Houston (1988), pp. 607–614.
63. G. J. Glynn and S. R. Sutton, Synchrotron x-ray fluorescence analyses of stratospheric cosmic dust: New results for chondritic and nickel-depleted particles, in Proceedings of the 20th Lunar and Planetary Science Conference, Houston (1990), pp. 335–342.
64. C. Tuniz, K. W. Jones, M. L. Rivers, S. R. Sutton, and Sz. Török, *Environ. Sci. Technol.*, to be published.
65. F. W. Karasek and L. C. Dickson, *Science 237*:754 (1987).
66. H. Hagenmaler, M. Kraft, H. Brunner, and R. Haag, *Environ. Sci Technol. 21*:1080 (1987).
67. E. R. Altwicker, J. S. Schonberg, R. K. N. V. Konduri, and M. S. Milligan, *Hazardous Waste Hazardous Materials 7*(1):73 (1990).
68. H. S. Isaacs, A. J. Davenport, J. H. Cho, A. L. Hanson, and M. L. Rivers, in National Synchrotron Light Source Annual Report 1990, BNL 52272, 1991, p. 348.
69. B. P. Flannery, H. W. Deckman, W. G. Roberge, and K. L. D'Amico, *Science 237*:1439 (1987).
70. U. Bonse, Q. Johnson, M. Nichols, R. Nusshardt, S. Krasmaki, and J. H. Kinney, *Nucl. Instr. Methods Phys. Res. A245*:644 (1986); J. H. Kinney, Q. C. Johnson, R. A. Saroyan, M. C. Nichols, U. Bonse, R. Nusshardt, and R. Pahl, *Rev. Sci. Instr. 59*:196 (1988).

71. P. Spanne and M. L. Rivers, *Nucl. Instr. Methods Phys. Res. B24/25*:1063 (1987).
72. P. Spanne, Abstract of invited talk at Bioscience 90, Swedish National Convention of Bio-Scientists, Malmo, Sweden, April 23–26, 1990, BNL 44369, 1991; see also K. W. Jones, R. S. Bockman, B. M. Gordon, M. L. Rivers, A. J. Saubermann, G. Schidlovsky, and P. Spanne, in *XRF and PIXE Applications in Life Science* (R. Moro and R. Cesareo, Eds.), World Scientific Publishing, Singapore (1990), p. 163.
73. A. J. Saubermann, private communication (1989).
74. R. M. LoPachin, Jr., J. Lowery, J. Eichberg, J. B. Kirkpatrick, J. Cartwright, Jr., and A. J. Saubermann, *J. Neurochem. 51*(3):764 (1988).
75. R. M. LoPachin, Jr., V. R. LoPachin, and A. J. Saubermann, *J. Neurochem. 54*:320 (1990).
76. D. C. Koningsberger and R. Prinz (Eds.), *X-ray Absorption*, John Wiley & Sons, New York, 1988.
77. S. P. Cramer, O. Tench, M. Yocum, and G. N. George, *Nucl. Instr. Methods Phys. Res. A266*:586 (1988).
78. H. Tolentino, F. Baudelet, E. Dartyge, A. Fontaine, A. Lena, and G. Tourillon, *Nucl. Instr. Methods Phys. Res. A289*:307 (1990).
79. A. Iida, M. Takahashi, K. Sakurai, and Y. Gohshi, *Rev. Sci. Instrum. 60*:2458 (1989).
80. S. R. Sutton, K. W. Jones, B. Gordon, M. L. Rivers, and J. V. Smith, *Lunar and Planetary Science XXII*, Lunar and Planetary Institute, Houston (1991), p. 1365.
81. R. M. Bionta, E. Ables, O. Clamp, O. D. Edwards, P. C. Gabriele, K. Miller, L. L. Ott, K. M. Skulina, R. Tilley, and T. Viada, *Opt. Eng. 29*:576 (1990).
82. C. J. Sparks and G. E. Ice, X-ray microprobe-microscopy, in T. A. Carlson, M. O. Krause, and S. T. Manson, eds., *AIP Conference Proceedings 215: X-ray and Inner Shell Processes*, AIP, Knoxville, TN, 1990, pp. 770–786.
83. S. R. Sutton, M. L. Rivers, K. W. Jones, and J. V. Smith, *Synchrotron X-Ray Sources and New Opportunities in the Earth Sciences*, Workshop Report, ANL/APS-TM-3, p. 93, April 1988.
84. M. L. Rivers, Stephen R. Sutton, and B. M. Gordon, *Mat. Res. Soc. Symp. Proc. 143*:285 (1989).

9

Total Reflection XRF

Heinrich Schwenke and Joachim Knoth   *GKSS Forschungszentrum,*
Geesthacht, Germany

I. INTRODUCTION

Total reflection x-ray fluorescence analysis (TXRF) is a variant of energy-dispersive XRF, which differs from conventional XRF in two essential features:

1. The primary beam is incident to the specimen at a glancing angle less than or near to the critical angle at which total reflection of x-rays occurs.
2. The primary beam is incident on a plane, smooth surface, which serves either as sample support (thin-film analysis) or is itself the object to be examined (surface analysis).

In either case, the effect of external total reflection of x-rays is exploited to prevent the primary radiation from penetrating into the respective surface in the ordinary, uncontrolled manner.

   A description of an experimental TXRF setup was first published by Yoneda and Horiuchi [1] in 1971, and the method has been further pursued by Aiginger and Wobrauschek [2] since 1974. Even at the early stages, the capability of TXRF to detect very small amounts to the nanogram level was demonstrated.

   Subsequently, the geometry for thin-film analysis has undergone substantial technical development. However, the principle remains unchanged: a well-collimated beam from an x-ray tube is aimed in grazing incidence at an optical flat that serves as a carrier for samples placed on the surface in the form of a thin film. Because the intensity of an x-ray beam grazing an interface at a subcritical angle is drastically reduced inside the reflecting material, the scattered and fluorescence radiation contributed by the carrier is less than by any other optical geometry. After years of technical development, fluorescence spectra can now be obtained with such low background levels that picogram quantities can be detected in samples with favorable matrices.

In the case of surface analysis, the total reflection arrangement is used to achieve high selectivity with regard to the atoms in the near surface layer. This results from the low penetration depth of the primary beam under total reflection conditions.

In brief, the two branches of TXRF analysis exploit two quite distinct features of the total reflection phenomenon: the high reflectivity of the substrate for the analysis of films placed on an optical flat, and the low penetration depth of the primary radiation for surface analysis. Both effects are described later on the basis of the physics of x-ray reflection.

II. PHYSICS OF TOTAL REFLECTION OF X-RAYS

The effect of the total reflection mode of excitation in XRF can be characterized by means of only three parameters: the critical angle, the reflectivity, and the penetration depth. These fundamental parameters of total reflection are supplied by the physics of electromagnetic radiation. The complexity and the level of the theoretical tools needed for their description are determined by the extent to which the real properties of the interfaces and materials to be examined are considered.

Experimentally, total reflection of x-rays was first discovered by Compton in 1923 [3]. Since then many workers have investigated the total reflection phenomenon, both experimentally and theoretically. All energy regions have been covered, and many materials have been studied. Most investigations have been concerned with the determination of reflectivity and verification of the Fresnel equations relating reflectivity to the angle of glancing incidence. The Fresnel equations are based on an approximation of classical dispersion theory [4]. They have been derived assuming a perfectly flat and smooth interface between homogeneous media. Even though a real surface is, in general, rough on a microscopic scale and scatters x-rays as well as reflecting them, the experimental results have shown sufficient agreement for the particular application described here. Hitherto, all TXRF instruments have been designed and operated on the basis of the classical dispersion theory and the Fresnel equations.

A. Critical Angle

The total reflection of x-rays is based on the fact that the real part of the complex index of refraction of matter

$$n' = 1 - \delta - i\beta$$

is smaller than unity with $\delta \simeq 10^{-5}$–10^{-6} in the x-ray energy regime, whereas the refractive index of vacuum (and, for practical purposes, air) is $n' = 1$. β describes the attenuation of x-rays in matter. It provides the damping that is an essential feature of the grazing incidence arrangement and is given by

$$\beta = \frac{\mu^*\lambda}{4\pi} \tag{1}$$

where:

μ^* = linear absorption coefficient
λ = wavelength of the x-radiation

In the approximation of the classical dispersion theory for regions far from any absorption edge of the material, the variation of δ with x-ray energy is given by

$$\delta = \frac{n_e e^2 \lambda^2}{2\pi m c^2} \tag{2}$$

where:

n_e = electron density
e = electron charge
m = mass of the electron
c = velocity of light

The electron density n_e is described by the properties of the target material:

$$n_e = N_A \rho \frac{Z}{A} \tag{3}$$

where:

N_A = Avogadro's number
ρ = density
Z = atomic number
A = atomic weight

The critical glancing angle of total reflection is defined from Snell's law as

$$\cos \phi_c = 1 - \delta \tag{4}$$

where ϕ_c is that angle of incidence for which the angle inside the material is just zero.

Serial development expansion of the cosine in Equation (4) yields, for small ϕ_c,

$$\phi_c = \sqrt{2\delta} \tag{5}$$

By inserting Equations (2) and (3) into Equation (5) and substituting λ (Ångström) = 12.4/E (keV), the critical angle becomes

$$\phi_c = \frac{99.1}{E} \sqrt{\frac{Z\rho}{A}} \qquad \text{minutes of arc} \tag{6}$$

From a practical point of view, the critical angle gives an estimate of the range of angles in which total reflection occurs.

B. Reflectivity

The reflectivity R is defined as the ratio of the reflected to the incident intensity. With the aid of the critical angle, the variation of the reflectivity with ϕ is given by the Fresnel equation

$$
R = \frac{\{\sqrt{2}X - [(X^2 - 1)^2 + Y^2]^{1/2} + (X^2 - 1)^{1/2}\}^2 + [(X^2 - 1)^2 + Y^2]^{1/2} - (X^2 - 1)}{\{\sqrt{2}X + [(X^2 - 1)^2 + Y^2]^{1/2} + (X^2 - 1)^{1/2}\}^2 + [(X^2 - 1)^2 + Y^2]^{1/2} - (X^2 - 1)} \tag{7}
$$

where:

$$
X = \frac{\phi}{\phi_c} \quad \text{and} \quad Y = \frac{\beta}{\delta}
$$

The variation of the reflectivity as a function of the angle of incidence is displayed in Figure 1 for two materials of different density. The curves represent the typical shapes for materials with low (curve 1) and high damping (curve 2).

C. Penetration Depth

The penetration depth is described by

$$
z_p = \frac{\lambda}{4\pi\sqrt{\delta}\{[(X^2 - 1)^2 + Y^2]^{1/2} - (X^2 - 1)\}^{1/2}} \tag{8}
$$

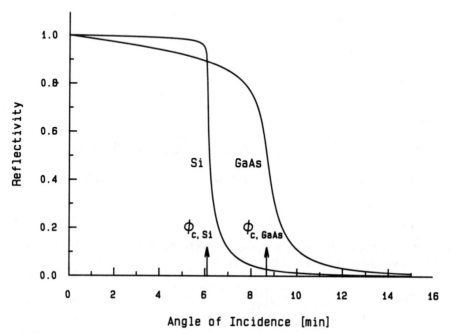

Figure 1 Calculated reflectivity for x-rays of 17.5 keV for Si ($\rho = 2.3$ g/cm³) and GaAs ($\rho = 5.3$ g/cm³).

where z_p is the distance measured normal to the interface at which the intensity of the refracted beam is reduced by the factor $1/e$. Figure 2 shows the penetration depth for the example of gallium arsenide as a function of the angle of incidence using a copper (curve 1) or a molybdenum anode (curve 2), respectively.

For special cases of the angle of incidence, Equation (8) is reduced to

$$z_p = \sqrt{\frac{mc^2}{16\pi N_A e^2}} \sqrt{\frac{A}{Z_\rho}} = 3.4 \times 10^{-7} \sqrt{\frac{A}{Z_\rho}} \qquad \text{for } \phi \to 0$$

$$z_p = \frac{1}{2} \sqrt{\frac{\lambda}{\pi \mu^*}} \qquad\qquad\qquad \text{for } \phi = \phi_c \qquad (8a)$$

$$z_p = \frac{\phi}{\mu^*} \qquad\qquad\qquad\qquad \text{for } \phi \gg \phi_c$$

In the case of $\phi \to 0$, it has been assumed that $\beta/\delta < 0.1$. This condition is commonly met even in cases of comparatively high damping (e.g. $\beta/\delta = 0.06$ for GaAs and 17.5 keV).

It can be seen from Equation (8a) that the penetration depth for very small glancing angles is independent of the energy of the primary beam. This is also shown in Figure 2, which is computed from Equation (8). The important role of z_p in surface analysis is discussed in more detail in Section V.A. A selection of critical angles, reflectivities, and penetration depths is given in Table 1 for various materials and two excitation energies.

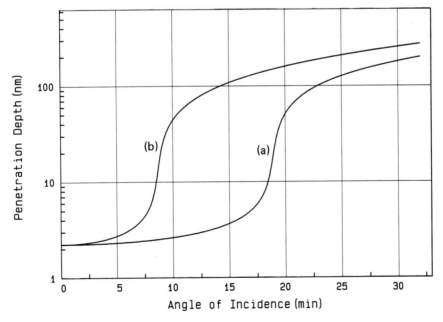

Figure 2 Calculated penetration depth for GaAs at an incident energy of 8 keV (a) and 17.5 keV (b).

Table 1 Critical Angles ϕ_c, Reflectivities R, and Penetration Depths z_p for Various Materials

Material	ρ (g/cm^3)	ϕ_c (minutes) (E_1/E_2)[a]	z_p (nm) $\phi = 3$ minutes (E_1/E_2)	z_p (nm) $\phi = \phi_c/2$ (E_1/E_2)	z_p (nm) $\phi = \phi_c$ (E_1/E_2)	R at $\phi = \phi_c/2$ (E_1/E_2)
Plexiglass	1.20	9.9/4.6	4.5/5.7	4.9/4.9	127.2/247.0	0.997/0.999
Glassy carbon	1.55	10.8/5.0	4.1/4.9	4.5/4.5	135.0/250.4	0.998/0.999
Quartz	2.20	12.9/5.9	3.4/3.8	3.8/3.8	40.1/84.3	0.985/0.997
Silicon	2.32	13.2/6.1	3.3/3.7	3.7/3.7	39.1/62.0	0.973/0.994
Aluminum	2.69	14.0/6.4	3.1/3.4	3.5/3.5	30.5/64.4	0.978/0.995
Titanium	4.54	17.8/8.2	2.4/2.6	2.7/2.7	11.6/23.2	0.907/0.976
Germanium	5.31	18.9/8.7	2.3/2.4	2.6/2.6	18.3/13.3	0.966/0.937
Steel	7.80	23.5/10.8	1.8/1.9	2.1/2.1	7.5/14.2	0.875/0.963
Tantalum	16.60	31.9/14.6	1.3/1.4	1.5/1.5	6.8/6.2	0.915/0.900
Platinum	21.41	36.0/16.6	1.2/1.2	1.4/1.4	5.4/4.9	0.897/0.878

[a] Energies of the incident x-rays: $E_1 = 8.04$ keV (Cu); $E_2 = 17.5$ keV (Mo).

III. DESIGN OF INSTRUMENTS

To obtain optimum limits of detection, which are defined according to IUPAC [5] as

$$x_L = \frac{\text{concentration}}{\text{peak area}} 3\sqrt{\text{background}}$$

the design considerations of a TXRF instrument cannot be limited to those that seek only to minimize background levels via the total reflection effect. It is even more important to maximize the analyte signal intensity. Because of the limited performance of the available x-ray tubes, solid angle losses should be minimized by close distances between tube anode and sample position or by focusing x-ray optics. A genuine focusing of the x-rays is not feasible with the existing limiting conditions, that is, a comparatively large anode (effective width of a 2 kW fine-focus anode $\simeq 40$ μm) and a sample diameter of about 10 mm. The alternative, an extremely compact construction, creates some difficulties because all parts must be carefully machined and adjusted to maintain control of the glancing angle of incidence within close tolerances.

The effects of the most important design parameters have been investigated employing a Monte Carlo technique. In this way the intensity of the primary beam and the distribution of angles at the sample position have been calculated along with the corresponding reflectivities and penetration depths as a function of those instrumental parameters that control the path of the x-rays. The most simple setup shown in Figure 3 is still used for surface analysis because it can be most easily fitted with equipment for angular displacement and scanning devices for large samples.

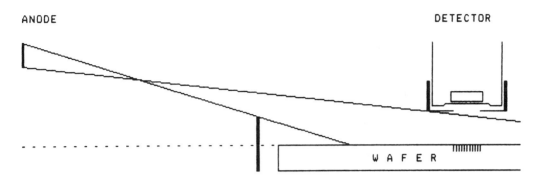

Figure 3   Arrangement for wafer surface examination.

In general, the arrangement outlined in Figure 4 is found to be beneficial. The primary radiation is first incident on the reflector, which in its turn directs the x-rays at the target position. The reflector may be planar or curved.

The results of Monte Carlo simulations of the available primary intensity are shown in Figure 5. The calculations are based on realistic instrumental data and take account of the divergence of the x-ray beam. Figure 6 illustrates that in the case of a curved mirror the increase in intensity (given by the integrated curve area) is obtained at the cost of an increase in the divergence of the primary beam.

Plane mirror optics, in contrast to the curved mirror case, do not influence either the intensity or the divergence of the primary beam, relative to direct incidence. Mirror optics, however, have an important advantage. This advantage results from the energy dependence of the reflectivity, shown for four distinct incident angles and two reflecting materials in Figure 7. Because of the sharp decrease in the reflectivity at a critical energy, a mirror in the x-ray path serves as an effective low-pass filter, cutting off all x-rays with energies exceeding an adjustable boundary energy. Moreover, the low-pass filtering effect of x-ray reflection permits the operation of the x-ray tube in its most effective mode, at its maximum voltage, because the bremsstrahlung hump, which normally occurs, is cut off by the inserted mirror. The benefit of the low-pass filtering effect is dem-

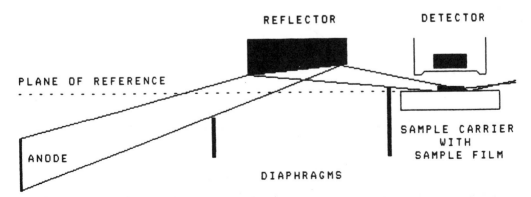

Figure 4   Arrangement with a mirror acting as low-pass filter of the primary radiation.

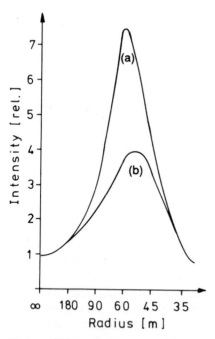

Figure 5 Effect of the radius of curvature of a mirror on the available x-ray intensity at the sample spot. The curves display the benefit of a curved mirror compared to a plane reflector ($r = \infty$) for two anode widths (curve a = 15 μm; curve b = 40 μm) calculated for a typical TXRF instrument [25].

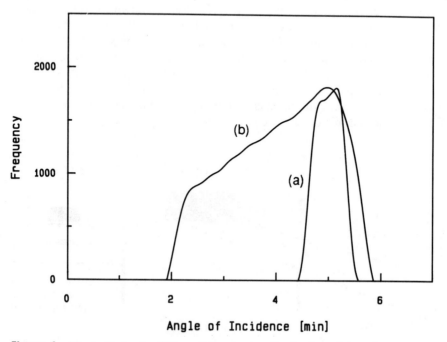

Figure 6 Monte Carlo simulation of the angular distribution of the primary x-rays at the sample spot in the case of (1) a plane and (2) a curved mirror ($r = 50$ m).

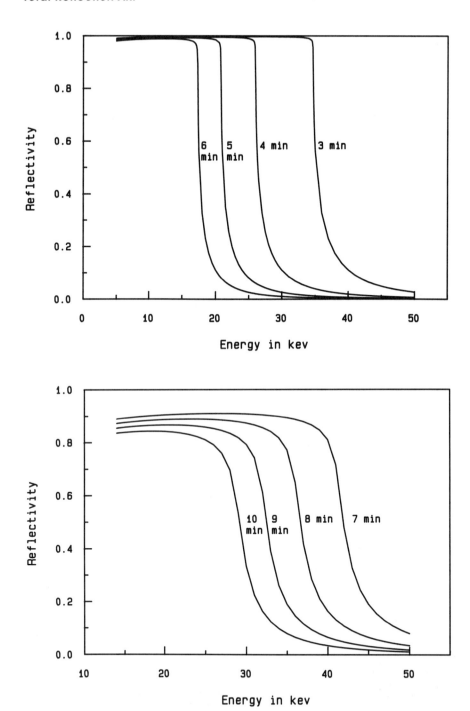

Figure 7   Reflectivity versus energy for (a) quartz and (b) platinum as a function of the incident angle.

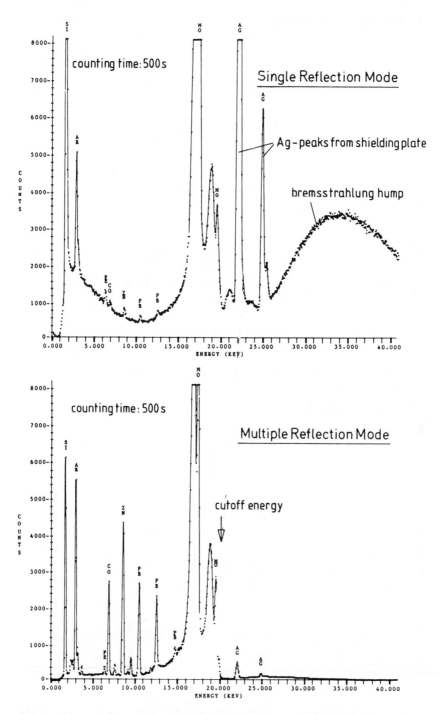

Figure 8 Low-pass filtering effect (lower spectrum) of a mirror in the path of the x-rays demonstrated with a sample of 1 ng Co, Zn, and Pb.

onstrated in Figure 8. The zigzag path of the x-rays additionally facilitates technical measures to eliminate stray and scattered radiation.

IV. ANALYSIS OF SMALL QUANTITIES ON REFLECTING CARRIERS

TXRF is mainly applied to the analysis of fluid samples in the form of solutions or dispersions. The scope of the work covers a wide field ranging from environmental research and monitoring [6–11], mineralogy [12–15], oceanography [16,17] to biology and medicine [18–20]. As a result, numerous sample matrices have been handled. A review of the applications and intercomparisons with other analytical methods may be found in References 21–24. As for AAS or ICP-AES, solid materials are normally digested before analysis employing one of many suitable methods, whereas fluids are diluted to appropriate concentration levels. Figure 9 shows a choice of matrix separation and sample preparation steps suited to TXRF analysis of rain, river, and sea waters, as well as of sediments, particulate matter, manganese nodules or crust, and mussel tissue. Finally, aliquots from 1 to 100 μl are placed on the center of the sample carrier and allowed to dry. The film-like residue, which ranges from picograms to ~10 μg in mass and covers an area of about 5 mm in diameter, is ready for analysis. Elements that tend to be volatile, such as mercury or arsenic, should be converted into more stable chemical forms by adding a minute amount of a complexing agent (e.g., a dithiocarbamate solution) to the drop on the surface of the sample carrier before the evap-

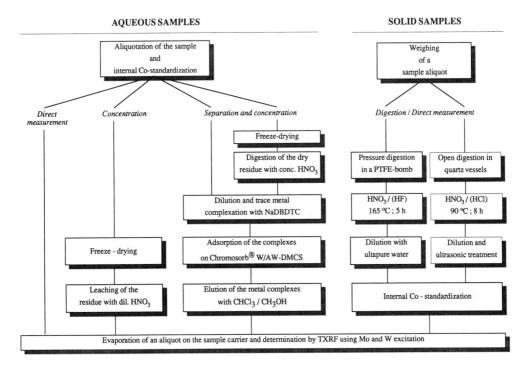

Figure 9   Typical sample preparation procedures for TXRF.

oration of the solvent. In any case, because of the low masses involved (less than 10 μg), the samples on the substrate may be regarded as "thin films" in XRF terms.

A. Detection Limits

With favorable matrices, TXRF provides x-ray spectra for a broad range of elements characterized by both high sensitivity and low background compared to conventional energy-dispersive XRF. The background generated by the sample support can be assessed from the absorbed portion of the primary radiation. One obtains for the background of a total reflecting carrier,

$$B_T = CI_0 \sin \phi (1 - R) \tag{9}$$

where C is a constant and $I_0 \sin \phi$ describes the component of the incident intensity I_0 normal to the interface. The term $1 - R$ is that fraction of the primary radiation that is not reflected and therefore available to inelastic processes. Hence, the product $\sin \phi (1 - R)$ describes the "energy transfer" into the substrate. Values of the energy transfer for different materials are given in Table 2. Figure 10 shows the energy transfer into a quartz substrate as a function of the angle of incidence. According to Equation (9), this quantity describes the angular dependence of the background when the incident angle goes beyond about 5 minutes of arc. It shows in addition that the smallest possible angle should be sought. In practice, angles of incidence below 2 minutes are not feasible. Because of an additional uncertainty of the incident angle of about 1 minute, the recommended operating angle amounts to 3 minutes of arc.

In normal XRF, the employment of thin foils is widespread to avoid the unacceptably strong background contribution from a massive sample support.

Table 2 Energy Transfer $\sin \phi (1 - R)$ into Various Materials at Different Angles of Incidence and X-ray Energies

Material	Energy transfer ($\times 10^{-5}$)		
	$\phi = 2$ minutes $(E_1/E_2/E_3)^a$	$\phi = 5$ minutes $(E_1/E_2/E_3)$	$\phi = 10$ minutes $(E_1/E_2/E_3)$
Plexiglass	0.05/0.03/16.4	0.38/>100/>100	>100/>100/>100
Glassy carbon	0.04/0.02/0.11	0.25/40.4/>100	2.32/>100/>100
Quartz	0.24/0.12/0.12	1.63/1.36/>100	9.37/>100/>100
Silicon	0.42/0.21/0.17	2.80/2.20/>100	15.5/>100/>100
Aluminum	0.33/0.17/0.13	2.17/1.55/>100	11.4/>100/>100
Titanium	1.10/0.61/0.35	7.00/4.61/>100	31.4/>100/>100
Germanium	0.37/1.54/0.93	2.38/11.2/>100	10.7/>100/>100
Steel	1.14/0.71/0.39	7.16/4.87/>100	30.0/42.9/>100
Tantalum	0.56/1.44/0.90	3.52/9.30/8.36	14.5/45.3/>100
Platinum	0.61/1.58/0.96	3.80/10.0/7.81	15.4/45.6/>100

[a] Energies of the incident x-rays: $E_1 = 8.04$ keV, $E_2 = 17.5$ keV, and $E_3 = 40$ keV.

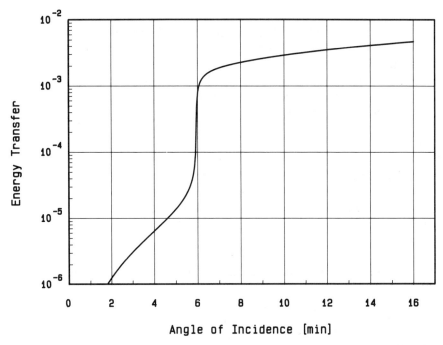

Figure 10 Background caused by a quartz support (energy transfer) as a function of the angle of incidence.

The background B_F produced by a foil is given by

$$B_F = CI_0 \sin \phi (1 - e^{-\mu^* d/\sin \phi}) \tag{10}$$

where

μ^* = linear absorption coefficient of the foil
d = thickness of the foil

Equation (10) simply considers the angular dependence of the absorption of the primary radiation by a foil.

To compare the background of both methods, typical operating values are inserted into Equations (9) and (10). For TXRF one obtains for a quartz substrate with $\phi = 3$ minutes of arc and $R = 0.9965$ (17.5 keV):

$$B_T = (3 \times 10^{-6})CI_0 \tag{9a}$$

For normal XRF employing an extremely thin Mylar foil, one obtains with $\phi = 45°$, $\mu^* = 1$ cm^{-1} (17.5 keV), and $d = 1$ μm,

$$B_F = 10^{-4}CI_0 \tag{10a}$$

Comparing Equation (9a) with Equation (10a), it follows that the limits of detection (x_L) are lower by a factor of about 6 in favor of TXRF compared to a foil as substrate, because the x_L varies with the square root of the background. Moreover, the primary intensity available for excitation of the sample is multiplied by $1 + R \approx 2$, because the reflected beam is additionally effective. This estimation thus results in a factor of $6 \times 2 = 12$ in favor of TXRF, even compared to an ultrathin foil as sample support. Additionally, Equation (9) indicates that the background reduction in TXRF has nothing to do with the low penetration depth but is caused by only the impaired energy transfer through the interface.

The effective background reduction in TXRF is demonstrated in Figures 11 and 12. They show measured spectra of 1 ng Ni and Bi each of 1 ng Mo and Cd, respectively. Detection limits of 2.0, 2.3, 5.6, and 9.5 pg, respectively, have been achieved with 1000 s counting time.

Figure 13 shows the detection limits as a function of the atomic number obtained by commercially available instruments [25] for elements ranging in atomic number from 16 through the remainder of the periodic table. The detection limits are measured on standard solutions containing elements selected to avoid peak overlap. Results are below 20 pg for about 40 elements. Because a detection gap would exist either in the 40–60 or in the 60–70 region of atomic numbers, this instrument permits excitation by a choice of two x-ray tubes with different anode materials. The arrangement consists of a compact block containing two reflection lines with independently adjusted optics aimed at the sample film. One selection is a tungsten tube operated at a voltage of 50 kV for the excitation of cadmium and neighboring elements through the bremsstrahlung continuum. Alternatively, it can be operated at 25 kV combined with an appropriate filter to improve the excitation conditions for the light elements through the W $L\alpha$ line (Fig. 13).

Thus far, detection limits have been given in terms of mass, because this appears to be the most definitive measure. In most cases, however, concentrations are required instead of amounts. Detection limits expressed in picograms must

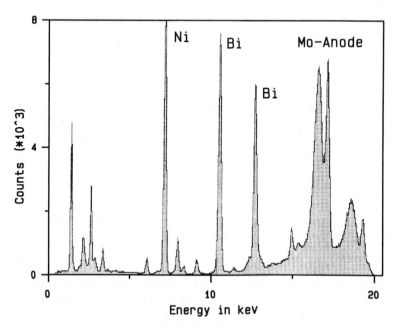

Figure 11 Spectrum of a 1 ng Ni and Bi sample using a Mo tube (40 mA, 50 kV; counting time = 1000 s). The detection limits are Ni 2.0 and Bi 2.3 pg.

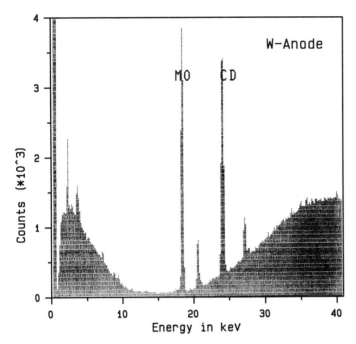

Figure 12 Spectrum of a 1 ng Mo and Cd sample using a W tube (40 mA, 50 kV; counting time = 1000 s). The detection limits are Mo 5.6 and Cd 9.5 pg.

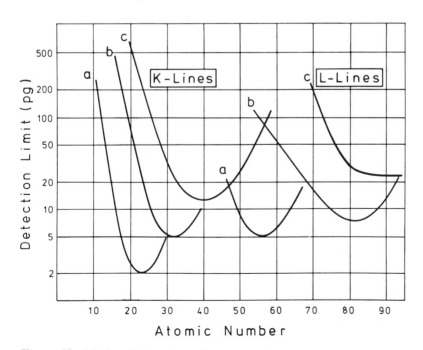

Figure 13 Limits of detection using a Mo tube (curve b) or a W tube (curve a, W-L excitation; curve c, 50 kV bremsstrahlung), respectively.

be converted into the corresponding concentrations by division by the sample volume placed on the sample carrier (e.g., 5 pg → 5 pg/50 μl = 0.1 ppb).

The detectability of elements with atomic numbers below 12 or so is affected by poor fluorescence yield, the well-known limitations of energy-dispersive XRF, that is, poor peak resolution, and other detector problems. It is additionally complicated by the Si fluorescence peak generated by the quartz substrate most frequently employed as a sample carrier. Other materials, especially glassy carbon, combined with a vacuum chamber and a special Si(Li) detector help to extend the range of detectable elements down to oxygen. Detection limits between 108 ng for oxygen and 2.3 ng for magnesium have been reported [26].

B. Sample Carriers

The demands on appropriate TXRF sample carrier materials are many. The material should have the following characteristics:

- Machinable to a perfectly flat and smooth surface
- Immune to aggressive chemicals and mechanical stresses
- Free of fluorescence lines over the energy range of interest
- Free of contamination
- Hydrophobic
- Available at reasonable price
- High reflectivity under operating conditions

The energy transfer, which accounts for the background produced by the carrier (see Sec. IV.A), is listed in Table 2 for a variety of materials and operating conditions. Currently, no material is known that fulfills all these requirements. Quartz and pure silicon are suitable for elements ranging in atomic number above 16. Besides quartz, glassy carbon [15], germanium [27], and plexiglass [28] have been used as sample carriers. At present glassy carbon appears to be the most promising material when the silicon K lines of quartz would interfere with light elements, although there still exist some problems with purity and surface quality.

Nonhydrophobic surfaces, such as quartz, must be rinsed with a silicone solution (e.g., available from Serva, Heidelberg, Germany; Catalog No. 35,130) and subsequently dried at 100°C for about 1 h to keep aqueous samples in position on the carriers. For organic solutions, such as alcohol and chloroform, an additional device is needed for sample preparation [16].

The sensitivity down to the picogram range sometimes makes cleaning of the sample carrier critical, particularly in routine operations when large numbers of carriers are in use. The following cleaning procedure, consisting of five steps, has proven to be effective:

1. Mechanical removal of sample residues from the previous analysis using a tissue or brush
2. Rinsing with water or, in the case of organic residuals, with acetone
3. Gentle boiling for 2 h in a detergent bath (e.g., RBS 50 from Carl Roth GmbH, 7500 Karlsruhe, Germany) in groups of about 25 carriers
4. Rinsing with 0.1 N ultrapure nitric acid
5. Rinsing with ultrapure water and subsequent drying

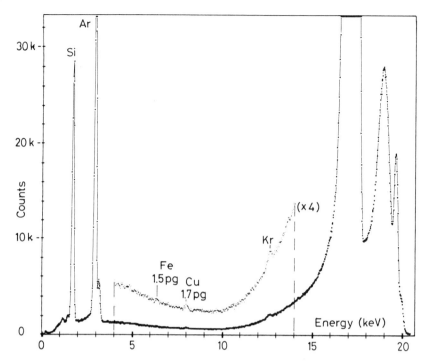

Figure 14   Spectrum of a cleaned sample carrier using a Mo tube (counting time = 10,000 s, residual contamination of Fe and Cu below 2 pg).

Figure 14 is a spectrum of a cleaned carrier. The residual contamination is reduced to less than 2 pg for the transition metals. The spectrum demonstrates that the cleaning method is effective. Moreover, it proves the absence of intrumental peaks. Both requirements must be met to take full advantage of the sensitivity of the method.

C. Standardization

Thin films are the most favorable form to present samples to primary x-rays because they require no corrections for the matrix absorption and enhancement effects of the fluorescence radiation. The measured intensity depends only on fluorescence yield, geometric parameters of the instrument, and detector efficiency for the elements of interest. All these parameters are either constant or remain stable over long periods. The slope of the calibration curve does not depend on the properties of the sample, thus facilitating the quantitation of all detected elements in any sample provided that the amount or concentration of one of these elements is known. This prerequisite can be met by spiking the sample solution with a well-defined quantity of an element that is detectable and not expected in the sample and hence that may serve as an internal standard.

The use of internal standardization in quantitative TXRF analysis of filmlike samples is recommended. Attempts to quantify elements directly (via a counting rate–mass relation for each fluorescence peak) can only be successful in limited cases characterized by extremely small amounts of nearly matrix-free samples. Quantitation with external standards is also feasible with near surface layers, for which the necessary corrections can be calculated exactly (see Sec. V.A).

A calibration curve is established by means of a number of multielement standard solutions, each containing about five elements in varying combinations. A larger number of elements in one standard solution is not recommended for standardization because of possible spectral overlap. The elemental compositions of the standard solutions should be chosen in such a manner that each standard solution contains at least one element common to another standard solution. Because of the smooth slope of the calibration curve, missing elements can be interpolated to a certain degree.

The calibration procedure should be carefully conducted during the installation phase of the instrument. Once established, the calibration remains valid as long as major components [e.g., tube anode or the Si(Li) detector] remain unchanged.

The primary radiation, in contrast to the fluorescence radiation, is normally attenuated by the matrix. The path length of the primary rays in the sample film is increased by a factor of about 1000 compared to conventional thin-film XRF as a result of the extremely low incident angle (Fig. 15). The effect of the attenuation of the primary radiation by the sample is normalized to the internal standard, provided that the standard element is homogeneously distributed in the sample.

For residual matrix, internal standardization helps to provide the accuracy of TXRF. There is, however, no equivalent means by which the detection limits can be preserved from deterioration by the impact of the matrix.

D. Influence of the Matrix

Apart from spectral interferences, the detection limits of TXRF depend considerably on the properties of the sample. The detection limits are raised by increasing the concentration of matrix simply because the matrix atoms contribute to the background radiation. Provided that the fluorescence spectra are acquired at the maximum count rates allowable for the analyzing system (e.g., by adjusting the

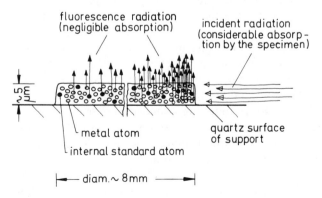

Figure 15 Sample excitation.

tube current) and that the count rate originating from the trace elements remains low compared to the total count rate, two conditions that are normally met in trace element analysis, there is a linear relation between matrix concentration and detection limit.

The strong effect of the matrix on the performance of TXRF is demonstrated in Figure 16. The analysis of nickel in a NaCl matrix indicates that the detection limit is a linear function of the NaCl content ranging over five orders of magnitude. This means that TXRF analysis can approach the fundamental limits of energy-dispersive XRF to a high degree. Instrumental contributions to scatter are largely eliminated, except for small residual effects revealed by the slope of the curve at very low concentrations. In the case of conventional XRF analysis, the detection limits are determined by effects other than matrix-induced effects up to comparatively high concentrations. This is demonstrated by the upper curve (broken line) in Figure 16. The curve is based on recently published data [29] obtained with a conventional XRF instrument specially optimized for trace element analysis, which also uses the thin-film technique.

The increase in the minimum detectable concentrations from 0.02 to 0.03 ppb in matrix-free samples to about 15 ppb in 1% NaCl solution could give the impression that the detection limits obtained in TXRF are more sensitive to the matrix than in conventional XRF. In fact, the matrix affects the sensitivity of any energy-dispersive XRF instrument equally. At low concentrations, however, the scatter

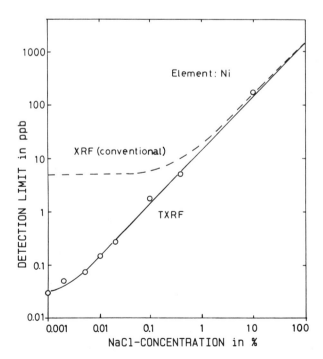

Figure 16 Effect of the residual matrix on the detection limit. The detection limits were determined using Ni standard solutions with increasing NaCl concentration. Water was removed by evaporating.

caused by the matrix may be concealed by the scattered radiation from the sample support.

E. Special Sample Preparation Techniques

Numerous TXRF analyses have been conducted [6–24] using sample preparation techniques that were not really specific, but basically the same as for AAS or ICP-AES. There are, however, special cases in which sample preparation takes advantage of the particular features of TXRF.

Two of these features, the capability of analyzing small samples and the advantage of an inert sample carrier, can be demonstrated using as an example the analysis of fine roots of trees. These fine roots have lengths of 1 mm and weights as low as about 50 μg. They are prepared as follows. A single fine root is weighed using a precision balance, dissolved in 100 μl nitric acid, and spiked with an internal standard. An aliquot of about 20 μl solution is placed on a quartz plate and allowed to dry. Further improvement in detection limits can be achieved by low-temperature ashing of the sample film directly on the quartz support using a microwave-generated oxygen plasma, a procedure that removes the organic matrix. Figure 17 shows the spectra of such a sample film along with analysis results, before and after plasma ashing. The reliability of the technique is demonstrated by the agreement in several trace element concentrations (see Fig. 17) and its effectiveness by the evident background reduction, which improves the precision and enables the detection of additional elements.

An additional example has been published [14,30] that demonstrates how the properties of the sample support can be utilized to allow chemical reactions in microliter volumes to be conducted on the carriers. The procedure is briefly described as follows. Up to 100 μl solution loaded with alkali and alkaline-earth salts, for example, is placed on the quartz carriers previously rendered hydrophobic by means of a silicone solution. The drop on the carrier is spiked with about 5 μl of a 1% sodium dibenzyldithiocarbamate solution. After allowing the sample to dry for about 30 minutes, the plate is rinsed with ultrapure water. The insoluble metal carbamates, including the internal standard, remain fixed on the hydrophobic surface, whereas the soluble matrix is dissolved and removed by the water. The resulting specimen gives substantially improved detection limits compared to the unprocessed film.

The properties of the sample support combined with the high detection performance of TXRF are also utilized in the analysis of thin sections of tissues [20] as well as in the effective analysis of air dust particles. In the latter case the carrier plates were used as targets for each stage of a size-separating sampler of airborne particles [31]. Figure 18 shows photographs of airborne particles with different sizes collected on TXRF sample carriers. With a sampling time of 10 minutes, a detection limit for lead in the 1–2 μm fraction of 0.5 ng/m^2 was obtained.

These examples illustrate the special capabilities of TXRF when only minute amounts of sample are available or permitted (e.g., radioactive samples). The performance can be further improved whenever suitable sample preparation steps can be performed on the carrier itself.

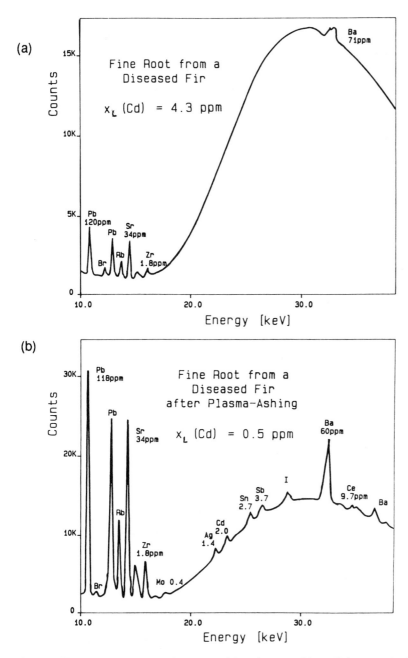

Figure 17 Influence of matrix removal by plasma ashing of the sample directly on the sample support.

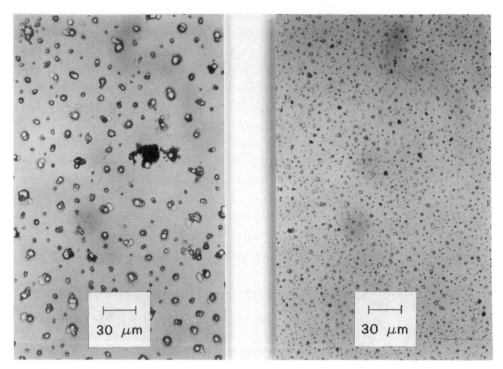

Figure 18 Airborne particles directly collected on TXRF quartz carriers. Shown here are two of five impactor stages.

V. SURFACE ANALYSIS

An application of TXRF that will possibly gain growing importance is the non-destructive element analysis of near surface layers. The extremely low penetration depth in the nanometer range under total reflection conditions can be exploited to concentrate the intensity of the primary x-ray beam in the uppermost surface layer of flat, polished samples.

One of the problems that can be tackled by this technique is the determination of metal traces in the surface of silicon wafers [32]. Surface contamination in wafers is known to affect the performance and yield of integrated circuits. For this reason, the first commercially available TXRF instrument for surface analysis was especially constructed for the examination of silicon wafers up to 150 mm in diameter [33]. The design of the TXRF surface analyzer is along the same lines as pointed out in Section III, except for changes chiefly related to the comparatively large size of the wafers to be examined. Therefore, similar detection limits were found [33], for example, 2.5 pg for nickel corresponding to 9×10^{10} atoms/ cm^2 or 70 ppm of a monoatomic layer of silicon atoms.

A. Quantitation

External standards providing known element concentrations in nanometer layers are difficult to prepare. Even if they were available, they would be of limited benefit because even small differences between the standard and the sample still require the full transformation procedure of the measured values. This is outlined here.

The internal standards normally used for quantitation in TXRF can also be ruled out, because a particular near surface layer cannot be doped with known concentrations of an element before analysis. Therefore, one must approach the problem in an indirect way. In a first step, the instrumental parameters are determined by measuring standard solutions as described in Section IV.C. Subsequently, the properties of the particular sample are taken into account. Fortunately, the effect of these properties on the fluorescence signal are amenable to calculation. Thus, quantitation is based on the application of the classical dispersion theory combined with a thin-film calibration of the instrument.

First, we consider the fluorescence intensity I_F of a thin-film sample on a reflecting surface as a function of the angle of incidence ϕ:

$$I_F = C_i I_0 A_s (1 + R(\phi)) \tag{12}$$

with

C_i = instrumental calibration factor for element i
I_0 = intensity of the primary beam
A_s = atoms per unit area
R = reflectivity according to Equation (7)

Equation (12) describes the fact that the effective primary intensity immediately above the reflecting interface is approximately doubled because the incoming as well as the reflected radiation excites the sample. Figure 19 shows the change in

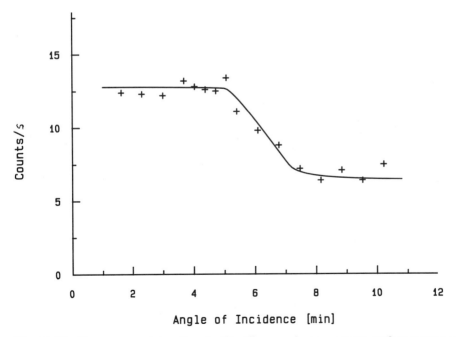

Figure 19 Fluorescence intensity of a film-like sample on a quartz surface versus angle of incidence. $E = 17.5$ KeV. (+ measured values; — calculated curve).

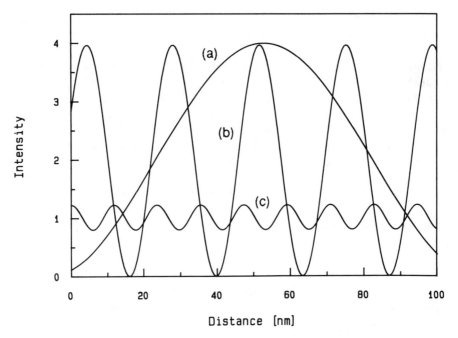

Figure 20 Intensity of the primary radiation as a function of the distance above a quartz surface. (Angle of incidence: a = 1 minute; b = 5 minutes; c = 10 minutes).

the counting rate of a filmlike sample with respect to ϕ according to Equation (12), along with measured data.

It should be noted that Equation (12) actually quantifies the fluorescence radiation only as long as the sample raised above the interface by more than about 100 nm, for example (quartz substrate, 17.5 keV). This is because the primary intensity above the interface builds up standing waves (Fig. 20) due to interferences, which can be simulated using the complex notation of the reflectivity [see Sec. V.C, Eq. (20)]. In fact, investigations report on the standing wave phenomenon being exploited to locate the perpendicular position of atoms above a surface embedded in extremely thin films [34]. In film samples of regular thickness (>1 µm), however, the radiation is normally averaged over the sample and thus yields the fluorescence intensity according to Equation (12).

At low angles of incidence, where R is nearly unity, the product $C_i I_0$ can easily be determined by rearranging Equation (12) as

$$C_i I_0 = \frac{I_F}{A_s 2} \tag{12a}$$

because all quantities on the right-hand side are either known or measurable.

In contrast to the atoms on the surface, atoms below the interface are excited differently. In this case the primary intensity $I(z)$ below the interface is given by

$$I(z) = I_0(1 - R(\phi))K(\phi)e^{-z/z_P(\phi)} \tag{13}$$

where:

I_0 = intensity of the primary beam above the interface as defined in Equation (12)

K = compression factor (see later)

z = penetration normal to the interface

z_p = value of z at which the intensity of the refracted beam is reduced to $1/e$ [see Eq. (8)]

This formulation of Equation (13) has been chosen to give a vivid description of the effects:

- The term $1 - R$ illustrates that only that portion of the primary radiation, not reflected at the interface, is available for inelastic processes in the surface, such as fluorescence.
- The compression factor K describes the change in the density of the flux lines of the refracted field compared to the incoming beam. Changes in the field strength arise in the total reflection case, because the flux lines of the refracted beam are concentrated, or sometimes diluted, in the near surface layer compared to normal incidence.
- Finally, the exponential function describes the decline in the intensity as a function of the distance from the interface.

Elementary geometrical considerations result in a simple expression for K. Using b and b^* from Figure 21 as measures for the radiation densities of the primary and the refracted beam, K is defined as

$$K = \frac{b}{b^*}$$

Taking into account that the absorption length of the refracted rays in matter

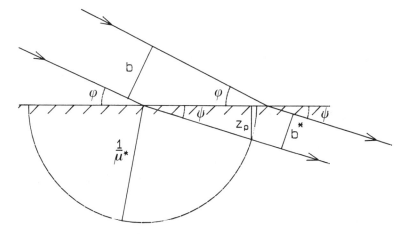

Figure 21   Refraction of x-rays in the total reflection case.

characterized by a linear absorption coefficient μ^* is $1/\mu^*$, it follows for small angles that

$$\frac{b}{b^*} = \frac{\phi}{\psi} \quad \text{and} \quad \psi = \mu^* z_p$$

hence,

$$K = \frac{\phi}{\mu^* z_p} \tag{14}$$

As expected, the compression factor K approaches unity for $\phi \gg \phi_c$, because with increasing ϕ the penetration depth becomes

$$z_p \rightarrow \frac{\phi}{\mu^*}$$

The intensity of the primary radiation in the near surface layer under total reflection conditions is derived elsewhere [35,36] in more detail.

The corresponding fluorescence intensity from a thin surface layer (no absorption of the fluorescence radiation) collected by the detector is given by

$$I_F = C_i \int_0^\infty A_v(z) I(z) \, dz \tag{15}$$

where $A_v(z)$ = concentration profile (number of atoms per cm^3).

Integration of Equation (15) is accomplished for the simple but practically important case of a homogeneous distribution A_v in a layer of thickness H, described by

$$A_v(z) = A_v = \text{constant} \quad \text{for} \quad 0 \le z \le H$$
$$A_v(z) = 0 \quad \text{for} \quad z > H$$

Hence $A_s = A_v H$ for atoms per unit area corresponding to Equation (12).

By inserting Equation (13) in Equation (15), one obtains after integration

$$I_F = C_i I_0 A_s (1 - R) K \frac{z_p}{H} (1 - e^{-H/z_p}) \tag{16}$$

For an infinitely thick layer ($H \rightarrow \infty$), Equation (16) becomes

$$I_F^\infty = C_i I_0 A_v (1 - R) K z_p \tag{17}$$

The calculated fluorescence intensity of an infinite thick layer is displayed in Figure 22 according to Equation (17). The angular dependence of the intensity originates from R, z_p, and K [see Eqs. (7), (8), and (14)].

The fluorescence intensity I_F^∞ of a homogeneous, infinitely thick sample irradiated under total reflection conditions can also be derived directly by more general considerations as

$$I_F^\infty = I_0 \phi (1 - R) C_i A_v \int_0^\infty e^{-\mu^* s} \, ds \tag{17a}$$

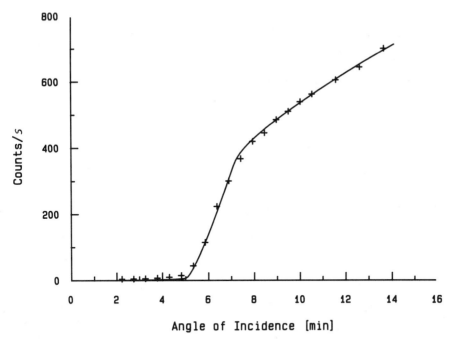

Figure 22 Fluorescence intensity of Si versus angle of incidence for a silicon wafer (+ measured values; — calculated curve).

consisting of the following terms:

$I_0\phi$ = geometrical reduction in the primary radiation on the surface

$1 - R$ = portion of the primary radiation available below the interface

$\int_0^\infty e^{-\mu^* s}\, ds = 1/\mu^*$ = attenuation inside the material, integrated over the path s of the refracted x-rays

The relation (17a) is identical with (17), if K is replaced by $\phi/\mu^* z_p$ according to Equation (14).

In the case of a homogeneous distribution of the atoms i in a layer of a thickness H, as described by Equation (16), this relation can be written as

$$A_s = \frac{I_F}{C_i I_0 f} \tag{18}$$

with

$$f = (1 - R)K\frac{z_p}{H}(1 - e^{-H/z_p}) \tag{19}$$

where I_F is the measured value, and the product $C_i I_0$ is known from calibration measurements applying Equation (12a). The function f depends essentially on the concentration profile and the incident angle. It can be calculated by applying Equations (7) and (8).

By looking at the fluorescence signal as a function of the angle of incidence, a principal distinction can be made between particles above and contamination below the interface (Fig. 23). For the determination of a contamination below the interface, A_s, by measuring the fluorescence intensity I_F, it follows from Equation (19) that additional information is needed, compared to the analysis of samples situated above the interface: the angle of incidence at which the measurement is made, together with some knowledge about the concentration profile of the atoms normal to the surface.

The effective angle of incidence ϕ is supplied by the current instrumentation [33] to an accuracy of about 0.1 minute of arc absolute. Hence, one of the requirements for a calculation of f is fulfilled.

Not as yet considered, however, is, in Equation (19), the angular spread of the primary beam. Because of the low-energy transfer into the surface (see Table 2), one of the main requirements for the design of a TXRF surface analyzer is a short distance between tube anode and sample spot, to avoid solid angle losses. Therefore, a substantial angular divergence must be accepted. The divergence has been calculated by a Monte Carlo simulation on the basis of data that describe a particular, commercially available instrument [33]. The calculations result in the introduction of an effective form function f^*, which replaces f in Equation (19). For practical purposes f^* should be implemented in the quantitation software that belongs to the TXRF instrumentation. The effect of the angular spread on the fluorescence signal from a near surface layer of 2 nm is displayed in Figure 24.

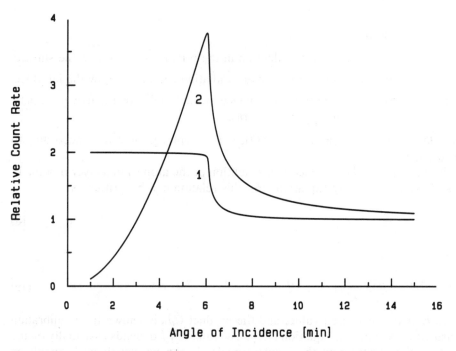

Figure 23 Comparison of the calculated fluorescence signals from the same number of atoms per unit area in near surface layers, which are differently situated with respect to the interface (1 = above interface; 2 = below interface).

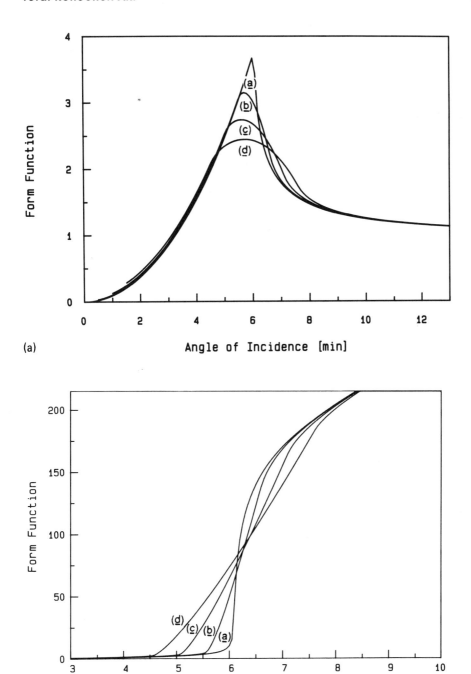

Figure 24 Form function of a near-surface layer (a) with various beam divergences: a = no divergence; b = 1 minute; c = 2 minutes; d = 3 minutes) and (b) for an infinitely thick layer.

Figure 25 demonstrates that an effective form function $f*$, which comprises the instrumental divergence of 2 minutes of arc, fits the experimental data satisfactorily.

At first glance, obtaining information concerning concentration profiles seems to pose a major problem. Fortunately, it turns out that in the case of thin surface layers, the form function is not influenced by the details of the profiles, only by the layer thickness (Fig. 26), which may, for example, be supplied by the physics of the sample manufacturing process. Moreover, within certain limits, the layer thickness can be checked by the results themselves. Figure 27 demonstrates that only one profile, which is characterized by a specific layer thickness, finally leads to a stable value of the surface loading that does not depend on the angle of incidence at which the respective value was obtained. Thus the TXRF method not only satisfies the main objectives but also provides additional information regarding the distribution of the atoms in the near surface layer.

B. Results

The strong and characteristic dependence of the fluorescence radiation on the angle of incidence also offers an opportunity to examine the reliability of the method.

Comparisons between calculated and experimental data covering a broad range of angles around the critical angle are a crucial test for the validity of the theoretical basis. Hence the comparisons between theoretical and measured data shown for a filmlike sample on a surface (Fig. 19), for the silicon signal of a silicon

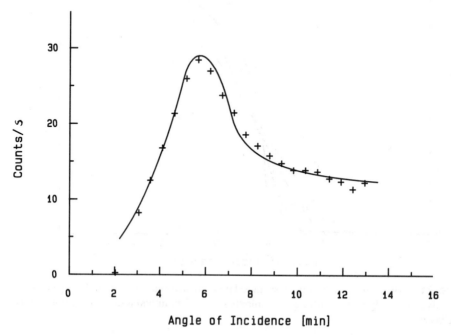

Figure 25   Fluorescence intensity of Ni in the surface of a silicon wafer versus angle of incidence (+ measured values; — calculated curve).

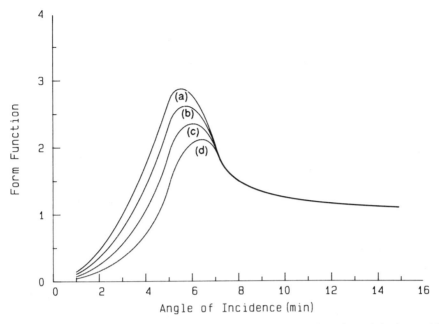

Figure 26 Form function versus angle of incidence as a function of the layer thickness (a, 0.1 nm; b, 2 nm; c, 5 nm; d, 10 nm).

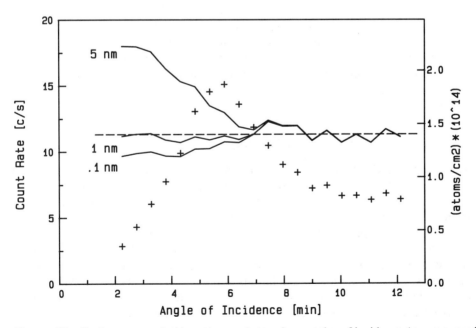

Figure 27 Surface contamination measured at various angles of incidence (+, measured values). The final result is obtained (best fit = 11.3 counts/s $\triangleq$ 1.4 $\times$ 10^{14} atoms/cm^2) by applying the form function f^* based on a layer thickness of 1 nm.

Table 3 Determination of Copper Concentrations in the Surface of Wafers[a]

	Wafer 1	Wafer 2	Wafer 3
TXRF	7.4×10^{12}	5.0×10^{13}	2.2×10^{15}
RBS	8.3×10^{12}	5.0×10^{13}	2.6×10^{15}

[a] All values are given in atoms/cm^2.

wafer (Fig. 22), and the nickel contamination in the near surface layer of a silicon wafer (Fig. 25) can serve as verifications of the technique, as can the results shown in Figure 27. This figure demonstrates with respect to concentration profiles that arbitrary assumptions that are not in line with reality can be perceived and sorted out during data evaluation.

To compare TXRF results with those obtained by Rutherford ion backscattering spectrometry (RBS), special test wafers with a well-defined metal coverage in the submonoatomic layer scale were prepared by argon ion sputtering. The bias on the pure metal sputter target was properly chosen to attain an implanted depth of less than 2 nm below the silicon interface. The test wafers and the RBS measurements were made by GeMeTec (München, Germany). As shown in Table 3, the results of both methods agree within 10–20%.

Compared to TXRF, the RBS method using nitrogen ions yields in general a somewhat better detection limit, at the expense, however, of element resolution; for example, Fe and Ni or Cu and Zn cannot be determined separately because of overlap effects.

Atomic absorption spectrometry and neutron activation analysis, which are also employed for the determination of impurities in silicon wafers, reach detection limits down to 10^8 atoms/cm^2 for selected elements. They need, however, an etching procedure to separate the near surface layer from the substrate. Therefore, TXRF provides unique features for elemental surface analysis: nondestructiveness and simplicity of operation as well as survey and quantitation capability of metals at the 10^{11} atoms/cm^2 level.

C. Multilayers and Depth Profiles

The fluorescence radiation that originates in a surface is described in general terms in Section V.A by Equation (15). Technically, this formulation covers perpendicular variations in the refractive index as well as changes in analyte concentrations given by a z dependence of A_v. In practice, however, Equation (15) can only be directly applied in a few special cases because even matrix effects are neglected. Moreover, integration problems may arise, and above all it must be regarded that the Fresnel relations are confined to homogeneous media and sharp interfaces. Therefore, model calculations of realistic near surface regions are usually conducted using stratified structures; any perpendicular inhomogeneities are approximated by a properly chosen sequence of homogeneous layers.

To calculate the radiation field in multilayers, the Fresnel equation given in Section II.B [Eq. (7)] must be substituted by a system of complex equations [Eq. (20)]. These relations cover the reflectivity at the boundary between arbitrary media in any stratified arrangement, in contrast to Equation (7), which is confined to the transition at the interface between a vacuum and an infinitely thick medium. In addition, this complex form considers the phase relations of the primary radiation. Therefore, it includes interference effects that occur when two or more layers contribute to reflection.

In this notation the relative intensity, which is reflected from layer $j + 1$ into layer j at the $j/j + 1$ interface in a stratified structure, is obtained from the complex Fresnel reflection coefficient $r_{j,j+1}$:

$$|r_{j,j+1}|^2 = \left| \frac{v_j - v_{j+1}}{v_j + v_{j+1}} \right|^2 \tag{20}$$

where

$$v_j = \sqrt{\phi^2 - 2\delta_j - 2i\beta_j}$$

with

ϕ = incident angle in vacuum
δ_j = δ of layer j [see Equation (2)]
β_j = β of layer j [see Equation (1)]

Based on Equation (20), different algorithms have been applied to compute the primary intensity at any position in a multilayer system. These techniques are described elsewhere [34,36,37].

Some typical results of these kind of model calculations, which additionally comprise the fundamental parameter method for interelement corrections, are given in Figures 28 and 29.

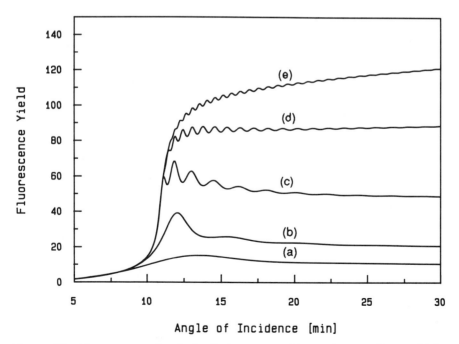

Figure 28 Fluorescence signal of a Co layer on a silicon substrate. (Layer thickness: a = 10, b = 20, c = 50, d = 100, e = 150 nm.)

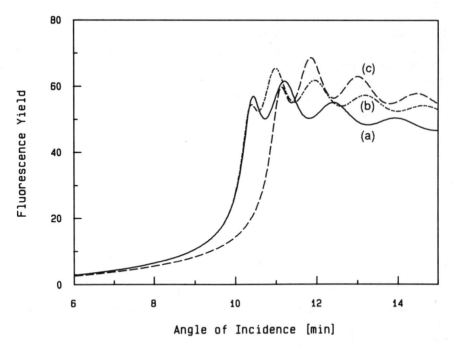

Figure 29 Effect of the density on the fluorescence signal of a Co layer on a silicon substrate: (a) p = 7 g/cm³, d = 50 nm; (b) p = 7 g/cm³, d = 57 nm; (c) p = 8 g/cm³, d = 50 nm.

In Figure 28, the effects of the layer thickness on the fluorescence signal of metal layers on silicon wafers are simulated as a function of the angle of incidence. The oscillations are caused by interferences of the primary radiation in the metal layer between the incoming and the reflected beam. Their period is determined by the layer thickness. In addition, Figure 29 also shows that even combinations of density and layer thickness that yield the same number of metal atoms per unit surface (curves b and c) may show considerable differences in their fluorescence behavior. The expected effects appear to be strong and specific enough to permit deducing density and thickness from measured data by inversion of multilayer calculations [38].

To check the capability of the TXRF technique for depth profiling, the measured fluorescence intensity of a rough silicon surface was compared with results obtained using the multilayer model. The method used for converting the roughness into a density profile is illustrated in Figure 30. A rough surface is treated as a stratified structure characterized by layers, which show increasing density from zero on top of the rough zone to the bulk density on the bottom. The density of each layer is given through its corresponding space filling. Figure 31 shows the fluorescence yield as a function of the incident angle for the unpolished reverse side of a silicon wafer. Consistency of measured and computed data was obtained assuming an averaged top-to-valley roughness of 4 μm. This agrees, within the margins of uncertainty, with independent determinations, which resulted in 3.6 and 4.2 μm.

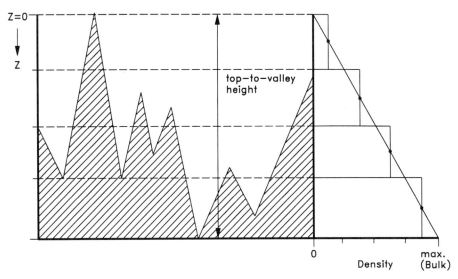

Figure 30   Model used for the simulation of a rough surface.

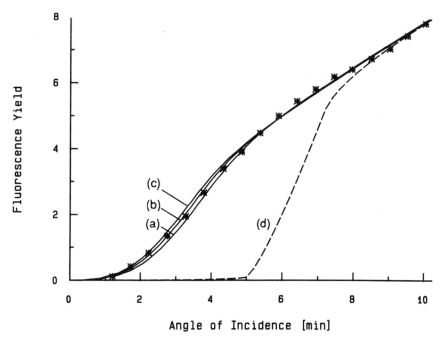

Figure 31   Determination of roughness by comparison with model calculations. *Measured values; curve a = 3 μm, curve b = 4 μm, curve c = 5 μm, curve d = 1 nm (top-to-valley, averaged).

VI. ROLE OF TXRF

The total reflection XRF technique is, despite its close kinship to conventional energy-dispersive XRF, quite different with respect to operation and performance. There is therefore much to be said for TXRF as a self-reliant branch of x-ray analysis. In fact, it is a technique that provides complementary capabilities for the employment of XRF analysis in fields that in practice have been reserved for other techniques. In surface spectroscopy specifically, it is at present the only available method that allows absolute concentration determinations to be made. Although the results obtained to date already look quite promising, the total reflection method, when compared to other XRF techniques, is just at the beginning of its development. Future work should cover instrumental and methodical investigations, as well as new applications in trace and surface analysis.

REFERENCES

1. Y. Yoneda and T. Horiuchi, *Rev. Sci. Instr. 42*:1069 (1971).
2. H. Aiginger and P. Wobrauschek, *Nucl. Instr. Methods 114*:157 (1974).
3. A. H. Compton, *Philos. Mag. 45*:1121 (1923).
4. A. H. Compton and S. K. Allison, *X-Rays in Theory and Experiment,* D. Van Nostrand Company, Princeton, NJ (1935).
5. Reprint of IUPAC document, *Spectrochim. Acta 33*:241 (1978).
6. W. Michaelis, A. Prange, and J. Knoth, in *Instrumentelle Multielement-Analyse* (B. Sansoni, Ed.), VCH-Verlagsgesellschaft, Weinheim (1985), p. 269.
7. W. Michaelis, J. Knoth, A. Prange, and H. Schwenke, *Adv. X-ray Anal. 28*:75 (1985).
8. R. P. Stössel and A. Prange, *Anal. Chem. 57*:2880 (1985).
9. W. Michaelis, *Techn. Mitteilungen 79*:266 (1986).
10. R.P. Stössel and W. Michaelis, Environmental Contamination, in 2nd International Conference, Amsterdam (1986), pp. 85–89.
11. A. Prange, J. Knoth, R.P. Stössel, H. Böddeker, and K. Kramer, *Anal. Chim. Acta 195*:275 (1987).
12. W. Junge, J. Knoth, and R. Rath, *N. Jb. Miner. Abh. 147*:169 (1983).
13. U. Hentschke, W. Junge, and R. Rath, *N. Jb. Miner. Abh. 152*:113 (1985).
14. W. Gerwinski and D. Goetz, *Fresnius Z. Anal. Chem. 327*:690 (1987).
15. A. von Bohlen, R. Eller, R. Klockenkämper, and G. Tölg, *Anal. Chem. 59*:2551 (1987).
16. A. Prange, A. Knöchel, and W. Michaelis, *Anal. Chim. Acta 172*:79 (1985).
17. A. Prange and K. Kremling, *Mar. Chem. 16*:259 (1985).
18. S. Batsford, M. Schwerdtfeger, R. Rohrbach, J. Knoth, C. Cambiaso, A. Vogt, and R. Kluthe, in *Histidine III: Laboratory and Clinical Aspects* (P. Fürst and R. Kluthe, Eds.), Wissenschaftliche Verlagsgesellschaft, Stuttgart (1986), p. 81.
19. R. Eller and G. Weber, *Fresenius Z. Anal. Chem. 328*:492 (1987).
20. A. von Bohlen, R. Klockenkämper, H. Otto, G. Tölg, and B. Wiecken, *Int. Arch. Occup. Environ. Health 59*:403 (1987).
21. W. Michaelis, H.-U. Fanger, R. Niedergesäss, and H. Schwenke, in *Instrumentelle Multielementanalyse* (B. Sansoni, Ed.), VCH Verlagsgesellschaft, Weinheim (1985), p. 693.
22. H. Schwenke, J. Knoth, and H. Böddeker, Totalreflexions-Röntgenfluoreszenzanalyse, in First Workshop, GKSS/E/61 (1986), p. 55.
23. A. Prange, *Spectrochim. Acta 44B*:437 (1989).

24. A. Prange and H. Schwenke, *Adv. X-ray Anal. 32*:209 (1989).

25. EXTRA II, TXRF with Multiple Total Reflection, Rich. Seifert & Co., Bogenstrasse 41, D-2070 Ahrensburg, Germany.

26. H. Aiginger, P. Wobrauschek, and C. Streli, Totalreflexions-Röntgenfluoreszenzanalyse, in First Workshop, GKSS/E/61 (1986), p. 20.

27. P. Wobrauschek and H. Aiginger, *Spectrochim. Acta 35B*:607 (1980).

28. M. Schmitt, P. Hoffman, and K. H. Lieser, *Fresenius Z. Anal. Chem. 328*:594 (1987).

29. B. Rastegar, F. Jundt, A. Gallmann, F. Rastegar, and M. J. F. Leroy, *X-ray Spectrom. 15*:83 (1986).

30. J. Knoth and H. Schwenke, *Fresenius Z. Anal. Chem. 294*:273 (1979).

31. B. Schneider, *Spectrochim. Acta 44B*:519 (1989).

32. P. Eichinger, H. J. Rath, and H. Schwenke, in *Semiconductor Fabrication: Technology and Metrology,* ASTM STP 990 (D. C. Gupta, Ed.), American Society for Testing and Materials, 1988.

33. XSA 8000 X-Ray Surface Analyzer, Atomika Technische Physik GmbH, Postfach 450 135, D-8000 München, Germany.

34. M. J. Bedzyk, G. M. Bommarito, and J. S. Schildkraut, *Phys. Rev. Lett. 62*:1376 (1989).

35. M. Blochin, *Physik der Rötgenstrahlen,* VEB Verlag Technik, Berlin, Germany (1957), p. 205.

36. L. G. Parratt, *Phys. Rev. 95*:359, (1954).

37. A. V. Andreev, A. G. Michette, and A. Renwick, *J Modern Optics 35*:1667 (1987).

38. E. Spiller, *Rev. Phys. Appl. 23*:1687 (1988).

10

Polarized Beam X-ray Fluorescence

Richard W. Ryon   *Lawrence Livermore National Laboratory, Livermore, California*

John D. Zahrt   *Los Alamos National Laboratory, Los Alamos, New Mexico*

I. INTRODUCTION

In this chapter, we explore the sources of spectral background and show how polarization effects can yield enhanced detection thresholds. We see that the stationary arrangement of components used in energy-dispersive x-ray fluorescence (EDXRF) is ideally suited for geometrical configurations that exploit polarization phenomena to reduce spectral background and thereby improve signal-to-noise ratios.

Observed spectral background is caused by several interactions of radiation with system components, the specimen, and the detector. A principal cause of background is the scatter of source radiation by the specimen into the detector. The scattered radiation adds directly to the background under analyte lines when broad-band primary radiation is used to excite fluorescence. Even when monochromatic radiation is used for excitation, the scatter of this primary radiation adds indirectly to the background because of incomplete charge collection in the detector. In addition, primary radiation carries with it fluorescence from the x-ray tube anode contaminants and collimator materials, and the scattered radiation causes fluorescence of collimator materials between the specimen and detector and of the gold contact layer on the surface of the detector. Scattered radiation also causes low-energy background due to the residual electron kinetic energy when Compton scatter from the detector itself occurs. All these sources of spectral noise and interference can be reduced by minimizing the scatter of source radiation into the detector by using polarized radiation to excite fluorescence. An additional advantage of this technique in trace element analysis is that the intensity of the source radiation can be increased, thereby proportionally increasing the intensity of analyte lines without exceeding the count rate limitations of the detection electronics due to the counting of unwanted scattered source radiation.

Polarized radiation can lead to dramatic improvements in signal-to-noise ratios and detection thresholds, but other sources of background remain. These residual sources of background include some source radiation due to finite solid angles required in real spectrometers, Klein-Nishina limitations to the obtainable polarization at higher energies, bremsstrahlung produced by photoelectrons in the specimen, and detector limitations, such as escape and sum peaks associated with analyte lines and Compton scatter in the detector. We explore some of these limitations and their magnitudes later in this chapter.

Polarized x-rays may be produced by various interactions of radiation with matter [1], or the source may be intrinsically polarized, as in synchrotron radiation. When using conventional x-ray tubes, radiation caused by a single collision of an electron with the anode is polarized. Examples are bremsstrahlung from thin targets and radiation near the maximum energy for thick targets. X-rays passing through crystals by Borrmann diffraction are also polarized. However, scattering of x-ray tube radiation from suitable materials has proven to be the most promising for self-contained EDXRF facilities. Polarization of visible light by scattering may be observed by looking at the sky with a Polaroid filter. Glass lasers are often cut at the Brewster angle of 0.317π (57°) to polarize the emitted light. With x-rays, the index of refraction is very nearly unity, so the corresponding angle required for nearly complete polarization is $\pi/2$.

A decade after the discovery of x-rays by Roentgen in 1895, Barkla [2] demonstrated that this newly discovered radiation could be polarized by scattering, thus supporting the hypothesis that x-rays are electromagnetic radiation (that is, their wave nature). A few years later, Friedrich et al. [3] demonstrated x-ray diffraction, which also substantiated the electromagnetic hypothesis.

Thomson found in 1933 [4], based on the classical electromagnetic theory, that the intensity of radiation scattered by a free electron is

$$I_s + I_0 \frac{1}{r^2} \left(\frac{e^2}{m_0 c^2} \right)^2 \sin^2 \alpha \tag{1}$$

where

I_0 = the intensity of the incident beam at the x-ray tube window
I_s = intensity of the scattered beam at distance r
e = charge of the electron (1.6×10^{-19} C)
m_0 = rest mass of the electron (9.11×10^{-31} kg)
c = speed of light (3.00×10^8 m/s)
r = distance to the point of observation (m)

and where α is the angle between the scattering direction $\vec{k}$ and the direction of acceleration of the electron. The acceleration vector is perpendicular to the direction of propagation of the incident radiation $\vec{k}_0$ and parallel to the electric field vector $\vec{E}_0$ of the incident radiation. If $\vec{k}_0$ is in the z direction and $\vec{k}$ is in the xz plane, we have the situation as shown in Figure 1. For standard x-ray sources, such as x-ray tubes, $\vec{E}_0$ of the photons incident on the scattering electron is random in direction (but always perpendicular to the direction of propagation) so that on

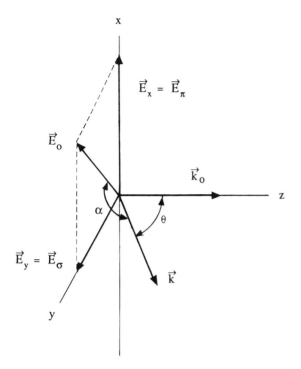

Figure 1   Geometry for Thomson scattering. $\vec{E}_\pi$ is in the xz scatter plane and $\vec{E}_\sigma$ is perpendicular to it. θ is the angle between the direction of the incident beam and the direction of the scattered beam in the xz plane. α is the angle between the electron acceleration and the direction of scatter.

the average

$$\langle E_x^2 \rangle = \langle E_y^2 \rangle = \frac{1}{2} E_0^2 \tag{2}$$

To obtain the scattered intensity at a general scattering angle θ in the xz plane, we can simply decompose the incident $\vec{E}_0$ into components and sum their individual intensity contributions. Now $\vec{E}_y$ causes the electron to oscillate in the y direction so that the angle $\alpha = \pi/2$ and $\sin^2 \alpha = 1$. $\vec{E}_x$ causes the electron to vibrate in the x direction, so the angle $\alpha = \pi/2 + \theta$ and $\sin^2 \alpha = \cos^2 \theta$, where θ is the indicated scattering angle.* Combining these considerations of the dependencies of E_x and E_y on the scattering angle α and recalling that the intensity of a wave is equal to the square of its amplitude ($I = E_0^2 = E_x^2 + E_y^2$), we obtain from Equations (1) and (2) the result

$$I_s = I_0 \frac{1}{(2r^2)} \left(\frac{e^2}{m_0 c^2} \right)^2 (1 + \cos^2 \theta) \tag{3}$$

From small angle solid geometry, the quantity r^2 in Equation (3) is equal to $dA/d\Omega$, where dA is the surface area of the expanding wavefront of solid angle $d\Omega$

---

* $\vec{E}_y$ is often denoted E_σ, σ or its Latin equivalent s standing for the German *senkrecht*, meaning perpendicular. Likewise, $\vec{E}_x$ is often denoted E_π; $\pi = p$ for parallel in German or English.

at a distance r from the point of scatter. Therefore,

$$\frac{I_s}{I_0} = \frac{1}{2}\left(\frac{e^2}{m_0 c^2}\right)^2 (1 + \cos^2 \theta) \frac{d\Omega}{dA} \tag{4}$$

If the intensity of the scattered radiation is summed over the surface area dA, we have the fraction of the incident photon flux, which is scattered by the electron into the solid angle $d\Omega$ in the direction θ. We therefore have

$$\frac{I_s}{I_0} dA = \frac{1}{2}\left(\frac{e^2}{m_0 c^2}\right)^2 (1 + \cos^2 \theta) \, d\Omega \equiv \frac{d\sigma_e}{d\Omega} \, d\Omega \tag{5}$$

where $d\sigma_e/d\Omega$ is defined as the differential cross section for scatter by a single electron and is given by

$$\frac{d\sigma_e}{d\Omega} = \frac{1}{2} r_0^2 (1 + \cos^2 \theta) \tag{6}$$

where $r_0 = (e^2/m_0 c^2) = 2.82 \times 10^{-15}$ m, the classical electron radius. The total cross section σ_e is found by integrating Equation (6) over all solid angles:

$$\sigma_e = \int_0^{4\pi} \frac{d\sigma_e}{d\Omega} \, d\Omega = \frac{1}{2} r_0^2 \int_0^{\pi} (1 + \cos^2 \theta) 2\pi \sin \theta \, d\theta = \frac{8}{3} \pi r_0^2 \tag{7}$$

This type of scattering is often referred to as Rayleigh scattering but is more appropriately called Thomson scattering.[†] Note that, relative to the core atomic electrons, the valence electrons are "free" in the Thomson sense. Corrections for electrons that are neither core nor valence requires the atomic structure factor, which is described later.

Compton scattering is often taken to have the same cross section as Thomson scattering, but Thomson cross sections are only low-energy approximations. At higher energies, the inclusion of quantum mechanical considerations leads to the Klein-Nishina formulation [5], presented in Chapter 1, Equation (54), which is repeated here:

$$\frac{d\sigma_{KN}}{d\Omega} = \frac{1}{2} r_0^2 \frac{1 + \cos^2 \theta}{[1 + \gamma(1 - \cos \theta)]^2} \left[1 + \frac{\gamma^2 (1 - \cos \theta)^2}{(1 + \cos^2 \theta)[1 + \gamma(1 - \cos \theta)]} \right] \tag{8}$$

where γ = incident photon energy/$m_0 c^2$ = $1.96 \times 10^{-3} \times$ incident photon energy (keV). What we present in Equation (8) is the collisional cross section, which is also the approach of most other authors. Because of the decrease in energy upon Compton scattering, we need the energy scattered in a particular direction as a

---

[†] Rayleigh, in 1871, concerned himself with scattering from dielectric spheres in the long-wavelength limit and derived (also from classical electromagnetic theory)

$$\frac{d\sigma}{d\Omega} = k^4 a^6 \left| \frac{\epsilon - 1}{\epsilon + 2} \right| \frac{1 + \cos^2 \theta}{2}$$

where $k = 2\pi/\lambda$ and a is the radius of a uniform isotropic dielectric sphere with dielectric constant ϵ. This has a large wavelength (energy) dependence; Thomson scattering from free electrons is wavelength (energy) independent. Unfortunately, some authors have looked only at the angle dependence of Equation (3) and termed it Rayleigh scattering. This is a fundamental misnomer with regard to the physics of x-ray scattering.

fraction of the incident intensity [6]. Thus, the differential Klein-Nishina scattering cross section is $d\sigma^s_{KN}/d\Omega = (v'/v_0)(d\sigma_{KN}/d\Omega)$. The wavelength shift due to Compton scattering is $\lambda' - \lambda_0 = h(1 - \cos\theta)/m_0 c$; converting to frequency by the relationship $\lambda v = c$, we obtain $v'/v_0 = 1/[1 + \gamma(1 - \cos\theta)]$, so that

$$\frac{d\sigma^s_{KN}}{d\Omega} = \frac{1}{2}r_0^2 \frac{1 + \cos^2\theta}{[1 + \gamma(1 - \cos\theta)]^3}\left\{1 + \frac{\gamma^2(1 - \cos\theta)^2}{(1 + \cos^2\theta)[1 + \gamma(1 + \gamma(1 - \cos\theta)]}\right\}$$

(9)

The fractional background reduction, even at a scattering angle of exactly $\pi/2$, is not quite zero because of the Klein-Nishina limitation. This limitation to achievable polarization is significant when exciting K-line fluorescence of heavy elements [7]. As γ approaches zero in Equation (9), the Klein-Nishina cross section approaches the classical Thomson value in Equation (6). If the small Klein-Nishina effect at lower energies is ignored, it follows that unpolarized incident electromagnetic radiation scattered through an angle of $\theta = \pi/2$ is linearly polarized, with only E_σ surviving. In Barkla's experiment [8], a source of x-rays is incident on a scatterer S_1, as shown in Figure 2. The scattered rays from S_1 intercepted by S_2 are nearly linearly polarized as the angle $S_0 - S_1 - S_2$ is $\pi/2$. The electrical field vector is perpendicular to the $S_0 - S_1 - S_2$ plane. Radiation from S_1 scattered by S_2 into I_1 also undergoes $\pi/2$ scattering, but it is in the same plane as the first scattering and no further annihilation takes place. However, scattering of the radiation from S_1 by S_2 into I_2 (also through $\pi/2$) annihilates all

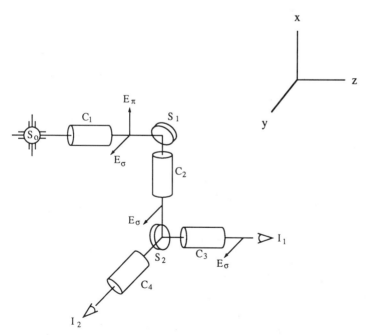

Figure 2 Schematic geometry for polarized beam XRF. S_0 is the source, S_1 is the polarizing scatterer, S_2 is the specimen, and I_2 is the intensity at the detector. I_1 is the intensity at an alternate, in-plane position for the detector, which does not eliminate source radiation. The measured degree of polarization is $(I_1 - I_2)/(I_1 + I_2)$.

remaining radiation. Thus I_1 receives radiation but I_2 does not. This discussion is idealized: the measured value of I/I_0 depends on the tightness of collimation of the paths S_0 to S_1, S_1 to S_2, S_2 to I_1, and S_2 to I_2, and also on the materials and thicknesses of S_1 and S_2.

Because of the nature of the polarization experiments, namely, the desirability of a constant $\pi/2$ scattering angle, the polarization of x-rays played little role in XRF (except for quantitative corrections) until the advent of EDXRF (but see Ref. 9). With the advent of EDXRF and the fixed geometry associated with it, it became possible to make good use of the polarization technique to significantly reduce the background radiation at the detector (which in unpolarized systems is predominantly scattered radiation from the source). The effectiveness of this technique is dramatically shown in Figure 3.

Several different kinds of polarized x-ray systems are in use. Systems that polarize the continuum, or bremsstrahlung, are discussed in Section II. We refer to these as Barkla scatterers, which includes both Thomson and Compton scat-

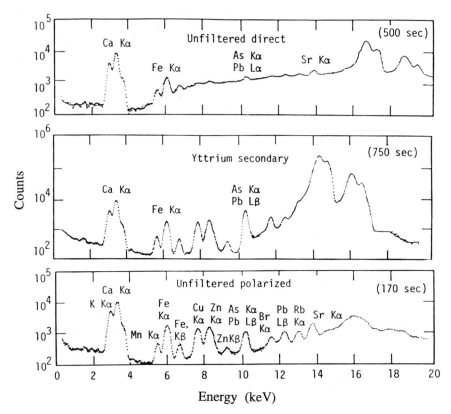

Figure 3 Comparison of excitation methods. The spectra are (top) unpolarized, direct excitation by radiation from molybdenum anode x-ray tube; (middle) yttrium secondary excitation; (bottom) polarized excitation by scatter from boron carbide. Each spectrum was accumulated until the Ca $K\alpha$ peak had a maximum of 10^4 counts. The counting rate was adjusted by first increasing the tube current to its maximum value and then increasing the collimator apertures to give a total count rate of approximately 2×10^4 counts per second.

tering. There are beam systems in orthogonal geometry and also systems that use curved or cylindrical scatterers and hence make use of a continuous fan of beams. Collimation effects and multiple-layer scatterers are discussed in Section II.

Because Bragg diffraction is one form of coherent scattering, it follows that diffracting a beam of x-rays through $2\theta_{\text{Bragg}} = \pi/2$ results in a monochromatic beam of plane-polarized x-rays. These can be used to cause analyte fluorescence; with the proper positioning of the detector one can obtain a very clean spectrum. This technique was pioneered by Aiginger et al. [10]. A crystal can be bent or bent and ground (Johann and Johansson crystal focusing optics, respectively), so that again one can use a fan of beams to increase the intensity at the sample. These ramifications make up the content of Section III.

In Section IV we note the very exciting area of synchrotron radiation, which is both very brilliant and highly polarized. In Section V we present a number of analytical applications in which researchers have successfully exploited the benefits of polarized radiation. In the conclusion, Section VI, we present our assessment of the present state of affairs in the field of polarized XRF.

Appendices VI, VII, and VIII in Chapter 1 provide x-ray cross sections for scattering and photoelectric absorption that may be used to calculate the efficiency of various materials that may be used as Barkla polarizers. The appendix in this chapter lists crystals that may be appropriate as Bragg diffraction polarizers.

II. BARKLA SYSTEMS
A. Orthogonal Systems

Proof-of-principle experiments in the early 1970s [11,12], using single collimators to define the three orthogonal beams (Fig. 2), demonstrated improved signal-to-noise ratios, but at severely reduced intensities. A few years later, it was demonstrated [13] that the way to maximize analyte intensities while minimizing background is to open the apertures of the collimators until maximum system throughput is achieved while maintaining the orthogonal beam geometry. Such nonzero beam divergence causes a small decrease in the degree of polarization, which is more than compensated for by the increase in x-ray flux. Detection limits are improved in comparison to either direct excitation or secondary fluorescers [13,14].

EDXRF is ideally suited to measuring a wide range of elements simultaneously because all photon energies are separately binned. When the analytical goal is to exploit this capability of measuring many elements with good sensitivity and low detection limits, a polarized continuum (or bremsstrahlung) source is preferred. This is in contrast to nonpolarized sources, for which the increased sensitivity due to efficient excitation is offset by the increased background under the peaks.

From the discussion in Section I of the idealized polarization of a random source of x-rays, it was seen that E_π can be completely eliminated when the beam is scattered at an angle of exactly $\pi/2$ and that E_σ can likewise be eliminated when the beam is scattered a second time at an angle of exactly $\pi/2$ relative to the first scatter plane. However, to achieve these exact angles, the beams must be highly collimated, approaching zero divergence. This means that the beam cross section must approach zero, and therefore the intensity likewise must approach zero. Clearly, some compromise must be made in a practical system. If one examines the system, a way becomes clear.

There are three scattering angles and three divergences about these angles to be considered that determine the amount of source radiation scattered through the system to the detector (referring to Fig. 2):

$$\theta_1 \pm \omega_1 = \text{first scattering angle} \pm \text{deviation in the } xz \text{ plane}$$

$$\theta_2 \pm \omega_2 = \text{second scattering angle} \pm \text{deviation in the } xy \text{ plane} \qquad (10)$$

$$\theta_3 \pm \omega_3 = \text{angle between the planes} \pm \text{deviation in this angle}$$

where the ω are ½ the range of the planar divergences about the θ. In the geometry for maximal polarization, all the θ are $\pi/2$, all the axes are mutually perpendicular, and the ω approach zero. When examining the scatter events, it is more convenient to think in terms of the collimator angles. These angles are defined as the arctangent of the ratio of the diameter to length for each tubular collimator:

$$\omega_{c1} = \text{source to polarizer collimator divergence}$$

$$\omega_{c2} = \text{polarizer to specimen collimator divergence} \qquad (11)$$

$$\omega_{c3} = \text{specimen to detector divergence}$$

The divergences about the scattering angles in Equations (10) are the convolution of the collimator angles in the definitions (11) and are approximated by

$$\omega_1 \simeq (\omega_{c1}^2 + \omega_{c2}^2)^{1/2} \qquad \text{in the } xz \text{ plane}$$

$$\omega_2 \simeq (\omega_{c2}^2 + \omega_{c3}^2)^{1/2} \qquad \text{in the } xy \text{ plane} \qquad (12)$$

$$\omega_3 \simeq (\omega_{c1}^2 + \omega_{c3}^2)^{1/2} \qquad \text{between the planes}$$

In practice, the collimators are made to be of similar dimensions; the subscripts in the definitions can thus be dropped, so that we have

$$\text{Average collimator divergence} = \pm\omega_c \qquad (13)$$

$$\text{Average deviation about scattering angle} = \pm\omega = (2\omega_c^2)^{1/2} \qquad (14)$$

The divergences through the collimators are small for a system with a fairly high degree of polarization. We can therefore use series approximations with only the first terms for the trigonometric function in Equation (1). That is, the average value of $\sin^2 \pm \omega = 1/3\omega^2$ and the average value of $\sin^2(\pi/2 \pm \omega) = 1 - \frac{1}{3}\omega^2$. We can likewise neglect higher order terms in ω in subsequent arithmetic.

We first consider the intensity component values along the three orthogonal axes of the coordinate system after the beam passes through the first collimator and before it strikes the polarizer. Because the ratio of collimator diameter to length is a finite value, the electrical vectors project onto all three coordinate axes. Along the x and y axes, the values are $\langle E_x^2 \rangle = \langle E_y^2 \rangle = \frac{1}{2}E_0^2 \cos^2 \omega$. Along the z axis, contributions are made due to tipping of both E vectors, and the intensity is $\langle E_z^2 \rangle = 2(\frac{1}{2}E_0^2) \sin^2 \omega$. When these projections are averaged over the possible range of angles $\pm\omega_c$ and the intensities are corrected for the solid angle, we find that the magnitudes of the electrical vectors before scattering at the first scatterer are

$$\langle E_x^2 \rangle_0 = \frac{1}{2} E_0^2 (1 - \frac{1}{3}\omega_c^2) \frac{\Omega_c}{4\pi}$$

$$= \frac{1}{2} E_0^2 (1 - \frac{1}{3}\omega_c^2) \frac{\omega_c^2}{4} \qquad (15)$$

$$\langle E_y^2 \rangle_0 = \frac{1}{2} E_0^2 (1 - \frac{1}{3}\omega_c^2) \frac{\omega_c^2}{4}$$

$$\langle E_z^2 \rangle_0 = \frac{1}{2} E_0^2 (\frac{2}{3}\omega_c^2) \frac{\omega_c^2}{4}$$

The intensities at the second scatterer are found by applying the Thomson scatter equation [Eq. (1)] to the intensity vector projections on the coordinate axes [Eqs. (15)] and correcting for the solid angles. When scattered in the x direction, the intensities are

$$\langle E_x^2 \rangle_1 = \frac{1}{2} E_0^2 r_0^2 (\frac{1}{3}\omega^2) \left(\frac{\omega_c^2}{4} \right)^2$$

$$\langle E_y^2 \rangle_1 = \frac{1}{2} E_0^2 r_0^2 (1 - \frac{1}{3}\omega_c^2) \left(\frac{\omega_c^2}{4} \right)^2 \qquad (16)$$

$$\langle E_z^2 \rangle = \frac{1}{2} E_0^2 r_0^2 (\frac{2}{3}\omega_c^2) \left(\frac{\omega_c^2}{4} \right)^2$$

Similarly, if the beam is scattered a second time in a direction perpendicular to the first scatter plane, that is, along the y axis, the electrical vector magnitudes at the end of the third collimator become

$$\langle E_x^2 \rangle_{\text{perp}} = \frac{1}{2} E_0^2 (r_0^2)^2 \frac{1}{3}\omega^2 \left(\frac{\omega_c^2}{4} \right)^3$$

$$\langle E_y^2 \rangle_{\text{perp}} = \frac{1}{2} E_0^2 (r_0^2)^2 \frac{1}{3}\omega^2 \left(\frac{\omega_c^2}{4} \right)^3$$

$$\langle E_z^2 \rangle_{\text{perp}} = \frac{1}{2} E_0^2 (r_0^2)^2 \frac{2}{3}\omega_c^2 \left(\frac{\omega_c^2}{4} \right)^3 \qquad (17)$$

$$= \frac{1}{2} E_0^2 (r_0^2)^2 \frac{1}{3}\omega^2 \left(\frac{\omega_c^2}{4} \right)^3$$

The intensity of the scattered radiation after scattering twice at mutually perpendicular angles is the sum of the three parts in Equations (17), which is approximately

$$I_{\text{perp}} = \langle E^2 \rangle_{\text{perp}} = \frac{1}{2} E_0^2 (r_0^2)^2 \omega^2 \left(\frac{\omega_c^2}{4} \right)^3 \qquad (18)$$

The quantity $\frac{1}{2}\omega^2$ is the fraction of the original intensity, uncorrected for solid angles and the interaction cross section, which is scattered into the detector. With very tight orthogonal geometry, this quantity would be zero and there would be no background (and, of course, no fluorescence signal either).

If the second scatter were parallel to the first scatter plane (along the z axis) instead of perpendicular to it, the electrical vectors are similarly found to be

$$\langle E_x^2\rangle_{\text{paral}} = \frac{1}{2} E_0^2(r_0^2)^2 \, \tfrac{1}{3}\omega^2 \left(\frac{\omega_c^2}{4}\right)^3$$

$$\langle E_y^2\rangle_{\text{paral}} = \frac{1}{2} E_0^2(r_0^2)^2(1 - \tfrac{1}{3}\omega_c^2) \left(\frac{\omega_c^2}{4}\right)^3$$

$$= \frac{1}{2} E_0^2(r_0^2)^2(1 - \tfrac{1}{6}\omega^2) \left(\frac{\omega_c^2}{4}\right)^3$$

$$\langle E_z^2\rangle_{\text{paral}} = \text{nil}$$

$$\tag{19}$$

Summing the three terms for the in-plane scatter given in Equations (19), the parallel scatter intensity is

$$I_{\text{paral}} = \langle E^2\rangle_{\text{paral}} = \frac{1}{2} E_0^2(r_0^2)^2(1 + \tfrac{1}{6}\omega^2) \left(\frac{\omega_c^2}{4}\right)^3 \tag{20}$$

The system polarization as determined by geometry is defined as

$$P_g = \frac{I_{\text{paral}} - I_{\text{perp}}}{I_{\text{paral}} + I_{\text{perp}}}$$

$$= 1 - 2\omega^2 = 1 - 4\omega_c^2 \tag{21}$$

Let us now turn to the question of whether a polarized beam system can compete with other means of excitation, considering that extra collimation is required to produce the polarization effect. We consider three methods of excitation: direct, secondary fluorescer, and polarizer. Spectral background is considered to arise from scattered radiation, and the "noise" is the statistical fluctuation in this quantity, namely, the square root of the scattered radiation.

For direct excitation, the scattered source radiation is the sum of the quantities in Equations (16), assuming $\pi/2$ geometry. If the constants are ignored so that the functionality is emphasized, the scattered radiation at the detector is

$$I_{\text{scat,dir}} = I_{0,sx}K(1 + \omega_c^2) \left(\frac{\omega_c^2}{4}\right)^2 \tag{22}$$

where K is used here and in what follows simply as a proportionality constant. The fluorescence signal [15] is given by

$$I_{\text{fluor,dir}} = I_{0,sx} \frac{K_A}{(\mu/\rho)} W_A(1 - e^{-\mu/\rho\rho T} \frac{d\Omega}{4\pi} \tag{23}$$

where

$\begin{aligned}
I_{0,sx} &= \text{beam intensity impinging on the sample, s}^{-1} \\
&= I_0(\omega_c^2/4) \text{ in this case of directed excitation from Equation (15)} \\
K_A &= \text{proportionality constant consisting of fundamental parameters for} \\
&\quad \text{the excitation of analyte } A \\
W_A &= \text{weight fraction of the analyte} \\
\mu/\rho &= \text{mass absorption coefficient, cm}^2/\text{g} \\
\rho &= \text{density, g/cm}^3 \\
T &= \text{specimen thickness, cm} \\
d\Omega/4\pi &= \text{fractional solid angle of the exit collimator, } = \omega_c^2/4
\end{aligned}$

Collecting the constants,

$$I_{fluor,dir} = I_0 K \left(\frac{\omega_c^2}{4}\right)^2 \tag{24}$$

With a secondary fluorescer, the primary radiation is largely eliminated by photoelectric absorption in the heavy metal secondary. Furthermore, the fluorescer can be placed in the orthogonal beam geometry, further reducing source radiation by polarization. However, the characteristic lines from the fluorescer contribute to the background. By combining the considerations of Equations (5) and (24), the secondary fluorescer radiation that reaches the detector is

$$I_{scat,sec} = I_0 K \left(\frac{\omega_c^2}{4}\right)^3 \tag{25}$$

The fluorescence from the specimen is proportional to the radiation incident upon it [Eq. (24)], to the fluorescence efficiency [Eq. (23)], and to the fractional solid angle:

$$I_{fluor,sec} = I_0 K \left(\frac{\omega_c^2}{4}\right)^3 \tag{26}$$

For the polarized system, the background is Equation (18), restated here in simplified form:

$$I_{scat,pol} = \langle E^2 \rangle_{perp} = I_0 K 2\omega_c^2 \left(\frac{\omega_c^2}{4}\right)^3 \tag{27}$$

That is, the source intensity is attenuated by the three solid angles as for the secondary fluorescer, with an additional attenuation factor of $2\omega_c^2$ due to the polarization. The excitation beam is the same as the background with direct excitation, that is, Equation (22), which is substituted into the expression for fluorescence, Equation (24):

$$I_{fluor,pol} = I_0 K (1 + \omega_c^2) \left(\frac{\omega_c^2}{4}\right)^3 \tag{28}$$

The signal-to-noise ratios are estimated by

$$\frac{S}{N} = \frac{I_{fluor}}{I_{scat}^{1/2}} \tag{29}$$

which are therefore

$$\left(\frac{S}{N}\right)_{dir} = I_0^{1/2} K_{dir} \omega_c^2$$

$$\left(\frac{S}{N}\right)_{sec} = I_0^{1/2} K_{sec} \omega_c^3 \tag{30}$$

$$\left(\frac{S}{N}\right)_{pol} = I_0^{1/2} K_{pol} \omega_c^2$$

for the three excitation schemes.

For the same source intensity I_0, the signal-to-noise ratios for direct, secondary fluorescer, and polarized excitation are proportional to ω_c^2, ω_c^3, and ω_c^2, respectively. To the first order, there is nothing gained using a polarized source, and indeed something is lost using a secondary fluorescer, if the source intensity is fixed. However, if the source intensity is increased to compensate for the collimation and scatter efficiency losses, a gain is achieved. By comparing Equations (22), (25), and (27), we can see how much the source intensity can be increased. If the maximum system counting rate is achieved with direct excitation, then I_0 for the secondary fluorescer can be increased by approximately $1/(\omega_c^2/4)$ and likewise I_0 can be increased by approximately $1/(2\omega_c^2\omega_c^2/4)$ using a polarizer. The signal-to-noise ratios in Equations (30) are then improved by the square root of the increase in intensity and become

$$\left(\frac{S}{N}\right)_{\text{dir}} = K_{\text{dir}}\omega_c^2$$

$$\left(\frac{S}{N}\right)_{\text{sec}} = K_{\text{sec}}\omega_c^2 \qquad\qquad (31)$$

$$\left(\frac{S}{N}\right)_{\text{pol}} = K_{\text{pol}}\omega_c^0 = K_{\text{pol}}$$

The relationships for the signal-to-noise ratios in Equations (30) suggest in all cases that the best results are obtained by opening up the collimation system to obtain the highest usable count rate. The approximate analysis in Equations (31) also shows that the polarized system can yield the best signal-to-noise ratios and detection limits if the source intensity is adquate to overcome the geometrical and scatter losses. The proportionality constants in Equations (31) could be replaced by the actual physical parameters. Experiments have verified these general conclusions.

Experiments have been performed to compare the various geometrical arrangements for energy-dispersive fluorescence measurements [13]. An unfiltered Mo anode x-ray tube was used, which could be operated at a maximum of 2500 W. In the comparisons, the current was adjusted so that the counting rate was at its maximum, that is, approximately 20,000 cps input to the amplifier. If the maximum power rating for the tube was reached before the maximum counting rate was achieved, the collimator divergences were increased to give the maximum counting rate. The results are shown graphically in Figure 3. When a boron carbide polarizer was used, the detection limits for a thick, pressed specimen of NBS Orchard Leaves were 2–4.5 times lower in comparison to direct excitation, and about <1–3 times lower than when using the optimum secondary fluorescer, yttrium. The case of the yttrium secondary is notable, because the detection limits are marginally better when the analyte absorption edge is just below the Y $K\alpha$ line. At lower energies, at which the bremsstrahlung excitation becomes important for the polarized system, the detection limits are lower than with the secondary fluorescer, which is the next best alternative. The conclusion is that secondary fluorescers, particularly in the orthogonal geometry [16,17], are very good for a narrow range of elements, but that the polarized system is better for multielement analysis.

The specimen itself causes the amount of scatter to increase above that calculated from the system geometry. This increase is due to multiple scatter depolarization. This effect was investigated experimentally using a highly collimated system, with the results shown in Figure 4. The geometrically defined polarization was about 96%. A boron carbide scatterer with a thickness of 0.13 cm was used, which contributes very little to multiple scatter depolarization. Since bulk specimens limit the observable polarization to somewhat above 90%, the system collimation can be opened up a corresponding amount with little effect on the signal-to-background ratio but with a large effect on the counting rate.

Polarization losses have also been calculated for both single and double scatter. Monte Carlo techniques can be used, but Gaussian quadrature [18,19] is much more efficient for simple geometries. It is found that scatterer material outside the area of direct view of the collimators scatters radiation back into the collimation system, thereby contributing significantly to multiple scatter depolarization. Therefore, shaving off material not seen by both entrance and exit collimators improves the degree of polarization.

Tolokonnikov [20] suggested a novel means of increasing the excitation beam intensity in an orthogonal system. If the scatterer S_1 in Figure 2 is thin, a secondary fluorescer can be placed a short distance behind it. The high-energy continuum from the source passes through the scatterer, where it excites fluorescence from the secondary target. The secondary target becomes a source of x-rays, some of which scatter off S_1 and become polarized. Model calculations show that the intensity of the characteristic lines from the secondary source may be comparable in magnitude to those from the x-ray tube anode. Thus, enhanced characteristic lines are obtained for excitation while minimizing the high-energy continuum.

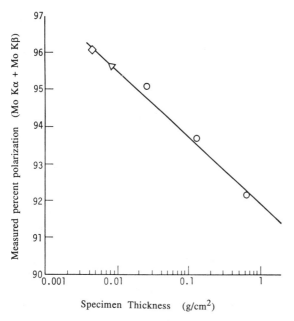

Figure 4   Measured polarization as a function of specimen thickness. The specimen is NBS Standard Reference Material 1571, Orchard Leaves. Polarization decreases as the specimen mass increases due to multiple scatter depolarization.

B. Multiple Layer Scatterers

As we have indicated before, if a single element is the object of an analysis, a monochromatic source gives the lowest detection limits. However, a polychromatic source is desirable when analyzing for a broad range of elements. If the polychromatic source is to be polarized by scattering, high-Z scatterers absorb the low-energy portion of the incident spectrum whereas low-Z scatterers do not interact effectively with the high-energy components of the spectrum. In both cases this limits the intensity over a broad range of energies.

It has been proposed [21] that a layered scatterer be used, with a low-Z material backed by a high-Z material. The effectiveness of these ideas was borne out by experiment, as shown in Figure 5. It should be noted that here Teflon is "high" Z whereas paraffin is "low" Z. These results also correlate nicely with simple model calculations as shown in Figure 6. The mathematical model is

$$P = \frac{3}{8\sqrt{2\pi}d} \left(\sigma_{1s} \int_0^{T_1} \frac{d}{\sqrt{2} - z} e^{-2\sqrt{2}\sigma_{1t}z} \, dz \right.$$
$$\left. + e^{-2\sqrt{2}\sigma_{1t}T_1} \sigma_{2s} \int_{T_1}^{d/\sqrt{2}} \frac{d}{\sqrt{2} - z} e^{-2\sqrt{2}\sigma_{2t}(z-T_1)} \, dz \right) \quad (32)$$

where

P = probability of scatter
d = thickness of the scatterer divided into two layers, 1 and 2
T_1 = thickness of layer 1
T_2 = thickness of layer 2
σ_{is} = scatter cross section of layer i (i = 1 or 2)
σ_{it} = total cross section of layer i (i = 1 or 2)

On the other hand, experiments have shown that pure B_4C is better for all analytes than any bilayer, such as B_4C-Teflon. At first surprising, these results were traceable to the comparisons of scattering and photoelectric linear cross sections of B_4C and Teflon. An imaginary system [21] that changed its cross section as a function of depth was modeled, and the finding was made that a linear variation was better than a step or quadratic variation. This result suggests that approximately equal thicknesses of low-Z and high-Z materials will yield the best broad-band system. This is borne out by Figure 5.

C. Cylindrical Systems

Because the scattering x-ray cross section for most materials is so low and because it is desirable to use a tightly collimated system, the time required to accumulate good counting statistics with polarized beam XRF may be excessive. However, it is possible to scatter from materials shaped into cylinders or cylinder segments. The x-rays reaching the sample would then constitute a broad fan beam, and the intensity should rise (roughly as the ratio of surface areas) with a concomitant decrease in analysis time. Figure 7 is a schematic sketch of such a device. Cylindrical devices have been constructed, but none have yet proven to be to be successful. Very careful attention must be given to design details to eliminate scattered and fluorescent radiation from the device components.

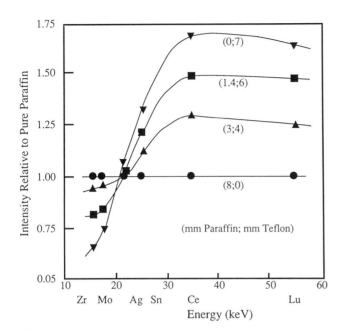

Figure 5 Experimental intensities for six analyte $K\alpha$ lines for various paraffin-Teflon bilayer scatterers, relative to pure paraffin. The best average sensitivity is measured when the two layers are roughly the same thickness. The primary x-ray source was a tungsten anode tube operated at 150 kV.

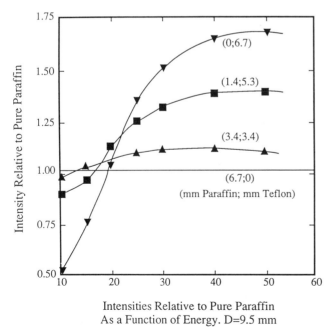

Intensities Relative to Pure Paraffin
As a Function of Energy. D=9.5 mm

Figure 6 Theoretical intensities for seven analyte $K\alpha$ lines for various paraffin-Teflon bilayer scatterers, relative to pure paraffin. The model used is given in Equation (32).

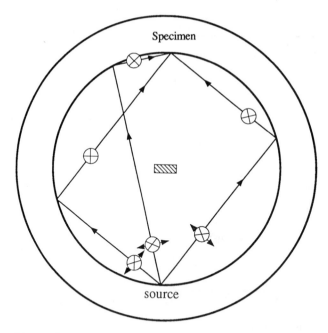

Figure 7   A cylindrical scatterer. The plain symbol represents electrical field oscillations perpendicular to the page. The symbol with arrow represents electrical field oscillations in the plane of the page. The detector would be located above the specimen along a line perpendicular to the page.

A cylindrical scatterer, such as that shown in Figure 7, has been modeled mathematically [22,23]. The modeling indicates that (1) the source and the sample should not sit on the cylinder wall but be slightly recessed to compensate for the finite mean free path of the x-rays in the scattering material; (2) the amount of double-scattered radiation reaching the sample is large; and (3) the amount of unpolarized radiation is large in double scatter. Items 2 and 3 may be involved in the poor performance of experimental instuments. The modeling also suggest that a baffle added along the line from source to sample will diminish double scatter significantly.

Cylindrical polarizers are being investigated in Scheer's group at the University of Bremen and in Wobrauschek's group at the Atomic Institute in Vienna. Results are beginning to appear from the group in Bremen [24,25]. They find a large scattering effect above 35 keV. They also find that a baffle on the source to sample line has little influence on the spectrum, in apparent contrast to the model calculations.

III. BRAGG SYSTEMS

A. Orthogonal System

Diffraction offers an excellent means of obtaining both monochromatic and polarized radiation. Synthetic multilayer materials that have very high reflectivities can have d spacings chosen virtually at will to give diffraction at 2θ of $\pi/2$ for energies up to about 0.90 keV (corresponding to the present minimum achievable d spacing of about 10 Å). For higher energies, crystalline materials must be used. Possible materials are given in the appendix.

In 1974, Aiginger et al. [10] reported using a copper crystal cut along the (113) plane to diffract and polarize a beam of Cu $K\alpha$ x-rays. The Bragg angle for this system is $45°01'39''$, and the intensity of the beam was reported as "magnitudes greater than that of a scattered beam from polycrystalline or any other matter."

The orthogonal triaxial geometry for Bragg diffraction polarization is similar to that for Barkla scattering, shown in Figure 2. What is important here is the physics of the diffraction process and the parameters of the diffracting crystal. There are two theories of x-ray diffraction; the wave kinematic (K theory) and the dynamic (D theory). The K theory is valid for small crystals or mosaic blocks. When absorption and/or the interaction between incident and diffracted beams becomes important, the D theory must be used. In Zachariasen's notation [26], the K theory gives

$$\frac{I_H}{I_0} = \left(\frac{e^2}{m_0 c^2 V}\right)^2 \frac{1 + \cos^2 2\theta_B}{2} |F_H|^2 \lambda^3 \frac{\delta V}{\sin 2\theta} \tag{33}$$

where

$$
\begin{aligned}
I_H/I_0 &= \text{the integrated intensity compared to the incident beam} \\
\theta_B &= \text{the Bragg diffraction angle} \\
F_H &= \sum_{n=1}^{N} f_n e^{2\pi i(hu_n + kv_n + lw_n)} \\
u_n, v_n, w_n &= \text{fractional coordinates} \\
&= 1/ax_n, \, 1/by_n, \, 1/cz_n \text{ of the } n\text{th atom in the unit cell} \\
a, b, c &= \text{unit cell lengths} \\
f_n &= \text{atomic scattering factor} \\
V &= \text{unit cell volume} \\
\delta V &= \text{volume of the crystal}
\end{aligned}
$$

and other symbols have their usual meanings. The atomic form factor f_n is approximately half-Gaussian in shape with argument $\sin \theta/\lambda$ and $f_n(0) = Z$, the number of electrons on atom n. To maximize I_H/I_0 we thus desire V to be small and F_H to be large. Making F_H large often means taking $\sin \theta/\lambda$ to be small, which from

$$\frac{1}{d^2} = \frac{h^2 + k^2 + l^2}{a^2} = \frac{4 \sin^2 \theta}{\lambda^2} \tag{34}$$

for cubic crystals implies low values for the Miller indices (h, k, l). Due consideration of the imaginary exponential in Equation (33) must be taken into account for each case. A list of possible diffracting crystals for various x-ray tube characteristic lines are given in the appendix.

For example, consider the Cu (113)–Cu $K\alpha$ system where $\lambda = 1.542$ Å and $\sin \theta/\lambda = 0.459$ Å^{-1}, so $f_{Cu} = 14.1$. Copper is face-centered cubic with atoms at $(0, 0, 0)$, $(\frac{1}{2}, \frac{1}{2}, 0)$, $(\frac{1}{2}, 0, \frac{1}{2})$, and $(0, \frac{1}{2}, \frac{1}{2})$, so the structure factor is $F(113) = 14.1\{1 + 1 + 1 + 1\} = 56.4$. The unit cell length is 3.61 Å, so $V = 47.0 \times 10^{-24}$ cm^3; $e^2/m_0 c^2 = 2.82 \times 10^{-13}$ cm, $\lambda^3 = 3.66 \times 10^{-24}$ cm^3, and taking

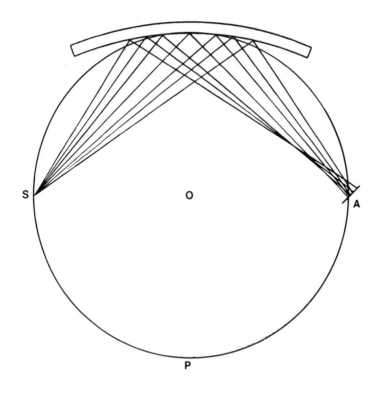

JOHANN

Figure 8 **(a)** The Johann curved crystal device. All the rays from the source S arrive arrive at a small, but defocused, region A. S and A are on the ends of a diameter of a circle (known as the Rowland circle) centered on O.

(arbitrarily) $\delta V = 0.001 \text{ cm}^3$, we have

$$\frac{I_H}{I_0} = \left(\frac{2.82 \times 10^{-13} \text{ cm}}{47.0 \times 10^{-24} \text{ cm}^3}\right)^2 \frac{1}{2} (56.4)^2 (3.66 \times 10^{-24} \text{ cm}^3) \, 10^{-3}$$

$$\approx 2.5 \times 10^{-4}$$

Real crystals usually have greater integrated intensities than estimated by the theory of ideal crystals as a result of the greater diffraction line width arising from the crystal mosaic structure. Nonetheless, the diffraction process may be no more efficient than Barkla scattering processes in depositing photons on the sample. The tradeoff is very high intensity in a narrow energy band for diffraction (good to excite one or two elements) versus lower intensity over a broad energy band for Barkla scattering (good for the excitation of many elements).

Detection limits using the copper system for Co (1.06 ng), Fe (1.1 ng), Mn (1.3 ng), and Cr (1.3 ng) have been reported [10]. With Cr $K\alpha$ radiation and Mg (1120) crystal, detection limits for Ca (3 ng) and Cl (14 ng) were also determined. Other sources and detection limits have been published [27,28].

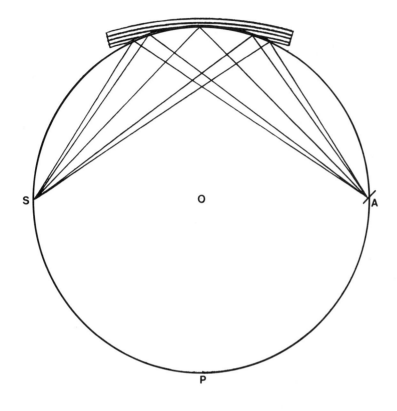

JOHANSSON

Figure 8 **(b)** The Johansson curved and ground crystal device. There is no defocusing as in a. The crystal is bent to the radius $2R$ and ground to the radius R of the Rowland circle.

B. Curved Crystals

To increase the intensity at the sample, one may bend (Johann) or bend and grind (Johansson) the crystals. Figure 8 shows the respective geometries with some ray tracing. With a flat crystal, the angular acceptance is essentially the rocking curve, generally $\geq 0.01°$ θ. With a bent crystal, an angular acceptance of 5–10° is readily conceivable, with an increase in intensity proportional to the size of the fan beam. Aiginger and Wobrauschek [27] report an intensity gain of 3.1 using Cu (113)–Cu $K\alpha$ in Johann geometry compared to the flat crystal. Calculations for Johann and Johansson geometries have been performed by Zahrt [29,30]. Although the calculations use estimated values of a mosaic distribution function and its full width at half-maximum, the reported gain is also calculated. Furthermore, gains of up to about 10-fold may be possible by proper preparation of the diffracting crystal.

The difficulties with using Bragg diffracted polarized x-rays center on two issues. First, it is somewhat difficult to obtain the diffracting crystal and cut it in such a manner that the diffracting planes are parallel to the surface plane. (This is likely the reason that there has been no rush to determine the mosaic parameters of the existing diffracting crystals.) Once the crystal is obtained, the imperfections should be optimized in such a manner as to maximize the integrated intensity at the sample. Second, the Barkla scatter systems are broad-band systems, whereas the Bragg diffraction systems are excellent for those elements that have K or L edges in a limited region below the energy of the diffracted beam [31].

IV. SYNCHROTRON RADIATION

Synchrotron radiation is intrinsically highly polarized because of its production by the radial acceleration of electrons in a plane. Other advantages include high flux, high brilliance, and tunable energy with very narrow bandwidth when the beamline is equipped with a double-crystal monochromator as is typically the case. High brilliance enables the use of highly collimated beams for such applications as scanning x-ray microscopy, the analysis of very small specimens, and other applications, such as very high resolution tomography [32]. Since the excitation energy can be selected to a bandwidth of 1 part in 10^4, it is possible to selectively excite elements that would otherwise suffer from spectral overlap. With respect to polarization, radiation from the 54-pole wiggler beam line at the Stanford Synchrotron Radiation Laboratory (SSRL) over a 2×2 mm area is approximately 95% [33].

V. RESULTS AND EXAMPLES

There is a dearth of real XRF applications of polarization. The early papers gave only peak-to-background ratios to show the reduction in background. More recent papers give sensitivities and detection limits for one type of specimen or another; however, these values are more for comparisons between polarized and unpolarized sources and are not necessarily state of the art since the systems were prototypes.

Polychromatic synchrotron radiation, when operating at a ring energy of 3.1 GeV, has given a detection limit as low as 1.0 ppm for iron in NBS SRM 1571 Orchard Leaves, when extrapolated to a beam current of 60–80 mA and a measurement time of 1000 s [34]. A standard Mo target x-ray tube whose output was polarized by scattering from boron carbide yielded similar detection limits: 0.84 ppm for iron, 18 ppm for potassium, and 0.30 ppm for strontium [13]. The similar results obtained from these two very different x-ray sources may be attributed to multiple-scatter depolarization in the specimens, which limits the lowest level of detection when analyzing bulk materials. Substantially lower detection limits are obtained when the synchrotron source is monochromatized and the specimen is thinner. Researchers using the 54-pole wiggler beam line at the Stanford Synchrotron Radiation Laboratory tuned to 17 keV obtained the following detection limits for cellulose of mass thickness 30 mg/cm^2: 0.038 ppm for chromium, 0.019 ppm for Zn, and 0.013 ppm for selenium when extrapolated to a 1000s counting time. An in-depth discussion of synchrotron-induced x-ray fluorescence analysis is given in Chapter 8.

At least one laboratory is using a standard polarized x-ray source for routine analyses for such materials as ash [35] and aircraft engine oils [36]. A 3000 W tungsten tube is used, with the output polarized by scatter from amorphous graphite. The system is collimated to give nearly 100% polarization, which is probably not quite optimum. In briquetted ash specimens with an areal density of 0.16 g/cm^2, the 3σ, 1000 s detection limits are 3–8 ppm for elements with Z between 26 and 42, 11–27 ppm for elements with Z greater than 42, and about 270 ppm for sulfur.

When the analysis is of thin deposits or small specimens, the problem of specimen depolarization is eliminated and synchrotron radiation can be fully utilized. Bragg-polarized Cu $K\alpha$ radiation has given such detection limits as 3 ng for manganese [10]. This is a very fine result for a prototype laboratory source, but the synchrotron radiation just mentioned gave detection limits at the subpicogram level for small area specimens on a thin substrate [33].

Various studies have been made with high-energy x-rays and γ-rays. One investigator, using a 200 kV x-ray tube, achieved about 90% of the Klein-Nishina polarization for radiation between 27 and 100 keV [6]. Polarization was produced by scattering from iron and aluminum. Another group used a similar 160 kV x-ray tube, polarized by scattering from boron carbide, to measure iodine in fluids [37]. They concluded that a 10-fold improvement in assay time could be achieved relative to using an ^{241}Am source (60 keV). Another group used a ^{57}Co source (122 keV) to investigate the measurement of heavy elements for medical studies. The investigators observed only a 50% reduction in background [38], which they attributed to inadequate shielding. There have been other successful in vivo measurements of heavy metals using polarized x-ray fluorescence [39].

VI. CONCLUSION

The published work on polarization for XRF analysis has been done in universities and research laboratories, but nothing has been published by researchers employed by instrument manufacturers. Since most analysts use commercial systems, the technique is not at present widely used. The instruments that can be purchased perform quite well, if not quite optimally. There is therefore little incentive in the commercial world to make changes in successful instrumentation. Instruments that use excitation by secondary fluorescers are arranged in an approximation to orthogonal geometry, thereby helping to suppress continuum radiation from reaching the detector. The principal drawback of polarization was alluded to in Section II.A, namely the need for sufficient power in the primary x-ray source to overcome the geometrical and scatter efficiency losses (even though there is evidence that this is not an insurmountable problem [14]). A higher power x-ray source translates into a more expensive spectrometer system.

The simplest expedient to improving Barkla polarizers was suggested by Dzubay et al. in the first widely distributed paper on polarization for XRF [11], namely, the use of a cluster of collimators rather than a single tubular collimator. Very fine leaded glass capillary bundles, such as those used in microchannel plate detectors, are obtainable from commercial sources. Nobody has yet done this experiment. Incremental improvements can also be expected by constructing polarizers to minimize multiple-scatter depolarization and the use of multiple-layer polarizers. The other advance that could be made is to use a curved polarizer.

This is not such a simple undertaking, but with careful attention to design details, an x-ray tube with polarized output may yet be constructed.

Polarized x-ray sources have repeatedly been demonstrated to give improved detection limits. There are many applications in which the reduced background from a polarized source can be useful. Typically, these are low-level measurements in bulk, low-Z specimens, such as organic materials, water, or glass fusion disks. Even in high-Z specimens in which the counting rate is dominated by the fluorescence of major elements, a reduced background can be appreciated when measuring the minor elements or when making the most precise measurements on the major elements.

APPENDIX: CRYSTALS THAT DIFFRACT SELECTED CHARACTERISTIC WAVELENGTHS AT $2\theta = \pi/2 = 90°$

The tables were prepared by Quintin C. Johnson, Chemistry and Materials Science Department, Lawrence Livermore National Laboratory, Livermore, CA 94550.

Good candidate materials for use in a polarized beam XRF experiment ideally have several desirable characteristics. The first criterion is a material that has a reflection of high diffraction power near 90° 2θ for the characteristic wavelength of the anode material used. In addition, the material should be stable, easy to orient and cut to the desired orientation, obtainable at reasonable cost, and perhaps nontoxic.

One way to find materials that diffract at the required angle is to search the JCPDS powder diffraction data base of approximately 45,000 patterns. The search was made for wavelengths equal to or longer than Cu $K\alpha$ for materials with strong reflections near 90°. This approach cannot be used for shorter wavelengths because the JCPDS data base was collected, for the most part, using copper radiation. Therefore, this data base generally does not contain useful information for reflections appropriate for short wavelengths. When the data base was first interrogated, many matches were found for obscure intermetallic alloys. The Common Phase subfile was therefore used with the expectation that the selected materials might actually be obtainable commercially. The results for Cu $K\alpha$ are contained in Table 1.

A second search was made, with the requirement for strong reflections eliminated, but with the requirements that the material must be easy to obtain, stable, and low in cost and have a reflection within 2° of 90° 2θ. This approach makes it much easier to find reasonable suggestions for materials, but no determination of the relative intensities of the reflections has been made. Tables 2 and 3 present some materials for the characteristic wavelengths of common anode materials.

For short-wavelength anodes the number of choices are many. One of the structures, garnet, is very common and can be altered chemically. It is suggested that the chemistry be tailored to obtain the ideal matching of the 90° reflection for the characteristic line of interest. For example, the readily obtainable single-crystal GGG (gadolinium gallium garnet) has a cell constant of approximately 12.383 Å. Other members of the garnet family have cell constants ranging from 11.5 to 13.7 Å. The technology of garnet synthesis and single crystal growth are

Table 1 Materials Selected from JCPDS Common Phase File for Cu $K\alpha$ Radiation

JCPDS	Material	hkl	d	I/I_0	2θ (degrees)
3-1215	LiAl	(440)	1.125	60	86.4
		(531)	1.074	60	91.6
25-0549	MnS_2	(250)	1.132	60	85.8
		(521)	1.114	60	87.5
		(440)	1.078170	70	91.2
16-0160	MnF_2 (α)	(910)	1.097	70	89.2
		(803)	1.063	70	92.9
		(515)	1.042	80	95.3
6-0697	Ni_3C	(112)	1.1277	100	86.2
2-1126	$Cs_2NaBi(NO_2)_6$	(628)	1.094	100	89.5
35-1011	$Cs_2NaBiCl_6$	(628)	1.062	100	93.0

Table 2 Materials with Reflections Within $90 \pm 2°$ 2θ for Various Wavelengths

Anode	Wavelength	Selected reflections within 88–92°		
Scandium	3.03114	MgO (200) (both are outside range)		
Chromium	2.28962	NaCl (111)	Ge, GaAs (222)	
Iron	1.93597	CaF_2 (400)	GGG (048), (248)	
Cobalt	1.78892	Cu (220)	MgO (311)	CaF_2 (331)
		NaCl (331)	Diamond (220)	Si (331)
		Ge, GaAs (420)	GGG (many)	
Copper	1.54051	Cu (311)	GGG (many)	NaCl (511), (333)
		Diamond (311)	Si (422)	Ge, GaAs (511), (333)
Zirconium	0.78588	Cu (533)	LiF (711), (551), (640)	
		Au (640)	MgO (642), (731), (553)	
		Diamond (026)	Si (448)	
		NaCl, CaF_2, Ge, GaAs, GGG (many)		
Molybdenum	0.70926	Cu (711), (551), (640)	LiF (800)	
		Au (800), (733), (820), (644)	Diamond (117), (155)	
		NaCl (11,1,1), (775), (880)		
		Si, GaF_2, Ge, GaAs, GGG (many)		
		Mo (611)* HOP-C Graphite (00.14)* Mg(24.3)*		
Rhodium	0.613245	Cu (337), (028), (446)	LiF (248), (466)	
		Au (466), (139) MgO (139), (448)		
		Diamond (337), (446)		
		CaF_2, NaCl, Si, Ge, GaAs, GGG (many)		
Silver	0.559363	Cu (119), (357), (248)		
		Diamond (048), (119), (357)		
		Lif, Au, MgO, CaF_2, NaCl, Si, Ge, GaAs, GGG (many)		

Source: From P. Wobrauschek and H. Aiginger, *Adv. X-ray Anal.* 28:71 (1985).

Table 3 References for Selected Materials

Material	Space group	Cell constants		Reference
		a_0	c_0	
Cu	$Fm3m$	3.61529		J. Appl. Phys. *36*, 2864 (1965)
LiF		4.0262		Acta Cryst. *8*, 36 (1955)
Au		4.0786		J. Appl. Cryst. *1*, 123 (1968)
MgO		4.2128		J. Appl. Cryst. *1*, 246 (1968)
CaF_2		5.46305		NBS US Monogr. 25, *21* (1984)
NaCl		5.6238		Acta Cryst. *A26*, 655 (1970)
Diamond	$Fd3m$	3.56		
Si		5.43088		NBS US Monogr. 25, *13*, 35 (1976)
GaAs		5.6532		Acta Cryst. *21*, 290 (1966)
Ge		5.6576		J. Appl. Phys. *23*, 330 (1952)
GGG	$Ia3d$	12.3829		Acta Chem. Scand. *A37*, 203 (1983)
Al_2O_3	$R-3c$	4.75848	12.9932	Z. Krist. *125*, 377 (1967)
$CaCO_3$		4.9900	17.002	Acta Cryst. *18*, 689 (1965)
$BaTiO_3$	$P4mm$	3.9945	4.0335	Acta Cryst. *A26*, 336 (1970)
SiO_2 (α)	$P3221$	4.731	5.280	Acta Cryst. *B35*, 550 (1979)

very well established as a result of commercial interest in certain compositions for bubble memories, lasers, and electronic devices.

A third method of searching for materials was used for the longer wavelengths. It is generally possible to find some material with a reflection close to 90°. However, it is perhaps desirable to have a low index reflection. A search of the Crystal Data database of approximately 115,000 materials was therefore made for the space group $Fm3m$ (like LiF) and a cell size appropriate for the wavelength. This method is probably of limited value since the number of materials found is small and most are not expecially good materials. Table 4 is therefore included to show the results of this type of search.

Table 4 Materials Selected for Low Index of Reflections

Anode	Ideal a_0	Reflection	Material	Cell constant
Scandium	3.71	(111)	Cu	3.615
			Cu_3Pt	3.702
	4.28	(200)	MgO	4.213
			CoO	4.263
			ZnO	4.280
			FeO	4.29
Chromium	2.80	(111)	No matches	
	3.24	(200)	No matches	
	4.58	(220)	ZrN	4.578

REFERENCES

1. R. Howell and W. Pickles, *Nucl. Instr. Methods 120*:187 (1974).
2. C. G. Barkla, *Philos. Trans. R. Soc. (Lond.) 204A*:467 (1905). Also, C. G. Barkla, *Proc. R. Soc. (Lond.) 77*:247–255 (1906).
3. W. Friedrich, P. Knipping, and M. von Laue, *Ann. Phys. 41*:971–988 (1913).
4. J. J. Thomson and G. Thomson, *The Conduction of Electricity Through Gases,* 3rd Ed., Cambridge University Press, Cambridge, 1933.
5. O. Klein and Y. Nishina, *Z. f Phys. 52*:853 (1928).
6. R. D. Evans, Compton effect, in *Handbuch der Physik*, Vol. 34, Springer-Verlag, Berlin, 1958.
7. R. B. Strittmatter, *Adv. X-ray Anal. 25*:75 (1982).
8. A. H. Compton and C. F. Hagenow, *J. Optical Soc. Am. 8*:487 (1924).
9. K. Champion and R. Whittem, *Nature 199*:1082 (1963).
10. H. Aiginger, P. Wobrauschek, and C. Brauner, *Nucl. Instr. Methods 120*:541 (1974). Also in *Measurement, Detection and Control of Environmental Pollutants*, IAEA, Vienna, 1976.
11. T. G. Dzubay, B. V. Jarrett, and J. M. Jaklevic, *Nucl. Instr. Methods 115*:297 (1974).
12. R. Howell, W. Pickles, and J. Cate, Jr., *Adv. X-ray Anal. 18*:265 (1974).
13. R. W. Ryon, *Adv. X-ray Anal. 20*:575 (1977).
14. A. A. Ter-Saakov and M. V. Glebov, *Atomnaya Énergiya 58*:260 (1984).
15. Eugene P. Bertin, *Principles and Practice of X-ray Spectrometric Analysis*, 2nd Ed., Plenum Press, New York, 1975, p. 117.
16. P. Standzenieks and E. Selin, *Nucl. Instr. Methods 165*:63 (1979).
17. K. M. Bisgård, J. Laursen, and B. Schmidt Nielson, *X-ray Spectrom. 10*:17 (1981).
18. J. D. Zahrt and R. W. Ryon, *Adv. X-ray Anal. 24*:345 (1981).
19. S. Chandrasekhar, *Radiative Transfer*, Dover, New York, 1960.
20. I. A. Tolokonnikov, *Atomnaya Énergiya 61*:224 (1985).
21. J. D. Zahrt and R. W. Ryon, *Adv. X-ray Anal. 29*:435 (1986).
22. J. D. Zahrt and R. W. Ryon, *Adv. X-ray Anal. 27*:505 (1984).
23. J. D. Zahrt, *Adv. X-ray Anal. 27*:513 (1984).
24. H. Liegmahl, University of Bremen, private communication, 1988.
25. P. Valtink, University of Bremen, private communication, 1988.
26. W. H. Zachariasen, *Theory of X-ray Diffraction in Crystals*, Dover, New York, 1967.
27. H. Aiginger and P. Wobrauschek, *J. Radioanal. Chem. 61*:281 (1981).
28. P. Wobrauschek and H. Aiginger, *Adv. X-ray Anal. 28*:69 (1985).
29. J. D. Zahrt, *Adv. X-ray Anal. 26*:331 (1983).
30. J. D. Zahrt, *Nucl. Instr. Methods A242*:558 (1986).
31. R. W. Ryon, J. D. Zahrt, P. Wobrauschek, and H. Aiginger, *Adv. X-ray Anal., 25*:63 (1982).
32. V. B. Baryshev, G. N. Kulipanov, and A. N. Skrinsky, *Nucl. Instr. Methods Phys. Res. A246*:739 (1986).
33. R. Giauque, J. Jaklevic, and A. Thompson, *Adv. X-ray Anal. 28*:53 (1985). Also, *Anal. Chem. 58*:940 (1986).
34. C. J. Sparks, Jr., in *Synchrotron Radiation Research* (H. Winick and S. Doniach, Eds.), Plenum, New York, 1980.
35. W. E. Maddox, *Adv. X-ray Anal. 27*:519 (1984).
36. W. E. Maddox, *Adv. X-ray Anal. 29*:497 (1986).
37. L. Kaufman and D. Shosa, *IEEE Trans. Nuclear Sci.* NS-24(1):525 (1977).
38. J. Dutton, et al., *Adv. X-ray Anal. 28*:151 (1985).
39. J. O. Christoffersson, *In Vivo Elemental Analysis in Occupational Medicine Using X-ray Fluorescence*, thesis, Malmo, Sweden, 1986.

11

Particle-Induced X-ray Emission Analysis

Willy Maenhaut   *University of Gent, Gent, Belgium*

Klas G. Malmqvist   *University of Lund and Lund Institute of Technology, Lund, Sweden*

I. INTRODUCTION

In 1970, Johansson et al. [1] demonstrated that the bombardment of a specimen with protons of a few MeV (megaelectron volts) gives rise to the emission of characteristic x-rays and that this can form the basis for a highly sensitive elemental analysis. This landmark paper formed the starting point of the x-ray emission analysis technique, which became known as particle-induced x-ray emission analysis (PIXE). Other names and acronyms, such as ion-induced x-ray emission (IIX) and charged-particle x-ray fluorescence (CPXRF), have also been proposed or are still occasionally used to indicate this technique, but PIXE has become the standard acronym for workers in the field. Although PIXE is sometimes considered a variant of x-ray fluorescence (XRF), such a classification is not correct in a strict sense, since the technique does not rely on excitation of the sample by x-rays. Instead, heavy charged particles, that is, protons, α particles, or heavy ions, are used in PIXE to create inner-shell vacancies in the atoms of the specimen. As in XRF and electron probe microanalysis (EPMA), the characteristic x-rays produced by deexcitation of the vacancies can be measured by either a wavelength-dispersive or an energy-dispersive detection system. However, although the two detection systems are employed in both XRF and EPMA, an energy-dispersive spectrometer with a Si(Li) detector is almost exclusively used in PIXE.

The incident charged-particle beams in PIXE are invariably generated by particle accelerators. For the great majority of PIXE work, protons of 1–4 MeV, which can be produced by small accelerators (e.g., Van de Graaff accelerators or compact cyclotrons), are employed. Such small accelerators are also used in other analytical techniques that utilize ions of a few MeV/u, such as Rutherford backscattering spectrometry (RBS), nuclear reaction analysis (NRA), charged-particle activation analysis (CPAA), and accelerator mass spectrometry (AMS). Because of their common use of ion beams, these techniques and PIXE are often jointly referred to as ion beam analysis (IBA) techniques. Furthermore, since the same incident particle type and energy may be used in several of the techniques, the simultaneous analysis of a sample by PIXE and some other IBA techniques (particularly RBS and NRA) is often feasible, so that the elemental coverage may be increased down to the very light elements and/or information on the depth distribution may be obtained.

Compared to x-rays, protons or other heavy charged particles have the advantage that they can be focused by electrostatic or (electro)magnetic lenses and may be transported over large distances without loss of beam intensity. As a result, incident fluence densities (expressed as number of impinging particles per square centimeter and per second) are generally much higher in PIXE than in ordinary tube-excited XRF. Moreover, focusing of particle beams down to micrometer sizes is possible, so that PIXE allows an analysis with high spatial resolution. The microbeam variant of PIXE has become known as micro-PIXE, whereas the common variant, which makes use of a millimeter-sized beam, is now often referred to as macro-PIXE. Focusing to micrometer sizes is also possible with electrons and has given rise to the powerful EPMA technique, but heavy charged particles have the clear advantage that they give rise to much lower continuum background intensity in the x-ray spectrum. As a result, the relative detection limits (micrograms per gram) are typically two orders of magnitude better in micro-PIXE than in EPMA.

After the initial experiment by Johansson et al. [1], the favorable characteristics of PIXE were rapidly realized by many researchers, particularly within the nuclear physics community. Its applicability and potential for solving numerous trace element analytical problems were extensively examined and abundantly demonstrated. Besides the traditional bombardment in vacuum, external beam approaches (with the specimen either in the laboratory air or in a nitrogen or He atmosphere) were also attempted and were found to be useful, particularly in examining delicate and/or large objects. The progress of PIXE over the years is exemplified by the proceedings of the five international conferences exclusively dedicated to the PIXE technique [2–6] and its applications. By now, PIXE has evolved into a rather well-used and mature technique, as demonstrated by the increasing numbers of research papers in which PIXE provided the analytical results and by the recent publication of the first textbook on the technique [7]. For a comprehensive treatment of PIXE and its applications, a reading of that book is highly recommended.

II. INTERACTIONS OF CHARGED PARTICLES WITH MATTER, CHARACTERISTIC X-RAY PRODUCTION, AND CONTINUOUS PHOTON BACKGROUND PRODUCTION

A. Interaction of Charged Particles with Matter

1. Slowing of Charged Particles in Matter: Stopping Power

When a beam of heavy charged particles of a few MeV/amu penetrates into matter, it loses its energy gradually with depth, until it is finally stopped. The energy loss occurs mainly through inelastic Coulombic encounters with bound electrons, and in contrast to the case of electron beams, the direction of travel of an ion beam is scarcely altered during the slowing process.

The stopping power $S(E)$ of an ion with energy E is defined as the energy loss per unit mass thickness traversed:

$$S(E) = -\frac{1}{\rho}\frac{dE}{dx} \tag{1}$$

where ρ is the density of the stopping material and x the distance. As defined here, $S(E)$ is expressed in units of keV/g/cm^2.

Numerous experimental measurements of stopping powers are available. They formed the data base to fit the parameters of semi-empirical equations [8,9], which are now commonly used to obtain the stopping powers for all elements of the periodic table. For the energy range of 1–4 MeV, which is most important in PIXE, the accuracy of the values calculated with the semi-empirical equations is estimated at 1–2%.

The stopping power for compounds or more complex matrices is obtained from those of the constituent elements through the Bragg-Kleemann additivity rule:

$$S_{\text{matrix}}(E) = \sum_{i=1}^{n} w_i S_i(E) \tag{2}$$

where w_i and $S_i(E)$ are the mass fraction and stopping power of constituent element i.

The total pathlength R of an ion may easily be obtained by integration of the stopping powers:

$$R = \int_{E_0}^{0} -\frac{dE}{S(E)} \tag{3}$$

where E_0 is the incident ion energy. Although the total pathlength is larger than the projected range, the difference between the two is smaller than 1% for incident protons of a few MeV.

2. Inner-Shell Vacancy Creation: Ionization and X-ray Production Cross Sections

Many of the Coulombic interactions between protons or heavier ions and matter result in the ejection of inner-shell electrons. It is these interactions and their cross sections that are of importance in PIXE. Three basic theoretical approaches have been used to calculate the cross sections for inner-shell vacancy creation: the binary encounter approximation (BEA), the semiclassical approximation (SCA), and the plane wave Born approximation (PWBA).

The PWBA model, which applies perturbation theory to a transition from an initial state (plane wave projectile and bound atomic electron) to a final state (plane wave projectile and ejected continuum electron), has been most elaborated, particularly by Brandt and coworkers. These authors incorporated a series of modifications in the model to correct for its inherent approximations, and this resulted in the so-called ECPSSR treatment [10,11] of K- and L-shell ionization cross sections. The ECPSSR treatment deals with the deflection of the projectile due to the nuclear Coulomb field (C), perturbation of the atomic stationary states (PSS) by the projectile, relativistic effects (R), and energy loss (E) during the collision.

Cohen and Harrigan [12] used the ECPSSR model to produce an extensive tabulation of K- and L-subshell ionization cross sections for most target elements and for protons and helium ions between 100 keV and 10 MeV. As in the original ECPSSR version [11], they employed nonrelativistic hydrogenic wave functions to describe the atomic electrons. More elaborate relativistic Dirac-Hartree-Slater (DHS) wave functions within a Brandt and Lapicki formalism [10] were used by Chen and Crasemann [13,14] to produce K-, L-, and M-shell ionization cross sections for protons of a few selected energies from 100 keV to 5 MeV and for a narrow range of selected target elements.

For K-shell ionization with protons, the cross sections as predicted by the ECPSSR and other theories were thoroughly compared with experimental data by Paul and coworkers [15–18]. Although no theory emerges that will predict the experimental data within a few percent for all target elements and energies [18], it is generally agreed [7,19] that the tables of both Cohen and Harrigan [12] and Chen and Crasemann [13,14] are adequate for most K-shell proton PIXE work. For the case of the L ionization, the situation is much less favorable, however, as discussed in detail by Johansson and Campbell [7], Campbell [20], and Cohen [19].

For practical purposes, one requires x-ray production cross sections for individual x-ray lines, that is σ_p^X, with p the x-ray line used for analysis. The x-ray production cross sections are related to the ionization cross sections through the following equations. For K x-rays,

$$\sigma_{Kp}^X = \sigma_K^I \omega_K \left(\frac{\Gamma_{Kp}}{\Gamma_K} \right) \tag{4}$$

where the index K refers to the K shell, σ_K^I is the K-shell ionization cross section, ω_K the fluorescence yield, and Γ_{Kp}/Γ_K the fractional radiative width for line p.

For L x-rays,

$$\sigma_{Li,p}^X = \sigma_{Li}^X \left(\frac{\Gamma_{Li,p}}{\Gamma_{Li}} \right) \tag{5}$$

where Li (with $i = 1, 2, 3$) refers to the ith L-subshell, σ_{Li}^X is the x-ray production cross section for the subshell Li whose vacancy is filled up by the radiative transition, and $\Gamma_{Li,p}/\Gamma_{Li}$ the fractional radiative width. The x-ray production cross sections σ_{Li}^X are related to the L-subshell ionization cross sections σ_{Li}^I through

$$\sigma_{L1}^X = \sigma_{L1}^I \omega_1 \tag{6}$$

$$\sigma_{L2}^X = (\sigma_{L2}^I + \sigma_{L1}^I f_{12})\omega_2 \tag{7}$$

$$\sigma_{L3}^X = [\sigma_{L3}^I + \sigma_{L2}^I f_{23} + \sigma_{L1}^I (f_{13} + f_{12}f_{23} + f_{13}')]\omega_3 \tag{8}$$

where ω_i denotes the fluorescence yield and f_{ij} the Coster-Kronig probability, and f_{13}' is the small radiative intrashell vacancy transfer probability.

Calculation of the x-ray production cross section from the ionization cross sections thus involves additional atomic parameters, that is, fluorescence yields and fractional radiative widths in the case of the K shell and fluorescence yields, Coster-Kronig yields, and fractional radiative widths for the L shell. As to the K fractional radiative widths, either the experimental data of Salem et al. [21] or the theoretical values of Scofield [22] that are derived from Dirac-Hartree-Fock (DHF) calculations are considered accurate data bases. The K fluorescence yields are usually taken from Krause [23]. These data are quite accurate for atomic numbers above $Z = 20$, but the situation for the lighter elements is less clear, as discussed by Maenhaut and Raemdonck [24] and Paul [18]. For the case of the L x-rays, Cohen [19] with Clayton [25] suggests employing the fractional radiative widths of Salem et al. [21] and the Krause [23] fluorescence and Coster-Kronig yields, and Cohen and Harrigan used this approach to convert their table of ECPSSR L ionization cross sections [12] into a very useful table of production cross sections for up to 16 individual L x-ray lines [26]. Campbell [20], however, advocates the use of the DHF radiative widths of Scofield [27] and of the DHS fluorescence and Coster-Kronig yields of Chen et al. [28] in combination with the tabulated DHS L ionization cross sections of Chen and Crasemann [13,14].

A very practical alternative to tabulated theoretical ionization and x-ray production cross sections, particularly with computer calculations in mind, are parameterized or analytical formulas obtained by fitting polynomial expressions to theoretical or empirical cross-sectional data. The equations derived by Johansson and Johansson [29] have been widely used by PIXE analysts in the past, but it is now known that they progressively underpredict cross sections with increasing Z of the target element. More accurate formulas, in which similar functions were often used as in the Johansson and Johansson equations, have in recent years been presented by Paul [30], Cohen and Clayton [25], and Miyagawa et al. [31].

For the proton energy range of 1–4 MeV typically used in PIXE, the ionization (and x-ray production) cross sections increase with increasing proton energy and decrease with increasing atomic number of the target atom. This is illustrated in Figure 1, which displays theoretical K- and total L-shell ionization cross sections [12] for selected target elements. The very steep fall in the K ionization cross section with Z is particularly notable.

3. Elastic Encounters

Protons or other heavy charged particles that pass near atomic nuclei can be scattered elastically, that is, without causing nuclear or atomic excitation. The cross section for the elastic scattering is strongly dependent on the scattering angle and on the atomic number (Z) of the scattering nuclide; it decreases with increasing angle and increases with increasing Z. The energy of the scattered

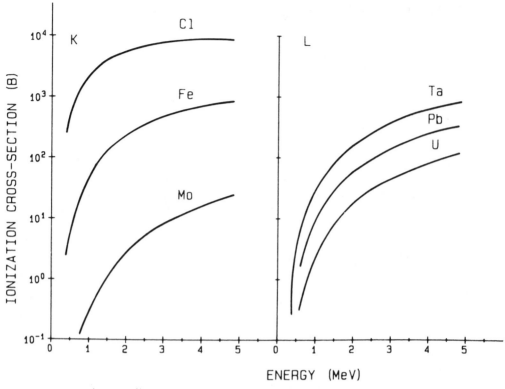

Figure 1 K- and L-shell ionization cross sections in barns (1 barn = 10^{-24} cm^2) as a function of proton energy and target atom. The values are theoretical ECPSSR predictions [12]. (Reproduced from Ref. 7 by permission of John Wiley & Sons, Ltd.)

particles also increases with Z. By measuring the elastically scattered particles, information can be obtained on the elemental composition of the sample and on the distribution of the elements with depth. This has given rise to the widely used technique of Rutherford backscattering spectrometry [32], which can often be elegantly employed in combination with PIXE to provide complementary infor-mation. On the other hand, the energetic scattered particles can also cause prob-lems in PIXE. Indeed, when they enter the Si(Li) detector, their large energy is deposited there, and this gives rise to serious resolution deterioration and other electronic problems in many x-ray pulse-processing systems [33]. A solution to this problem is to interpose a light element absorber (preferably Be) between the specimen and the Si(Li) detector (e.g., Ref. 24), but this invariably results in a loss of sensitivity for the light elements.

4. Nuclear Reactions

Besides elastic encounters between incident particles and target nuclei, various inelastic interactions or nuclear reactions are possible. They include (p, γ), $(p, p'\gamma)$ and $(p, \alpha\gamma)$ reactions when protons are used as incident particles. The cross sections for these reactions vary in a rather irregular way with target nuclide and with incident particle energy. They generally increase with increasing energy but may exhibit intense resonance peaks at particular energies. Also, because of the Coulomb barrier, the cross sections are smaller for the heavier target elements than for lighter ones. By detecting the promptly emitted γ-rays or charged particles for these nuclear reactions, elemental analysis and depth profiling of certain elements is possible. The analytical technique that employs these possibilities is referred to as nuclear reaction analysis. An introduction to this technique can be found in the book by Deconninck [34]. The technique is also briefly presented in Section VII.B.

5. Other Interactions

As indicated, the interactions of charged particles and matter occur mainly through inelastic Coulombic encounters with bound electrons. This results in electron excitation and ionization but also in secondary phenomena that contribute to the continuous photon background in the PIXE spectrum. Another interaction that contributes to this background is projectile bremsstrahlung. All these interactions are discussed in some detail here.

B. Continuous Photon Background Production

1. Electron Bremsstrahlung

The characteristic x-ray lines in a PIXE spectrum are superimposed on a continuum background that has a strong resemblance to that observed in EPMA. In both cases, electron bremsstrahlung is the major background component. Although this originates from primary electron interactions in EPMA, secondary phenomena are at its origin in PIXE. In the mid-1970s, Folkmann and various coworkers made very significant contributions to our understanding of the continuum background in PIXE (e.g., Refs. 35–37). This work was later complemented and extended by Ishii and Morita [38–40]. It is by now clear that the electron bremsstrahlung in PIXE originates essentially from three processes, that is, quasi-free electron bremsstrahlung (QFEB), secondary electron bremsstrahlung (SEB), and atomic bremsstrahlung (AB). SEB is formed by a two-step process: the incident particle first ejects an electron from a target atom, and the secondary electron is subsequently scattered in the Coulomb field of a target nucleus, thus producing the bremsstrahlung. The photon spectrum of SEB is characterized by an "end-point" energy $T_m = 4m_e E_p/M_p$, with m_e and M_p the electron and projectile masses and E_p the projectile energy. Above T_m the intensity of SEB decreases rapidly. QFEB is emitted when an electron of a target atom is scattered by the Coulomb field of the projectile (this is a process in the projectile frame). The QFEB end-point energy T_r is equal to $T_m/4$. The process AB occurs when a bound target electron is excited to a continuum state by the projectile and, returning to its original state, emits a photon. The relative contributions of QFEB, SEB, and AB to the electron bremsstrahlung background are shown schematically in Figure 2. As can be seen from Figure 2, AB predominates in the high-energy part of the spectrum (i.e., for photon energies above T_m), whereas QFEB becomes the prevailing component at low photon energies (below T_r). As both T_r and T_m increase linearly with increasing projectile energy E_p, the PIXE spectrum (which

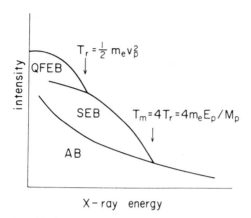

Figure 2   Relative contributions of QFEB, SEB, and AB to the electron bremsstrahlung background. For an explanation of the acronyms and symbols, see text. (Reproduced from Ref. 39 by permission of Elsevier Science Publishers.)

typically extends from 0 to about 20–30 keV) has a quite different appearance depending upon the value of E_p. Besides E_p, the matrix composition of the target also plays a critical role in both the shape and intensity of the electron bremsstrahlung background. As illustrated in Figure 3, the background becomes more intense with increasing Z of the target.

The electron bremsstrahlung is emitted anisotropically and is lower at forward and backward angles than at 90°. For this reason, the Si(Li) detector is frequently positioned at an angle of 135° in an experimental PIXE chamber.

2. Projectile Bremsstrahlung

While passing through the target, the incident charged particles may be decelerated in the Coulomb field of atomic nuclei and thereby give rise to the emission of projectile bremsstrahlung (PB). Although this process is negligible within the context of the projectile energy loss, it contributes to some extent to the continuum background in the PIXE spectra. For a detailed treatment on PB in PIXE we refer to the papers by Folkmann and coworkers [35–37]. As can be seen from Figure 3, the PB is much less important than the electron bremsstrahlung and becomes only significant at photon energies above 10–20 keV.

3. γ-Ray Background

The prompt γ-rays emitted as a result of nuclear reactions between the projectiles and the target atoms are generally quite energetic and are therefore far outside the 0–30 keV energy range typically observed in PIXE. However, through Compton interactions of these γ-rays with the Si(Li) detector, a wide spectrum of electron energies is generated, and this in turn gives rise to a slowly varying continuum in the PIXE spectrum. This Compton scattering background is not predictable in the sense that the bremsstrahlung background is predictable, since it does not depend on the matrix composition of the target alone but rather on the presence of particular elements with large cross sections for nuclear reactions yielding γ-rays. Na and particularly F are examples of elements that are often responsible

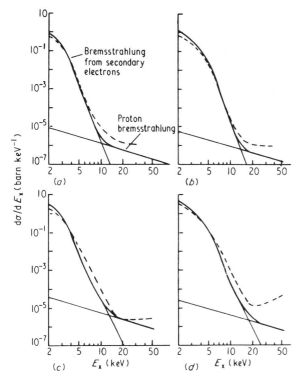

Figure 3 Background radiation spectra at 90° expressed as differential cross-sectional dependence on photon energy E_x for thin elemental targets and for two proton energies: (a) 2 MeV on carbon; (b) 3 MeV on carbon; (c) 2 MeV on aluminium; and (d) 3 MeV on aluminium. The full curves are calculated and the dashed curves measured. (Reproduced from Ref. 35 by permission of Elsevier Science Publishers.)

for a noticeable or even significant γ-ray background. A large fraction of the γ-ray background in the PIXE spectrum may originate from nuclear reactions that do not take place in the specimen but instead in the beam collimators, Faraday cup, or various other parts of the PIXE irradiation chamber [even the x-ray absorber placed in front of the Si(Li) detector]. As discussed in Reference 7 and briefly indicated in Section III.B.1, the materials for the collimators and various other parts of the chamber should therefore be selected with care.

4. Other Sources of Background

In many cases, intense characteristic x-ray peaks are present in the PIXE spectrum. This is particularly true when the matrix consists of element(s) with atomic number above $Z = 11$. These intense peaks and their pileup peaks seriously hamper (or even preclude) the detection of other elements with characteristic x-rays of similar energy. Furthermore, incomplete charge collection in the Si(Li) detector and other processes [7] have the effect that each x-ray peak exhibits a low-energy tail, which is roughly a flat shelf with height up to 1% of the peak height. Consequently, tails associated with very intense peaks form a substantial component of the total background. Because of the intense peaks and associated tails for higher Z matrices, and because in addition the electron bremsstrahlung intensity increases with increasing Z of the matrix, PIXE is much more suitable for analyzing trace elements in light-element matrices than in heavy-element matrices.

When heavy ions are used as projectiles, other processes than those discussed thus far also contribute to the continuum background. According to Folkmann [37], radiative electron capture and molecular orbital x-rays are important.

III. INSTRUMENTATION

A. Accelerators

The ion beams used for PIXE analysis are produced in an accelerator. The energy range required, 1–4 MeV/u, means that relatively small accelerators are sufficient. Electrostatic accelerators or cyclotrons are suitable sources [41]. Electrostatic accelerators are most commonly used. They may be of a regular or modified Van de Graaff type, in which a high-voltage terminal is charged by a belt or a metal chain. A second type of electrostatic accelerator uses a high-frequency voltage multiplication stage to charge the terminal, the ions being accelerated in the electrostatic field created by the terminal. Cyclotrons have been employed at some laboratories. These are particularly suitable for PIXE at higher ion energies [42–44] and in certain combinations of PIXE with complementary techniques, such as particle elastic scattering analysis (PESA) or forward α scattering (FAST) [45].

In most analytical situations in which PIXE is applied, the choice of accelerator is governed by access to a particular machine rather than by free choice. When a dedicated accelerator can be selected, however, a single-ended or tandem electrostatic machine of 2–3 MV is usually chosen. Such accelerators may also favorably be employed for a combined use of ion implantation and ion beam analysis, including PIXE.

Certain types of analysis, such as analyses by microbeam techniques, impose special requirements on ion source brightness, energy stability, beam emittance, and minimum scattering of the ion beam. Such requirements may play an important role in selecting an accelerator.

It is often stated that the need for an accelerator in PIXE is a major drawback for reasons of both technical complexity and economy. Although such objections contain some truth, it should be pointed out that the prices of small electrostatic accelerators are of the same order as those for complex "traditional" analytical equipment, such as mass spectrometers or electron microscopes, and that a comparable technical skill is required of the operator. It is therefore unjustified to rule out the PIXE method for such reasons.

B. Beam Handling and Target Chambers

1. Macro-PIXE in Vacuum

Commercially available vacuum chambers used as scattering chambers in nuclear physics experiments are generally not well suited for PIXE. Their shapes and dimensions hamper obtaining large solid angles for detection and introducing multispecimen holders and sample changers. As a consequence, few existing PIXE chambers are of commercial origin. There are exceptions to this rule, however, and certain chambers have been built entirely from commercially available components.

Figure 4 shows the design and principal components of a hypothetical "typical" setup for macro-PIXE analysis. In the following, the details of such a system are outlined.

The ion beam emerging from the accelerator first passes through an analyzing magnet, which sorts out the ions of the correct mass and velocity, and is then normally focused by electrostatic or magnetic quadrupole lenses onto the specimen.

A typical experimental facility for macro-PIXE analysis employs a system for producing a homogenous beam so that the specimen is evenly irradiated. This is required for quantitative analysis of heterogeneous samples. There are various methods of achieving this:

1. Focusing the beam onto a scattering foil, which distributes the ions homogeneously over the cross section selected by a diaphragm. The foil material should be able to withstand the ion beam for long times. It is often made of thin metal, such as gold, which is an excellent scatterer with low stopping power for the ions.
2. Rastering the beam over the sample by electrostatic deflection, for example. This does not give a genuinely homogeneous distribution and may therefore produce high instantaneous count rates in heterogeneous samples. On the other hand, this method has the advantage of relatively low loss of beam intensity, whereas such a loss may be more than 90% in the case of method 1.
3. After the initial alignment of the beam, straightforward symmetrical defocusing by the quadrupole magnets.

After homogenization, the beam is defined by a pair of collimators (diaphragms), which are normally circular with diameters between 1 and 10 mm and placed approximately 100–200 mm apart, just before the entrance to the irradiation chamber. The collimator material should preferably withstand bombardment by the ions over long periods and produce low background in the PIXE spectra. This can be achieved by using a material containing nuclei with low cross sections for

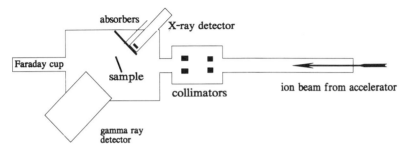

Figure 4 Typical experimental arrangement for routine PIXE, with detectors for x-rays and γ-rays. Normally, the collimators can be changed between 1 and 10 mm diameter, and the x-ray absorber holder would contain 5–10 different filters.

producing γ-rays at the ion energy interval in question. Common choices of collimator material are tantalum and graphite. To measure the beam fluence on the target accurately, the collimators should be grounded and electrically shielded from the irradiation chamber in which the target is placed, and secondary electrons generated at these collimators should be prevented from entering the target chamber. A target holder defines the proper position of the sample during analysis. In a typical setup for PIXE analysis, in vacuo, several samples are simultaneously inserted into the vacuum, so that a high sample throughput is facilitated. The normal slide frame format (5×5 cm^2) is becoming a standard target frame in most laboratories, and often 36–80 target frames are contained in vacuum. The samples are irradiated at an angle suitable to detect the x-rays, and often the detector views the sample from a backward angle of 135° or more, since this produces higher peak-to-background ratios in the PIXE spectra (see Sect. II.B.1).

In contrast to energy-dispersive XRF (ED-XRF) spectra, PIXE spectra have most of their x-ray intensity in the low-energy region, so that the spectral shape can advantageously be modulated by placing an x-ray absorber between sample and detector. The aim of absorbers invariably is to reduce or eliminate unwanted continuum background and/or intense x-ray peaks and their associated pileup peaks and at the same time allow bombardments at higher beam intensities, so that the elements of interest can be measured in shorter bombardment times and with fewer spectral interferences. The absorbers are usually made from organic material (e.g., Mylar) or from light-element metal foils (Be and Al). Plain absorbers are employed if complete elimination of the low-energy part of the spectrum is desired. In many cases, however, it is preferable to allow a certain fraction (e.g., a few percent) of the low-energy x-rays to pass onto the detector. This can elegantly be realized by resorting to pinhole absorbers (usually called "funny filters" within the PIXE community). Also, more sophisticated designs consisting of a combination of various layers with different thicknesses and pinhole diameters have been used [46]. Figure 5 illustrates the effect of some absorbers on the appearance of the PIXE spectrum for a biological reference material [47]. In the acquisition of the top spectrum a funny filter, consisting of a 52 μm thick Be foil and a 324 μm Mylar filter with a 5.54% hole, was placed in front of the detector, and even at a beam current of 10 nA the count rate exceeded 2000 cps. For acquiring the bottom spectrum of Figure 5, a 660 μm Mylar absorber was interposed between specimen and detector. This had the effect of allowing a beam current of 150 nA, and the count rate remained below 1000 cps. If one wants to analyze samples with high concentrations of elements whose lines severely interfere with nearby lines of the elements of interest, it is useful to resort to "bandpass" absorbers. For instance, in the PIXE analysis of steel a chromium absorber selectively suppresses the strong Fe K lines [48]. Other examples in which such complex absorbers are advantageous are the analysis of metal alloys in archeology, for example, of bronzes [49], and various analyses in material studies.

When one is interested in measuring the light elements ($Z < 13$) by PIXE, only a very thin x-ray absorber may be used, and for the very light elements, a windowless detector or a detector with an ultrathin polymer window [50] would even be advisable. However, as indicated in Section II.A.3, bombardment of a sample with charged particles also gives rise to backscattered particles, and a fraction of these may penetrate into the Si(Li) detector and cause problems.

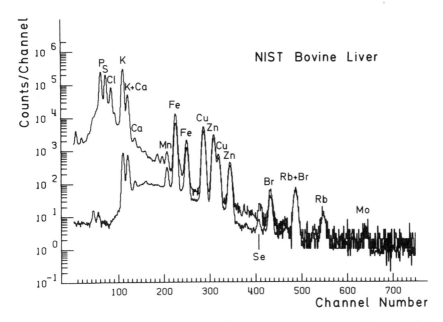

Figure 5 PIXE spectra for a 5 mg/cm^2 NIST Bovine Liver specimen. Incident proton beam of 2.4 MeV. Top spectrum: taken with funny filter (see text); beam current 10 nA, x-ray count rate 2200 cps, and preset charge 20 μC. Bottom spectrum: taken with a 660 μm Mylar absorber; beam current 150 nA, count rate 900 cps, and preset charge 200 μC. The ordinate scale applies to the top spectrum. All marked peaks are K lines ($K\alpha$ and $K\beta$). Most of the unmarked peaks in the top spectrum are sum peaks. (Reproduced from Ref. 47.)

Hence, it is advisable to use an absorber that is sufficiently thick to stop all scattered particles, but this hampers or precludes the detection of those light elements. According to Musket [51], placing a magnetic deflection trap between the sample and the detector crystal may be a viable alternative to an absorber for removing the scattered ions. However, the installation of such system also results in a lower solid angle of detection.

The Si(Li) detectors used for PIXE are the same as those in EDXRF. They typically have a sensitive area from 10 to 80 mm^2, but this area is often reduced by inserting an x-ray collimator in front of the detector. The aim of such collimator is to minimize the low-energy peak tailing that results from incomplete charge collection at the edge of the detector crystal.

The amplifiers and pulse processors in energy-dispersive x-ray spectroscopy require large time constants for optimum energy resolution. Unfortunately, large time constants also imply that pulse pileup already becomes a serious problem at relatively low count rates. It is therefore common practice to incorporate an electronic pileup rejector and its associated dead-time correction circuitry in the pulse-processing chain. Such a solution reduces the throughput of signals to the analyzer; that is, it decreases the ratio of output signal count rate to beam current, and this is a drawback when analyzing samples that deteriorate during bombardment. To minimize beam-induced damage during PIXE analysis, an on-demand beam excitation system is recommended. Such a system employs a set of deflection plates placed in the beam path upstream of the sample; when an x-ray event is detected, the beam is deflected immediately and then held off until the event has been processed [52]. It should be indicated here that pileup peaks (or sum peaks) are not entirely eliminated by pileup rejectors or on-demand beam excitation, since such systems have a finite pulse pair-resolving time or beam-switching time.

In PIXE of thin specimens, the ions pass through and are dumped in a Faraday cup (see Fig. 4) for charge integration. To minimize the γ-ray background originating from nuclear reactions in the Faraday cup, long cups in which the particles are collected far from the detector are preferable. Furthermore, the escape of secondary electrons from the cup as well as the entrance of secondary electrons from the specimen should be prevented by placing a negatively biased ring in front of the cup. The charge integration itself is accomplished by connecting the Faraday cup to a sensitive current integrator/digitizer. When the samples are thick enough to stop the ions, the beam current must be measured either on the whole irradiation chamber or through some indirect approach. One possible method is to place a thin foil in the beam path upstream of the target and to measure the intensity of the particles scattered from it with a surface barrier detector [53]. Another, related method uses a beam chopper, for example, a thin metal strip, which periodically passes through the beam [54].

The bombardment of an insulating thick (or semithick) specimen in vacuum generally gives rise to charge build-up, and the specimen may reach a positive potential of up to several tens of kV before breakdown and sparking. The high potential accelerates electrons up to tens of keV, and as a consequence a huge bremsstrahlung background is produced in the PIXE spectrum. The peak-to-background ratios then decrease significantly [55]. To avoid this, various methods can be used:

1. Increased pressure in the chamber [55]
2. Thin carbon coating of the specimen [56]
3. Placing a thin carbon foil just in front of the sample [57]
4. Spraying with electrons from an electron gun [55]
5. The use of strong permanent magnets [58]

These methods either avoid the buildup of a high potential by conduction of the positive charge or by producing electrons to neutralize the positive charge or avoid the effects from the electrons impinging on the sample (magnets).

In addition to the equipment used for quantitative PIXE analysis, target chambers also often include equipment necessary for the complementary IBA techniques presented in Section VII, such as surface barrier detectors for measuring charged particles or Ge detectors for measuring prompt γ-rays.

2. Nonvacuum Macro-PIXE

As has already been indicated, it is sometimes advantageous to use atmospheric pressure or moderate vacuum instead of high vacuum during analysis [59]. The advantages are that the heat conductivity is increased and the target temperature decreased, the charge is conduced from the target, and the vacuum requirements are less, thus making it easier to design a low-cost chamber. Because the accelerator requires a good high vacuum, the ion beam must be extracted into the moderate vacuum or atmospheric pressure region through a thin exit foil. The beam eventually deteriorates this foil, and therefore its material must be carefully selected. The best choice is the polyimide foil Kapton, which withstands high intensities and a high radiation dose before mechanical breakdown. Supporting the foil by a carbon grid and direct flow of liquid nitrogen-cooled helium gas allows the use of high beam intensities over extended time intervals [60]. The chamber gas is normally helium or nitrogen. In addition to its better cooling properties, helium produces less bremsstrahlung than nitrogen. Some x-ray detector windows are not leakproof to helium, however, and this has disastrous consequences for the cryostat vacuum unless the detector is separated from the chamber gas by an additional window.

The choice of chamber atmosphere is determined by the objectives of the analysis and by the samples to be analyzed. A rather low pressure of helium gas, for example, suffices to improve the heat conductivity and to reduce thermal losses of elements or compounds [61]. On the other hand, an external beam in air is needed when large objects are to be analyzed without sectioning or subsampling. This is very important in archeology and art science, for example. The disadvantages of bombardment in air are that there is a danger of sample oxidation during irradiation and interfering argon x-ray lines are present in the spectrum. The strong argon lines may also be used to monitor the beam fluence, however, which is otherwise difficult to do in nonvacuum PIXE. In fact, one generally relies on some indirect method of beam current measuring [53,54], although with special precautions it is possible to measure the beam current directly [62].

The advantages of using higher pressure during irradiation have been demonstrated in various applications. For example, in studies on volatile organic compounds, using a combination of PIXE and complementary techniques (PESA), it was shown that significant losses of certain constituents occurred when bombarding in high vacuum, whereas the losses were insignificant for bombardments of similar beam intensity in helium at a pressure of 100 torr [61].

3. Nuclear Microprobes

In nuclear microprobe or micro-PIXE analysis, the particle beam is collimated and/or focused down to dimensions in the range of 1–50 μm. With the regular equipment used in accelerator-based ion beam analysis, beam sizes down to typically a few tenths of a mm in diameter are easily obtained, but to produce a genuine microbeam, specially designed equipment is required.

The simplest way of producing a microbeam is to employ a pinhole collimator [63]. For very small collimator sizes, however, the beam intensity obtainable is much too low for practical use. In addition, a substantial fraction of the ions are scattered at the edge of the collimator, and this gives rise to a halo (with an intensity of several percent of the beam current) around the central beam. In most micro-PIXE systems, collimation is therefore combined with an electrostatic or magnetic demagnification system, such as quadrupole doublets, triplets, or quadruplets. The initial collimation in an object aperture normally provides a beam with dimensions of 10–100 μm, and this beam is then demagnified by a factor of

5–25. The best systems currently available are able to produce a spatial resolution of about 1 μm at the specimen while maintaining an ion current that is useful for PIXE analysis (>100 pA). Figure 6 shows the layout of a nuclear microprobe. Although the exciting particles form the only difference between micro-PIXE and energy-dispersive EPMA, much better peak-to-background ratios and, consequently, much lower detection limits are obtainable by micro-PIXE. This is illustrated in Figure 7, where x-ray spectra for a biological specimen obtained by both techniques are compared. The much better peak-to-background ratios, particularly for the heavier trace elements ($Z > 20$), justify the use of the much more complex analytical equipment of the nuclear microprobe.

As shown in Figure 6, the components of a nuclear microprobe are as follows:

1. An accelerator, preferably with a very bright ion source
2. A precision collimator
3. Magnetic quadrupoles for focusing
4. A scanning system to raster the beam over the sample, as in a scanning electron microscope

The detection system is in principle identical to that for macro-PIXE but usually includes surface barrier detectors for scattered particles and for particles emitted in nuclear reactions [64]. Sometimes a detection system for secondary electrons

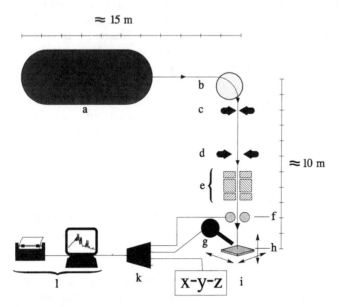

Figure 6 Typical experimental arrangement for a nuclear microprobe: (a) electrostatic accelerator; (b) analyzing magnet; (c) precision collimator that forms the object (10–100 μm) for imaging in the lens system (e); (d) collimator for restricting the beam cone to the acceptance angle of the lens; (h) x-y-z translator sample stage in the irradiation chamber; (f) sweeping system for rastering the beam over the specimen; (k) front-end computer and other electronics for controlling the sample movement (i), the beam sweeping, and the data acquisition with the x-ray detector (g) and other detectors; (l) host computer and peripherals used for elemental mapping and for displaying of the results.

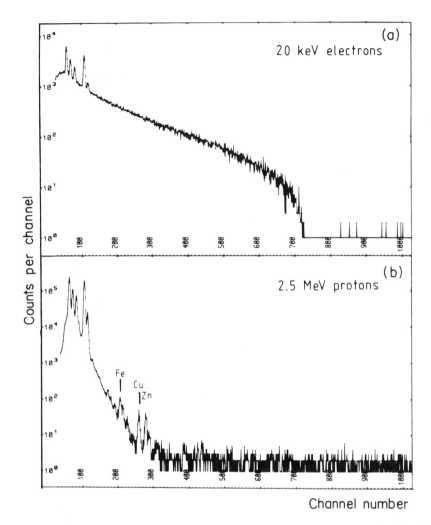

Figure 7 X-ray spectra from a thin biological specimen (human brain) obtained with electron (a) and nuclear microprobe (b) excitation. Note the very large difference in peak-to-background ratio, particularly from about 3 keV (channel 100) up. (Reproduced from Ref. 7 by permission of John Wiley & Sons, Ltd.)

[65] is also added, so that an image of the specimen can be made as in the scanning electron microscope. Facilities for accurate positioning of the beam on the specimen area of interest are also required. Hence, an optical viewing system with high magnification and good resolution is needed, as well as a precision sample holder controlled by stepping motors with micrometer accuracy.

Various types of object collimators can be used. The collimator design developed by the Heidelberg group [66] minimizes the contribution of scattered ions to the specimen and is employed in several micro-PIXE setups. Other laboratories prefer to use fixed apertures of the same design as for electron probes [67]. A new design, which has been shown to give very low edge scattering, uses precision-polished cylinder surfaces to define the ion beam [68]. The precision parts of the collimators are normally protected against beam damage and excessive heating by adding a slightly larger precollimator in which most of the energy is dissipated. The spacing of the variable collimators may be controlled manually by micrometer screws, but remote control using piezoelectric crystals or stepping motor-driven micrometer screws is preferred.

Each configuration of magnetic quadrupoles has its advantages and drawbacks. The complex Russian quadruplet, which was used in the first micro-PIXE system at Harwell [69], produces a completely symmetrical image but is difficult to align mechanically. Because of its simplicity, several laboratories use the doublet configuration, although it requires rectangular collimation to obtain a symmetrical image. When more than two quadrupoles are used, two magnets are connected in series to the same current supply. Whatever configuration is used, the current supplies should be very stable ($<10^{-4}$) to avoid beam aberrations.

The image size at the specimen is easily determined from first-order calculation. For small apertures, however, the first-order calculation does not suffice and second- and third-order calculations are required. The lens aberrations can be calculated if the details of the configuration are known. For more information on this subject, the reader is referred to Grime and Watt [70], which provides a comprehensive compilation of various magnetic quadrupole systems. It should be noted here, however, that among the intrinsic aberrations, the chromatic aberrations (i.e., second-order aberrations that depend on the angle and momentum of the ion beam) and third-order, spherical (angular) aberrations dominate in most nuclear microprobes. It is therefore essential to have a stable accelerator that delivers an ion beam of very constant energy and angle (emittance). Other important factors are the mechanical precision (to reduce parasitic aberrations) and particularly rotational defects in the magnetic fields (these can severely distort the image).

Special demagnification designs are worth mentioning here. A first example is the achromatic lens system with both magnetic and electrostatic elements, as described by Tapper and Nielsen [71]. This system has the advantages of eliminating chromatic aberrations and being less dependent on accelerator stability. Other demagnification systems make use of octupole magnets to reduce spherical aberrations [72]. Magnetic superconducting solenoids [73] and plasma lenses [74] have also been employed.

To allow full use of its powerful analytical capabilities, a nuclear microprobe setup should include a scanning system for rastering the ion beam over the specimen surface. The scanning system could be installed before focusing (predeflection), but such a system deflects the beam out of the optical axis and therefore significantly increases the aberrations. By proper selection of the position of deflection, however, the beam can be made to pass through the optical center of the lens system so that the aberrations are minimized. To avoid the aberrations of predeflection systems, the scanning system should be installed after the demagnification lenses. Because of the short distance between lens and specimen, however, scanning systems using electrostatic deflection plates require a high electrical field (with concomitant risk of electrical discharges) to obtain a sufficiently large deflection amplitude. The alternative is to employ magnetic scanning. By using scanning coils with ferrite cores, a reasonably large amplitude and a scanning frequency of more than 5 kHz can be obtained [75]. Such systems provide a good compromise between scanning speed and amplitude. The scanning system is usually computer controlled and connected to the data acquisition system (see Sec. III.C.2).

The detailed design of the irradiation chamber for the nuclear microprobe is beyond the scope of this chapter. We therefore limit ourselves to giving some recommendations for the essential components. The sample-positioning system may be a commercially available x-y-z precision translator as designed for scanning electron microscopes. The sample holder should preferably take many samples to avoid the need for frequent opening of the vacuum chamber. The microscope used for viewing the specimen should have a magnification of $\times 200\text{-}400$ and be equipped with a zoom lens. In addition to the Si(Li) detector, surface barrier detectors should be entered in the forward and backward directions. Such detectors are needed for extending the elemental coverage and for specimen thickness determinations by RBS or PESA. The specimen thickness is required for quantification in thin and semithick samples (see Secs. III.B and III.C). The specimen thickness may also be provided by other techniques than RBS or PESA, for example by measuring the energy loss in each specimen pixel, as is done in scanning transmission microscopy (STIM) [76]. Finally, it is recommended that secondary electron detectors be installed to allow imaging of the specimen surface in a way similar to scanning electron microscopy.

The high vacuum is maintained by direct pumping with oil diffusion, turbo-molecular, or cryo pumps. Although they are very efficient, the last two types may transmit vibrations to the demagnification lens system and to the specimen holder. The focusing of small beams is adversely affected by these or other vibrations. It is therefore common to place the whole microprobe system on a very rigid optical bench. The accurate positioning of the optical elements on this bench is realized by high-precision mechanical controls. Furthermore, to avoid the effects of the earth's magnetic field and stray fields from surrounding equipment, beam tubes are sometimes shielded by μ-metal foils [77].

As is the case for macro-PIXE, micro-PIXE may also be done under nonvacuum conditions. Because of scattering of the beam by the gas, however, nonvacuum micro-PIXE is only feasible for moderately small beam sizes (20–100 μm). After collimation and/or focusing, the beam is passed to the nonvacuum region through a pinhole or an exit foil. In the first approach, the high vacuum in the beam line is maintained by means of differential pumping. If the spatial resolution requirements are not too high, the nonvacuum micro-PIXE technique is rather straightforward and simple to use. It can be applied to examine large samples or sensitive art objects, such as bronze figures and ancient documents, and may provide unique information [49].

C. Data Acquisition

1. Macro-PIXE

The data acquisition systems used in PIXE have a great similarity to the EDXRF data acquisition systems discussed in Chapter 3. For x-ray detection, a high-resolution solid-state detector is virtually always employed. This could be either a Si(Li) or HPGe detector, but the former is highly preferable in most PIXE work. Indeed, for the energy region from 0 to about 20–30 keV, Si(Li) detectors offer a better energy resolution and, particularly important, a much lower escape peak probability than Ge detectors. The higher detection efficiency of the latter for energies above 20 keV is not very useful in 1–4 MeV PIXE because of the very rapid decrease in K ionization cross sections with increasing Z (see Sec. II.A.2). On the contrary, the higher detection efficiency of Ge may rather be a disadvantage, as the background contribution resulting from Compton-scattered γ-rays is larger than with Si. The great majority of PIXE chambers contain only one Si(Li) detector for x-ray detection. However, to improve the sensitivity for the heavier elements while retaining the capability of measuring the light elements during the same bombardment, the use of two Si(Li) detectors has been advocated [78]. The second Si(Li) detector, used for measuring the heavier elements, is provided with a thick absorber to cut down the high count rate from the light elements and has a larger solid angle of detection than the first Si(Li). As an alternative to such a second Si(Li), a Ge x-ray detector is well worth considering. In addition to the Si(Li) detector(s) for x-rays, PIXE chambers generally also contain detectors for the complementary IBA techniques, that is, surface barrier detectors for scattered particles or for particles resulting from nuclear reactions and a Ge detector for measuring prompt γ-rays. Hence, the acquisition system must include several analog-to-digital converters. Furthermore, since the spectral intensity in PIXE is proportional to the number of incident particles, the measurements are generally carried out for a preset charge (or some parameter related to it in case of indirect beam current measurement) instead of for a preset live time. The charge (or related parameter) is usually measured by an external counter, and this unit forces the data collection to stop when its preset is reached. For acquiring the spectra, either a personal computer (PC)-based multichannel analyzer (MCA) or a classical MCA may be used, but the latter should be interfaced to a computer so that the spectra can be saved on disk and evaluated by appropriate computer programs.

2. Micro-PIXE

The data acquisition in scanning nuclear microprobe analysis is more complex than in macro-PIXE and is therefore invariably controlled by computers. As in macro-PIXE, signals from various detectors must be handled, but in addition the positional information must be dealt with. Two main principles are employed: on-line display of elemental maps and event-by-event acquisition with off-line sorting of data. Sometimes a combination of both is used. Elemental maps on-line are obtained by setting energy windows for the characteristic x-ray lines of interest (and for background regions) and reading out the count rate within each window for each position of the beam. In its simplest form, the map can be produced by a storage oscilloscope operating in the X-Y mode with its intensity modulated by the intensity of the x-ray line selected. In another approach, which can map several elements simultaneously, the line intensity values (with or without background subtraction) for each beam position (pixel) are stored in a computer, and maps are generated on-line with intensity modulation by gray or color codes. This technique gives good feedback, so that one can concentrate the analysis on the more interesting regions of the specimen. This is quite important in micro-PIXE analyses, since they are often very time consuming. However, a disadvantage of on-line elemental mapping is that truly quantitative information about the elemental concentrations is not provided.

In the event-by-event or list-mode analysis, data acquisition is handled by the same computer that controls beam scanning [79]. When an event (x-ray, scattered particle, γ-ray, or secondary electron) is registered in a detector, the computer is triggered, and the detector label, the energy of the radiation, and the coordinates of the pixel where the event occurred are recorded on disk or magnetic tape. The

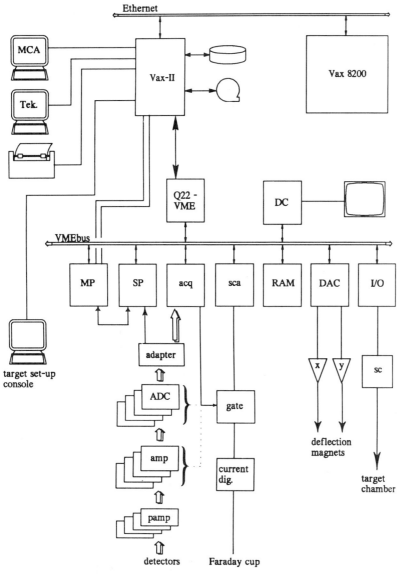

Figure 8 A VME-based data acquisition and beam scanning system; acq, acquisition modules; ADC, analog-to-digital converter; amp, amplifier; DAC, digital-to-analog converter; DC, display controller; I/O, digital input/output module; MCA, multichannel analyzer; MP, master processor (first processor unit); pamp, preamplifier; RAM, memory board; sc, stepping motor controller; sca, scaler board; SP, slave processor (second processor unit). (Reproduced from Ref. 79 by permission of Elsevier Science Publishers.)

same procedure is repeated for each event detected. It is also possible to record the dead-time losses and the accumulated charge for each pixel. The data obtained for all events can be sorted or analyzed off-line in any selected manner. Spectrum evaluation programs of various kinds may be applied to spectra generated from the data, so that quantitative results of high accuracy may be obtained. However, it must be kept in mind that detailed off-line processing of all the recorded data is very time consuming, and it is therefore not recommended for general use.

A powerful, low-cost system for list-mode data acquisition in scanning microprobe analysis can be constructed by using an industrial standard 32-bit modular system (VME) both to control the beam and to acquire and sort data. The schematic diagram of such a VME-based system, as described by Lövestam [79], is presented in Figure 8. The modular VME computer is linked to a host computer (a μVAX-II) via an appropriate interface between the VME bus and the Q bus of the μVAX. The μVAX is equipped with a 500 Mbyte Winchester disk for fast data storage over long time intervals. The digital-to-analog converters, analog-to-digital converters, and stepping motor controls are connected to the VME bus. An assembler coded software package is used to set up the system and to perform user-friendly analysis. The results are calculated by the host computer and presented on a graphic terminal as color- or gray-coded intensity maps of selected elements or of normalized results. It is also possible to display the signals from charged particle surface barrier and secondary electron detectors.

IV. QUANTITATION, DETECTION LIMITS, ACCURACY, AND PRECISION

A. Analysis of PIXE Spectra

Once a PIXE spectrum has been acquired, the first step in the quantitation is the extraction of the net peak intensities for the elements of interest. This task is similar to that to be carried out in all other x-ray emission analysis techniques with energy-dispersive detection, and we therefore refer to Chapter 4 for a detailed discussion of the subject. By far the most common spectrum analysis approach in PIXE is to model the spectrum by an analytical function. This function includes modified Gaussians to describe the characteristic x-ray peaks and a polynomial or exponential polynomial to represent the underlying continuum background [80]. An alternative to analytical background modeling is to use some kind of mathematical background removal method [81].

Because of the low continuum background in PIXE (particularly compared to energy-dispersive EPMA), the range of peak heights in a PIXE spectrum can be up to five to six orders of magnitude. This leads to PIXE spectra that often exhibit fine details, such as escape and sum peaks, and low-energy tailing for intense peaks. Whereas escape peaks and low-energy tailing may also be quite important in EDXRF spectra, sum peaks tend to be a minor problem in the latter, because the most intense peaks are generally located in the upper part of the spectrum. In PIXE, however, the most intense peaks are usually situated in the lower part of the spectrum (this results from the fact that cross sections increase with decreasing Z), and sum peaks therefore show up when high count rates are used during spectrum acquisition. An illustration of the importance of sum peaks

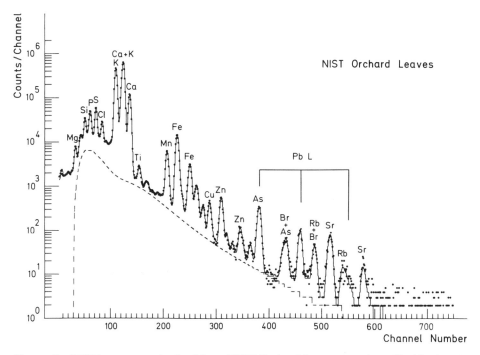

Figure 9 PIXE spectrum obtained for a NIST Orchard Leaves specimen [incident proton energy 2.4 MeV, funny filter in front of Si(Li) detector, x-ray count rate 2200 cps]. The marked peaks are K lines, except where indicated otherwise. The unmarked peaks between Fe $K\beta$ and Cu $K\alpha$ are sum peaks. The dots represent the experimental spectrum, the curve through the dots is the modeled spectrum, as obtained from a non-linear least-squares fit, and the dashed line is the modeled background (Reproduced from Ref. 47).

is given in Figure 9. The presence of the many fine details in the PIXE spectra places stringent requirements on the spectrum model. As far as the modeling of the sum peaks is concerned, this is generally done by representing them by a single pileup element, according to an approach first proposed by Johansson [82]. Despite the many fine details, accurate modeling of PIXE spectra is quite feasible, as was demonstrated in an intercomparison exercise of five different PIXE spectrum analysis programs [80]. It is also illustrated by the good agreement between modeled and experimental spectrum in Figure 9.

B. Quantitation for Thin Specimens

When protons of 1–4 MeV are used in PIXE, elements with Z up to about 50 are generally determined through their K x-rays (typically the $K\alpha$ line), and the heavier elements are measured through their L x-rays ($L\alpha$ line). The basis for a quantitative analysis is that there is a relationship between the net area of an element's characteristic K or L x-ray line in the PIXE spectrum and the amount of element present in the sample. For proton bombardment and an indefinitely thin specimen (by this is meant a specimen that is sufficiently thin that matrix effects become negligible), the relation is given by

$$Y_p(Z) = \frac{N_0 \sigma_{pZ}^X(E_0) \epsilon_p N C_Z \rho t}{A_Z \sin \theta} \tag{9}$$

where $Y_p(Z)$ is the number of counts in a characteristic x-ray line p of the analyte element with atomic number Z, N_0 Avogadro's number, $\sigma_{pZ}^X (E_0)$ the x-ray production cross section for line p at the incident proton energy E_0, ϵ_p the absolute detection efficiency (including the solid angle) for x-ray line p, N the number of incident protons, C_Z the concentration of the analyte element in the specimen, ρ the specimen density, t the specimen thickness, A_Z the atomic mass of the analyte element, and θ the angle between the incident proton beam and the specimen surface. In formulating Equation (9), it is implicitly assumed that the specimen is uniform and that the beam size is smaller than the specimen area. For a formulation of the situation in which the beam size is larger than the specimen area, the reader is referred to the book by Johansson and Campbell [7].

In deriving the analyte concentration from its x-ray yield, several approaches are possible [7]. One can solve Equation (9) for C_Z and thus employ the absolute or fundamental parameter method. This requires accurate knowledge of all the parameters involved. The most critical parameters are the x-ray production cross sections and the absolute detection efficiency. The accuracy of the x-ray production cross sections was addressed earlier (see Sec. II.A.2). The absolute detection efficiency of a Si(Li) detector has been the subject of numerous research papers. Its determination no longer poses problems for photon energies in the 5–30 keV region [83], in which use can be made of accurate long-lived radionuclide standards and the relative detection efficiency can be accurately modeled. As can be concluded from recent papers on this subject [84–87], however, there is still no easy way to calibrate a Si(Li) detector in the 1–5 keV energy gap. The method that at present offers the best solution to this problem is based on the use of an [55]Fe source in combination with secondary fluorescers [84,86].

Because of the difficulties with the absolute quantitation method, many PIXE workers prefer to rely on a relative approach, and they calibrate their experimental PIXE set up using thin-film standards (e.g., Refs. 24, 88, and 89). This method yields so-called thin-target sensitivities $k_p(Z)$, which combine several of the quantities of Equation (9):

$$k_p(Z) = \frac{N_0 \sigma_{pZ}^X(E_0) \epsilon_p}{A_Z \sin \theta} \tag{10}$$

The units of $k_p(Z)$ are x-ray counts per unit proton charge (usually μC) and per μg/cm^2.

Both the absolute and relative quantitation methods generally require knowledge of the specimen mass thickness if the results are to be obtained as concentrations in the specimen material. As discussed by Johansson and Campbell [7], there are a variety of ways to determine the specimen thickness. They include direct weighing, thickness measurement via ancillary photon transmission measurements, thickness measurement via energy loss of transmitted protons, and energy loss determination by means of a beam stop.

The requirement for knowing the specimen mass thickness can be avoided by spiking the sample with a known amount of an internal standard element before specimen preparation. Such spiking is easily done for liquid samples and is also feasible when one deals with powdered solid materials but is of course impossible in the nondestructive analysis of solid samples. Another advantage of spiking is that the number of incident particles (beam fluence or preset charge) need not be measured accurately. Indeed, when a spike is used, quantitation involves division of Equation (9) for the analyte element by the same equation for the internal standard, so that the number of incident particles N as well as the mass thickness ρt cancel out.

C. Quantitation for Specimens of Intermediate Thickness and for Infinitely Thick Specimens

In practice, specimens are rarely thin enough that matrix effects are entirely negligible. For example, in 2.5 MeV proton PIXE of a 0.5 mg/cm² thick U.S. National Institute of Standards and Technology (NIST) Bovine Liver specimen, and when basing the analysis on the $K\alpha$ x-ray line, the matrix correction factor is 1.03–1.05 for the elements K to Sn, and it increases strongly with decreasing Z for the light elements (e.g., it is 1.1 for S, 1.2 for Si, and 1.5 for Mg).

For specimens of intermediate thickness and for infinitely thick specimens (the latter are specimens that are thicker than the particle range), Equation (9) must be replaced by

$$Y_p(Z) = \frac{N_0 \epsilon_p N C_Z}{A_Z} \int_{E_0}^{E_f} \frac{\sigma_{pZ}^X(E) T_p(E)}{S(E)} \, dE \tag{11}$$

where E_0 and E_f are the incident proton energy and the energy of the protons after passage through the target ($E_f = 0$ for an infinitely thick specimen), respectively, E the proton energy, $T_p(E)$ the transmission of the x-rays from successive depths in the specimen, and $S(E)$ the matrix stopping power. $T_p(E)$ is itself given by

$$T_p(E) = \exp\left(\frac{-\mu_p \sin \theta}{\sin \phi} \int_{E_0}^{E} \frac{dE}{S(E)}\right) \tag{12}$$

with μ_p the mass attenuation coefficient for line p in the sample matrix, and ϕ the angle between the specimen surface and the specimen-detector axis (i.e., the x-ray takeoff angle).

It should be noted here that the relation between the analyte element line intensity $Y_p(Z)$ and the concentration C_Z, as expressed by Equations (11) and (12), does not include secondary or tertiary fluorescence enhancement effects. A detailed treatment of these effects was given in a recent paper by Campbell et al. [90]. Although enhancement effects are less pronounced in PIXE than in XRF and are in fact often negligible, secondary fluorescence should be accounted for when the analyte elements are lighter than the matrix elements. For example, in 3 MeV thick target PIXE of stainless steel (with θ and ϕ both equal to 45°), the Cr $K\alpha$ intensity is raised by about 50% as a result of the secondary fluorescence from the Fe K x-ray lines [90].

As for thin specimens, several quantitation approaches are possible for thick specimens. If one relies on the fundamental parameter approach and thus solves Equations (11) and (12) for C_Z, other parameters are needed in addition to those already required for the thin-specimen case. These additional parameters are μ_p and $S(E)$, that is, the mass attenuation coefficient for line p in the sample matrix and the matrix stopping power. The values of those parameters for the sample matrix can be obtained from those for the matrix constituents by employing Bragg's additivity rule, as already indicated in Section II.A.1 for the matrix stopping power. Such calculations require knowledge of the matrix composition and data bases for the mass attenuation coefficients in the various elements and for the elemental stopping powers. The stopping power data base and its accuracy were dealt with in Section II.A.1. The problem of selecting an accurate data base for the mass attenuation coefficients is the same as in all other x-ray emission techniques and is not discussed here. The matrix elemental composition places the major burden on the calculations and usually contributes most to the uncertainty in the calculated μ_p and $S(E)$ values for the sample matrix. For heavier element matrices, in which all matrix elements are detected in the PIXE spectrum, iterative procedures can be applied to obtain the matrix composition, but for light-element matrices, one must resort to a priori information (e.g., obtained by other techniques), or certain assumptions must be made (e.g., that the elements are present as oxides). In any case, this problem of matrix composition is common to all x-ray emission analysis techniques. When specimens of intermediate thickness are analyzed, the transmitted proton energy E_f (or rather the energy loss $E_0 - E_f$) is also needed for evaluating the integral in Equation (11). This implies knowledge (or determination) of the specimen mass thickness, because the energy loss is related to the latter through the matrix stopping power. Alternatively, the energy loss can be measured experimentally. It is evident that any error in the specimen mass thickness (or in the experimental energy loss) is also transmitted to the value of C_Z. The magnitude of this error transmission increases with decreasing specimen thickness, and ultimately a given relative error in the specimen mass thickness produces an identical relative error in the value of C_Z, as is in fact also the situation for the infinitely thin specimens just discussed. The error transmission from the matrix composition, from the data bases, and for intermediately thick specimens also from the specimen mass thickness can be much reduced by the use of an internal standard element. Indeed, the error transmitted in the integral of Equation (11) is to a large extent in the same sense for the analyte and the spike, so that a significant error reduction occurs when dividing the two integrals.

An alternative to the pure fundamental parameter quantitation approach is to make use of experimental thin-target sensitivities $k_p(Z)$, as defined by Equation (10), so that Equation (11) can be written as

$$Y_p(Z) = \frac{k_p(Z)NC_Z}{\sigma_{pZ}^X(E_0)/\sin\theta} \int_{E_0}^{E_f} \frac{\sigma_{pZ}^X(E)T_p(E)}{S(E)}\, dE \tag{13}$$

By solving this equation for C_Z, one basically uses a relative method (relative to thin-film standards), but the correction for matrix effects is made by a fundamental parameter approach. As in the relative quantitation method for thin specimens, this mixed approach requires no knowledge of the absolute detection efficiency or of the radiative transition probabilities and fluorescence yields [the latter two parameters cancel out in the ratio of the x-ray production cross sections in Eq. (13)], but ionization cross sections and Coster Kronig yields are still required.

However, the division of the x-ray production cross sections also has the effect that the impact of the Coster Kronig yields is marginal and that for ionization cross sections essentially only their dependence upon proton energy is needed, which has a much smaller uncertainty than the absolute value of the ionization cross section.

In the analysis of infinitely thick specimens, one can also utilize experimental thick-target calibration factors instead of relying on the fundamental parameter approach or on experimental thin-target sensitivities. The thick-target calibration factors incorporate the integral of Equation (11) and are usually expressed in x-ray counts per μC and per $\mu g/g$. They are commonly derived from PIXE measurements on samples with known trace element composition (standards). In a strict sense, the thick-target factors are only valid for the analysis of unknown samples with identical (matrix) composition as the standards, but in practice, some variability in composition can be tolerated or corrected for. The necessary correction factor is in this case the ratio of the integral of Equation (11) for the standard to the corresponding integral for the unknown.

As discussed in Johansson and Campbell [7], still other quantitation approaches are possible, for example, making use of thick single-element standards.

Before closing this section on quantitation for semithick and infinitely thick specimens, it should be warned that Equations (11) through (13) are in a strict sense only valid for perfectly flat homogeneous samples, and that, for specimens made up of particulate material, the particle size should be as small as possible (ideally below 1 μm when quantifying light elements, such as Na through Si). For a discussion of the importance of surface roughness effects in PIXE, we refer to the paper by Cookson and Campbell [91]. As far as the study of particle size effects in PIXE is concerned, Jex et al. [92] recently presented a theoretical description of these effects for monolayers of spherical particles, but unfortunately no studies have been made thus far of the effects for multilayers of particulate material.

D. Detection Limits in Thin- and Thick-Target PIXE

As in other spectrometric techniques, the detection limits in PIXE are determined by the sensitivity (calibration) factors, on the one hand, and by the spectral background intensity where the analyte signal (x-ray line) is expected, on the other hand. Various definitions for the limit of detection (x_L) were proposed in the early years of PIXE [7], but it is now general practice to define x_L as that amount (or concentration) of analyte element that gives rise to a peak area equal to three times the standard deviation (square root) of the background intensity N_B in the spectral interval of the principal x-ray line. The spectral interval for integration of N_B is invariably defined in terms of the full-width at half-maximum (FWHM) of the principal x-ray line, but regions of one, two, and three FWHMs have been used in the PIXE literature; this range of choice introduces a variation of 1.7 in detection limits deduced from the same data set.

For thin specimens, the relationship between line intensity $Y_p(Z)$ and analyte concentration C_Z was given by Equation (9). If we represent the probability for the production of continuum background radiation per unit of x-ray energy by σ_B and if we further assume (to keep the formulation simple) that the background originates from a single matrix element with atomic mass A_B, the following relation can be written for the background intensity N_B:

$$N_B = \frac{N_0 \sigma_B(E_0) n\text{FWHM} \epsilon_p N \rho t}{A_B \sin \theta} \tag{14}$$

where nFWHM indicates the spectral interval used for summation of the background, and all other symbols have the same meaning as in Equation (9). By setting $Y_p(Z)$ equal to $3N_B^{1/2}$ and solving Equations (9) and (14) for the x_L value of C_Z, one obtains

$$x_L(Z) = \frac{3A_Z}{\sigma_{pZ}^X(E_0)} \sqrt{\frac{\sigma_B(E_0) n\text{FWHM} \sin \theta}{N_0 A_B \epsilon_p N \rho t}} \tag{15}$$

In thus appears that x_L is proportional (or inversely proportional) to the square root of the experimental parameters FWHM, ϵ_p, N, and the specimen mass thickness. Hence, to optimize x_L, a detector with very good resolution must be used, and rather obviously, the solid angle of detection should be made as large as possible, but improvement in this parameter is limited by the area of the detector and by the fact that the detector can only get to about 2 cm from the specimen. Much more flexibility is provided by the number of incident protons N (preset charge), which can be increased either by a longer measurement time or by an increase in beam current.

Both theoretical calculations and experimental measurements have been performed to obtain x_L values for thin specimens (e.g., Refs. 37, 29, and 40). In most of this work one adopted (or employed) a light-element matrix (typically carbon or an organic polymer) and a specimen mass thickness of 1 or 0.1 mg/cm^2. Johansson and Johansson [29] produced a very useful contour plot of x_L values as a function of incident proton energy and atomic number of the analyte element for the case of a 0.1 mg/cm^2 thick carbon matrix. Their plot, reproduced here in Figure 10, was based on experimental measurements of the continuum background, and it was further assumed that elements with atomic number up to about 50 are determined through their $K\alpha$ x-ray line and the heavier elements through their $L\alpha$ line. As can be seen in Figure 10, there is a valley of optimum detection limits for both the K and L cases, with the best K x_L values (less than 0.5 μg/g) obtained at lower proton energy than the best L x_L values (0.5–1 μg/g). Furthermore, within either the K or L case, the bombarding energy for optimum detection limits depends upon the atomic number of the analyte elements of interest, with higher bombarding energies favoring the heavier elements. Selection of the energy should thus be made with the objective of the analysis in mind, but in practice some compromise is necessary. Johansson and Johansson [29] concluded from their contour plot that the optimum proton energy is about 2 MeV for the analysis of trace elements in biological and environmental samples. Such bombarding energy also has the advantage that the x_L values show rather little variation (about one decade only) for the analyte elements with Z between 15 and 90. More recently, Ishii and Morita [40] produced a contour plot similar to that of Johansson and Johansson [29], but they based it solely on theoretical calculations and adopted a pure oxygen matrix (which was considered representative for biological samples). The conclusion of this study was that the best detection limits were obtained

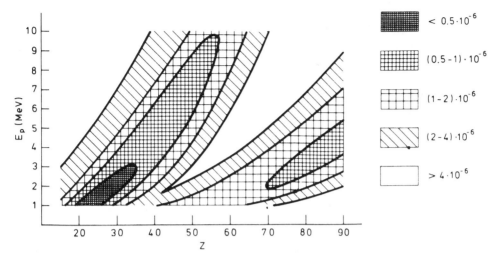

Figure 10 Contour plot of the limit of detection (x_L) as a function of incident proton energy and atomic number of the analyte element for the case of a 0.1 mg/cm^2 thick carbon matrix. Experimental conditions: detector FWHM 165 eV, solid angle of detection 38 msr, collected charge 10 μC. The background interval selected for calculating x_L was equal to one FWHM. (Reproduced from Ref. 29 by permission of Elsevier Science Publishers.)

with about 3 MeV protons. Considering these investigations on x_L values, it is thus no surprise that most PIXE laboratories employ proton energies in the range of 2–3 MeV. Several laboratories actually use consistently the same bombarding energy (e.g., 2.5 MeV) for their typical applications. Much higher proton energies (e.g., 40 MeV) are also useful for certain applications, however, particularly for measuring the rare earth elements (REE) in geological samples [42–44]. At such a high beam energy, the K ionization cross sections for the REE are sufficiently high that one can base their analysis on the K x-rays and thereby eliminate the very severe line overlap problems encountered when the REE are determined through their L x-rays. Detection limits of 5 μg/g or better have been reported for the REE in 40 MeV proton PIXE [42].

The detection limits in Figure 10 are expressed in relative units (concentrations), but they are easily converted into absolute units (masses) by multiplying by the specimen mass probed by the beam. Because this mass is invariably small (e.g., for a 0.1 mg/cm^2 thick specimen as in Figure 10 and a typical beam size of 0.2 cm^2, the probed mass is 0.02 mg), it follows that the absolute detection limits are in the range of 10–100 pg for the case of 2 MeV protons and the experimental parameters of Figure 10.

As indicated by Equation (15), the concentration limit of detection improves with the square root of the mass thickness. However, this relation remains valid only as long as the thin-specimen criterion holds. Because of the existence of matrix effects (i.e., the decrease in σ_{pZ}^X and in the cross sections for bremsstrahlung production with decreasing proton energy and the attenuation of the characteristic x-rays and continuum radiation by the sample matrix), x_L rapidly approaches its optimum value. In general, the improvement in x_L is quite limited above a few mg/cm^2. Depending upon the origin of the continuum background and the energy dependence of its production cross section, x_L may actually deteriorate somewhat beyond a certain specimen thickness. Such situations may occur for analyte lines in the spectral region in which prompt γ-radiation forms the major background source (typically above about 10 keV).

Whereas x_L studies for thin specimens have concentrated on light-element matrices, in similar investigations for infinitely thick targets, heavy-element matrices were also considered. Teesdale et al. [93] conducted a comprehensive experimental study of x_L values for 1–5 MeV proton bombardment of pure single-element matrices of carbon, aluminum, silicon, titanium, iron, germanium, molybdenum, silver, tin, ytterbium, and lead. They paid particular attention to the choice of appropriate x-ray filters for suppression of the matrix characteristic x-ray lines and their pileup peaks and to the choice of the optimum proton energy. It was found that increase in energy up to 3 MeV is profitable, but that a further increase confers only small benefits. Increasing absorber thicknesses were suggested for increasing atomic number of the matrix up to $Z = 40$. Under these conditions and using a beam charge of 1 μC, the x_L values were a few μg/g for the light-element matrices (C and Al) and 10–50 μg/g for the intermediate matrices Ti through Ge.

By using an optimized PIXE setup and a preset charge of 100 μC (which corresponds to a 15 minute bombardment with 100 nA current), x_L values down to a few tenths of a μg/g may be obtained for light-element matrices. The detection limit, expressed as concentration in the original sample, can be further improved by a physical or chemical separation of the material of interest or by a chemical separation of the analyte element(s) from the bulk of the sample. Drying or freeze drying is an obvious preconcentration step for natural water samples, but also for biological tissues (which typically contain 80% water). For natural waters with high mineral content, such as seawater, chemical preseparation schemes are advantageous, and detection limits of 1 ng/L (or 1 pg/g) have been obtained by such approach [94]. For dried biological tissues, further preconcentration may be realized by resorting to high- or low-temperature ashing. As demonstrated by Pallon and Malmqvist [95] and Maenhaut et al. [96], however, the gain in detection limits remains limited to a factor of about 2.

In this entire discussion on x_L values, it was assumed that the PIXE analyses are carried out in a vacuum chamber. In external beam or nonvacuum PIXE (in air or in a helium or nitrogen atmosphere), poorer x_L values are expected because of the background contribution from interactions in the beam exit foil material and in the air or chamber gas and, for the light elements, also because of the substantial attenuation of their soft x-rays by the same gases. However, practical x_L values in nonvacuum PIXE appear to be rather comparable to the x_L values in vacuum PIXE, at least for analyte elements with atomic number above 25 [97].

E. Precision and Accuracy in Thin- and Thick-Target PIXE

As in any other analytical technique, high precision and accuracy should be aimed for in PIXE. It is therefore essential that careful attention is given to all stages of the analysis. These include sample and specimen preparation, specimen bombardment, spectral data processing, quantification, and correction for matrix effects. For a discussion of the critical facets in the various stages, the reader is referred to the specific section dealing with each stage. As far as the specimen bombardment stage is concerned, it should be added here that one should be aware of the danger of radiation- or heat-induced losses during PIXE bombardment. Such losses are particularly feared for volatile analyte elements (e.g., the halogens, S, As, Se, and Hg) and, in the case of organic or biological specimens, also for certain matrix elements (mainly H and O). The current density applied during analysis plays a major role, and the danger for losses is therefore more severe in micro-PIXE than in macro-PIXE. In any case, it should be determined which irradiation conditions are safe for a particular application. More information on this subject can be found in a tutorial paper by Maenhaut [47] and in research papers of Cholewa and Legge [98] and Themner et al. [99].

The reproducibility (precision) of an entire PIXE analytical procedure (including the contribution from sample processing and specimen preparation) can be examined by preparing several specimens from the same material, subjecting these to PIXE, and calculating a standard deviation (s) from the spread in the results obtained. Under optimum conditions, this standard deviation should be the same as that expected on the basis from counting statistics alone (s_c). However, when the percentage standard deviation from counting statistics ($\%s_c$) approaches values smaller than about 1–2%, differences between s and s_c are often unavoidable because of the limitations in sample and specimen homogeneity. It should indeed be realized that, even in macro-PIXE of infinitely thick specimens, the probed sample mass is at most a few mg, so that only ng amounts of analyte elements are actually examined for concentration levels of a few μg/g. Particularly in the analysis of biological, geological, and atmospheric aerosol samples, a $\%s$ of 1–2% is often the ultimate practical limit of precision. Such precisions were obtained by Maenhaut et al. [100], for example, in PIXE analysis of biological reference materials.

The accuracy of a PIXE procedure should be evaluated by analyzing (certified) reference materials or through comparisons with other analytical techniques. Ultimately, the accuracy will depend on the extent of spectral interferences and matrix effects and on how well these can be controlled or corrected for. Several PIXE analyses of reference materials and other accuracy investigations have been reported in the literature, and selected studies dealing with the analysis of trace elements in biological, environmental, and geological samples were reviewed by Maenhaut [101]. As an example of an accuracy investigation on biological materials, the study by Maenhaut et al. [100] can be cited. A total of 18 elements were measured in up to 14 (certified) reference materials, and from a comparison of the PIXE results with the reference values (when available), it was concluded that the accuracy was better than 5%. As far as atmospheric aerosol samples are concerned, a fine intercomparison with other techniques was presented by the PIXE group from the University of Marburg [102,103]. This group participated in an intercomparison program that involved the analysis of 250 aerosol filters by means of various techniques at different laboratories. The PIXE results for S, V, Cr, Mn, Zn, Se, Cd, Sb, and Pb were plotted against the average values of the different techniques, and most data points followed the ideal 1:1 slope quite closely. An example of an accuracy investigation for the case of geological samples is the study by Carlsson [104]. This author used thick-target PIXE and particle-induced prompt γ-ray emission analysis (PIGE) to analyze a geological standard, which was prepared and thoroughly calibrated at the University of Lund. The PIXE/PIGE data deviated less than 3.3% from the reference values for the elements Na, Al, Si, K, Ca, Fe, Rb, and Sr.

V. SAMPLE COLLECTION AND SAMPLE AND SPECIMEN PREPARATION FOR PIXE ANALYSIS

A. General

In this section general aspects of sample collection (sampling) and sample and specimen preparation are discussed. Methods or procedures that apply only to samples of a specific type (e.g., biological and environmental) are touched upon in Section VI, which deals with the applications of PIXE. Furthermore, the present section discusses aspects that are of general importance in both macro- and micro-PIXE. Points that are relevant for micro-PIXE only are dealt with in Section V.B.

PIXE can in principle be applied to any type of sample. Considering that the bombardments are normally done in vacuum, however, it is evident that the technique is more suitable for analyzing solids than liquids. PIXE analysis of liquids normally involves some preconcentration by drying (which can be as simple as drying a drop of the liquid on a suitable backing film) or some other physical or chemical separation of the analyte elements from the liquid phase. As far as the analysis of solids is concerned, it should be kept in mind that even in macro-PIXE of infinitely thick samples the mass actually probed by the beam is at most a few mg. Determination of the bulk composition of a solid sample without preliminary sample preparation is therefore possible only for samples that are homogeneous in all three dimensions. Hence, before deciding to analyze a sample by PIXE, it should be carefully examined whether the analytical problem is not much better solved by some other technique. The analysis of liquid samples and the bulk analysis of large heterogeneous samples are clearly problems for which PIXE may not be most appropriate. However, as concluded in a recent paper [105] in which PIXE and various nuclear and atomic spectrometric techniques were compared, there are numerous analytical problems for which PIXE is the most suitable technique or at least among the more suitable. Examples are the multielemental analysis of milligram-sized samples consisting of a light-element matrix (e.g., biomedical and atmospheric aerosol samples), the nondestructive analysis of millimeter-sized areas on a large sample or of thin superficial layers on a bulk sample, and various problems that require sensitive analysis with high spatial resolution.

When it has been decided to tackle an analytical problem by PIXE, full use should be made of the inherent characteristics of the technique, particularly of its nondestructive and instrumental character. Therefore, if possible, sampling should be done in such a way that subsequent sample preparation can be avoided or kept to a strict minimum. The collection of suspended atmospheric particulate material (atmospheric aerosols) is a good example of when such strategy should be adopted. For unique samples or samples of high commercial or historical value, sample preparation or subsampling may even not be allowed because the sample generally must be returned unaltered after the analysis. Examples of such samples are historical documents, various objects of art, and extraterrestrial dust particles.

In many situations, however, some sample preparation is required. This may vary from simple cleaning of the sample (to remove surface contamination), polishing (to eliminate surface roughness effects), and powdering (to homogenize the sample and to reduce the particle size), to digestion or physical or chemical preconcentration or separation. Furthermore, the last step in the sample preparation usually consists of preparing specimens that are suitable for PIXE bombardment. Such specimen preparation may involve depositing a drop of a liquid (e.g., of an acid digest) or a few mg of powdered material on a clean, strong substrate film (for thin and intermediate specimens) or pressing a certain amount of sample into a pellet (for infinitely thick specimens). The prepared specimen is often mounted or held in a target frame (e.g., a 25 mm diameter plastic ring or a square target frame that fits in a standard 35 mm slide tray holder). Overall, the sample and specimen preparation procedures in PIXE are quite similar to those in the other x-ray emission analysis techniques. As far as the backing films for thin and semithick specimens are concerned, an additional requirement in PIXE is that such films be able to withstand irradiation by the particle beam. More detailed information on specimen backing films for PIXE is given by Johansson and Johansson [29] and Russell et al. [106].

Considering that PIXE analyses often aim at measuring µg/g levels of trace elements and that the absolute amounts of analyte elements actually examined are then in the ng region or below, contamination control is very important. Hence, acid-cleaned plastic (e.g., polyethylene or Teflon) or quartz containers and tools should be employed during sampling and sample processing. The chemicals, acids, and water used in sample preparations (e.g., for digestion or dissolution) should be of high purity. Also, all critical manipulations should be done in a clean bench with laminar airflow. When applying thin-specimen procedures, realistic blank specimens should always be prepared. This should be done by applying the same procedures and using the same substrate films as for the actual sample specimens. Another point of concern is potential loss of analyte elements during sample storage, sample processing, and specimen preparation. During storage of aqueous samples, analyte elements may be deposited on the container walls. Perhaps more important is that some analyte elements (e.g., the halogens, S, As, Se, Hg, and Pb) may be volatilized by drying of the sample at elevated temperatures, particularly in certain sample preparation methods (e.g., in high- and low-temperature ashing or in acid digestions in open vessels).

For more information about general aspects of sample and specimen preparation for PIXE, the recent tutorial paper by Mangelson and Hill [107] can be recommended. This paper also provides a fine overview of the various physical and chemical methods of sample preparation.

B. Specimen Preparation for Micro-PIXE

Because of the similarity between micro-PIXE and EPMA, the specimen preparations techniques developed for EPMA are generally also applicable in micro-PIXE. However, the difference in ionizing particles (typically 10–20 keV electrons in EPMA versus MeV protons in PIXE) has the effect that the depth probed in the analysis is significantly greater in micro-PIXE. For example, this depth amounts to several tens of micrometers for 3 MeV protons. To obtain meaningful results, the specimen should be homogeneous throughout the depth analyzed, and for optimum use of the spatial capability of the nuclear microprobe, the specimen thickness should preferentially be of the same order as the size of the microbeam. On the other hand, the specimen must be sufficiently thick (0.5–1.5 mg/cm^2) to obtain a high x-ray yield. It should also be realized that the support material (backing film or other support material) may cause interference in the x-ray spectrum, and it should therefore be selected with care.

The actual specimen preparation is to a large extent determined by the material to be studied. For biological materials, cryosectioning of frozen samples followed by freeze-drying and mounting of the material on a thin clean plastic foil is often the method of choice. For minerals, the specimens may consist of thin, polished, or ion-milled disks mounted on a glass plate or an electron microscope grid. Finally, the preparation of thin or semithick specimens should be done in such a way that the sample mass examined in each pixel can be determined. This mass thickness is required to allow expressing the results as elemental concentrations (see Secs. IV.B and IV.C).

VI. APPLICATIONS

The applicability of PIXE to various analytical problems has been amply demonstrated in many publications. Furthermore, numerous studies have already been carried out in which PIXE provided part or all of the requested trace element concentration data. This section presents a brief selection of the applications of PIXE. Many more examples can be found in the proceedings of the various international PIXE conferences [2–6] and in the recent book on PIXE [7].

A. Biomedical Samples

Most biological tissue is built up of organic compounds, so that its matrix elements do not give rise to characteristic x-ray lines in the PIXE spectrum. Furthermore, the electron bremsstrahlung background is lower for an organic matrix than for a matrix of heavy elements. Biomedical samples are therefore ideal samples for trace element determinations by PIXE. The elements of interest in biological materials are either "essential" minor or trace elements (e.g., K, Ca, Mn, Fe, Cu, Zn, and Se) or "toxic" trace elements (e.g., Cd and Pb). A comparison of the detection limits in PIXE (see, e.g., Fig. 10) with the levels of the minor or trace elements in biological tissues (e.g., Ref. 108) reveals that most elements of interest in physiology or pathology, with a few exceptions, such as Cd, can suitably be determined by PIXE in most tissue types. For a more thorough discussion of the applicability of PIXE in the biomedical field, we refer to some recent reviews [109,110].

1. Sample and Specimen Preparation

The sample preparation of biomedical samples for PIXE analysis depends on the type of sample, its composition, the information looked for (bulk concentrations or spatially resolved data), the elements of interest, and the mode of irradiation (vacuum or nonvacuum). For nuclear microprobe analyses, special requirements apply, as indicated in Section V.B.

Many elements exhibit sufficiently high concentrations in biological material that a simple physical sample preparation method may suffice to obtain the requested concentration data. Several of the purely physical sample preparation methods are discussed in detail by Mangelson and Hill [107]. These include drying or freeze-drying, homogenizing and pulverizing, and cutting of thin sections. In some cases, however, particularly for natural levels of toxic elements and for levels of some essential elements in certain tissue types, preconcentration by destruction of the organic matrix or some other chemical preconcentration or preseparation is required. This may be performed in various ways [107]. The most common methods are (1) wet digestion in acids, either in open or closed vessels, (2) dry ashing in an oven, (3) low-temperature ashing in a plasma asher, and (4) biochemical separation techniques.

In wet digestion the organic sample matrix is decomposed by concentrated strong acids, normally nitric and/or hydrochloric acid. Because the acid must be removed by evaporation before PIXE analysis, the use of hygroscopic acids (i.e., sulfuric, perchloric, and phosphoric acid) should definitely be avoided. Moderate heating in an oven or microwave-assisted digestion significantly increases the decomposition rate. The digestion rate may be further increased by performing the digestion in a closed pressurized vessel (e.g., a Teflon bomb). Such a procedure has the additional advantage that losses of volatile analyte elements are greatly reduced. Unfortunately, losses of certain elements (particularly of the halogens) may still occur while opening the vessel and during drying of the digest on the backing film while preparing the PIXE specimen.

Dry ashing provides a greater mass reduction factor than wet ashing, with high-temperature ashing even better in this respect than low-temperature ashing. However, the very severe risk of losing volatile analyte elements may prohibit the use of a high temperature. The more complex low-temperature ashing in oxygen plasma is therefore often preferred, but this technique also involves a serious risk of loss [96,47].

Biochemical separation techniques are quite useful, particularly in studies in which the exact biological role and/or the molecular association or exact molecular site of the trace element are of interest. An example of this in which the properties of PIXE are fully exploited is the biological monitoring of environmental exposure when appropriate indices for heavy metals are to be established and the metal is located at particular sites in special proteins.

After the physical or chemical sample preparation, or even when no sample preparation is used, specimens for the actual PIXE bombardment must generally be prepared, as indicated in Section V.A. When preparing specimens by pipetting a drop of a liquid (e.g., of an acid digest) on a backing film, the uniformity of the deposit can be enhanced by pretreatment of the film [111,112] by adding lecithin or some other suitable additive to the liquid before pipetting [107,113] and by rapid drying of the pipetted solution (e.g., by placing the targets in a vacuum desiccator), thereby favoring the formation of fine crystallites [47].

2. Examples

The application examples presented here were chosen somewhat arbitrarily. However, their selection was guided with the aim of demonstrating the particular potential of the PIXE technique. Furthermore, most examples were taken from the medical field.

a. Dermatological Samples

PIXE has contributed significantly to our knowledge of trace element levels in hair, skin, and nails, in both normal and diseased states. Hair strands have long been popular study objects for PIXE and other trace element analysis techniques. They are easily collected, and their elemental composition is supposed to reflect the individual's state of health or possible exposure to toxic metals. Several characteristics of PIXE can be favorably used when analyzing hair samples, that is, the high absolute sensitivity combined with high lateral resolution, the multielemental capability, and the fact that PIXE allows nondestructive and direct analysis. By employing a millimeter-sized beam, the distribution of elements along the hair strand can be determined, thereby revealing the time of intake or exposure of the elements. Furthermore, by scanning a nuclear microbeam over cross sections of hair from different stages of development, very interesting information can be obtained about the incorporation routes of elements in hair [114]. The capability of micro-PIXE to perform quantitative analysis on a micrometer scale is also very useful in studies of normal and pathological human epidermis [115]. Reproducible elemental distributions were observed for Zn, Fe, and Ca in normal skin, as well as deteriorated distributions of Fe in paralesional psoriatic skin.

b. Neurobiology

Trace and minor element distributions in the normal and diseased central nervous system are of great interest both from a basic physiological viewpoint and for clinical use. As shown in several papers, macro- and microbeam PIXE analyses can provide very important information in this field.

Duflou et al. [116] used macro-PIXE in an extensive study of the elemental composition of up to 50 different structures of normal human brain. It was found that there are reproducible and significant differences between various structures. The same authors subsequently applied macro-PIXE to examine elemental alterations in pathological brain samples [117]. In a study on malignant brain tumors [118], the multielement data set, as obtained by PIXE, was examined by a multivariate statistical method, and this provided a prognostic tool to estimate the survival time of patients. Furthermore, both macro- and micro-PIXE measurements revealed significant changes in elemental composition for the tumor front of gliomas.

Another important area of research in neurology and neuropathology is that of neurodegenerative diseases, such as Alzheimer's disease. Traditional electron microprobe measurements have indicated that the senile plaques observed in Alzheimer's cases contain high levels of Al and Si. However, the sensitivity of the electron probe is insufficient to measure other trace elements that may aid in understanding the etiology of this serious mental state. Some preliminary analyses of plaques have been done with the micro-PIXE technique at a resolution of 1 μm, and these analyses revealed extremely high local concentrations of titanium [119].

c. Mineralized Tissue

Various types of mineralized or calcified tissue can be directly examined by PIXE with virtually no sample preparation. The basic aspects of the PIXE analysis of a Ca-rich matrix were discussed by Ahlberg and Akselsson [120]. Möller et al. [121] developed a method for cleaving teeth without touching the fresh surfaces, and they analyzed the specimens in vacuum with a 1 mm proton beam. The nuclear microprobe was also successfully applied, for example to samples from the inner ear [122], the femur [123], and teeth [124].

d. Body Fluids and Single Cells

Because of the relative ease of collection, some body fluids (e.g., blood, urine, and saliva) have become very valuable for diagnostic purposes. The elemental profiles of such body fluids are often used as health indices and indicators of environmental or occupational exposure. However, it should be noted here that there may be quite substantial short-term variations in the trace element profiles of certain body fluids (e.g., of urine), so that interpretation of the elemental data is not always easy. Also, the detection limits of PIXE are too high to measure the natural levels of most trace elements in body fluids, and preconcentration is therefore often required. The application of PIXE or other x-ray techniques for the bulk analysis of such samples is generally justified only if a multielement analysis is indeed needed.

However, some elements that are of great medical relevance, such as Cu and Zn, are easily measured by PIXE. The ratio between both can be used as an index of the whole-body status in some pathological states, such as Wilson's disease. Starting from a whole-blood sample, or from a plasma or serum sample, the Cu/Zn ratio can be determined within minutes by PIXE [125].

By using special preconcentration techniques it is possible to measure the natural levels of Ni and several other metals in urine [126]. For analyzing cerebrospinal fluid by PIXE, a specimen preparation method was developed that consists of making very thin self-supporting pellets [127]. The capability of PIXE for analyzing small sample masses was fully exploited in the study of Pallon et al. [128] on the association of trace elements with blood serum proteins. These authors used gel filtration to separate a serum sample into 100 fractions and analyzed each fraction by macro-PIXE. The results are displayed in Figure 11.

The application of nuclear microprobe analysis to single cells [129,130], such as red or white blood cells, can provide unique elemental information. In a study on the link between diseases and elemental profiles, it was found that there are very large fluctuations in the profiles of the individual cells [130]. Such a phenomenon would have remained obscured by a "collective" analysis of many cells.

B. Atmospheric Aerosols

The multielement analysis of airborne particulate material (atmospheric aerosols) has been a very popular and highly successful application of PIXE since the early days of the technique. Actually, the pioneering paper of Johansson et al. [1] presented a PIXE spectrum of such material. A more recent spectrum, obtained from the bombardment of an urban aerosol sample [80], is shown in Figure 12.

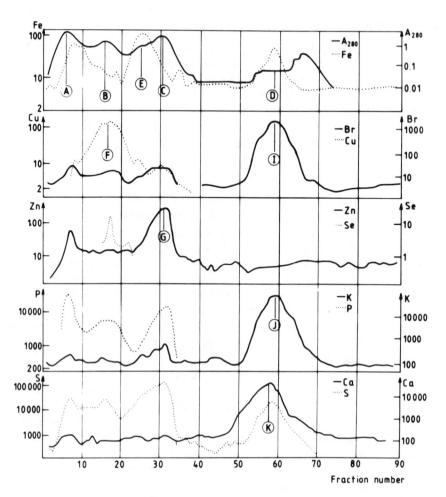

Figure 11 Elemental concentrations (in ng per fraction) in blood serum after separation by gel filtration [128]. The horizontal scale is roughly a logarithmic scale of decreasing molecular size. The curve A_{280} (UV absorbance at 280 nm) denotes the total protein content. (A) IgM and α2-macroglobulin; (B) IgG, IgA, and ceruloplasmin; (C) albumin; and (D) proteins of low molecular weight. The curves for Fe, Cu, and Zn have peaks at E, F, and G, which correspond to transferrin, ceruloplasmin, and albumin, respectively. Se is mainly associated with the protein peak containing IgG, IgB, and ceruloplasmin. (Reproduced by permission of the Humana Press, Inc.)

Considering that atmospheric aerosols are often collected as a thin sample layer on some thin filter or substrate film, that such samples can be analyzed nondestructively by PIXE without sample preparation, that the sample matrix consists of light elements, and that a 5–10 minute bombardment suffices to detect up to 20 elements, including interesting anthropogenic elements, such as S, V, Ni, Cu, Zn, As, and Pb, the analysis of aerosol samples forms almost an ideal application of PIXE. Compared to EDXRF, which shares several favorable characteristics, PIXE offers sensitivities (expressed as characteristic x-ray count rate per μg element actually exposed to the analysis) that are typically at least one

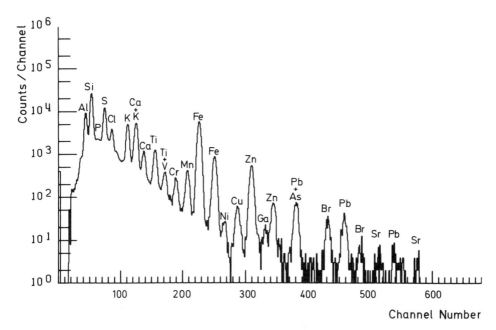

Figure 12 PIXE spectrum of an urban aerosol sample. (Reproduced from Ref. 80 by permission of Elsevier Science Publishers.)

order of magnitude better [131], so that it requires much less sample mass and thus allows the use of compact samplers with high time and size resolution. Another advantage of PIXE over EDXRF is that it can be complemented with other ion beam analysis techniques [45,132], so that the light elements (H, C, N, and O) that make up most of the aerosol mass can be measured as well.

1. Sampling Devices and Collection Surfaces

Sampling of atmospheric aerosols for chemical analysis is usually carried out by means of filters, cascade impactors, cyclones, or a combination of these devices (e.g., Refs. 133–135). Similarly to XRF, samplers that collect the particulate material on the surface of a filter or substrate film are also most suitable for PIXE. Although it is highly preferable for EDXRF that the particles are present as a uniform layer, PIXE can also easily handle nonuniform samples, such as those collected by single-orifice cascade impactors.

Collection of the aerosol by filtration is done in total filter samplers (high-volume or low-volume samplers) but also in certain devices that fractionate the aerosol into two size fractions, such as the stacked filter unit [136] and the dichotomous sampler or virtual impactor [137,138]. Membrane or fibrous filters may be used as filter materials, but the former, in particular Nuclepore polycarbonate filters, are preferred for analysis by PIXE. With fibrous filters, and in fact also with certain membrane filters, a large fraction of the aerosol particles penetrates into the filter material, so that cumbersome corrections for matrix effects are required [139]. For certain applications Teflon membrane filters may be preferable to Nuclepore filters. Unfortunately, their high fluorine content gives rise to a pronounced prompt γ-ray background in the PIXE spectrum. Furthermore, some Si(Li) detection systems, when exposed to high rates of such prompt γ-rays, may exhibit resolution deterioration and other electronic problems similar to those encountered when scattered particles penetrate the detector.

Total filter samplers, stacked filter units, and dichotomous samplers allow collecting samples that may be analyzed by a variety of techniques including PIXE, but they do not make full use of the favorable characteristics of PIXE, in particular its small mass requirement. Several PIXE groups have therefore spent considerable time and effort in designing innovative samplers that combined several of the following features: light weight, battery-powered, automated, and providing good time and size resolution. A detailed overview of the various samplers is given in References 132, 140, and 141. We restrict ourselves here to discussing some samplers that have been extensively and successfully used in actual sampling campaigns. Nelson and coworkers [142–144] developed the linear streaker and its successors, the circular and two-stage circular streaker. In these devices, a Nuclepore polycarbonate filter surface is continuously moved over a sucking orifice, so that the particulate material is collected as a linear or circular streak on the filter material. The rate of movement is typically 1 mm/h, so that subsequent analysis of the Nuclepore strip with a 2 mm wide beam provides a time resolution of 2 h. In the two-stage streaker version, particles larger than 2.5 μm equivalent aerodynamic diameter (EAD) are collected by impaction on a suitable thin-sub-

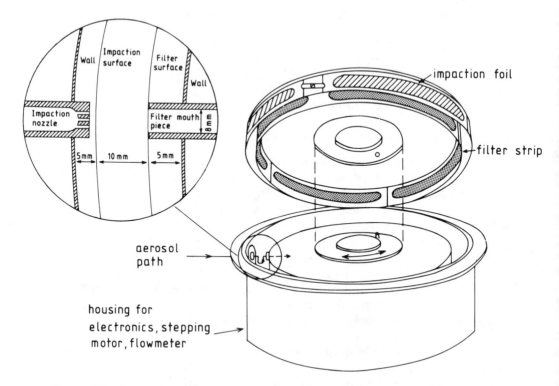

Figure 13 Automatic, microcomputer-controlled, two-stage aerosol sampler especially designed for sample collection for subsequent PIXE analysis. (Reproduced with permission from Ref. 145. Copyright 1985 American Chemical Society.)

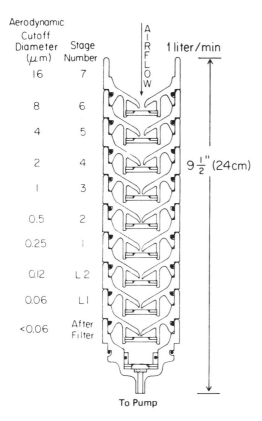

Aerodynamic Cutoff Diameter (μm) / Stage Number

Aerodynamic Cutoff Diameter (μm)	Stage Number
16	7
8	6
4	5
2	4
1	3
0.5	2
0.25	1
0.12	L 2
0.06	L 1
<0.06	After Filter

AIRFLOW 1 liter/min

9½" (24 cm)

To Pump

Figure 14 Cross section of the 1 L/min, single-orifice, Battelle-type cascade impactor, as modified and commercialized by PIXE International Corporation, Tallahassee, FL.

strate film (e.g., Mylar). The stationary aerosol monitor (SAM) developed by the University of Lund [145] and shown in Figure 13 is essentially based on the same concept as the two-stage streaker but incorporates several refinements, such as microprocessor control. To obtain finer size resolution than with the samplers just mentioned, the 1 L/minute single-orifice Battelle-type cascade impactor, as modified and commercialized by PIXE International Corporation [144], can be used. This unit, shown in Figure 14, differentiates the aerosol in up to 10 size fractions, provides good size resolution in the submicrometer range, and can be operated from small battery-powered pumps. A more sophisticated design that combines both good time and size resolution is the DRUM impactor, developed by the University of California at Davis [146]. The collection surfaces in cascade impactors are typically thin polyester (Mylar) or polycarbonate (Kimfol) films. To reduce particle bounce-off effects during sampling, the films are commonly coated with petrolatum or paraffin.

2. Examples

As indicated in a recent review [132] on the status of PIXE in aerosol research, the technique can equally well be applied to samples from research dealing with health effects and samples from studies that involve welfare. Numerous examples of both types of applications are available. In fact, the samples analyzed by PIXE range from those collected indoors (e.g., in work environments) or near specific pollution sources, to samples dealing with urban or regional air pollution problems, to samples collected in areas as remote as the geographical South Pole.

The PIXE research on work environment aerosols up to about 1983 was reviewed by Malmqvist [147]. Among the more recent work, that carried out by Annegarn et al. [148] provides a fine example of the usefulness of PIXE in this area. These authors employed streakers, cascade impactors, and stacked filter units to collect aerosol samples emitted during various underground gold-mining operations (drilling, blasting, and clearing). The multielement PIXE data sets were examined by multivariate statistical techniques, and this resulted in the identification of five source processes contributing to the aerosol variability. A surprising finding was that Cl-containing particles derived from groundwater made up a large fraction of the aerosol mass. In some samples the Cl mass even exceeded the mass of the quartz-related element Si.

Particulate matter emitted from the stack of oil-fired and coal-fired power plants was examined by the Milan PIXE group [149]. The samples were collected with a 1 L/minute cascade impactor, and their analysis by PIXE resulted in detailed size distributions for several elements. The same Milan group also made an extensive study of the pollution aerosol at four sites in the city of Milan and in several Sicilian towns [150]. A variety of samplers, including total Nuclepore filter samplers, stacked filter units, and cascade impactors, were employed for the aerosol collection. The annual trend (seasonal variability) of several anthropogenic elements, including S, V, Ni, and Pb, were examined, associations (correlations) between the various elements were investigated, and detailed size distributions were given for S, V, Mn, Ni, Cu, Zn, Pb, and some other elements. PIXE has been employed in several other urban aerosol studies in the Western world [7], but it is increasingly used for urban studies in other areas as well. Research projects carried out in various cities in Brazil [151] and in Beijing, China [152], are good examples of such studies.

A great deal of the PIXE effort has dealt with regional aerosol problems. To study such problems, large sampling networks were set up that involved the continuous and simultaneous operation of aerosol collection devices at various locations, so that the cost effectiveness of PIXE-compatible samplers and PIXE analysis could be fully exploited. Groups at the Florida State University (FSU) and at the University of California—Davis have done pioneering work in this area. Among many other things, the FSU group studied the relation between humidity, temperature, and sulfate aerosol formation in the eastern United States [153]. The Davis group established networks in the state of California [154], in the western United States [155], and in most major national parks in the United States [156]. The objective of the last network was to investigate the causes of visibility impairment at various national parks. Stacked filter units with a coarse Nuclepore polycarbonate filter and a fine Teflon filter were used for the aerosol collection, and the analytical work relied heavily on the complementary IBA technique FAST in addition to PIXE. It was found that the fine mass (which is responsible for the visibility degradation) was dominated by ammonium sulfate in the eastern United States, whereas in the western United States, fine soil and organics were also important aerosol components.

PIXE is also very useful in examining the long-range transport of anthropogenic and natural atmospheric trace elements, in assessing their sources or source processes, and in studying various other aspects of global atmospheric chemistry. The Lund group has a long tradition of examining the long-range transport from various European countries to Sweden [157–160]. The PIXE data sets obtained were carefully examined by classifying the samples according to air mass history and by applying multivariate statistical techniques, particularly SIMCA. By this approach pollution aerosol signatures for large source regions could be established. Furthermore, extensive studies were carried out on the release of particulate material by the vegetation in the Amazon basin [161,162]. Most of the aerosol collections were done with stacked filter units, and the coarse- and fine-particle PIXE data sets obtained were examined by absolute principal factor analysis. This procedure allowed extraction of the biogenic component and determining its elemental composition. The composition of particulate material over the oceans and the long-range transport of continentally derived materials have also been the subject of extensive investigations. One such study involved the use of cascade impactor samplers provided with battery-powered pumps for collecting samples from a sailboat on a voyage from Belgium to Tahiti [163,164]. In a follow-up study of the aerosol composition above the tropical and equatorial Pacific [165], stacked filter units were used in addition to cascade impactors, and the PIXE analyses were complemented by PESA. An important finding of this study was that the non–sea-salt sulfate levels could be quantitatively related to the levels of its biogenic precursor dimethylsulfide. PIXE analyses of Arctic aerosol samples, with the aim of improving our understanding of the phenomenon of Arctic haze, were carried out by Heidam [166], Heintzenberg et al. [167], Maenhaut et al. [168], and Li and Winchester [169], among others. In most of this work, elegant use was made of receptor modeling by principal components analysis and/or chemical mass balance techniques. Antarctic aerosol studies involving analysis by PIXE were also done (e.g., Refs. 170 and 171).

C. Other Applications in the Earth Sciences

1. Aquatic Systems

Liquid samples are not very suitable for direct analysis by PIXE. By resorting to appropriate preconcentration methods, however, it becomes possible to measure even ultratrace levels of elements in such difficult samples as seawater. In one preconcentration method described [94], a complexing agent is added to a 100 ml water simple, and the metal complexes formed are then collected by shaking with activated carbon and filtering it off. The carbon can be analyzed by PIXE after pressing it into a pellet, or further preconcentration may be done by digesting it in a small volume of strong acid and pipetting the digest on a thin carbon foil, so that a suitable specimen is obtained for PIXE analysis. Figure 15 shows the PIXE spectrum from such a specimen prepared from normal seawater. In another sample preparation approach, the water is sprayed into a fine aerosol and dried, and the particles formed are collected by inertial impaction on a thin substrate suitable for PIXE analysis [172]. This method provides good detection limits for waters with low levels of dissolved salts, such as rainwater, and was shown to give results

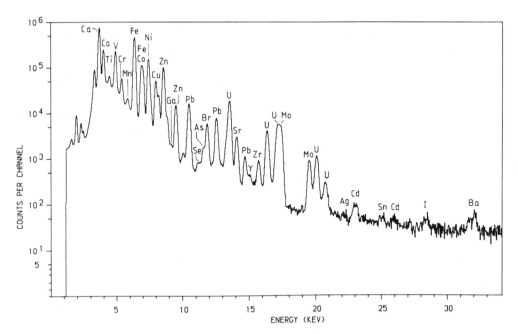

Figure 15 PIXE spectrum of a sample prepared by preconcentration from sea water. (Courtesy of S. A. E. Johansson.)

that are in good agreement with more traditional trace element analysis techniques [173]. With both these sample preparation methods, the special ability of PIXE to allow trace element determinations in small samples can be fully utilized.

Beside the dissolved elements, the suspended phase and sediments may also be of interest in aqueous systems. By a simple sample preparation method, consisting of filtration of the particulate material on a thin filter (e.g., a Nuclepore polycarbonate filter) and subsequent drying, samples are obtained that are ready for direct bombardment by PIXE. Analysis of particulate matter by most common optical spectrometric techniques, such as atomic absorption spectrometry (AAS) and inductively coupled plasma atomic emission spectrometry (ICPAES), requires dissolution. Furthermore, the practical detection limits (expressed in μg/g solid material) are significantly poorer in ICPAES than in PIXE, and AAS is essentially a single-element technique [105]. The potential of PIXE for analyzing solid matter in water and sediment cores and for examining temporal variations along the cores was demonstrated by Respaldiza et al. [174].

Somewhat special water samples are ice cores drilled from glaciers or inland ice. Such samples are collected to investigate variations in their composition and trace element content with depth (e.g., along the core) and, by doing so, to obtain information on past environmental or climatic changes or to assess the impact of major volcanic eruptions. PIXE offers the possibility of analyzing ice cores with mm resolution, so that the seasonal and annual elemental variation can be studied in great detail. Samples for PIXE may be prepared by placing the core on a thin film and subjecting it to freeze-drying [175].

2. Mineral Prospecting

Special methods with potentially large applicability in mineral prospecting have been developed around the PIXE technique. One such method makes use of the fact that minute amounts of particulate matter are transported with geogas from great depths in the earth crust upward to the surface. Collection of the matter for subsequent analysis by PIXE is simply done by exposing very thin foils for some time to the geogas stream [176]. The usefulness of this method has been amply demonstrated by trial experiments at sites with known mineralizations, and the method is now commercially available. As an alternative to PIXE, total-reflection XRF (see Chap. 9) could possibly also be applied, but the particulate material would then better be collected on total reflectors (e.g., quartz disks).

In another method, the drill cores obtained during ore prospecting are analyzed directly in air without sample preparation. An extracted ion beam is used, and the x-rays and γ-rays produced are measured. By computer control it is possible to scan tens of meters of drill core per hour and to generate on-line elemental distributions [176]. Ion beam analysis (PIXE/PIGE) has significant advantages over competitive methods for such samples. For example, compared to on-line XRF, more elements can be detected and accurately quantified.

3. Microbeam Analysis

Because of its unique ability to provide quantitative trace element information on a micrometer scale, micro-PIXE offers great potential for mineral prospecting and for improving our understanding of basic geological processes. Actually, the nuclear microprobe could serve as an excellent complement to the electron microprobe in these areas. The capabilities of micro-PIXE are increasingly recognized by scientists in mineralogy and geology, and some major research programs have already been initiated [56,177].

The mineral sample preparation techniques used for other microscopic techniques (e.g., for the electron microprobe) are to a large extent also suitable for micro-PIXE. Great care should be taken to reduce surface roughness, however, as otherwise accurate quantitative data cannot be obtained. For realizing a smooth surface, grinding or polishing is normally used, but in applying such procedures for PIXE, contamination control is much more stringent than for the electron microprobe. Hence, it is recommended that diamond powder be used for polishing, because this contains only carbon. Furthermore, to take full advantage of the spatial resolution of the microbeam, it is necessary to prepare specimens that are less than a few μm thick. This can be done using an "ion-milling" technique.

The complex matrix composition of many geological specimens means that accurate correction for projectile slowing and x-ray attenuation is quite complicated. Also, the PIXE spectra of such specimens are often quite complex, so that special evaluation codes are required [178]. Finally, it should be realized that, because of the relatively large range of protons in matter, underlying layers of possibly quite different composition may also contribute to the PIXE spectrum for certain specimens.

The micro-PIXE analysis can be done either point by point or by performing a two-dimensional scan over a selected area. The latter method provides fine

562   **Maenhaut and Malmqvist**

maps for the major and minor elements but is normally not sensitive enough at the trace element level. The two-dimensional image can be used for selecting interesting spots for subsequent point-by-point analysis. Figure 16 shows an elemental map obtained by scanning over a 4×4 mm^2 area of a garnet [179]. In this section of the garnet Sr can be seen to be concentrated at the grain boundaries; in another section it was almost homogeneously distributed in the grains. In a

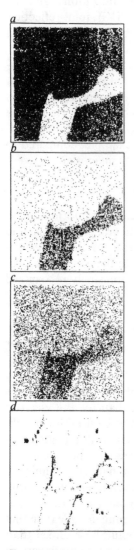

Figure 16 Elemental x-ray maps of the distribution of Ni (a), Cr (b), Mn (c), and Sr (d) in a 4×4 mm^2 thin section of a coarse garnet. Dark areas indicate high concentrations of the elements. Sr is concentrated at the grain boundaries; the other elements are homogeneously distributed within the grains. (Reproduced from Ref. 179 by permission of *Nature*. Copyright 1984 MacMillan Magazines, Ltd.)

study of liquid inclusions in minerals [180], elegant use was made of the fact that a MeV ion beam penetrates deeper in matter than an electron microprobe beam, and the composition of the liquid under a top layer of the mineral specimen could be determined. Meteorites [181], ore minerals [57,176], and zircons [182] are other examples of geological samples analyzed by micro-PIXE.

As indicated in Section III.B.3, micro-PIXE can also be done in an external beam setup in air, and this possibility is particularly useful for analyzing some geological specimens. Accurate elemental maps for areas of several hundreds of mm^2 can be produced by mounting the specimen, which may be very large, on a simple x-y plotter table controlled by normal beam-scanning electronics [183].

D. Applications in the Arts and Archeology

One application of PIXE that has been unexpectedly successful is the analysis of archeological and art objects. The status and possibilities of PIXE and other ion beam analysis techniques in this field were assessed in a topical meeting, and its proceedings are recommended for detailed information [184]. Recognition of the potential of PIXE for art investigations has led to the installation of an ion beam analysis facility in the scientific laboratory of the Louvre museum in Paris [185], which is the first such facility within an art museum.

The multielement character of PIXE and the speed of analysis are very useful in provenance studies of pottery, for example. By analyzing tens or hundreds of objects and applying multivariate statistical techniques to the data obtained, it is possible to discern groupings within the artifacts. Such a study on African potsherds revealed population movements and trade routes [186]. In a similar investigation, obsidians from Australia were analyzed by a combination of PIXE and PIGE [187].

Many archeological and art objects are unique and very valuable and must be returned unaltered after the analysis. Furthermore, some objects are quite sensitive to heat or radiation. These are all reasons for resorting to an external beam facility. Nonvacuum ion beam analysis is less destructive because of the enhanced convective cooling, and it also allows the analysis of large objects that would not fit in a vacuum chamber. That PIXE can offer high spatial resolution in addition to its external beam capability is also of great use for some studies, and several semimicroprobes (or miniprobes or milliprobes) have already been constructed with problems in archeology and arts as their major application. One such milliprobe, set up in the University of California–Davis, has been used in large-scale studies on ancient documents. Both the Gutenberg Bible [188] and the "Vinland" map [189] were investigated in great detail to reveal the printing technique and the authenticity, respectively. At Lund, an external semimicroprobe was used to examine portions of handwriting on 2000-year-old papyrus. By scanning the microbeam, maps of several elements were produced for regions with faded writing. The multielement data were then treated by a multivariate statistical technique to improve the contrast, and by doing so, a two-dimensional pattern was obtained that revealed the actual letters [183]. The external "milliprobe" in combination with PIXE and PIGE has also been used to check the authenticity of oil paintings [190].

E. Materials Analysis

Materials analysis is a vast field of application for ion beam analysis techniques. Rutherford backscattering spectrometry (RBS) in particular is very commonly used for examining surface layers and determining elemental depth distributions in various materials. PIXE generally serves only as a complementary technique, and the situation will likely remain so in the future. Since most matrices in this area are made up of elements with relatively high atomic number, the characteristic x-rays of the matrix elements show up in the PIXE spectra, thereby worsening the detectability of the trace elements. When the matrix is rather light, however, the use of PIXE alone could be very well justified, as in investigations on polymers and some semiconductor materials. It is worth noting that special ion beam facilities using PIXE as an important tool have been implemented for materials analysis in such completely different places as the Louvre (see earlier discussion) and the hot cell department in the Centre D'Etudes Nucléaires at Saclay, where the setup will be used for the direct analysis of highly radioactive material [191].

1. Solid-State Physics and Electronics

Since PIXE is essentially a surface-oriented technique, it is natural to apply it to problems of surface phenomena, for instance corrosion and erosion. However, the multielemental character of PIXE also makes the technique useful for the panoramic analysis of bulk impurities that are present at levels of a few $\mu g/g$ or higher. By exploiting the ultimate lateral resolution of micro-PIXE, it is even possible to study elemental distributions in integrated electronic devices, for example to diagnose errors in the manufacturing process [192].

   PIXE has been used in combination with RBS and channeling to measure impurities in GaAs single crystals [193]. The yield from interactions requiring close encounters between the projectiles and the atoms or nuclei depends, in single crystals, on the direction of travel of the ions through the crystal. Under channeling conditions, that is, when the projectiles are "guided" along crystal axes or crystal planes, the interaction probability is decreased. By comparing the yields of x-rays and scattered particles obtained for channeling and for random orientation, information on the crystal structure and on interstitial atoms may be obtained.

2. Other Materials

The surface character of PIXE makes the technique very suitable for examining catalytical material and for studying corrosion processes. Also, pure metals can be successfully analyzed with reasonable detection limits for the analyte elements when one resorts to a suitable x-ray filter for suppressing the characteristic lines of the matrix element [48,49]. The difficulties associated with thick-target PIXE analysis of high-Z elements may be circumvented by dissolving the material in acids and preparing thin specimens [194]. The recent interest in high-temperature superconductor materials based on ceramics with oxides of rare earth elements, barium, and copper has led to the use of ion beam analysis for characterizing such materials [195]. In this application a combination of PIXE, NRA, and RBS was very useful. Other examples of materials that have been examined by PIXE and complementary IBA techniques are fission reactor [196] and fusion reactor [197] materials, ancient bronzes, and ion-implanted steel.

Considering the growing use of small accelerators and dedicated ion beam facilities in industry, it is anticipated that PIXE and other IBA techniques will be increasingly applied for research in material science in the future [198].

VII. COMPLEMENTARY ION BEAM ANALYSIS TECHNIQUES

On several occasions in this chapter, it has already been indicated that the MeV ion beams employed for PIXE are also very useful for other ion beam analysis techniques. In this section, the various IBA techniques are briefly presented. In particular, it is shown through examples of applications that the other IBA techniques can provide information that is complementary to that obtainable by PIXE (e.g., by extending the elemental coverage down to hydrogen) and that such analyses may often be done simultaneously with the PIXE analysis. For comprehensive accounts of the various complementary IBA techniques we refer to a number of textbooks [32,34,199–201] and the proceedings of the biennial IBA conferences (see Refs. 202 and 203 for the last two conferences in the series).

A. Elastic Scattering Spectrometry and Related Techniques

Elastic scattering spectrometry involves the detection of either the elastically scattered incident particles or the elastically recoiled target nuclei. The first approach is used in such techniques as Rutherford backscattering spectrometry, Rutherford forward scattering spectrometry (RFS), particle elastic scattering analysis, and forward α scattering. Although, strictly speaking, the terms RBS and RFS imply that the scattering process is purely Coulombic (i.e., according to the Rutherford law), they are normally often used in a broader sense and also include non-Rutherford scattering. Also, both forward scattering and backward scattering are often referred to by the single acronym RBS. RBS is a very prominent technique in materials analysis, in which one generally applies 2 MeV α particles and large scattering angles. With such particles deviations from the Rutherford cross section are small for all target elements. In RBS with protons of a few MeV, as used for PIXE, quite substantial deviations occur for the lighter target elements (up to about Ti) because of nuclear interference. As the incident particle energy increases, the deviations become greater for both protons and α particles. The energy of the scattered particles depends on the scattering angle, the energy and mass of the incident ion, and the mass of the target element. When the specimen is not infinitely thin, the scattered particle energy also depends on the interaction depth. Indeed, as a consequence of inelastic Coulombic encounters with bound electrons (see Sec. II.A.1), incident particles lose energy as they penetrate into the specimen and the scattered particles in turn lose energy on their way out to the detector. Elastic scattering spectrometry therefore allows us to obtain information on both the elemental composition of a sample and the distribution of the elements with depth, which is a very important asset for materials research. However, the dependence of scattered particle energy on the interaction depth also has the effect that RBS spectra, in contrast to PIXE spectra, may contain broadened peaks (in the case of semithick samples), consist of a staircase of nearly rectangular shelves (for infinitely thick specimens of homogeneous composition), or exhibit a more complicated appearance. Furthermore, even for infinitely thin specimens, the scatter peaks from analyte elements (nuclei) of neighboring mass become less and less resolved with increasing mass. For complex specimens, such as aerosol samples or biomedical or geological samples, RBS can only provide information on the lighter elements (up to about Si), and extraction of the requested concentrations is generally difficult. The reasons for the successful application of RBS for examining medium and heavy analyte elements in materials research is that substantial a priori information about the specimen is generally available and only a very limited number of medium- or high-Z constituents are present in detectable concentrations.

To improve the elemental mass discrimination (and at the same time ease the spectral analysis) in elastic scattering spectrometry of the light elements in atmospheric aerosol samples, Nelson and coworkers (e.g., Ref. 204) resorted to 16 MeV protons as incident particles and called this variant of the technique PESA. Figure 17 shows a PESA spectrum obtained for an urban aerosol sample collected on a thin polystyrene backing. To raise the scattering cross sections and also to allow measuring hydrogen, Cahill and coworkers [45] prefer forward scattering and the use of 30 MeV α particles as incident ions (FAST). The disadvantages of both 16 MeV PESA and 30 MeV FAST is that these techniques are not applied concurrently with the PIXE measurements but imply a separate ion bombardment; and also, higher energy accelerators are required. Other groups have therefore developed PESA techniques that are more compatible with PIXE and with 1–4 MeV accelerators. Martinsson [205] performed careful investigations on the de-

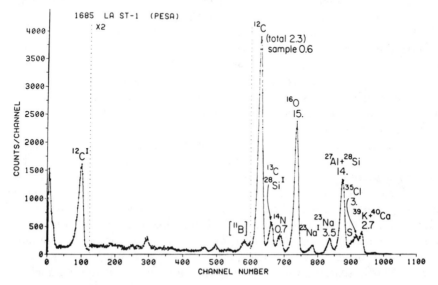

Figure 17 PESA spectrum of an urban aerosol sample. The superscript I for certain isotopes indicates that the peak results from inelastic scattering. The numbers above the peaks denote the concentration of the element in $\mu g/m^3$ air. For carbon, both the gross and blank corrected concentrations are listed. (Reproduced from Ref. 204 by permission of Elsevier Science Publishers.)

pendence of the scattering cross section on incident proton energy and detection angle, with the measurement of H, C, N, and O in thin aerosol samples in mind, and he proposed employing 3.58 MeV incident protons with detectors set up at 170° (for C, N, and O) and at an angle between 29° and 59° (for H). PESA and FAST have been extensively applied for measuring total carbon and the other light elements in various atmospheric aerosol studies. As collection substrate for the particulate material, Teflon was generally used. Andreae and coworkers [165,206–208] employed 16 MeV proton PESA for analyzing marine aerosol samples and related the levels of total carbon to those of soot (black) carbon and of other aerosol constituents. As indicated in Section VI.B.2, the Davis national park study [156] relied heavily on the use of FAST in addition to PIXE to elucidate the causes of visibility impairment.

Whereas elastic scattering spectrometry is a very useful tool in aerosol analysis, its applicability as a complement to macro-PIXE for analyzing biological and geological samples is more limited. As far as biological samples are concerned, RBS and PESA are generally only employed to obtain information on the mass thickness of the specimen and/or to determine the beam fluence. These types of applications are especially useful in microprobe investigations, in which the local mass thickness of the microtome slice must be measured to convert the x-ray intensities from the PIXE spectrum into concentration values [209,210].

As indicated, the depth profiling capability of RBS is an important asset for materials research. This characteristic is also very useful in investigations related to art and archeology. For example, RBS can be employed to measure depth profiles in samples in which surface uniformities may be of high significance, as in patina, corrosion and surface segregation processes, glass aging, and polishing procedures [211].

For very light analyte elements, in particular hydrogen, it is advantageous to detect the elastically recoiled target nuclei instead of the scattered projectiles. The technique using this approach is elastic recoil detection (ERD). Because of the kinematics of the scattering process, the particle detector must be placed in the forward direction in ERD, and it therefore implies that thick specimens must be bombarded at a glancing angle. Similarly to RBS, ERD is particularly useful in materials research [212], but it has also been employed in other applications, such as for depth profiling of hydrogen in obsidians (volcanic glasses) [213] with the aim of dating them.

Besides the elastically scattered particles, the unscattered transmitted beam may also be employed to extract valuable information on thin specimens. Indeed, by placing a surface barrier downstream from the sample and measuring the energy loss of the incident particles, the local mass thickness may be derived. Because of the small beam currents required, this approach is particularly useful in nuclear microprobe work, where it has given rise to the technique of scanning transmission ion microscopy (STIM) [214,215], which may be employed for high-resolution imaging of a specimen. By judicious optimization it is possible to obtain images with a spatial resolution of better than 100 nm [216].

B. Nuclear Reaction Analysis

Nuclear reaction analysis is based on the detection of the prompt γ-rays or prompt particles emitted as a result of nuclear reactions between the incident beam and the target nuclei. As indicated in Section II.A.4, the cross sections for such reactions vary in a rather irregular way with target nuclide and with incident particle energy. When using light ion beams of only a few MeV, nuclear reaction cross sections are only important for light- and medium-weight target nuclei. Of the two forms of NRA, the one in which the prompt γ-rays are detected is by far the most common. It is usually referred to as particle-induced γ-ray emission analysis (PIGE), but other names and acronyms, such as PIGME, are also sometimes employed. The prompt γ-ray measurement has the advantage over the detection of promptly emitted charged particles that it allows greater flexibility in the experimental setup, and for thick samples, the PIGE spectra are much less complicated and far more easily analyzed than prompt particle spectra.

Proton-induced γ-ray emission analysis, which employs (p, γ), $(p, p' \gamma)$, or $(p, \alpha\gamma)$ reactions, lends itself easily to concurrent use with PIXE for virtually all specimen types. Moreover, it is able to provide good detection limits for several elements that are not accessible by PIXE, so that it is a truly complementary technique. Räisänen [217] carried out extensive investigations on the applicability of PIGE for analyzing thick biomedical specimens, and reported that, under favorable conditions, the detection limits are down to the sub–microgram per gram level for Li, B, F, and Na, about 10–25 μg/g for N, Mg, and P, and 100–300 μg/g for C, O, and Cl. Unfortunately, however, the optimum proton energies for all these elements are not the same. Moreover, the detection limits in PIGE depend strongly on the sample composition and may be much worse when some of the elements for which the sensitivity of the method is highest (e.g., Li, B, F, and Na) are present in elevated concentrations.

The most popular application of PIGE on biomedical samples is unquestionably the determination of fluorine. Either the reaction $^{19}F(p, p' \gamma)^{19}F$ or $^{19}F(p, \alpha\gamma)^{16}O$ may be used for this purpose, but the former offers about five times better detection limits [217]. Particularly favorite study objects for PIGE fluorine analyses are teeth. For example, Kim et al. [218] measured several elements, including fluorine, in limpet teeth by a combination of PIGE and PIXE. PIGE measurements of fluorine, and sometimes also other elements, in human teeth were carried out by several researchers (e.g., Refs. 124 and 219–221). In several of these studies a proton microprobe was used, so that it became possible to determine the fluorine profile across the entire thickness of the enamel layer [124] and across precarious and artificially induced lesions [220,221]. Several additional examples of the application of PIGE to biomedical samples, including examples not related to fluorine and/or teeth, can be found in Reference 109.

PIGE is also very useful as a complementary technique to PIXE in the analysis of atmospheric aerosol samples. Several groups have examined the optimum bombarding energy and/or presented procedures or systems for measuring several light elements in such samples by PIGE [222–224]. Asking et al. [223] conclude that two proton energies, 2.64 and 2.96 MeV, are most suited for measuring Na in thin (<0.25 mg/cm^2) aerosol samples, and they report a detection limit of 100 ng/cm^2. To smooth out the variations in the cross-sectional curves, Boni et al. [224] advocate spreading out the beam energy, and they propose an incident proton beam of 3.2–3.5 MeV, with rectangular energy distribution. The detection limits obtained with their setup are 1 ng/cm^2 for Li, 3 ng/cm^2 for F, 25 ng/cm^2 for Na, and of the order of 100 ng/cm^2 for B, Mg, Al, Si, and P.

Simultaneous PIGE/PIXE analyses of thick geological samples have been done by Clark et al. [225] and by Carlsson [104] with Akselsson [46]. Carlsson used PIGE for measuring Li, F, Na, Mg, and Al, whereas the heavier elements were determined by PIXE.

Archeological specimens are often examined by a combination of PIGE and PIXE. As mentioned in Section VI.D, such a combination was used for provenancing obsidian artifacts [187]. The determination of 11 elements by PIXE and 3 by PIGE provided a very good characterization of the samples, as concluded by applying principal components analysis on the data set obtained. Tuurnala et al. [190] applied external "milliprobe" PIGE/PIXE to check the authenticity of oil paintings. A relatively high beam energy of 4 MeV was used to raise the penetration depth, and Na, Mg, and Al were measured by PIGE. Still other examples of applications of PIGE in art and archeology are given by Demortier [226]. For analyses in this field, the depth profiling capacity of PIGE is also quite useful.

Although PIGE is almost exclusively performed with proton beams, deuterons are more commonly used as incident particles when the nuclear reaction analysis is based on the detection of the promptly emitted charged particles. Consequently, this form of NRA is less suitable for concurrent use with PIXE. Furthermore, as indicated earlier, prompt particle spectra are often much more complex than PIGE spectra, thus making the technique less attractive for routine use.

As far as biomedical materials are concerned, NRA with particle detection can be used for measuring nitrogen (an indicator of protein content). For example, Gönczi et al. [227] applied the $^{14}N(d, p)^{15}N$ nuclear reaction, with 6 MeV deuterons, to measure the nitrogen depth profile in 1000 individual wheat grains. The nitrogen distributions showed striking correlations with parameters describing the nitrogen content of the fertilizer, the time of harvesting, the grain position in a head, and the analyzed variety. The (d, p) reactions on ^{12}C and ^{14}N were used in combination with deuteron-induced x-ray emission (DIXE) measurements to determine the N/C and S/N ratios along a single hair [228]. Some general aspects of NRA with particle detection using protons as incident particles were discussed by Räisänen [217].

As an example of the application of (d, p) reactions for analyzing atmospheric aerosol samples, the work by Braga Marcazzan et al. [150] can be cited. In this work, oxygen, nitrogen, and carbon were measured in particulate matter collected on a silver filter. In nuclear microprobe investigations on meteorites and cosmic dust particles, a (d, p) reaction was used to measure the carbon content [229]. The applicability of particle detection NRA in art and archeology was addressed by Amsel et al. [211] and Demortier [226].

Similarly to RBS, NRA (including PIGE) is particularly useful in materials analysis. In such applications, the depth profiling capacity of the two variants of NRA (e.g., by making use of strong resonance peaks in the cross-sectional curves) is often quite valuable. As indicated in Section VI.E, however, PIXE, if applied at all, generally serves as the complementary technique, whereas the essential information is obtained by RBS or NRA. For these reasons, the applications of NRA (including PIGE) to materials research are not discussed here. Instead, the reader is referred to the textbooks and proceedings mentioned at the beginning of this section.

C. Ion Beam Thermography

In biomedical and environmental research, considerably more insight may often be gained by determining the chemical species or chemical association of minor or trace elements than merely their elemental concentrations. For example, to understand the source and transformation processes responsible for particulate nitrogen in the atmosphere, it is of interest to find out what fraction of total nitrogen is present as ammonium, nitrate, or some other N-containing species and whether the ammonium is associated with the nitrate or with sulfate, for example. To obtain such information with IBA techniques and IBA-compatible aerosol samplers, Martinsson and Hansson [230] developed the technique of ion beam thermography (IBT). This technique consists of an ingenious use of controlled heating of the aerosol sample, whereby its temperature is gradually raised, with concurrent PIXE and PESA analysis. In principle, simultaneous PIGE measurements are also possible. To reduce or eliminate vacuum- or beam-induced losses of compounds or analyte elements from the sample, Martinsson and Hansson [230] performed the measurements at prevacuum pressure ($p < 10^{-2}$ mbar) or in a 10 mbar He atmosphere. Protons of 3.58 MeV were selected as incident particles. H, C, N, and O were measured by PESA and the elements with $Z > 14$ by PIXE. Heating of the sample occurred by passing an electrical current through the sample substrate, which consisted of a 0.8 μm thick aluminum foil. The temperature of the sample was monitored by determining the change in relative resistivity (defined as the ratio of the resistance of the heated sample to that

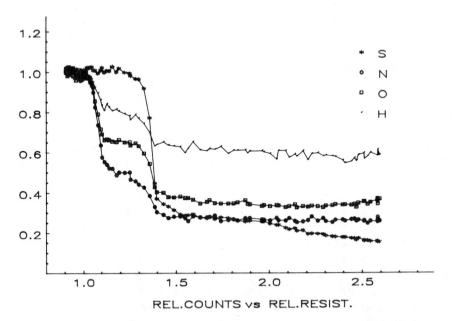

REL.COUNTS vs REL.RESIST.

Figure 18 Thermograms from a mixture of NH_4NO_3 and $(NH_4)_2SO_4$. Incident proton current 70 nA; beam charge $\triangle C$ for each data point 1 μC; chamber pressure prevacuum. Amounts inferred: $m(N) = 4.1$ μg/cm^2, $m(O) = 4.3 + 3.6$ μg/cm^2, $m(S) = 1.8$ μg/cm^2, $m(NH_4NO_3) = 7.2$ μg/cm^2, $m((NH_4)_2SO_4) = 7.4$ μg/cm^2. (Reproduced from Ref. 230 by permission of Elsevier Science Publishers.)

of the unheated sample). To achieve time resolution in the data acquisition so that the course of sample deterioration during thermographic treatment could be followed, a special data acquisition program was developed. The IBT results are presented as a set of thermograms, one for each element, which show the amount of element remaining in the sample as the temperature increases. From these thermograms, the chemical compounds and their concentrations are inferred. Figure 18 shows the set of thermograms obtained for a mixture of equal amounts of NH_4NO_3 and $(NH_4)_2SO_4$. The first 10 data points, at relative resistivities between 0.9 and 1, were taken without heating the sample and are listed in chronological order. The two compounds are clearly separated in the O thermogram, where the high-temperature edge corresponds to the S edge. In the N thermogram a slight interference between NH_4NO_3 and the initial dissociation of $(NH_4)_2SO_4$ is observed. The elemental amounts inferred from the thermograms are given in the legend to Figure 18 and are in good agreement with those expected on the basis of stoichiometry.

The especially interesting features of IBT are that chemical compounds (not ions) are determined, that it is a multicompound technique, that it requires no pretreatment of the sample, and that some compounds may be measured down to trace levels under favorable conditions.

VIII. CONCLUSIONS

PIXE is undoubtedly an invaluable addition to the broad arsenal of x-ray emission techniques. The major advantages of PIXE are its multielement character (as in XRF and EPMA, all elements from Na to U can in principle be measured), the high sensitivity (relative detection limits down to 0.1 $\mu g/g$; absolute detection limits down to 10^{-12} g in macro-PIXE and down to 10^{-15} g in micro-PIXE), the smooth variation in the relative detection limit with atomic number of the analyte element, the ability to analyze tiny samples (1 mg or less), the speed of the analysis (1–10 minute bombardment time per specimen), the possibility for automation, and the fact that it is often nondestructive. Compared to conventional EDXRF, PIXE offers relative detection limits (in $\mu g/g$) often one order of magnitude better, it allows one to analyze smaller sample masses, and it is faster. Another favorable feature of PIXE is that it can be complemented with other IBA techniques, such as elastic scattering spectrometry, NRA, and PIGE, so that a simultaneous measurement of the light elements (Li, B, C, N, O, and F) is feasible. Finally, and certainly not least, the microbeam variant of PIXE offers the possibility of spatially resolved analyses with high resolution (down to 1 μm or better) and high sensitivity. As already indicated, the relative detection limits are typically two orders of magnitude better in micro-PIXE than in EPMA.

The most serious drawbacks of PIXE are that it requires a MeV particle accelerator and that commercial PIXE apparatus are not readily available (the great majority of PIXE laboratories have built their own setup, and commercial systems are usually custom-built). Other limitations of the technique, which are also shared by XRF, are that it suffers from spectral interferences, that matrix effects must be accounted for, and that it does not allow the direct measurement of ultratrace elements that are present at ng/g levels. Furthermore, the sub–microgram per gram detection limits in PIXE are only obtained when the sample matrix consists of light elements, as in biomedical samples. For heavier element matrices, the detection limits are worsened because of the increased continuum background and the presence of intense x-ray lines from the matrix elements in the PIXE spectrum.

Considering that many alternative techniques for trace element analysis are currently available, it is of interest to compare the characteristics and capabilities of PIXE with those of the other techniques and to assess for what sample types and/or problems PIXE is most appropriate. Maenhaut [105] recently made such an evaluation for macro-PIXE and compared it to several nuclear and atomic spectrometric techniques selected because they are most competitive with PIXE (from the viewpoint of both sample type and elemental coverage) and/or because they are widely applied or experience impressive growth. The techniques were nuclear activation analysis, mainly instrumental neutron activation analysis (INAA), x-ray fluorescence, including total-reflection XRF (TXRF) and synchrotron radiation XRF (SRXRF), atomic emission spectrometry (AES), atomic absorption spectrometry, atomic fluorescence spectrometry (AFS), and atomic mass spectrometry, in particular inductively coupled plasma mass spectrometry (ICPMS). Table 1 presents an intercomparison taken from Reference 105. With regard to the price of the instrument, it should be stressed that the small accelerator required in a PIXE setup is seldom used for PIXE alone but generally serves a wide variety of other IBA techniques. It is therefore more relevant to express the cost of PIXE on the basis of price per sample analyzed, and an estimate of $30/sample for a multielement analysis has been published [231]. As indicated in Table 1, PIXE, conventional XRF, and INAA lend themselves better to the analysis of solids than of liquids; TXRF, ICPMS, and the optical spectrometric techniques ICPAES, AAS, and AFS are better suited for the analysis of liquids. The analysis of solid samples by the latter techniques generally requires dissolution, decomposition, or extraction procedures, which always involve the danger of extraneous additions, losses, or incomplete dissolution of the elements of interest. A strong point of PIXE is that it allows the nondestructive, multielement analysis of solids with little or no sample preparation. This characteristic is also shared by conventional XRF, however, so that the latter is generally the technique of choice. When the sample matrix consists of elements with $Z > 11$ (as in geological samples) and when enough sample is available, there is indeed little reason for resorting to PIXE, because it offers no better detection limits than XRF in such cases. That PIXE can be combined with other IBA techniques, so that a simultaneous measurement of the light elements is possible, would then essentially be the only motivation for its use. However, when only mg amounts or less of solid sample are available or when one is interested in measuring the composition of millimeter-sized areas on a large sample or of thin superficial layers on a bulk sample, PIXE is virtually the only nondestructive multielement technique available. The analysis of small aerosol deposits, as collected by compact samplers that provide good time or size resolution or both, is an application area in which PIXE has virtually no competition. Since PIXE and related IBA techniques allow the analysis of delicate samples without visible damage and without introducing radioactivity, they can also very advantageously be used in studies in art and archeology or for analyzing other unique samples of which subsampling is im-

Table 1 Summary of Some Characteristics of PIXE and Other Analytical Techniques for Bulk Trace Element Analysis

Technique	Price of instrument[a]	Detection limit[b] (μg/g or ng/ml)	Spectral interference	Matrix effects	Multielement	Preferred sample type
PIXE	(++++)	0.2–3	High	Medium	Yes	Solid
XRF[c]	+/++	1–10	High/low	Medium	Yes	Solid
TXRF	++	0.2	High	Medium	Yes	Liquid[d]
INAA	(+++++)	0.001–1	Low	Low	Yes	Solid
ICPAES	++	1–30	High	Medium	Yes	Liquid
ETAAAS[e]	+	0.01–0.2	Medium	High	No	Liquid
LIFETA[f]	+++	0.001	Low	High	No	Liquid
ICPMS	+++	0.03–0.1	High	High	Yes	Liquid

[a] Code for price of instrument: (+) less than $100,000, (++) from $100,000 to $250,000, (+++) from $250,000 to $500,000, and (++++) more than $500,000; as indicated in the text and in References 105 and 131, the instrument cost is not a very relevant measure in PIXE and INAA.

[b] Detection limits are in μg/g for INAA, XRF, and PIXE and in ng/ml for the other techniques, including TXRF.

[c] The indications on this line before each slash apply to EDXRF; those after each slash to WDXRF.

[d] The dried residue of the liquid on a totally reflecting substrate is subjected to analysis.

[e] ETAAAS denotes electrothermal atomization AAS.

[f] LIFETA denotes laser-induced atomic fluorescence spectrometry with electrothermal atomization.

possible. As far as the analysis of biomedical samples is concerned, PIXE experiences an increasing competition from other techniques. However, the fact that PIXE is able to provide sub–microgram per gram detection limits for milligram-sized biological samples remains an important asset, and furthermore, it allows the essential elements Mn and Se to be measured in many tissues whereas they are generally near or below the detection limit in conventional XRF and TXRF.

Comparisons of micro-PIXE with competing techniques have also been made. In such evaluation, Nobiling [232] included various nondestructive and destructive techniques, all based on photon or particle excitation, including EPMA, laser microprobe mass analysis (LAMMA), and ion micromass analysis (IMMA). Particular attention was paid to resolution, sensitivity, and analytical depth. It was concluded that the nuclear microprobe (i.e., combination of micro-PIXE with RBS and NRA) offers the widest applicability for nondestructive quantitative analysis.

REFERENCES

1. T. B. Johansson, K. R. Akselsson, and S. A. E. Johansson, *Nucl. Instr. Meth.* *84*:141 (1970).
2. S. A. E. Johansson (Ed.), Proc. Int. Conf. on Particle Induced X-ray Emission and Its Analytical Applications, *Nucl. Instr. Meth. 142*:1 (1977).
3. S. A. E. Johansson (Ed.), Proc. 2nd Int. Conf. on Particle Induced X-ray Emission and Its Analytical Applications, *Nucl. Instr. Meth. 181*:1 (1981).
4. B. Martin (Ed.), Proc. 3rd Int. Conf. on Particle Induced X-ray Emission and Its Analytical Applications, *Nucl. Instr. Meth. B3*:1 (1984).
5. H. A. Van Rinsvelt, S. Bauman, J. W. Nelson, and J. W. Winchester (eds.), Proc. 4rd Int. Conf. on Particle Induced X-ray Emission and Its Analytical Applications, *Nucl. Instr. Meth. B22*:1 (1987).
6. R. D. Vis (Ed.), Proc. 5th Int. Conf. on Particle Induced X-ray Emission and Its Analytical Applications, *Nucl. Instr. Meth. B49*:1 (1990).
7. S. A. E. Johansson and J. L. Campbell, *PIXE: A Novel Technique for Elemental Analysis*, Wiley, Chichester, 1988.
8. H. H. Andersen and J. F. Ziegler, *The Stopping and Ranges of Ions in Matter*, Vol. 3, *Hydrogen: Stopping Powers and Ranges in All Elements*, Pergamon Press, New York, 1977.
9. J. F. Ziegler, J. P. Biersack, and U. Littmark, *The Stopping and Ranges of Ions in Solids*, Vol. 1, Pergamon Press, New York, 1985.
10. W. Brandt and G. Lapicki, *Phys. Rev. A20*:465 (1979).
11. W. Brandt and G. Lapicki, *Phys. Rev. A23*:1717 (1981).
12. D. D. Cohen and M. Harrigan, *At. Data Nucl. Data Tables 33*:255 (1985).
13. M. H. Chen and B. Crasemann, *At. Data Nucl. Data Tables 33*:217 (1985).
14. M. H. Chen and B. Crasemann, *At. Data Nucl. Data Tables 41*:257 (1989).
15. H. Paul, *Z. Phys. D4*:249 (1987).
16. H. Paul and J. Muhr, *Phys. Rep. 135*:47 (1986).
17. H. Paul and J. Sacher, *At. Data Nucl. Data Tables 42*:105 (1989).
18. H. Paul, *Nucl. Instr. Meth. B42*:443 (1989).
19. D. D. Cohen, *Nucl. Instr. Meth. B49*:1 (1990).
20. J. L. Campbell, *Nucl. Instr. Meth. B31*:518 (1988).
21. S. I. Salem, S. L. Panossian, and R. A. Krause, *At. Data Nucl. Data Tables 14*:91 (1974).

22. J. H. Scofield, *Phys. Rev. A9*:1041 (1974).
23. M. O. Krause, *J. Phys. Chem. Ref. Data 8*:307 (1979).
24. W. Maenhaut and H. Raemdonck, *Nucl. Instr. Meth. B1*:123, (1984).
25. D. D. Cohen and E. Clayton, *Nucl. Instr. Meth. B22*:59 (1987).
26. D. D. Cohen and M. Harrigan, *At. Data Nucl. Data Tables 34*:393 (1986).
27. J. H. Scofield, *Phys. Rev. A10*:1507 (1974); see also Erratum in *Phys. Rev. A12*:345 (1975).
28. M. H. Chen, B. Crasemann, and H. Mark, *Phys. Rev. A24*:177 (1981).
29. S. A. E. Johansson and T. B. Johansson, *Nucl. Instr. Meth. 137*:473 (1976).
30. H. Paul, *Nucl. Instr. Meth. B3*:5 (1984); see also Erratum in *Nucl. Instr. Meth. B5*:554 (1984).
31. Y. Miyagawa, S. Nakamura, and S. Miyagawa, *Nucl. Instr. Meth. B30*:115 (1988).
32. W. -K. Chu, J. W. Mayer, and M. -A. Nicolet, *Backscattering Spectrometry*, Academic Press, New York, 1978.
33. D. W. Mingay, *X-ray Spectrom 12*:52 (1983).
34. G. Deconninck, *Introduction to Radioanalytical Physics*, Elsevier Scientific, Amsterdam, 1978.
35. F. Folkmann, C. Gaarde, T. Huus, and K. Kemp, *Nucl. Instr. Meth. 116*:487 (1974).
36. F. Folkmann, J. Borggreen, and A. Kjeldgaard, *Nucl. Instr. Meth. 119*:117 (1974).
37. F. Folkmann, in: *Ion Beam Surface Layer Analysis*, Vol. 2 (O. Meyer, G. Linker, and F. Käppeler, Eds.), Plenum, New York, 1976, p. 239.
38. K. Ishii and S. Morita, *Phys. Rev. A30*:2278 (1984).
39. K. Ishii and S. Morita, *Nucl. Instr. Meth. B22*:68 (1987).
40. K. Ishii and S. Morita, *Nucl. Instr. Meth. B34*:209 (1988).
41. W. Scharf, *Particle Accelerators: Applications in Technology and Research*, Research Studies Press Ltd., Taunton, England, 1989.
42. J. J. G. Durocher, N. M. Halden, F. C. Hawthorne, and J. S. C. McKee, *Nucl. Instr. Meth. B30*:470 (1988).
43. J. S. C. McKee, G. R. Smith, Y. H. Yeo, K. Abdul-Retha, D. Gallop, J. J. G. Durocher, W. Mulholland, and C. A. Smith, *Nucl. Instr. Meth. B40/B41*:680 (1989).
44. M. Peisach and C. A. Pineda. *Nucl. Instr. Meth. B49*:10 (1990).
45. T. A. Cahill, Y. Matsuda, D. Shadoan, R. A. Eldred, and B. H. Kusko, *Nucl. Instr. Meth. B3*:263 (1984).
46. L. -E. Carlsson and K. R. Akselsson, *Nucl. Instr. Meth. 181*:531 (1981).
47. W. Maenhaut, *Scanning Microsc. 4*:43 (1990).
48. M. S. Ahlberg, K. R. Akselsson, D. Brune, and J. Lorenzen, *Nucl. Instr. Meth. 123*:385 (1975).
49. C. P. Swann, *IEEE Trans. Nucl. Sci. NS-30*:1298 (1983).
50. Quantum X-ray detector, Kevex Corp., 1987.
51. R. G. Musket, *Nucl. Instr. Meth. B15*:735 (1986).
52. K. G. Malmqvist, E. Karlsson, and K. R. Akselsson, *Nucl. Instr. Meth. 192*:523 (1982).
53. I. V. Mitchell, K. M. Barfoot, and H. L. Eschbach, *Nucl. Instr. Meth. 168*:233 (1980).
54. V. N. Volkov, V. B. Vykhodets, I. K. Golubkov, S. M. Klotsman, P. V. Lerkh, and V. A. Pavlov, *Nucl. Instr. Meth. 205*:73 (1983).
55. M. Ahlberg, G. Johansson, and K. Malmqvist, *Nucl. Instr. Meth. 131*:377 (1975).
56. L. J. Cabri, J. L. Campbell, J. H. G. Laflamme, R. G. Leigh, J. A. Maxwell, and J. D. Scott, *Can. Mineralogist 23*:133 (1985).
57. M. A. Chaudri and A. Crawford, *Nucl. Instr. Meth. 181*:31 (1981).
58. D. W. Mingay and E. Barnard, *Nucl. Instr. Meth. 157*:537 (1978).
59. E. T. Williams, *Nucl. Instr. Meth. B3*:211 (1984).

60. M. Hyvonen-Dabek, J. Räisänen, and J. T. Dabek, *J. Radioanal. Chem 63*:163 (1982).
61. B. G. Martinsson, *Nucl. Instr. Meth. B22*:356 (1987).
62. C. W. Wookey and J. L. Rouse, *Nucl. Instr. Meth. B18*:303 (1987).
63. P. Horowitz and L. Grodzins, *Science 189*:795 (1975).
64. J. A. Cookson, *Nucl. Instr. Meth. 165*:477 (1979).
65. H. Kneis, B. Martin, R. Nobiling, B. Povh, and K. Traxel, *Nucl. Instr. Meth. 197*:79 (1982).
66. R. Nobiling, Y. Civelekoglu, B. Povh, D. Schwalm, and K. Traxel *Nucl. Instr. Meth. 130*:323 (1975).
67. G. J. F. Legge, D. N. Jamieson, P. M. J. O'Brien, and A. P. Mazzolini, *Nucl. Instr. Meth. 197*:85 (1982).
68. B. Fischer, *Nucl. Instr. Meth. B30*:284 (1988).
69. J. A. Cookson, A. T. G. Ferguson, and F. D. Pilling, *J. Radioanal. Chem. 12*:39 (1972).
70. G. W. Grime and F. Watt, *Beam Optics of Quadrupole Probe-Forming Systems*, Adam Hilger, Bristol, 1984.
71. U. A. S. Tapper and B. R. Nielsen, *Nucl. Instr. Meth. B44*:219 (1989).
72. D. N. Jamieson and G. J. F. Legge, *Nucl. Instr. Meth. B34*:411 (1988).
73. C. J. Maggiore, *Scanning Electron Microsc. 1*:439 (1980).
74. R. Booth and H. W. Lefevre, *Nucl. Instr. Meth. 151*:143 (1978).
75. U. A. S. Tapper, N. E. G. Lövestam, E. Karlsson, and K. G. Malmqvist, *Nucl. Instr. Meth. B28*:317 (1988).
76. J. C. Overley, R. C. Connolly, G. E. Sieger, J. D. Macdonald, and H. W. Lefevre, *Nucl. Instr. Meth. 218*:43 (1983).
77. D. Heck, personal communication, 1989.
78. U. Wätjen, H. Prins, R. Van Bijlen, and E. Louwerix, *Nucl. Instr. Meth. B49*:78 (1990).
79. N. E. G. Lövestam, *Nucl. Instr. Meth. B36*:455 (1989).
80. J. L. Campbell, W. Maenhaut, E. Bombelka, E. Clayton, K. Malmqvist, J. A. Maxwell, J. Pallon, and J. Vandenhaute, *Nucl. Instr. Meth. B14*:204 (1986).
81. J. L. Campbell, *Scanning Microsc. 3*:449 (1989).
82. G. I. Johansson, *X-ray Spectrom. 11*:194 (1982).
83. J. L. Campbell and P. L. McGhee, *Nucl. Instr. Meth. 248*:393 (1986).
84. C. A. Baker, C. J. Batty, and S. Sakamoto, *Nucl. Instr. Meth. A259*:501 (1987).
85. J. M. Jaklevic, J. T. Walton, R. E. McMurray, Jr., N. W. Madden, and F. S. Goulding, *Nucl. Instr. Meth. A266*:598 (1988).
86. I. Orlic, W. J. M. Lenglet, and R. D. Vis, *Nucl. Instr. Meth. A276*:202 (1989).
87. M. Pajek, A. P. Kobzev, R. Sandrik, R.A. Ilkhamov, and S. H. Khusmudurow, *Nucl. Instr. Meth. B42*:346 (1989).
88. G. I. Johansson, J. Pallon, K. G. Malmqvist, and K. R. Akselsson, *Nucl. Instr. Meth. 181*:81 (1981).
89. I. Borbely-Kiss, E. Koltay, S. Laszlo, G. Szabo, and L. Zolnai, *Nucl. Instr. Meth. B12*:496 (1985).
90. J. L. Campbell, J. -X. Wang, J. A. Maxwell, and W. J. Teesdale, *Nucl. Instr. Meth. B43*:539 (1989).
91. J. A. Cookson and J. L. Campbell, *Nucl. Instr. Meth. 216*:489 (1983).
92. D. G. Jex, M. W. Hill, and N. F. Mangelson, *Nucl. Instr. Meth. B49*:141 (1990).
93. W. J. Teesdale, J. A. Maxwell, A. Perujo, J. L. Campbell, L. Van Der Zwan, and T. E. Jackman, *Nucl. Instr. Meth. B35*:57 (1988).
94. E. -M. Johansson and S. A. E. Johansson, *Nucl. Instr. Meth. B3*:154 (1984).
95. J. Pallon and K. G. Malmqvist, *Nucl. Instr. Meth. 181*:71 (1981).

96. W. Maenhaut, L. De Reu, and J. Vandenhaute, *Nucl. Instr. Meth. B3*:135 (1984).
97. J. Räisänen, *X-ray Spectrom. 15*:159 (1986).
98. M. Cholewa and G. J. F. Legge, *Nucl. Instr. Meth. B40/41*:651 (1989).
99. K. Themner, P. Spanne, and K. W. Jones, *Nucl. Instr. Meth. B49*:52 (1990).
100. W. Maenhaut, J. Vandenhaute, and H. Duflou, *Fresenius Z. Anal. Chem 326*:736 (1987).
101. W. Maenhaut, *Anal. Chim. Acta 195*:125 (1987).
102. E. Bombelka, F. -W. Richter, H. Ries, and U. Wätjen, *Nucl. Instr. Meth. B3*:296 (1984).
103. U. Wätjen, E. Bombelka, F. -W. Richter, and H. Ries, *J. Aerosol Sci. 14*:305 (1983).
104. L. -E. Carlsson, *Nucl. Instr. Meth. B3*:206 (1984).
105. W. Maenhaut, *Nucl. Instr. Meth. B49*:518 (1990).
106. S. B. Russell, C. W. Schulte, S. Faiq, and J. L. Campbell, *Anal. Chem. 53*:571 (1981).
107. N. F. Mangelson and M. W. Hill, *Scanning Microsc. 4*:63 (1990).
108. I. V. Iyengar, W. E. Kollmer, and H. J. M. Bowen, *The Elemental Composition of Human Tissues and Body Fluids*, Verlag Chemie, Weinheim, 1978.
109. W. Maenhaut, *Nucl. Instr. Meth. B35*:388 (1988).
110. K. G. Malmqvist, *Nucl. Instr. Meth. B49*:183 (1990).
111. N. F. Mangelson, D. J. Eatough, N. L. Eatough, L. D. Hansen, M. W. Hill, M. L. Lee, L. R. Phillips, M. E. Post, B. E. Richter, and J. F. Ryder, *IEEE Trans. Nucl. Sci. NS-28*:1378 (1981).
112. H. Duflou, W. Maenhaut, and J. De Reuck, *Biol. Trace Elem. Res. 13*:1 (1987).
113. J. L. Campbell, W. J. Teesdale, and R. G. Leigh, *Nucl. Instr. Meth. B6*:551 (1985).
114. A. J. J. Bos, C. C. A. H. van der Stap, V. Valkovic, R. D. Vis, and H. Verheul, *Sci. Total Environ. 42*:157 (1987).
115. K. G. Malmqvist, B. Forslind, K. Themner, G. Hyltén, T. Grundin, and G. M. Roomans, *Biol. Trace Elem. Res. 12*:297 (1987).
116. H. Duflou, W. Maenhaut, and J. De Reuck, *Neurochem. Res. 14*:1099 (1989).
117. H. Duflou, W. Maenhaut, and J. De Reuck, in: *Trace Element Analytical Chemistry in Medicine and Biology*, Vol. 5 (P. Brätter and P. Schramel, Eds.), Walter de Gruyter, Berlin, 1988, p. 483.
118. L. Salford, U. A. S. Tapper, A. Brun, and K. G. Malmqvist, in: *Development of a Nuclear Microprobe and Its Application to Neurobiology*, U. A. S. Tapper, Thesis, LUTFD2/(TFKF-1012), Department of Nuclear Physics, Lund University, Lund, Sweden, 1989, p. 89.
119. F. Watt and G. Grime, unpublished results, 1990.
120. M. Ahlberg and R. Akselsson, *Int. J. Appl. Radiat. Isot. 27*:279 (1976).
121. B. Möller, L. -E. Carlsson, G. Johansson, K. G. Malmqvist, L. Hammarström, and M. Berlin, *Scand. J. Work Environ. Health 8*:267 (1982).
122. J. Krmpotic-Nemanic, V. Valkovic, and C. Nemanic, *Acta Otolaryngol. (Stockh.) 99*:466 (1985).
123. U. Lindh, *Anal. Chim Acta 150*:233 (1983).
124. I. D. Svalbe, M. A. Chaudhri, K. Traxel, C. Ender, and A. Mandel, *Nucl. Instr. Meth. B3*:648 (1984).
125. W. Maenhaut, L. De Reu, U. Tomza, and J. De Roose, in: *Nuclear Methods in Environmental and Energy Research*, J. R. Vogt (ed.), Fourth International Conference, Columbia, Missouri, CONF-800433, 1980, p. 378; Report available from the National Technical Information Service, U.S. Department of Commerce, Springfield, Virginia 22161.
126. P. Pakarinen, A. -K. Ekholm, and J. Pallon, *Nucl. Instr. Meth. B49*:241 (1990).

127. J. Räisänen, M. Hyvönen-Dabek, R. Lapatto, and J. T. Dabek, *Appl. Radiat. Isot.* *38*:373 (1987).

128. J. Pallon, P. Pakarinen, K. Malmqvist, and K. R. Akselsson, *Biol. Trace Elem. Res.* *12*:401 (1987).

129. G. J. F. Legge, *Nucl. Instr. Meth. B3*:561 (1984).

130. E. Johansson and U. Lindh, *Biol. Trace Elem. Res. 12*:309 (1987).

131. W. Maenhaut, in: *Control and Fate of Atmospheric Trace Metals* (J. M. Pacyna, and B. Ottar, Eds.), Kluwer Academic, Dordrecht, The Netherlands, 1989, p. 259.

132. T. A. Cahill, *Nucl. Instr. Meth. B49*:345 (1990).

133. P. J. Lioy and M. J. Y. Lioy (Eds.), *Air Sampling Instruments for Evaluation of Atmospheric Contaminants*, 6th ed., Am. Conf. Gov. Ind. Hyg., Cincinnatti, 1983.

134. C. H. Murphy, *Handbook of Particle Sampling and Analysis Methods*, Verlag Chemie, Deerfield Beach, FL, 1984.

135. K. R. Spurny (Ed.), *Physical and Chemical Characterization of Individual Airborne Particles*, Ellis Horwood, Chichester, 1986.

136. N. Z. Heidam, *Atmos. Environ. 15*:891 (1981).

137. T. G. Dzubay and R. K. Stevens, *Environ. Sci. Technol. 9*:663 (1975).

138. B. W. Loo, J. M. Jaklevic, and F. S. Goulding, in: *Fine Particles* (B. Y. H. Liu, Ed.), Academic, New York, 1976, p. 311.

139. K. Kemp, in: *X-ray Fluorescence Analysis of Environmental Samples* (T. G. Dzubay, Ed.), Ann Arbor Science, Ann Arbor, MI, 1977, p. 203.

140. K. R. Akselsson, *Nucl. Instr. Meth. B3*:425 (1984).

141. H. J. Annegarn, T. A. Cahill, J. P. F. Sellschop, and A. Zucchiatti, *Phys. Scripta 37*:282 (1988).

142. J. W. Nelson, G. G. Desaedeleer, K. R. Akselsson, and J. W. Winchester, *Adv. X-ray Anal. 19*:403 (1976).

143. J. W. Nelson, in: *X-ray Fluorescence Analysis of Environmental Samples* (T. G. Dzubay, Ed.), Ann Arbor Science, Ann Arbor, MI, 1977, p. 19.

144. S. Baumann, P. D. Houmere, and J. W. Nelson, *Nucl. Instr. Meth. 181*:499 (1981).

145. H. -C. Hansson and S. Nyman, *Environ. Sci Technol. 19*:1110 (1985).

146. O. G. Raabe, D. A. Braaten, R. L. Axelbaum, S. V. Teague, and T. A. Cahill, *J. Aerosol Sci. 19*:183 (1988).

147. K. G. Malmqvist, *Nucl. Instr. Meth. B3*:529 (1984).

148. H. J. Annegarn, A. Zucchiatti, J. P. F. Sellschop, and P. Booth-Jones, *Nucl. Instr. Meth. B22*:325 (1987).

149. P. Bacci, E. Caruso, G. M. Braga Marcazzan, P. Redaelli, C. Sabbioni, and A. Ventura, *Nucl. Instr. Meth. B3*:522 (1984).

150. G. M. Braga Marcazzan, E. Caruso, E. Cereda, P. Redaelli, P. Bacci, A. Ventura, and G. Lombardo, *Nucl. Instr. Meth. B22*:305 (1987).

151. M. H. Tabacniks, C. Orsini, and P. Artaxo, *Nucl. Instr. Meth. B22*:315 (1987).

152. J. W. Winchester and M. -T. Bi, *Atmos. Environ. 18*:1399 (1984).

153. J. W. Winchester and A. C. D. Leslie, in: *Heterogeneous Atmospheric Chemistry* (D. R. Schryer, Ed.), Geophys. Monograph Ser., Vol. 26, Am. Geophys. Union, Washington, D. C., 1982, p. 250.

154. R. G. Flocchini, T. A. Cahill, D. J. Shadoan, S. J. Lange, R. A. Eldred, P. J. Feeney, G. W. Wolfe, D. C. Simmeroth, and J. K. Suder, *Environ. Sci. Technol. 10*:76 (1976).

155. R. G. Flocchini, T. A. Cahill, M. L. Pitchford, R. A. Eldred, P. J. Feeney, and L. L. Ashbaugh, *Atmos. Environ. 15*:2017 (1981).

156. R. A. Eldred, T. A. Cahill, and P. J. Feeney, *Nucl. Instr. Meth. B22*:289 (1987).

157. H. Lannefors, H. -C. Hansson, and L. Granat, *Atmos. Environ. 17*:87 (1983).

158. B. G. Martinsson, H. -C. Hansson, and H. O. Lannefors, *Atmos. Environ. 18*:2167 (1984).

159. E. Swietlicki, H. -C. Hansson, and B. G. Martinsson, *Nucl. Instr. Meth. B22*:264 (1987).
160. E. Swietlicki, H. -C. Hansson, B. Svantesson, and L. Asking, in: *European Source Region Identification of Long Range Transported Ambient Aerosol Based on PIXE Analysis and Related Techniques*, E. Swietlicki, Thesis, LUTFD2/(TFKF-1015), Department of Nuclear Physics, Lund University, Lund, Sweden, 1989, p. 79.
161. P. Artaxo and C. Orsini, *Nucl. Instr. Meth. B22*:259 (1987).
162. P. Artaxo, H. Storms, F. Bruynseels, R. Van Grieken, and W. Maenhaut, *J. Geophys. Res. 93*:1605 (1988).
163. W. Maenhaut, A. Selen, P. Van Espen, R. Van Grieken, and J. W. Winchester, *Nucl. Instr. Meth. 181*:399 (1981).
164. W. Maenhaut, H. Raemdonck, A. Selen, R. Van Grieken, and J. W. Winchester, *J. Geophys. Res. 88*:5353 (1983).
165. H. Raemdonck, W. Maenhaut, and M. O. Andreae, *J. Geophys. Res. 91*:8623 (1986).
166. N. Z. Heidam, *Atmos. Environ. 15*:1421 (1981).
167. J. Heintzenberg, H. -C. Hansson, and H. Lannefors, *Tellus 33B*:40 (1982).
168. W. Maenhaut, P. Cornille, J. M. Pacyna, and V. Vitols, *Atmos. Environ. 23*:2551 (1989).
169. S. -M. Li and J. W. Winchester, *J. Geophys. Res. 95*:1797 (1990).
170. B. A. Bodhaine, J. J. Deluisi, J. M. Harris, P. Houmere, and S. Bauman, *Tellus 38B*:223 (1986).
171. P. Artaxo, F. Andrade, and W. Maenhaut, *Nucl. Instr. Meth. B49*:383 (1990).
172. H. -C. Hansson, E. -M. Johansson, and A. -K. Ekholm, *Nucl. Instr. Meth. B3*:158 (1984).
173. H. -C. Hansson, A. -K. Ekholm, and H. B. Ross, *Environ. Sci. Technol. 22*:527 (1988).
174. M. A. Respaldiza, M. Garcia-Leon, and G. Madurga, *J. Trace Microprobe Tech. 6*:87 (1988).
175. H. -C. Hansson, personal communication, 1990.
176. K. G. Malmqvist, H. Båge, L. -E. Carlsson, K. Kristiansson, and L. Malmqvist, *Nucl. Instr. Meth. B22*:386 (1987).
177. S. H. Sie, C. G. Ryan, D. R. Cousens, and W. L. Griffin, *Nucl. Instr. Meth. B40/41*:690 (1989).
178. C. G. Ryan, D. R. Cousens, S. H. Sie, and W. L. Griffin, *Nucl. Instr. Meth. B49*:271 (1990).
179. D. G. Fraser, F. Watt, G. W. Grime, and J. Takacs, *Nature 312*:352 (1984).
180. A. U. Horn and K. Traxel, *Fortschr. Mineralogie 65*:80 (1987).
181. T. M. Benjamin, C. J. Duffy, C. J. Maggiore, P. S. Z. Rogers, D. S. Woolum, D. S. Burnett, and M. T. Murrell, *Nucl. Instr. Meth. B3*:677 (1984).
182. M. A. Lucas, T. C. Hughes, and C. D. McKenzie, *Nucl. Instr. Meth. 191*:34 (1981).
183. N. E. G. Lövestam and E. Swietlicki, *Nucl. Instr. Meth. B43*:104 (1989).
184. C. Lahanier, G. Amsel, C. Heitz, M. Menu, and H. H. Andersen (Eds.), Proc. Int. Workshop on Ion Beam Analysis in the Arts and Archaeology, *Nucl. Instr. Meth. B14*:1 (1986).
185. M. Menu, T. Calligaro, J. Salomon, G. Amsel, and J. Moulin, *Nucl. Instr. Meth. B45*:610 (1990).
186. M. Peisach, *Nucl. Instr. Meth. B14*:99 (1986).
187. P. Duerden, E. Clayton, J. R. Bird, and D. D. Cohen, *Nucl. Instr. Meth. B14*:50 (1986).
188. B. H. Kusko, T. A. Cahill, R. A. Eldred, and R. N. Schwab, *Nucl. Instr. Meth. B3*:689 (1984).

189. T. A. Cahill, R. N. Schwab, B. H. Kusko, R. A. Eldred, G. Möller, D. Deutschke, D. L. Wick, and A. S. Pooley, *Anal. Chem.* 59:829 (1987).
190. T. Tuurnala, A. Hautojärvi, and K. Harva, *Nucl. Instr. Meth.* B14:70 (1986).
191. C. Engelmann, *Nucl. Instr. Meth.* B49:33 (1990).
192. J. S. Williams, J. C. McCallum, and R. A. Brown, *Nucl. Instr. Meth.* B30:480 (1988).
193. R. S. Bhattacharya and P. P. Pronko, *Appl. Surface Sci.* 18: 1 (1984).
194. I. Roelandts, G. Robaye, G. Weber, and J. M. Delbrouck, *Fresenius Z. Anal. Chem.* 320:541 (1985).
195. G. Demortier, F. Bodart, G. Deconninck, G. Terwagne, Z. Gabelica, and E. G. Derouane, *Nucl. Instr. Meth.* B30:491 (1988).
196. J. W. Macmillan, P. M. Pollard, and F. C. W. Pummery, *Nucl. Instr. Meth.* B15:394 (1986).
197. B. L. Doyle, R. T. McGrath, and A. E. Portau, *Nucl. Instr. Meth.* B22:34 (1987).
198. J. L. Duggan and I. L. Morgan (Eds.), Proc. 10th Int. Conf. on the Application of Acclerators in Research and Industry, Denton, 1988, *Nucl. Instr. Meth.* B40/41:1 (1989).
199. J. F. Ziegler (Ed.), *New Uses of Ion Accelerators*, Plenum Press, New York, 1975.
200. J. P. Thomas and A. Cachard (Eds.), *Material Characterization Using Ion Beams*, Plenum Press, New York, 1978.
201. J. R. Bird and J. S. Williams (Eds.), *Ion Beams for Materials Analysis*, Academic Press, Sydney, 1989.
202. H. J. Annegarn, J. K. Basson, T. E. Derry, R. W. Fearick, J. P. F. Sellschop, and J. I. W. Watterson (Eds.), Proc. 8th Int. Conf. on Ion Beam Analysis, *Nucl. Instr. Meth.* B35:205 (1988).
203. J. F. Ziegler, P. J. Scanlon, W. A. Lanford, and J. L. Duggan (Eds.), Proc. 9th Int. Conf. on Ion Beam Analysis, *Nucl. Instr. Meth.* B45:1 (1990).
204. J. W. Nelson and W. J. Courtney, *Nucl. Instr. Meth.* 142:127 (1977).
205. B. G. Martinsson, *Nucl. Instr. Meth.* B15:636 (1986).
206. M. O. Andreae and W. R. Barnard, *Nucl. Instr. Meth.* 181:383 (1981).
207. M. O. Andreae, *Science* 220:1148 (1983).
208. M. O. Andreae, T. W. Andreae, R. J. Ferek, and H. Raemdonck, *Sci. Total Environ.* 36:73 (1984).
209. D. Heck and E. Rokita, *Nucl. Instr. Meth.* B3:259 (1984).
210. K. Themner and K. G. Malmqvist, *Nucl. Instr. Meth.* B15:404 (1986).
211. G. Amsel, C. Heitz, and M. Menu, *Nucl. Instr. Meth.* B14: 30 (1986).
212. B. L. Doyle and P. S. Piercy, *Appl. Phys. Lett.* 34:811 (1979).
213. R. Pretorius, M. Peisach, and J. W. Mayer, *Nucl. Instr. Meth.* B35:478 (1988).
214. R. M. Sealock, D. N. Jamieson, and G. J. F. Legge, *Nucl. Instr. Meth.* B29:557 (1987).
215. R. M. S. Schofield, H. W. Lefevre, J. C. Overley, and J. D. MacDonald, *Nucl. Instr. Meth.* B30:398 (1988).
216. G. S. Bench and G. J. F. Legge, *Nucl. Instr. Meth.* B40/41:655 (1989).
217. J. Räisänen, *Biol. Trace Elem. Res.* 12:55 (1987).
218. K. -S. Kim, J. Webb, D. J. Macey, and D. D. Cohen, *Nucl. Instr. Meth.* B22:227 (1987).
219. G. S. Hall and E. Navon, *Nucl. Instr. Meth.* B15:629 (1986).
220. I. D. Svalbe, M. A. Chaudhri, K. Traxel, C. Ender, and A. Mandel, *Nucl. Instr. Meth.* B3:651 (1984).
221. G. E. Coote and I. C. Vickridge, *Nucl. Instr. Meth.* B30:393 (1988).
222. G. Robaye, J. M. Delbrouck-Habaru, I. Roelandts, G. Weber, L. Girard-Reydet, J. Morelli, and J. P. Quisefit, *Nucl. Instr. Meth.* B6:558 (1985).
223. L. Asking, E. Swietlicki, and L. M. Garg, *Nucl. Instr. Meth.* B22:368 (1987).

224. C. Boni, E. Caruso, E. Cereda, G. M. Braga Marcazzan, and P. Redaelli, *Nucl. Instr. Meth. B40/41*:620 (1989).
225. P. J. Clark, G. F. Neal, and R. O. Allen, *Anal. Chem. 47*:650, (1975).
226. G. Demortier, *Spectroscopy 4*(6):35 (1989).
227. L. Gönczi, R. Didriksson, B. Sundqvist, and M. A. Awal, *Nucl. Instr. Meth. 203*:577 (1982).
228. L. Varga, I. Demeter, and Z. Szökefalvi-Nagy, *Nucl. Instr. Meth. B3*:357 (1984).
229. R. D. Vis, C. C. A. H. van der Stap, and D. Heymann, *Nucl. Instr. Meth. B22*:380 (1987).
230. B. G. Martinsson and H. -C. Hansson, *Nucl. Instr. Meth. B34*:203 (1988).
231. S. A. E. Johansson, *Fresenius Z. Anal. Chem. 324*:635 (1986).
232. R. Nobiling, *Nucl. Instr. Meth. B14*:142 (1986).

12

Electron-Induced X-ray Emission

John A. Small   *National Institute of Standards and Technology, Gaithersburg, Maryland*

I. INTRODUCTION

Several microanalytical instruments are currently being used to analyze microscopic and submicroscopic regions of samples. The schematic diagram in Figure 1 outlines the generic features of a microanalytical instrument. Primary radiation is focused to form a beam about 1 μm or smaller in size. The interaction of the primary beam with the specimen results in the emission of secondary radiation, which is then analyzed with a spectrometer system to provide information on the structure and composition of the sample. The electron probe is one of the instruments used for the analysis of microscopic dimains. It incorporates electron-induced x-ray emission as the basis for analysis.

In 1951, Castaing [1] developed the first successful electron probe microanalyzer and outlined the fundamental physical concepts of quantitative analysis. The electron microprobe he developed made use of a focused beam of electrons to excite x-rays from a microscopic domain on a sample surface.

In conventional electron microprobes, the electron beam is generated from either a tungsten filament or a lanthanum hexaboride (LaB_6) electron source. The filament or LaB_6 source acts as the cathode in the electron gun and is maintained at a negative potential of -1 to -400 keV (Fig. 2). In the system employing a tungsten filament, the electron beam is produced by the thermal emission of electrons from the "hairpin" filament, which consists of tungsten wire bent in a V shape. The tip of the V is approximately 200 μm in diameter (see Fig. 3). The

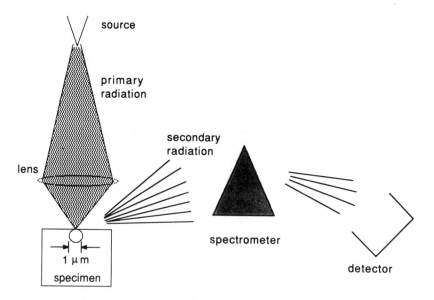

Figure 1   A generic microanalytical instrument.

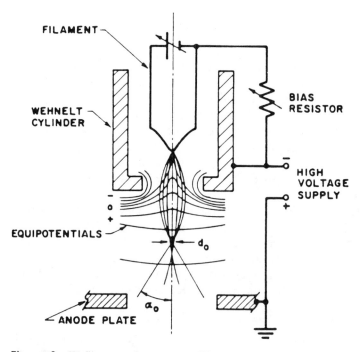

Figure 2   W filament electron gun [2].

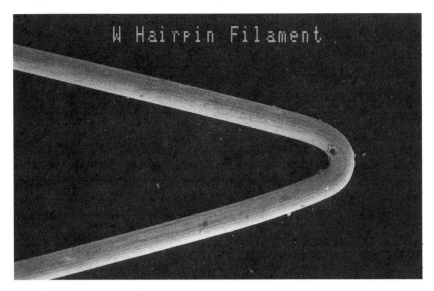

Figure 3 Photomicrograph of a W filament.

tungsten wire is directly heated to a temperature of 2400–2600°C, which results in an electron emission current density of about 10 A/cm^2.

In systems that employ the LaB_6 electron source, the cathode consists of a rod of solid LaB_6 milled on one end to a tip with a diameter of approximately 10 μm. The LaB_6 source is heated to a temperature of about 1680°C. At this temperature, the emission current density for the LaB_6 source is of the order of 100 A/cm^2, or a factor of 10 greater than the current density for the tungsten filament. The higher current density of the LaB_6 enables the use of smaller primary beams for a given current than the tungsten filament. In the past the LaB_6 emission source found only limited use in electron probe microanalysis (EPMA) primarily because of instability in the emission current. It is only in the last few years that the electron emission from the LaB_6 sources has been stable enough to allow the analyst to perform high-quality x-ray analysis.

After emission, the electrons are focused through an initial crossover by the presence of the Wehnelt cap that surrounds the filament (see Fig. 2). The cap is biased several hundred volts negative compared to the filament, which creates an immersion field that focuses the electrons to a crossover with a diameter d_0 of approximately 50 μm. After passing through the initial crossover, the electrons are accelerated by an anode plate biased from 1 to 400 keV positive with respect to the cathode. Next, the initial electron crossover or spot is demagnified by a series of apertures and electron optical lenses, including both condenser and objective lenses. The final probe diameter used to interrogate the sample is dependent on the filament material, emission current, and acceleration voltage, as shown in Figure 4. A schematic of the modern electron probe microanalyzer is shown in Figure 5. A detailed description of electron optics and electron probe formation is given in References 2 and 3.

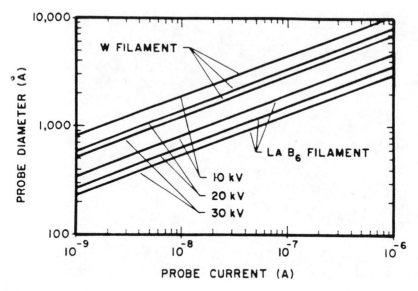

Figure 4 Probe size as a function of filament material, emission current, and acceleration voltage [2].

In classical electron probe microanalysis, the characteristic x-rays emitted as a result of the primary electron beam interaction with the atoms in the specimen are analyzed by wavelength-dispersive spectrometry (WDS) and/or energy-dispersive spectrometry (EDS) [3]. The various characteristic x-ray intensities can then be used to determine elemental compositions. The volume of the specimen

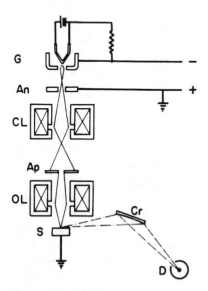

Figure 5 An electron probe: G = electron gun, An = anode plate, CL = condenser lens, Ap = column aperture, OL = objective lens, S = specimen, Cr = diffraction crystal, and D = x-ray detector [3].

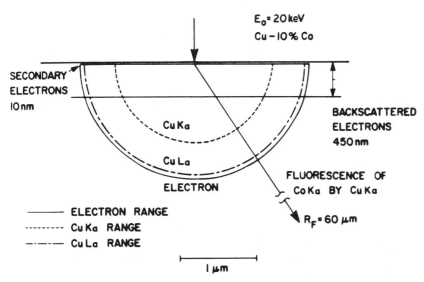

Figure 6 Relative dimensions of the primary electron beam, the electron excitation and emission volumes, and the x-ray emission volumes in a Cu target. (From Ref. 4, p. 121.)

that is excited depends on the specimen composition and the energy of the primary electron beam. Since the absorption path lengths for x-rays are considerably greater than those of the secondary electrons used for electron imaging, the spatial resolution for x-ray microanalysis is of the order of 1–2 μm, compared to 3–5 nm for electron imaging. This is shown schematically in Figure 6, where the x-ray emission volume is compared to the electron beam diameter of 1–5 nm. Specimens that are inhomogeneous at dimensions below the x-ray resolution cannot be readily analyzed by conventional microprobe analysis. These samples represent a special class of samples that are best analyzed in the analytical electron microscope as thin films (see Sec. III).

II. QUANTITATIVE ANALYSIS

As mentioned previously, the basic concepts for quantitative electron probe microanalysis were first introduced by Castaing in his thesis, which was published in 1951 [1]. Castaing proposed that quantitative elemental analysis could be carried out in the electron probe by comparing the x-ray intensity generated from a given element i in an unknown to the x-ray intensity of i generated in a standard containing a known amount of the element. The ratio of the intensity of i in the sample to i in a pure element standard, Equation (1), was referred to as the k ratio by Castaing and forms the basis for quantitative analysis:

$$\frac{I_i^{sam}}{I_i^{std}} = k_i \tag{1}$$

In Equation (1), the intensities must be corrected for background and dead-time differences by the methods described in the chapter on x-ray detection. Ideally,

$$k_i = \frac{C_i}{\sum C_a} = C_i \quad \text{since} \quad \sum C_a = 1 \tag{2}$$

where C refers to the elemental concentrations in wt% and the subscript a refers to all the elements in the sample. This equation only applies to a system in which the sample and the standard are identical and have been measured under identical experimental conditions. In practice, as the similarities between the sample and the standard decrease, Equation (2), even as an approximation, fails and a series of corrections must be applied to the k ratio to obtain an accurate quantitative analysis. The corrections that must be applied to the k ratio include the following:

1. The atomic number correction for the differences between the electron scattering and penetration in the sample and the standard
2. The absorption correction for the difference in the absorption of the x-rays as they pass through the sample or standard before reaching the detector
3. The fluorescence correction for the fluorescence of x-rays by the characteristic and continuum x-rays generated in the sample by the primary electron beam

These corrections are applied to the various k ratios as part of theoretical or empirical correction procedures to obtain quantitative results.

A. Theoretical Corrections for Electron Probe Microanalysis
1. ZAF Corrections

In classical electron probe analysis (ZAF method) the various corrections are supplied to the k ratios as multiplicative terms:

$$C_i = k_Z k_A k_F k_i \tag{3}$$

where the terms k_Z, k_A, and k_F refer to the atomic number, absorption, and fluorescence corrections, respectively. In the remainder of this section, each of the corrections is discussed separately.

a. Atomic Number Correction k_Z

The atomic number correction in electron microprobe analysis is applied to the k ratio to compensate for the difference between the electron retardation and electron backscattering in the sample and standard. Both the electron retardation and backscattering are dependent on the average atomic number of the sample $\overline{Z}$, defined in Equation (4). Therefore, any difference between the average atomic number of the sample and the standard should be addressed by this correction. As a general rule, ignoring the effects of the atomic number correction results in an underestimation of the concentration of high-Z elements in low-Z matrices and the overestimation of concentrations of low-Z elements in high-Z matrices [4]. The average atomic number $\overline{Z}$ for the sample is given by

$$\overline{Z} = \sum_i C_i Z_i \tag{4}$$

The general formulation of k_z for element i is given in Equation (5):

$$(k_Z)_i = \frac{R_i^{sam} \int_{E_c}^{E_0} (Q_i/S_i^{sam}) \, dE}{R_i^{std} \int_{E_c}^{E_0} (Q_i/S_i^{std}) \, dE} \tag{5}$$

where the R and the S terms refer to the electron backscattering and the electron stopping power, respectively, and Q is the ionization cross section. The limits on the integral are from the incident electron energy E_0 to the critical excitation energy E_c for the x-ray line of interest.

i. Electron Stopping Power. The electron stopping power S is defined in Equation (6) as the energy lost per unit electron pathlength in material of density ρ.

$$S = -\frac{1}{\rho} \frac{dE}{dX} \tag{6}$$

The most commonly used term for S is the approximation from Bethe [5] and Bethe and Ashkin [6], which assumes a continuous function for the electron energy loss.

$$S = 78,500 \frac{Z_i}{A_i} \frac{1}{E} \ln \frac{1.166E}{J_i} \tag{7}$$

The value for the mean ionization potential J in Equation (7) is not directly measured, and several different expressions are used in the literature for the calculation of J. Various literature values for J are listed in Table 1 (from Heinrich [3]) and are plotted as a function of Z in Figure 7. The various expressions for J all yield similar results for elements above $Z = 10$, with relatively large discrepancies between expressions for $Z < 10$. The Berger-Seltzer expression for J is currently the most widely used in quantitative analysis procedures.

$$\frac{J}{Z} = 10.04 + 8.25 \, e^{-Z/11.22}$$

Zeller [13]. Several models in the literature have also been used to calculate Q, all of which have the general form

$$Q = C \frac{\ln U}{U E_c^2} \tag{8}$$

where U is the overvoltage defined as E_0/E_c.

Table 1 Different Values for J

Equation in eV	Reference
$J/Z = 13.5$	Block [7]
$J/Z = 11.5$	Wilson [8]
$J/Z = 9.76 + 58.82Z^{-1.19}$	Berger and Seltzer [9]
$J/Z = 9.0(1 + Z^{-0.67}) + 0.03Z$	Springer [10]
$J/Z = 12.4 + 0.02Z$	Heinrich and Yakowitz [11]
$J/Z = 14.0[1 - e^{-0.1Z}] + 75.5/Z^{0.13Z} - Z/(100 + Z)$	Duncumb et al. [12]

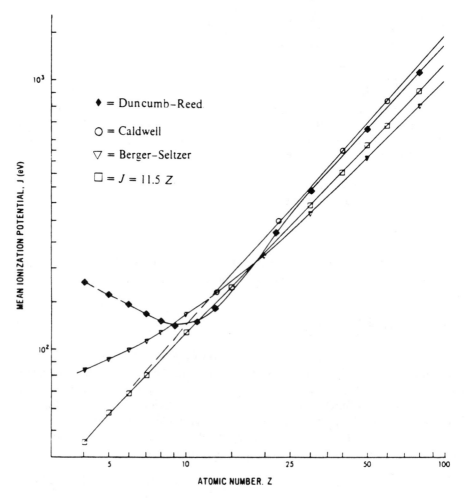

Figure 7   Different formulations for the mean ionization potential J plotted as a function of the atomic number Z [2].

Heinrich and Yakowitz [11], however, showed that the difference in the various models result in "negligible changes in the final elemental concentrations" for elements with $Z > 10$.

Duncumb and Reed [14] simplified the integration in Equation (5), eliminating the need for the numerical integration of Equation (5) and the evaluation of Q. In their procedure, they assumed that the values for $1/S\ dE$ were constant for the sample and standard over the electron range used in electron probe analysis and therefore could be removed from the integral. As a result of this assumption, the integration of Q is unnecessary since it appears in both the numerator and denominator and is the same for the unknown and the standard. In addition, from the work of Thomas [15], the average energy $(E_0 + E_c)/2$ can replace the integration of dE. The stopping power factor can then be expressed as S_{ij}, which is the stopping power for element i in element j:

$$S_{ij} = \frac{Z_i}{A_j(E_0 + E_c)} \ln \frac{583 (E_0 + E_c)}{J_j} \tag{9}$$

where A is atomic weight, i refers to the measured element, and j refers to all elements in the specimen, including i.

Equation (9) is the most commonly used formulation for microanalysis. Duncumb and Reed [14] showed from experimental work that the stopping power for a multielement specimen can be expressed as a weighted sum of the stopping power factor for each element.

$$S_i = \sum_0^j C_j S_{ij} \tag{10}$$

where S_i is the stopping power for element i and C_j is the weight fraction of element j.

In addition to Equation (9), alternative formulations for S were also proposed by Philibert and Tixier [16] and by Love et al. [17]. Philibert and Tixier derived an exact analytical solution to Equation (6) that defines S in terms of a logarithmic integral:

$$\frac{1}{S} = \frac{1}{m} \left[U_0 - 1 - \frac{\ln w}{w} \langle Li(wU_0) - Li(w) \rangle \right] \tag{11}$$

where

$$w = 1.166 \frac{E_c}{\bar{J}}$$

$$m = \sum C_i \frac{Z_i}{A_i}$$

$$\bar{J} = \sum C_i \frac{Z_i}{A_i/m}$$

$$U_0 = E_0/E_c$$

The logarithmic integral Li of a variable Y expressed as an infinite series equals

$$Li(Y) = \ln | \ln Y | + \sum_{F=1}^{\infty} \frac{(\ln Y)^F}{FF!} + D \tag{12}$$

where D is Euler's constant, which equals 0.577.

Love and Scott noted that the Bethe expression [5] for $dE/d\rho s$ is valid only if $E \gg J$ and as a result modified the Bethe expression "to give better limiting behavior as E approaches J." Their work results in the following formulation for S:

$$\frac{1}{S} = \frac{\{1 + 16.05(\bar{J}/E_c)^{1/2}[(U_0^{1/2} - 1)/(U_0 - 1)]^{1.07}\}}{\sum C_i Z_i/A_i} \tag{13}$$

ii. Electron Backscatter Factor. The electron backscatter factor R in Equation (5) is defined as

$$R = 1 - \frac{I_b}{I_t} \tag{14}$$

where I_b is the x-ray intensity lost to backscatter electrons and I_t is the x-ray intensity if no electrons are backscattered.

The fraction of electrons that are backscattered from a sample is known as the electron backscatter coefficient η, and is given by Equation (15).

$$\eta = \int_0^1 \frac{d\eta}{dw} \, dw \tag{15}$$

where w is the ratio of the energy of the backscattered electron E_b to the beam energy E_0 (i.e., E_b/E_0) and η is the backscatter coefficient. An empirical expression for η, Equation (16), was obtained by Reuter [18] from a fit to Heinrich's data [19]:

$$\eta = -0.0254 + 0.016Z - 1.86 \times 10^{-4}Z^2 + 8.3 \times 10^{-7}Z^3 \tag{16}$$

The number of ionizations generated in a sample by an electron with energy E is given for a characteristic x-ray line with critical excitation energy E_c by

$$\int_E^{E_c} \frac{Q}{dE/d\rho s} \, dE \tag{17}$$

I_b can then be obtained by multiplying Equations (15) and (17):

$$I_b = \int_{w_0}^1 \frac{d\eta}{dw} \int_E^{E_c} \frac{Q}{dE/d\rho s} \, dE \, dw \tag{18}$$

where the integration limit for η is $w_0 = E_c/E_0$ since electrons with energies less than E_c cannot excite the x-rays of interest.

Similarly, I_t can be calculated from Equation (19) with integration limits of E_0 to E_c:

$$I_t = \int_{E_0}^{E_c} \frac{Q}{dE/d\rho s} \, dE \tag{19}$$

Finally, substituting Equations (18) and (19) into (14) results in the following formulation for R:

$$R = 1 - \frac{\int_{w_0}^1 d\eta/dw \int_E^{E_c} Q/(dE/d\rho s) \, dE \, dw}{\int_{E_0}^{E_c} Q/(dE/d\rho s) \, dE} \tag{20}$$

Several tabulations of R (including those by Duncumb and Reed [14], Green [20], and Springer [21] have been made for pure elements as a function of Z and U. Duncumb and Reed produced a table of R values for various elements and several different overvoltage values. Their values were determined indirectly from Bishop's [22] measurements of the energy distributions of backscattered electrons and were in agreement with the direct measurements of R made by Derian and Castaing [23]. Figure 8 shows a plot of the Duncumb-Reed R values versus Z at different overvoltages.

Duncumb [24] derived an algebraic expression for R in terms of w_q and Z from his calculated values. Equation (21) is still used in many analytical procedures.

$$
\begin{aligned}
R = 1 &+ (-0.581 + 2.162w_q - 5.137w_q^2 + 9.213w_q^3 \\
&- 8.619w_q^4 + 2.962w_q^5) \times 10^{-2}Z + (1.609 - 8.298w_q \\
&+ 28.791w_q^2 - 47.744w_q^3 + 46.540w_q^4 - 17.676w_q^5) \\
&\times 10^{-4}Z^2 + (5.400 + 19.184w_q - 75.733w_q^2 + 120.050w_q^3 \\
&+ 110.700w_q^4 + 41.792w_q^5) \times 10^{-6}Z^3 + (-5.725 \\
&- 21.645w_q + 88.128w_q^2 - 136.060w_q^3 + 117.750w_q^4 \\
&- 42.445w_q^5) \times 10^{-8}Z^4 + (2.095 + 8.947w_q - 36.510w_q^2 \\
&+ 55.694w_q^3 - 46.079w_q^4 + 15.851w_q^5) \times 10^{-10}Z^5
\end{aligned}
\tag{21}
$$

where $w_q = E_c/E_0$.

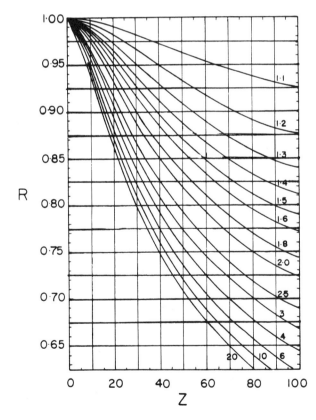

Figure 8 Duncumb backscatter correction factor R versus Z and overvoltage [3].

Yakowitz et al. [25] obtained a simplified expression for R from a fit of the Duncumb and Reed values:

$$R_{ij} = R'_1 - R'_2 \ln (R'_3 Z_j + 25) \tag{22}$$

where:

$$R'_1 = 8.73 \times 10^{-3} U^3 - 0.1669 U^2 + 0.9662 U + 0.4523$$
$$R'_2 = 2.703 \times 10^{-3} U^3 - 5.182 \times 10^{-2} U^2 + 0.302 U - 0.1836$$
$$R'_3 = \frac{0.887 U^3 - 3.44 U^2 + 9.33 U - 6.43}{U^3}$$

In this expression, i represents the element being measured and j represents the elements in the specimen including i. R_{ij} is therefore the backscatter correction for element i in the presence of element j.

Myklebust [26] further simplified the expression for R:

$$R = 1 - 0.0081512Z + 3.613 - 10^{-5}Z^2 \tag{23}$$
$$+ 0.009582Ze^{-U} + 0.00114E_0$$

This expression, which is used in the current versions of the FRAME analysis procedure [25], represents a fit of the R values obtained from the National Bureau of Standards (NBS) Monte Carlo program [27]. Figure 9 from Myklebust [28] shows the behavior of various R values from different authors as a function of Z for selected $K\alpha$ x-ray lines.

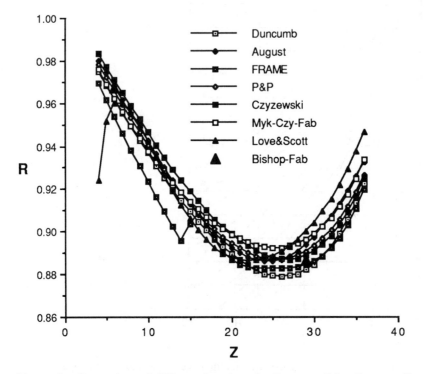

Figure 9 Comparison of different functions for R versus Z for $K\alpha$ x-ray lines [26].

In a multielement system, the factor R for element i can be calculated from Equation (24), which was proposed by Duncumb and Reed [14]:

$$R_i = \sum_j C_j R_{ij} \tag{24}$$

Recently, Myklebust and Newbury [29] compared Monte Carlo results to results from Equation (23) for a 10% Cu-Au alloy to determine the accuracy of mass concentration averaging in multielement targets. Their results indicate that Equation (24) is valid for the alloy studied.

iii. Evaluation of k_Z. The results of an analysis for copper in 2 wt% Cu-Al alloy was used by Goldstein et al. [30] to demonstrate the magnitude of the atomic number correction. The authors used both pure elements and a 46% Al-Cu alloy as standards for the analysis. The results, given in Table 2, indicate that the atomic number correction for this analysis is as high as 16% for the pure element standards and is reduced by 8% when the alloy is used as the standard.

b. Absorption Correction k_A

The primary electron beam generates x-rays at varying depths within the sample. As a result, the x-rays must pass through that portion of the specimen that lies between the x-ray generation point and the detector before they escape the sample and are measured. As shown in Figure 10, the distance A-B is referred to as the absorption pathlength since a percentage of the generated x-rays are absorbed, interacting with specimen atoms before escape. The effect of this absorption is an attenuation of the x-ray intensity measured by the detector. The magnitude of the attenuation is dependent on the composition of the specimen, and a correction must be considered when the sample and the standards used for the analysis are dissimilar.

In 1951, Castaing [1] described the characteristic x-ray intensity (without absorption) generated in a layer of thickness dz, at a depth z below the specimen surface, in a sample of density ρ as

$$dI = \phi(\rho z)\, d(\rho z) \tag{25}$$

Table 2 k_Z Correction for Copper in a 2 wt% Copper in Aluminum Alloy

	15 keV
Standard	k_Z (Cu)
Elemental Cu	1.16
Alloy	1.08

	30 keV
Standard	k_Z (Cu)
Elemental Cu	1.11
Alloy	1.05

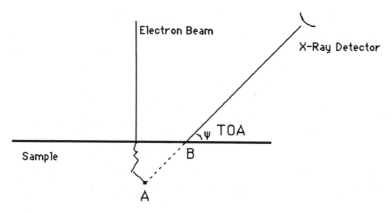

Figure 10   X-ray absorption pathlength.

where $\phi(\rho z)$ is the distribution of characteristic x-rays as a function of depth (density × distance, mg/cm²) in the sample. A typical curve for Cu $K\alpha$ radiation is shown in Figure 11. The total generated x-ray intensity for a given line can be obtained by integrating the area under the curve over the entire x-ray range:

$$I = \int_0^\infty \phi(\rho z)\, d(\rho z) \tag{26}$$

The introduction of x-ray absorption into Equation (26) results in the following expression for the x-ray intensity I'. In this expression, I' refers to the x-ray intensity after absorption.

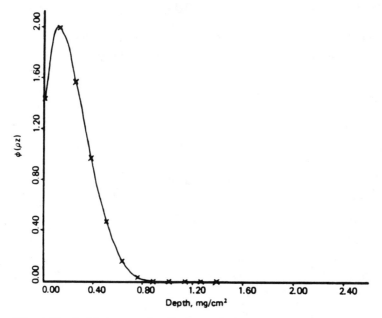

Figure 11   A $\phi(\rho z)$ curve for Cu $K\alpha$ radiation. (From Ref. [4], p. 309.)

$$I' = \int_0^\infty \phi(\rho z)e^{-(\mu/\rho)\rho z \, \csc \Psi} \, d(\rho z) \tag{27}$$

where $\rho z \csc \Psi$ is the absorption pathlength for the x-rays in the specimen, Ψ is the detector emergence angle, and μ/ρ is the mass absorption coefficient of the specimen for the characteristic line of the element of interest.

The absorption term for EPMA is referred to as $f(\chi)$ or f_p, where $\chi = \mu/\rho \csc \Psi$ and $f(\chi) = I'/I$. From Equations (26) and (27), $f(\chi)$ can be expressed in terms relating to the specimen as

$$f(\chi) = \frac{\int_0^\infty \phi(\rho z)e^{-(\mu/\rho)\rho z \, \csc \Psi} \, d(\rho z)}{I = \int_0^\infty \phi(\rho z) \, d(\rho z)} \tag{28}$$

The absorption correction for EPMA can then be expressed as

$$K_A = \frac{f(\chi)_{\text{std}}}{f(\chi)_{\text{sam}}} \tag{29}$$

The basic formulation of the $f(\chi)$ term was derived by Philibert [31]. For the derivation, he considered the number of ionizations produced in a layer of element i with thickness $d(z)$ at some depth z within a specimen:

$$f(\chi) = \frac{1 + \phi(0)h\chi/[(4 + \phi(0)h)\sigma]}{(1 + \chi/\sigma)\{1 + [h/(1 + h)]\chi/\sigma\}} \tag{30}$$

In this equation, $\phi(0)$ is the surface ionization function, σ is Lenard's constant for a given incident electron energy, and h is given by

$$h = 1.67 \times 10^{-6} \frac{A}{Z^2} \sigma E_0^2 \tag{31}$$

Noting that $f(\chi)$ was not sensitive to either $\phi(0)$ or σ, Philibert simplified Equation (31) by setting $\phi(0) = 0$ and $h = 1.2A/Z^2$. This results in the following expression for $f(\chi)$:

$$\frac{1}{f(\chi)} = \left(1 + \frac{\chi}{\sigma}\right)\left(1 + \frac{h}{1 + h}\frac{\chi}{\sigma}\right) \tag{32}$$

Duncumb and Shields [32] modified σ to take into account that electrons with energies less than E_c cannot generate x-rays from the line associated with E_c. They proposed the following expression for σ:

$$\sigma = \frac{2.39 \times 10^5}{E_0^{1.5} - E_c^{1.5}} \tag{33}$$

Heinrich [33] fit experimental values of $f(\chi)$ from Green [34] and proposed the following formulation for σ:

$$\sigma = \frac{4.5 \times 10^5}{E_0^{1.65} - E_c^{1.65}} \tag{34}$$

The most common form of the absorption term found in the various analytical schemes is Equation (32), with $h = 1.2A/Z^2$ and Heinrich's σ. This form of Equation (32) is often referred to as the Philibert-Duncumb-Heinrich equation [35].

Heinrich et al. [36] empirically derived a simplified absorption term based on experimental data and the Philibert equation. In the formulation of this term, it was noted that the compositional dependence in Philibert's h term was small compared to the scatter in the available experimental data. The resulting formulation for $f(\chi)$, given in Equation (35), is referred to by Heinrich as the quadratic model. In this equation, the compositional dependence in the h term has been eliminated, making the model independent of the atomic weight and atomic number of the target.

$$\frac{1}{f(\chi)} = (1 + 1.2 \times 10^{-6}\gamma\chi)^2 \tag{35}$$

where γ is the quantity $E_0^{1.65} - E_c^{1.65}$.

In Equation (32), h is dependent on target composition and must be averaged for the various elements in multielement targets. A compositionally weighted average for h [Eq. (36)] is generally employed in analytical procedures:

$$h_t = \sum_j C_j h_j \tag{36}$$

In addition, the mass absorption coefficient μ/ρ for the characteristic line of element i in a multielement target is the weighted sum over all elements in the target, including i:

$$\left(\frac{\mu}{\rho}\right)_i^{sam} = \sum_j C_j \left(\frac{\mu}{\rho}\right)_i^j \tag{37}$$

where $(\mu/\rho)_i^j$ is the mass absorption coefficient for the line of element i in element j and C_j is the concentration of element j.

The calculation of x-ray absorption from Equation (32) is most accurate when the value of $f(\chi)$ is greater than 0.7. Toward this end, it is important in setting up an experiment to minimize the degree of x-ray absorption by

1. Minimizing the absorption pathlength by selecting a low overvoltage E_0/E_c.
2. Selecting high-energy x-ray lines that have low absorption cross sections. It is important to note that selecting a high-energy x-ray line requires a high acceleration potential and therefore increases the absorption for elements in the sample that have relatively low energy analytical lines.
3. Measuring the intensities at the highest possible x-ray emergence angle (which typically is fixed for any given instrument).

c. Fluorescence Correction k_F

The characteristic fluorescence correction is necessary when the analysis involves the following conditions:

1. The characteristic x-ray peak from element j at energy E_j is greater than the critical excitation potential $E_{c,i}$ for element i.
2. The energy difference $E_j - E_{c,i} < 0.5$ keV.

When these conditions exist, the characteristic line of element j excites the characteristic line of element i in the specimen, as shown in Figure 12. This fluorescence results in an increased intensity for element i that must be taken into account in obtaining a quantitative analysis.

Since the excitation in this case is caused by x-rays, not electrons, generation of the secondary x-rays originates at a greater depth within the sample than if they were generated by electrons. Since the absorption pathlength for x-rays is shorter at higher emergence angles, the magnitude of the fluorescence radiation is proportional to the x-ray emergence angle.

The basic formulation of the fluorescence correction is given in Equation (38):

$$F_i = \frac{\left(1 + \sum_j I^f_{i,j}/I_i\right)_{std}}{\left(1 + \sum_j I^f_{i,j}/I_i\right)_{sam}} \tag{38}$$

The factor $I^f_{i,j}/I_i$ is the ratio of the x-ray-excited intensity for the characteristic line of element i to the electron-excited intensity. The summation over j is required since the total correction must be summed over all the elements in the specimen.

The most common used fluorescence correction factor is Equation (39), which was developed by Reed [37]:

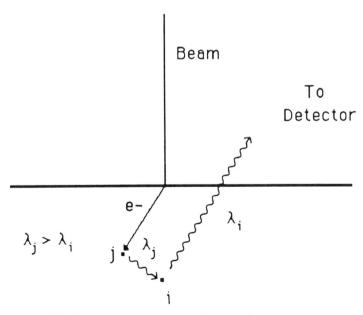

Figure 12   Secondary fluorescence in a specimen.

$$\frac{I^f_{i,j}}{I_i} = C_j Y_0 Y_1 Y_2 Y_3 P_{i,j} \tag{39}$$

where C_j is the concentration of element j.

$$Y_0 = 0.5 \frac{r_i - 1}{r_i} \omega_j \frac{A_i}{A_j} \tag{40}$$

where r_i is the absorption jump ratio of element i defined as the ratio of μ/ρ on the high-energy side of the absorption edge to μ/ρ on the low-energy side; the factor $(r_i - 1)/r_i$ is 0.88 for K-line fluorescence and 0.74 for L-line fluorescence; ω_j is the fluorescent yield of element j; and A_i and A_j are the atomic weights for elements i and j.

$$Y_1 = \left(\frac{U_j - 1}{U_i - 1}\right)^{1.67} \tag{41}$$

where U_i and U_j and E_0/E_c for elements i and j.

$$Y_2 = \frac{\mu/\rho^j_i}{\mu/\rho^j_{sam}} \tag{42}$$

where the numerator is the mass absorption coefficient for the absorption of x-rays from element j in element i, and the denominator is the mass absorption coefficient for the absorption of x-rays from element j in the specimen.

$$Y_3 = \frac{\ln(1 + t)}{t} + \frac{\ln(1 + k)}{k} \tag{43}$$

where

$$t = \left[\frac{(\mu/\rho)^i_{sam}}{(\mu/\rho)^j_{sam}}\right] \csc \Psi$$

and

$$K = \frac{3.3 \times 10^5}{[E_0^{1.65} - (E_c^i)^{1.654}](\mu/\rho)^j_{sam}}$$

Finally, P_{ij} is a factor related to the x-ray lines associated with both the excited element i and the exciting agent j. The values of P_{ij} for the different combinations of K and L x-rays are as follows:

Element i excited	Element j agent	P_{1j}
K	K	1.0
L	L	1.0
K	L	0.24
L	K	4.76

Table 3 Fluorescence of Fe $K\alpha$ by Ni in 10 wt% Fe-90 wt% Ni Alloy

Ψ (degrees)	E_0 (keV)	$I^f_{\text{Fe-Ni}/I_{\text{Fe}}}$	F_{Fe}
52.5	15	0.263	0.792
15.5	15	0.168	0.856
52.5	30	0.346	0.743
15.5	30	0.271	0.787

In practice, when calculating the fluorescence correction from Equation (38), the standard is either a pure element or there is no significant fluorescence of element i by other elements in the standard. For this situation, Equation (38) reduces to

$$F_i = \frac{1}{\left(1 + \sum_j I^f_{i,j}/I_i\right)_{\text{sam}}} \tag{44}$$

which is the most common form of the correction found in analytical procedures. Goldstein et al. [38] demonstrated the magnitude of the fluorescence correction with the analysis of a 10 wt% Fe-90 wt% Ni alloy. The results of the analysis are summarized in Table 3 and show that the intensity of the fluoresced iron $I^f_{\text{Fe-Ni}}/I_{\text{Fe}}$ ranges from 16.8 to 34.6% of the observed iron x-ray intensity. The magnitude of the correction is lower at the smaller accelerating potential and detector take-off angle.

In addition to the secondary excitation by characteristic x-rays, x-rays can also be excited by the continuum x-rays produced in the sample. The continuum x-rays are the result of the deceleration of beam electrons in the Coulombic field of the specimen atoms. This radiation forms an x-ray background that is slowly varying with energy and ranges from E_0 to zero. Although there has been some recent work by Heinrich [39] to simplify the calculations for determining continuum fluorescence, the basic formulation for the added intensity from the continuum fluorescence I_c derived by Henoc [40]:

$$I_c = f(\overline{Z}, \omega, r, \mu/\rho, \Psi) \tag{45}$$

The calculation of the continuum fluorescence is relatively complicated, involving integration over the range of E_0 to E_c for each element in the specimen .Since the correction can be as large as 2–4%, it should be included for highest accuracy. Myklebust et al., investigating the continuum fluorescence, determined the following experimental conditions for which a correction for the fluorescence by the continuum is required [41]:

$f(\chi) > 0.95$

$C_i < 0.5$

$\overline{Z}$ from the standard much different from $\overline{Z}$ of the sample

These requirements translate to the analysis of a small amount of an element with a high-energy x-ray line, such as zinc, in a light matrix, such as carbon. Heinrich compared the fluorescence intensities caused by the x-ray continuum to the intensities generated by the primary electron beam for a series of elements at different electron beam energies (Fig. 13) [42].

2. Analysis Method Based on Integration of $\phi(\rho z)$ Curves

Another correction procedure, known as $\phi(\rho z)$, combines, in principle, the atomic number and absorption corrections already mentioned. Several researchers [43,44] have made experimental measurements of x-ray depth distributions, $\phi(\rho z)$ curves, in a large number of targets including pure elements, alloys, and oxides. In addition, other researchers [45] have made Monte Carlo calculations of x-ray depth distributions.

The increased knowledge regarding the behavior of these curves for different materials has made it possible to obtain empirically a correction based on the integration of $\phi(\rho z)$ curves. This procedure is particularly attractive for the analysis of low-energy x-ray lines, where $f(\chi)$ is much less than 0.7 and the accuracy of the ZAF method is low. In this procedure, $\phi(\rho z)$ refers to the depth distribution of x-rays in a bulk sample normalized to the x-ray intensity produced in an infinitely thin film of the same composition. The analysis of specimens using the $\phi(\rho z)$ correction method is dependent on the derivation of an accurate expression that describes the experimental $\phi(\rho z)$ curves and the degree to which that expression can be universally applied to systems with unknown $\phi(\rho z)$ curves. As pointed out by Armstrong [46], the use of $\phi(\rho z)$ expressions for the correction of x-ray absorption requires only that a given expression produce the correct shape for the $\phi(\rho z)$ curve. He cautions against the use of a $\phi(\rho z)$ expression for the atomic number correction since this requires "that the thickness of the tracer films and the normalizing thin films in the $\phi(\rho z)$ experiments for different matrices be known to a high degree of accuracy." Additional problems also occur with the atomic number correction because of the limited $\phi(\rho z)$ data available on multielement targets.

Packwood and Brown [47] proposed a modified Gaussian function to describe the shape of $\phi(\rho z)$ curves. The Gaussian was centered at the sample surface and was modified by a transient function that was introduced to model the near-surface x-ray distribution. In the model given in Equation (46) the α term relates to the

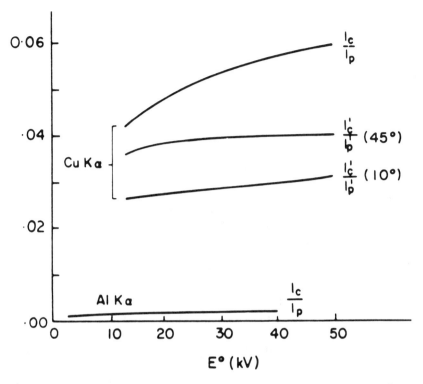

Figure 13 Comparison of x-ray intensities fluoresced by the x-ray continuum to the intensities generated by the primary electron beam [3].

width of the Gaussian and γ relates to the amplitude. The β term in the transient is related to the slope of the curve in the near-surface region, that is, the rate at which the focused electron beam is randomized through scattering in the target. The y intercept $\phi(0)$ is related to the surface ionization potential:

$$\phi(\rho z) = [\gamma e^{-\alpha^2(\rho z)^2}]\left[1 - \left(\frac{\gamma - \phi(0)}{\gamma}\right)e^{-\beta\rho z}\right]$$

$$\underbrace{\phantom{[\gamma e^{-\alpha^2(\rho z)^2}]}}_{\text{Gaussian term}} \quad \underbrace{\phantom{\left[1 - \left(\frac{\gamma - \phi(0)}{\gamma}\right)e^{-\beta\rho z}\right]}}_{\text{Transient term}}$$

(46)

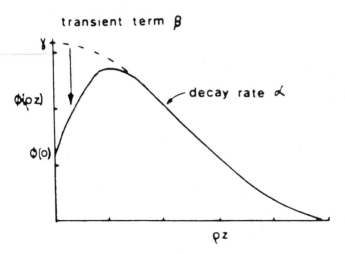

Figure 14 Modified Gaussian function for fitting $\phi(\rho z)$ curves [44].

Figure 14 is a plot of the modified Gaussian function showing the influence of the various parameters α, β, γ, and $\phi(0)$. The α, β, and γ parameters are expressed in terms of several different experimental terms, including specimen composition, incident electron energy, atomic number, and atomic mass. Since the development of Equation (46), various researchers have modified the original Brown and Packwood expressions to optimize the fit for different sets of experimental data [48–50].

In 1984, Pouchou and Pichoir [51] proposed a $\phi(\rho z)$ model, PAP, consisting of a pair of intersecting parabolas. For this model, the curve is defined by the following parameters:

- The surface ionization function $\phi(0)$
- The depth at which the maximum in the $\phi(\rho z)$ curve occurs R_m
- The x-ray range R_x
- The area under the $\phi(\rho z)$ curve, that is, the total generated intensity

The equation for the first parabola nearest the target surface is given in Equation (47) and the second parabola by Equation (48):

$$\phi_1(\rho z) = A_1(\rho z - R_m)^2 + B_1 \tag{47}$$

$$\phi_2(\rho z) = A_2(\rho z - R_x)^2 \tag{48}$$

where the parameters A_1, A_2, and B_1 are functions of R_m, R_x, and $\phi(0)$ and R_c is the crossover point of the parabolas. The equation for R_c is a function of the generated x-ray intensity I, which is the integral of the area under the $\phi(\rho z)$ curve, Equation (26). Figure 15 is the plot of the double-parabola curve with the various parameters located on the curve [44]. Pouchou and Pichoir [52] recently introduced a simplified version of their model based on Equation (49):

$$\phi(\rho z) = Ae^{-\alpha\rho z} + (B\rho z + \phi(0) - A)e^{-\beta\rho z} \tag{49}$$

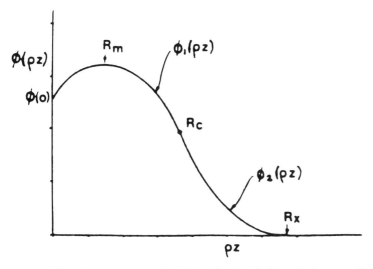

Figure 15 PAP double-parabola function for fitting $\phi(\rho x)$ curves [44].

The coefficients A, B, α, and β can be determined from the following parameters:

$$\int_0^\infty \phi(\rho z)\, d\rho z = \frac{A}{\alpha} + \frac{\phi(0) - A}{\beta} + \frac{B}{\beta^2}$$

$\phi(0) =$ surface ionization

$$\overline{R} = \text{mean ionization depth} = \frac{A}{\alpha^2} + \frac{\phi(0) - A}{\beta^2} + \frac{2B}{\beta^2}$$

Initial slope $= B - \alpha A - \beta(\phi(0) - A)$

Love et al. [53] proposed a $\phi(\rho z)$ method in which they introduced a separate atomic number correction. With the separate treatment of the atomic number correction, the $\phi(\rho z)$ curve is used only for target absorption. As mentioned, target absorption requires only that the shape of the curve is correct, not the absolute height. The authors proposed a quadrilateral profile (Fig. 16) to fit the shape of the curve. The quadrilateral is defined by the y intercept $\phi(0)$, the position and amplitude of the peak $(\rho z_m, \phi(m))$, and the x-ray range ρz_r. From the quadrilateral model, the analytical expression for $f(\chi)$ can be written as

$$f(\chi) = 2[A]^{-1}(B + C) \tag{50}$$

where

$$A = [(\rho z_r - \rho z_m)(\rho z_m + h(\rho z_r))\chi^2]$$

$$B = [-e^{-\chi \rho z_m} + he^{-\chi \rho z_r} + \chi(\rho z_r - \rho z_m) - h + 1]$$

$$C = \frac{e^{-\chi \rho z_m}(\rho z_r - h\rho z_r) + h\rho z_r - \rho z_r}{\rho z_m}$$

and

$$h = \frac{\phi(m)}{\phi(0)}$$

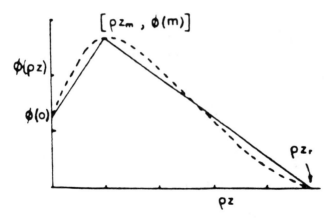

Figure 16 Quadrilateral function for fitting $\phi(\rho z)$ curves [44].

To constrain the quadrilateral model when $f(\chi)$ values exceed 0.5, the authors expressed both ρz_m and ρz_r in terms of the mean depth of x-ray generation, ρz [54]. In conjunction with the quadrilateral fit to the $\phi(\rho z)$ curve for target absorption, they use the atomic number correction of Love et al. [17].

B. Analytical Correction Procedures

1. ZAF Procedures

The simple relationship describing the ZAF correction procedure, Equation (3), gives the impression that the concentration of a given element can be calculated directly from the k ratio. This is not the case, however, since the different ZAF correction factors are dependent on the composition of the sample, including the element of interest. For a simple binary system, the analyst could construct a calibration curve relating measured x-ray intensity to concentration. This technique does not apply to more complicated systems, and the primary method used in modern ZAF procedures is an interactive method.

A general flowchart for an interactive ZAF method is shown in Figure 17. The measured k ratio is used as the initial estimate of composition. The different ZAF corrections are then applied to the initial estimate of concentration to obtain the corresponding estimate of k. Next the hyperbolic iteration, Equation (52), which is described in Section II.C, is used to obtain a value for α. The α value is then used with the measured k ratio to obtain a new estimate of concentration, which is then normalized and the iteration continued. Convergence with the hyperbolic function is rapid and seldom requires more than three interations. The practical lower limit of detection for quantitative x-ray microanalysis is of the order of 0.1% by weight for an element homogeneously distributed in an infinitely thick, polished specimen. This limit is primarily the result of the x-ray excitation cross sections and the signal-to-noise ratio of the x-ray detection systems. The accuracy of conventional probe analysis is shown in Figure 18, from Goldstein et al. [55], which is a histogram of the relative errors from 264 analyses of binary metal alloys. It is important to note that the distribution for other matricies, such as oxides, may be different.

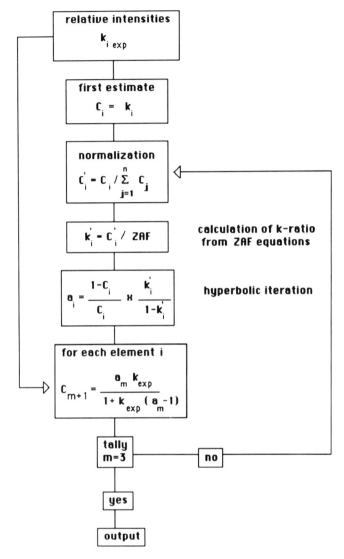

Figure 17 A ZAF analysis procedure [3].

2. φ(ρz) Procedures

A flowchart for the φ(ρz) procedures is shown in Figure 19. The percentage root mean square (RMS) errors for the comparison of various φρz models to experimental data bases are given in Table 4 from Love and Scott [44]. In general, for elements with atomic numbers >11, the RMA errors are less than 5%. For elements with atomic numbers <11, the RMS errors are generally greater than 4% and are, in many cases, dependent on the choice of mass absorption coefficients used in the calculations.

In addition to the Love and Scott work, Armstrong [59] compared the results from the analysis of silicates and oxides for various ZAF, Monte Carlo, and φ(ρz)

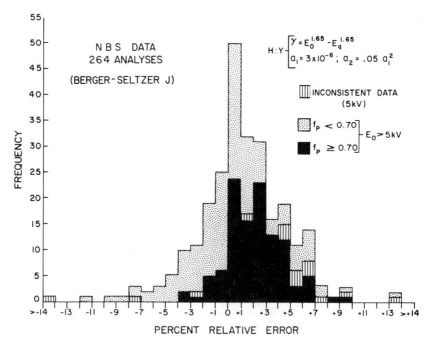

Figure 18 Relative errors for conventional electron probe analysis [3].

Table 4 % RMS Errors for $\phi\rho z$ Models

		Models			
$Z > 11$ data base	Assessor	Bastin	Bastin II	PAP	L&S
Sewell et al. [56] (554 measured)	Love and Scott	3.7	4.6	3.7	2.9
Love et al. [57] (430 measured)	Bastin	5.46	—	—	—
Bastin et al. [49] (627 measured)	Bastin		2.99		4.33
Pouchou and Pichoir	Pouchou and Pichoir			2.67	
Oxides and fluorides [56] (94 measured)	L&S Henke		10.7	10.3	4.9
	L&S Heinrich		8.1	5.8	6.2
Carbides [58] (117 measured)	L&S Henke		15.0	15.5	18.8
	L&S Heinrich		12.8	11.6	18.4

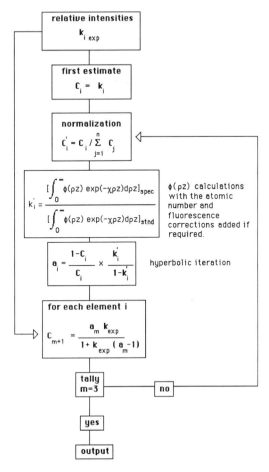

Figure 19 A $\phi(\rho z)$ analysis procedure.

methods. His results indicate that several of the methods did not provide accurate analyses on silicate and oxide minerals. In addition, he found, in general, that the $\phi(\rho z)$ methods that were optimized for oxide systems performed the best.

C. Empirical Approach to Quantitative Analysis

In 1964, Ziebold and Ogilvie [60] developed a quantitative analysis procedure for emitted x-rays based on Castaing's third approximation for generated x-rays [1]. In this empirical procedure, Ziebold and Ogilvie assume that the relationship between the k ratio and concentration, expressed in Equation (2), can be expanded to include real samples by the introduction of an efficiency factor α:

$$k_i = \frac{\alpha_i C_i}{\sum \alpha_j c_j} \tag{51}$$

Ziebold and Ogilvie expressed the relationship for a binary system in the form

$$\frac{C}{1 - C} \frac{1 - k}{k} = \alpha \tag{52}$$

which is called the hyperbolic approximation since the plots of the calibration curves described by Equation (52) are hyperbolas. Rearranging Equation (52) so that the dependent variable is expressed as either C/k or k/C results in

$$\frac{C}{k} = \alpha + (1 - \alpha)C \tag{53}$$

or

$$\frac{k}{C} = \frac{1}{\alpha} + \left(1 - \frac{1}{\alpha}\right) k \tag{54}$$

Equations (53) and (54) indicate a linear relationship between C/k and C and between k/C and k, so that a plot of C/k versus C or k/C versus k are straight lines. The slope of the line is positive if α is greater than 1 and negative if α is less than 1. Experimental confirmation of the hyperbolic approximation is shown in Figure 20, from Goldstein et al. [61], which is a plot of k/C versus k for experimental measurements on Ag $L\alpha$ x-rays from four different Ag-Au alloys at

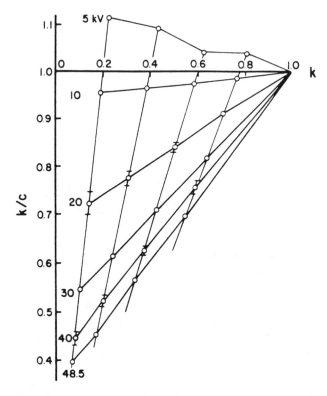

Figure 20 k/C versus k for Ag-Au alloys at various electron beam voltages from 5 to 48.5 kV [61].

several different beam energies. Except for the 5 keV line, the plots for the different alloys as predicted by the hyberbolic approximation are straight lines.

The concentrations C_i and C_j, for elements i and j in an unknown binary, can be determined from Equation (52) if the appropriate values for α_i, α_j, k_i, and k_j are known. In theory, the k values do not present a problem since they are experimentally measured. Determination of α_i and α_j, however, requires the analysis of a standard binary containing the same elements as the unknown. In the procedure for determining the α values, k_i and k_j are obtained for the standard binary by comparison to pure element standards of i and j. Since the values of C_i and C_j in the standard binary are known, it follows that α_i and α_j can be calculated from Equation (51). In the case of a binary, Equation (55) can be used to calculate the concentration for element i in the unknown binary:

$$C_i = \frac{k_i \alpha_{i,j}}{1 + k_i(\alpha_{i,j} - 1)} \tag{55}$$

where $\alpha_{i,j} = \alpha_j/\alpha_i$ and $\alpha_{j,i} = \alpha_i/\alpha_j$.

The values for α are specific for a given experimental setup, and if the analysis parameters are changed, the α values must be recalculated. As long as the analyst is concerned with measurements on a given binary system under constant conditions, then Equation (55) provides a very rapid and easy method for the determination of elemental composition with an accuracy that is comparable to that of the full ZAF procedure.

In addition to their work on binary systems, Ziebold and Ogilvie also derived the analytical expression for a ternary system [60]. The basic equation for the ternary remains [Eq. (55)], except that the combined α factor for element 1 in a ternary compound containing elements 1, 2, and 3 is given by the expression

$$\frac{\alpha_{1,2,3} = \alpha_{1,2}C_2 + \alpha_{1,3}C_3}{C_2 + C_3} \tag{56}$$

An important aspect of the ternary derivation is that the combined α factor for the ternary system in Equation (56) can be expressed as a combination of the individual α factors for the different binaries.

Bence and Albee [62], building on the work of Ziebold and Ogilvie, extended the application of the hyperbolic approximation to multielement systems containing six to eight elements. They were interested in obtaining a rapid, accurate procedure for the analysis of specimens of geological origin. In such specimens, it was not practical for time reasons to apply the full ZAF corrections, although the advancements in computers since 1964 has changed this situation considerably. The general formula for element n in a system of n components is given as

$$C_n = k_n \beta_n \tag{57}$$

where

$$\beta_n = \frac{k_1 \alpha_{n1} + k_2 \alpha_{n2} + k_3 \alpha_{n3} + \cdots + k_n \alpha_{nn}}{k_1 + k_2 + k_3 + \cdots + k_n}$$

In Equation (57), the α_{n1} factor refers to the α factor for element n in a binary of element 1 and element n. Similarly, k_1 refers to the intensity ratio of element

1 in the specimen to the element 1 in the standard. In practice, the analyst is required to measure $n(n-1)$ values for α, which requires $n(n-1)$ binary standards.

The flowchart for the empirical method of quantitative analysis is shown in Figure 21. The analyst, having determined the needed α values, measures the appropriate k values for the unknown and calculates the β term from Equation (57). The β values are then used to calculate a first approximation of the elemental concentrations. From Equation (57), a second set of β values are determined with the approximate concentrations used for the ks. The iteration loop is continued until the difference in successive β terms is below a predetermined limit.

As mentioned, the analysis of multielement systems requires $n(n-1)$ standards. Since the large number of standards for the determination of α factors in a complex system may not be available, a combination ZAF-α factor approach has been developed. In this procedure, the analyst calculates the necessary α factors by assuming an appropriate composition and then running the ZAF pro-

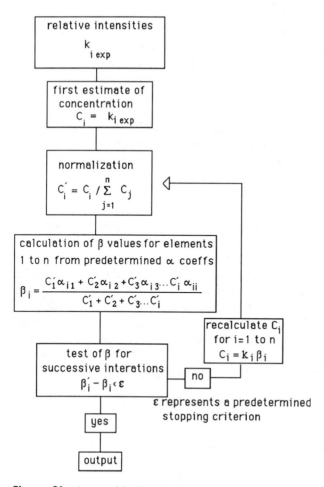

Figure 21   An empirical analysis procedure.

cedure to backcalculate the corresponding k ratios. Finally, the α factors can be determined from Equation (52). Albee and Ray [63] used this procedure to determine the α factors for 36 elements relative to simple oxides. Laguitton et al. [64] introduced additional corrections along with more elaborate determinations of the α factors. It is important that the analyst realize that the total uncertainty in the combined ZAF-α factor procedure includes the uncertainties of both the empirical and the ZAF methods.

The empirical method of analysis is most accurate when the difference between the sample and standards is mainly the result of x-ray absorption. The procedure is less accurate when there are large differences in atomic number and is least accurate when there is a significant contribution from fluorescence. Given these limitations, the empirical procedure works best for the analysis of such materials as geological specimens and oxide systems, which have relatively low mean atomic numbers and minimal secondary fluorescence.

D. Use of Quantitative Schemes with Archival Standards

In conventional electron probe analysis, the measurement of standards in conjunction with the analysis of an unknown eliminates or at least minimizes the need to correct the measured x-ray intensities for changes in electron dose and spectrometer calibration and efficiency.

Although the measurement of standards immediately before or after the measurement of an unknown provides the analyst with the most accurate results, improvements in the stability of electronics and the introduction of the Si-Li x-ray detector have made it possible to store a series of standard measurements and use these measurements over a long period of time for analysis. This procedure permits the analyst to reduce the time pr analysis since standards need only be measured infrequently. In theory, if the analyst can return the instrument to the identical conditions under which the standards were measured or correct the unknown intensities for any instrumental changes, then the standards need only be measured once. In practice, this is often not possible and the time between standard measurements is dependent on the stability of the instrument and the x-ray detector. It is also important to note that the standard measurement are applicable only for a particular accelerating voltage and beam current. Changes resulting from beam current or dead-time measurements are relatively minor since these systems are quite stable and the parameters easily measured. In addition, the corrections are linear over the entire x-ray energy range. If the standards and the unknowns are measured at two different beam voltages, then additional uncertainties are introduced as a result of the voltage conversion. Of greater importance are shifts in the efficiency of the x-ray detection systems that can be caused by changes in detector window thickness as a result of contamination or, in the case of crystal spectrometers, changes in room temperature and atmospheric pressure.

One method that can be used to minimize such effects as detector window shift or calibration shifts is to measure one reference standard spectrum at the same time as the unknown. This reference standard can then be compared, over the energy range used for the analysis, to its counterpart in the archival spectra to determine the magnitude of any shifts that have occurred as a result of long-term changes in either the electron probe or the x-ray spectrometer systems. These corrections can then be applied to the unknown spectra and the analysis performed with the archival standards. An alternative but comparable procedure would be to determine elemental ratios by normalizing the various compositions to the element measured both times. In this way, the instrumental effects cancel out for both the standards and the unknown. The analysis of samples with archival standards should approach the accuracy of the conventional method provided the analyst invests the time necessary to very accurately measure the initial standards and the conditions under which they were measured and uses this same care in measuring the spectra from the unknowns. It is also important that care be taken to ensure that sample preparation procedures, such as sample coating, be accurately controlled.

E. Standardless Analysis

A third analytical procedure for quantitative analysis is standardless analysis. In this procedure, which is used primarily for the analysis of EDS detector results, one or more x-ray intensities that have been calculated from first principles are used as the standards for the analysis of unknowns [71].

Russ [65] developed a standardless procedure in which the generated intensity for K-line emission is given by Equation (58):

$$I_g = p\omega \frac{N}{A} R \int_{E_0}^{E_c} \frac{Q}{dE/d\rho s} \, dE \tag{58}$$

where p is the transition probability, ω is the fluorescence yield, N is Avogadro's number, R is the backscatter factor, A is the atomic weight, and Q is the ionization cross section. In this procedure, the Thompson-Widdington law [66] is used to describe the rate of energy loss by the electrons and Q is the Bethe [5] cross section as modified by Green and Cosslett [67]. The integration of Equation (58) results in Equation (59) for I_g:

$$I_g = Dp\omega \frac{N_{el}R}{Ab} f(\chi)T(U_0 \ln U_0 - U_0 + 1) \tag{59}$$

where T is a term to correct for absorption of x-rays in the dead layer and window of the EDS detector and D is a constant. N_{el} is equal to 2 for K-shell ionization, 8 for L-shell, and 18 for M-shell. The fluorescence yield is determined from Equation (60) by Wentzel [68]:

$$\omega = \frac{Z^4}{a + Z^4} \tag{60}$$

where a is 10^6 for K-shell ionization, 10^8 for L, and 7.5×10^8 for M.

Nasir [69] developed a standardless procedure similar to that of Russ but substituted the Bethe energy loss law for the Thomas-Widdington energy loss term in Equation (58). I_g can then be expressed as

$$I_g = D'p\omega \frac{N}{A} C_a \frac{U_0 \ln U_0 - U_0 + 1}{\sum c_i Z_i/A_i \ln(1.166\overline{E}/J_i)} \tag{61}$$

where $\overline{E}$ is $(E_0 + E_c)/2$, and D' is a constant. Nasir, assuming that the denominator in Equation (61) is the same for all elements, proposed that the concentration ratio for elements a and b in a sample can be calculated from the equation

$$\frac{C_a}{C_b} = \frac{I_m^a}{I_m^b} \frac{p_b \omega_b A_a}{p_a \omega_a A_b} \frac{U_0^b \ln U_0^b - U_0^b + 1}{U_0^a \ln U_0^a - U_0^a + 1} \frac{f(\chi)_a}{f(\chi)_b} \frac{T_a}{T_b} \frac{F_a}{F_b} \tag{62}$$

where I_m^a and I_m^b are the measured intensities for elements a and b and F and $f(\chi)$ are the fluorescence and absorption corrections, respectively. By using elemental ratios, such unknown variables as detector solid angle, beam dose, and detector efficiency are normalized out.

The principal use of this technique is in the analysis of samples for which at least one of the required standards is not available. The analyst, in performing standardless analysis, makes the assumption that the electron beam and x-ray detector system are very accurately calibrated and that the various parameters are known and stable.

Over the past decade, the advancement in electronics used for the control and measurement of the electron dose reaching the sample has improved to a level such that the electron dose is probably not the significant factor in the uncertainty associated with standardless analysis. The main limitation is in the calibration, efficiency, and stability of the x-ray detection systems. Standardless analysis with wavelength-dispersive spectrometers is not recommended because a large number of parameters that cannot be readily controlled or measured have a pronounced effect on the x-ray intensities measured. These parameters include room temperature and pressure and energy calibration.

The introduction of solid-state Si-Li x-ray detectors has greatly improved the long-term stability of x-ray measurements and has made standardless analysis realistic, provided high accuracy is not required. Any changes in the x-ray detector parameters affect the accuracy of an analysis [70]. These changes include, among others,

1. The increase in effective window thickness due to oil and/or ice contamination
2. Microphonics in the detector due to ice accumulation in the liquid nitrogen Dewar
3. Electronic ground loops

In addition, if first-principle calculations are used for an analysis, then the detector efficiency, including the solid angle, must be known accurately.

If archival standards are used, then the analyst needs to compare the current standard to earlier standards to determine if there are any changes in the effective window thickness and/or detector resolution. An example of the accuracy for standardless analysis is shown in Table 5 (from Ref. 71), which lists the results for a stainless steel analysis by the Russ method. In general, the results are of the same magnitude as those that can be expected from a ZAF procedure on intensities measured with an EDS detector.

Table 5 Standardless Analysis of Stainless Steel NBS SRM #348

Element	Certified wt%	Calculated wt%	Absolute error (%)
Al	0.23	0.15	−0.08
Si	0.54	0.49	−0.05
Ti	2.24	2.26	+0.02
V	0.25	0.21	−0.04
Cr	14.54	14.95	+0.41
Mn	1.48	1.46	−0.02
Fe	53.30	53.66	+0.36
Ni	25.80	24.87	−0.93
Cu	0.22	0.35	−0.13
Mo	1.30	1.62	+0.32

III. QUANTITATIVE ANALYSIS OF THIN SPECIMENS

The quantitative analysis of thin films or small particles with dimensions less than about 0.5 μm represents a unique class of analysis that must be considered separately from conventional probe analysis. Unlike conventional electron-opaque samples, thin films and small particles are best analyzed in analytical electron microscopes by high-energy electron beams with accelerating potentials in excess of 80 keV. Under these circumstances, there is minimal electron backscattering and the electrons lose very little of their energy in the sample. As a result, the atomic number correction k_z can be neglected and the generated x-ray intensity from a characteristic line of element a can be given by [72]

$$I_a^* = KC_a\omega_a Q_a a_a \frac{t}{A_a} \qquad (63)$$

where

C_a = weight fraction of element a
ω_a = fluorescence yield for the analytical line of element a
a_a = fraction of total K-, L-, or M-line intensity measured
A_a = atomic weight of element a
Q_a = ionization cross section of the analytical line of element a
t = film thickness
K = constant

In addition to a simplified generation expression, if the thin film or particle can be approximated as an infinitely thin film, the effects of x-ray absorption and fluorescence can be neglected. This condition is often referred to as the thin-film approximation and makes it possible to express the relationship between the measured and the generated x-ray intensities for element a by

$$I_a = I_a^* \epsilon_a \qquad (64)$$

where ϵ_a is an efficiency factor related to the overall efficiency of the Si(Li) detector for the detection of x-rays from element a. Since the determination of sample thickness at each analytical point is impractical and the value of ϵ_a is not constant, analysis schemes for thin films involve the measurement of elemental ratios in which the relative concentration of one element to another can be expressed by the equation

$$\frac{C_a}{C_b} = k_{ab} \frac{I_a}{I_b} \tag{65}$$

The sensitivity factor approach to analysis is common to many analytical techniques, and the factor k_{ab} in Equation (65) is referred to as the Cliff-Lorimer factor or k_{ab} factor. It is related to the generation and efficiency terms for element b in proportion to those for element a [73]. In many analytical schemes, the analyst determines, from binary standards, a set of k_{ab} factors. Then, for a binary system containing unknown concentrations of elements a and b C_a and C_b can be determined from Equation (64) and the knowledge that $C_a + C_b = 1$. In ternary and higher order systems, the relative concentrations for the various elements can be determined, if the k_{ab} factors are known, as combinations of various binaries as shown in the following equations for a ternary system:

$$\frac{C_a}{C_b} = k_{ab} \frac{I_a}{I_b} \tag{66}$$

$$\frac{C_c}{C_b} = k_{cb} \frac{I_c}{I_b} \tag{67}$$

$$C_a + C_b + C_c = 1 \tag{68}$$

In analytical electron microscopy, particularly in analyses related to geological specimens, the convention is to express the k_{ab} factors relative to silicon: $k_{a\mathrm{Si}}$. In this format they are referred to as simply k factors and are related to the k_{ab} factors as

$$k_{ab} = \frac{K_{a,\mathrm{Si}}}{K_{b,\mathrm{Si}}} = \frac{K_a}{K_b} \tag{69}$$

Figure 22 shows the $k_{a\mathrm{Si}}$ values from the work of Wood et al. [74] and Schreiber and Wims [75].

In addition to the k factors relative to silicon, several researchers, particularly those involved in metallurgy, have found it useful to report the k factors relative to iron. Table 6 lists the $k_{a\mathrm{Fe}}$ for the K and L lines of several elements [72].

It is also possible to calculate k factors from Equation (63). A detailed discussion of the equations used to calculate the various terms in Equation (63) is given in Reference 72. Figures 23 and 24 from Reference 72 show a comparison of the calculated and measured k factors for K- and L-line radiation from various elements.

Recently, Sheridan determined a large set of experimental $k_{a\mathrm{Si}}$ factors for a wide range of $K\alpha$, $L\alpha$, and $M\alpha$ lines [81]. The measurements were performed on a series of submicrometer particles ground from NBS multielement research glasses. After applying an absorption correction, the experimental results were compared to theoretical $k_{a\mathrm{Si}}$ factors determined with several different ionization cross sections.

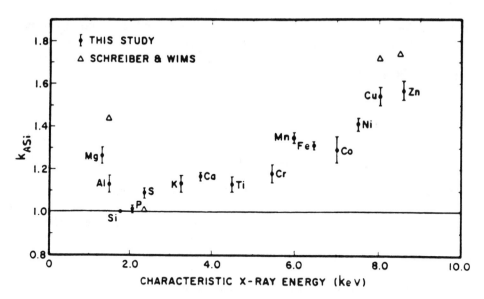

Figure 22 Measured $k_{a\text{Si}}$ factors for thin-film analysis [72].

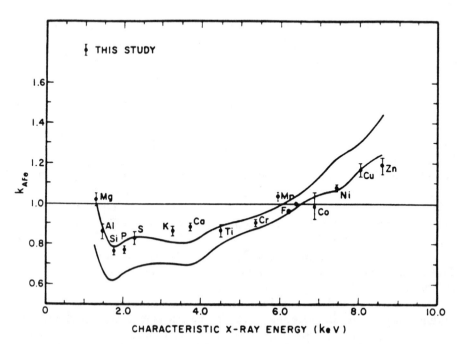

Figure 23 Calculated and measured k factors for K-line radiation from several elemental thin films [72].

Table 6 Experimental $k_{a\text{Fe}}$ Factors for Various Element

Element	Lorimer et al. [76,77] (100 keV)	McGill and Hubbard [78] (100 keV)	Wood et al. [79] (120 keV)
K Lines			
Na	2.46		
Mg	1.23 ± 0.08	1.16	0.96 ± 0.03
Al	0.92 ± 0.08	0.8	0.86 ± 0.04
Si	0.76 ± 0.08	0.71	0.76 ± 0.004
P			0.77 ± 0.005
S			0.83 ± 0.03
K	0.79	0.77	0.86 ± 0.014
Ca	0.81 ± 0.05	0.75	0.88 ± 0.005
Ti	0.86 ± 0.05		0.86 ± 0.03
Cr	0.91 ± 0.05		0.90 ± 0.006
Mn	0.95 ± 0.05		1.04 ± 0.025
Fe	1.0		1.0
Co	1.05		0.98 ± 0.06
Ni	1.14 ± 0.05		1.07 ± 0.06
Cu	1.23 ± 0.05		1.17 ± 0.03
Zn	1.24		1.19 ± 0.04
Nb			2.14 ± 0.06
Mo	3.38		3.80 ± 0.09
Ag	6.65		9.52 ± 0.03
L lines	Wood et al. [79] (120 keV)		Goldstein et al. [80] (100 keV)
Sr[a]	1.21 ± 0.06		—
Zr[a]	1.35 ± 0.01		—
Nb[a]	0.90 ± 0.06		—
Ag[a]	1.18 ± 0.06		1.04
Sn	2.21 ± 0.07		2.39
Ba	—		2.18
W	—		2.43
Au	3.10 ± 0.09		3.27
Pb	—		4.14

[a] *k* factors for these elements are for combined *L*α and *L*β lines.

As the thickness of the sample increases, the electron transparency of the film or particle decreases, eventually reaching a thickness for which the sample no longer conforms to the "thin-film criteria." Under these circumstances, corrections for x-ray absorption and fluorescence must be included in the analysis scheme. According to Goldstein et al. [80], the limits for the failure of the thin-film criterion are an x-ray absorption >3% and/or fluorescence >5%. For these

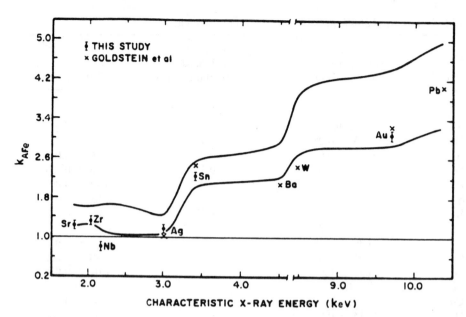

Figure 24 Calculated and measured k factors for L-line radiation from several elemental thin films [72].

samples, Equation (62) can be expanded to include absorption and fluorescence:

$$\frac{C_a}{C_b} = k_{ab} \frac{I_a}{I_b} F_{abs} \frac{1}{1 + (F_{fl})} \tag{70}$$

where F_{abs} is the absorption factor given in Equation (71) and F_{fl} is the ratio of the fluoresced to primary intensity I_a/I_0 given in Equation (72). The absorption factor was derived by Goldstein from the work of Tixier and Philibert [82] and Konig [83]:

$$F_{abs} = \left[\frac{(\mu/\rho)^a_{sam}}{(\mu/\rho)^b_{sam}} \right] \left(\frac{1 - e^{-\chi^a(\rho t)}}{1 - e^{-\chi^b(\rho t)}} \right) \tag{71}$$

$$\frac{I_a}{I_0} = C_b \omega_b \frac{r_a - 1}{r_a} \frac{1}{A_b} \frac{A_a}{A_b} \left(\frac{\mu}{\rho} \right)^b_a \frac{[E_c]^a}{[E_c]^b} \frac{\ln(U)^b}{\ln(U)^a} \frac{\rho t}{2} \tag{72}$$

$$\times \left[0.932 - \ln \left(\frac{\mu}{\rho} \right)^b_{sam} \rho t \right] \sec \psi$$

where r_a is the absorption jump ratio for element a and A_a and A_b are the atomic numbers for a and b.

The equations for absorption and fluorescence involve the measurement of the film or particle thickness and the calculation of the mass absorption coefficients for elements a and b in the specimen. As a result, it is necessary to measure the thickness at each analysis location and to calculate the concentrations in an iteration loop similar to that used for conventional ZAF schemes.

IV. QUANTITATIVE ANALYSIS OF PARTICLES AND ROUGH SURFACES

In classical electron probe analysis schemes employing either a ZAF or Bence-Albee approach, both sample and standard must be infinitely thick with respect to the penetration of the electron beam and have flat surfaces. By controlling sample dimensions and shape, the corrections for the interaction of the electron beam with the sample and the subsequent x-ray emission can be calculated from simple geometrical relationships. In the quantitative analysis of particles and samples with rough surfaces, the shape and thickness of the specimen cannot be controlled or in many cases measured. The difficulties in quantitative analysis of particles or rough surfaces result from three different effects that influence the generation and measurement of x-rays from these samples [84].

The first effect is the result of the finite size (mass) of the sample. The mass effect is related to the elastic scattering of the electrons and is strongly affected by the average atomic number of the sample. The mass effect is important when the sample thickness or particle size is smaller than the range of the primary electron beam so that a fraction of the beam escapes the sample before exciting x-rays. This is shown in Figure 25, where the majority of the primary electron trajectories terminate within the boundries of the larger sphere but very few terminate within the boundries of the smaller sphere. As the size of the sample decreases, the mass of material from which x-rays are excited drops and results in a reduction in x-ray intensities from the specimen compared to a bulk sample of the identical composition. The mass effect can be demonstrated by comparing the x-ray emissions from a bulk target to the emissions from a particle of the same composition. This effect can be seen in Figure 26, which represents the Ba $L\alpha$ x-ray intensity from particles normalized to the intensity from a bulk material of the same composition plotted versus particle diameter. The energy of the Ba x-rays is 4.47 keV, which is high enough that the absorption effects are minimal. The mass effect is demonstrated by the decrease in the intensity measured on the

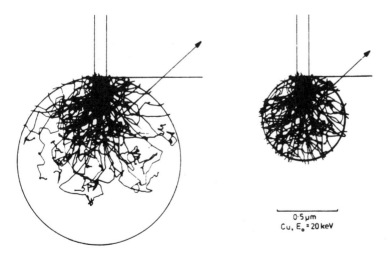

Figure 25   Effect of particle mass on the intensity of measured x-rays [84].

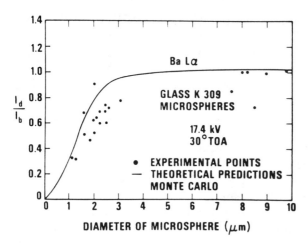

Figure 26 Normalized Ba $L\alpha$ x-ray intensity plotted as function of particle diameter [84].

particles compared to the bulk for particles less than 3 μm in diameter. The net result of not correcting for the mass effect in the analysis of particles less than about 3 μm in size is an underestimation of the composition for all elements analyzed in the sample.

The second effect that must be corrected for is the absorption effect. In most analyses of specimens with nonstandard configurations, the x-ray emergence angle and therefore the absorption pathlength cannot be as predicted accurately as for polished specimens. The magnitude of this effect is largest when there is high absorption, as is typically the case for soft x-rays from such elements as Al or Si that have energies less than 2 keV. The difference between the absorption pathlength in a sample with a nonstandard shape and that in a bulk flat sample can result in widely different values of emitted x-ray intensities. For the particle shown in Figure 27, the pathlength *A-B* in the particle is less than the pathlength *A-C* in the polished sample. The pathlength *A*-D* in the particle, however, is greater than the pathlength *A-C* in the polished sample. The result of the varying absorption pathlengths is shown in Figure 28, which shows the Si and Al *K*α x-ray intensities from particles normalized to the intensity from a bulk material of the same composition plotted versus particle diameter. For these particles, the

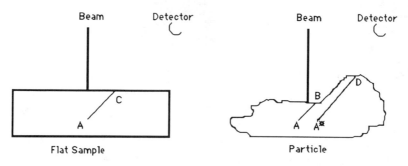

Figure 27 X-ray absorption in a particle compared to absorption in a bulk material [84].

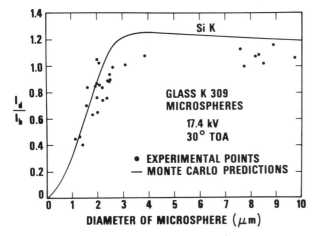

Figure 28 Normalized Si $K\alpha$ x-ray intensity plotted as a function of particle diameter [84].

absorption pathlengths in the particles are less than the bulk material, resulting in a higher emitted x-ray intensity from the particles compared to bulk. This effect is detectable for the lower energy x-ray lines, such as that of aluminum at 1.49 keV, which are highly absorbed.

The third effect is caused by the fluorescence of x-rays by either the continuum or other characteristic x-rays. Because x-ray absorption coefficients in solids are relatively small compared to electron attenuation, secondary x-ray fluorescence occurs over a much larger volume than primary electron excitation. In bulk samples and standards, the x-rays for the most part remain in the specimens. In the case of particles, however, the particle volume may be only a small fraction of the x-ray excitation volume. As a result, the exciting x-rays excite relatively few x-rays before leaving the particle. In those samples in which the fluorescence is important (see section on fluorescence), the effect of comparing a particle to a bulk standard may be significant. This effect is shown in Figure 29, which is a plot of the range for Ni $K\alpha$ x-rays causing fluorescence of Fe $K\alpha$ x-rays in a Ni-Fe alloy.

The net effect of not correcting for the secondary excitation of x-rays in particle analysis is as follows:

1. An underestimation of the concentration for elements that have a significant contribution to their characteristic line or lines from excitation by other characteristic x-rays
2. In the case of continuum fluorescence, an underestimation of the concentrations for all elements, particularly those with higher energy lines, that are excited by the higher energy, longer range continuum

A. Normalization

One of the simplest methods for the quantitative analysis of particles is to normalize to 100% the concentrations, determined with bulk standards, from a con-

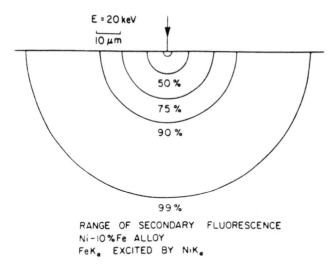

E = 20 keV

10 μm

50 %

75 %

90 %

99 %

RANGE OF SECONDARY FLUORESCENCE
Ni-10%Fe ALLOY
FeKₐ EXCITED BY NiKₐ

Figure 29   Range of secondary x-ray fluorescence in a Ni-Fe alloy [84].

ventional procedure like ZAF [85]. The analyst, in selecting this method of correction, makes the assumption that the x-ray absorption and fluorescence are the same for the particle as for a bulk specimen and that the mass effect is the same for all elements. In practice, normalization of results is most effective for the correction of the mass effect since the decrease in intensity as a function of particle size is nearly the same for all elements. Figure 30 shows that the different elemental curves merge together for particle diameters less than 2 μm. Since this procedure does not accurately compensate for absorption and fluorescence effects, the most accurate results are obtained on particle systems that meet the following conditions:

1. Systems for which all the analytical lines for the elements are above 4 keV, where the absorption effects are minimal.
2. If any of the analytical lines are below about 4 keV in energy, then the lines for all the elements should be as close together in energy as possible so that the matrix absorption is approximately the same in all cases.
3. Systems for which there is no significant fluorescence.

Table 7 lists the results from the analysis of lead silicate glass particles. The first set of results is taken from the analysis of the Pb $M\alpha$ line at 2.3 keV, which is close in energy to the Si $K\alpha$ line at 1.74 keV. Since the absorption and mass corrections are similar for these two lines, the lead and silicon concentrations are in good agreement with the true values. The second set of results was determined by analyzing the Pb $L\alpha$ line at 10.6 keV. In this case, the two analytical lines have very different energies and the particle absorption effect is not similar in magnitude. As expected, the errors associated with this analysis are considerably higher than those associated with the Pb $M\alpha$ analysis.

One of the major disadvantages of the normalization of results from a bulk analysis procedure is that the analyst cannot determine, by obtaining an analysis

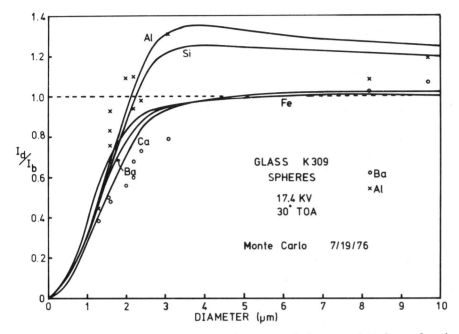

Figure 30 Normalized x-ray intensities from several elements plotted as a function of particle diameter [84].

Table 7 Analysis of Lead Silicate Glass K-229 by Normalization of ZAF Results[a]

Analysis	Si wt%	% error	Pb wt%	% error
Analysis done with Pb $M\alpha$ line (meets conditions)				
1	0.155	+10.7	0.620	−4.6
2	0.144	+2.9	0.643	−1.0
3	0.136	−2.7	0.658	+1.2
4	0.138	−1.1	0.653	+0.5
5	0.127	−9.0	0.675	+3.8
6	0.170	+22	0.588	−9.5
7	0.137	−2.5	0.657	+1.1
Analysis done with Pb $L\alpha$ line (does not meet conditions)				
1	0.134	−4.5	0.663	+1.9
2	0.177	+26.3	0.578	−11
3	0.159	+14	0.612	−5.8
4	0.166	+18	0.602	−7.4
5	0.017	+88	0.894	+37
6	0.100	−29	0.731	+12.4
7	0.157	−12.3	0.616	−5.2

[a] Nominal composition: Si = 0.140; Pb = 0.650.

total of less than 100%, the presence of any unanalyzed elements, such as those with atomic numbers less than 9.

B. Particle Standards

The analyst can use a conventional analysis scheme and substitute particle standards for the normal polished standards [86]. In this procedure, the assumption is that the particle effects, particularly the absorption effect, will be approximately the same for the sample and standard. This assumption is reasonably valid provided the sample and standard are close in composition and shape and the particle diameter is above about 2 μm. Below 2 μm, as shown in Figure 30, any difference in size and shape between unknown and standards is critical since a small change in effective diameter results in a large change in x-ray intensity.

C. Geometrical Modeling of Particle Shape

This method for the analysis of particles was developed by Armstrong and Buseck [87,88]. It is based on the determination of a simple geometrical shape or combination of shapes, such as square or pyramid, which define the boundaries of the particle of interest. The various particle effects are then calculated for the chosen geometrical shapes defining the shape of the particle. The particle mass effect (loss of primary x-ray production due to electron transmission, sidescatter, and backscatter) and the absorption effect are based on Equation (73). The emitted primary x-ray intensity for element A, corrected for mass and absorption, I_a can be written as

$$I_a = a_a \iiint \phi_a e^{-\mu_a \rho g} \, dx \, dy \, dz \tag{73}$$

where a_a is a term reflecting various constants, such as detector efficiency, that are not dependent on particle geometry, ϕ_a is the probability of x-ray production, μ_a is the mass absorption coefficient, g is the absorption pathlength in the particle for the x-ray, and x, y, and z are the coordinates for x-ray production in a given shape.

From Equation (73), the procedure backcalculates the x-ray production from the particle to an appropriate value for an infinitely thick sample and for which all primary electrons remain in the sample and it has the same composition as the unknown. The various mechanisms responsible for x-ray loss are corrected as follows:

- Electron transmission. The amount of primary radiation lost as a result of electron transmission through the particle and sidescatter is calculated from a modified form of an expression developed by Reuter [18] to calculate relative x-ray production in thin films. In the modified form, the expression can be used to calculate x-ray production as a function of position within a particle.
- Electron backscatter. The expression of Duncumb and Reed [24] is used to calculate the loss of x-rays as a result of electron backscatter.
- Electron sidescatter. The loss of primary x-rays from the sidescatter of elec-

trons is minimized by rastering or defocusing the electron beam over the entire particle area, by using an overvoltage less than 1.5, and by expressing the concentrations as a ratio.

X-ray absorption. The correction for x-ray absorption is done by setting the integration limits for the absorption pathlength to the limits determined for the geometrical shape or shapes that define the overall particle. The shapes that have been included in the program include rectangular prism, tetragonal prism, right triangular prism, square pyramid, hemisphere, and sphere.

The x-ray intensity from the secondary fluorescence of characteristic x-rays is calculated from Armstrong and Buseck [89], with the integration limits adjusted to calculate the probability of secondary fluorescence at a given location within a particle [Eq. (67)].

$$I_{fa} = \sum_i \sum_j b_{ij} \iiint\!\!\!\iiint \phi_{ij} \tan \theta\, e^{-\mu_a \rho g} e^{\mu_{ij}\rho(\sec \theta)s}\, ds\, d\theta\, d\xi\, dx\, dy\, dz \qquad (74)$$

where s, θ, and ξ are cylindrical coordinates defining the travel of the exciting x-ray to the point of excitation (see Fig. 31), b_{ij} is the collection of constants and terms that are not dependent on particle size or shape, and $\mu_{i,j}$ is the mass absorption coefficient for the absorption of j-line x-rays of element i in the matrix. The boundaries that are determined for the various geometrical shapes defining the particle are used to calculate the absorption pathlengths for the x-rays. Results from this procedure are given in Table 8 for anorthite oxide, rhodonite oxide, and pyrite [88]. In all cases, the analyses are in good agreement with the known composition and have standard deviations less than $\pm 6\%$ relative.

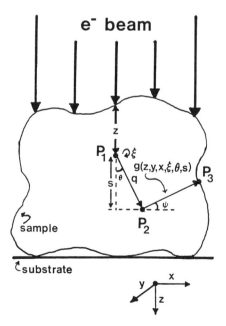

Figure 31 Armstrong's secondary fluorescence correction for particles [89].

Table 8 Results from the Analysis of Mineral Particles by the Geometrical Modeling Method

	Actual composition (wt%)	Determined (wt%)
Anorthite oxide[a]		
CaO	18.9	19.0 ± 0.8
Al₂O3	35.7	35.6 ± 1.2
SiO_2	44.3	44.4 ± 1.0
Rhodonite oxide[b]		
MnO	35.5	35.5 ± 1.1
FeO	13.0	13.1 ± 0.8
CaO	4.1	4.0 ± 0.2
SiO_2	46.9	46.8 ± 0.8
Pyrite element[c]		
FeS	46.6	46.4 ± 1.0
S	53.4	53.6 ± 1.0

[a] Average of 122 analyses.
[b] Average of 100 analyses.
[c] Average of 140 analyses.

D. Peak-to-Background Ratios

A fourth method for the quantitative analysis of particles was developed by Small et al. [90] and Statham and Pawley [91]. This method, derived from work on biological specimens by Hall [92], is based on the following observation: to a first approximation, the ratio of a characteristic x-ray intensity to the continuum intensity of the same energy for a flat, infinitely thick target is equivalent to the ratio from a particle or rough surface of the same composition.

In the form of an equation, this observation can be expressed as

$$\left(\frac{I}{B}\right)_{particle} = \left(\frac{I}{B}\right)_{bulk} \tag{75}$$

where I is the background-corrected peak intensity and B is the continuum intensity for the same energy window as the peak. It is assumed that the spatial distribution for characteristic x-ray excitation is identical to the distribution for continuum x-ray excitation. As a result, the effects of particle shape and size on measured x-ray intensity are the same for the continuum and the characteristic x-rays. It therefore follows that by taking the ratio of the two intensities, the particle mass and absorption effects cancel.

In the procedure developed by Small et al. [93], Equation (75) is rearranged and the peak intensities for the particle or rough surface are scaled up to values similar to a bulk material of the identical composition:

$$I^*_{particle} = I_{bulk} = \frac{I_{particle}B_{bulk}}{B_{particle}} \tag{76}$$

The values of I^* for each element in the unknown can then be used as input for a standard quantitative analysis scheme. In practice, a bulk material does not exist for a particle of unknown composition and the value for B_{bulk} at any given characteristic peak energy must be estimated as part of the analysis scheme. In the procedure, B_{bulk} is determined from Equation (77) and for each characteristic peak energy must be summed over all elements in the unknown.

$$B_{bulk} = \sum_i \frac{F(\overline{Z}_{bulk})}{F(\overline{Z}_{std})} B_{std} \frac{f(\chi)_{bulk}}{f(\chi)_{std}} \tag{77}$$

The $F(\overline{Z})$ terms describe the continuum intensity as a function of the concentration-weighted average atomic number for the hypothetical bulk material and standards. For the simplest case, Kramers' relationship [94], the first part of Equation (77) would be $\overline{Z}_{bulk}/\overline{Z}_{std}$, where the current estimation of concentration from the iteration loop in the ZAF procedure is used for the calculation of $\overline{Z}_{bulk}$. B_{std} is the measured continuum intensity for a given standard at the energy of the analyzed x-ray. For multielement standards, B_{std} must be multiplied by the weight fraction of the element of interest. Once the values for B_{bulk} are determined, they can be used in Equation (76) to obtain the first estimates for I^*. A set of k ratios can then be calculated and used as input to the ZAF routine. The set of concentrations from each iteration is used to calculate new values of B_{bulk}, and the sequence is repeated until successive iterations are within a predetermined limit. The block diagram showing the interfacing of Equation (76) with the ZAF procedure FRAME C [41] is shown in Figure 32.

Various types of mineral particles have been analyzed with the peak-to-background method, FRAME P [95]. The results of these analyses are reported in Table 9 along with the result from the conventional ZAF routine, FRAME C. In all cases, the analyses with the peak-to-background routine are within 10% and usually

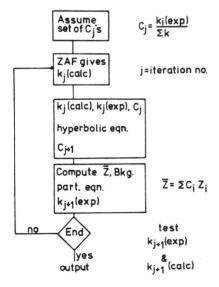

Figure 32 The FRAME P particle analysis procedure [84].

Table 9 Analysis of Mineral Particles by the Peak-to-Background (P-B) Method[a]

	Talc, $Mg_3[Si_4O_{10}](OH)_2$				FeS_2				ZnS_2			
	Wt% Mg	% Error	Wt% Si	% Error	Wt% S	% Error	Wt% Fe	% Error	Wt% S	% Error	Wt% Fe	% Error
Nominal Normalized	19.3		29.8		53.4		46.6		32.9		67.1	
FRAME C P-B	19.4	+0.5	29.7	−0.3	52.7	−1.3	47.3	+1.5	35.4	+7.8	64.5	−3.8
FRAME P	18.5	−4.0	29.7	−0.3	52.9	−0.9	46.4	−0.4	36.0	+9.0	67.7	−0.7

[a] Results are the average from seven analyses.

Table 10 f_{ab} Values for K-961 Microspheres

Diameter (μm)	Mg	Al	Si	P	K	Ca	Ti	Mn	Fe
2	1	1.24	1.74	1.72	2.43	3.12	3.01	3.77	3.06
3	1	1.36	1.68	1.57	2.44	3.46	3.13	2.56	2.90
4	1	1.30	1.66	1.95	2.63	3.36	3.16	3.29	3.21
6	1	1.48	1.83	1.43	2.96	3.56	3.14	3.81	3.06
9	1	1.27	1.51	1.74	2.72	3.17	3.06	3.28	3.33
Average	1	1.33	1.69	1.68	2.64	3.33	3.10	3.34	3.11
σ, %	0	6.8	7.1	12	8.3	5.7	2.2	15	5.1

better than 5% relative error of the stoichiometric values. In contrast, the errors with the conventional ZAF routine range from 7.9% for S in ZnS to 47% for Mg and Si in talc. In addition, the standard deviations for individual measurements are less for the peak-to-background routine than they are for the conventional ZAF routine.

In this case, f_{ab} is a correction factor that should have a minimum dependence on particle size and can be calculated or determined empirically from standards. Table 10 lists the f_{ab} values for five particles of K-961 glass, the concentration of which is listed in Table 11. These results show that the f_{ab} values do not exhibit any noticeable trend with particle size, and the standard deviation of the f_{ab} values is less than 8% relative for elements with a concentration greater than 1 wt%. For the elements Mn and P, with concentrations less than 1%, the standard deviation in f_{ab} is small considering the relatively poor counting statistics.

In the analysis of particles, the assumption that the generation volumes for characteristic and continuum x-rays are identical is valid only for particles larger than about 2 μm in diameter. Below this size, the anisotropic generation of the

Table 11 Nominal Composition of Glass K-961

Element	Wt%
Na[a]	2.97
Mg	3.02
Al	5.82
Si	29.9
P	0.22
K	2.49
Ca	3.57
Ti	1.20
Mn	0.32
Fe	3.50

[a] Na was not included in the results because of poor statistics and high ion mobility.

Table 12 FRAME P Analysis of Fracture Surfaces

	Actual wt%	FRAME C wt% Normalized	Relative error	FRAME P wt%	Relative error
SRM 482 60% Au-40% Cu					
Location 1					
Au	60.3	49.6	−18	58.0	−4
Cu	39.6	50.4	27	44.0	11
Location 2					
Au		29.1	−52	52.0	−14
Cu		70.8	79	41.0	3.5
SRM 482 80% Au-20% Cu					
Location 1					
Au	80.1	73.8	−8	76.9	−4
Cu	19.8	26.2	32	19.1	−3.5
Location 2					
Au		69.3	−13	76.7	−4.2
Cu		30.6	55	20.1	1.5
SRM 483 Fe-3.22% Si					
Location 1					
Fe	96.8	97.0	0.2	100	3.3
Si	3.22	2.9	−8.2	3.2	−0.3
Location 2					
Fe		96.4	−0.4	97.7	0.9
Si		3.6	11	3.5	7.4

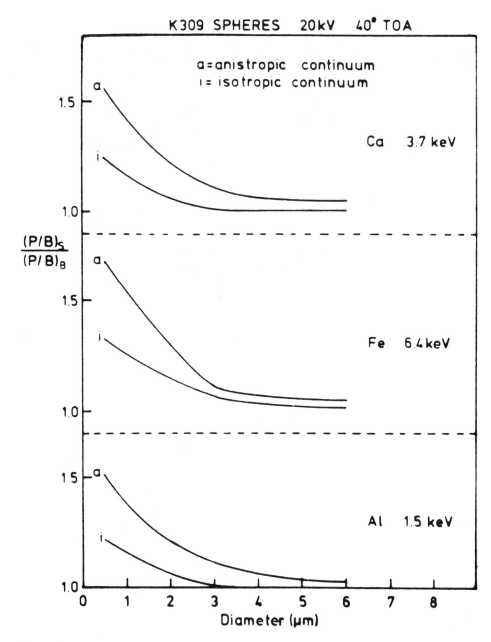

Figure 33 Peak-to-background ratios from glass spheres normalized to bulk ratios showing the effects of isotropic and anisotropic cross sections for continuum generation [84].

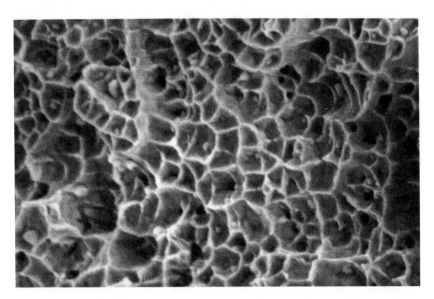

Figure 34 Electron micrograph of a Au-Cu fracture surface [94].

continuum results in a significantly different excitation volume for the continuum compared to the isotropically generated characteristic x-rays. This effect is shown in Figure 33, where isotropic and anisotropic cross sections have been used to calculate peak-to-background ratios from K-309 glass particles normalized to bulk glass. The composition of K-309 is 7.9% Al, 18.7% Si, 10.7% Ca, 13.4% Ba, and 10.5% Fe. These plots show that the introduction of an anisotropic cross section for the continuum results in significantly higher peak-to-background ratios for the smaller particles. As a result, it is necessary to introduce a correction for aniso-tropic generation of the continuum for quantitative analysis with peak-to-back-ground procedures [83].

Results from the FRAME P analysis of rough surfaces show a similar improve-ment in accuracy with the peak-to-background method compared to conventional ZAF. Table 12 lists the results of analyses on fracture surfaces of various Au-Cu alloys and a Fe and 3.22% Si alloy, all of which are certified NBS Standard Reference Material Microprobe Standards. A micrograph of one of the fracture surfaces is shown in Figure 34. The use of the peak-to-background method for the analysis of rough surfaces leads to significant improvement in accuracy and precision.

In the peak-to-background method developed by Statham and Pawley [91], the peak-to-background ratio from a given element is compared to the ratio from a second element:

$$\frac{C_a}{C_b} = f_{ab} \frac{(P/B)_a}{(P/B)_b} \tag{78}$$

V. SPATIALLY RESOLVED X-RAY ANALYSIS

A. *x-y* Mapping

In spatially resolved x-ray analysis, the position of the electron beam on the sample is coupled to the output of an x-ray spectrometer. The x-ray spectrometer in turn is coupled to the output of the display on a cathode ray tube (CRT), photographic plate, or computer memory such that when the beam interrogates a point on the sample, the output from the x-ray spectrometer is displayed or stored at the corresponding point on the storage medium. In this way, multiple analyses can be taken on a sample and the spatial relationship between each analysis location is retained. The analyst can then manipulate the stored data and construct an image of the sample based on its elemental composition.

One of the earliest methods of obtaining spatially resolved analysis was to deflect the analog scan of an electron beam in a single direction, usually *x*, across the specimen surface and simultaneously record and display the output from a wavelength-dispersive (WD) x-ray spectrometer on a CRT. In this way, a line profile for the intensity of a chosen element can be overlain on an electron or target current image of the sample. An example is shown in Figure 35 [96].

The first method of x-ray analysis to display elemental information in both the *x* and *y* directions was the x-ray dot map or x-ray area scan [97]. This procedure, like the line scan, initially made use of the analog scan of the electron

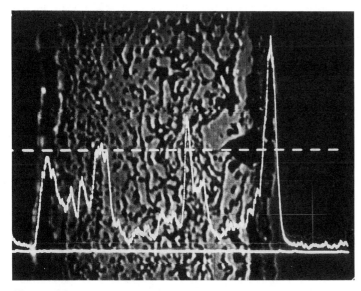

Figure 35 Line profile of S *K*α x-ray intensity overlain on a target current image of a specimen from a failed pressure vessel.

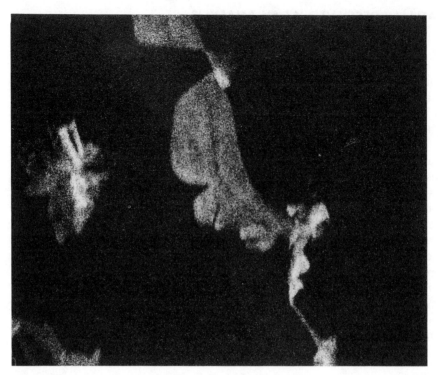

Figure 36   X-ray dot map of a grain boundary migration in a Zn-Cu system [98].

beam. As the beam was scanned slowly over the sample, the x-ray intensity of a given element as determined by a spectrometer was compared to a preselected threshold value. If the x-ray counts were above the threshold value, then a signal is sent to the recording CRT on the instrument, which records a full-intensity spot at that location. The entire sample area selected by the instrument's magnification is scanned in this manner, and the resulting x-ray dot map is constructed. Figure 36 shows the result of such an analysis of grain boundary migration of Zn in a Zn-Cu system. In this procedure, only the presence or absence of a given element, as determined by the preselected threshold value, can be displayed. Minimal information is provided that reflects the amount or concentration of that element present. In addition to displaying black and white maps, the analyst could produce color overlay of up to three different elements by separately photographing the dot maps using red, green, and blue filters.

More recent instruments, which are interfaced with computer systems, make it possible for the analyst to store the various elemental dot maps in the computer memory or on peripheral storage devices, such as floppy disks. The images can then be retrieved at a later date and displayed in various colors on an appropriate monitor.

In the past few years, the computer interfacing of instruments has advanced beyond the utilization of the computer primarily as a data collection and mass storage device. In the newer instruments, the computer is fully integrated, providing digital control of the electron beam or sample stage. In an instrument with digital beam control, the electron beam is deflected such that it interrogates the sample in a two-dimensional array consisting of $n \times n$ spot analyses, where n is usually 16–256. In an instrument with stage control, the electron beam is static and the stage is stepped in the array pattern. Each beam or stage position in the two-dimensional array corresponds to a picture element, pixel, on a map of the specimen. By storing WDS or EDS spectra at each position, the analyst can obtain a full quantitative analysis at each pixel. Once the results from the analyses are stored in the array, the variations in elemental compositions can be coupled to the gray scale or color output of a computer monitor. The completed array of pixels provides the analyst with a spatially resolved map of the elemental composition across the specimen. In addition, since the array is stored in the computer, the analytical information can be processed in several ways. For example, a row of pixels can be selected, providing the analysts with elemental x-ray profiles across the sample in any direction or pattern desired. This information can be displayed as a simple lien plot or can be used to construct contour maps of a given x-ray distribution.

The uncertainty associated with quantitative analysis obtained for compositional mapping is essentially identical to the uncertainty associated with conventional analysis obtained at the same electron dose. For maps in which high accuracy is required on minor constituents, the dwell time per pixel and or beam current must be sufficiently large to provide the appropriate x-ray intensity in the peak of interest. As a result, compositional maps under these circumstances may require several hours to accumulate and care must be taken to ensure instrument stability over the required time interval.

As previously mentioned, both EDS and WDS analysis systems can be used for compositional mapping. WDS detection of x-rays, however, is superior to EDS detection for compositional mapping, particularly of samples requiring the analysis of constituents at the minor and trace levels [99]. One of the most important advantages of WDS detection is that the pulse-processing time for WDS is at least an order of magnitude shorter than that for EDS. WDS detectors have a limiting count rate of about 10^5 cps compared with a limiting count rate of 10^4 for the EDS detectors. In addition, the WDS systems use diffraction crystals so that only a narrow energy band of x-rays is detected at any given time. The energy is adjustable so that most of these counts are the characteristic counts from the element of interest. This count rate compares to a limiting rate of about 10^2–10^3 cps in the peak of interest on an EDS detector since the limiting count rate of 10^4 is distributed over the entire energy range of the spectrum.

In addition to its higher limiting count rate, the WDS detector has a factor of 10 higher peak-to-background ratio than the EDS detector. Taking into account the higher limiting count rate and peak-to-background ratio of the WDS detector, the WDS detector has a lower detection limit of about 100 ppm compared to a lower limit of 1% or greater for mapping on the EDS system.

One limitation of the WDS systems compared to EDS systems is in the mapping of samples that contain a large number of elements of interest. In WDS systems, the number of elements mapped during a given digital scan is limited to the number of WDS spectrometers. Typically in EDS systems, all elements with characteristic x-ray energies above about 1 keV can be detected simultaneously.

Two important aspects that must be considered in the interpretation of compositional maps are the counting statistics and resulting uncertainties associated with the various point analyses at each pixel location. Under normal mapping conditions, it is impractical to obtain the level of accuracy associated with conventional probe analysis, where 100 s or more is used for data accumulation.

Marinenko et al. showed that for a 128×128 map at a magnification below $\times 500$, each pixel represents an area of about 2 μm in diameter on the specimen surface [99]. This means that the spot size of the beam can be of the order of 1 μm in diameter and carry a current of about 1 μA. At this current density, the limiting count rate of 10^5 cps can be obtained on pure element samples. Assuming a dwell time per pixel of 0.1 s, a 128×128 map would require 1600 s to accumulate and have 10^4 counts per pixel. From Poisson statistics, the 1σ counting uncertainty per pixel is 1%.

When multielement standards are required, the dwell time per pixel must be increased to obtain the same counting statistics since each element is present at less than 100%. In general, dwell times of about 0.4 s are sufficient for multielement standards and require less than 2 h per 128×128 map. As in conventional analyses, the standard maps can be archived and used indefinitely, provided the instrument conditions are constant.

In the analysis of samples in which one or more of the constituents are at the minor or trace level (1% or less by weight), digital maps may require 10 h or more to accumulate. A 10 h scan on a 128×128 map translates to 2 s dwell time per pixel. If the element of interest is at the 1% level by weight, then the x-ray intensity in the peak would be 2000 counts, assuming a limiting count rate of 10^5 cps as stated earlier. The associate counting uncertainty is 2.2% for 1σ. The actual uncertainty is larger than this because of fluctuations in the intensity of the x-ray background that must be taken into account for quantitative analysis. In general, it should be possible to obtain 2σ confidence intervals of 10% or better for constituents at the 1% weight concentration in an unknown sample.

The power of compositional mapping can be seen in Figure 37, which is a compositional map of the same Zn grain boundary migration shown in Figure 36. The brighter areas on the map correspond to a Zn concentration of 10% by weight. The darker labeled area corresponds to a Zn concentration of 0.2% by weight.

One of the major difficulties in obtaining quantitative x-ray maps with crystal spectrometers is defocusing of the diffraction crystal as the electron beam is moved off the crystal axis. The magnitude of the defocusing effect is proportional to the distance of the beam from a point centered on the optical focus of the electron probe. The defocusing is most severe at the low magnifications of $\times 200$–800 routinely used in digital x-ray maps. Corrections for the defocusing fall into four categories [99].

The first method to avoid spectrometer defocusing is to employ, as mentioned earlier, the digital movement in the microscope stage rather than the electron beam. In this situation, the electron beam position remains static at the optical focus of the diffraction crystal and the sample is moved in a raster pattern under the beam. For WDS mapping, it is critical that the sample remain at the focus of the optical microscope during the x-y movement of the stage. This procedure has been used successfully by Mayr and Angeli [100]. The success of this method, since it involves the mechanical movement of the stage, is dependent on the

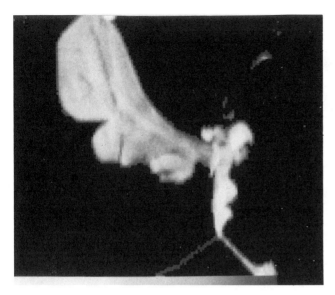

Figure 37 Composition x-ray map of the same grain boundary area shown in Figure 36 [98].

reproducibility of the stage motion. If the reproducibility of the stage movement is poor, then the accuracy of the quantitative maps is poor except for extremely low magnifications. In more modern instruments, which have optical encoding and a guaranteed stage positioning of 0.1 μm, the results from moving the stage are comparable to those with other correction methods.

The second method used to correct for spectrometer defocusing is to move or rock the diffraction crystal in synchronization with the raster of the electron beam [101]. By slightly rocking the crystal, the entire sample area under study is maintained at the focus of the x-ray crystal. The problems associated with crystal rocking are similar to those involving stage motion. Since beam rocking requires mechanical movement of the crystal, the results are dependent on the reproducibility of the crystal-rocking mechanism. This procedure was described for multiple spectrometers by Swyt and Fiori in [102].

The third correction method is to collect a series of standard maps in conjunction with the collection of a map from an unknown. If the standard and the unknown are collected under the same experimental conditions, then the standard maps can be used to construct k ratios at each pixel location, thus normalizing the effects of spectrometer defocusing. Although this method does not require mechanical movement, it does require that the instrument remain stable over the period of time necessary to collect all the maps from the standards and unknown. This is particularly important since a pixel-by-pixel comparison of the unknown to the standard is made, which requires a constant crystal orientation with respect to the beam raster. Unless great care is taken during analysis, the use of archival standard maps is not recommended because of the possibility of introducing large errors as a result of spectrometer drift. This procedure was used by Marinenko et al. [99] for the analysis of the Au-Ag standard reference materials with the

Figure 38 Compositional maps from Au-Ag alloys showing the effects of spectrometer defocusing [98]. (Upper left) pure Ag; (upper right) Au20-Ag80; (lower left) Au60-Ag40; and (lower right) Ag80-Ag20. The analytical line is the Ag $L\alpha$.

results shown in Figures 38 and 39. Figure 38 presents the uncorrected maps from the alloys and the pure silver standard, which show the characteristic banding from the defocusing of the spectrometers. Figure 39 contains the pure silver map and the subsequent alloy maps, which have been corrected for defocusing by normalization to the standard map. The normalized maps show very little gra-

Figure 39 Compositional maps, corresponding to those in Figure 39, of the Au-Ag alloys corrected for defocusing by normalization to a standard map [98].

dation or structure in the image gray levels, indicating that the artifacts from defocusing have been removed. Table 13 contains a comparison of the average composition obtained from the digital map and the results from a conventional point analysis. In all cases, the average composition from each map compares quite well with the conventional point analysis as well as with the NIST-certified values.

The fourth method is to model the spectrometer defocusing. Marinenko et al. [103] observed that the defocusing artifacts, that is, banding, were equivalent to an intensity profile from a wavelength scan across the elemental peak of interest. This is shown in Figure 40, which compares a wavelength line scan across the Cr $K\alpha_{1,2}$ peaks to a line profile taken from a Cr map perpendicular to the defocusing bands. The similarity of these two traces implies that the intensity can be measured in the center band of a map, and from a model of the line profile, the intensity at any other pixel location can be calculated. This procedure eliminates the need for measuring a standard map and greatly reduces the time required for obtaining a quantitative x-ray map.

Figure 41 is a schematic showing the relationship between the x-rays emitted from a defocused point on the sample and the corresponding angular deviation from the line of focus for a vertical crystal spectrometer. The electron probe is designed such that the maximum x-ray intensity obtained on a crystal corresponds to point A in Figure 42, which is centered on the electron optical axis with the aid of an optical microscope. The distance between point A and the crystal S_0 is

$$S_0 = 2R \sin \theta \tag{79}$$

where R is the spectrometer radius and θ is the Bragg angle. At a defocused point, point B, which is a distance ΔS from the optical axis, the angular deviation from the exact Bragg angle is $\Delta\theta$. Given that $\Delta\theta$ is small, less than about 0.01 rad, which is the case for most mapping applications, it can be approximated as

$$\Delta\theta = \sin \Delta\theta = \frac{\Delta s'}{S_0} \tag{80}$$

Table 13 Quantitative Analysis of Gold-Silver Alloys, SRM 481, Concentrations in W%

Alloy	Element line	Certified value	Point beam analysis[a]	Digital mapping
Au20-Ag80	Ag L	77.58	77.33 ($\pm$ 0.4%)[b]	77.63
	Au M	22.43	21.93 ($\pm$ 2.6%)	20.81
	Au L	22.43	22.69 ($\pm$ 2.3%)	22.94
Au60-Ag40	Ag L	39.92	39.44 ($\pm$ 0.5%)	39.96
	Au M	60.05	59.59 ($\pm$ 2.4%)	57.59
	Au L	60.05	61.26 ($\pm$ 2.3%)	60.87
Au80-Ag20	Ag L	19.93	19.59 ($\pm$ 0.6%)	19.72
	Au M	80.05	80.56 ($\pm$ 2.5%)	77.26
	Au L	80.05	81.09 ($\pm$ 2.0%)	80.30

[a] Excitation potential = 20 kV, Faraday cup current = 38 nA, point beam, five randomly selected samplings averaged for each alloy.
[b] One relative standard deviation of a single measurement is in parentheses.

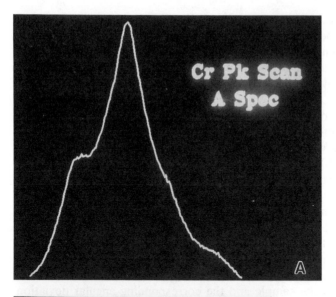

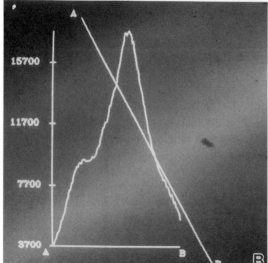

Figure 40 Comparison of a WDS line scan across the Cr $K\alpha$ peak to a line profile taken from a Cr map perpendicular to the defocusing bands of the spectrometer [102]. (A) Line scan across the Cr $K\alpha$ peak with a LiF crystal. (B) Cr $K\alpha$ intensity profile, from the compositional map, taken along a series of pixels perpendicular to the maximum intensity band.

where $\Delta S' = \Delta S \sin \Psi$ for a vertical spectrometer and Ψ is the x-ray takeoff angle for the electron probe.

In calculating the background value for a given pixel in the compositional map, it is necessary to determine ΔS, which is the distance of the pixel from the maximum focus line of the spectrometer. The orientation and equation of the line

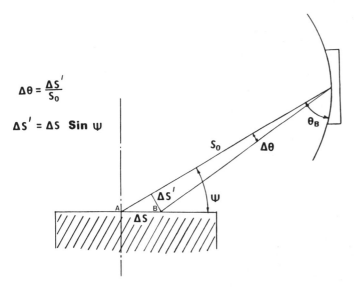

$$\Delta\theta = \frac{\Delta s'}{S_0}$$

$$\Delta s' = \Delta s\ \sin\psi$$

Figure 41 Relationship between the x-rays emitted from a defocused point on the sample and the angular deviation from the line of focus in a vertical crystal spectrometer [102].

of maximum focus for a given spectrometer can be defined by the coordinates of two pixel points, (x_1, y_1) and (x_2, y_2), which lie on the line. The orientation of the line can be determined from any raw intensity map by increasing the image threshold, as shown in Figure 42. The line has the general form, $A * x + B * y + C = 0$, and it follows that ΔS for any pixel point, represented by coordinates (x_3, y_3) in the x-ray map, can calculated from Equation (81):

$$\Delta s = \frac{-D * x_3 + Y_3 - E}{\pm(D^2 + 1)^{1/2}} \tag{81}$$

where $D = (y_2 - y_1)/(x_2 - x_1)$ and $E = y_1 - D * x_1$.

ΔS is positive for points located above the line of focus and negative for points below the line of focus. ΔS as defined in Equation (81) is defined in units of pixel elements, which must be converted to centimeters to obtain the linear distance for the calculation of Δs. For this purpose, the scaling factor F is used, which is defined in terms of the magnification M, the linear dimension of the CRT display L, and the number of pixel points in the matrix N:

$$F = \frac{L}{M * N} \tag{82}$$

$\Delta\theta$ can then be expressed as

$$\Delta\theta = \Delta s' * \frac{F}{S_0} \tag{83}$$

The accuracy of this procedure is demonstrated in Figure 43. Figure 43A is an experimentally measured intensity profile taken perpendicular to the focal axis,

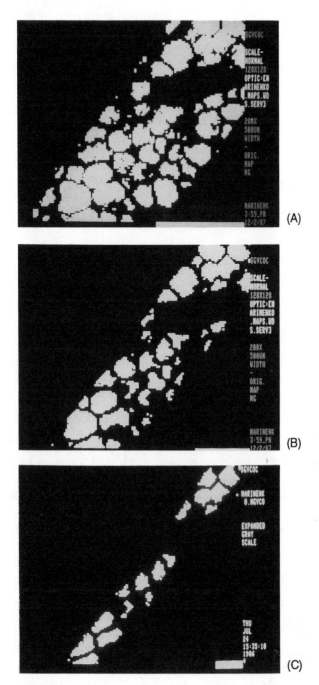

Figure 42 Line of maximum x-ray intensity on a crystal spectrometer determined by increasing the image threshold. The threshold is low in the top image and high in the bottom image [102].

Figure 43 Comparison of experimental and modeled x-ray intensity profiles for Ti $K\alpha$ x-rays [102]. (A) Experimental x-ray intensity profile for Ti $K\alpha$ overlain on the corresponding Ti map. Magnification $\times 448$, $E_0 = 20$ keV. (B) Modeled intensity profile. (C) Profiles from A and B shown together.

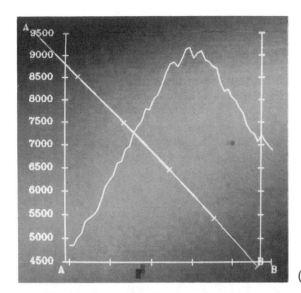

(A)

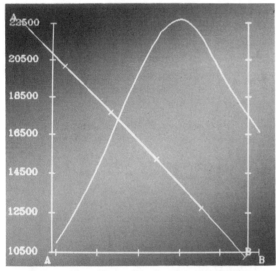

(B)

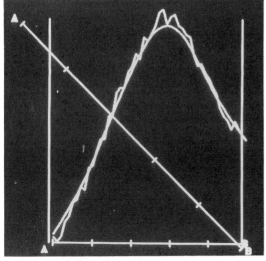

(C)

and Figure 43B is the modeled profile along the same axis. The intensity profiles shown together in Figure 43C are identical within the limitations of the experiment.

Background correction procedures 3 and 4 are magnification limited. At magnifications of about ×150 or less, the deviation in intensity from the edge to the center of a map is about a factor of 2 and cannot be accurately corrected by these methods. For very low magnification maps for which quantitative results are important, correction procedures 1 and 2 are preferred.

B. Composition-Composition Histograms

The final aspect of electron probe analysis considered here is composition-composition histograms. Compositional maps provide the analyst with a visual method of interpreting the results from the elemental analysis of some 4000 individual points on a sample [104–106]. The analyst, by associating color and intensity, is able to relate spatially the various elements and their compositions within the analyzed region of the sample. One problem in the interpretation of compositional maps is that it is often difficult to visualize the compositional ranges and resulting interelement correlations from a color composite image, particularly for minor or trace constituents. An alternative method of displaying the analytical information is in the form of a composition-composition histogram (CCH) [106]. The CCH provides the analyst with an image that can be used to interpret the numerical relationship between the various components in the sample.

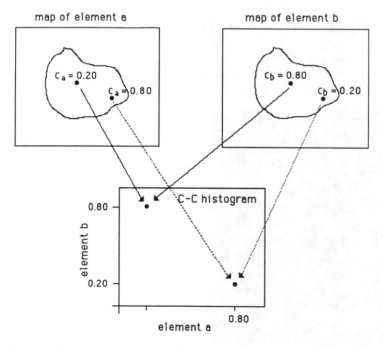

Figure 44 The basis of a concentration-concentration histogram (CCH).

Figure 44 is a schematic diagram describing the CCH. The concentration of element a at each pixel in the compositional map of a is associated with the corresponding pixel and concentration of element b in the map of b. This is represented by the top two blocks in Figure 44. The associated concentrations for a and b at each pixel in the compositional image are then plotted in the CCH, bottom block, as a single point in a scatter diagram. Associations between elements are visible in the CCH as features, such as lines or areas, that have a

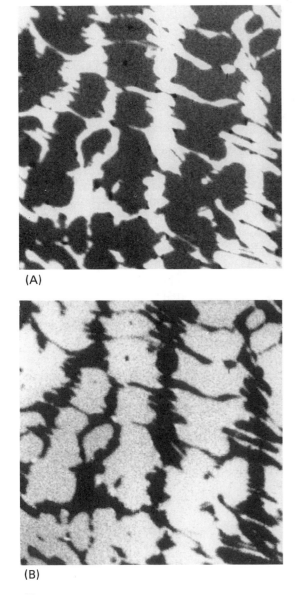

(A)

(B)

Figure 45 Compositional maps for an aligned Cu-Ti eutectic [105]. (A) Cu compositional map. (B) Ti compositional map.

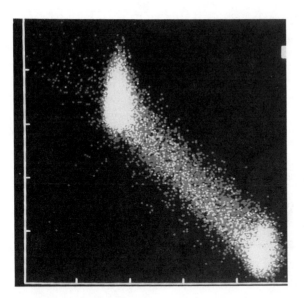

Figure 46 CCH for the Cu-Ti aligned eutectic shown in Figure 46 [105].

detectable density of points above background. In addition, the overlapping of multiple image pixels at one CCH point can be coupled to the intensity of the recording CRT and displayed in an appropriate gray level or color scale.

An example of the type of information conveyed by the CCH is shown in Figures 45 and 46, which are the elemental and CCH for an aligned Cu-Ti eutectic alloy. The CCH (Fig. 47) shows two lobes that have a high density of pixels and correspond to the two different phases of the alloy [106].

In addition to the two lobes in Figure 46, there is a detectable region of points between the lobes that has a lower pixel density compared to the density in the lobes. This lower density region with 1–2 pixels per point forms the linear strip connecting the two main lobes.

A useful feature of the CCH is the ability to relate a given region or feature on the CCH to the corresponding pixels in the compositional map. The outline of the high-Cu lobe of the CCH is shown in Figure 47A, along with the corresponding pixels highlighted as the bright areas on the compositional map (Fig. 47B). Figures 48 and 49 are the corresponding sets for the Ti lobe and the connecting strip. This display method allows the analyst to readily determine that the connecting line between the two lobes corresponds to the boundary regions between the Ti and Cu phases.

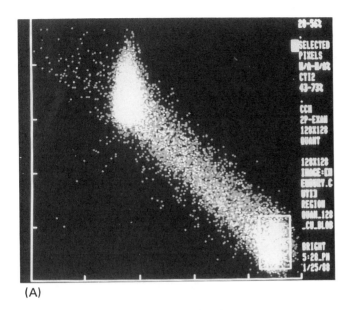

(A)

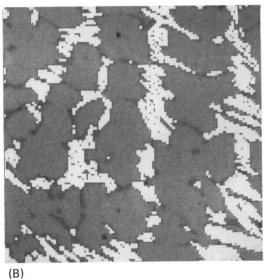

(B)

Figure 47 CCH and corresponding map areas for the high-Cu region of the eutectic [105]. (A) Outline of the high-Cu region of the CCH. (B) Pixels in the compositional map that correspond to the high-Cu region.

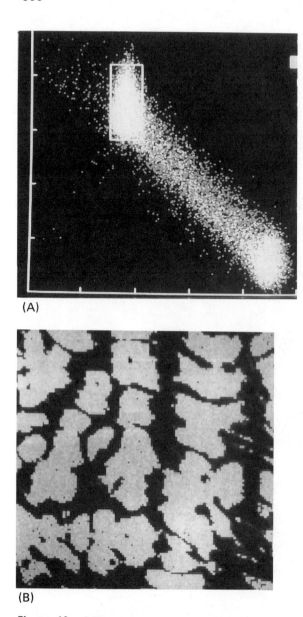

(A)

(B)

Figure 48   CCH and corresponding map areas for the high-Ti region of the eutectic [105].
(A) Outline of the high-Ti region of the CCH. (B) Pixels in the compositional map that
correspond to the high-Ti region.

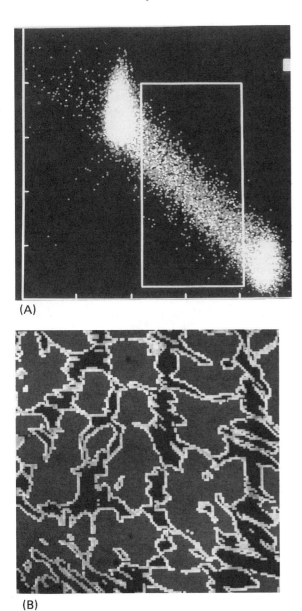

(A)

(B)

Figure 49 CCH and corresponding map areas for the region of the eutectic containing both Cu and Ti [105]. (A) Outline of the connecting pixels between the high-Cu and high-Ti lobes. (B) Pixels in the compositional map that correspond to the connecting strip.

REFERENCES

1. R. Castaing, Ph. D. Thesis, University of Paris, 1951.
2. J. I. Goldstein, H. Yakowitz, D. E. Newbury, E. Lifshin, J. W. Colby, and J. R. Coleman, *Practical Scanning Electron Microscopy*, Plenum Press, New York, 1975.
3. K. F. J. Heinrich, *Electron Beam X-ray Microanalysis*, Van Nostrand-Reinhold, New York, 1981.
4. J. I. Goldstein, D. E. Newbury, P. Echlin, D. C. Joy, C. E. Fiori, and E. Lifshin, *Scanning Electron Microscopy and X-ray Microanalysis*, Plenum Press, New York, 1981, pp. 315–322.
5. H. Bethe, *Ann. Phys. (Leipzig)* 5:325 (1930).
6. H. A. Bethe and J. Ashkin, *Experimental Nuclear Physics*, Wiley, New York, 1953, p. 252.
7. F. Bloch, *Z. Phys. 81*:363 (1933).
8. R. R. Wilson, *Phys. Rev. 60*:749 (1941).
9. M. J. Berger and S. M. Seltzer, in *NRC Publication 1133*, National Academy of Sciences, Washington, D.C., 1964, p. 205.
10. G. Springer, *Neues. Jahrb Fuer Mineral. Monatsh. 9/10*:304 (1967).
11. K. F. J. Heinrich and H. Yakowitz, *Mikrochim. Acta 1*:123 (1970).
12. P. Duncumb, P. K. Shields-Mason, and C. DaCasta, in *5th Int. Cong. on X-Ray Optics and Microanal.*, Springer-Verlag, Berlin, 1969, p. 146.
13. C. Zeller, cited by J. Ruste and M. Gantois, *J. Phys. D. Appl. Phys. 8*:872 (1975).
14. P. Duncumb, and S. J. B. Reed, in *Quantitative Electron-Probe Microanalysis*, NBS Special Pub. 298, Depart. of Commerce, 1968, p. 133.
15. P. M. Thomas, *Br. J. Appl. Phys. 14*:397 (1963).
16. J. Philibert and R. Tixier, in *Quantitative Electron-Probe Microanalysis*, NBS Special Pub. 298, Dept. of Commerce, 1968, p. 53.
17. G. Love, M. G. Cox, and V. D. Scott, *J. Phys. D., Appl. Phys. 11*:7 (1978).
18. W. Reuter, in *6th Int. Conf. X-ray Optics and Microanal.*, Univ. Tokyo Press, Japan, 1972, p. 121.
19. K. F. J. Heinrich, in *4th Int. Cong. on X-ray Optics and Microanal.*, Hermann, Paris, 1966, p. 1509.
20. M. Green, in *3rd Int. Cong. on X-ray Optics and Microanal.*, Academic, New York, 1963, p. 361.
21. G. Springer, *Mikrochim. Acta 3*:587 (1966).
22. H. Bishop, in *4th Int. Cong. on X-ray Optics and Microanal.*, Hermann, Paris, 1966, p. 153.
23. J. D. Derian and R. Castaing, in *4th Int. Cong. on X-ray Optics and Microanal.*, Hermann, Paris, 1966, p. 193.
24. P. Duncumb, in *Electron Beam X-ray Microanalysis*, Van Nostrand-Reinhold, New York, 1981, p. 249.
25. H. Yakowitz, R. L. Myklebust, and K. F. J. Heinrich, *FRAME: An On-Line Correction Procedure for Quantitative Electron Probe Microanalysis*, NBS Tech. Note 796, Dept. of Commerce, Washington, D.C., 1973.
26. R. L. Myklebust, *J. Phys. 45(Suppl. 2)*:C2-41 (1984).
27. R. L. Myklebust, D. E. Newbury, and H. Yakowitz, in *Use of Monte Carlo Calculations in Electron Probe Microanalysis and Scanning Electron Microscopy*, NBS Special Publ. 460, 1975, p. 105.
28. R. L. Myklebust, in *Electron Probe Quantitation*, K. F. J. Heinrich et al (eds.), Plenum Press, New York, 1991, pp. 177–191.
29. R. L. Myklebust and D. E. Newbury, in *Microbeam Analysis—1988*, D. E. Newbury (ed.), San Francisco Press, San Francisco, CA, 1988, pp. 261–262.

30. J. I. Goldstein, D. E. Newbury, P. Echlin, D. C. Joy, C. E. Fiori, and E. Lifshin, *Scanning Electron Microscopy and X-ray Microanalysis*, Plenum Press, New York, 1981, p. 321.
31. J. Philibert, in *3rd Int. Cong. on X-ray Optics and Microanal.*, Academic, New York, 1963, p. 379.
32. P. Duncumb and P. K. Shields, *The Electron Microprobe*, John Wiley, New York, 1966, p. 284.
33. K. F. J. Heinrich, NBS Technical Note 521, Dept. of Commerce, Washington, D.C., 1969.
34. M. Green, Ph.D. Thesis, Cambridge University, 1962.
35. J. I. Goldstein, D. E. Newbury, P. Echlin, D. C. Joy, C. E. Fiori, and E. Lifshin, *Scanning Electron Microscopy and X-ray Microanalysis*, Plenum Press, New York, 1981, p. 312.
36. K. F. J. Heinrich, H. Yakowitz, and D. L. Vieth, in *Proc. 7th Natl. Conf. Electron Probe Analysis Soc.*, San Francisco, CA, 1972, paper 3.
37. S. J. B. Reed, *Br. J. Appl. Phys. 16*:913 (1965).
38. J. I. Goldstein, D. E. Newbury, P. Echlin, D. C. Joy, C. E. Fiori, and E. Lifshin, *Scanning Electron Microscopy and X-ray Microanalysis*, Plenum Press, New York, 1981, pp. 323–325.
39. K. F. J. Heinrich, in *Microbeam Analysis—1987*, D. Joy (ed.), San Francisco Press, San Francisco, CA, 1987, p. 24.
40. J. Henoc, in *Quantitative Electron-Probe Microanalysis*, NBS Special Pub. 298, Dept. of Commerce, 1968, p. 197.
41. R. L. Myklebust, C. E. Fiori, and K. F. J. Heinrich, NBS Technical Note 1106, 1979.
42. K. F. J. Heinrich, *Electron Beam X-ray Microanalysis*, Van Nostrand-Reinhold, New York, 1981, p. 331.
43. V. D. Scott and G. Love, X-ray absorption correction, in *Quantitative Electron-Probe Microanalysis*, John Wiley, Chichester, 1983, pp. 175–176.
44. G. Love and V. D. Scott, in *Microbeam Analysis—1988*, D. E. Newbury (ed.), San Francisco Press, San Francisco, CA, 1988, p. 247.
45. K. F. J. Heinrich, D. E. Newbury, and R. L. Myklebust, *Microbeam Analysis—1988*, 1988, pp. 273–276.
46. J. T. Armstrong, in *Microbeam Analysis—1988*, San Francisco Press, San Francisco, CA, 1988, 239–246.
47. R. H. Packwood and J. D. Brown, *X-ray Spectrom. 10*:138–146 (1981).
48. G. F. Bastin, F. J. J. Van Loo, and H. J. M. Heijligers, *X-ray Spectrom. 13*:91–97 (1984).
49. G. F. Bastin, H. J. M. Heijligers, and F. J. J. Van Loo, *Scanning 8*:45–67 (1986).
50. J. T. Armstrong, *Microbeam Analysis—1982*, K. F. J. Heinrich (ed.), San Francisco Press, San Francisco, CA, 1982, pp. 175–177.
51. J. L. Pouchou and F. Pichoir, *J. Phys. C2(Suppl. 2)*:17–20 (1984).
52. J. L. Pouchou and F. Pichoir, in *Microbeam Analysis—1988*, San Francisco Press, San Francisco, CA, 1988, pp. 315–318.
53. G. Love, D. A. Sewell, and A. D. Scott, *J. Phys.* C2(Suppl. 2):21–24 (1984).
54. V. D. Scott and G. Love, in *Quantitative Electron Probe Analysis*, NIST Special Publ., Dept. of Commerce, Washington, D.C., (in press).
55. J. I. Goldstein, D. E. Newbury, P. Echlin, D. C. Joy, C. E. Fiori, and E. Lifshin, *Scanning Electron Microscopy and X-ray Microanalysis*, Plenum Press, New York, 1981, p. 330.
56. D. A. Sewell, G. Love, and V. D. Scott, *J. Phys. D18*:1245–1267 (1985).
57. G. Love, M. G. C. Cox, and V. D. Scott, *J. Phys. D9*:7–13 (1976).

58. G. F. Bastin and H. J. M. Heijligers, *Quantitative Electron Probe Microanalysis of Carbon in Binary Carbides*, Technical Report, University of Technology, Eindhoven, 1985.

59. J. T. Armstrong, in *Microbeam Analysis—1988*, San Francisco Press, San Francisco, CA, 1988, pp. 239–246.

60. T. O. Ziebold and R. E. Ogilvie, *Anal. Chem. 36*:322 (1964).

61. J. I. Goldstein, D. E. Newbury, P. Echlin, D. C. Joy, C. E. Fiori, and E. Lifshin, *Scanning Electron Microscopy and X-ray Microanalysis*, Plenum Press, New York, 1981, p. 333.

62. A. E. Bence and A. Albee, *J. Geol. 76*:382 (1968).

63. A. L. Albee and L. Ray, *Anal. Chem. 42*:1408 (1970).

64. D. Laguitton, R. Rousseau, and F. Claisse, *Anal. Chem. 47*:2174 (1975).

65. J. C. Russ, in *Proc. 9th Ann. Conf. Microbeam Analysis Soc.*, Ottawa, Ontario, 1974, paper 22.

66. R. Widdington, *Proc. R. Soc. A86*:360 (1912).

67. M. Green and V. E. Cosslett, *Proc. Phys. Soc. 78*:1206 (1961).

68. G. Wentzel, *Z. Phys. 43*:524 (1927).

69. M. J. Nasir, *J. Microsc. 108*:79 (1976).

70. C. E. Fiori and D. E. Newbury, in *Scanning Electron Microsc.*, O. Johani (ed.), Scanning Electron Microscopy, Inc., Chicago, IL, 1978, pp. 401–421.

71. V. P. Scott and G. Love, in *Quantitative Eletron-Probe Microanalysis*, John Wiley, Chichester, 1983, pp. 254–258.

72. J. I. Goldstein, D. B. Williams, and G. Cliff, in *Principles of Analytical Electron Microscopy*, Plenum Press, New York, 1986, p. 161.

73. G. Cliff and G. W. Lorimer, *J. Microsc. 103*:203 (1975).

74. J. E. Wood, D. B. Williams, and J. I. Goldstein, *J. Micros. 133*:255 (1984).

75. T. P. Schreiber, and A. M. Wims, in *Microbeam Analysis—1981*, R. H. Geiss (ed.), San Francisco Press, San Francisco, CA, 1981, p. 317.

76. G. W. Lorimer, S. A. Al-Salman, and G. Cliff, in *Inst. Phys. Conf. Ser. No. 36*, Inst. of Physics, Bristoll and London, 1977, p. 369.

77. G. W. Lorimer, G. Cliff, and J. N. Clark, in *Developments in Electron Microscopy and Analysis*, Academic Press, London, 1976, p. 153.

78. R. L. McGill and F. H. Hubbard, in *Quantitative Microanalysis with High Spatial Resolution*, Metals Soc., London, Book 277, 1981, p. 30.

79. J. E. Wood, B. B. Williams, and J. I. Goldstein, in *Quantitative Microanalysis with High Spatial Resolution*, Metals Soc., London, Book 277, 1981, p. 24.

80. J. I. Goldstein, J. L. Costley, G. W. Lorimer, and S. J. B. Reed, in *Scanning Electron Microsc. I*, O. Johani (ed.), Scanning Electron Microscopy, Inc., Chicago, IL, 1977, p. 315.

81. P. J. Sheridan, *J. Elec. Microprobe Tech. 11*:41 (1989).

82. R. Tixier and J. Philibert, in *Proc. 5th Int. Cong. X-ray Optics and Microanal.*, Springer-Verlag, Berlin, 1969, p. 180.

83. R. Konig, in *Electron Microscopy in Mineralogy*, Springer-Verlag, Berlin, 1976, p. 526.

84. J. A. Small, in *Scanning Electron Microsc. I*, O. Johani (ed.), Scanning Electron Microscopy, Inc., Chicago, IL, 1981, pp. 447–461.

85. F. W. Wright, P. W. Hodge, and C. G. Langway, *J. Geophys. Res. 68*:5575–5587 (1963).

86. E. W. White, P. J. Denny, and S. M. Irving, in *The Electron Microprobe*, John Wiley and Sons, New York, 1966, p. 791–804.

87. J. T. Armstrong and P. R. Buseck, *Anal. Chem. 47*:2178–2192 (1975).

88. J. T. Armstrong, *Scanning Electron Microsc. I*:455–467 (1978).

89. J. T. Armstrong and P. R. Buseck, in *Proc. 12th Ann. Conf. Microbeam Anal. Soc.*, Boston, MA, 1977, pp. 42A–42F.

90. J. A. Small, K. F. J. Heinrich, C. E. Fiori, R. L. Myklebust, D. E. Newbury, and M. F. Dilmore, *Scanning Electron Microsc. II*:807–816 (1979).

91. P. J. Statham and J. B. Pawley, in *Scanning Electron Microsc. I*, O. Johani (ed.), Scanning Electron Microscopy, Inc., Chicago, IL, 1978, pp. 469–478.

92. T. A. Hall, in *Quantitative Electron-Probe Microanalysis*, NBS Special Pub. 298, Dept. of Commerce, 1968, pp. 269–299.

93. J. A. Small, K. F. J. Heinrich, D. E. Newbury, and R. L. Myklebust, in *Scanning Electron Microsc. II*, O. Johani (ed.), Scanning Electron Microscopy, Inc., Chicago, IL, 1979, pp. 807–816.

94. H. A. Kramers, *Philos. Mag. 46*:836 (1923).

95. J. A. Small, D. E. Newbury, and R. L. Myklebust, in *Microbeam Analysis—1979*, P. E. Newbury (ed.), San Francisco Press, San Francisco, CA, 1979, pp. 243–246.

96. K. F. J. Heinrich, *Electron Beam X-ray Microanalysis*, Van Nostrand-Reinhold, New York, 1981, p. 540.

97. V. E. Cosslett and P. Duncumb, *Nature 177*:1172 (1956).

98. H. Yakowitz and K. F. J. Heinrich, *J. Res. Nat. Bur. Stds. A73*:113 (1969).

99. R. B. Marinenko, R. L. Myklebust, D .S. Bright, and D. E. Newbury, *J. Microsc. 145(2)*:207–223 (1987).

100. M. Mayr and A. Angeli, *X-ray Spectrom. 14*:89–98 (1985).

101. K. F. J. Heinrich, *Electron Beam X-ray Microanalysis*, Van Nostrand-Reinhold, New York, 1981, pp. 521, 542.

102. C. R. Swit and C. E. Fiori, in *Microbeam Analysis—1986*, A. Romig and W. Chambers (eds.), San Francisco Press, San Francisco, CA, 1986, p. 482.

103. R. B. Marinenko, R. L. Myklebust, D. S. Bright, and D. E. Newbury, *J. Micros. 155(2)*:183–198 (1988).

104. M. Prutton, M. M. El Gomati, and C .G. Walker, *Anal. Elec. Microsc.* San Francisco Press, San Francisco, CA, 1987, pp. 304–310.

105. R. Browning, *Anal. Elec. Microsc.*, San Francisco Press, San Francisco, CA, 1987, pp. 311–316.

106. D. S. Bright, D. E. Newbury, and R. B. Marinenko, in *Microbeam Analysis—1988*, D. E. Newbury (ed.), San Francisco Press, San Francisco, CA, 1988, pp. 18–24.

13

Sample Preparation for XRF

Jasna Injuk and René E. Van Grieken   *University of Antwerp, Antwerp, Belgium*

I. INTRODUCTION

Technological advances in x-ray fluorescence (XRF) instrumentation, for both the wavelength-dispersive and energy-dispersive modes, have given the spectroscopist the means to accommodate virtually all types of sample materials. Higher degrees of analytical accuracy and precision and lower detection limits can be achieved. Refinements in sample presentation methods have been made, and in many instances, new systems and accessory sample preparation equipment have been introduced. Sample preparation is frequently more time consuming than the actual analysis step, which has been shortened dramatically. This is particularly true with powdered solid sample materials, as a result of the need to change the physical state of specimens, for example for reduced particle size differences and inhomogeneities to insignificant levels. Solution samples, in most cases, simply require transfer to an appropriate device for containment and presentation to the instrument, at least when the analytes are not present at trace levels. Metal alloys are often sufficiently homogenous to be measured in their natural, compact state; a flat polished surface is sufficient.

The quality of sample preparation in XRF is as important as the quality of the experimental techniques and strongly influences the final result of the quantitative analysis. As a rule, the literature should always be consulted before starting any analytical method development, specimen preparation should be avoided as much as possible or kept to a strict minimum, and simple physical sample preparation methods, like drying, freeze-drying, homogenizing, pulverizing, and cutting of thin sections, should be considered if feasible. Any sample preparation method further requires contamination control, concern about losses of analyte elements (particularly halogens), and always simultaneously, blank specimen preparation. An adequately prepared specimen must be representative of the material, homogeneous, with a flat and smooth surface, and, when possible, thick enough to comply with the requirements of an infinitely thick specimen. The preparation must be simple, accurate, rapid, reproducible, and of low cost.

A wide variety of sample types may be analyzed by x-ray spectrometry, including solids, liquids, and even gases; hence, a wide variety of sample preparation techniques is required. The use of XRF techniques in the analysis of gases is not widespread, although it has been demonstrated, for example by Knöchel et al. [1] in synchrotron research, in which they analyzed gaseous nitrogen in the presence of small amounts of xenon. Special sample cups closed with appropriate windows (polymer films) can be designed for this purpose but are, as far as we are aware, not commercially available. In our opinion the analysis of gaseous substances will be of increasing importance in the near future. However, because of the lack of definite development, gas analysis is not discussed further here.

The following discussion thus deals with the preparation of solid and liquid samples for XRF. More emphasis is placed on more recently developed techniques. Some special preparation techniques, relevant for particular x-ray analysis types, have been discussed to some extent in previous chapters (SXRF, TXRF, PIXE, and EPMA in Chaps. 8, 9, 11, and 12, respectively). Reference should also be made to excellent chapters on sample preparation for XRF in earlier books [2–5].

II. SOLID SAMPLES

Determination of the bulk composition of solid samples, without sample preparation, is possible for samples that are homogeneous in all three dimensions, with a flat surface. In many cases, solid samples may be machined to the shape and dimensions of the sample holder. For large objects, small portions of the unfinished bulk material may be cut so that a smaller surface can be properly prepared for analysis. The reproducibility of the positioning of such small samples is critical. Often, it is found useful to make a wax mold of the part that will fit into the sample holder. Using the mold as a positioning aid, other identical samples may be reproducibly placed in the spectrometer. This technique is especially useful for small manufactured parts [3].

A. Metallic Specimens

Metallic specimens are usually prepared for x-ray analysis as solid disks by conventional methods of cutting, milling, and grinding (for hard alloys). A major source of error in the analysis of solid samples is improper surface preparation [2]. The surface finishing is accomplished by polishing with an abrasive, such as alumina, boron carbide, silicon carbide, or diamond dust, depending on the hardness of the material [4]. Most metallic specimens must be polished since the analyte intensity usually depends on the surface roughness ("shielding effect") [6,7].

In general, the surface roughness should not exceed the pathlength that would cause 10% absorption of the measured radiation [5]. As an illustration, Table 1 shows some 10% absorption (90% transmission) pathlengths for common situations in the analysis of metallic specimens. It is clear that most metallic specimens require some surface finishing or polishing. For example, the measurement of copper in aluminum would require a 7.7 μm surface finish, which could not be obtained even with 600 grit paper. On the other hand, mechanical polishing may present problems for soft malleable multiphase alloys or metals, since polishing may cause smearing or embedding of the abrasive in the material. This state of affairs is especially common for such samples as Pb, Al, or alloys containing those two metals. In such multiphase alloys, the fluorescent intensities in the softer phases increase and those in the harder phase decrease, thus contributing to the analytical error [2,3]. To avoid such difficulties, such techniques as etching and electropolishing may be used. However, such methods are expensive and time consuming.

Moreover, even for specimens that require prolonged surface polishing, it is important to consider not only the nature of the grooves and surface profile but also the orientation of the sample and its grooves in relation to the direction of the incident x-rays [8]. Polishing grooves on the surface of the sample may have a serious effect on the measured intensity of low-energy x-rays. This can be examined by repetitive measurement of the intensity of a sample after 45° or 90° rotation. The use of sample spinner reduces this effect. Thus large specimens are usually rotated during measurement. If this is not possible, then the specimens should be oriented in such a manner that the grooves are parallel to the plane formed by incident and emergent radiation. Several mathematical models have been developed to determine the relationship between surface roughness and emergent analyte line intensity [8,9].

Figure 1 shows a plot of fluorescent intensity versus groove size for the elements Al, Fe, Cu, Pb, and Mo [8]. It can be seen that for Al $K\alpha$ x-rays, a groove size of 50 μm results into a 20% intensity reduction, whereas for Fe $K\alpha$ x-rays, the reduction in intensity is practically negligible.

When conventional milling and polishing procedures are not applicable to the specimen (e.g., in the case of turnings), these may be pressed into pellets using

Table 1 Mass Absorption Coefficients and 90% Transmission Pathlengths for Some Cases

Measured radiation	Average matrix composition	μ Matrix $(cm^2\ g^{-1})$	ρ $(g\ cm^{-3})$	$x_{90\%}$ (μm)
Mg $K\alpha$	Fe	5430	7.87	0.02
Ti $K\alpha$	Cr	144	7.19	1.02
Cu $K\alpha$	Al	50	2.70	7.7
Cu $K\alpha$	U	306	18.7	0.18
Zr $K\alpha$	U	57	18.7	0.98

Intensity

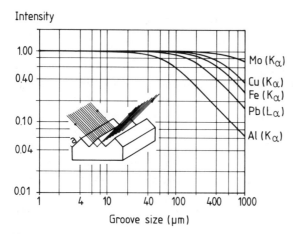

Figure 1   Change in relative fluorescent intensity of massive samples as a function of groove size *a* for different elements, calculated for pure elements (From Ref. 8. Reprinted by permission of A. Hilger, Ltd.)

a hydraulic press at 250–310 MPa. At such high pressures, most turnings are compacted to solid specimens with a fairly smooth surface for analysis.

Irregularly shaped bits of metallic samples can also be prepared for analysis by embedding them in a special wax resin (e.g., acrylic resin and methyl acrylic resin) before analysis [10]. Once mounted and set, the resin block needs to be ground and polished with, for example, silicon carbide abrasive and diamond paste progressively from 120 grit to 1 μm size.

Samples whose surface is not representative of the bulk composition, for example because of smearing at grinding or because of a corrosion layer, can be treated by chemical etching. For some archeological materials the etching procedure is superfluous or can be detrimental since many inclusions are dissolved during etching. For bronze alloys the recommended etching reagent is a solution of alcoholic ferric chloride (120 ml C_2H_5OH, 30 ml HCl and 10 g $FeCl_3$), whereas for brass alloys a solution of aqueous ammonium persulfate [100 ml H_2O and 10 g $(NH_4)_2S_2O_8$] is preferred. Ancient metals etch very fast: only a 5–10 s etching time is required.

Sometimes, a complete dissolution step is necessary, as for irregular metallic samples. A number of metals and alloys that are otherwise very difficult to deal with can be brought into solution in aqua regia (the mixture of one part concentrated HNO_3 with three parts concentrated HCl) with gentle heating to 60°C. The effectiveness of this reagent is due to the complexing power of the chloride ion and the catalytic effect of molecular Cl_2 and NOCl and of the chloride ion present during the reaction. The reagent is more effective if it is first allowed to stand for 10–20 minutes before use. Because of the speed of the reaction, aqua regia is used to attack a number of metals (e.g., palladium, platinum, antimonum, and germanium) and alloys, including steels, molybdenum, rhodium, iridium, ruthenium, and high-temperature alloys. Iron and vanadium ores and crude phosphates can be dissolved by this mixture, as well as some sulfides (e.g., pyrites and copper sulfide ores), although a part of the sulfur is lost as H_2S. Titanium remains passively resistant. HNO_3 and HCl are often used together in other proportions. So-called inverted aqua regia (mixture 3:1) is used to oxidize sulfur and pyrites.

The mixture of HNO_3 and HF acids is particularly useful for dissolving metallic silicon, titanium, niobium, tantalum, zirconium, hafnium, tungsten, rhenium, and tin, together with their alloys, and it may also be used for various carbides and nitrides, uranium and tungsten ores, sulfide ores, and many silicates [11]. For the determination of niobium, molybdenum, tungsten, iron, and titanium in tantalum by thin-film x-ray spectrometry (XRS), Eddy and Balaes [12] dissolved the metal in HF/HNO_3, extracted the tantalum by reversed-phase extraction chromatography using tributyl phosphate, and either spotted the sample solution directly on a filter paper or coprecipitated the trace elements with indium. The detection limits were 0.05–0.7 µg, depending on the element being measured. Samples heated to fumes with $HNO_3/HF/HClO_4$ mixtures lose selenium and chromium completely and mercury, arsenic, germanium, tellurium, rhenium, osmium, and ruthenium to some extent [13]. Semiconductors, PbTe and GeTe, can be dissolved by concentrated HNO_3 mixed with 10% sodium oxalate solution (1:2) [14].

Sometimes, less common dissolution reagents are used in XRF. Trace levels of zirconium, molybdenum, hafnium, and tungsten in tantalum metal could be determined after removing the matrix elements with diantipyrylmethane and collecting the impurity elements on anion-exchange paper [15].

B. Powdered Specimens

Powdered specimens may be analyzed in loose form or as a pressed pellet. Also, they may be fused with a flux (see Sec. III). The fused product may be reground and compressed or cast as a disk; experience indicates that the latter technique is the best compromise. For specimens that do not form self-supporting pellets or do not present surface properties amenable to polishing and finishing, the best approach in preparing the samples for analysis is to convert them into fine powders. In quantitative analysis, most correction algorithms (as discussed in Chap. 5) assume homogeneous samples, so the powders must be ground finely enough to ensure conditions for homogeneous dense materials. Typical bulk samples pulverized before analysis are ores, minerals, refractory materials, and freeze-dried biological tissue. Frequently ground sample materials are comprised of individual particles or crystallites that are dissimilar in composition, density, hardness, configuration, size, and distribution and exert a different influence on analyte line intensities. These effects are easily rectified by admixing the powdered sample material with an additive selected for its ability to aid in the grinding, blending, or briquetting process. Generally, 1–5 wt% additive to 10 g sample is sufficient for many troublesome sample materials. The main disadvantages of powder specimen preparation is that surface effects strongly influence low atomic number analysis.

Heterogeneity effects in loose powders are often associated with segregation of particles due to specific gravity differences and with texture effects due to particle orientation, as in micas and surface roughness effects. A qualitative interpretation of the origin of the heterogeneity effects was proposed by Claisse [16–18]. The particle size effects can be a major source of error in the analysis of heterogeneous materials and loose powders since they may cause large variations in the fluorescent intensities of the analyte [9,19,20]. Furthermore, these effects become more pronounced and more difficult to avoid as the atomic number of the fluorescent element decreases [21]. Determination of elements using low-energy x-irradiation may lead to errors of as much as 50% arising from particle size effects.

Mathematical methods of correction for particle size effects have been developed, but with few exceptions (see Chap. 6, Sec. IV.B) they are not practically useful because the particle size distribution in the sample is required but not known [8,22,23]. Most reports conclude that optimized operating conditions are necessary for the analysis of powdered specimens.

1. Grinding

In the reduction of a bulk solid to a fine powder, grinding or milling is usually employed with attendant problems arising as a result of variations in the particle sizes of the various components of the powder as well as contamination from the grinding material. Particularly for the trace element level, precautions should be taken by suitable choice of containers (agate, alundum, silicon carbide, boron carbide, tungsten carbide, and others) and by proper adding and mixing of binders or matrix diluents. Geological material contamination is negligible for agate grinding elements and is serious for only a few elements, such as Ag, Cr, Cu, Zn, and Pb, when grinding is done with corundum. When typical biological materials are ground, however, the contamination from corundum grinding elements is absolutely intolerable for most trace elements and permissible only for abundant constituents, such as Na, Mg, P, S, Cl, K, Ca, Rb, and Sr. Agate grinding elements may introduce into biological material important contamination by Ti, V, Cr, Mn, Fe, and Pb, but many other elements can probably be determined safely [24].

Variations in particle sizes may be minimized by grinding the specimen with a mixture of SiC and Al_2O_3 with ether for 30 minutes [25]. Wet grinding with ethyl or isopropyl alcohol and diethyl ether increases grinding efficiency and homogenization. For porous materials, such as platinum, the pore structure can be destroyed by calcination in an air stream at 800°C and grinding to a fine powder [26,27]. Grinding the specimen to a mean particle diameter of 50 μm is recommended to attenuate particle size variations. Grinding times of 4 h [28] and even 15 h [29] have been shown to be necessary for obtaining reproducible results in the analysis of rocks. The addition of a few milligrams of a grinding aid like sodium stearate greatly reduces the time required to attain small particle diameters. The effect of the grinding time or particle size on the fluorescent intensities of the elements Si, S, K, Ca, and Fe in coal samples prepared by the powder technique (5 g sample was ground in a tungsten carbide rotary swing mill together with 1 g boric acid binder and 100 mg sodium stearate grinding aid) is shown in Figure 2a [30]. From this figure it is clear that a grinding time of about 6 minutes will reduce the particle size in a coal sample to the order of 50 μm, but no further increase in intensity is attained for longer grinding times.

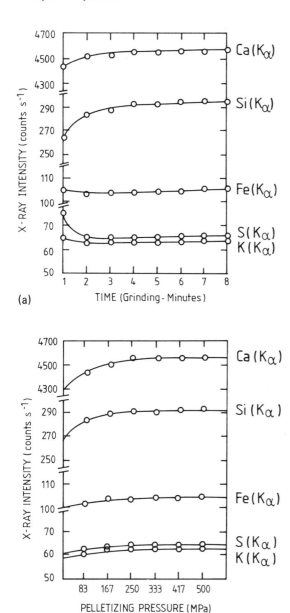

(a)

(b)

Figure 2 (a) Grinding time versus intensity. (From Ref. 30. Reprinted by permission of Plenum Press.) (b) Pelletizing pressure versus intensity. (From Ref. 30. Reprinted by permission of Plenum Press.)

2. Analyzing Loose Powders

As powders, the specimens may be analyzed in loose form in special specimen cups or holders or dusted on Scotch tape before being presented for irradiation. Another approach is to make a slurry by mixing a small amount of the fine powder with a solution of 2–5% (wt/vol) of nitrocellulose in a mg L^{-1} acetate. The resultant slurry is supported on a microscope slide or some other suitable substrate [3]. A simpler approach is to take a small portion of the finely powdered sample and shake it in a beaker with water, thereby forming a fine suspension. The suspended particles are then deposited on a Millipore or Nuclepore filter material as a thin layer using a vacuum pump [10]. In this way, a rather homogeneous sample is prepared for analysis, but soluble analyte elements or those adsorbed reversibly at the particle surfaces may be lost in this process. Therefore, apolar organic solvents like hexane are sometime used to prepare the suspension. An alternative procedure involves mixing the samples ($\sim$1 g) with a few milliliters of double-distilled water and grinding, either in a micronizing mill for 1 minute or manually with a corundum pestle and mortar. From the resulting suspension, a 0.5 ml fraction is immediately pipetted onto a Mylar foil. Careful evaporation at 80°C results in a target containing about 2 mg cm^{-2} of sample material. In this case, microscopic photographs showed the sample to be quite homogeneous with a grain size below 10 μm [31,32].

3. Pelletizing

Briquettes or pressed powders yield better precision than loosely packed powder samples and are relatively simple and economical to prepare. There are various methods of making pellets with or without a binding substance. In many cases all that is needed is a hydraulic press and a suitable die set, which includes a die body, base, a plunger, and two polished metal disks. These disks may be made of an alloy like tungsten carbide, which makes them rather useful for pressing hard or abrasive materials. In most cases the ensemble is fitted with O ring vacuum seals and knockout rings for extraction of the sample pellet after pressing.

Materials that do not make stable pellets or do not bind on their own require the addition of a binder or backing material. In this method, the backing material is poured into the die to form a thin uniform layer. The specimen powder is poured on this layer and leveled. Suitable pressure is then applied to form a pellet. This technique is particularly useful when the amount of specimen is small (less than 1 g). Most binders are composed of light elements that scatter x-rays and tend to increase the background. Therefore it is essential that the minimum amount of binder needed should be used, to avoid excessive background that may interfere with the analysis of light elements and of elements in trace amounts. The most common binders are starch, detergent powders, stearic acid, boric acid, lithium carbonate, and chromatography-grade cellulose [33]. A few grams of the starch or cellulose are added to and mixed thoroughly with the powdered specimen before pressing. Other effective binders are graphite, ethyl and methylcellulose, and polyvinyl alcohol. Polyvinyl acetate is particularly good for forming stable pellets from soils or powdered rocks. Good analytical results have been obtained with pellets containing only few drops of binder in diluted solution. For most powdered material preparations, a mixture containing 5 ml liquid binder (e.g., a 10% solution of a polymeric binding ingredient in methylene chloride) and 10 g sample is sufficient to produce slurries of adequate consistency for subsequent wet grinding and to form briquettes. After the grinding process is complete, the solvent should be evaporated under an infrared lamp with occasional stirring of the slurry to expedite drying. With a new type of sample, experimentation is usually required.

Other methods consist of pouring the powdered specimen into a thin-walled aluminum cup and pressing. In this case there is no need for a binder and the process is fast and easy. An exhaustive description of briquetting procedures is given in several references [2–5].

The effect of the compaction or pelletizing pressure on the fluorescent intensities for the same coal sample, as presented in Section II.B.1, is displayed in Figure 2b [30]. The resulting coal powders were pelletized with a boric acid backing at 250 MPa and placed in the spectrometer for fluorescence intensity measurement, and the optimum pelletizing pressure was determined. Light elements, such as Si and Ca, show higher dependence on pressure than Fe, for example. Another notable feature is that S and K have less dependence on the pressure that Si or Ca, showing that there is a significant different in the particle sizes of these elements in the coal samples. Increasing the packing density (by increasing the pelletizing pressure) greatly reduces these variations, and stable analyte intensities are obtained for pressures ranging from 250 to 410 MPa.

III. FUSED SPECIMENS

The fusion specimen preparation method is one of the most successful techniques that almost completely overcomes any of the common problems in x-ray spectrometric analysis. It was first proposed by Claisse in 1956 for the transformation of aggregates into glass disks [34]. Essentially the fusion procedure consists of heating a mixture of sample and flux at a high temperature, for times ranging from 5 to 30 minutes, so that the flux melts and reacts with the sample or dissolves it. Agitation during fusion is necessary for the production of homogeneous glass [35]. The final product after cooling is a one-phase glass. To prepare a 3 cm diameter glass disk, a total weight of 7 g, sample plus flux, is required. The dissolution or decomposition of a portion of the sample by a flux and the production of a homogeneous glass eliminates the particle size and mineralogical effects entirely [33,36]. The additional advantages of the fusion technique are the possibility of decreasing or compensating the matrix effects by proper specimen dilution or by adding heavy absorbers and internal standards and the possibility of preparing standards of desired composition. The most severe disadvantages to the fusion techniques are the time and material costs involved, the dilution of the elements, which can result in an order of magnitude reduction in x-ray intensity, and high detection limits, particularly for low atomic number elements ($\sim$600 μg g^{-1}). Considerable practice is normally required before quality specimens can be produced routinely.

A. Fluxes and Additives

Various fluxes have been developed to dissolve a wide variety of sample materials and yield mechanically stable disks [37]. Selection of the appropriate fusion flux is related to the composition and properties of the specimen. Low-temperature fusion may be done using potassium pyrosulfate. More common are the glass-forming fusions with lithium borate, lithium tetraborate, or sodium tetraborate. When the analysis of light elements is concerned, it is better to use lithium fluxes instead of sodium because of their lower mass absorption coefficients and, therefore, lesser effect on the intensity of the low-energy x-rays. A variety of flux additive recipes are reported for various sample types in Table 2. Flux-to-sample ratios range from 1:1 to 10:1.

Lithium-based borates are a favorite for fusion because the resulting melt is rather more fluid than that of sodium-based borate fluxes. The high viscosity of sodium tetraborate fluxes is a disadvantage because the resultant melt tends to "wet" or stick to the crucibles, thus necessitating frequent cleaning of the crucibles.

Lithium metaborate has been shown to have superior fluidity and dissolution properties for specimens with a high content of silica, such as silicate rocks [38]. Very low sample-to-flux dilution ratios can be obtained as a result of the fluidity of the metaborate, and the attendant increase in background due to the scatter from the low atomic number flux is appreciably lowered by addition of a heavy absorber [38,39]. Matrix modification by addition of a heavy absorber, such as

Table 2 Examples of Fluxes and Applications

Flux base	Flux composition	Properties	Application
Lithium metaborate	$LiBO_2$ and $LiBO_2$ + $Li_2B_4O_7$ (4:1)	Good mechanical properties, low x-ray absorption	Acid oxides (SiO_2, TiO_2); silica-alumina refractories
Lithium tetraborate	$Li_2B_4O_7$	Beads crack easily, low x-ray absorption	Basic oxides (Al_2O_3); metal oxides; alkali, alkaline earth oxides; carbonates; cements
Sodium, potassium carbonate, and their mixtures	Na_2CO_3, K_2CO_3	Unsuitable for glass beads	Silicates
Sodium hydrogen sulfate, sodium pyrosulfate	$NaHSO_4$, $Na_2S_2O_4$		Nonsilicate minerals (chromates, ilmenite)
Sodium tetraborate	$Na_2B_4O_7$	Low viscosity of melt	Metal oxides; rocks, refractories; bauxites
Sodium metaphosphate	$NaPO_3$		Various oxides (MgO, Cr_2O_3)

$K_2S_2O_7$, BaO, CeO_2, $BaSO_4$, or La_2O_3, has a buffering effect; this makes the analyte intensity linearly dependent on the concentration, thus making quantization rather easy since artificial standards may be used with a minimum of matrix correction procedures [37,40,41].

Slags from copper-smelting processes may be analyzed after dissolution in sodium tetraborate combined with lanthanum oxides as a heavy absorber and sodium nitrate as an oxidant to prevent attack of the crucible by elemental copper [35].

In the fluxing of slags from blast furnaces, a mixture ratio of 3:1 of sodium tetraborate to sodium carbonate is preferred. Normally, the metals are first converted to their oxides before dissolution in the mixture [37].

B. Fusion Procedures

For most x-ray applications, a 9:1 ratio of flux to sample is fused in a muffle furnace, over a gas burner, or with automatic fusion equipment at a temperature up to 1100°C until the mixture is completely molten and homogenized. Occasional swirling of the melt ensures homogeneity, dissolution, and removal of bubble occlusion [38–42]. For very fluid melts, however, agitation is not necessary as the convection and heat gradients are sufficient to cause homogeneous mixing of the sample and flux [37].

Several automatic fluxes have been developed, such as the Claisse fluxer [2], which heats and agitates the melt for a preset time with a predetermined flame size without operator attendance [43]. Others, like PERL'X, or SOL'X (Philips Materials, Automatic Sample Preparation Machines) are more sophisticated and embody operation under microprocessor control. These devices have reliable operation as well as reproducibility, and their use has been documented by several workers in analytical chemistry [44,45].

Fusion in borate and similar fluxes requires temperatures in excess of the melting point to ensure viscosity and mixing of the resulting melt. Usually the flux is heated at about 900°C or just until it melts. The temperature is then raised to about 1100°C and maintained at this level for 10–15 minutes.

For samples that are difficult to dissolve, a small increment of an oxidizing agent, 2–10% of total weight (e.g., BaO, CeO_2, KNO_3, or $LiNO_3$), may be added to the melt to expedite dissolution. Samples containing organic matter should be preignited for several hours at approximately 450°C before fusion. For samples lacking glass-forming constituents (e.g., alkaline earths), at least 25% of total weight of SiO_2, Al_2O_3, or GeO_2 should be added for better mechanical stability of the glass disks. Also, for inadequate quantities of sample, the balance of weight is adjusted by adding SiO_2.

After fusion, the melt is poured immediately into a graphite, Pt, or Pt-Au mold. It is common practice to have an intermediate annealing stage, in which the fresh bead is kept at a temperature of about 200°C for 2 minutes [38–43]. This may not be necessary, especially if the heating or melting is done in a furnace or induction heater, since the crucible and mold can be maintained at the same temperature and the melt can be poured without suffering from thermal shock or other effects deleterious to its mechanical properties. Addition of non-wetting substances, like KI, LiI, CsI, NH_4I, NaCl, NaBr, and LiBr, in a small concentration (less than 1% of total weight) prevents the tendency of a molten glass to wet the crucible. As a result, transfer from the crucible to the mold is more complete, removal of the glass disk from the mold is easier, and cleaning of the crucible is not necessary. In practice, solutions of these salts are prepared in advance, and one or two drops are placed on the top of the mixture before heating [2]. Grinding the sample to a grain size of approximately 150 μm before fluxing quickens the dissolution of the sample appreciably, especially in the case of refractory materials [46,47].

Fused samples may be cast into glass disks and formed into briquettes. Because of their robust nature, glass disks can be stored for a long time, without deterioration, in desiccators or other moisture-free environments. The borates are slightly hygroscopic, and prolonged exposure of the disk to the atmosphere changes the surface properties. Slight polishing with an abrasive (100, 200, 400, or 600 grit sizes on silicon carbide) and cleaning of the analytical surface with alcohol is usually sufficient to render the disks fit for qualitative or quantitative analysis. In extreme cases, the disk can be recast in a few minutes. Cast sample disks may be inscribed for identification and stored for future use.

Trace element determinations after fluxing in a borax melt are possible, as demonstrated by Luke [42] using special disk-casting techniques. In this procedure, the molten flux is poured onto a hot aluminum plate kept at 225°C and then quenched and flattened to a thickness of 1.6 mm using a 13 cm^2 aluminum plate with a wood-tipped handle. With this preparation, the minimum amount of metal that can be determined ranges from 1 μg for Ni to 10 μg for Sc or Ca.

Detection limits of better than 50 μg g^{-1} for Zn, Pb, Rb, and Sr traces in rock specimens have been demonstrated by using proportions of 2:2:1 of lithium metaborate, lanthanum oxide, and the sample, respectively. Only mathematical corrections for interelement effects are required for quantization of the major elements [39,40]. However, the determination of trace elements without special disk-casting techniques and low flux-to-specimen dilutions is rather dubious, since most trace elements can easily be diluted below their detection limits [41].

During the fusion process, the sample and the flux lose weight through volatile evolvement. If the percentage loss for both the flux and the sample is identical, no correction is generally necessary. If they differ the flux-to-sample ratio is imbalanced and weight loss correction must be calculated. During the fusion preparation of geological materials, the loss of volatile constituents as bound water, moisture water, or CO_2, for example, from the flux-rock mixture gives rise to higher apparent concentrations of constituents in the product glass disk. This problem can be corrected mathematically [9,47] if the weight loss on ignition (LOI) is less than 20% or by adding SiO_2 to compensate for the loss [48]. For carbonate rocks, the method developed by King and Vivit [49] is worth mentioning. First, they ground the samples into a fine powder and then dried them at 105°C for 2 h. Then, about 2 g of each sample was accurately weighed, placed in a platinum crucible, and calcined in a preheated muffle furnace at 925°C until constant weight was achieved. The weight loss was recorded as LOI. Finally, the samples were fused into glass disks with lithium tetraborate at 1100°C using a 10:1 flux-sample ratio. Matrix effects are corrected empirically using Lachance-Traill correction algorithms.

Analysis of bauxites [50], cements [51], and rocks [52] are good illustrations of the potential of the fusion techniques. The observed deviations are small and include variations in sample preparation as well as errors in matrix effect corrections.

IV. LIQUID SAMPLES

Liquid specimens represent nearly ideal targets for XRS. Liquids are perfectly homogeneous, obviously no particle size effects are present, there are no problems of surface texture, matrix effects are reduced or eliminated by dilution, matrix absorption is very low, and standards can be prepared straightforwardly. Blanks for measurement of background and for evaluating possible contamination are usually prepared easily. Analysis of liquid samples may be accomplished by presenting the sample in a cup with a thin Mylar or polyethylene bottom directly to the x-ray system. Solutions are particularly suited to on-line process control, since the liquid can be arranged to continuously flow through the sample chamber of the spectrometer.

The major problems arising with liquids are usually high x-ray scatter background resulting in poor detection limits, evaporation of the solvent, and bubble formation during the analysis, which causes a change in analyte line intensity. The liquid solution technique is especially unsatisfactory for low atomic number elements whose radiation must pass through the window of the sample cup. Also, there is always a danger of leakage with corrosive solvents, such as acids. Liquids are unsuitable for vacuum irradiation but can be analyzed in a He atmosphere. Heating during irradiation may cause chemical reactions or ionization of liquid due to interaction with the radiation, for example, presenting many problems that are either absent or minimal in solid samples. Many inorganic specimens (e.g., minerals, ores, rocks, cements, and refractories) are too heterogeneous for direct examination even if they are finely ground. Accurate analysis requires that they be transformed to liquid or solid solution.

Typical detection limits of conventional direct energy-dispersive XRF (EDXRF) with 30 minutes counting time are in the ppm range [53], certainly not satisfactory for most natural water analysis applications. Consequently, it is necessary to remove the matrix to improve the detection limits. Preconcentration is defined as an operation or process as a result of which the ratio of the concentration or the amount of microcomponents to the matrix increases. It also reduces matrix effects and so thus improves the analytical detection limit and enhances the accuracy of the results, as well as facilitating calibration procedures [54]. However, preconcentration can be time consuming, involves risks of contamination, and may not collect simultaneously all species of an element. In principle any preconcentration and separation method developed for any analytical technique could be used in combination with XRS, but multielement preconcentration leading to solid thin targets is ideal for XRS. In an extensive review of the analytical and environmental literature [55] covering nearly 200 references relevant to preconcentration for XRS of water between 1967 and 1982, the reported merits and drawbacks of all procedures were compared. At present, still about 40 articles per year appear on XRS for water analysis. XRS is ideally suited for the analysis of particulate matter and suspended sediments in natural water, although it is not very frequently used for this purpose. Suspension can be collected by filtering a known volume of water through a membrane filter, and sediments can easily be stirred up and sampled in the same way. The major analytical problems are then only those pertaining to particle size effects and to corrections for x-ray absorption.

In comparing different filter media for the EDXRS analysis of particulate matter in seawater, Vanderstappen and Van Grieken [56] found Nuclepore membranes (1.1 mg cm^{-2}) and ultrathin Millipore-THWP membranes (1.1 mg cm^{-2}) to be preferable over common HAWP-Millipore filters (5.5 mg cm^{-2}) or Whatman 41 cellulose filters. For the Nuclepore membranes, interference-free detection limits of 5–10 ng cm^{-2} for Mn, Ni, Cu, Zn, As, and Br, 10–20 ng cm^{-2} for Ti, V, and Cr, and around 50 ng cm^{-2} for K and Ca were quoted for 2000 s counting and Mo secondary fluorescer excitation; for the Whatman filters, values about two times higher were found.

A. Physical Preconcentration Methods

Physical preconcentration methods are quite suitable for dilute aqueous samples like rainwater and, in fact, are the only practical possibility for the analysis of sewage, wastewaters, or similar samples. Several approaches are available to remove liquid matrix physically.

The most obvious and straightforward method for preconcentrating ions from solution is perhaps evaporation of the solvent. In the evaporation residue, all nonvolatile elements are collected quantitatively, irrespective of their speciation in water, and the risk of contamination is minimal.

A large water sample can be directly evaporated or freeze-dried, and the evaporation residue can be pelletized, possibly after mixing with an organic binder to reduce matrix effect variations. Freeze-drying of, for example, 250 ml wastewater on 100 mg graphite followed by grinding and pelletizing of the residue leads for EDXRF to typical detection limits of 5 μg L^{-1} for many elements with accuracies around 10% [57]. An alternative approach for similar samples with low salinity is the so-called vapor filtration procedure [42,58]. In vapor filtration, the sample is presented to a container whose bottom surface is permeable to water vapor but not the liquid phase. This surface is subsequently exposed to vacuum, and the dissolved solids are deposited on the bottom surface, which may be a membrane, such as a Millipore or Nuclepore filter. The filter is then presented for analysis. In combination with particle-induced x-ray emission (PIXE), typical detection limits are in the range of 0.1–3 μg L^{-1}. This method suffers from several disadvantages: it is time consuming, samples with high salinity or hardness cannot be preconcentrated efficiently, and incomplete recovery of the residual from the container invariably leads to analytical errors. The formation of finite crystals (rather common in this method), even from very dilute solutions, can lead to large and unknown x-ray absorption effects, resulting in microscopically inhomogeneous residues that involve rather problematical matrix corrections [54].

Another procedure, which allows evaporation of larger samples, includes impregnation of filter paper with sample solution, such as spotting a 1.5 ml water sample on a Whatman 41 cellulose filter paper provided with a wax ring 29 mm in diameter and evaporating the water by passing an unheated air stream from underneath. The hydrophobic wax ring reduces the spreading of the solution in the filter and any differential chromatographic effects. The detection limits were found to be below 50 μg L^{-1} for 2000 s analyzing time with optimal secondary target excitation [57].

Other approaches that give a more homogeneously distributed deposit on a filter paper include the nebulization technique [59] and the multidrop spotting technique [60]. Occasionally, water samples can be sprayed as a fine aerosol, dried, and presented to the x-ray spectrometer [61].

In general, the more advantageous detection limits of PIXE and its predominant use in nuclear physics laboratories, where chemical manipulations are often unpopular, have resulted frequently in a combination of PIXE with a simple water pretreatment step consisting of pipetting one drop of water onto a thin Formvar or Mylar carrier and evaporating at 50°C. The reported detection limits are around 10 μg L^{-1}.

Total-reflection XRF (TXRF) allows direct water analysis by simply spotting a drop on a reflecting substrate. The detection limits obtained are very impressive, particularly for highly diluted media. For example, for Ni, the detection limits of conventional XRF and TXRF do not differ much in a solution containing 1% NaCl or more. Yet, with only 0.001% NaCl present, for example, the detection limits for Ni for TXRF (around 0.05 μg L^{-1}) are two orders of magnitude below those for conventional XRF [62].

It seems, however, that the most suitable techniques for the screening of water when extreme sensitivity and accuracy are of paramount importance probably include some form of chemical preconcentration.

B. Chemical Preconcentration Methods

Chemical preconcentration methods may be grouped into three broad classes:

1. (Co)precipitation
2. Ion exchange
3. Chelation and sorption immobilization

These methods have been applied to various types of water samples [54,58,63–68]. They often involve a final stage in which the analyte precipitates or complexes, for example, are deposited on membrane filters that are then subjected to irradiation and analyzed.

1. (Co)precipitation Methods

Since the early days of analytical chemistry, much effort has been devoted to finding precipitation methods, especially for the separation of particular elements or element groups. However, for an inherently multielement technique like XRF, selective precipitation is usually not necessary. Nonspecific multielement reagents are more appropriate, and therefore, coprecipitation and broad-spectrum precipitation techniques have often been combined with XRF.

Probably the most popular (co)precipitation reagents are the carbamates in view of the low solubility of their metal chelates [54]. In his pioneering work on the "Coprex" method, Luke [69] established the conditions for the preconcentration of many trace metals using sodium diethyldithiocarbamate (NaDDTC) at a certain pH and in the presence of a suitable metal ion spike serving as both a carrier and an internal standard. NaDDTC was afterward applied for XRF by many authors [70–78]. Kessler and Vincent [79] improved the detection limits by filtering the NaDDTC/hydroxide precipitate on a filter area of 2.5 mm diameter using a highly collimated XRF setup.

At concentrations below 10 μg L^{-1}, ammonium pyrrolidine dithiocarbamate (APDC) seems preferable over DDTC for quantitative recoveries. Ulrich and Hopke [80] compared several preconcentration methods and found that APDC at pH 4 was the best nonspecific reagent, allowing quantitative recoveries for a dozen elements in water, independent of the alkaline ion content. Elder et al. [81] pulverized precipitation at pH 2 with a fresh APDC solution for Cu, Hg, and Pb, but they observed that the recoveries of Fe and Zn were seriously depressed in natural waters.

Dibenzyldithiocarbamate (DBDTC) was proposed by Lindner et al. [82] and later recommended after a comparison by Ellis et al. [83] because its very low solubility eliminates the need for a metal carrier. Quantitative recoveries were found for Mn, Fe, Co, Ni, Cu, Zn, Se, Sb, Hg, Tl, Ag, Cd, and Pb, with detection limits around 1 μg ml^{-1} for 100 ml samples. More recently, DBTC was used to recover trace amounts of Mo in water at pH 3, for concentrations of the order of 10 μg L^{-1} [84]. The addition of 0.5 mg Cr as a carrier enabled Saitoh et al. [85] to determine quantitatively less than 5 μg vanadium in 500 ml seawater at pH 4 with DBDTC.

Another attractive reagent for some water types is 1-(2-pyridylazo)-2-naphthol (PAN). PAN is very soluble in hot water and ethanol but not in cold water. Cocrystallizations of at least 15 cations with PAN are thus simply be achieved by adding 20 mg PAN, dissolved in ethanol, to a 2 L water sample at 70°C, and after cooling and filtering of the precipitate, EDXRF detection limits of 0.5 μg L^{-1} are realistic [86].

Coprecipitation with hydrated iron and aluminum hydroxide have classically been used in analytical chemistry but seldom in combination with XRF. Addition of 10 mg L^{-1} of Fe at pH 9 allow collection of Co, Ni, Cu, Zn, and Pb in the precipitate in a predictable way, that is, with high and constant distribution coefficients, and led to typical detection limits of 0.4 μg L^{-1} for EDXRF [64].

Various other reagents have been used for multielement (co)precipitation in XRF; many earlier references are quoted by Van Grieken [54].

Precipitation has also been used extensively for the specific preconcentration of one or a few elements before XRF. Only a few examples are mentioned here. Lebedev et al. [87] determined rubidium in seawater in the 2–4 μg concentration range using coprecipitation with Ni$_3$K$_2$[Fe(CN)$_6$]$_2$ at pH 3.5. The determination of the Cl$^-$, Br$^-$, and I$^-$ ions in fresh snow was reported by Yamamoto [88]. The snow was melted and acidified to 0.9 M in HNO$_3$. After cooling, Ag(NO$_3$) solution was added and the precipitated silver halides were collected on membrane filters for analysis. No significant interference effects were observed, and detection limits of 0.1–0.5 μg were obtained. For spring waters, Tanaka et al. [89] determined arsenic and antimony and reported detection limits of 0.3 and 6.1 μg, respectively. They first reduce the As(V) and Sb(V) to As(III) and Sb(III) with KBr and HCl at 80°C for 1 h. The As and Sb were coprecipitated at pH 9 with a solution of $ZrOCl_2$. Becker et al. [65] used a variant of the technique to determine microgram amounts of arsenic in a salt matrix. The dissolved As was precipitated by addition of a 7% solution of the monosodium salt of EDTA at pH 10. Good recoveries ranging from 94–102% were reported. The minimum amount of As that could be detected using this approach was 0.14 μg. By careful choice of the pH speciation is possible. Trace amounts of vanadium(V) and (IV) can be quantitatively recovered from water at pH 1.8 and 4.0, respectively, by reaction with NaDDTC. However, it must be pointed out that quantitative recovery is only possible if the interfering elements Fe, Co, Ni, Zn, and Pb do not exceed a total concentration of 100 μg [67].

It should be emphasized that although many (co)precipitation enrichment procedures have been proposed for XRF, most studies failed to check systematically the performance of their procedure for natural waters, which may contain high levels of alkali and alkaline earth ions and of organics, leading to a variable speciation of the analytes.

2. Ion-Exchange Methods

About half the literature on preconcentration for XRS of water deals with ion exchangers and ion-collecting filters, yet common cation and anion exchangers are of limited use for preconcentrating trace elements from natural waters because these resins operate on an ion-association basis and are not very selective, and the abundant alkali and alkaline ions can compete well in the preconcentration process with the transition metals that are usually to be determined. Still, the acid or basic resins may be applied for sample with a limited alkali and alkaline earth ion content.

Common chelating ion-collecting resins like Dowex Al or Chelex-100 (Bio-Rad Laboratories), with iminodiacetate functional groups, are an even better choice. They were earlier used in combination with XRS by several authors [90,91]. More recently, Brykina et al. [92] used Chelex-100 resin columns to capture Zn, Pb, Cd, and Hg from river water and obtain detection limits of 0.2–0.8 μg L^{-1} by XRS analysis of the pelletized resin. Voutsa et al. [93] recovered 80% of Cu, Zn, and Cd by passing a water sample of pH 9.0 through an Amberlite XAD-4 chelating column before XRS analysis. An interesting application of separation and preconcentration with chelating resins is speciation of the chemical state of the analyte ion [94]. Cr(IV) and Cr(III) can be separated and preconcentrated by using a mixture of Ag1-X8 in (Cl^- form), AG5OW-X8 (Na^+ form), and Chelex-100 resin beads. The AG1-X8 separates out the Cr(IV) ions; Cr(III) is selectively adsorbed on the remaining two resins at enrichment factors of 1000 and 500 for Cr(IV) and Cr(III), respectively. The loaded resin beads are then spread on an adhesive film (Scotch tape) and presented for XRF analysis.

For more efficient preconcentrations from natural waters, however, the ion-exchange substrate should not show any affinity at all for alkali and alkaline earth ions and be selective toward transition metals. Several materials have been developed for this purpose. Leyden and Luttrell [95] prepared polyamine-polyurea resin columns from tetraethylenepentamine and toluene diisocyanate and used 100 mg aliquots to preconcentrate Ni, Cu, and Zn quantitatively from 4 L seawater of natural pH and to determine these trace metals by wavelength-dispersive XRS (WDXRS) down to 0.1–0.3 $\mu g\ L^{-1}$ concentrations. Burba and Lieser [96] immobilized 1-(2-hydroxyphenylazo)-2-naphthol or Hyphan on cellulose powder and determined Fe, Zn, Pb, and especially Cu and U in diverse water samples, down to 0.3 $\mu g\ L^{-1}$ for U. Leyden and coworkers [97] intensively studied and used the silylation reaction to immobilize chelating amine functional groups onto glass beads or silica gel; they used N-β-aminoethyl-γ-aminopropyltrimethoxysilane, γ-aminopropyltrimethoxysilane, and their dithiocarbamate derivatives and obtained capacities of 0.5–0.9 mEq g^{-1} and detection limits around 1 $\mu g\ L^{-1}$.

All preconcentration procedures based on ion-exchange resin columns share some drawbacks for combination with XRS analysis: the sensitivity is not optimal because the preconcentration factor (ratio of original to final sample weight) is typically around 5000 only and after column preconcentration the ion exchanger should be homogenized and pelletized before XRS analysis.

Ion-collecting filters offer an attractive alternative for preconcentration in XRS: such loaded filters are ideal thin homogeneous targets with low Z that can be presented directly to the XRF instrument without preparation. Much effort has been devoted to preconcentration by ion-collecting filters for XRS, but unfortunately, most earlier research was focused on the ion-collecting membrane types that are least interesting for environmental water samples, such as cation-exchange papers SA-2, with sulfonic acid functional groups, containing approximately 50% Amberlite IR-120 resin and 50% cellulose and with an exchange capacity of 2 mEq g^{-1}. By 1977, a dozen publications, all listed in Reference 54, had already been devoted to these ion-collection filters for XRS analysis. However, they have a significant affinity for the alkali and alkaline earth ions, and hence, the collection of transition metal ions is not quantitative for large sample volumes of natural water, but for small sample volumes, their preconcentration factor is too low to yield environmentally relevant detection limits. SA-2 filters can successfully be applied only when the alkali and alkaline earth content of the water samples is very low, as in rainwater, or in combination with ion-exchange columns [98].

Hyphan filters, prepared by immobilizing 1-(2-hydroxyphenylazo)-2-naphthol on short-fibered cellulose powder and pelletizing the resulting material into thin layers, are much more suitable: high recoveries of trace metals are possible from large volumes of water a pH 7, and the detection limits for EDXRF are typically around 1 $\mu g\ L^{-1}$ [99]. The influence of alkali and alkaline earth ions is much smaller but still not negligible. The heavy metals ions Co, Cu, Zn, Fe, and Pb could easily be recovered from water by filtration through a Hyphan filter at pH 7.5 [100,101]. Selective analysis of Co alone could be performed after washing the Hyphan-loaded filter with 3 M HCl. The washing removes all metals except Co. It appeared that 0.4 $\mu g\ L^{-1}$ of Co could be determined in water containing 1 mg L^{-1} or more Fe, without significant interference.

Theoretical considerations indicated that 2,2-diaminodiethylamine, often called diethylenetriamine (DEN), would well satisfy the requirements for preconcentrating filters in XRS. Therefore these filters were also studied in much detail [102] after their synthesis was optimized [103]. At a filtration rate below 1.5 ml $min^{-1}\ cm^{-2}$ and at a pH > 6, recoveries of 90–100% were obtained for many ions, without significant interferences. Anions could also be preconcentrated by DEN filters, at least from dilute solutions. Practical detection limits were around 0.5 μg L^{-1} and often lower; the accuracy and precision were both around 10% for higher concentration levels. The major drawback of these filters is that their synthesis is not trivial and requires many precautions.

Lately, no ion-collecting papers of interest for XRS preconcentrations have become commercially available, and the interest in this field seems to have faded somewhat.

3. Chelation and Sorption Immobilization Methods

Although liquid-liquid extraction is undoubtedly the most popular preconcentration method in atomic absorption spectrometry, this method is not often applied to XRS because a subsequent evaporation step is necessary.

Reversed-phase techniques in which organic chelating agents are immobilized on a solid support seem more interesting. Knapp et al. [104] chelated simultaneously eight transition metals in large water volumes of pH 4 with NaDDTC, which was then adsorbed on Chromosorb W-DMCS columns and subsequently eluted with 2 ml chloroform onto a filter paper. For 100 ml samples, detection limits to 0.1 μg could be obtained. Also, APDC complexes of transition metals have been immobilized from 100 μl sample solvents on siliconized quartz reflectors [105]. Even for seawater, detection limits at the 20 pg or 0.2 μg L^{-1} level could be then obtained by total-reflection XRF for numerous elements.

An alternative approach is to chelate the trace metals in large volumes of water by APDC, DDTC, or DBDTC, adsorb the complexes on a Chromosorb column, and elute them in a small volume of $CHCl_3/CH_3OH$ directly on a polished carrier, evaporate, and measure by total-reflection XRF [106]. The high accuracy, freedom of matrix effects, multielement character, and excellent detection limits have been proven by several workers in recent years.

Braun et al. [107] determined phenyl mercury, methyl mercury, and inorganic mercury after preconcentration on DDTC-loaded polyurethane foam disks. The disks were prepared by adding a 4% solution of DDTC in $CHCl_3$ to provide a 20% concentration of the reagent on the polyurethane disk and addition of 0.2 ml dinonyl phthalate plasticizer. The various types of mercury were then recovered from the sample by shaking 25 ml specimen for 1 h with the polyurethane disks, after adjustment of the sample pH to 5. The reported extraction efficiencies for Hg in the concentration range 0.2 and 2 μg L^{-1} were between 88 and 100%.

Immobilization on activated carbon has also been used extensively. Activated carbon is known to be a good adsorber for organic and colloidal material and hence also for the species of trace metals that are bound to natural organic and colloidal matter. Free ions are not quantitatively adsorbed on activated carbon, but addition of a chelating agent converts them into an adsorbable form. Addition of a chelating compound and a subsequent adsorption step onto activated carbon should thus collect both originally free and colloidal and organic trace metal species. Several groups have exploited this idea, for example for combination with XRS. Vanderborght and Van Grieken [108] considered 8-quinolinol as a particularly suitable multielement chelating agent. The optimized preconcentration procedure consists of adding 10 mg 8-quinolinol per liter of water sample at pH 8 (either adding solid 8-quinolinol and heating the sample to 60°C or adding a 10%

8-quinolinol solution in acetone), adding 100 mg precleaned activated carbon, and filtering off the suspension. Quantitative recoveries with enrichment factors near 10,000 were achieved for about 20 ions from various media, more or less independently of the alkali and alkaline earth content. The influence of naturally occurring humic material was also studied [109]. Johansson and Akselsson [110] used oxine and APDC chelation and activated carbon adsorption in combination with PIXE analysis to obtain typical detection limits in the range 0.02–2 μg L^{-1} for many elements in brackish and distilled water.

Adsorption on activated carbon has also been used for single-element preconcentration before XRS, for example for Mo in seawater after chelation with NH$_4$-tetramethylenedithiocarbamate [111]. Robberecht and Van Grieken [112] selectively determined Se(IV) in diverse water samples to 0.05 μg L^{-1} with a precision of 6% by reducing Se(IV) with ascorbic acid to elemental Se and subsequent adsorption onto activated carbon, which was filtered off and analyzed by EDXRS. Total Se was determined after reduction of both Se(IV) and Se(VI) by refluxing with thioureum and H$_2$SO$_4$. These procedures were applied to various types of natural waters [113].

Many other approaches for XRS analysis of liquid samples and environmental waters have been published, but it seems that for many applications in this field, XRS can, in practice, not compete very well with the more recent inductively coupled plasma emission and mass spectrometry techniques.

V. BIOLOGICAL SAMPLES

The determination of inorganic compounds in biological and environmental materials at trace and ultratrace levels has now become a major aspect of various diagnostic and monitoring programs. For this reason, there is an increasing interest in applying PIXE, XRF, and especially TXRF to determine trace elements reliably at relevant levels [114–116]. The minimum detectable concentration in an organic matrix is usually assumed to be of the order of 10^{-6}–10^{-7} for PIXE and XRF and 10^{-8} for TXRF [117]. To determine some of the minor trace elements it is necessary to reduce the detection limit. For certain setups detection limits to 10^{-8} can be achieved through a reduction in the spectral background and by an increase in concentration via various physical or chemical preconcentration methods [118]. Sample preparation procedures to remove the matrix are of extreme importance and range from simple to sophisticated techniques. Some procedures are designed to combine parts of the sample into a homogeneous target representative of a whole sample; others are designed to separate a sample into its physical or chemical components, such as blood separation into various cellular and subcellular fractions [119,120]. Additionally, when biochemical separation procedures are used, the amount of substances available for analysis is often quite small. Hence, high sensitivity in absolute terms is of great importance for the analysis of biomedical samples [121].

A number of review papers describe sample preparation techniques of biological materials in much detail. They usually contain a detailed description of the sample preparation technique used [122–125]. Here we give only a brief outline of the various alternatives that exist for sample preparation, with emphasis on more recent approaches.

A. Physical Methods of Sample Preparation

In some cases hardly any preparation is needed. Examples of human and animal tissues that lend themselves more or less to direct irradiation in XRS are bone, teeth, hair, and nails. When dealing with various botanical hard tissues, such as leaves, wood, bark, or tree rings, the sample can be analyzed directly, too, or after very limited sample preparation, such as drying, cutting, or polishing. An adequate procedure for x-ray microanalysis is to embed such samples in a resin and preform the sections (5–30 μm) with a microtome. The thicker sections are self-supporting and can be mounted on a target frame; thinner sections can be placed on one of the backing foils (e.g., Kapton, Hostaphan, Mylar, polyethylene, Teflon, Kinfol, Selectron, or Nuclepore) and subsequently mounted on a frame. The main disadvantage of direct bombardment is that no internal standard can be added and the matrix effects can be considerable. Hair is a typical example of tissue that can advantageously be analyzed directly [126–128].

In many cases, simply drying is the only operation needed. This is particularly true for TXRF, SXRF, and PIXE, because of their low detection limits. Also, conventional XRF can determine several trace elements simultaneously in serum [129]. In the freeze-drying (lyophilization) approach, the specimen is kept in vacuum at liquid nitrogen temperature for at least 1 day. The freeze-drying technique has the additional advantage that the mass of the specimen is reduced by a factor of about 5. This reduces the spectrum background and improves the sensitivity. Lyophilization of biological material removes 80–90% of the water content. After drying or lyophilization, the residue can be pulverized and pressed into a pellet for XRF analysis.

Most human and animal tissues are very soft. Accordingly, they must first be stabilized and strengthened before sections can be cut. Maenhaut et al. [130] freeze-dried human and animal tissue and pulverized at liquid nitrogen temperatures. A small amount of the powder was fixed to a thin membrane with a solution of 1% polystyrene in benzene, making a target of intermediate thickness for PIXE.

Frozen organic tissue can be cut with a microtome, and if the sections are not too thin, they can be dried and irradiated as self-supporting targets. Another possibility is to support them on a thin plastic foil. Freeze-cutting can be applied to biomaterials of very different origins. To quantify the results, it is sufficient to add an internal standard before the analysis and to determine the dry mass of the sections after analysis. The powdered material can be prepared as a suspension and freeze-cut in sections of around 200 μg.

B. Chemical Methods of Sample Preparation

If even better sensitivity is required, the only possibility is to remove the organic matrix by dry ashing or wet digestion. High-temperature ashing (600°C) has been little used because of concomitant trace element losses, which may be minimized by the addition of an ashing aid, such as a sulfate. The matrix reduction factor is usually around 10 [131]. However, many examples of partial ashing at a temperature below 450°C have been reported. Havranek and coworkers [132] recommended the use of partial ashing of hair (1 g) in a muffle furnace at 200°C. The cool ash was homogenized in a mechanical agate mortar, and a 0.3 g portion was treated with few drops of polystyrene solution in chloroform and pressed at 15 kPa to form a pellet (~80 mg cm^{-2}). Standards were prepared by applying a multielement solution to a pellet, which is then dried at room temperature, homogenized, and re-pressed. A simple technique for the analysis of urine samples is evaporation followed by partial decomposition at 220–450°C. The resulting dry residue is then weighed, mixed, and pelletized. The same procedure can easily be adapted for radioactive samples, since minimal liquid sample handling is required and the pellet can be readily sealed between thin plastic films in a sandwich geometry. To avoid partial losses of S and Se, addition of acids is required (1 ml 15 M HNO_3 or H_2SO_4 to 25 ml sample) [133].

More often, dry ashing implies the use of a low-temperature oxygen plasma oven. Dry-ashed samples can be pelletized or deposited (few mg) on a substrate foil. It is more common to dissolve the residual salts in a small volume of acid (e.g., nitric acid) and then pipette 10–100 μl of the solution onto a backing foil to be dried and analyzed. By ashing biological materials, considerable background reduction and consequent improvement in detection limits can be achieved, but the mass reduction (between 2 and 5) depends upon the specimen type. If the specimen contains considerable amounts of inorganic matter, such as sodium chloride or calcium carbonate, these salts are the source of most of the background. Removing the organic matrix then gives only a small increase in sensitivity. Ashing is rather time consuming, and the gain in sensitivity is not always worth the effort. Even for low-temperature ashing, such elements as Se and Br are lost and Cd is partially lost [130].

The other alternative is wet digestion with an oxidizing agent, such as nitric acid, sometimes in combination with a small amount of sulfuric acid or hydrogen peroxide. The drawbacks to this technique are the risks of contamination and incomplete digestion, yet it is most frequently employed for solubilizing and homogenizing biological materials. Several workers have employed acid digestion of biological materials by adding weighed quantities to a suitable acid mixture at atmospheric pressure in open or half-open vessels. To minimize losses of analyte elements, however, it is advisable to perform the digestion in a closed vessel at elevated pressure. Such pressure digestions are generally done in so-called Teflon bombs in a regular oven. Even then, partial losses of such elements as Cl and Br may occur. A few microliters of the solution resulting from acid digestion are usually deposited on thin target backings and allowed to dry in preparation for analysis. The matrix reduction factor is usually between 5 and 4 [130]. Some alkaline digestion reagents, like tetramethylammonium hydroxide (TMAH), brought human autopsy tissues into homogeneous solution but increased the matrix by a factor of 1.5 [119].

Low-temperature preparation methods, such as cryofixation, cryoultramicrotomy, cryotransfer, and freeze-drying of biological materials, have recently been compared [134]. Particularly, cryofixation must be applied carefully to the particular biological problem to be investigated.

The use of gas-phase attack to dissolve biological materials is rare, despite the success of this technique with siliceous materials. Thomas and Smythe describe a simple all-glass apparatus for vapor-phase oxidation of up to 90% plant material with HNO_3 in 5–6 minutes [135]. Addition of $HClO_4$ ensured fast and complete oxidation, and the presence of HNO_3 during the final $HClO_4$ oxidation step eliminated any danger of explosion. Klitenick et al. used the same technique with a simplified pressurized polytetrafluoroethylene (PTFE) digestion vessel for the determination of Zn in brain tissue [136].

Microwave acid digestion has been used in a variety of sample preparations [137,138] since it was first reported 16 years ago [139]. There are now hundreds of units designed specifically for laboratory microwave digestion in use worldwide for this application. These systems are designed to overcome the deficiencies of home units, which include lack of cavity ventilation to remove acid fumes, poor resistance of the system to acid fumes, coarse power control, and short lifetime of the magnetron. The review by De la Guardia et al. [140] outlines modifications that can be made to domestic microwave ovens to overcome some of these problems. A complete line of metal-free vessels is designed specifically for treating both inorganic or organic samples with strong acids or alkalis under heat and pressure by using microwave energy. These high-pressure digestion vessels, manufactured for trace and ultratrace analysis of difficult-to-digest samples, are able to safely withstand up to 120 MPa pressure. The tetrafluorometoxil (TFM) type of digestion vessel showed by far the least memory effects of previous digestions. Two heating cycles of 30 minutes, each with concentrated nitric acid, are sufficient to reduce the blank background below the ppb range [141]. For the perfluoroalcoholoxil (PFA) vessels, a 1 day vapor procedure with nitric acid helps to reduce the blank background [142].

The combination of microwave and pressure digestion allows quantitative recovery of elements that may be volatilized in open digests (like Ge, Se, Hg, Pt, Os, and Ir), as well as the recovery of rare trace metals (Pt, Os, and Ir), to the ppb level in the solid sample even in the extremely difficult digestion of organic-rich material.

The potential hazards involved in the use of microwave heating techniques are associated with difficulties in the determination of the exact internal pressure in the vessel. A rough but often misleading value can be estimated from standard vapor pressure curves. With such acids as hydrochloric, hydrofluoric, and sulfuric acid and aqua regia, which are commonly used with inorganic samples, gases are released from the solution at elevated temperatures and the resultant pressure is a function of the temperature and the initial loading density. Sample decomposition products can contribute to the total internal pressure of the system. In addition to the partial pressure of the acids, reaction products like CO_2 rapidly increase the pressure in vessels. For microwave digestion, concentrated (70%) nitric acid (e.g., 2.5 ml to 0.1 g dry weight of the sample) is the recommended aid for organic materials. This can be diluted (5:1) with water. Nitric-sulfuric acid mixtures are not recommended. Digestion with perchloric acid can be dangerous and must not be used. A number of explosions resulting from overpressurizing microwave digestion bombs have been reported [143,144]. Hence, for safety's sake, it is recommended to start with small amounts of samples and acids and with short digestion times, to increase these variables only if necessary, and to obtain a preliminary estimate of the effective heating rate by heating a small amount of the sample and its digestion medium in an open Teflon cup and observing the time required to bring the medium to boiling.

For a basic understanding of microwave acid digestion theory, including safety guidelines and dissolution methods for geological, metallurgical, botanical, biological, food, and other samples, we recommend the professional reference book edited by Kingston and Jassie [145].

For the microwave acid digestion procedure of certified biological reference material (Hay V-10, IAEA), 500 mg powdered specimen was carefully heated in 4 ml concentrated HNO_3 and 1 ml H_2O_2 in several (three to five) steps, each for 1 minute, at 300 W. After each heating step, the vessel was cooled and the pressure released. Heating was then repeated for 7 minutes at 300 W and 2 minutes at 600 W. The total digestion time, including time for cooling the vessels, was around 30 minutes [141]. The reported analytical results confirm certified values. For the oxidation of marine biological reference material Tort-1, a 1:1 volume ratio HNO_3/ H_2O_2 was found to be optimal [146]. Digestion times of less than 1 minute were used for the microwave digestion of NIST standard reference material Oyster Tissue (SRM 1566a) and Bovine Liver (SRM 1577a) in a closed PFA bomb with HNO_3 [147]. Recently, the potential of microwave acid digestions of animal and plant tissue for PIXE analysis was examined [148]. The digestion experiments were carried out with a microwave oven normally used for domestic purposes. A set of eight certified reference materials were digested with high-purity 14 M HNO_3 (1 ml to 100 mg dry weight of sample) in closed Teflon vessels. Concentrated digestion solutions are required if one wants to minimize the amount of solution to be pipetted during target preparation and still have enough biological material on the target. The vessel was filled with vegetable oil (7.5 ml to 100 mg of sample), which absorbs the excess microwave power and prevents damage to the antenna. Targets were prepared by pipetting 10 µl digestion solution onto a 1.5 µm thick Kimfol polycarbonate film. The film was treated with 14 M HNO_3 and a 0.05% polyvinylpyrrolydone solution to make it hydrophilic. The targets were dried in a vacuum desiccator. The results indicate an accuracy better than 5% and a matrix reduction factor of 5.

An important consideration for the selection of digestion methods for the determination of trace elements in biological materials is the required digestion time. The wet digestion methods are considerably more rapid, requiring about 2–3 h, compared with at least 8 h for complete decomposition by dry-ashing methods. Generally, the required decomposition time increases with increasing amount of sample digested. Hence, the analysis time available is a major factor influencing the choice of a particular decomposition method for the reliable determination of trace elements. A recent comparison of various techniques for the sample preparation of biological materials showed comparable accuracy and precision, but the major differences were in the speed: slurry preparation took 5 minutes, microwave dissolution 45 minutes, dry ashing 24 h, and wet digestion 36 h [149].

Chemical preconcentration methods have often been invoked before XRF analysis, both on biological fluids and on the solutions obtained by ashing or digesting tissue. For example, cell fluid and blood serum produce a relatively thick layer of low-Z matrix during the evaporation process. This matrix causes a high scattering contribution and consequently considerably impairs the lower limits of detection. This matrix can be removed by pretreating either by acid-digestion in a pressure vessel or by low-temperature oxygen plasma ashing.

There are many interesting biomedical applications, but the available space does not allow more than few typical and recent examples to be mentioned.

Precipitation and coprecipitation can be advantageous preconcentration methods. For example, a very low concentration of Cr (0.3 ng ml^{-1}) in plasma was determined by complexation with APDC and extraction with methyl isobutyl ketone. After evaporation of organic solvent, the residue was dissolved in acid and deposited on a thin polycarbonate foil [150]. Eltayeb and Van Grieken [151] digested 0.1 g hair samples with a 1:5 mixture of $HClO_4$ and HNO_3 in a Teflon bomb and coprecipitated the trace elements with Y carrier at pH 3–4 using APDC to obtain detection limits in the range of 0.4–2.2 μg g^{-1} for Fe, Ni, Cu, Zn, and Pb. Cesaril et al. [152] used Te as internal standard and coprecipitant in the preconcentration method for a PIXE study of Se in serum of patients with liver cirrhosis. Iodine in rat thyroids was determined following extensive preparation; homogenized samples were treated with NaCl, HNO_3, and $AgNO_3$ and filtered and the precipitate dried. A disk punched from the filter was placed between x-ray films and analyzed by XRF [153].

The simultaneous determination of Br, Cu, Fe, and Zn in human serum by TXRF was proposed by diluting (1:1) samples with water containing Y as internal standard and drying [117].

Jing and Lihua [154] pelletized a mixture of 1 g hair, 0.1 g CaO, and 20 g cellulose and determined Cr, Cu, Fe, Ni, Mn, Se, Sr, and Zn with a detection limit of 0.5 μg g^{-1}. Methods of removing external metal contamination in hair samples for environmental monitoring have recently been developed [155].

Various ion-exchange resins have been proposed for preconcentration of trace elements from biological samples before XRF. A reduction in the detection limits by a factor of 10 can be accomplished by separating the trace elements from the main electrolytes with a chelating ion-exchange column of cellulose-Hyphan [156]. Urine analyses have been reported using ion-exchange methods [157]. Proteins can also be isolated by gel electrophoresis [158].

C. Sample Preparation for Analysis with Spatial Resolution

Sample preparation for microanalysis with spatial resolution has been developed extensively and well known in electron microprobe laboratories for many years, several books on this subject have been published [159–165]. The inherent problems related to EPMA are discussed briefly in Chapter 12. Micro-PIXE, Synchrotron-induced XRF (SXRF), and the emerging milli- or microversion of XRF will certainly be applied increasingly in this field in the near future.

First, the time at which the tissue is obtained is of crucial importance. Postmortem changes affect the redistribution of mobile ions, leading to an increase in the cellular concentrations of sodium, chloride, and calcium and to a decrease in magnesium and potassium [166]. Thus, there should be a minimal delay in obtaining the sections of tissue or cells from the organism under investigation.

In microanalysis, sample preparation is of utmost importance. Unfortunately, the environment inside a microprobe is not compatible with the continued existence of living specimens. For example, the high vacuum requires that tissue water be either withdrawn (by chemical or physical methods) or immobilized (by freezing). However, all available methods of preparing biological specimens for microprobe analysis involve a degree of restrictive compromise. The specimen must meet the often conflicting requirements of possessing a chemical integrity at the time of analysis that bears a known relationship to that in vivo, and it must exhibit interpretable structural information.

Adaptations of the "wet chemical techniques" routinely used for ultrastructural studies by transmission electron microscopy can in some cases be used for microanalytical purposes. The sample is chemically fixed so that cellular processes are arrested and the cell contents are immobilized. After removal of the tissue water by immersion in an organic solvent, the dehydrated specimen is infiltrated with a suitable resin for sectioning purposes. Detailed procedures and theoretical discussions may be found in the books by Glauert [167] and Hayat [168].

Because of the serious limitations in the more conventional ambient-temperature methods, however, interest is shifting more and more toward the technically more demanding cryoprocedures in which the chemical and structural integrity of the tissue is more faithfully preserved. An extensive review of cryobiological techniques can be found in the book by Steinbrecht and Zierold [169].

VI. ATMOSPHERIC PARTICLES

In air, the heavy metals of interest are almost exclusively in the particulate phase, and very large preconcentration factors are achieved straightforwardly by filtration of large volumes. Homogeneously loaded thin filters are nearly ideal targets for XRS. Therefore, XRS is invoked very frequently in many air pollution studies and monitoring networks. However, there are several important pitfalls to be avoided and choices to be made before reliable results can be expected.

Some serious problems stem primarily from particle size effects and x-ray absorption effects due to the filter media ("filter penetration effect"). If the particle size distribution is known for a particular element, an appropriate correction is possible. The major limitation remains that measuring the particle size distribution is not easy. The absorption of x-rays by the collection medium depends on the particle depth distribution, which can be calculated more or less by measuring the fluorescent intensities from the front and back of the loaded filter. Attenuation from the particles themselves is calculated after evaluating the mass absorption coefficient by transmission measurements. Assuming an exponential function for the depth distribution, the average penetration depth of the element is calculated [170]. On a thin filter medium (e.g., Millipore and Nuclepore), the x-ray absorption effects are usually negligible for the energy range above 5 keV. One can usually assume that the particles stay on the surface. For the lower energy range, which includes such elements as silicon, phosphorus, sulfur, and chlorine, and for small particle elements, like sulfur and lead, the correction is necessary. The effect is also much more pronounced for thicker cellulose fiber filters (Whatman 41) that collect aerosols in depth. The variability of the effects can be reduced by measuring the aerosol filter in sandwich geometry [171]. This only implies folding the loaded aerosol filter inward for XRS measurement. For such a sampling procedure, a larger filter area should be used since the area presented to the primary beam is halved. Although the x-ray absorption effect by particle penetration in the filter is increased, the advantages of this measuring geometry are a reduction in the possible filter heterogeneity effects, simpler and more accurate absorption corrections (the dependence of the x-ray absorption on the generally unknown deposition depth is averaged by assuming that all the particles are on the surface and therefore in the center of the sandwich), and increased sensitivity for high-Z elements.

The back-front and sandwich methods are accurate when the x-ray absorption cross section of the media is not excessively high. The variations in the portion of particulates on the surface affects the accuracy of the correction more than the variation in the depth distribution [172].

By using a dichotomous sampler, the filter penetration correction for the x-ray measurements can simply be assumed to be negligible. With these sampling systems, particles are separated into those smaller and larger than approximately 2 μm in diameter. The smaller particles exhibit fewer particle size effects, and x-ray spectrometric determination of even low atomic number elements is possible. The aerosol size distribution could be also obtained by the use of cascade impactor samplers, which fractionate airborne particles according to their aerodynamic diameter.

If particle size and matrix effects are expected to be severe, a fusion of the air particulate sample with lithium tetraborate can be employed [173].

Another problem that should be of concern is the possibility of losing volatile elements or unstable compounds from the target. This may be more severe in PIXE than in x-ray-induced XRS since a high vacuum must be used and the charged particle beam can be very intense and highly focused. Negative errors of 50% may occur when sulfur is determined by PIXE as a result of volatility losses but also chemical reactions leading to significant sulfur loss when 2 MeV protons, 18 MeV α particles, and other types of high-energy excitation radiation are used [174]. Gases like chlorine and bromine might be adsorbed at the filter material during aerosol sampling. Bromine compounds deposited in trace amounts can rapidly be lost when exposed to x-ray radiation. Therefore, the samples are sealed with adhesive tape and analysis times are limited to a few minutes to reduce losses to about 20% [175].

The selection of a particular filter type is the result of a compromise among many factors that include cost, availability, collection efficiency, the requirements of the analytical procedures, and the ability of the filter to retain its filtering properties and physical integrity under the ambient sampling conditions. The increasing variety of commercially available filter media sometimes makes the choice more difficult but also increases the possibility of a selection that satisfies all important criteria.

Before selecting a filter for a particular application, the blank count of the filter or background level of the material to be analyzed must be determined. All filters contain various elements as major, minor, and trace constituents, and the filter medium of choice for analyzing a particular element must be one with little or no background level for the analyte.

In interpreting filter efficiency data, it is important to consider that the test data are usually based on the efficiency of a "clean filter." For most filters, the collection efficiency increases with the accumulation of solid particles on the filter surfaces. The resistance to flow also increases with increasing loading, but usually at a much slower rate. A practical implication is that even with reliable published filter efficiency and aerosol size distribution data, it is not possible to know precisely what the collection efficiency of a filter will be for a given sampling interval. The filter efficiency data can provide only an estimate of the minimum collection efficiency. The actual collection efficiency is usually higher.

The collection efficiency of Whatman 41 is lowest for 0.264 μm particles at a face velocity of 15 cm s^{-1}. It increases for both larger and smaller particles and is more than 95% for all sizes at face velocities above 100 cm s^{-1}. It has the advantages of low cost, high mechanical strength, and high purity. This filter is fabricated with an organic binder, which makes it unsuitable for analytical work involving organic solvents, and the hygroscopy of the paper necessitates weighing under controlled humidity conditions to determine the filter load. For 10 μm Teflon filters (type LC), the collection efficiencies for 0.003–0.1 μm particles at low face velocities are in the 60–65% range; for 5 μm (type LS) filters, they are in the 80–85% range. For higher velocities and large particles, the efficiencies are higher. For 0.5 μm (type FH) and 1 μm (type FA) filters, the collection efficiencies were >99.99% under all conditions tested [176].

Very suitable for aerosol collection before XRS are Nuclepore membranes. They are very thin (1 mg cm^{-2}) and pure and essentially collect the aerosol at the surface only, but they are quite expensive. Cellulose fiber filters, such as Whatman 41, are thicker (8.5 mg cm^{-2}) and hence produce a larger x-ray scatter background, resulting in higher detection limits. For size-fractionated collection, stacked filter units with two subsequent membranes with different size cutoffs can very well be used, or dichotomous samplers, or single-orifice impactors with their stages coated with a thin and removable polymer film. Glass fiber filters are in most cases more expensive and have poorer mechanical properties than cellulose papers and contain higher inorganic impurity levels. They also have many advantages, that is, reduced hygroscopicity, ability to withstand higher temperatures, and higher collection efficiencies. These properties, combined with the ability to make benzene, water, and nitric acid extracts from particulates collected on them, led to the selection of a high-efficiency glass fiber filter as the standard collection medium for high-volume samplers and organic analyses.

Filters may also be used for nonaerosol atmospheric components, such as reactive gases. Filter materials may be impregnated with a reagent reactive to the gas that will be trapped chemically. Sampling is accomplished by passing atmospheric gases through a treated filter using carefully controlled conditions; for example, a filter treated with ferric ion solution is used for absorbing H_2S. The excess iron can be rinsed from the filter, whereas the precipitated Fe_2S_3 remains. The sulfur can be determined directly, or indirectly by measuring the iron x-irradiation. For such determinations of atmospheric components, the development of suitable standards is necessary. Some standards for aerosols are commercially available [177].

Many air-monitoring agencies around the world now invoke tube-excited XRS for the analysis of loaded air filters (e.g., Sartorius filters) of dichotomous sampler deposits. Radioisotope-excited EDXRS is increasingly used in portable instruments together with Geiger counters or scintillation detectors for the *in situ* monitoring of few elements in the work environment [178]. Gilfrich and Birks designed a portable vacuum WDXRS analyzer for *in situ* measurements of sulfur and other elements in aerosols [179]. The high analytical sensitivity of PIXE permits the use of small aerosol samplers, like 1 L min^{-1} six-stage cascade impactors, in combination with battery-operated vacuum pumps. Thus aerosol sampling can be easily done in remote areas [180].

The listing of the literature on advances and applications of XRS techniques to air pollution studies includes more than 500 publications that have appeared in the last few years. A large part of the basic sampling methodology of the airborne particulate material together with numerous examples of applications is given in Chapter 11. Also, a wide range and different types of samples, excitation strategies, and detection modes were described earlier [181].

VII. SAMPLE SUPPORT MATERIALS

In addition to the actual preparation of the sample, it must be remembered that precise positioning of the sample in the spectrometer is critical to quantitative determinations, except when internal standards are used. In most commercial instruments the size of the sample is determined by the size of the holder. An exception is in the case of custom-built instruments tailored for specific applications, with large sample chambers for direct nondestructive analysis of large objects, such as archaeological specimens and objects of artistic or esthetic value. Most sample holders on commercial instruments are cylindrical, with diameters of 5.1 cm, and allow a maximum sample thickness of about 4 cm. The designs vary slightly for sample-up sample-down spectrometer geometry. For small samples, masks are provided that are fixed to the bottom of the cup or placed inside it. In general, the bottom surfaces of specimen holders are covered with a thin film.

Thin-film samples are ideal for x-ray spectrometric analysis. The x-ray intensity of an infinitely thin sample is proportional to the mass of the element on the film, and the spectral intensities are free of interelement and mass absorption coefficient effects. In practice, however, perfect thin-film samples are difficult to encounter. To be "infinitely thin" to most x-rays, the specimen must be of the order of 10–200 μ, thick.

A thin-film substance and its thickness should be selected that provide the greatest degree of transmittance, particularly for low concentration levels and low-energy photons. The materials must also be reasonably free of impurities and with a high "degradation resistance." Degradation resistance represents the ability of a thin-film material to safely retain a specimen in an XRF cup during preparation and analysis. This includes resistance to chemical attack, thermal softening, tearing, and stretching. Some of these properties for the commonly used thin films are listed in Table 3.

The effect of thin-film thickness on analyte line transmittance for various substances is illustrated in Figure 3 [182]. Most commercial polymeric thin-film materials permit transmission better than 90% for photons of 3 keV or more. In

Table 3 Degradation Resistance Properties of Some Commonly Used Thin-Film
Materials

Thin-film substances	Contaminants	Poor degradation resistance for
Mylar	P, Ca, Zn, Sb	Acids (HCl, HNO$_3$), mineral; strong
Polypropylene	Al, Si, Ti, Cu, Fe	Acids, oxidizing, concentrated; aqua regia
Polyethylene	—	Acids, oxidizing, concentrated; alcohols, esters, ketones
Polycarbonate	—	Acids, oxidizing, concentrated; alcohols; alkalines; esters, ketones; hydrocarbons, aliphatic; hydrocarbons, aromatic; oils, mineral, vegetable, animal
Polystyrene	—	Esters, ketones; hydrocarbons, aromatic; hydrocarbons, aliphatic; oils, mineral, vegetable
Kapton	—	Acids, mineral, strong; alkalines
Formvar	—	Acids

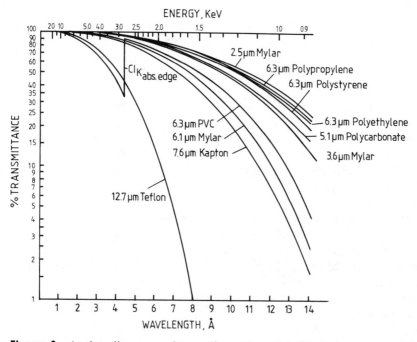

Figure 3 Analyte line transmittance for various thin-film substances and thicknesses.
(From Ref. 182. Reprinted by permission of *American Laboratory*.)

the low-keV region, absorption effects predominate. Teflon prohibits 50% analyte line transmittance at 3 keV and therefore is limited to use with the more energetic analyte lines. Polyvinyl chloride (PVC) exhibits a sharp discontinuity in the percentage transmittance correlation to the analyte line appearing at 2.8 keV (K absorption edge for chlorine) and unacceptable resistance to chemical attack. Mylar possesses good properties with respect to degradation resistance and percentage of transmittance, but it has the drawback of inherent detectable trace impurities.

More important thin-film types are platings and coatings on various substrates, and this rapid, on-line measurement of film thickness and composition analysis is of increasing importance in the electronics industry.

REFERENCES

1. A. Knöchel, W. Petersen, and G. Tolkiehn, *Nucl. Instrum. Methods 208*:659 (1983).
2. R. Tertian and F. Claisse, *Principles of Quantitative X-ray Spectrometry*, Marcel Dekker, New York, 1981, p. 373.
3. E. P. Bertin, *Introduction to X-ray Spectrometric Analysis*, Plenum, New York, 1978, p. 402.
4. E. P. Bertin, *Principles and Practice of X-ray Spectrometric Analysis*, Plenum, New York, 1970, p. 176.
5. R. Jenkins, R. W. Gould, and D. Gedcke, *Quantitative X-ray Spectrometry*, Marcel Dekker, New York, 1981, p. 373.
6. K. Togel, *Zerstorungsfreie Materialprufung*, R. Oldenbourg, Munich, 1961.
7. R. Jenkins and P. W. Hurley, *Proceedings of XII Colloquium Spectroscopicum Internationale*, Hilger and Watts, London, 1965, p. 444.
8. R. O. Müller, *Spectrochemical Analysis by X-ray Fluorescence*, Adam Hilger, London, 1972, p. 72.
9. P. Berry, T. Furuta, and J. Rhodes, *Adv. X-ray Anal. 12*:612 (1969).
10. F. Wybenga, *X-ray Spectrom. 8*:182 (1979).
11. R. Bock, *A Handbook of Decomposition Methods in Analytical Chemistry*, T.&A. Constable, Edinburgh, 1979, p. 199.
12. B. T. Eddy and A. M. E. Balaes, *X-ray Spectrom. 17*:195 (1988).
13. G. Z. Tölg, *Anal. Chem. 190*:161 (1962).
14. V. Fano and L. Zanotti, *Anal Chim. Acta 72*:419 (1974).
15. H. Knote and V. Krivan, *Anal. Chem. 54*:1858 (1982).
16. F. Claisse, *Norelco Rep. 4*:3 (1957).
17. F. Claisse, *Norelco Rep. 4*:95 (1957).
18. F. Claisse and C. Samson, *Adv. X-ray Anal. 5*:335 (1972).
19. J. Rhodes and C. Hunter, *X-ray Spectrom. 10*:107 (1972).
20. B. Holynska and A. Markowicz, *X-ray Spectrom. 11*:117 (1982).
21. K. W. Madlem, *Adv. X-ray Anal. 9*:441 (1966).
22. C. A. Volborth, *Am. Mineral. 49*:634 (1964).
23. H. A. Liebhafsky and P. D. Zemany, *Anal. Chem. 28*:455 (1956).
24. R. Van Grieken, R. Van de Velde, and H. Robberecht, *Anal. Chim. Acta 118*:137 (1980).
25. I. Adler and J. Axelrod, *Anal. Chem. 27*:1002 (1955).
26. J. Lincoln and E. Davix, *Anal. Chem. 31*:1317 (1959).
27. J. Mitchell, *Anal. Chem. 30*:1894 (1958).
28. A. A. Chodos and C. G. Engel, *Adv. X-ray Anal. 4*:401 (1961).

29. W. J. Campbell and J. W. Thatcher, *Adv. X-ray Anal. 2*:313 (1959).
30. B. D. Wheeler, *Adv. X-ray Anal. 26*:457 (1983).
31. R. Van Grieken, L. Van'T Dack, C. Costa Dantas, and H. Da Silveira Dantas, *Anal. Chim. Acta 108*:93 (1979).
32. L. Sauer, D. Van der Ben, and R. Van Grieken, *X-ray Spectrom. 8*:159 (1979).
33. G. Frechette, J. C. Hebert, P. T. Thinh, R. Rousseau, and F. Claisse, *Anal. Chem. 51*:957 (1979).
34. F. Claisse, *Norelco Rep. 4(3)*:95 (1957).
35. R. LeHouillier and S. Turmel, *Anal. Chem. 46*:734 (1974).
36. S. Banajee and B. Olsen, *Appl. Spectrosc. 32*:576 (1978).
37. D. Stephenson, *Anal. Chem. 41*:966 (1969).
38. K. Norrish and J. Hutton, *Geochim. Cosmochim. Acta 33*:431 (1968).
39. J. Hutton and S. Elliot, *Chem. Geol. 29*:1 (1980).
40. R. H. Dow, *Adv. X-ray Anal. 25*:117 (1981).
41. P. Pella, G. Tao, A. Dragoo, and J. Epp, *Anal. Chem. 57*:1752 (1985).
42. C. Luke, *Anal. Chem. 35*:1553 (1963).
43. B. Schroeder, G. Thompson, and M. Sulanowska, *X-ray Spectrom. 9*:198 (1980).
44. S. Turmel and C. Samson, *X-ray Spectrom. 13*:13 (1984).
45. S. Banajee and B. Olsen, *Adv. X-ray Anal. 18*:317 (1974).
46. C. Luke, *Anal. Chem. 35*:57 (1963).
47. R. LeHouillier, S. Turmel, and F. Claisse, *Adv. X-ray Anal. 20*:459 (1976).
48. P. K. Harvey and D. M. Taylor, *X-ray Spectrom. 2*:33 (1958).
49. B. S. King and D. Vivit, *X-ray Spectrom. 17*:85 (1988).
50. R. Tertian, *X-ray Spectrom. 4*:52 (1975).
51. G. Frechette, J. C. Hebert, T. P. Thinh, R. Rousseau, and F. Claisse, *Anal. Chem. 51*:957 (1979).
52. A. García-Cervigón Bellón, *Est. Geol. Madrid 33*:433 (1977).
53. R. Van Grieken, K. Bresseleers, J. Smits, B. Vanderborght, and M. Vanderstappen, *Adv. X-ray Anal. 19*:435 (1976).
54. R. Van Grieken, *Anal. Chim. Acta 143*:3 (1982).
55. M. S. Cresser, J. Armstrong, J. Dean, M. H. Ramsey, and M. Cave, *J. Anal. At. Spectrom. 6*:1R (1991).
56. M. Vanderstappen and R. Van Grieken, *Fresenius Z. Anal. Chem. 282*:25 (1976).
57. J. Smits and R. Van Grieken, *Anal. Chim. Acta 88*:97 (1977).
58. F. Rickey, K. Mueller, P. Simms, and B. Michael, in: *X-ray Fluorescence Analysis of Environmental Samples* (T. G. Dzubay, Ed.), Ann Arbor Science, Ann Arbor, MI, 1977, p. 135.
59. R. D. Giauque, R. B. Garret, and L. Y. Goda, in: *X-ray Fluorescence Analysis of Environmental Samples* (T. G. Dzubay, Ed.), Ann Arbor Science, Ann Arbor, MI, 1977, p. 153.
60. D. C. Camp, J. A. Cooper, and J. R. Rhodes, *X-ray Spectrom. 3*:47 (1974).
61. H. C. Hansson, E. M. Johansson, and A. K. Ekholm, *Nucl. Instr. Meth. Phys. Res. B3*:158 (1984).
62. A. Prange and H. Schwenke, *Adv. X-ray Anal. 32*:211 (1989).
63. E. Bruninx and A. Van Ebergen, *Anal. Chim. Acta 109*:419 (1979).
64. R. Chakravorty and R. Van Grieken, *Int. J. Environ. Anal. Chem. 11*:67 (1982).
65. N. Becker, V. Mcrae, and J. Smith, *Anal. Chim. Acta 173*:361 (1985).
66. T. Andrew, D. Leyden, W. Wegscheider, B. Jablonski, and W. Bochnar, *Anal. Chim. Acta 142*:73 (1982).
67. K. Hirayama and D. Leyden, *Anal. Chim. Acta 188*:1 (1986).
68. X. Shan, J. Tie, and G. Xie, *J. Anal. At. Spectrom. 3*:259 (1988).
69. C. Luke, *Anal. Chim. Acta 41*:237 (1968).

70. S. Brüggerhoff, E. Jackweth, B. Raith, A. Stratmann, and B. Gonsior, *Fresenius Z. Anal. Chem. 311*:252 (1982).

71. T. Florkowski, B. Holynska, and J. Niewodniczanski, in: *Nuclear Techniques in Environmental Pollution*, IAEA, Vienna, 1971, p. 335.

72. B. Holynska and K. Bisiniek, *J. Radioanal. Chem. 31*:159 (1976).

73. N. Moriyama, K. Kimata, and S. Andou, *Adv. X-ray Anal. Jpn. 5*:93 (1972).

74. R. Perry and R. J. Young, *Handbook of Air Pollution Analysis*, Wiley, New York, 1977.

75. H. Sasuga, A. Abe, T. Nakamura, E. Asada, and T. Aota, *Adv. X-ray Anal. 25*:63 (1982).

76. T. Tanoue, H. Nara, and S. Yamaguchi, *Adv. X-ray Anal. Jpn. 11*:69 (1979).

77. T. Tanoue, H. Nara, and S. Yamaguchi, *Adv. X-ray Anal. Jpn. 11*:81 (1979).

78. H. Watanabe, J. Berman, and D. S. Russell, *Talanta 19*:1363 (1972).

79. J. E. Kessler and S. M. Vincent, *Pittsburgh Conf. Anal. Chem. Appl. Spectrosc.*, March, 1972, paper 70.

80. M. M. Ulrich and P. K. Hopke, *Res. Dev. 28*:34 (1977).

81. J. F. Elder, S. K. Perry, and F. P. Brady, *Environ. Sci. Technol. 9*:1039 (1975).

82. H. R. Lindner, H. D. Seltner, and B. Schreiber, *Anal. Chem. 50*:896 (1978).

83. A. T. Ellis, D. E. Leyden, W. Wegscheider, B. B. Jablonski, and W. B. Bodnar, *Anal. Chim. Acta 142*:89 (1982).

84. I. Watanabe, N. Shibata, Y. Kose, and M. Takahashi, *Nippon Kagaku Kaishi 8*:1079 (1986).

85. Y. Saitoh, A. Yoneda, Y. Maeda, and T. Azumi, *Bunseki Kagasku 33*:412 (1984).

86. M. G Vanderstappen and R. Van Grieken, *Talanta 25*:653 (1978).

87. V. Lebedev, V. Alvares, A. Krasnyanskii, T. Shurupova, and V. Golubtsov, *Vestri. Mosk. Univ. Serz. Khim. 28*:174 (1987).

88. T. Yamamoto, *Bunseki Kagaku 36*:592 (1987).

89. S. Tanaka, M. Nakamura, and Y. Hashimoto, *Bunseki Kagaku 36*:114 (1987).

90. T. Flokowski, B. Holynska, and S. Piorek, in: *Measurement, Detection and Control of Environmental Pollutants*, IAEA, Vienna, 1976, p. 213.

91. F. Clanet and R. Deloncle, *Anal. Chim. Acta 117*:343 (1980).

92. G. Brykina, A. Stefanov, O. Okuneva, N. Alekseeva, V. Alekseev, and S. Nikitin, *Zh. Anal. Khim. 39*:1750 (1984).

93. D. Voutsa, C. Samara, K. Tytianos, and T. Kovimtzis, *Fresenius Z. Anal. Chem. 330*:596 (1988).

94. A. Prange and A. Knochel, *Anal. Chim. Acta 172*:79 (1985).

95. D. E. Leyden ad G. H. Luttrell, *Anal. Chem. 47*:1612 (1975).

96. P. Burba and K. H. Lieser, *Fresenius Z. Anal. Chem. 286*:191 (1977).

97. D. E. Leyden, G. H. Luttrell, W. K. Nonidez, and D. B. Werho, *Anal. Chem. 48*:67 (1976).

98. H. Kingston and P. A. Pella, *Anal. Chem. 53*:223 (1981).

99. P. Burba and K. H. Lieser, *Fresenius Z. Anal. Chem. 297*:374 (1979).

100. H. James and P. Lin, *Int. Lab. 12*:44 (1982).

101. P. Coetzee and K. Leiser, *Fresenius Z. Anal. Chem. 323*:257 (1986).

102. J. Smits and R. Van Grieken, *Anal. Chem. 52*:1479 (1980).

103. J. Smits and R. Van Grieken, *Angew. Makromol. Chem. 72*:105 (1978).

104. G. Knapp, B. Schreiber, and R. W. Frei, *Anal. Chim. Acta 77*:293 (1975).

105. J. Knoth and H. Schwenke, *Fresenius Z. Anal. Chem. 294*:273 (1979).

106. A. Prange, A. Knöchel, and W. Michaelis, *Anal. Chim. Acta 172*:79 (1985).

107. T. Braun, N. Abbas, S. Török, and Z. Szökefalor-Nagy, *Anal. Chim. Acta 16*:277 (1984).

108. B. M. Vanderborght and R. Van Grieken, *Anal. Chem. 49*:311 (1977).

109. B. M. Vanderborght and R. Van Grieken, *Int. J. Environ. Anal. Chem. 5*:221 (1978).
110. E. M. Johansson and K. R. Akselsson, *Nucl. Instrum. Methods 181*:221 (1981).
111. H. Monien, R. Bovenkerk, K. P. Krings, and D. Rath, *Fresenius Z. Anal. Chem. 300*:363 (1980).
112. H. J. Robberecht and R. Van Grieken, *Anal. Chem. 52*:449 (1980).
113. H. Robberecht, R. Van Grieken, D. Vanden Berghe, M. Van Sprundel, and H. Deelstra, *Sci. Total Environ. 26*:165 (1983).
114. G. A. Quamme, *Scanning Microsc. 2*:2195 (1988).
115. E. A. Maier, M. L. Sargentini-Maier, F. Rastegar, C. Christope, C. Ruch, R. Heimburger, and M. J. F. Leroy, *Fresenius Z. Anal. Chem. 331*:58 (1988).
116. W. Maenhaut, *Nucl. Instrum. Methods B35*:388 (1988).
117. C. T. Yap, *Appl. Spectrosc. 42*:7/1250 (1988).
118. J. L. Campbell, S. B. Russell, S. Faiq, C. W. Schulte, R. W. Ollerhead, and R. R. Gingerich, *Nucl. Instrum. Methods 181*:285 (1981).
119. N. F. Mangelson and M. W. Hill, *Nucl. Instrum. Methods 181*:143 (1981).
120. G. Weber, G. Robaye, J. M. Delbrouck, I. Roelandts, O. Dideberg, P. Bartsch, and M. C. De Pauw, *Nucl. Instrum. Methods 168*:551 (1980).
121. K. S. Kim, J. Webb, D. J. Macey, and D. D. Cohen, *Nucl. Instrum. Methods B22*:227 (1987).
122. H. M. Crews, D. J. Halls, and A. Taylor, *J. Anal. At. Spectrom. 5*:75R (1990).
123. N. F. Mangelson and M. W. Hill, *Scanning Microsc. 4*:63 (1990).
124. J. R. Bacon, A. T. Ellis, and J. G. Williams, *J. Anal. At. Spectrom. 5*:243R (1990).
125. W. Maenhaut, *Scanning Microsc. 4*:43 (1990).
126. H. Kubo, *Phys. Med Biol. 26*:867 (1981).
127. T. Y. Toribara, D. A. Jackson, W. R. French, A. C. Thomson, and J. M. Jaklevic, *Anal. Chem. 54*:1844 (1982).
128. Sz. Török, P. Van Dyck, and R. Van Grieken, *X-ray Spectrom. 13*:27 (1984).
129. H. Robberecht, R. Van Grieken, J. Shani, and S. Barak, *Anal. Chim. Acta 136*:285 (1982).
130. W. Maenhaut, L. De Reu, and J. Vandenhaute, *Nucl. Instrum. Methods B3*:135 (1984).
131. H. Rudolph, J. K. Kliwer, J. J. Kraushaar, R. A. Ristinen, and W. R. Smythe, Elemental analysis by proton and x-ray induced characteristic x-rays, in: Proceedings of the 18th Annual ISA Analysis Instrumentation Symposium, San Francisco, CA, 1972, p. 151.
132. E. Havranek, A. Bumbalova, and M. Harangozo, *J. Radioanal. Nucl. Chem. 135*(5):321 (1986).
133. K. K. Nielson and D. R. Kalkwarf, Multielement analysis of urine using energy dispersive x-ray fluorescence, in: *Electron Microscopy and X-ray Applications* (P. A. Russell and A. E. Hutchings, Eds.), Ann Arbor Science, Ann Arbor, MI, 1978. p. 31.
134. R. Wroblewski, J. Wroblewski, and G. M. Roomans, *Scanning Microsc. 1*:1225 (1987).
135. A. D. Thomas and L. E. Smythe, *Talanta 20*:469 (1973).
136. M. A. Klitenick, C. J. Frederickson, and W. I. Manton, *Anal. Chem. 55*:921 (1983).
137. E. M. Skelly and F. T. di Stefano, *Appl. Spectrosc. 42*:1302 (1988).
138. H. M. Kingston and L. B. Jassie, *Anal. Chem. 58*:2534 (1986).
139. A. Abu-Samra, J. S. Morris, and S. R. Koirtyohann, *Anal. Chem. 47*:1475 (1975).
140. M. De la Guardia, A. Salvador, J. L. Burguera, and M. Burguera, *J. Flow Injection Anal. 5*:121 (1988).
141. T. Noltner, P. Maisenbacher, and H. Puchelt, *Spectroscopy 5*(4):49 (1990).
142. G. Knapp, *Int. J. Environ. Anal. Chem. 22*:71 (1985).

143. S. A. Matthes, R. F. Farrelli, and A. J. Mackie, *Tech. Prog. Rep.—U.S. Bur. Mines*, No. 120, 1983.

144. L. A. Fernando, W. D. Heavner, and C. C. Gabrielli, *Anal. Chem. 58*:511 (1986).

145. H. M. Kingston and L. B. Jassie, *Introduction to Microwave Sample Preparation, Theory and Practice*, ACS Reference Book Series (H. M. Kingston and L. B. Jassie, Eds.), American Chemical Society, Washington, D.C. 1988.

146. H. Matusiewicz, R. E. Sturgeon, and S. S. Berman, *J. Anal. At. Spectrom. 4*:323 (1989).

147. R. A. Stripp and D. C. Bogen, *J. Anal. Toxicol. 13*:57 (1989).

148. T. Pinheiro, H. Duflou, and W. Maenhaut, *Biol. Trace Elem. Res. 26–27*:589 (1989).

149. N. J. Miller-Ihli, *J. Res. Natl. Bur. Stand. (U.S.) 93*:350 (1988).

150. M. Simonoff, Y. Llabador, G. N. Simonoff, M. R. Boisseau, and M. F. Lorient Rodaut, *J. Radioanal. Nucl. Chem. Lett. 94*:297 (1985).

151. M. A. H. Eltayeb and R. Van Grieken, *J. Radioanal. Nucl. Chem. Articles 131*(2):331 (1989).

152. M. Cesaril, A. M. Stanzial, G. B. Gabrielli, F. Capra, L. Zenari, S. Galassini, G. Moschini, Q. I. Niang, and R. Corrocher, *Clin. Chim. Acta 182*:221 (1989).

153. J. Pavel, M. Fille, and U. Frey, *Fresenius Z. Anal. Chem. 331*:51 (1988).

154. W. Jing and J. Lihua, *Huaxue Fence 24*:366 (1988).

155. L. Raghupathy, M. Harada, H. Ohno, A. Naganuma, N. Imura, and R. Doi, *Sci. Tot. Environ. 77*:141 (1988).

156. H. M. Crews, D. J. Halls, and A. Taylor, *J. Anal. At. Spectrom. 5*:75R (1990).

157. M. Agarwal, R. B. Bennett, I. G. Stump, and J. M. D'Auria, *Anal. Chem. 47*:924 (1975).

158. Z. Szökefalvi-Nagy, I. Demeter, Cs. Bagyinka, and K. L. Kovacs, *Nucl. Instrum. Methods B22*:156 (1987).

159. T. Hall, P. Echlin, and R. Kaufmann, (Eds.), *Microprobe Analysis as Applied to Cells and Tissues*, Academic Press, London, 1974.

160. J. P. Revel, (Ed.), *Science of Biological Specimen Preparation for Microscopy and Microanalysis*, SEM, AFM O'Hare, Chicago, IL 60666-0507, 1984.

161. P. Echlin and P. Galle, (Eds.), *Biological microanalysis, J. Microsc. Biol. Cell. 22*(2–3):121, (1975).

162. S. J. B. Reed (Ed.), *Electron Microprobe Analysis*, University Press, Cambridge, 1975.

163. D. A. Erasmus (Ed.), *Electron Probe Microanalysis in Biology*, Chapman and Hall, London, 1978.

164. M. A. Hayat (Ed.), *X-ray Microanalysis in Biology*, University Park Press, Baltimore, 1980.

165. M. A. Hayat (Ed.), *Principles and Techniques of Electron Microscopy, Biological Applications*, Vol. 1, *Sample Preparation*, Van Nostrand-Reinhold, New York, 1970.

166. T. von Zglincki, M. Rimmler, and H. J. Purz, *J. Microsc. 141*:79 (1985).

167. A. M. Glauert, *Fixation, Dehydration and Embedding of Biological Specimens: Practical Methods in Electron Microscopy*, Vol. 3 (A. M. Glauert, Ed.), North-Holland, Amsterdam, 1974.

168. M. A. Hayat, *Principles and Techniques in Electron Microscopy: Biological Applications*, 3rd Ed., Macmillan, London, 1989.

169. R. A. Steinbrecht and K. Zierold (Eds.), *Cryotechniques in Biological Electron Microscopy*, Springer-Verlag, Berlin, 1987.

170. F. Adams and R. Van Grieken, *Anal. Chem. 47*:1767 (1975).

171. R. Van Grieken and F. Adams, *X-ray Spectrom. 4*:190 (1975).

172. D. W. Davis, R. L. Reynolds, G. C. Tsou, and L. Zafoute, *Anal. Chem. 49*:1990 (1977).
173. P. A. Pella, K. E. Lorber, and K. F. J. Heinrich, *Anal. Chem. 50*:1268 (1978).
174. L. D. Hansen, J. F. Ryder, N. F. Mangelson, M. W. Hill, K. J. Faucette, and D. J. Eatough, *Anal. Chem. 52*:821 (1980).
175. B. H. O'Connor, G. C. Kerrigan, and E. P. Hinchliff, *X-ray Spectrom. 6*:83 (1977).
176. M. Lippman, Filter media and filter holders for air sampling, in: *Air Sampling Instruments for Evaluation of Atmospheric Contaminants*, 5th Ed., American Conference of Governmental Industrial Hygienists, Cincinnati, OH, 1978.
177. D. E. Leyden, *Fundamentals of x-ray Spectrometry as Applied to Energy Dispersive Techniques*, Tracor Xray, Mountain View, CA, 1984.
178. C. Von Alfthan, P. Rautala, and J. R. Rhodes, *Adv. X-ray Anal. 23*:27 (1980).
179. J. V. Gilfrich and L. S. Birks, *Portable Vacuum X-ray Spectrometer: Instrument for On-site Analysis of Airborne Particulate Sulfur and Other Elements*, EPA Report 600/7-78-103, Research Triangle Park, NC, 1978.
180. W. Maenhaut, H. Raemdonck, A. Selen, R. Van Grieken, and J. W. Winchester, *J. Geophys. Res. 83*:5353 (1983).
181. R. Van Grieken and J. J. LaBrecque, in *Trace Analysis, Vol. 4*, (J. F. Lawrence, Ed.), Academic Press, Inc., Florida, 1985, p. 101.
182. M. J. Solazzi, *Am. Lab. 17*(11):3 (1985).

Index

Bold numbers refer to pages where a term is explained or discussed in more detail.

Aberration of lens system, **99**
Absorber, 17, 19, 115, 185, 234, 253,
    315, 522, 528–529, 536, 546,
    665–667, 675
Absorption, x-ray, 9, **17**, **19–21**, 32, 77,
    84, 91, 107, 131, 153, 159–161,
    167, 172, 185, 219, 224, 228,
    233–234, 252–254, 257, 296, 306,
    311–312, 323, 326, 329, 343, 360,
    361, 365–366, 368, 372, 374, 376,
    378, 387–389, 397, 399, 428, 454,
    465, 501, 507, 561, 588, 595–598,
    603, 605, 613–616, 619, 624–627,
    659, 682
    edge, **9**, 19–20, **21**, 25, **27**, 28–29, 33,
      77, 93, 99, 112, 114, 133, 170, 173,
      175, 185, 219, 222, 224, 253–254,
      297–299, 312, 320–321, 342–345,
      365–366, 384, 387–389, 403, 424,
      427, 443, 502, 599
    fine structure, **21**, 27
    jump ratio, **20**, 296, 298, 320–321, 363,
      599, 619
    path length, 587, 595, 597–599, 622,
      626–627

Absorption, coefficients, **17**, 19, 21,
    25–27, 28, 30–31, **69–70**, **76**,
    76–77, 95, 112, **120**, 122, **130**, **142**,
    160, 186, 234, 296–297, 311–312,
    315, 320–321, 339, 342–351, 353,
    362–366, 384, 403, 412, 419, 423,
    434, 443, 455, 465, 478, 500–501,
    541–542, 597–598, 600, 607, 620,
    623, 626–627, 666, 682, 685
    calculation, **19**, 342–343
Absorption correction, x-ray, 5, **91**,
    228–229, 233–234, 341, 343,
    344–350, 588, **595**, 597, 601, 615,
    617, 670, 683
    factor, **340**, 342, 345–346, 348, 350
Absorption-enhancement effects (*see
    also* Enhancement), 30–31, 77,
    233, 257, 297–299, 621–622, 626,
    668, 670, 682–683, 687
Accuracy, 83, **131**, 134, 137, 145, 162,
    177–178, 196, 204–205, 300, 310,
    319, 322, 333, 349, 365, 377, **390**,
    391, 402, 407, 470, 480, 519, **538**,
    540, 542, **546**, 547, 601, 603, 606,
    611, 614, 634, 639, 643, 657, 669,
    671, 675